**WITHDRAWN
UTSA LIBRARIES**

Handbook of Applied Spatial Analysis

Manfred M. Fischer • Arthur Getis
Editors

Handbook of Applied Spatial Analysis

Software Tools, Methods and Applications

Editors

Professor Manfred M. Fischer
Vienna University of Economics
and Business
Institute for Economic Geography
and GIScience
Nordbergstraße 15/4/A
1090 Vienna
Austria
manfred.fischer@wu.ac.at

Professor Arthur Getis
San Diego State University
Department of Geography
5500 Campanile Drive
San Diego, CA 92182-4493
USA
arthur.getis@sdsu.edu

ISBN 978-3-642-03646-0   e-ISBN 978-3-642-03647-7
DOI 10.1007/978-3-642-03647-7
Springer Heidelberg Dordrecht London New York

Library of Congress Control Number: 2009940922

© Springer-Verlag Berlin Heidelberg 2010

This work is subject to copyright. All rights are reserved, whether the whole or part of the material is concerned, specifically the rights of translation, reprinting, reuse of illustrations, recitation, broadcasting, reproduction on microfilm or in any other way, and storage in data banks. Duplication of this publication or parts thereof is permitted only under the provisions of the German Copyright Law of September 9, 1965, in its current version, and permission for use must always be obtained from Springer. Violations are liable to prosecution under the German Copyright Law.

The use of general descriptive names, registered names, trademarks, etc. in this publication does not imply, even in the absence of a specific statement, that such names are exempt from the relevant protective laws and regulations and therefore free for general use.

Cover design: WMXDesign GmbH, Heidelberg, Germany

Printed on acid-free paper

Springer is part of Springer Science+Business Media (www.springer.com)

Preface

The *Handbook* is written for academics, researchers, practitioners and advanced graduate students. It has been designed to be read by those new or starting out in the field of spatial analysis as well as by those who are already familiar with the field. The chapters have been written in such a way that readers who are new to the field will gain important overview and insight. At the same time, those readers who are already practitioners in the field will gain through the advanced and/or updated tools and new materials and state-of-the-art developments included.

This volume provides an accounting of the diversity of current and emergent approaches, not available elsewhere despite the many excellent journals and textbooks that exist. Most of the chapters are original, some few are reprints from the *Journal of Geographical Systems*, *Geographical Analysis*, *The Review of Regional Studies* and *Letters of Spatial and Resource Sciences*. We let our contributors develop, from their particular perspective and insights, their own strategies for mapping the part of terrain for which they were responsible. As the chapters were submitted, we became the first consumers of the project we had initiated. We gained from depth, breadth and distinctiveness of our contributors' insights and, in particular, the presence of links between them.

The chapters were rigorously refereed blindly by the contributors to this volume. Referee reports were sent to each author and changes made accordingly. We supervised this process to guarantee that authors received reviews that would be useful for finalizing their chapters. The soundness of the comments and ideas have contributed immensely to the quality of the *Handbook*. Fortunately, we were dealing with truly exemplary scholars, the most distinguished and sophisticated representatives of the fields of inquiry.

We thank the contributors for their diligence, not only in providing extremely thoughtful and useful contributions, but also in meeting all deadlines in a timely manner and in following stringent editorial guidelines. Moreover, we acknowledge the generous support provided by the Institute for Economic Geography and GIScience, Vienna University of Economics and Business. Thomas Seyffertitz greatly assisted in keeping the project well organized. Last but not at least, we have benefitted greatly from the editorial assistance he and Ingrid Divis provided. Their expertise in handling several word processing systems, formatting, and indexing, together with their care and attention to detail, helped immeasurably.

August 2009　　　　　　　　　　　　　　　　　　　Manfred M. Fischer, Vienna
　　　　　　　　　　　　　　　　　　　　　　　　　Arthur Getis, San Diego

Contents

Preface   *v*

Introduction   *1*
Manfred M. Fischer and Arthur Getis

PART A   GI Software Tools

A.1   Spatial Statistics in ArcGIS
Lauren M. Scott and Mark V. Janikas

     A.1.1   Introduction   *27*
     A.1.2   Measuring geographic distributions   *28*
     A.1.3   Analyzing patterns   *30*
     A.1.4   Mapping clusters   *33*
     A.1.5   Modeling spatial relationships   *35*
     A.1.6   Custom tool development   *38*
     A.1.7   Concluding remarks   *39*
     References   *40*

A.2   Spatial Statistics in SAS
Melissa J. Rura and Daniel A. Griffith

     A.2.1   Introduction   *43*
     A.2.2   Spatial statistics and SAS   *43*
     A.2.3   SAS spatial analysis built-ins   *44*
     A.2.4   SAS implementation examples   *45*
     A.2.5   Concluding remarks   *51*
     References   *51*

A.3   Spatial Econometric Functions in R
Roger S. Bivand

     A.3.1   Introduction   *53*
     A.3.2   Spatial models and spatial statistics   *55*
     A.3.3   Classes and methods in modelling using R   *57*
     A.3.4   Issues in prediction in spatial econometrics   *60*
     A.3.5   Boston housing values case   *65*
     A.3.6   Concluding remarks   *68*
     References   *69*

A.4	GeoDa: An Introduction to Spatial Data Analysis	
	Luc Anselin, Ibnu Syabri and Youngihn Kho	
	A.4.1 Introduction	73
	A.4.2 Design and functionality	76
	A.4.3 Mapping and geovisualization	78
	A.4.4 Multivariate EDA	80
	A.4.5 Spatial autocorrelation analysis	82
	A.4.6 Spatial regression	84
	A.4.7 Future directions	86
	References	87
A.5	STARS: Space-Time Analysis of Regional Systems	
	Sergio J. Rey and Mark V. Janikas	
	A.5.1 Introduction	91
	A.5.2 Motivation	92
	A.5.3 Components and design	92
	A.5.4 Illustrations	98
	A.5.5 Concluding remarks	109
	References	111
A.6	Space-Time Intelligence System Software for the Analysis of Complex Systems	
	Geoffrey M. Jacquez	
	A.6.1 Introduction	113
	A.6.2 An approach to the analysis of complex systems	115
	A.6.3 Visualization	116
	A.6.4 Exploratory space-time analysis	117
	A.6.5 Analysis and modeling	119
	A.6.6 Concluding remarks	122
	References	123
A.7	Geostatistical Software	
	Pierre Goovaerts	
	A.7.1 Introduction	125
	A.7.2 Open source code versus black-box software	127
	A.7.3 Main functionalities	128
	A.7.4 Affordability and user-friendliness	131
	A.7.5 Concluding remarks	132
	References	133
A.8	GeoSurveillance: GIS-based Exploratory Spatial Analysis Tools for Monitoring Spatial Patterns and Clusters	
	Gyoungju Lee, Ikuho Yamada and Peter Rogerson	
	A.8.1 Introduction	135
	A.8.2 Structure of GeoSurveillance	137

	A.8.3	Methodological overview	*138*
	A.8.4	Illustration of GeoSurveillance	*142*
	A.8.5	Concluding remarks	*148*
	References		*149*

A.9 Web-based Analytical Tools for the Exploration of Spatial Data
Luc Anselin, Yong Wook Kim and Ibnu Syabri

	A.9.1	Introduction	*151*
	A.9.2	Methods	*152*
	A.9.3	Architecture	*158*
	A.9.4	Illustrations	*163*
	A.9.5	Concluding remarks	*170*
	References		*171*

A.10 PySAL: A Python Library of Spatial Analytical Methods
Sergio J. Rey and Luc Anselin

	A.10.1	Introduction	*175*
	A.10.2	Design and components	*177*
	A.10.3	Empirical illustrations	*180*
	A.10.4	Concluding remarks	*191*
	References		*191*

PART B   Spatial Statistics and Geostatistics

B.1 The Nature of Georeferenced Data
Robert P. Haining

	B.1.1	Introduction	*197*
	B.1.2	From geographical reality to the spatial data matrix	*199*
	B.1.3	Properties of spatial data in the spatial data matrix	*204*
	B.1.4	Implications of spatial data properties for data analysis	*208*
	B.1.5	Concluding remarks	*214*
	References		*214*

B.2 Exploratory Spatial Data Analysis
Roger S. Bivand

	B.2.1	Introduction	*219*
	B.2.2	Plotting and exploratory data analysis	*220*
	B.2.3	Geovisualization	*224*
	B.2.4	Exploring point patterns and geostatistics	*229*
	B.2.5	Exploring areal data	*236*
	B.2.6	Concluding remarks	*249*
	References		*250*

B.3 Spatial Autocorrelation
Arthur Getis

B.3.1	Introduction	*255*
B.3.2	Attributes and uses of the concept of spatial autocorrelation	*257*
B.3.3	Representation of spatial autocorrelation	*259*
B.3.4	Spatial autocorrelation measures and tests	*262*
B.3.5	Problems in dealing with spatial autocorrelation	*272*
B.3.6	Spatial autocorrelation software	*274*
References		*275*

B.4 Spatial Clustering
Jared Aldstadt

B.4.1	Introduction	*279*
B.4.2	Global measures of spatial clustering	*280*
B.4.3	Local measures of spatial clustering	*289*
B.4.4	Concluding remarks	*297*
References		*298*

B.5 Spatial Filtering
Daniel A. Griffith

B.5.1	Introduction	*301*
B.5.2	Types of spatial filtering	*303*
B.5.3	Eigenfunction spatial filtering and generalized linear models	*312*
B.5.4	Eigenfunction spatial filtering and geographically weighted regression	*313*
B.5.5	Eigenfunction spatial filtering and geographical interpolation	*315*
B.5.6	Eigenfunction spatial filtering and spatial interaction data	*316*
B.5.7	Concluding remarks	*317*
References		*317*

B.6 The Variogram and Kriging
Margaret A. Oliver

B.6.1	Introduction	*319*
B.6.2	The theory of geostatistics	*319*
B.6.3	Estimating the variogram	*321*
B.6.4	Modeling the variogram	*327*
B.6.5	Case study: The variogram	*331*
B.6.6	Geostatistical prediction: Kriging	*337*
B.6.7	Case study: Kriging	*344*
References		*350*

Part C Spatial Econometrics

C.1 Spatial Econometric Models
   James P. LeSage and R. Kelley Pace

C.1.1	Introduction	355
C.1.2	Estimation of spatial lag models	360
C.1.3	Estimates of parameter dispersion and inference	365
C.1.4	Interpreting parameter estimates	366
C.1.5	Concluding remarks	374
	References	374

C.2 Spatial Panel Data Models
   J. Paul Elhorst

C.2.1	Introduction	377
C.2.2	Standard models for spatial panels	378
C.2.3	Estimation of panel data models	382
C.2.4	Estimation of spatial panel data models	389
C.2.5	Model comparison and prediction	399
C.2.6	Concluding remarks	403
	References	405

C.3 Spatial Econometric Methods for Modeling Origin-Destination Flows
   James P. LeSage and Manfred M. Fischer

C.3.1	Introduction	409
C.3.2	The analytical framework	410
C.3.3	Problems that plague empirical use of conventional spatial interaction models	416
C.3.4	Concluding remarks	431
	References	432

C.4 Spatial Econometric Model Averaging
   Olivier Parent and James P. LeSage

C.4.1	Introduction	435
C.4.2	The theory of model averaging	436
C.4.3	The theory applied to spatial regression models	440
C.4.4	Model averaging for spatial regression models	444
C.4.5	Applied illustrations	450
C.4.6	Concluding remarks	458
	References	459

C.5 Geographically Weighted Regression
   David C. Wheeler and Antonio Páez

C.5.1	Introduction	461
C.5.2	Estimation	462
C.5.3	Issues	467

	C.5.4	Diagnostic tools	469
	C.5.5	Extensions	472
	C.5.6	Bayesian hierarchical models as an alternative to GWR	474
	C.5.7	Bladder cancer mortality example	477
	References		484
C.6	Expansion Method, Dependency, and Multimodeling *Emilio Casetti*		
	C.6.1	Introduction	487
	C.6.2	Expansion method	488
	C.6.3	Dependency	493
	C.6.4	Multimodeling	496
	C.6.5	Concluding remarks	501
	References		502
C.7	Multilevel Modeling *S.V. Subramanian*		
	C.7.1	Introduction	507
	C.7.2	Multilevel framework: A necessity for understanding ecological effects	509
	C.7.3	A typology of multilevel data structures	510
	C.7.4	The distinction between levels and variables	511
	C.7.5	Multilevel analysis	512
	C.7.6	Multilevel statistical models	513
	C.7.7	Exploiting the flexibility of multilevel models to incorporating 'realistic' complexity	521
	C.7.8	Concluding remarks	523
	References		524

Part D The Analysis of Remotely Sensed Data

D.1	ARTMAP Neural Network Multisensor Fusion Model for Multiscale Land Cover Characterization *Sucharita Gopal, Curtis E. Woodcock and Weiguo Liu*		
	D.1.1	Background: Multiscale characterization of land cover	529
	D.1.2	Approaches for multiscale land cover characterization	530
	D.1.3	Research methodology and data	532
	D.1.4	Results and analysis	534
	D.1.5	Concluding remarks	540
	References		541
D.2	Model Selection in Markov Random Fields for High Spatial Resolution Hyperspectral Data *Francesco Lagona*		
	D.2.1	Introduction	545

	D.2.2	Restoration, segmentation and classification of HSRH images	549
	D.2.3	Adjacency selection in Markov random fields	550
	D.2.4	A study of adjacency selection from hyperspectral data	554
	D.2.5	Concluding remarks	560
	References		561

D.3 Geographic Object-based Image Change Analysis
Douglas Stow

	D.3.1	Introduction	565
	D.3.2	Purpose of GEOBICA	566
	D.3.3	Imagery and pre-processing requirements	568
	D.3.4	GEOBIA principles	569
	D.3.5	GEOBICA approaches	571
	D.3.6	GEOBICA strategies	572
	D.3.7	Post-processing	575
	D.3.8	Accuracy assessment	576
	D.3.9	Concluding remarks	578
	References		579

Part E Applications in Economic Sciences

E.1 The Impact of Human Capital on Regional Labor Productivity in Europe
Manfred M. Fischer, Monika Bartkowska, Aleksandra Riedl, Sascha Sardadvar and Andrea Kunnert

	E.1.1	Introduction	585
	E.1.2	Framework and methodology	586
	E.1.3	Application of the methodology	592
	E.1.4	Concluding remarks	595
	References		596

E.2 Income Distribution Dynamics and Cross-Region Convergence in Europe
Manfred M. Fischer and Peter Stumpner

	E.2.1	Introduction	599
	E.2.2	The empirical framework	601
	E.2.3	Revealing empirics	608
	E.2.4	Concluding remarks	622
	References		623
	Appendix		626

E.3 A Multi-Equation Spatial Econometric Model, with Application to EU Manufacturing Productivity Growth
Bernard Fingleton

E.3.1	Introduction	629
E.3.2	Theory	630
E.3.3	Incorporating technical progress variations	632
E.3.4	The econometric model	637
E.3.5	Model restriction	639
E.3.6	The final model	642
E.3.7	Concluding remarks	644
References		645
Appendix		647

Part F Applications in Environmental Sciences

F.1 A Fuzzy *k*-Means Classification and a Bayesian Approach for Spatial Prediction of Landslide Hazard
Pece V. Gorsevski, Paul E. Gessler and Piotr Jankowski

F.1.1	Introduction	653
F.1.2	Overview of current prediction methods	655
F.1.3	Modeling theory	658
F.1.4	Application of the modeling approach	666
F.1.5	Concluding remarks	679
References		680

F.2 Incorporating Spatial Autocorrelation in Species Distribution Models
Jennifer A. Miller and Janet Franklin

F.2.1	Introduction	685
F.2.2	Data and methods	687
F.2.3	Results	691
F.2.4	Concluding remarks	697
References		699

F.3 A Web-based Environmental Decision Support System for Environmental Planning and Watershed Management
Ramanathan Sugumaran, James C. Meyer and Jim Davis

F.3.1	Introduction	703
F.3.2	Study area	704
F.3.3	Design and implementation of WEDSS	705
F.3.4	The WEDSS in action	712
F.3.5	Concluding remarks	715
References		716

Part G Applications in Health Sciences

G.1 Spatio-Temporal Patterns of Viral Meningitis in Michigan, 1993-2001
Sharon K. Greene, Mark A. Schmidt, Mary Grace Stobierski and Mark L. Wilson

G.1.1	Introduction	721
G.1.2	Materials and methods	723
G.1.3	Results	725
G.1.4	Concluding remarks	730
References		734

G.2 Space-Time Visualization and Analysis in the Cancer Atlas Viewer
Dunrie A. Greiling, Geoffrey M. Jacquez, Andrew M. Kaufmann and Robert G. Rommel

G.2.1	Introduction	737
G.2.2	Data and methods	739
G.2.3	Results	742
G.2.4	Concluding remarks	750
References		751

G.3 Exposure Assessment in Environmental Epidemiology
Jaymie R. Meliker, Melissa J. Slotnick, Gillian A. AvRuskin, Andrew M. Kaufmann, Geoffrey M. Jacquez and Jerome O. Nriagu

G.3.1	Introduction	753
G.3.2	Data and methods	755
G.3.3	Features and architecture of Time-GIS	757
G.3.4	Application	759
G.3.5	Concluding remarks	765
References		766

List of Figures	769
List of Tables	779
Subject Index	785
Author Index	793
Contributing Authors	805

Introduction

Manfred M. Fischer and *Arthur Getis*

1 Prologue

The fact that the 2008 Nobel Prize in Economics was awarded to Paul Krugman indicates the increasing attention being given to spatially related phenomena and processes. Given the growing number of academics currently doing research on spatially related subjects, and the large number of questions being asked about spatial processes, the time has come for some sort of summary statement, such as this *Handbook*, to identify the status of the methods and techniques being used to study spatial data. This Handbook brings together contributions from the most accomplished researchers in the area of spatial analysis. Each was asked to describe and explain in one chapter the nature of the types of analysis in which they are expert. Clearly, having only one chapter to explain, for example, exploratory spatial data analysis or spatial econometric models, is a daunting task, but the authors of this book were able to summarize the key notions of their spatial analytic fields and point readers in directions that will help them to better understand their data and the techniques available to them.

Whether or not spatial analysis is a separate academic field, the fact remains that in the last twenty years spatial analysis has become an important by-product of the interest in and the need to understand georeferenced data. The current interest in environmental sciences is a particular stimulant to the development of new and better ways of analyzing spatial data. Environmental studies have become either an important subfield of or a major thrust in such fields as ecology, geology, atmospheric sciences, sociology, political science, economics, urban planning, epidemiology, and the field that sometimes characterizes itself as the archetype environmental science, geography. There is no shortage of articles in the applied journals of these fields where the analysis of spatial data is central.

Many researchers are busy developing techniques for the study of georeferenced data, and many more use spatial analytic tools. Following the adage 'Necessity is the mother of invention,' very often the developers are also the users. Thus, we see that in the fields mentioned above, new and tantalizingly imaginative techniques have been created for analytic purposes. Most often, but not exclusively, however, the fundamental principles for spatial analysis come from mathematics, statistics, and econometrics. Applied spatial scientific studies require use of probability and statistics and, for model development, the techniques of the econometricians and geostatisticians.

Since the practical nature of spatial analysis is the driving force for the field's development, it was inevitable that creating software would be a major activity of spatial analysts. Unlike most compendia, where principles are laid out first, followed by applications and notes on software, the editors of this Handbook placed software tools first. Some of the very best innovative techniques for spatial analysis come from the wide variety of software packages discussed in Part A. As the reader will deduce from perusing the table of contents, we approach spatial analysis as a series of surveys of what is available for the practical user. We want it to be possible for anyone new to the field to find relevant ideas and techniques for his/her research. In addition, our goal is to have seasoned researchers find new ideas or key references from unfamiliar spatial analytic fields. The fact that not everything available for spatial analysis is included in the discussions that follow has more to do with the background, research interests, and points of view of the editors than it does with space limitations.

Not unusual to academic research is the disciplinary boundaries surrounding some of the types of work being done in spatial analysis. For example, economists have a record of being reluctant to look at literature outside of their own field. It usually takes a strong societal interest in a given problem to encourage disciplinarians to consider, or become conversant, with other literatures. Although less true in a field such as spatial analysis, many are unwilling to get involved with names and ideas outside of their immediate research area. Fortunately, spatial analysis is the type of field that tends to break down those barriers. Especially with the development of GISystem software, user friendly software packages, National Institute of Health and National Science Foundation summer institutes in the US, interdisciplinary conferences and meetings, and internet activity, spatial analysis is taking on an ecumenical flavor. The difficulty that remains is the need for researchers to become familiar with the language of spatial analysis, the spatial point of view, and the techniques of those working on similar problems, but in other fields. We hope that this Handbook enhances the interdisciplinary nature of this field.

The history of spatial analysis is noteworthy for its genesis in a number of different fields nearly simultaneously. Much of the development has been based on the types of data characteristic of the particular research being done in the respective fields. For example, geologists and climatologists tend to study continuous data. Economists and political scientists pay a great deal of attention to time series data. Geographers, anthropologists, and sociologists are especially fond of point and area (choropleth) data. Transportation planners favor network data. Many environmentalists use remotely sensed spatial data. The data-driven emphasis of spatial analysis helped to create specialized 'schools of thought' on spatial analysis methodologies. Our view is that in recent years these schools are being opened to include ideas and methods from other schools. We believe, too, that in the future the field of spatial analysis will become less discipline oriented as the need for interdisciplinary research teams becomes a greater part of the research landscape. For example, no longer is it possible for a microbiologist or an epidemiologist alone to solve problems of disease transmission. Researchers well versed in the nuances of continuous or discrete spatial data must become members of the team. Moreover, the epidemiologist must be conversant with the techniques of analysis used to solve a disease transmission problem.

In the following section we briefly outline what may be called the points of view of the various schools of thought. Our goal, of course, is to have readers better understand how others approach spatial data. In this Handbook, these areas of interest are described, explained, and demonstrated.

2 Schools of thought on spatial analysis methodologies

Exploratory spatial data analysis (ESDA) is the extension of a Tukey-type data exploration (see Tukey 1977) to georeferenced data. ESDA represents a preliminary process where data and research results are viewed from many different vantage points, one of which is the display of data on maps. The power of computers to summarize and visualize large sets of georeferenced data has helped to stimulate the creation of amazingly evocative procedures for data manipulation. Science has always emphasized the need for high quality data and for researchers to have an informed sense of what problems may be in the offing once data are subjected to rigorous study. Computer programmers in a number of different fields have now made it possible to view data in a myriad of ways.

Of particular interest is GI software that allows for the mapping of data, making measurements on the mapped data, identifying weaknesses in the data, correcting incorrect data or data placed in incorrect locations, producing summary measures of data, manipulating point data into surfaces, viewing these surfaces from many different angles, and, if the data are time related, viewing data changes over time. The summary measures are the usual histograms and box plots, but the ability of the programs to, for example, identify a data outlier on a map at the same time as one views the location of the outlier in a histogram, in a cumulative distribution function, and in a three dimensional scatter diagram that can be viewed from any angle, gives ESDA a powerful role to play in much research.

Our view is that much of ESDA is used prior to the model building phase of research, but interestingly enough, some new techniques of ESDA act as model builders by allowing us to see how variables relate to one another in space. The field of data visualization, especially as related to maps, is just beginning to make an impact on research. There is a need to more closely unite those working on new techniques for data visualization with the actual needs of the various spatially-oriented fields of study.

The software discussed in Part A of the Handbook gives researchers an idea of the many tools and functions available for them to engage in ESDA. At one time it was anathema for many 'purists' to engage in exploratory work when preparing their data for analysis. The goal was to statistically test a model that was a direct descendant of well-documented theory. Now, awareness of all that is available in the software stimulates us to create final models only after performing a good deal of exploration and experimentation. In a sense, ESDA and EDA represent a new wave of research methodology. The traditional six steps of hypothesis guided inquiry – problem, hypothesis, sampling distribution, test, results, decision – has been expanded to a seventh step, data exploration, but instead of squeezing data exploration between two of the former steps, exploration is now represented at nearly all stages of analysis.

Spatial Statistics. The roots of spatial statistics go back to Pearson and Fisher, but their modern manifestation is mainly due to Whittle, Moran, and Geary. The field is indebted to Cliff and Ord for explicating, extending, and making their work socially relevant. From Cliff and Ord's papers and books of the late 1960s to the early 1980s comes the basic outline of what constitutes spatial statistics (see, for example, Cliff and Ord 1973, 1981). It probably is a stretch to call this area a school of thought, but the vast number of researchers who look to spatial autocorrelation statistics, for example, indicates a strong interest area. The point is that spatial statistics is also a part of ESDA, spatial econometrics, and remote sensing analysis, and to a lesser extent, geostatistics. One might ask the question, how can we model spatially varying phenomena without testing patterns on maps? The process of creating hypotheses and testing map patterns gives spatial statistics its raison d'être. Because of space limitations, this Handbook cannot cover in any detail all of the types of issues that spatial statistics practitioners address.

As a field, spatial statistics is concerned with map-related problems. Geometrically, one can think of point, line, and area patterns as well as mixtures of these three as the fundamental elements that are included in the use and study of spatial statistics. What is crucial, of course, is that these points, lines, and areas represent real world phenomena. How these phenomena pattern themselves and interact with one another has come to be an important element of scientific inquiry.

This Handbook reviews the fundamental knowledge required of the user of spatial statistics. Users are found in all of the social and environmental sciences and, to a lesser extent, the physical sciences. Hypotheses include conjectures about the mapped patterns of diseases and crime, the pattern of residuals from regression, the tendency for some phenomena to cluster or disperse, the differences among patterns, the spatial relationship between a given observation and other designated observations, and perhaps most important, how defined points, lines, and areas, interact with one another, either statically or over time and space.

Since the field's inception, certain particular problems have given rise to new statistical tests and routines. For example, the large data sets that began to emerge in the 1980s required researchers to find ways to reduce data redundancy or to subdivide regions into smaller units for statistical analysis. Eventually, the focused spatial tests developed and popularized in the 1990s became widely used, especially because spatial cluster analyses have come to depend on them. The ability of computers to create interaction data between all members of a population or sample has given rise to large sample statistics like the K function of Ripley (see Ripley 1977). The fundamental patterns of Voronoi polygons have now been studied using algorithms capable of manipulating tessellations of area patterns. The same is true of networks of lines.

Two of the most promising areas of spatial statistical analysis are the creation of defensible spatial weights matrices and the employment of spatial filters, discussed in the chapters of Section B of this Handbook. These new techniques are designed to facilitate understanding of what may be called the nature of spatial effects in any spatial system of variables. In addition, work is proceeding on ways to better test hypotheses concerning pattern representation. These include such tests as false discovery rates and simulation routines that create sampling distributions on which tests can be carried out.

Spatial Econometrics. Since Jean Paelinck and Leo Klaassen's description of the field in 1979 and Luc Anselin's influential volume *Spatial Econometrics,* published in 1988, spatial econometrics has blossomed. Before those auspicious events, economists with a spatial bent, such as Walter Isard (see Isard 1960), had begun to study the spatial manifestation of economic activities. The models that Anselin classified as spatial lag models and spatial error models (among several others), while related to the well-established field of econometrics, have become the fundamental regression tools of the spatial econometrician.

Although not deeply ingrained into the thinking characteristic of the discipline of economics, the discipline of regional science has become the home for spatial econometrics practitioners. Judging from the number of researchers who are in daily contact with Anselin's website, this field is growing very rapidly. Today, such researchers originally educated in economics and/or geography, such as LeSage, Pace, Kelejian, Florax, and Rey, are expanding the field to Bayesian thinking, new spatial regression estimating techniques and tests, and time-space modeling.

An interesting and crucial overlap between spatial statistics and spatial econometrics is the need to apply spatial statistical tests in order to check for the validity of the assumption of spatial randomness among the residuals of spatial, and nonspatial diagnostic, models. Commonly the well known Moran's I statistic is used for testing purposes. In Anselin's GeoDa software and in LeSage's spatial econometrics toolbox, Moran's I and newly developed tests are prominent parts of the software capabilities.

A new and useful system of econometric study, described in this Handbook, is geographically weighted regression (GWR). The realization that the constant nature of regression coefficients seems to fly in the face of reality when a geographic system is being modeled, stimulated Fotheringham, Brunsdon, and Charlton to create a spatial econometric system that allows regression parameters to vary over space (see Fotheringham et al. 2002). The developers of GWR are continually improving the system to avoid some of the difficulties in dealing with georeferenced data. Related to, but in addition to GWR, are expositions in this Handbook on the expansion method and the new techniques of spatial hierarchical models.

Geostatistics. Evolving differently than the previous schools of thought is the field of geostatistics, which is outlined in this Handbook. Primarily as a way to describe and explain physical phenomena in a continuous spatial data environment, geostatistics is the principal methodology of analysis. From its roots in the

1950s as a way to predict gold ore quality to its current widespread use for the study of all manner of physical phenomena, including petroleum reserve locations, soil quality, and patterns of weather and climate, geostatistics has become a mainstay of most earth science departments both in the academy and in the business world.

The field includes both spatial data descriptive routines and sophisticated modeling. The major themes are the study of variograms and the use of predictive devices called kriging, named after the mining engineer, Krige (1951), who pioneered the techniques. Matheron (1963), and most recently Cressie (1993), have laid out the statistical principles on which the methodology is based.

Variogram analysis is based on the principle of intrinsic stationarity, that is, inherent in the nature of spatial effects is that as distance increases between observations on the same variable, variance will increase. The increasing variance continues with increasing distance until a particular distance is reached when the variance will equal the population variance. The semivariogram is a function represented in a diagram that shows the nature of this increasing function. Considered to be theoretical, the function is most often estimated from real world data. The large amount of software available for the study of geostatistics is one of the field's features. Some GISystem modules include many exploratory features as well as capabilities for sophisticated modeling.

The second area of study, mentioned above – kriging – is a series of techniques that allows for the prediction of variable values or multi-variable interactions at locations where no data are available. Thus, via the simultaneous equation systems of kriging, point data can be used to create surfaces where each location in the study area is represented by a point estimate of the true value at that point. Kriging creates map surfaces and error surfaces, that is, surfaces that represent the confidence level in spatial point estimates. The manner in which kriging is carried out ranges from relatively simple procedures (simple and ordinary kriging) to complex prediction systems (co-kriging and disjunctive kriging). Given the enormous number of calculations that must be performed, the techniques require large samples and high levels of computer power.

3 Structure of the handbook

This volume is not intended as a textbook or research monograph, nor does it attempt to cover the field of spatial analysis exhaustively, or in great depth. It does attempt, though, to provide a useful manual or guidebook to spatial analytic fields, and to offer a wide range of views on spatial analysis that may lead the reader to inquire more deeply into specific areas that are touched on herein. It is intended that this Handbook should be as accessible as possible, especially to those who are relatively unfamiliar with this area of work.

The material in this volume has been chosen to provide an accounting of the diversity of current and emergent models, methods, and techniques, not available elsewhere despite the many excellent journals and text books that exist. The international collection of authors was selected for their knowledge of a subject area, and their ability to communicate basic information in their subject area succinctly and accessibly.

The volume is structured as a series of parts ranging from software tools over spatial statistical and geostatistical approaches to spatial econometric models and techniques, and finally to applications in various domain areas. The parts are as follows:

- *Part A*: GI software tools,
- *Part B*: Spatial statistics and geostatistics,
- *Part C*: Spatial econometrics,
- *Part D*: The analysis of remotely sensed data,
- *Part E*: Applications in economic sciences,
- *Part F*: Applications in environmental sciences, and
- *Part G*: Applications in health sciences.

where the chapters in *Part D* to *Part G* represent in many ways an application of models, methods, and techniques discussed in the preceding chapters.

Part A: GI software tools

The focus of *Part A* is on GI software packages, from which some of the very best innovative techniques for spatial analysis come. This part is composed of ten contributions, viz:

- Spatial statistics in ArcGIS (Chapter A.1),
- Spatial statistics in SAS (Chapter A.2),
- Spatial econometric functions in R (Chapter A.3),
- GeoDa: An introduction to spatial data analysis (Chapter A.4),
- STARS: Space-time analysis of regional systems (Chapter A.5),
- Space-time intelligence system software for the analysis of complex systems (Chapter A.6),
- Geostatistical software (Chapter A.7),
- GeoSurveillance: A GIS-based exploratory spatial analysis tool for monitoring spatial patterns and clusters (Chapter A.8),
- Web-based analytical tools for the exploration of spatial data (Chapter A.9), and
- PySAL: A Python library of spatial analytical methods (Chapter A.10).

The first chapter, written by *Lauren M. Scott* and *Mark V. Janikas,* provides an overview of the tools in the ArcGIS spatial statistics toolbox, an extendible set of feature pattern analysis and regression analysis tools, specifically designed to work with spatial data. There are four core analytical toolsets: measuring geographic distributions, analysing patterns, mapping clusters, and modeling spatial relationships. The chapter not only provides an overview of the tools, but presents also application examples and references, and outlines strategies for extending ArcGIS functionality through custom tool development.

The next chapter, by *Melissa I. Rura* and *Daniel A. Griffith*, describes ways SAS has been used in the past for spatial statistical analyses. It covers recent work that explicitly includes spatial information and geographic visualization, and gives two SAS implementation examples, namely the calculation of Moran's I and the eigenvector spatial filtering spatial statistical technique. First, SAS's embedded spatial functionality is discussed in terms of function options and procedures like PROC VARIOGRAM and PROC MIXED. Next, SAS's GISystem module functionality, including map display and data import, is described. Then PROC NLIN-based spatial autoregressive code capabilities are discussed. Finally, two example implementations and their necessary input and output data are described. An example calculation of Moran's I is presented, and an implementation of eigenvector spatial filtering is described, in order to illustrate how customized SAS can be created to put spatial statistical techniques into practice. Several sources are summarized from which a user may download or look up freely available spatial statistical SAS implementations. This chapter seeks to show how the use of a mature statistical programming language like SAS can enable advanced spatial analysis.

Placing spatial econometrics and more generally spatial statistics in the context of an extensible data analysis environment such as R exposes similarities and differences between traditions of analysis. This can be fruitful, and is explored in Chapter A.3, written by *Roger S. Bivand*, in relation to prediction and other methods usually applied to fitted models in R. R is a language and environment for statistical computing and graphics, available as Free Software under the terms of the Free Software Foundation's GNU General Public License in source code form. It compiles and runs on a wide variety of UNIX platforms and similar operating systems (including Linux), Windows, and MacOS. Objects in R may be assigned a class attribute, including fitted model objects. Such fitted model objects may be provided with methods allowing them to be displayed, compared, and used for prediction, and it is of interest to see whether fitted spatial models can be treated in the same way.

Chapter A.4, by *Luc Anselin*, *Ibru Syabri*, and *Younghin Kho*, presents an overview of GeoDa^{TM}, a free software program intended to serve as a user-friendly and graphical introduction to spatial analysis for non-GIS specialists. It includes functionality ranging from simple mapping to exploratory data analysis, the visualization of global and local spatial autocorrelation, and spatial regression. A key feature of GeoDa is an interactive environment that combines maps with statistical graphics, using the technology of dynamically linked windows. A brief review of the software design is given, as well as some illustrative examples that highlight distinctive features of the program in applications dealing with public health, economic development, real estate analysis and criminology.

Space-Time Analysis of Regional Systems (STARS) is an open source software package designed for the dynamic exploratory analysis of data measured for areal units at multiple points in time. STARS consists of four core analytical modules: exploratory spatial data analysis; inequality measures; mobility metrics; spatial Markov. Developed using the Python object oriented scripting language, STARS lends itself to three main modes of use. Within the context of a command line interface (CLI), STARS can be treated as a package which can be called from within customized scripts for batch oriented analyses and simulation. Alternatively, a graphical user interface (GUI) integrates most of the analytical modules with a series of dynamic graphical views containing brushing and linking functionality to support the interactive exploration of the spatial, temporal and distributional dimensions of socioeconomic and physical processes. Finally, the GUI and CLI modes can be combined for use from the Python shell to facilitate interactive programming and access to the many libraries contained within Python. Chapter A.5, by *Serge J. Rey* and *Mark V. Janikas*, provides an overview of the design of STARS, its implementation, functionality and future plans. A selection of its analytical capabilities is also illustrated that highlight the power and flexibility of the package.

The development and implementation of software tools that account for both spatial and temporal dimensions, and that provide advanced visualization and space-time analysis capabilities is recognized as an important technological challenge in Geographic Information Science. Chapter A.6, written by *Geoffrey M. Jacquez*, provides an overview of space-time intelligence system (STIS) software that has been developed by BioMedware with funding from the National Institutes of Health. STIS is founded on space-time data structures for representing points, geospatial lifelines, polygons and rasters, and how they morph through time. Linked windows, cartographic and statistical brushing are time-enabled, as are visualizations including tables, maps, principal coordinate plots, histograms, scatterplots, variogram clouds, and box plots. Spatial weight relationships that change through time for points, geospatial lifelines and polygons include nearest neighbors, inverse distance, geographic distance, and adjacencies. These are used by advanced space-time analysis methods including clustering, regression (linear, logistic, Poisson, and step-wise), geographically-weighted regression, variogram models, kriging, and disparity statistics, among others. STIS allows researchers to span the analytical continuum for space-time data on one software platform, from visualization, animation, exploratory space-time data analysis, through hypothesis testing and modeling.

During the last two decades one has witnessed an increasing interest in the application of geostatistics to the analysis of space-time datasets. A critical issue for many novice users is the availability of affordable and user-friendly software that offer basic (for example, variogram estimation and modeling, kriging) and advanced (for example, non-parametric kriging, stochastic simulation) algorithms for geostatistical modeling. The chapter, by *Pierre Goovaerts*, presents a brief overview of the main geostatistical software, stressing their advantages and weaknesses in terms of flexibility and completeness. Concomitant with the growing range of geostatistical applications, the software market is expanding and nowadays fairly general software or add-on modules that are open source but have limited graphical capabilities coexist with highly visual commercial software that are often tailored to specific applications, such as 2D health data or 3D assessment of contaminated sites. In particular, when geostatistics is combined with classical statistical techniques, such as regression analysis for trend modeling, the user often will have to rely on several programs to accomplish the different steps of the analysis.

Chapter A.8, written by *Gyoungju Lee*, *Ikuho Yamada*, and *Peter Rogerson*, describes GeoSurveillance, a GIS-based exploratory spatial analysis tool for monitoring spatial patterns and clusters over time. During the past decade, significant methodological advances have been made in assessing geographic clustering and in searching for local spatial clusters based on diverse statistical models. Recently, prospective surveillance models have been proposed to detect spatial pattern changes over time quickly, in contrast with traditional retrospective tests. As frequent updates of spatial databases are now made possible on a regular basis with the rapid development of GISystems, the development of prospective methods for monitoring emerging spatial clusters of geographic events (for example, disease outbreak) has been facilitated. GeoSurveillance provides a statistical framework integrated with a GISystem platform, where both retrospective and prospective tests for spatial clustering can be carried out effectively. To demonstrate the program, illustrations are given for Sudden Infant Death Syndrome (SIDS) in North Carolina and breast cancer cases in the northeastern part of the US.

In the next chapter, *Luc Anselin*, *Yong Wook Kim,* and *Ibru Syabri* deal with the extension of internet-based geographic information systems with functionality for exploratory spatial data analysis. The specific focus is on methods to identify and visualize outliers in maps for rates or proportions. Three sets of methods are included: extreme value maps, smoothed rate maps and the Moran scatterplot. The implementation is carried out by means of a collection of Java classes to extend the Geotools open source mapping software toolkit. The web based spatial analysis tools are illustrated with applications to the study of homicide rates and cancer rates in US counties.

PySAL is an open source library for spatial analysis written in the object oriented language Python. It is built upon shared functionality in two exploratory spatial data analysis packages: GeoDA and STARS and is intended to leverage the shared development of these components. This final chapter of *Part A*, written by *Serge J. Rey* and *Luc Anselin*, presents an overview of the motivation behind and the design of PySAL, as well as suggestions for how the library can be used with other software projects. Empirical illustrations of several key components in a variety of spatial analytical problems are given, and plans for future development of PySAL are discussed.

Part B: Spatial statistics and geostatistics

This part of the Handbook shifts attention to spatial statistical and geostatistical approaches, methods and techniques, and includes the following chapters:

- the nature of georeferenced data (the Chapter B.1),
- exploratory spatial data analysis (Chapter B.2),
- spatial autocorrelation (Chapter B.3),
- spatial clustering (Chapter B.4),
- spatial filtering (Chapter B.5), and
- the variogram and kriging (Chapter B.6).

In the first chapter of *Part B, Robert Haining* identifies various types of georeferenced data but focuses attention on the spatial data matrix. He considers the relationship between it and the complex, continuous geographic reality from which it is obtained and the difficulties that need to be addressed in constructing a spatial data set for the purpose of undertaking practical spatial data analysis. The links between each of the stages involved in the construction of the data matrix and the properties of spatial data are described. The author continues to discuss the implications of these findings for the conduct of exploratory and confirmatory data analysis and for the interpretation of results. The chapter concludes by discussing the role of models in influencing the types of georeferenced data that are needed and the consequences for model inference.

The focus of Chapter B.2, written by *Roger S. Bivand*, is on exploratory spatial data analysis, an extension of exploratory data analysis geared especially to dealing with the spatial aspects of data. This chapter presents the underlying intentions of ESDA, and surveys some of the outcomes. It challenges the frequently drawn conclusion that ESDA can somehow replace proper modeling. Exploratory spatial data analysis remains a key step prior to the model building phase of research, but interestingly enough, some new techniques of ESDA act as model builders by allowing us to see how variables relate to one another in space.

A fundamental concept for the study of spatial phenomena is spatial autocorrelation. The concept has played a pivotal role in the development of the field of spatial analysis.

Chapter B.3, written by *Arthur Getis*, reviews the literature on spatial autocorrelation and explains its various representations. Most definitions of the concept concern the spatial relationships among realizations of a random variable. The uses of spatial autocorrelation are many, including its major role in testing for model mis-specification and for testing hypotheses concerned with spatial relationships. The cross product statistic, a fundamental spatial autocorrelation structure, is used to record the geometrical relationships and the variable relationships among the spatial units under study and to assess the degree of similarity between the two relationships. The spatial weights matrix represents the geometric relationships. Each matrix element records the spatial association among the spatial units under study. Many tests and indicators of spatial autocorrelation are available. Chief among these is Cliff and Ord's extension of Moran's spatial autocorrelation statistic. At the local scale, Getis and Ord's statistics and Anselin's LISA statistics enable researchers to evaluate spatial autocorrelation at particular sites. Also at the local level, geographically weighted regression is an entire system devoted to the study of stationarity in spatial relationships among variables by location. Many software packages are available for the study of various aspects of spatial autocorrelation, including exploratory, global, local, time-space, and spatial econometric.

Chapter B.4, by *Jared Aldstadt*, reviews techniques for spatial clustering analysis. Emphasis is placed on the most commonly used techniques and their direct precursors. Some attention is given to recently developed clustering routines. These techniques may not yet be in wide use, but they are relevant because they overcome deficiencies in existing methodologies. They also indicate the direction of current research. Following the path of development, global clustering indices are covered first, followed by local clustering techniques. When applicable, test statistics are presented in the general cross-product form. In this format the similarities between and distinguishing characters of the clustering statistics are apparent.

Chapter B.5, written by *Daniel A. Griffith*, directs attention to spatial filtering, a spatial statistical methodology that enables spatial autocorrelation effects to be accounted for while preserving conventional statistical model specifications. A spatial filter is a synthetic variate that is constructed from locational information independent of the thematic nature of affiliated georeferenced data, being based upon the underlying geographic configuration of the data georeferencing. The primary idea is that some spatial proxy variables extracted from a spatial relationship matrix are added as control variables to a standard statistical model specification. To date, four principal approaches to spatial filtering have been implemented: autoregressive linear operators (*à la* Cochrane-Orcutt prewhitening), Getis's G_i-based specification, linear combinations of eigenvectors extracted from either distance-based principal coordinates of neighboring matrices, or topology-based spatial weights matrices. Not only does spatial filtering allow a more detailed analysis of spatial autocorrelation effects for geographic distributions of attribute variables, but it also supports sounder geographically varying coefficients analyses, spatial interpolation, and the analysis of spatial autocorrelation effects in geographic flows data. Spatial filtering can be employed with both the normal probability model, and the entire family of probability models affiliated with generalized linear models.

The final chapter of *Part B*, by *Margaret Oliver,* shifts focus to the variogram and kriging, the two central techniques of geostatistics. The variogram describes quantitatively how a property changes as the separation between places increases. Its values are estimated from data for a set of separating distances or lags to give the experimental variogram. This may then be modeled by a limited set of mathematical functions. Methods of estimating the variogram and the models that are fitted most frequently in the earth sciences are described and illustrated with a case study of soil data. The parameters of the models fitted to the variograms are used with the data to predict by employing kriging techniques. Kriging is a best linear unbiased predictor; it provides predictions and estimates of errors at each prediction point. Kriging is now a generic term that embraces several types of kriging that have been developed to solve particular problems in prediction. The emphasis in this chapter is on ordinary kriging, which is the type of kriging most often used. Factorial kriging is also described because of its value when the variation has more than one spatial scale..

Part C: Spatial econometrics

Part C is concerned with estimation and testing problems encountered when attempting to implement regional economic models. The problems often are characterized by the difficulties associated with assessing the importance of spatial dependence and spatial heterogeneity in a regression setting. Seven chapters represent the diversity of spatial econometric approaches, methods and techniques:

- spatial econometric models (Chapter C.1),
- spatial panel data models (Chapter C.2),
- spatial econometric methods for modeling origin-destination flows (Chapter C.3),
- spatial econometric model averaging (Chapter C.4),
- geographically weighted regression (Chapter C.5),
- expansion method, dependency, and multimodeling (Chapter C.6), and
- multilevel modeling (Chapter C.7).

The first chapter, written by *James P. LeSage* and *R. Kelley Pace*, provides an introduction to spatial econometric models and methods in a cross-sectional context. The authors show how conventional regression models can be augmented with spatial autoregressive processes to produce models that incorporate simultaneous feedback between regions located in space, and discuss methods estimating these models that are useful when modeling cross-sectional regional observations. The authors conclude the chapter in showing that for models containing spatial lags of the explanatory or dependent variables, interpretation of the parameters becomes richer and more complicated than in a least squares regression context with independent observations. Interpretation of parameter estimates and inferences requires an interpretation based on a steady-state equilibrium view, where changes in the explanatory variables lead to a series of simultaneous feedbacks that produce a new steady-state equilibrium. Because of working with cross-sectional sample data, these model adjustments appear as if they are simultaneous. The authors argue that these spatial regression models can be viewed as containing an implicit time dimension.

Chapter C.2, written by *J. Paul Elhorst*, focuses on the estimation of the spatial fixed effects model and the spatial random effects model extended to include spatial error autocorrelation or a spatially lagged dependent variable, including the determination of the variance-covariance matrix of the parameter estimates. In addition, it deals with robust LM tests for spatial interaction effects in standard panel data models, the estimation of fixed effects and the determination of their significance levels, a test for the fixed effects specification against the random effects specification using Hausman's specification test, the determination of goodness-of-fit measures, and the best linear unbiased predictor when using these models for prediction purposes. Finally, it briefly discusses possibilities for testing for endogeneity of one or more of the explanatory variables and to include dynamic effects.

Spatial interaction models of the gravity type are used in conjunction with sample data on flows between origin and destination locations to analyse international and interregional trade, commodity, migration, and commuting patterns. The focus of Chapter C.3, by *James P. LeSage* and *Manfred M. Fischer*, is on problems that plague empirical implementation of conventional regression-based spatial interaction models and econometric extensions that have appeared in the literature. The new models replace the conventional assumption of independence between origin-destination flows with formal approaches that allow for spatial dependence in flow magnitudes. Particular emphasis is laid on discussing problems, such as efficient computation, spatial dependence in origin-destination flows, large diagonal flows matrix elements, and the zero flows problem.

Model specification decisions represent a source of uncertainty typically ignored in applied modeling when we conduct statistical inference regarding model parameters. Chapter C.4, written by *Olivier Parent* and *James P. LeSage*, discusses formal methods that can be used to incorporate model specification uncertainty into inferences about model parameters. The focus is on how this can be accomplished in the context of spatial regression models, with an applied illustration involving the relation between local government expenditures and population migration.

Chapter C.5, by *David Wheeler* and *Antonio Páez*, deals with geographically weighted regression (GWR), a local form of spatial analysis drawing from statistical approaches for curve fitting and smoothing applications. GWR is based on the idea of estimating local models using subsets of observations centered on a focal calibration point. Since its introduction in 1996, GWR rapidly captured the attention of many in spatial analysis for its potential to investigate non-stationary relations in regression models. The basic concepts of GWR have also been used to obtain local descriptive statistics and other spatially weighted models, such as for Poisson regression. GWR has been instrumental in calling attention to the existence of potentially complex spatial relationships in linear regression. At the same time, there have been a number of issues raised concerning the nature and range of applications of the method, including its application for formal statistical inference on regression relationships. The available evidence suggests that GWR is an effective tool for spatial interpolation, but that it is problematic for inferring spatial processes. Collinearity has been shown to exacerbate inferential issues in GWR, but diagnostic tools have been developed to highlight local collinearity. In addition, other available approaches are discussed, such as hierarchical Bayesian regression models.

Chapter C.6, by *Emilio Casetti*, shows that the expansion method can provide an avenue for remedying residual spatial dependence, and, moreover, that within a multimodel frame of reference the expansion method can be used to identify the correlates and determinants of spatial dependence. The expansion method is a technique for widening the scope of a simpler initial model by expansion equations that redefine some or all of the initial model's parameters into functions of contextual variables. By replacing the parameters of the initial model with their expansions a terminal model is produced that encompasses both the initial model and a specification of its contextual variation. An initial model that upon estimation and testing displays significant residual spatial autocorrelation can be often expanded into terminal models that upon estimation and testing display no significant autocorrelation. Thus, the expansion method may provide an avenue to remedy the problem of spatial dependence. Omitted variables can produce autocorrelated residuals. The variables added to a terminal model by expansions obviously do not appear in its initial model. If upon estimation and testing, significant autocorrelation is found in the initial model's residuals but not in the terminal model's residuals, it follows that the variables generated by expansions are what makes the difference. These results can be used to investigate which properties and attributes of the models are associated with the occurrence of spatial dependence.

The final chapter of *Part C*, by *S.V. Subramanian*, continues to discuss the concept of multilevel statistical models as it relates to understanding place effects and more generally contextual effects. The chapter begins by identifying what constitutes a multilevel data analysis followed by a discussion on how a range of data structures that are observed in the real word or due to sampling design can be accommodated within a multilevel framework. After laying down the substantive motivation to utilize multilevel methods, some key statistical models are specified with a description of the property of each of the model. In particular, multilevel models are contrasted with fixed effect models. Finally, the chapter closes with a discussion of the substantive as well as the technical advantages of using a multilevel modeling approach to statistical analysis.

Part D: The analysis of remotely sensed data

Part D deals with the analysis of remotely sensed data. Remote sensing is the acquisition and analysis of data about an object or area acquired from a device that is not in contact with the object or area. Most of the remote sensor devices are placed in earth-observing satellites and both high and low flying aircraft. Much of the spatial analysis that is carried out on the data must take into account the usually very large number of observations, sometimes in the billions, and the size of the fundamental observations (the pixels). Increasingly, spatial statistics has become an integral part of the remote sensing experience. The main issues facing researchers are that results differ by spatial scale and that typical study regions (landscapes) vary appreciably, even over short distances. The type of data sensed is usually values on the electromagnetic spectrum condensed into pixels of a particularly scale. A principal task is to aggregate refined data or select a sensor that will capture data at a scale appropriate to the problem being solved. Spatial variation is often modeled by covariance, variograms or fractals. Surfaces are constructed using Fourier transforms of the covariance. Variograms are often used to model topography, vegetation indices, and soil properties. GISystems and data based management systems provide the computing capability for organizing and storing what usually are very large data sets. Analysis is dependent on visualization techniques designed to extract information from the massive data sets. Issues of spatial sampling, especially with regard to spatial scales are an ongoing research question. *Part D* of the Handbook is made up of three major constituent chapters, viz:

- ARTMAP neural network multisensor fusion model for multiscale land cover characterization (Chapter D.1),
- model selection in Markov random fields for high spatial resolution hyperspectral data (Chapter D.2), and
- geographic object-based image change analysis (Chapter D.3).

Land cover characterization is one of the primary objectives in using and analyzing geospatial information gathered by remote sensing. Land cover characterization is essential for terrestrial ecosystem modeling and monitoring, as well as climate modeling and prediction. To improve estimates of proportions or mixtures of land cover at a global scale, it is necessary to exploit information from multiple sensors and develop models that explicitly handle scale effects in data fusion. In Chapter D.1, *Sucharita Gopal, Curtis Woodcock,* and *Weiguo Liu* present a framework for multisensor fusion using an ARTMAP neural network to extract sub-pixel information from coarser resolution imagery. The framework is applied to the extraction of the proportion of forest cover using an image pair-TM (30M) and MODIS (one K) imagery for a region of North Central Turkey. The ARTMAP neural network multisensor fusion model is compared to a conventional linear mixture model and shows its superiority in terms of estimation of sub-pixel class proportion. This research suggests that nonlinear mixture models hold considerable promise for land cover mapping using information from multiple sensors.

Chapter D.2, written by *Francesco Lagona,* implements Markov random fields, implemented for the analysis of remote sensing images to capture the natural spatial dependence between band wavelengths taken at each pixel, through a suitable adjacency relationship between pixels, to be defined *a priori*. In most cases several adjacency definitions seem viable and a model selection problem arises. A BIC-penalized pseudo-likelihood criterion is suggested which combines good distributional properties and computational feasibility for analysis of high spatial resolution hyperspectral images. Its performance is compared with that of the BIC-penalized likelihood criterion for detecting spatial structures in a high spatial resolution hyperspectral image for the Lamar area in Yellowstone National Park.

The objective of Chapter D.3, by *Douglas A. Stow*, is to provide an overview of the use of multi-temporal remotely sensed image data to map earth surface changes from an object-based perspective. An initiation of research activity on GEOBICA techniques for detecting, identifying, and/or delineating earth surface changes has occurred over the past five or six years. Such techniques may be referred to as geographic object-based image change analysis or GEOBICA. GEOBICA is based on quantitative spatial analytical methods and generates data sets that can support spatial analysis of geographic areas. The chapter provides background and details on: (i) reasons and purposes for conducting GEOBICA, (ii) image acquisition and pre-processing requirements and types of image data that are input to GEOBICA routines, (iii) image segmentation and segment-based classification, (iv) approaches to multi-temporal image analysis, (v) GEOBICA strategies, (vi) post-processing techniques, and (vii) accuracy assessment for object-based and land cover change maps.

Part E: Applications in economic sciences

The focus of *Part E* is on applications in economic sciences in general and regional economics in particular. Three chapters have been chosen to demonstrate the range of spatial analytical applications in economic research:

- the impact of human capital on regional labor productivity in Europe (Chapter E.1),
- income distribution dynamics and cross-region convergence in Europe (Chapter E.2), and
- a multi-equation spatial econometric model, with application to EU manufacturing productivity growth (Chapter E.3).

The focus of Chapter E.1, by *Manfred M. Fischer* and associates, is on the role of human capital in explaining labor productivity variation among 198 European regions. Human capital is measured in terms of educational attainment using data for the active population aged 15 years and older that obtained the level of tertiary education. The existence of unobserved human capital that is excluded from the model but correlated with the included educational attainment variable and most likely exhibiting spatial dependence motivates the use of a spatial regression relationship that is known as spatial Durbin model. The chapter outlines the model along with the associated methodology for estimating the impact of human capital on regional labor productivity, based upon LeSage and Pace's approach to calculating scalar summary measures of direct and indirect impacts, described in detail in Chapter C.1. A simulation approach with 10,000 random draws is used to produce an empirical distribution of the model parameters that are needed for computing measures of dispersion for the impact estimates. The results obtained shed some interesting light on the role given to human capital in explaining labor productivity variation among European regions. Based on the estimate for the direct impact, we can conclude that a ten percent increase in human capital will on average result in a 1.3 percent increase in the final period level of labor productivity. But this positive direct impact is offset by a significant and negative indirect impact producing a total impact that is not significantly different from zero.

Chapter E.2, written by *Manfred M. Fischer* and *Peter Stumpner*, presents a continuous version of the model of distribution dynamics to analyze the transition dynamics and implied long-run behavior of the EU-27 NUTS-2 regions over the period 1995-2003. It departs from previous research in two respects: *first*, by introducing kernel estimation and three-dimensional stacked conditional density plots as well as highest density regions plots for the visualization of the transition function and *second*, by combining Getis' spatial filtering view with kernel estimation to explicitly account for the spatial dimension of the growth process. The results of the analysis indicate a very slow catching-up of the poorest regions with the richer ones, a process of shifting away of a small group of very rich regions, and highlight the importance of geography in understanding regional income distribution dynamics.

In the next chapter, *Bernard Fingleton* uses a multi-equation spatial econometric model to explain variations across EU regions in manufacturing productivity growth based on recent theoretical developments in urban economics and economic geography. The chapter shows that temporal and spatial parameter homogeneity is an unrealistic assumption, contrary to what is typically assumed in the literature. Constraints are imposed on parameters across time periods and between core and peripheral regions of the EU, with the significant loss of fit providing overwhelming evidence of parameter heterogeneity, although the final model does highlight increasing returns to scale, which is a central feature of contemporary theory.

Part F: Applications in environmental sciences

With the focus on applications in environmental sciences, *Part F* includes three chapters that may illustrate the potential of spatial analysis in this domain area:

- fuzzy *k*-means classification and a Bayesian approach for spatial prediction of landslide hazard (Chapter F.1),
- incorporating spatial autocorrelation in species distribution models (Chapter F.2), and

- a Web-based environmental decision support system for environmental planning and watershed management (Chapter F.3).

In Chapter F.1, *Pece V. Gorsevski, Paul E. Gessler,* and *Piotr Jankowski* describe a robust method for spatial prediction of landslide hazard in roaded and roadless areas of forest. The method is based on assigning digital terrain attributes into continuous landform classes. The continuous landform classification is achieved by applying a fuzzy *k*-means approach to a watershed scale area before the classification is extrapolated to a broader region. The extrapolated fuzzy landform classes and datasets of road-related and non road-related landslides are then combined in a GISystem for the exploration of predictive correlations and model development. In particular, a Bayesian probabilistic modeling approach is illustrated using a case study of the Clearwater National Forest in central Idaho, which experienced significant and widespread landslide events in recent years. The computed landslide hazard potential is presented on probabilistic maps for roaded and roadless areas. The maps can be used as a decision support tool in forest planning involving the maintenance, obliteration or development of new forest roads in steep mountainous terrain.

Spatial analysis is one of the most rapidly growing areas in ecology. This is due in part to an increasing awareness among ecologists about the importance of spatial structure in ecological phenomena, as well as an expanding variety of spatial analysis tools. Species distribution models, used to quantify the distribution of a (plant or animal) species along environmental gradients, have become an important research focus in this area. These models generally ignore or attempt to remove spatial autocorrelation in the data. When explicitly included in the model, spatial autocorrelation can increase model accuracy and clarify the influence of other predictor variables. Chapter F.2, written by *Jennifer A. Miller* and *Janet Franklin,* develops presence/absence models for eleven vegetation alliances in the Mojave Desert with classification trees and generalized linear models (GLMs), and uses geostatistical interpolation to calculate spatial autocorrelation terms (autocovariates) used in the models. Results are mixed across models and methods, but in general, the autocovariate terms more consistently increase model accuracy for widespread alliances. GLMs tend to have higher accuracy in general.

Local governments often struggle to balance competing demands for residential, commercial and industrial development with imperatives to minimize environmental degradation. In order to effectively manage this development process on a sustainable basis, local planners and government agencies are increasingly seeking better tools and techniques. In Chapter F.3, *Ramanathan Sugumaran, James C. Meyer* and *Jim Davis* describe the development of a Web-based environmental decision support system, which helps to prioritize local watersheds in terms of environmental sensitivity using multiple criteria identified by planners and local government staff in the city of Columbia, and Boone County, Missouri. The development of the system involved three steps, the first was to establish the relevant environmental criteria and to develop data layers for each criterion, then a spatial model was developed for analysis, and lastly a Web-based interface with analysis tools using client-server technology. The system is an example of a way to run spatial models over the Web and represents a significant increase in capability over other WWW-based GI applications that focus on database querying and map display. The decision support system seeks to aid in the development of agreement regarding specific local areas deserving increased protection and the public policies to be pursued in minimizing the environmental impact of future development. The tools are also intended to assist ongoing public information and education efforts concerning watershed management and water quality issues for the City of Columbia (Missouri) and adjacent developing areas within Boone County, Missouri.

Part G: Applications in health sciences

Part G closes the *Handbook* with three chapters illustrating applications in health sciences:

- spatio-temporal patterns of viral meningitis in Michigan, 1993-2001 (Chapter E.1),
- space-time visualization and analysis in the Cancer Atlas Viewer (Chapter E.2), and
- exposure assessment in environmental epidemiology (Chapter E.3).

Viral meningitis results in an estimated 26-42 thousand hospitalizations in the US each year. The incidence of this and other diseases can be successfully understood and controlled by examining cases in terms of person, place and time, and exploring spatio-temporal patterns. Areas with high incidence may be targeted for heightened surveillance, education, and prevention efforts. In Chapter G.1, *Sharon K. Greene, Mark A. Schmidt, Mary Grace Stobierski,* and *Mark L. Wilson* applied spatial analytical techniques to investigate viral meningitis incidence in Michigan and clarify disease patterns. Specifically, viral meningitis cases from 1993 to 2001 were analysed using standard epidemiological methods, mapped with a GISystem, and then further analysed using spatial and temporal cluster statistics.

Chapter G.2, written by *Dunrie A. Greiling, Geoffrey M. Jacquez, Andrew M. Kaufmann,* and *Robert G. Rommel*, demonstrates the use of the Cancer Atlas Viewer – an example of a space-time information system as described in Chapter A.6 – by exploring colon patterns for African-American and white females and males in southeastern United States over the period 1970-1995. Specifically, the authors use data from the National Cancer Institute and assess changes in spatial patterns of mortality from colon cancer by examining trends in the local Moran and the Getis-Ord statistics, and the persistence of patterns over time.

A key component of environmental epidemiologic research is the assessment of historic exposure to environmental contaminants. The expansion of space-time databases, coupled with the need to incorporate mobility histories in environmental epidemiology, has highlighted the deficiencies of current software to visualize and process space-time information for exposure assessment. This need is most pressing in retrospective studies or large studies where collection of individual biomarkers is unattainable or prohibitively expensive, and models and software tools are required for exposure reconstruction. In diseases of long latency such as cancer, exposure may need to be reconstructed over the entire life course, taking into consideration residential mobility, occupational mobility, changes in risk behavior, and time changing maps generated from models of environmental contaminants. Chapter G.3, written by *Jaymie R. Meliker, Melissa J. Slotnick, Gillian A. AvRuskin, Andrew Kaufman, Geoffrey D. Jacquez,* and *Jerome O. Nriagu*, undertakes a modest attempt to apply Time-GIS software tools – as described in Chapter A.6 – for space-time exposure reconstruction, using data from a bladder cancer case control study in Michigan.

4 Outlook

The field of spatial analysis can be defined by the problems it attempts to solve. The problems emanate from the peculiarities of georeferenced data. Even if dealing with multidirectional data were the only problem of the spatial analyst, researchers in this field would have their hands full. Multidirectionality issues require insight into problems of dependency, heterogeneity, the meaning of clustering, what constitutes filtering, nonstationarity, scale differences, spatial sampling, and so on. Much of this Handbook is devoted to these issues. But, in addition, there are spatial problems having to do with boundaries, object size, zoning, redundancy, data transformations, representations, parameter estimation, and the design of appropriate tests.

Particular problems characterize the schools of thought mentioned above. ESDA is challenged by use of the technology and new types of scripts for display and manipulation of spatial data. Spatial statisticians are beginning to address the problem of multiple, simultaneous, spatially dependent tests. Spatial econometricians are translating some of the more traditional problems into a Bayesian framework. Geostatisticians are developing new models employing spatial-temporal data. The use of new sensors, especially high resolution instruments challenges remote sensing specialists to devise methods for the classification and study of land cover at a variety of scales.

The outlook for spatial analysis is one of promise for new and innovative solutions to all of these problems. Given these challenges, the fact that many of the substantive issues revolve around environmental concerns means that for the foreseeable future the field will grow. By bringing together the technical and substantive issues of spatial analysis into a computer aided statistical setting has served and will continue to serve to move the field forward quickly. Granting agencies usually support research concerning societal issues. If the economies of the world hold up, we expect that granting agencies around the world will continue to emphasize this field.

References

Aldstadt J (2009) Spatial clustering. In Fischer MM, Getis A (eds) Handbook of applied spatial analysis. Springer, Berlin, Heidelberg and New York, pp.279-300
Anselin L (1988) Spatial econometrics: methods and models. Kluwer, Dordrecht
Anselin L, Florax RJGM, Rey SJ (eds) (2004) Advances in spatial econometrics. Springer, Berlin, Heidelberg and New York
Anselin L, Kim YW, Syabri I (2009) Web-based analytical tools for the exploration of spatial data. In Fischer MM, Getis A (eds) Handbook of applied spatial analysis. Springer, Berlin, Heidelberg and New York, pp.151-173
Anselin L, Syabri I, Kho Y (2009) GeoDa: An introduction to spatial data analysis. In Fischer MM, Getis A (eds) Handbook of applied spatial analysis. Springer, Berlin, Heidelberg and New York, pp.73-89
Bivand RS (2009a) Exploratory spatial data analysis. In Fischer MM, Getis A (eds) Handbook of applied spatial analysis. Springer, Berlin, Heidelberg and New York, pp.219-254
Bivand RS (2009b) Spatial econometric functions in R. In Fischer MM, Getis A (eds) Handbook of applied spatial analysis. Springer, Berlin, Heidelberg and New York, pp.53-71
Casetti E (2009) Expansion method, dependency, and modeling. In Fischer MM, Getis A (eds) Handbook of applied spatial analysis. Springer, Berlin, Heidelberg and New York, pp.487-505
Cliff AD, Ord JK (1973) Spatial autocorrelation. Pion, London
Cliff AD, Ord JK (1981) Spatial processes. Models and applications. Pion, London
Cressie NAC (1993) Statistics for spatial data (revised edition). Wiley, New York, Chichester, Toronto and Brisbane
Elhorst JP (2009) Spatial panel data models. In Fischer MM, Getis A (eds) Handbook of applied spatial analysis. Springer, Berlin, Heidelberg and New York, pp.377-407
Fingleton B (2009) A multi-equation spatial econometric model, with application to EU manufacturing productivity growth. In Fischer MM, Getis A (eds) Handbook of applied spatial analysis. Springer, Berlin, Heidelberg and New York, pp.629-649
Fischer MM, Getis A (eds) (1997) Recent developments in spatial analysis. Springer, Berlin, Heidelberg and New York
Fischer MM, Nijkamp P (eds) (1993) Geographical information systems, spatial modelling and policy evaluation. Springer, Berlin, Heidelberg and New York
Fischer MM, Stumpner P (2009) Income distribution dynamics and cross-region convergence in Europe. In Fischer MM, Getis A (eds) Handbook of applied spatial analysis. Springer, Berlin, Heidelberg and New York, pp.599-628
Fischer MM, Scholten HJ, Unwin D (1996) Spatial analytical perspectives on GIS. Taylor and Francis, London
Fischer MM, Bartkowska M, Riedl A, Sardadvar S, Kunnert A (2009) The impact of human capital on regional labor productivity in Europe. In Fischer MM, Getis A (eds) Handbook of applied spatial analysis. Springer, Berlin, Heidelberg and New York, pp.585-597

Fotheringham AS, Brunsdon C, Charlton M (2002) Geographically weighted regression: the analysis of spatially varying relationships. Wiley, New York, Chichester, Toronto and Brisbane

Getis A (2009) Spatial autocorrelation. In Fischer MM, Getis A (eds) Handbook of applied spatial analysis. Springer, Berlin, Heidelberg and New York, pp.255-278

Goovaerts P (2009) Geostatistical software. In Fischer MM, Getis A (eds) Handbook of applied spatial analysis. Springer, Berlin, Heidelberg and New York, pp.125-134

Gopal S, Woodcock C, Liu W (2009) ARTMAP neural network multisensor fusion model for multiscale land cover characterization. In Fischer MM, Getis A (eds) Handbook of applied spatial analysis. Springer, Berlin, Heidelberg and New York, pp.529-543

Gorsevski PV, Gessler PE, Jankowski P (2009) A fuzzy k-means classification and a Bayesian approach for spatial prediction of landslide hazard. In Fischer MM, Getis A (eds) Handbook of applied spatial analysis. Springer, Berlin, Heidelberg and New York, pp.653-684

Greene SK, Schmidt MA, Stobierski MG, Wilson ML (2009) Spatio-temporal patterns of viral meningitis in Michigan, 1993-2001. In Fischer MM, Getis A (eds) Handbook of applied spatial analysis. Springer, Berlin, Heidelberg and New York, pp.721-735

Greiling DA, Jacquez GM, Kaufmann AM, Rommel RG (2009) Space-time visualization and analysis in the Cancer Atlas Viewer. In Fischer MM, Getis A (eds) Handbook of applied spatial analysis. Springer, Berlin, Heidelberg and New York, pp.737-752

Griffith DA (2003) Spatial autocorrelation and spatial filtering. Springer, Berlin, Heidelberg and New York

Griffith DA (2009) Spatial filtering. In Fischer MM, Getis A (eds) Handbook of applied spatial analysis. Springer, Berlin, Heidelberg and New York, pp.301-318

Haining RP (1990) Spatial data analysis in the social and environmental sciences. Cambridge University Press, Cambridge

Haining RP (2009) The nature of georeferenced data. In Fischer MM, Getis A (eds) Handbook of applied spatial analysis. Springer, Berlin, Heidelberg and New York, pp.197-217

Isard W (1960) Methods of regional analysis: An introduction to regional science. The MIT Press, Cambridge [MA] and London

Jacquez GM (2009) Space-time intelligence system software for the analysis of complex systems. In Fischer MM, Getis A (eds) Handbook of applied spatial analysis. Springer, Berlin, Heidelberg and New York, pp.113-124

Krige DG (1951) A statistical approach to some basic mine valuation problems on the Witwatersrand. J Chem Met Min Soc of South Africa 52(6):119-139

Lagona F (2009) Model selection in Markov random fields for high spatial resolution hyperspectral data. In Fischer MM, Getis A (eds) Handbook of applied spatial analysis. Springer, Berlin, Heidelberg and New York, pp.545-563

Lee G, Yamada I, Rogerson P (2009) GeoSurveillance: GIS-based exploratory spatial analysis exploratory spatial analysis tools for monitoring spatial patterns and clusters. In Fischer MM, Getis A (eds) Handbook of applied spatial analysis. Springer, Berlin, Heidelberg and New York, pp.135-149

LeSage JP (2004) The MATLAB spatial econometrics toolbox. URL: http://www.spatial.econometrics.com

LeSage JP, Fischer MM (2009) Spatial econometric methods for modeling origin-destination flows. In Fischer MM, Getis A (eds) Handbook of applied spatial analysis. Springer, Berlin, Heidelberg and New York, pp.409-433

LeSage JP, Pace RK (2009a) Introduction to spatial econometrics. Taylor and Francis, London

LeSage JP, Pace RK (2009b) Spatial econometric models. In Fischer MM, Getis A (eds) Handbook of applied spatial analysis. Springer, Berlin, Heidelberg and New York, pp.355-376

Matheron G (1963) Principles of geostatistics. Econ Geol 58(8):1246-1266

Meliker JR, Slotnick MJ, AvRuskin GA, Kaufmann A, Jacquez GM, Nriagu JO (2009) Exposure assessment in environmental epidemiology. In Fischer MM, Getis A (eds) Handbook of applied spatial analysis. Springer, Berlin, Heidelberg and New York, pp.753-767

Miller JA, Franklin J (2009) Incorporating spatial autocorrelation in species distribution. In Fischer MM, Getis A (eds) Handbook of applied spatial analysis. Springer, Berlin, Heidelberg and New York, pp.685-702

Oliver M (2009) The variogram and kriging. In Fischer MM, Getis A (eds) Handbook of applied spatial analysis. Springer, Berlin, Heidelberg and New York, pp.319-352

Ord JK (1975) Estimation methods for models of spatial interaction. J Am Stat Assoc 70(349):120-126

Ord JK, Getis A (1995) Local spatial autocorrelation statistics. Distributional issues and an application. Geogr Anal 27(4):287-306

Ord JK, Getis A (2001) Testing for local spatial autocorrelation in the presence of global autocorrelation. J Reg Sci 41(3):411-432

Paelinck JHP, Klaassen LH (1979) Spatial econometrics. Saxon House, Farnborough

Parent O, LeSage JP (2009) Spatial econometric model averaging. In Fischer MM, Getis A (eds) Handbook of applied spatial analysis. Springer, Berlin, Heidelberg and New York, pp.435-460

Rey SJ, Anselin L (2009) PySAL: A Python library of spatial analytical methods. In Fischer MM, Getis A (eds) Handbook of applied spatial analysis. Springer, Berlin, Heidelberg and New York, pp.175-193

Rey SJ, Janikas MV (2009) STARS: Space-time analysis of regional systems. In Fischer MM, Getis A (eds) Handbook of applied spatial analysis. Springer, Berlin, Heidelberg and New York, pp.91-112

Ripley BD (1977) Modeling spatial patterns. J Roy Stat Soc B 39(2):172-194

Rura MI, Griffith DA (2009) Spatial statistics in SAS. In Fischer MM, Getis A (eds) Handbook of applied spatial analysis. Springer, Berlin, Heidelberg and New York, pp.43-52

Scott LM, Janikas MV (2009) Spatial statistics in ArcGIS. In Fischer MM, Getis A (eds) Handbook of applied spatial analysis. Springer, Berlin, Heidelberg and New York, pp.27-41

Stow DA (2009) Geographic object-based image change analysis. In Fischer MM, Getis A (eds) Handbook of applied spatial analysis. Springer, Berlin, Heidelberg and New York, pp.565-582

Subramanian SV (2009) Multilevel modeling. In Fischer MM, Getis A (eds) Handbook of applied spatial analysis. Springer, Berlin, Heidelberg and New York, pp.507-525

Sugumaran R, Meyer JC, Davis J (2009) A Web-based environmental decision support system for environmental planning and watershed management. In Fischer MM, Getis A (eds) Handbook of applied spatial analysis. Springer, Berlin, Heidelberg and New York, pp.703-718

Tukey JW (1977) Exploratory data analysis. Addison-Wesley, Reading [MA]

Wheeler D, Paéz A (2009) Geographically weighted regression. In Fischer MM, Getis A (eds) Handbook of applied spatial analysis. Springer, Berlin, Heidelberg and New York, pp.461-486

Part A

GI Software Tools

A.1 Spatial Statistics in ArcGIS

Lauren M. Scott and *Mark V. Janikas*

A.1.1 Introduction

With over a million software users worldwide, and installations at over 5,000 universities, Environmental Systems Research Institute, Inc. (ESRI), established in 1969, is a world leader for the design and development of Geographic Information Systems (GIS) software. GIS technology allows the organization, manipulation, analysis, and visualization of spatial data, often uncovering relationships, patterns, and trends. It is an important tool for urban planning (Maantay and Ziegler 2006), public health (Cromley and McLafferty 2002), law enforcement (Chainey and Ratcliffe 2005), ecology (Johnston 1998), transportation (Thill 2000), demographics (Peters and MacDonald 2004), resource management (Pettit et al. 2008), and many other industries (see http://www.esri.com/industries.html). Traditional GIS analysis techniques include spatial queries, map overlay, buffer analysis, interpolation, and proximity calculations (Mitchell 1999). Along with basic cartographic and data management tools, these analytical techniques have long been a foundation for geographic information software. Tools to perform spatial analysis have been extended over the years to include geostatistical techniques (Smith et al. 2006), raster analysis (Tomlin 1990), analytical methods for business (Pick 2008), 3D analysis (Abdul-Rahman et al. 2006), network analytics (Okabe et al. 2006), space-time dynamics (Peuquet 2002), and techniques specific to a variety of industries (e.g., Miller and Shaw 2001). In 2004, a new set of spatial statistics tools designed to describe feature patterns was added to ArcGIS 9. This chapter focuses on the methods and models found in the Spatial Statistics toolbox.

Spatial statistics comprises a set of techniques for describing and modeling spatial data. In many ways they extend what the mind and eyes do, intuitively, to assess spatial patterns, distributions, trends, processes and relationships. Unlike traditional (non-spatial) statistical techniques, *spatial* statistical techniques actually use space – area, length, proximity, orientation, or spatial relationships – directly in their mathematics (Scott and Getis 2008).

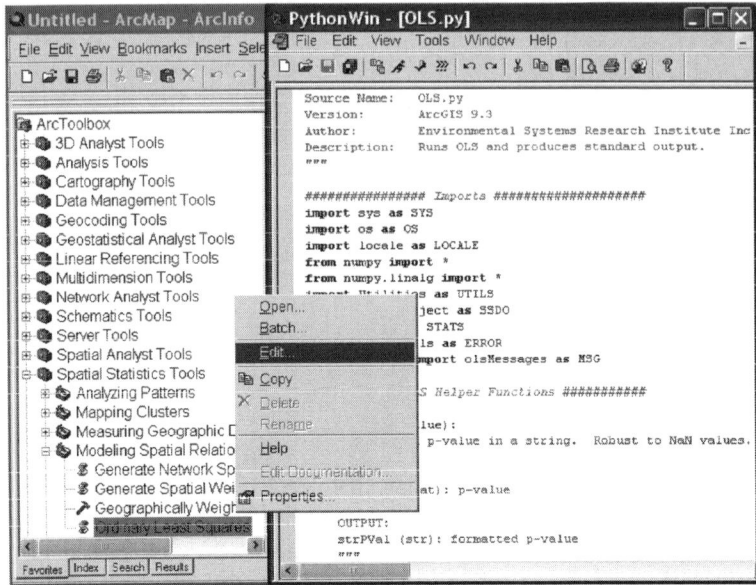

Fig. A.1.1. Right click on a script tool and select *Edit* to see the Python source code

By 2008 the Spatial Statistics toolbox in ArcGIS contained 25 tools. The majority of these were written using the Python scripting language. Consequently, ArcGIS users have access not only to the analytical methods for these tools, but also to their source code (see Fig. A.1.1).

The Spatial Statistics toolbox includes both statistical functions and general-purpose utilities. With the most recent release of ArcGIS 9.3, statistical functions are grouped into four toolsets: Measuring Geographic Distributions, Analyzing Patterns, Mapping Clusters, and Modeling Spatial Relationships.

A.1.2 Measuring geographic distributions

The tools in the Measuring Geographic Distributions toolset (Table A.1.1) are descriptive in nature; they help summarize the salient characteristics of a spatial distribution. They are useful for answering questions like:

- Which site is most accessible?
- Is there a directional trend to the spatial distribution of the disease outbreak?
- What is the primary wind direction for this region in the winter?
- Where is the population center?
- Which species has the broadest territory?

Table A.1.1. Tools in the measuring geographic distributions toolset

Tool	Description
Central feature	Identifies the most centrally located feature in a point, line, or polygon feature class
Directional distribution (standard deviational ellipse)	Measures how concentrated features are around the geographic mean, and whether or not they exhibit a directional trend
Linear directional mean	Identifies the general (mean) direction and mean length for a set of vectors
Mean center	Identifies the geographic center for a set of features
Standard distance	Measures the degree to which features are concentrated or dispersed around the geographic mean center

Even the simplest tool in the Spatial Statistics toolbox can be a powerful communicator of spatial pattern when used with animation. The mean center tool is a measure of central tendency; it computes the geometric center – the average X and average Y coordinate – for a set of geographic features. In Fig. A.1.2, the weighted mean center of population for the counties of California is computed every decade from 1910 to 2000. The center of population is initially located in the northern half of the state near San Francisco. Animation reveals steady movement of the mean center south, every decade, as population growth in Southern California outpaces population growth in the state's northern counties.

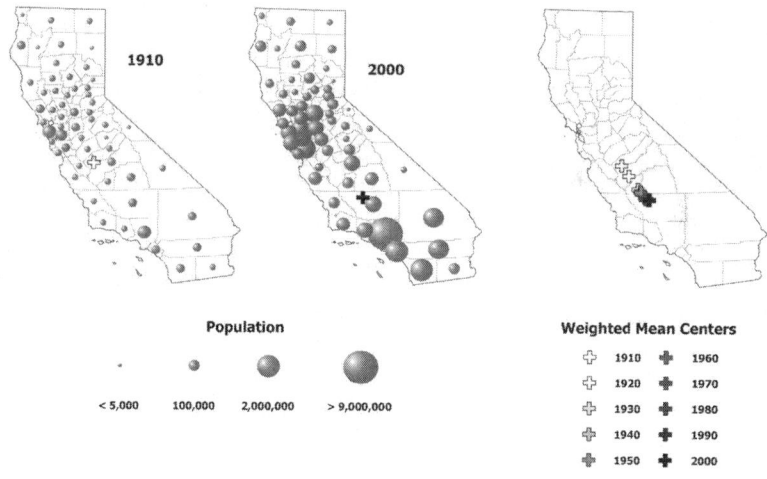

Fig. A.1.2. Weighted mean center of population, by county, 1910 through 2000

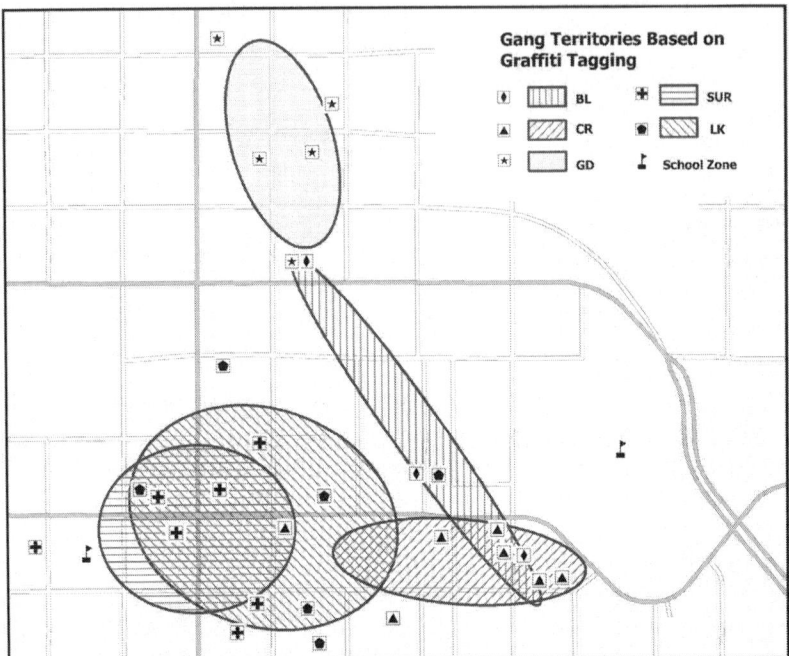

Fig. A.1.3. Core areas for five gangs based on graffiti tagging

The Standard Deviational Ellipse and Standard Distance tools measure the spatial distribution of geographic features around their geometric center, and provide information about feature dispersion and orientation. Gangs often mark their territory with graffiti. In Fig. A.1.3, a standard deviational ellipse is computed, by gang affiliation, for graffiti incidents in a city. The ellipses provide an estimate of the core areas associated with each gang's turf. The potential for increased gang-related conflict and violence is highest in areas where the ellipses overlap. By increasing the presence of uniformed police officers in these overlapping areas and around nearby schools, the community may be able to curtail gang violence. Mitchell (2005), Scott and Warmerdam (2005), Wong (1999) and Levine (1996) provide additional examples of applications for descriptive statistics like mean center, standard distance, and standard deviational ellipse.

A.1.3 Analyzing patterns

The Analyzing Patterns toolset (Table A.1.2) contains methods that are most appropriate for understanding broad spatial patterns and trends (Mitchell 2005). With these tools you can answer questions like:

- Which plant species is most concentrated?
- Does the spatial pattern of the disease mirror the spatial pattern of the population at risk?
- Is there an unexpected spike in pharmaceutical purchases?
- Are new AIDs cases remaining geographically fixed?

Consider the difficulty of trying to measure changes in urban manufacturing patterns for the United States over the past few decades. Certainly broad changes have occurred with globalization and the move from vertical integration to a more flexible and dispersed pattern of production. One approach might be to map manufacturing employment by census tract for a series of years, and then try to visually discern whether or not spatial patterns are becoming more concentrated or more dispersed. Most likely a range of scenarios would emerge. The Global Moran's I tool computes a single summary value, a z-score, describing the degree of spatial concentration or dispersion for the measured variable (in this case manufacturing employment). Comparing this summary value, year by year, indicates whether or not manufacturing is becoming, overall, more dispersed or more concentrated.

Similarly, viewing thematic maps of per capita incomes (PCR)[1] in New York for a series of years (see Fig. A.1.4), it is difficult to determine whether rich and poor counties are becoming more or less spatially segregated. Plotting the resultant z-scores from the Spatial Autocorrelation (Global Moran's I) tool, however, reveals decreasing values indicating that spatial clustering of rich and poor has dissipated between 1969 and 2002.

Table A.1.2. A summary of the tools in the analyzing patterns toolset

Tool	Description
Average nearest neighbor	Calculates the average distance from every feature to its nearest neighbor based on feature centroids
High/low clustering (Getis-Ord general G)	Measures concentrations of high or low values for a study area
Spatial autocorrelation (global Moran's I)	Measures spatial autocorrelation (clustering or dispersion) based on feature locations and attribute values
Multi-distance spatial cluster analysis (Ripley's K function)	Assesses spatial clustering/dispersion for a set of geographic features over a range of distances

The K function is a unique tool in that it looks at the spatial clustering or dispersion of points/features at a series of distances or spatial scales. The output from the K function is a line graph (see Fig. A.1.5). The dark diagonal line represents the expected pattern, if the features were randomly distributed within the study area. The X axis reflects increasing distances. The solid curved line represents the

---

[1] PCR is per capita income relative to the national average.

observed spatial pattern for the features being analyzed. When the curved line goes above the diagonal line, the pattern is more clustered at that distance than we would expect with a random pattern; when the curved line goes below the diagonal line, the pattern is more dispersed than expected. Based on a user-specified number of randomly generated permutations of the input features, the tool also computes a confidence envelope around the expected line. When the curved line is outside the confidence envelope, the clustering or dispersion is statistically significant.

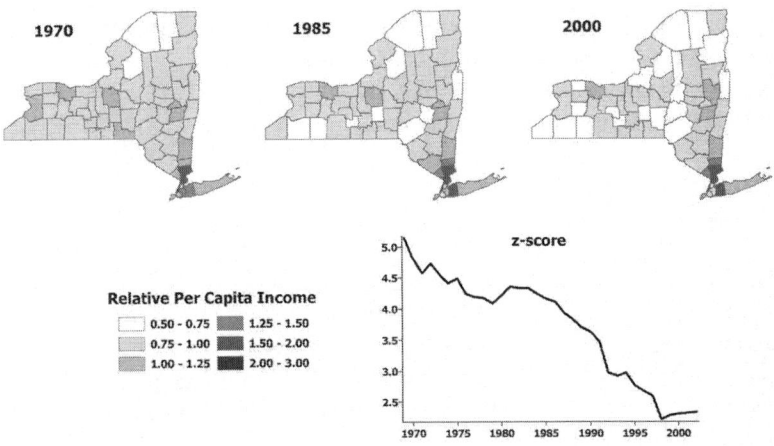

Fig. A.1.4. Relative per capita income for New York, 1969 to 2002

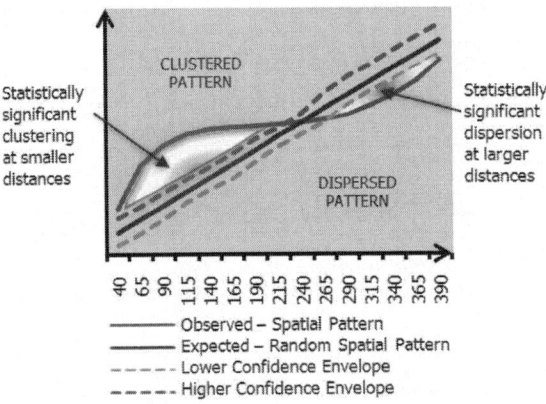

Fig. A.1.5. Components of the K function graphical output

The *K* function is useful for comparing different sets of features within the same study area, such as two strains of a disease or disease cases in relation to population at risk. Similar observed spatial patterns suggest similar factors (similar spatial processes) are at work. A researcher might compare the spatial pattern for a disease outbreak, for example, to the spatial pattern of the population at risk to help determine if factors other than the spatial distribution of population are promoting disease incidents. Wheeler (2007), Levine (1996), Getis and Ord (1992), and Illian et al. (2008) provide examples of additional applications for the tools in the Analyzing Patterns toolset.

A.1.4 Mapping clusters

The tools discussed above in the Analyzing Patterns toolset are global statistics that answer the question: *Is* there statistically significant spatial clustering or dispersion? Tools in the Mapping Clusters toolset (Table A.1.3), on the other hand, identify *where* spatial clustering occurs, and *where* spatial outliers are located:

- Where are their sharp boundaries between affluence and poverty in Ecuador?
- Where do we find anomalous spending patterns in Los Angeles?
- Where do we see unexpectedly high rates of diabetes?

In Fig. A.1.6, the Local Moran's *I* tool is used to analyze poverty in Ecuador. A string of outliers separate clusters of high poverty in the north from clusters of low poverty in the south, indicating a sharp divide in economic status.

Table A.1.3. A summary of the tools in the mapping clusters toolset

Tool	Description
Cluster and outlier analysis (Anselin's local Moran's *I*)	Given a set of weighted features, identifies clusters of high or low values as well as spatial outliers
Hot spot analysis (Getis-Ord G_i^*)	Given a set of weighted features, identifies clusters of features with high values (hot spots) and clusters of features with low values (cold spots)

The Hot Spot Analysis (Getis-Ord G_i^*) tool is applied to vandalism data for Lincoln, Nebraska in Fig. A.1.7. In the first map (left), raw vandalism counts for each census block are analyzed. The picture that emerges would not surprise local police officers. Most vandalism is found where most people and most overall crime are found: downtown and in surrounding high crime areas. Fewer cases of vandalism are associated with the lower density suburbs. In the second map (right), however, vandalism is normalized by overall crime incidents prior to analysis. Running the Hot Spot Analysis tool on this normalized data shows that

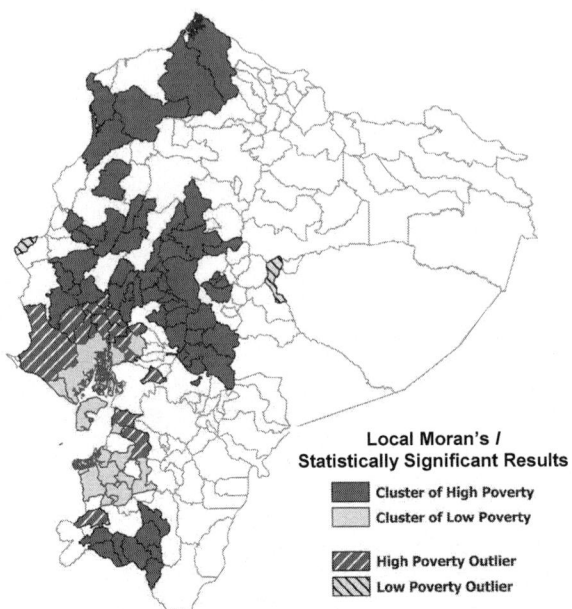

Fig. A.1.6. An analysis of poverty in Ecuador using local Moran's *I*

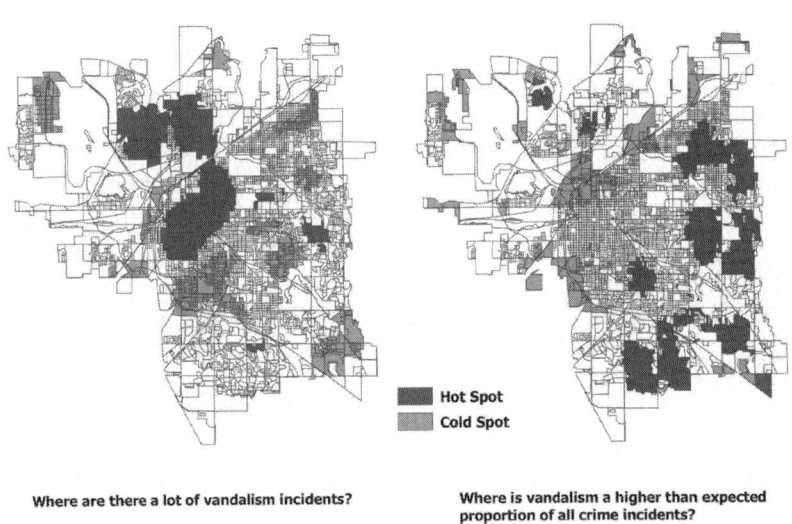

Fig. A.1.7. An analysis of vandalism hot spots in Lincoln, Nebraska using G_i^*

while Lincoln may have more incidents of vandalism in downtown areas, vandalism represents a larger proportion of total crime in suburban areas. Zhang et al. (2008), Jacquez and Greiling (2003), Getis and Ord (1992), Ord and Getis (1995), and Anselin (1995) provide additional applications for the tools in the Mapping Clusters toolset.

A.1.5 Modeling spatial relationships

The tools in the Modeling Spatial Relationships toolset (Table A.1.4) fall into two categories. The first category includes tools designed to help the user define a conceptual model of spatial relationships. The conceptual model is an integral component of spatial modeling and should be selected so that it best represents the structure of spatial dependence among the features being analyzed (Getis and Aldstadt 2004).

The options available for modeling spatial relationships include inverse distance, fixed distance, polygon contiguity (Rook's and Queen's case), k nearest neighbors, Delaunay triangulation, travel time and travel distance. Figure A.1.8 illustrates how spatial relationships change when they are based on a real road network, rather than on straight line distances.

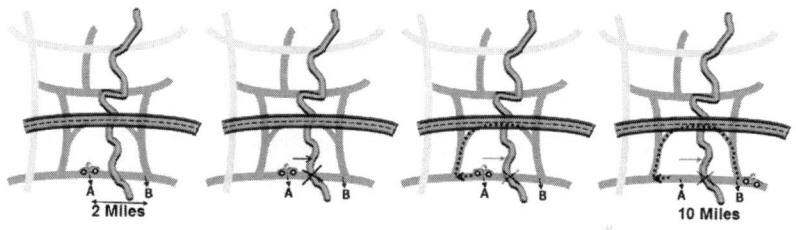

Fig. A.1.8. Traffic conditions or a barrier in the physical landscape can dramatically change actual travel distances, impacting results of spatial analysis

Table A.1.4. A summary of the tools in the modeling spatial relationships toolset

Tool	Description
Generate network spatial weights	Builds a spatial weights matrix file specifying spatial relationships among features in a feature class based on a Network dataset
Generate spatial weights matrix	Builds a spatial weights matrix file specifying spatial relationships among features in a feature class
Geographically weighted regression	A local form of linear regression used to model spatially varying relationships among a set of data variables
Ordinary least squares regression	Performs global linear regression to model the relationships among a set of data variables

Constructing spatial relationships prior to analysis generally results in improved performance, particularly within the context of larger datasets or when applied to multiple attribute fields. The spatial weights matrix files (*.swm) are sharable, reusable and can be directly edited within ArcGIS. Furthermore, options are available to facilitate both importing and exporting spatial weights matrix files from/to other formats (*.gal, *.gwt, or a simple *.dbf table).[2]

The second category of tools in the Modeling Spatial Relationships toolset includes ordinary least squares (OLS) (Woolridge 2003), and geographically weighted regression (GWR) (Fortheringham et al. 2002 and Chapter C.5). These tools can help answer the following types of questions:

- What is the relationship between educational attainment and income?
- Is there a relationship between income and public transportation usage? Is that relationship consistent across the study area?
- What are the key factors contributing to excessive residential water usage?

Regression analysis may be used to model, examine, and explore spatial relationships, in order to better understand the factors behind observed spatial patterns or to predict spatial outcomes. There are a large number of applications for these techniques (Table A.1.5).

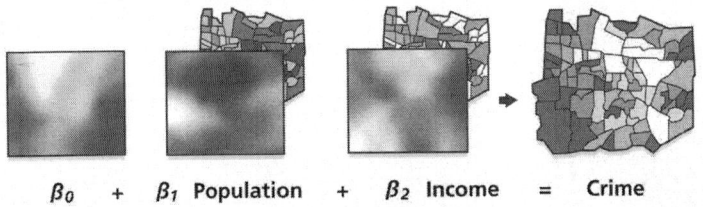

β_0 + β_1 Population + β_2 Income = Crime

Fig. A.1.9. GWR optionally creates a coefficient surface for each model explanatory variable reflecting variation in modeled relationship

OLS is a global model. It creates a single equation to represent the relationship between what you are trying to model and each of your explanatory variables. Global models, like OLS, are based on the assumption that relationships are static and consistent across the entire study area. When they are not – when the relationships behave differently in separate parts of the study area – the global model becomes less effective. You might find, for example, that people's desire to live and work close, but not too close, to a metro line encourages population growth: the relationship for being fairly close to a metro line is positive while the relation-

---

[2] See http://resources.esri.com/geoprocessing/ for a description and examples of exporting /importing *.swm files to *.gal and *.gwt formats.

ship for being right up next to a metro line is negative. A global model will compute a single coefficient to represent both of these divergent relationships. The result, an average, may not represent either situation very well.

Local models, like GWR, create an equation for every feature in the dataset, calibrating each one using the target feature and its neighbors. Nearby features have a higher weight in the calibration than features that are farther away. What this means is that the relationships you are trying to model are allowed to change over the study area; this variation is reflected in the coefficient surfaces optionally created by the GWR tool (see Fig. A.1.9). If you are trying to predict foreclosures, for example, you might find that an income variable is very important in the northern part of your study area, but very weak or not important at all in the southern part of your study area. GWR accommodates this kind of regional variation in the regression model.

Table A.1.5. A variety of potential applications for regression analysis

Application Area	Analysis Example
Public health	Why are diabetes rates exceptionally high in particular regions of the United States?
Public safety	What environmental factors are associated with an increase in search and rescue event severity?
Transportation	What demographic characteristics contribute to high rates of public transportation usage?
Education	Why are literacy rates so low in particular regions?
Market analysis	What is the predicted annual sales for a proposed store?
Economics	Why do some communities have so many home foreclosures?
Natural resource management	What are the key variables promoting high forest fire frequency?
Ecology	Which environments should be protected to encourage reintroduction of an endangered species?

The default output for both regression tools is a residual map showing the model over- and underpredictions (see Fig. A.1.10). The OLS tool automatically checks for muliticollinearity (redundancy among model explanatory variables), and computes coefficient probabilities, standard errors, and overall model significance indices that are robust to heteroscedasticity. The online help documentation for these tools provides a beginner's guide to regression analysis, suggested step by step instructions for the model building process, a table outlining and carefully explaining the challenges and potential pitfalls associated with using regression analysis with spatial data, and recommendations for how to overcome those potential problems.[3]

---

[3] See http://webhelp.esri.com/arcgisdesktop/9.3/index.cfm?TopicName=Regression_analysis_basics

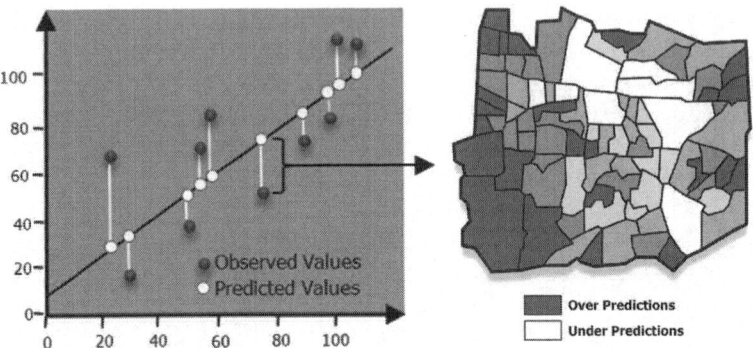

Fig. A.1.10. Default output from the regression tools is a map of model over- and under-predictions

A.1.6 Custom tool development

The tools in the Spatial Statistics toolbox were developed using the same methods and techniques that an ArcGIS user might adopt to create his/her own custom tools. They illustrate the extendibility of ArcGIS, and ESRI's commitment to providing a framework for custom tool development.

The simplest way to create a new tool in the geoprocessing framework is to use Model Builder to string existing tools together. The resultant model tool can then be exported to Python and further extended with custom code. In addition, any third party software package that can be launched from the DOS command line is an excellent custom tool candidate. Simply point to the executable for that software and define the needed tool parameters.

For software developers, the geoprocessing framework offers sophisticated options for custom tool development. Python script tools can be run 'in process', resulting in a cohesive interface that improves both performance and usability. Numerical Python (NumPy) provides an avenue to perform complex mathematical operations (Oliphant 2006), and is currently part of the ArcGIS software installation. Other Python libraries can be added as well. Perhaps the most logical extension is Scientific Python (SciPy),[4] which provides a host of powerful statistical techniques and works directly with NumPy. PySAL (a Python Library for Spatial Analytical Functions, see Chapter A.10), developed in conjunction with GeoDa (see Chapter A.4) and STARS (see Chapter A.5), is a crossplatform library of spatial analysis functions that may also provide opportunities for extending Arc GIS functionality.[5]

---

[4] http://www.scipy.org/

[5] See http://www.sal.uiuc.edu/tools/tools-sum/pysal and http://www.sal.uiuc.edu/tools-sum/pysal

Python works nicely with other programming languages, and this has resulted in several hybrid libraries including Rpy and PyMat, giving users access to the methods in R (see Chapter A.3 for spatial econometric functions in R) and in MatLab, respectively.[6] There are also a number of spatial data analysis add-on packages for R (Bivand and Gebhardt 2000) and a spatial econometrics toolbox for MatLab (LeSage 1999). Sample scripts demonstrating integration of ArcGIS 9.3 with R are available for download from the Geoprocessing Resource Center[7] (see Fig. A.1.11).

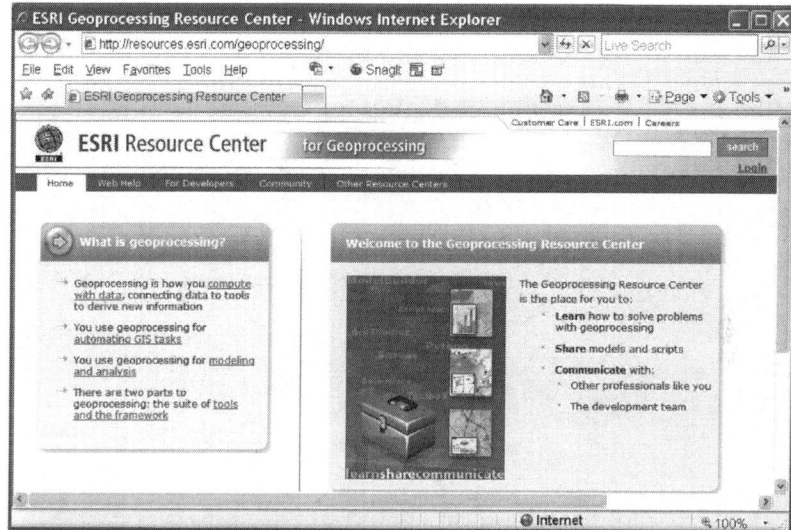

Fig. A.1.11. Geoprocessing Resource Center Web page

A.1.7 Concluding remarks

The Spatial Statistics toolbox provides feature pattern analysis and regression analysis capabilities inside ArcGIS where users can leverage, directly, all of its powerful database management and cartographic functionalities. The source code for these tools is provided inside a geoprocessing framework that encourages development and sharing of custom tools and methods. People and organizations developing custom Python tools can take advantage of existing libraries, documentation, sample scripts, and support from a worldwide community of Python software developers. The Geoprocessing Resouce Center (see Fig. A.1.11),

---

[6] See http://rpy.sourceforge.net/, http://www.r-project.org/, http://www.mathworks.com/, and http://claymore.engineer.gvsu.edu/~steriana/Python/pymat.html.

[7] http://resources.esri.com/geoprocessing/

launched in August of 2008, offers a platform for asking questions and getting answers, for sharing ideas, tools, and methodologies, and for participating in an ongoing conversation about spatial data analysis. The sincere hope is that this conversation will extend beyond the realm of academics, theoreticians, and software developers – that it will embrace the hundreds of thousands of GIS users grappling with real world data and problems – and that, as a consequence, this might foster new tools, new questions, perhaps even new approaches altogether.

References

Abdul-Rahman A, Zlantanaova S, Coors V (2006) Innovations in 3D geo information systems. Springer, Berlin, Heidelberg and New York
Anselin L (1995) Local indicators of spatial association: LISA. Geogr Anal 27(2):93-115
Anselin L, Syabri I, Kho Y (2009) GeoDa: an introduction to spatial data analysis. In Fischer MM, Getis A (eds) Handbook of applied spatial analysis. Springer, Berlin, Heidelberg and New York, pp.73-89
Bivand RS (2009) Spatial econometrics functions in R. In Fischer MM, Getis A (eds) Handbook of applied spatial analysis. Springer, Berlin, Heidelberg and New York, pp.53-71
Bivand RS, Gebhardt A (2000) Implementing functions for spatial statistical analysis using the R language. J Geogr Syst 2(3):307-317
Chainey SP, Ratcliffe JH (2005) GIS and crime mapping. Wiley, London
Cromley EK, McLafferty SL (2002) GIS and public health. Guilford, New York
Fotheringham SA, Brunsdon C, Charlton M (2002) Geographically weighted regression: the analysis of spatially varying relationships. Wiley, New York, Chichester, Toronto and Brisbane
Getis A, Aldstadt J (2004) Constructing the spatial weights matrix using a local statistic. Geogr Anal 36(2):90-104
Getis A, Ord JK (1992) The analysis of spatial association by use of distance statistics. Geogr Anal 24(3):189-206
Illian J, Penttinen A, Stoyan H, Stoyan D (2008) Statistical analysis and modeling of spatial point patterns. Wiley, London
Jacquez GM, Greiling DA (2003) Local clustering in breast, lung and colorectal cancer in Long Island, New York. Int J Health Geographics 2:3
Johnston CA (1998) Geographic information systems in ecology. Blackwell Science, Malden [MA]
LeSage JP (1999) Spatial econometrics using MATLAB. www.spatial-econometrics.com
Levine N (1996) Spatial statistics and GIS: software tools to quantify spatial patterns. J Am Plann Assoc 62(3):381-391
Maantay J, Ziegler J (2006) GIS for the urban environment. ESRI, Redlands [CA]
Miller HJ, Shaw S-L (2001) Geographic information systems for transportation: principles and applications. Oxford University Press, Oxford and New York
Mitchell A (1999) The ESRI guide to GIS analysis, volume 1: geographic patterns and relationships. ESRI, Redlands [CA]
Mitchell A (2005) The ESRI guide to GIS analysis, volume 2: spatial measurements and statistics. ESRI, Redlands [CA]
Okabe A, Okunuki K, Shiode S (2006) SANET: a toolbox for spatial analysis on a network. Geogr Anal 38(1):57-66

Oliphant T (2006) Guide to NumPy, Trelgol [USA]

Ord JK, Getis A (1995) Local spatial autocorrelation statistics: distributional issues and an application. Geogr Anal 27(4):287-306

Peters A, MacDonald H (2004) Unlocking the census with GIS. ESRI, Redlands [CA]

Pettit C, Cartwright W, Bishop I, Lowell K, Pullar D, Duncan D (eds) (2008) Landscape analysis and visualization: spatial models for natural resource management and planning. Springer, Berlin, Heidelberg and New York

Pick JB (2008) Geo-Business: GIS in the digital organization. Wiley, New York

Peuquet DJ (2002) Representations of space and time. Guilford, New York

Rey SJ, Anselin L (2009) PySAL: a Python library of spatial analytical methods. In Fischer MM, Getis A (eds) Handbook of applied spatial analysis. Springer, Berlin, Heidelberg and New York, pp.175-193

Rey SJ, Janikas MV (2009) STARS: Space-time analysis of regional systems. In Fischer MM, Getis A (eds) Handbook of applied spatial analysis. Springer, Berlin, Heidelberg and New York, pp.91-112

Scott L, Getis A (2008) Spatial statistics. In Kemp K (ed) Encyclopedia of geographic informations. Sage, Thousand Oaks, CA

Scott L, Warmerdam N (2005) Extend crime analysis with ArcGIS spatial statistics tools. ArcUser Magazine, April-June [USA]

Smith MJ, Goodchild MF, Longley PA (2006) Geospatial analysis. Troubador, Leicester

Thill J-C (2000) Geographic information systems in transportation research. Elsevier Science, Oxford

Tomlin DC (1990) Geographic information systems and cartographic modeling. Prentice-Hall, New Jersey

Wheeler D (2007) A comparison of spatial clustering and cluster detection techniques for childhood leukemia incidence in Ohio, 1996-2003. Int J Health Geographics 6(1):13

Wheeler D, Paéz A (2009) Geographically Weighted Regression. In Fischer MM, Getis A (eds) Handbook of applied spatial analysis. Springer, Berlin, Heidelberg and New York, pp.461-486

Wong DWS (1999) Geostatistics as measures of spatial segregation. Urban Geogr 20(7):635-647

Woolridge JM (2003) Introductory econometrics: a modern approach. South-Western, Mason [OH]

Zhang C, Luo L, Xu W, Ledwith V (2008) Use of local Moran's I and GIS to identify pollution hotspots of Pb in urban soils of Galway, Ireland. Sci Total Environ 398 (1-3):212-221

A.2 Spatial Statistics in SAS

Melissa J. Rura and *Daniel A. Griffith*

A.2.1 Introduction

From the abacus to the adding machine to the supercomputer, for centuries humans have used aids to enable mathematical computations. As the mathematical tabulations grew in complexity, so did the 'machines' that enabled more complex calculations. This in turn presented the problem of implementing beautifully written formulas in a form a computer 'aid' could understand. Today statistics specifically has a huge variety of software implementations available to choose from, some of which focus on a specific subdiscipline of statistics, while others encompass statistics more broadly. SAS Institute, as did many specialized software companies, evolved from an academic background in partnership with IBM, and its statistical package is used widely in statistics as well as a plethora of disciplines that rely on statistical results. Here we describe some of the ways SAS has been used in the past for spatial statistics, and some of the more recent additions made to explicitly include spatial information and geographic visualization, and give two SAS implementation examples, the calculation of Moran's I and the eigenvector spatial filtering spatial statistical technique.

A.2.2 Spatial statistics and SAS

SAS provides a programming language and components called procedures that perform data management functions as well as many different kinds of analyses. Combining the SAS language and its procedures allows a user to do tasks ranging from general-purpose data processing to highly specialized analysis, including accessing raw data files and data in external databases, managing data efficiently, analyzing data using descriptive statistics, multivariate techniques, forecasting and modeling, linear programming, customized analyses, and presenting data in reports and statistical graphics. Although in the past SAS did not include any strictly spatial statistical procedures, the computational mathematics of spatial statistics has been done for decades using statistical functions and procedures available in SAS. The reason for this is twofold. First, large spatial datasets, like census data or image pixels, often cause problems for software that hold the entire dataset in virtual memory. By integrating with both host (OS) file systems and a variety of third party DBMS products, SAS efficiently handles very large datasets. Second, the functions and procedures provided by SAS are flexible enough that models can be designed to allow researchers to include spatial information. Griffith (1993) implements spatial autoregressive models using SAS's nonlinear procedure PROC NLIN. The estimation is a nonlinear problem because: (i) the Jacobian term, which is a function of the spatial autocorrelation parameter, appears as a divisor of each regression model term; and, (ii) each regression coefficient also appears in a product term where it is multiplied by the spatial autocorrelation parameter.

The availability of SAS code for spatially informed models is found both in print (Moser 1987; Griffith 1993; Griffith et al. 1999), and on the web (Waller and Gotway 2004; Yiannakoulias 2008; Rura 2008; UCLA 2008), from a variety of authors across several disciplines. One can find freely available SAS code for the Moran Coefficient (i.e., Moran's I), the Geary Ratio (i.e., Geary's c), spatial autoregressive models, spatial random effects models, cluster detection, spatial diffusion, and much more. Regardless of the operating environment in which a particular version of SAS is running, the precision and algorithms do not change. SAS also creates log files that give a user feedback about what is happening inside procedures, including warnings about possible problems with model specification, convergence, and explanations of error messages.

A.2.3 SAS spatial analysis built-ins

Recently SAS has included many specifically geographical functions for mapping. The SAS/GIS and SAS/GRAPH software provide many mapping capabilities within SAS (see SAS/GIS 2008 and SAS/GRAPH 2008 for details about the software and its procedures). Also, SAS has implemented geostatistical procedures like PROC VARIOGRAM, which includes an option for computing Moran's I and Geary's c statistics using binary, row standardized or distance weights matrices, and PROC KRIGE2D which performs ordinary kriging in two dimensions. Also available is a spatially structured random effects intercept option in PROC MIXED based on a geostatistical semivariogram model, using a statement like *repeated/sub=intercept type=SP(EXP) (U V)*, where *EXP* is the exponential characterization of semivariance and (*U V*) are geographic coordinate pairs. A spatially structured random effects intercept also can be specified without this built-in geostatistical option, using, for instance, an eigenvector spatial filter specification by including selected eigenvectors in the model statement and specifying *random intercept/type=VC sub=ID*.

Probably the most powerful use of SAS in spatial statistics is the ability to modify existing procedures to include spatial information. PROC NLIN can be used with weights to estimate any valid semivariogram model using the output of PROC VARIOGRAM. A spatial regression can be specified a variety of different ways using the PROC NLIN procedure (see Griffith 1993 for details). And, PROC GENMOD can be used for generalized linear model spatial regression specifications by including eigenvector spatial filter proxy variables in a regression. Wang (2006) gives an example of wasteful commuting and sample SAS code using PROC LP to solve linear programming problems. These procedures, along with the flexibility of PROC IML, SAS's interactive matrix language, and the data step enable the customized programming of many standard quantitative geographical models, such as the Huff model, the Garin-Lowry model, and the doubly constrained gravity model.

SAS/GIS is an interactive Geographic Information System (GIS) within the SAS System that has an open data model, meaning the information stored in both the attribute and spatial datasets is accessible to users. SAS spatial datasets also must conform to the topological rules outlined by Boudriault (1987) and published by the American Society for Photogrammetry and Remote Sensing and the American Congress on Surveying and Mapping. These rules include topological completeness and topological geometric consistency. Spatial files failing to meet the topological criteria cause errors, alerting a user that quality control is necessary if spatial analysis is to be conducted. PROC GIS creates and maintains spatial datasets for use in SAS, and allows for batch accessibility to the GIS functionality. PROC MAPIMPORT can be used to import ESRI shapefiles into SAS. A user interface exists (found in the Solutions menu ► Analysis ► Geographic Information Systems), with an interactive GIS window (not supported on all platforms). This interface is not very intuitive, so producing sophisticated maps, although possible, is programmatically challenging. Mapping also can be done using SAS/GRAPH, which can be used to create four types of maps using PROC GMAP: two-dimensional choropleth maps and three-dimensional block, prism, and surface maps. SAS and ESRI also have partnered to create a bi-directional bridge between SAS data and analytical tools and the ERSI mapping environment. This bridge has been implemented by the U.S. Bureau of the Census to create school district demographics.

A.2.4 SAS implementation examples

Two examples of spatial statistical implementations within a GIS are presented here. Neither of these implementations takes advantage of built-in spatial functions within SAS, and both require only the base SAS license. The first example is the calculation of Moran's I, a straightforward computation and the creation of a Moran scatterplot. The second example is an implementation of eigenvector spatial filtering that includes a user interface within ArcGIS, a widely used GIS software package, and a simple exchange file in order for mapping to be done in ArcGIS and statistical analysis to be done in SAS.

Moran's I. Moran's *I* is a common statistical diagnostic that tests for spatial autocorrelation in data. First proposed by Moran (1950), it is implemented in a variety of software packages, including, but not limited to R, Geoda, and ArcGIS. This statistic essentially computes a weighted Pearson product moment correlation of a variable against itself, where the weighting relates to the variable's spatial arrangement (see Chapter B.5 for more details). Moran's *I* allows for the investigation of correlation within a single variable due to the spatial relationship amongst its observations. The work flow shown in Fig. A.2.1 illustrates the steps involved in its calculation.

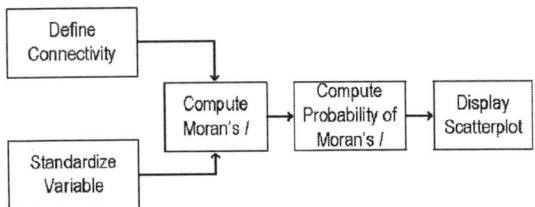

Fig. A.2.1. Moran's *I* workflow implemented in SAS

Initially, the variable of interest should be standardized (for example, the mean made equal to zero, and the standard deviation made equal to one); in SAS this can be done using PROC STANDARD or with simple calculations (that is, subtract the mean and divide by the standard deviation) in a DATA step. Next geographic connectivity should be defined for use as the weights. This is done by importing a neighbor file as in Table A.2.1. A connectivity matrix is created based on the weight information in the neighbor file. Next, Moran's *I* and the probability of Moran's *I* can be calculated. These calculations can be made either using PROC IML or inside a DATA step, depending on a programmer's preference for matrix or summation notation. Finally, using PROC GPLOT, a Moran scatterplot is displayed by plotting the variable information against the weighted variable information.

Tiefelsdorf and Boots (1995) show a relationship between Moran's *I* and the eigenvalues of $(I-11^T/n)C(I-11^T/n)$ (see Chapter B.5). After the calculation of the eigenvalues, a simple conversion calculates Moran's *I* again either in a data step or using PROC IML. This relationship is very useful, especially in the case where data have a massive number of observations. SAS code for both the traditional Moran's *I* calculation and eigenvalue conversion to Moran's *I* is available for download from Rura (2008).

Eigenvector spatial filtering with SAS and ArcGIS. An example of a spatial statistical model implemented in SAS is the technique of eigenvector spatial filtering. This spatial regression technique accounts for spatial autocorrelation in georeferenced variables by including map patterns based on, say topological connectivity as covariates in a model (see Chapter B.5 for details on this method). The map patterns are portrayals of eigenvectors extracted from a connectivity matrix of the underlying surface. This technique can be implemented in a tight coupling of SAS and ArcGIS.

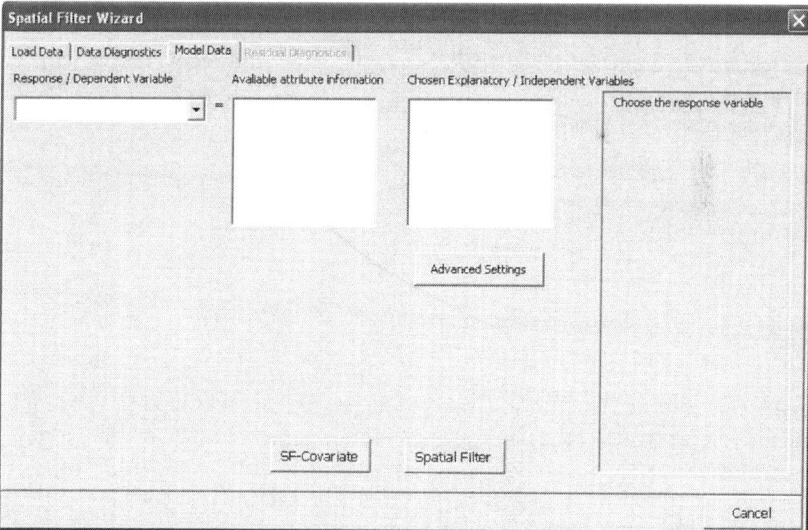

Fig. A.2.2. Visual Basic interface inside ArcGIS, the Load Data and Model Tabs; information is collected through this interface and sent to SAS programmatically for computation

Combining the established statistical procedures in SAS with the familiar interface of ArcGIS, a tool for creating eigenvector spatial filter models was created. This tool consists of a Visual Basic (VB) interface within ArcGIS that acts as a simple facade for SAS procedures executing in the background. The interface consists of four tabs: Load Data, Data Diagnostics, Model Data, and Residual Diagnostics (see Fig. A.2.2).

This tool, consisting of an open source ArcMap map file (for example, *.mdx) and a series of SAS program files (for example, *.sas), and can be downloaded from Rura (2008). The SAS programs can be run independently from ArcGIS to produce the spatial statistical outputs, but they include no mapping functionality within SAS.

Necessary data. Although eigenvector spatial filter models can be incorporated into a generalized linear model specification (and probably should be when data are counts or percentage), this implementation assumes that the given data can be modeled by a Gaussian-normal spatial linear regression. First, two pieces of information are necessary: the variable information and the geographic connectivity information. Generally, the attribute variable information is stored in a database. Since this program interacts with shapefiles, a Dbase4 (for example, *.dbf) file is assumed; but when running the program independently from ArcGIS, any database format supported by SAS (see SAS 2008 for a current listing of supported formats) can be used. The definition of the surface connectivity might be characterized in many ways, including contiguity rules or distance measures (see Cliff and Ord 1981; Griffith 1987). The definition of connectedness of a surface should be considered carefully by a user, and be justifiable in theoretical terms. This implementation only requires that the defined connectivity can be written into a file of the form shown in Table A.2.1, where each row is written in the following format: a unique ID, a region ID, a neighbor ID, and the weight associated with a connection.

Table A.2.1. Neighbor file format

ID	Region	Neighbor	Weight
1	1	3	1
2	1	2	1
3	2	1	1
4	2	4	1

When executed inside ArcGIS, this file should be comma-delimited (for example, *.csv), and can be created by default using a button in the tool. The default connectivity creates a first-order Queen's adjacency connectivity matrix, using the ArcGIS spatial query 'esriSpatailRelTouches' to query each region for neighbors. If topological problems exist with data (for example, slivers or unclosed polygons), mistakes will occur in the resulting neighbor file. The accuracy of any neighbor file should be checked. Once a neighbor file is created, considered reliable and loaded, an output path should be specified, where the statistical outputs from SAS are stored; a list of these outputs is found in Table A.2.2. Also, within this toolset is a set of geographic data diagnostic tools (not discussed here); some of these diagnostics require centroid locations. This centroid file can be specified as a point or a polygon file containing X and Y coordinate pairs in an ArcGIS attribute table, or when run in SAS, any table containing centroid values; these values are not necessary to specify a spatial filter model.

The model: Spatial filtering with eigenvectors in SAS. After data are collected, perhaps the response variable transformed, and a spatial regression model specified based upon data conceptualization and diagnostics, a Gaussian-normal linear spatial filter model can be computed. Within the interface in ArcGIS, a response variable and a set of explanatory variables are chosen from the attribute fields of a polygon file. The advanced button allows a user to set the adjusted Moran Coefficient (MC/MC_{Max}) threshold (see Chapter B.5 and Griffith 2003 for details) and the stepwise selection criterion for a model. Finally, the Spatial Filter button calls a function that initiates an instance of SAS, and the information input into ArcGIS by a user is read by SAS and used to calculate a spatial filter model. This interface is a convenient way for collecting the information used by SAS. The file read into SAS is comma-delimited, and includes the following information:

   Attribute, the file-path and filename for the variable information;
   Neighbor, the file-path and filename for the neighbor file;
   Response, the response variable name;
   Explain, the set of explanatory variable names, space delimited;
   Selection, the stepwise selection criterion;
   MCadj, the adjusted Moran Coefficient threshold value; and,
   SavePath, the file-path where all output information is to be saved.

An example of this file and other data formats can be downloaded from Rura (2008).

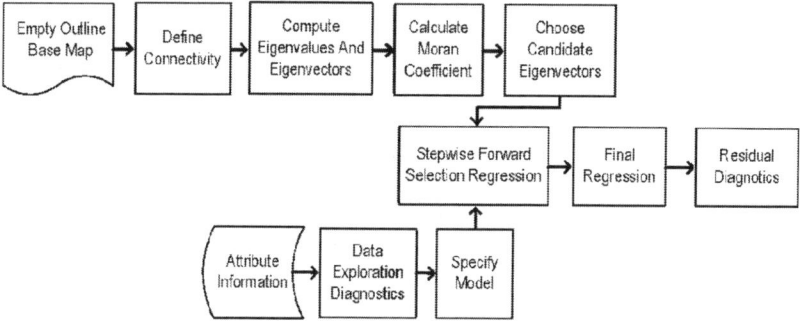

Fig. A.2.3. Eigenvector spatial filtering work flow implemented in SAS

The flow chart appearing in Fig. A.2.3 shows the work flow and the procedures used within SAS to create an eigenvector spatial filter model. First, a connectivity matrix is created from a neighbor file using PROC IML. This matrix is pre- and post-multiplied by the standard projection matrix, making the mean of all but one of the eigenvectors zero (see Griffith 2003 for details). Next, the sets of eigenvalues and eigenvectors are extracted from the connectivity matrix using the EIGEN function. In the case of positive spatial autocorrelation, a threshold set of eigenvectors containing positive spatial autocorrelation, called the candidate set, is chosen by a PROC SQL statement, using a user-defined minimum amount of positive spatial autocorrelation. This candidate set of eigenvectors is included in the Gaussian-normal linear regression model. The subsequent stepwise regression, which forces the chosen explanatory variables to remain in a model, is executed (for example, *PROC REG; Model Response = Explain CandidateEigenvectors /selection=stepwise sle= UserValue include = Number of Explain*), choosing those eigenvectors that are statistically significant, and that explain the most residual variation in the response variable. Then these eigenvectors are used to specify a final Gaussian-normal linear regression model that includes the explanatory variable and the selected eigenvectors. Finally, using the SAS *ods* system, the files in Table A.2.2 are exported to the file-path specified by a user, a Dbase4 file containing an ID, the spatial filter, and the model residuals is joined to the attribute table of the given polygon file, and the constructed spatial filter is mapped in ArcGIS.

Table A.2.2. Eigenvector spatial filtering for ArcGIS and SAS output file descriptions

Output	Description
Map	(SFmap.dbf) The response variable, the predicted values, and the residuals for each region are automatically joined to the associated polygon file (these are mapped in ArcGIS)
Initial data diagnostics	(Diagnostics.pdf) Univariate diagnostics, correlation diagnostics, and SAS experimental *ods* graphics for response and explanatory variables
Stepwise regression information	(StepwiseReg.pdf) SAS stepwise regression output (including all steps), univariate diagnostics of residuals, and output from regression of the observed on the predicted values
Final regression information	(FinalReg.pdf) SAS regression output, including ods graphs, univariate residual diagnostics, and the regression of the observed on the predicted values
Chosen eigenvectors	(SFChosenEV.dbf) Eigenvectors chosen to be included in the final spatial filter regression (an output file)
Coefficient values	(FinalCoef.dbf) The regression coefficient values for the intercept, covariates and each chosen eigenvector
Log	(SFCOVlog.pdf) The SAS log file

No SAS/ESRI bridge is used for this implementation. A text file containing information collected using the VB interface in ArcGIS is written to a predefined location and an instance of SAS is initiated. Next, SAS reads this file and uses the path and file names to automatically import the data into SAS. Then a series of sql statements results in the storage of variable names and data information. Finally, SAS executes the code and exports the output to a user-defined location. After the completion of this task, the SAS instance is closed, and the output table is automatically joined to the shapefile attribute table and mapped in ArcGIS.

A.2.5 Concluding remarks

The tools available in SAS for statistical analysis have been used for decades to analyze spatial statistical problems. Often this analysis is done through customization, although SAS now embeds spatial functionality into function options through its own GIS module. An implementation of eigenvector spatial filtering in SAS is described here in order to illustrate how customized SAS code can be created to put spatial statistical techniques into practice. This implementation uses standard statistical procedures and SAS functionality to estimate a spatial statistical model, expanding the already existing PROC NLIN-based spatial autoregressive code capabilities. The flexibility of SAS's statistical procedures, its stability with large datasets, and its ability to interface with many different software packages across a variety of platforms gives a researcher a large tool box from which to build upon in finding solutions to many different kinds of spatial statistical problems.

References

Boudriault G (1987) Topology in the TIGER File. In Chrisman N (ed) Proceedings of the Eighth International Symposium on Computer Assisted Cartography (Auto-Carto 8), Am Soc Photogramm Remote Sens and Am Congr Surv Mapp, Bethesda [MD], pp.258-263

Cliff AD, Ord JK (1981) Spatial processes: models and applications. Pion, London

Griffith DA (1987) Spatial autocorrelation: a primer. Association of American Geographers, Washington [DC]

Griffith DA (1993) Spatial regression analysis on the PC: spatial statistics using SAS. Association of American Geographers, Washington [DC]

Griffith DA (2003) Spatial autocorrelation and spatial filtering: gaining understanding through theory and scientific visualization. Springer, Berlin, Heidelberg and New York

Griffith DA (2009) Spatial filtering. In Fischer MM, Getis A (eds) Handbook of applied spatial analysis. Springer, Berlin, Heidelberg and New York, pp.301-318

Griffith DA, Layne LJ, Ord JK, Sone A (1999) A casebook for spatial statistical data analysis: A compilation of analyses of different thematic data sets. Oxford, New York

Moran PAP (1950) Notes on continuous stochastic phenomena. Biometrika 37(1):17

Moser EB (1987) The analysis of mapped spatial point patterns. In Smith P (ed) Proceedings of the 12th Annual SAS Users Group International Conference, SAS Institute, Cary [NC], pp.1141-1145

Rura MJ (2008) Web based storage and downloads. http://melissa.rura.us

SAS (2008) SAS customer support. http://support.sas.com/index.html

SAS Institute Inc (2008a) SAS/GIS 9.2: spatial data and procedure guide. SAS Institute Inc., Cary [NC]

SAS Institute Inc (2008b) SAS/Graph 9.2 Reference. SAS Institute Inc., Cary [NC]

Tiefelsdorf M, Boots B (1995) The exact distribution of Moran's I. Environ Plann 27(6): 985-999

UCLA: Academic Technology Services (2008) Introduction to SAS. http://www.ats.ucla.edu/stat/sas /notes2/

Waller L, Gotway C (2004) Applied spatial statistics for public health data, chapter appendix. Wiley. http://www.sph.emory.edu/ lwaller/WGindex.htm

Wang F (2006) Quantitative methods and applications in GIS. CRC Press (Taylor and Francis Group), Boca Raton [FL], London and New York

Yiannakoulias N (2008) SAS programming for spatial problems. http://www.ualberta.ca/ nwy/

A.3 Spatial Econometric Functions in R

Roger S. Bivand

A.3.1 Introduction

Developments in the R implementation of the S data analysis language are providing new and effective tools needed for writing functions for spatial analysis. The release of an R package for constructing and manipulating spatial weight, and for testing for global and local dependence during 2001 has been followed by work on functions for spatial econometrics (package spdep; the package may be retrieved from: http://cran.r-project.org). This chapter gives an introduction to some of the issues faced in writing this package in R, to the use of classes and object attributes, and to class-based method dispatch. In particular, attention will be paid to the question of how prediction should be understood in relation to the most commonly employed spatial econometrics simultaneous autoregressive models. Prediction is of importance because fitted models may reasonably be expected to be used to provide predictions of the response variable using new data – both attribute and position – that may not have been available when the model was fitted.

Class-based features are important because they encapsulate information about the data in a generic way, also when the data is given, for example, in form of a model formula, an object describing spatial neighbourhood relationships, or the results of fitting a model to data. This permits the flexible handling of subsetting, missing data, dummy variables, and other issues, based on existing classes that are extended to handle spatial econometrics functions. For the analyst, it is convenient if generic access functions can be applied to spatial analysis classes, such as making a summary or plotting a spatial neighbours' structure. The same applies to the use of model formulae, describing the model to be estimated, for a range of estimating functions. In this setting, a spatial linear model should build on the classes of the arguments of the underlying linear model. There should be no difference in the syntax of shared arguments between the a-spatial linear model, spatial econometrics models, or geographically weighted regression models, although of course function-specific arguments would be introduced.

It is also of interest to compare spatial econometric formulations with other related model structures, such as those for mixed effects models, and to explore other alternative approaches. These may include extensions to repeated measurements, to spatial time series, and to generalised linear models, although here the spatial case is often currently unresolved. However, the underlying classes are important in that their implementation may make the flexible extension of spatial analysis tools more or less difficult, and consequently may admit the quick prototyping of experimental new modelling techniques rather than hinder it.

It is clear that different disciplines and data analysis communities do not approach the writing of code, or the use of command line interfaces, in the same ways, and have varying expectations regarding the concerns of users. It is however arguable that language environments such as S, and its implementations R and S-PLUS, are instrumental in reducing barriers between users, who are not supposed to meddle with the software, but who can be expected to know about their data and methods, and developers. When these qualities of the S language environment for data analysis are coupled to free access to source code, opportunities for mutual peer-review and exchange between and among users and developers arise that are otherwise very difficult to create.

This chapter – after sketching the position of the R project and the spdep package – first reviews some open problems in spatial econometrics and then draws on the experience of the class/method mechanisms in S and R, including a discussion of the use of classes in spdep at present. This leads to an extended discussion of how prediction might be approached in spatial econometrics, since predict() is a method typically implemented for classes of fitted models. This is exemplified using the revised Harrison and Rubinfeld Boston house price data set, which is also distributed with spdep .

The R *project.* As summarised in brief in Bivand and Gebhardt (2000), R is a language and environment for statistical computing and graphics[1], and is similar to S. The S language is described and documented in Becker et al. (1988), Chambers and Hastie (1992), and more recently in Chambers (1998). There are differences between implementations of S: S-PLUS – which is a well-supported commercial product with many enhancements – manages both memory and data object storage in different ways from R. The chief syntactic differences are described in Ihaka and Gentleman (1996). Perhaps the most comprehensive introduction to the use of current versions of S-PLUS and R is Venables and Ripley (2002); a simpler alternative for R is Dalgaard (2002).[2]

R is available as source, and as binaries for Unix/Linux, Windows, and Macintosh platforms[3]. Contributed code is distributed from mirrored archives

---

[1] See also Bivand (2006) and Bivand et al. (2008).

[2] See also http://www.r-project.org/doc/bib/R-books.html for an up to date list of books of relevance to R .

[3] Both R itself and contributed packages may be downloaded from http://cran.r-project.org.

following control for adherence to accepted standards for coding, documentation and licensing. The contributed packages are distributed as source, and for some platforms – including Windows – as binaries, which can in addition be updated on-line using the `update.packages()` function within R. As usual in Free Software projects, there is no guarantee that the code does what it is intended to do, but since it is open to inspection and modification, the analyst is able to make desired changes and fixes, and if so moved, to contribute them back to the community, preferably through the package maintainer.

The spdep package. The current version of the spdep package is a collection of functions to create spatial weight matrix objects from polygon contiguities, from point patterns by distance and tessellations, for summarising these objects, and for permitting their use in spatial data analysis[4]; a collection of tests for spatial autocorrelation, including global Moran's *I*, Geary's *c*, Hubert-Mantel general cross product statistic, and local Moran's *I* and Getis-Ord *G*, saddlepoint approximations for global and local Moran's *I*; and functions for estimating spatial regression models. It contains contributions including code and/or assistance in creating code and access to legacy data sets from quite a number of spatial data analysts. Full details are in the licence file installed with the package. It is indeed central to the dynamics of free software/open source software projects such as R and its contributed packages, that communities are brought into being and fostered, leading where appropriate to collaborative development, and indeed to the replacement of code or class structures found by users in the community to be unsatisfactory or limiting.

A.3.2 Spatial models and spatial statistics

It often seems to be the case that spatial statistical analysis, including spatial econometrics, finds it challenging to give insight into general relationships guiding a data generation process. It is quite obvious that inference to general relationships from cross-section spatial data using a-spatial techniques raises the question of whether the locations of the observations in relation to each other should not have been included in the model specification. We now have quite a range of tests for examining these kinds of potential mis-specifications. We can also offer tools for exploring and fitting local and global spatial models, so that perhaps better supported inferences may be drawn for the data set in question, under certain assumptions.

These assumptions are not in general easy or convenient to handle, and constitute a major part of the motivation for further work on inference for spatial data generation processes. As Ripley (1988, p.2) suggests and Anselin (1988, p.9) confirms, they remove hope that spatial data are a simple extension of time series

---

[4] The treatment of spatial weight matrices has been discussed in greater length in Bivand and Portnov (2004).

to a further dimension (or dimensions). The assumptions of concern (Ripley 1988) here include those affecting the edges of our chosen or imposed study region, how to perform asymptotic calculations and how this doubt impacts the use of likelihood inference, how to handle inter-observational dependencies at multiple scales (both short-range and long-range), stationarity, and discretisation and support. Ripley (1988, p.8) concludes: '(T)he above catalogue of problems may give rather a bleak impression, but this would be incorrect. It is intended rather to show why spatial problems are different and challenging'.

Although many of these challenges are intractable in the point-process part of spatial statistics, more has been done to address them here. In particular, it has been recognised for some time that if we have a simple null hypothesis to simulate the spatial process model, we can generate exchangeable samples permitting us to test how well the model fits the data. As Ripley (1992) notes, an early example of this approach for the non-point-process case is the use of Monte Carlo simulation by Cliff and Ord (1973, pp.50-52). Substantial advances have also been taking place in geostatistics (Cressie 1993; Diggle et al. 1998). In addition, the implications of large volumes of data from remote sensing and geographical information systems, including data with differing support, have been recognised in a recent review by Gotway and Young (2002).

One of the characteristics of treatments of the statistical modelling of spatial data – especially lattice data – is that changes in techniques occur slowly, despite radical changes in data acquisition and computing speed. Haining's discussion of the research agenda twenty years ago (Haining 1981, pp.88-89), focusing on spatial homogeneity and stationarity, is taken up again by him ten years later (Haining 1990, pp.40-50), and remains relevant. Apart from the actual difficulty of the problems, it may be argued that exploring feasible solutions has been hindered by poor access to toolboxes combining both the specificity needed for handling spatial dependence between observations and general numerical and statistical functions. The coming first of SpaceStat (Anselin 1995), then James LeSage's Econometrics Toolbox for MATLAB[5], have created important opportunities, which the R spdep package attempts to follow up and build upon. In addition, code by Griffith (1989) for MINITAB, and by Griffith and Layne (1999) for SAS and SPSS has been made available. Finally, the spatial statistics module for S-PLUS provides additional and supplementary analytical techniques in a somewhat different form (Kaluzny et al. 1996).

To concentrate attention on the problem at hand, it may help to express the relationship between data and model in a number of parallel ways:

$$\text{data} = \begin{Bmatrix} \text{model} \\ \text{fit} \\ \text{smooth} \end{Bmatrix} + \begin{Bmatrix} \text{error} \\ \text{residuals} \\ \text{rough} \end{Bmatrix} \quad\quad (A.3.1)$$

---

[5] http://www.spatial-econometrics.com/

where our general grasp of the spatial data generation process on the data is incorporated in the first term on the right hand side, while the second term comprises the difference between this understanding and the observed data for our possibly unique region of study (Haining 1990, p.29 and p.51; cf. Hartwig and Dearing 1979, p.10; Cox and Jones 1981, p.140).

The model term may be made up of say fixed and random effects, of global and local smooths, of a-spatial and spatial component models, of trend surface and variogram model components, or of locally or geographically weighted parts. The distribution of the error term is assumed to be known, and should be such that as much as possible of the predictable regularity is taken up in the model.

In general, the model term should give a parsimonious description of the process or processes driving the data, and techniques used to choose between alternative models should take this requirement into account. It is also not necessarily the case that the model should be fitted using all of the data to hand; indeed many model forms may be compared by partitioning the available data into training and testing subsets. This position in fact reaches back to fundamental questions regarding the application of statistical estimation methods to spatial data, especially when the goals of such application may include inference, generalisation to a wider domain than the data used for calibration (Olsson 1970, Gould 1970). In particular, Olsson's comment that: 'If the ultimate purpose is prediction, then it also follows that specification of the functional relationships is more urgent than specification of the geometric properties of a spatial phenomenon' (Olsson 1968, p.131) continues to point up the question of what is being inferred to in spatial statistical analyses, also known as the geographical inference problem.

A.3.3 Classes and methods in modelling using R

Three main programming paradigms underly S: object-oriented programming, functional languages, and interfaces (Chambers and Hastie 1992, pp.455-480). Classes and methods were introduced to S at the time of this 1992 'White' book, and were not part of the 1988 'Blue' book (Becker et al. 1988) defining the fundamentals of the language. This step was, for practical reasons, incremental, and was intended to assist in the further development of modelling functions. For this reason, language objects may, but do not have to, have a class attribute – all objects may have attributes with name strings, and class is simply one such string with specific consequences for the way that functions in the system handle objects.

This established form of class and method use in S and hence R is the one which will be covered here. It should however be noted that a new class/method formalism has been introduced to S in the 1998 'Green' book (Chambers 1998), and is being introduced to R, as well as underlying S-PLUS 6.x. Programming using both styles of classes and methods is described in detail in Venables and Ripley (2000, pp.75-121). From the point of view of the user, however, the differences are either few or beneficial, and now require that each object shall have a class, and that each object of a given class shall have the same structure, requirements which were not present before.

The class/method formalisms in S have been adopted in the spirit of object-oriented programming, that evaluation should be data-driven. Functions for generic tasks, such as print(), plot(), summary(), or logLik(), are constructed as stubs that pass their own arguments through to the UseMethod(). In the following code snippets, > is the R command line prompt, entering the name of a function causes its body to be printed:

```
> print
function (x, ...)
UseMethod("print")
<environment: namespace:base>
```

Within UseMethod(), the first argument object is examined to see if it has an attribute named "class". If it does, and a function named, say, print."class"() exists, the arguments are passed to this function. If it has no class attribute, or if no generic function qualified with the class name is found, the object is passed to, say, print.default(). If we have estimated a spatial error model for the Columbus data set, and wish to display the log likelihood value of the object, we might do the following:

```
> COL.err <- errorsarlm(CRIME~INC+HOVAL,data = COL.OLD, nb2listw(COL.nb))
> class(COL.err)

[1] "sarlm"

> ll.COL.err <- logLik(COL.err)
> class(ll.COL.err)

[1] "logLik"
> ll.COL.err

'log Lik.' -183.3805 (df=5)
```

The model object COL.err has class sarlm, so the function used by method dispatch from logLik() is logLik.sarlm(), yielding a resulting object with class logLik. If an object with class logLik, is to be printed, UseMethod() will look for print.logLik(). As can be seen, this function expects the logLik object to be a scalar value, with an attribute named "df", the value of which is also printed.

```
> print.logLik
function (x, digits = getOption("digits"), ...)
```

```
{
    cat("'log Lik.' ", paste(format(c(x), digits = digits), collapse = ", "),
        " (df=", format(attr(x, "df")), ")", sep = "")
    invisible(x)
}
<environment: namespace:stats>
```

This brief example shows both the convenience of the class/method mechanism, and the reason for moving to the new style, since in the old style there are no barriers to prevent the class attribute of an object being changed or removed, nor are there any structures to ensure that class objects have the same properties. It could be argued that software code, and by extension the formalisms employed in writing software, such as class/method formalisms in object oriented programming described briefly above, are not of importance for advancing spatial data analysis.

A response to this position is that, for computable applications, abstractions and conjectures are enriched by being implemented in structured code, especially where the code is available, documented, and open to peer review, as in R and other community supported software projects and repositories. Further, formalisms such as class/method mechanisms also provide useful standards through which the assumptions and customs underlying computing practises may be exposed and compared. Finally, class/method mechanisms, in particular care in constructing classes, are associated with concern for data modelling as also understood for example in geographical information systems. In this case, it is important that classes support data types, structures, and metadata components adequately and in a robust way.

At present the key classes in spdep are written in the old style, and are "nb", "listw", "sarlm", and the generic class "htest" for hypothesis tests.[6] The first is for lists of neighbours, the second for sparse neighbour weights lists, and the third for the object returned from the fitting of SAR (simultaneous autoregressive) linear models of three types: lag, mixed, and error (corresponding to LeSage's sar(), sdm(), and sem() functions; there is no equivalent to his sac() function). The "htest" class is used to report the results of hypothesis tests, not least because `print.htest()` already existed, and conveniently standardised the displaying of test results.

The "sarlm" class is still under development, not least because writing methods leads to changes in components that need to be in the object itself, or can conveniently be computed at a later stage by functions such as `summary.sarlm()`, `logLik.sarlm()`, `residuals.sarlm()`, and so on. Migration to new-style classes will occur when the requirements have been refined following further exploration – old-style classes can be augmented without breaking existing code more easily than can new-style classes.

---

[6] While classes in spdep are still written in the old style, there are now many more of them, suiting the increasing number of model fitting functions and local indicators of spatial association; functions and methods in spdep use new style class objects defined in the sp package for spatial data, and in the Matrix package for sparse matrices.

The function that has prompted the most thought is however `predict.sarlm()`. Essentially all the fitted model classes in S (and R and its contributed packages) have methods for prediction, including prediction from new data. It is to this problem we will turn to show that class/method formalisms are more than a programming convenience, but also establish baselines for what analysts should expect from model fitting software.

A.3.4 Issues in prediction in spatial econometrics

Prediction may be subdivided into several similar kinds of tasks: calculating the fitted values when the values of the response variable observation are known and are those used in fitting the model, the same scenario, but when the predictions are not for observations used to fit the model, and finally predictions for observations for which the value of the response variable is unknown. Here we choose to measure the difference between the predicted and observed values of the response variable using the root mean square error of prediction. In the a-spatial linear model, predictions are a function of the fitted coefficients and their standard errors, and confidence intervals may be obtained using the fitted residual standard error. Extensions to the linear model can be furnished with prediction mechanisms in generally similar ways, although expressing standard errors and confidence intervals may become more difficult.

Work on filling in missing values (Bennett et al. 1984; Haining et al. 1989; Griffith et al. 1989) has not been followed up in the spatial econometrics literature, and was focused on the case when the position of an observation was known, but where one or more attribute values was missing (see also Martin 1990). This differs from prediction using new data where there is no contiguity between the positions of the data used to fit the model and the new data, where both the positions of the observations are new, and only explanatory variable values are available for making the prediction. Where contiguity between the data sets' positions is present, predicting missing values can be accommodated in the present approach; the main thrust of this literature has been to explore the consequences for parameter estimation of the absence of some data values. Given the provision noted by Martin (1984, p.1278) that data should be missing at random, it is not clear how to proceed when the new data adjoin the data used for fitting, for instance in one direction.

Trend, signal and noise. Prediction for spatial data may be seen as the core of geostatistics; most applications of kriging aim to interpolate from known data points to other points within or adjacent to the study area, or to other support. Interpolation of this kind also underlies the use of modern statistical techniques, such as local regression or generalised additive models among many others. As pointed out above, it is usual for prediction functions to accompany each new variety of fitted model object in S, not least because the comparison of prediction errors for in-sample and out-of-sample data give insight into how well models perform. Some model fitting techniques can be found to perform very well in relation to in-sample data, but do very poorly on out-of-sample data, that is, they are 'over-fitted'. While they may exhaust the training data, they will be very restricted to that particular region of data-space, and may perform worse than other, less 'over-fitted' models, on unseen test data.

The three terms: trend, signal and noise, are taken from Haining (1990, p.258), and the S-PLUS spatial statistics module (Kaluzny et al. 1996, pp.154-156), in which Haining's comment is followed up. In Haining (1990), the underlying linear model was a trend surface model, so that it was logical to partition the data into trend and noise

$$\underbrace{y}_{\text{data}} = \underbrace{X\beta}_{\text{trend}} + \underbrace{\varepsilon}_{\text{noise}} \qquad (A.3.2)$$

where $E[\varepsilon] = 0$ and $E[\varepsilon\varepsilon^T] = \sigma^2 I$. If we generalise this model to the error autoregressive form, we get

$$y = X\beta + u \qquad (A.3.3)$$

with $E[u] = 0$ and $E[uu^T] = V$. If we write $V = \sigma^2 LL^T$, and $L^{-1} = (I - \lambda W)$, we can rewrite the relationship

$$(I - \lambda W)y = (I - \lambda W)X\beta + \varepsilon \qquad (A.3.4)$$

$$\underbrace{y}_{\text{data}} = \underbrace{X\beta}_{\text{trend}} + \underbrace{\lambda W(y - X\beta)}_{\text{signal}} + \underbrace{\varepsilon}_{\text{noise}} \qquad (A.3.5)$$

To predict y, we could pre-multiply by $(I - \lambda W)^{-1}$

$$y = X\beta + (I - \lambda W)^{-1}\varepsilon \qquad (A.3.6)$$

which can yield the trend component, but for which the signal and noise components are combined. Cliff and Ord (1981, p.152, cf. pp.146-147) give $u =$

$\sigma(I - \lambda W)^{-1}\varepsilon$ as the simultaneous autoregressive generator from ε independent identically distributed random deviates, yielding $u \sim \mathcal{N}(0,V)$. If normality is assumed for ε, then u is multivariate normal. Here, predictions from error autoregressions are restricted to the trend component.

Kaluzny et al. (1996, pp.158-160) use Haining's results (1990, p.116) to suggest that a simulation of the unobservable autocorrelated error term may be used to attempt to predict the signal, but this necessarily depends on the assumption of normality. In the SAR case, they suggest computing $V = \sigma^2[(I - \lambda W)^T(I - \lambda W)]^{-1}$, next computing L as the lower triangular matrix of the Cholesky decomposition of V, and finally simulating u by $u = L\varepsilon$, where ε is a random deviate as above.

A further alternative based on work by Martin (1984, see also modifications by Haining et al. 1989; Griffith et al. 1989, and comment by Martin 1990) is to base the approximation of the unobservable autocorrelated signal on the projection of the residuals of the fitted process through a covariance matrix expressing the spatial dependence of the positions used to fit the model and the positions of the new data (using the spatial parameter from the fitted model). If the data used for fitting the model and the new data are not contiguous in position, this term is zero.

This alternative may be compared to the case of for time series with autocorrelated errors, since the estimate of the autoregressive coefficient is needed to make an estimate of the one-period forecast error (Stewart and Wallis 1981, pp.239-241; Johnson and NiDardo 1997, pp.192-193). Johnson and DiNardo term this the feasible forecast, and note that there is no closed form expression for the forecast variance in this case. Suppose we have $y_t = X_t^T\beta + u_t$, where $u_t = \lambda u_{t-1} + \varepsilon_t$. The same model can be written

$$y_t - \lambda y_{t-1} = X_t^T\beta - \lambda X_{t-1}^T \beta + \varepsilon_t. \tag{A.3.7}$$

Assuming λ known, β can be estimated, and substituting and rearranging, we can make a forecast of y_{t+1} by

$$\hat{y}_{t+1} = \underbrace{X_{t+1}^T\hat{\beta}}_{\text{trend}} + \underbrace{\lambda(y_t - X_t^T\hat{\beta})}_{\text{signal}} \tag{A.3.8}$$

for which the forecast variance is also available; the terms trend and signal here describe the non-autoregressive and the autoregressive components of the forecast by analogy with Haining's description. When we only have an estimate of λ, the feasible forecast becomes

$$\hat{y}_{t+1} = X_{t+1}^T \hat{\beta} + \hat{\lambda}(y_t - X_t^T \hat{\beta}) \quad \text{(A.3.9)}$$

that is the sum of products of the new x_{t+1} values and the $\hat{\beta}$ fitted using observations $1, \ldots, t$, plus $\hat{\lambda}$ times the residual at time t, representing the temporal dependency of the series, the forecast error for the one-step-ahead forecast.

Since t and $t+1$ are contiguous, it is possible to use the residual value from the fitted model in prediction in the time series case. In the simultaneous autoregressive spatial error model, when the new data positions coincide with, or are contiguous to, the positions of data used for fitting, it may be possible to calculate a signal component on the basis of the residuals of the fitted model and a rectangular matrix expressing the correlation structure of the original and new data positions. This approach has, however, not been attempted here, although Martin (1984, p.1279) provides a solution. To accommodate this, modifications to the current spatial weights list class in spdep are required, but have not yet been implemented. Consequently, for the simultaneous autoregressive error model, the prediction currently implemented in `predict.sarlm()` for the newdata case is the trend, and the signal is set to zero.

Haining's approach may be extended to the spatial lag model, in which dependence is not present in the error term, but rather in the dependent variable. Here we have

$$\underbrace{\hat{y}}_{\text{data}} = \underbrace{X\hat{\beta}}_{\text{trend}} + \underbrace{\hat{\rho}Wy}_{\text{signal}} + \underbrace{\varepsilon.}_{\text{noise}} \quad \text{(A.3.10)}$$

Rewriting, we have

$$(I - \rho W)y = X\beta + \varepsilon. \quad \text{(A.3.11)}$$

Once again, to predict y, we could pre-multiply by $(I - \rho W)^{-1}$

$$y = (I - \rho W)^{-1} X\beta + (I - \rho W)^{-1} \varepsilon. \quad \text{(A.3.12)}$$

The second term on the right hand side is equivalent to that in the error autoregressive case, and combines signal and noise components, while the first term combines trend and signal components.

As a first approximation, the `predict.sarlm()` function assumes that the trend can be expressed by $X\hat{\beta}$, and part of the signal by $\hat{\rho}W(I-\hat{\rho}W)^{-1}X\hat{\beta}$. The rationale is that if

$$(I - \hat{\rho} W)y = X \hat{\beta} \tag{A.3.13}$$

$$\hat{y} = (I - \hat{\rho}W)^{-1} X \hat{\beta} \tag{A.3.14}$$

then the signal may be approximated by

$$\hat{\rho} W \hat{y} = \hat{\rho}W(I - \hat{\rho}W)^{-1} X \hat{\beta}. \tag{A.3.15}$$

While this yields an estimate of part of the signal component, it is not complete, for new data missing the part combined with the noise component. This is clearly less than adequate, and more work is required here, as with the completely missing signal component for the error model.

Finally, it has been assumed that the weight matrix used for fitting the model is furnished with attributes detailing its construction: whether it is row standardised, and which type of underlying binary or general neighbourhood representation has been used (contiguity, distance, triangulation, k-nearest neighbours, etc.). Consequently, in predicting from new data, it is expected that the new attribute data will be accompanied by a suitable spatial weight list. This is not used in the error model predictions, but is used for the lag model, in the approximation to the part of the signal component described above.

Even if prediction for new data is as yet less well grounded, the partition of spatial model fitted values into trend and signal allows us to use alternative diagnostic plots. Examples of such plots for the data set discussed in Section A.3.5 below are shown in Fig. A.3.1. Tracts lying in towns in Boston city are distinguished in the plot, since their patterns seem to indicate different behaviour both in relation to the a-spatial trend, and the spatial autoregressive error signal. It may be remarked that the fit of the spatial error model (AIC = −506.85) is better than that of the spatial lag model (AIC = −496.02), than the a-spatial linear model (AIC = −283.96), but worse than the mixed spatial lag model (AIC = −543.23). The full results may be obtained by executing `example(boston)` after loading spdep into R[7], in which the sphere of influence row standardised weighting scheme is also presented.

---

[7] Copy and paste commented out lines from the help page to the console.

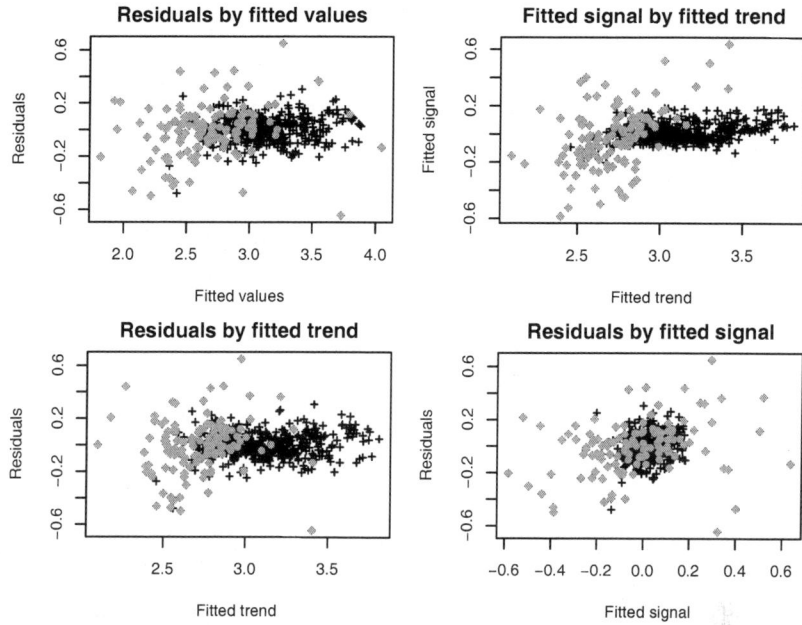

Fig. A.3.1. Boston tract log median house price data: plots of spatial autoregressive error model fit components and residuals for all 506 tracts; tracts in towns in Boston plotted with a grey •

A.3.5 Boston housing values case

The data set chosen here is that described by Gilley and Pace (1996), a revision of the Harrison and Rubinfeld Boston hedonic house price data, relating median house values to a range of environmental and social variables over 506 tracts. It is chosen because it is easily available, it has been used in a range of spatial econometric studies, including particularly LeSage's online materials on spatial econometrics[8]. The original data set is also featured as one of a corpus of machine learning datasets[9], and as such is well suited to applications such as the present. Most use of this dataset in machine learning research also seems to ignore the spatial nature of the data. Here, two prediction settings will be used.

In the first, the data are divided into northern and southern parts at UTM zone 19 northing 4,675,000m (dividing the tracts into two almost equal groups, with the dividing line running through the Boston city tracts). The data frame is subsetted by a logical variable expressing whether the centre point of the tract is north or

---

[8] http://www.rri.wvu.edu/WebBook/LeSage/etoolbox/index.html
[9] ftp://ftp.ics.uci.edu/pub/machine-learning-databases

south of the dividing line. The spatial weights used are constructed using the sphere of influence approach based on a triangulation of the UTM zone 19 projected tract centres, subsetted using the same north/south logical variable. An ordinary least squares model was fitted to each of the parts of the city, and predictions were made with the data used for fitting the models, and then using the model fitted on the southern data with the northern data, and vice-versa. The same procedure was repeated for the spatial lag model, the spatial error model, and the spatial mixed model (the spatial lag model augmented with the spatial lags of the explanatory variables – also known as the spatial Dublin model).

Although it can be seen from Fig. A.3.2 that the spatial models are better fitted to the data, the cross-predictions are no better than, and often worse than those for the a-spatial linear model (lm). The linear model gives the best prediction of the southern median house prices using the fitted coefficient values from the northern data. At least part of the reason for this is that the fits of the models, both a-spatial and spatial coefficient values, differed between the two parts of the metropolitan area, suggesting that spatial regimes and/or non-stationarity are present. This could be held to justify the abandonment of methods not accommodating this lack of stability in parameter estimates across the chosen data set, for example by comparing the fit of a geographically weighted regression with the baseline model. This will, however, not be pursued here, although some indication is given of the specific behaviour of Boston city tracts is given in Fig. A.3.1.

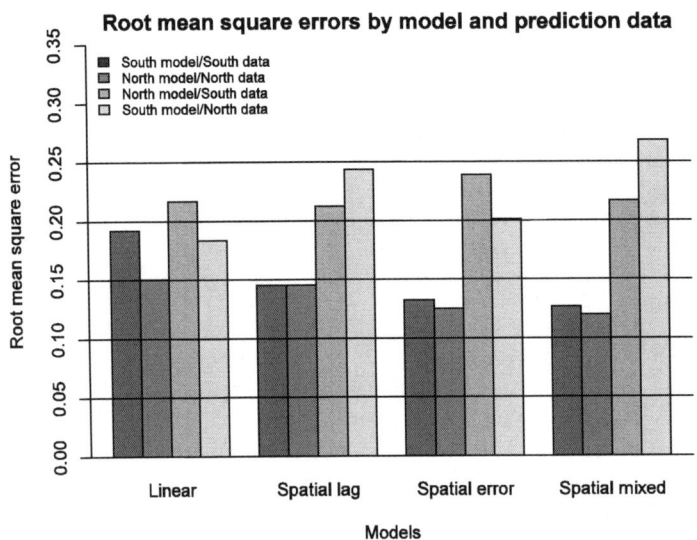

Fig. A.3.2. Comparison of model prediction root mean square errors for four models divided north/south, Boston house price data

In the second approach, 100 samples of 250 in-sample tracts were chosen, leaving 256 tracts out-of-sample. The samples were replicated in order to get a feeling for the variations in predictions which could result. Here, the spatial weight matrices were prepared for each data set as row standardised schemes for the six nearest neighbours of each tract centre (UTM zone 19). In addition, use was made of the gam() function in package mgcv to fit a generalised additive model (see Kelsall and Diggle 1998 for a similar use of GAM). In this specification, the model fitted was:

$$y = X\beta + s(\text{lon}, \text{lat}) + \varepsilon \qquad (A.3.16)$$

where $s(\text{lon}, \text{lat})$ is a smoothing function using a penalised thin plate regression spline basis in 12 dimensions to incorporate spatial dependence. Alternative modern statistical fitting techniques could have been used, and here the joint smoothing of longitude and latitude was chosen after inspecting the results of smoothing each of them and their interaction separately. Although such fitting techniques are not typically used in spatial econometric analyses, it may be of interest to compare prediction results across such analysis-community boundaries. It can be noted that GAM predictions in the first setting, with the data set divided into Northern and Southern parts, were very poor when predicting for new data.

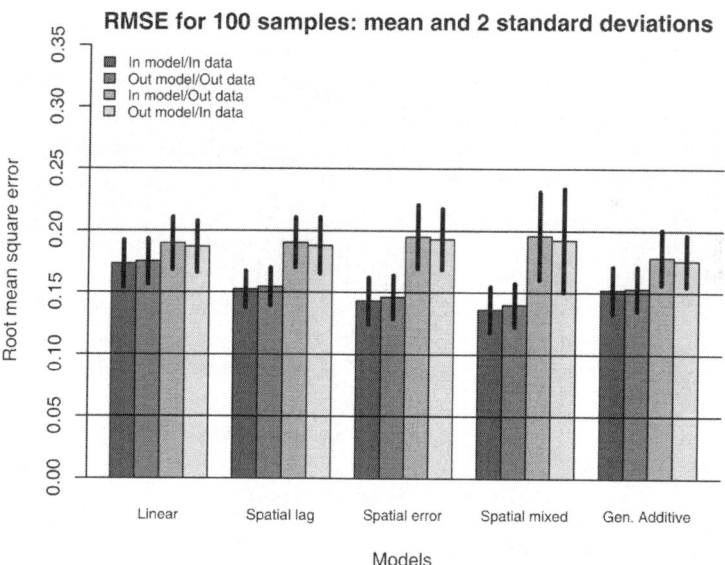

Fig. A.3.3. Comparison of model prediction root mean square errors means and standard deviations for 100 random samples of 250 in-sample tracts and 256 out-of-sample tracts, for five models, Boston house price data

Figure A.3.3 reinforces the results of testing model predictions after dividing Boston into two parts. The linear model (lm) has the least satisfactory fit within the sample from which the model was fitted, but performs as well or better than all the spatial econometrics models when predicting for other data than those used to fit the model. The mixed spatial lag model (the Common Factor model) does best in predicting on the training set the data it was fitted with, but worst on the test-excluded data. This may be taken as an indication of over-fitting, capturing too much of the specificity of the spatial dependencies of the training data set. The performance of the generalised additive model is better than that of the linear model both on the training and the test data sets, despite the 'black-box' nature of the specification of the spatial pattern in this case as penalised thin plate regression spline.

A.3.6 Concluding remarks

Among the opportunities and challenges posed by trying to implement spatial econometric techniques in R in the spdep package have been issues raised by the object-oriented data-driven approach implicit in classes and methods. So far, old style classes and methods have been used for spatial neighbour objects, spatial weights objects, and for spatial simultaneous autoregressive model objects. Many of the methods usually accompanying fitted model objects are simple to write, but `predict.sarlm()` revealed areas of spatial econometrics which perhaps have received little attention hitherto. The current implementation does however need to be augmented to handle situations in which the dependencies between the locations of observations from which the model to be used for prediction was fitted, and the locations of new data observations, can be represented as a correlation structure of some kind, thus better capturing the signal component.

It does seem that Haining's partitioning of the fitted values of spatial models is of interest in itself, as indicated by the diagnostic plots in Fig. A.3.1. It may well be that such diagnostic plots, perhaps dynamically linked to maps, will help us in establishing which further misspecification problems are present in our spatial models, shifting focus from criticising the mis-specification of a-spatial models to trying to construct spatial models with better properties. Haining's proposals for more general regression diagnostics for models in which spatial dependence is present do not seem to have as yet met with the acceptance they deserve (Haining 1990, 1994). Prediction for new data and new spatial weight matrices is a challenge for legacy spatial econometric models, raising the question of what spatial predictions should look like. Can for example spatial econometrics models be recast as mixed effects models, since as Pinheiro and Bates (2000) show, spatial correlation structures can 'plugged' into such models?

A further consequence of examining fitted model classes and methods, in particular with regard to prediction, is to question whether we need to fit models on very large data sets. Can we not rather fit and refine them on smaller data sets and predict or interpolate to larger data sets? Housing values are not infrequently the subject of analysis, and would perhaps be an attractive target for prediction. An advantage of fitting on moderate sized data sets, maybe training sets from larger data collections, is that the use of sparse matrix techniques in some circumstances would become unnecessary. Standard errors of prediction remain open.

It also seems that a relaxation of single data set fitting of spatial econometrics models may also help to lower barriers between geostatistics and legacy spatial econometrics models when using distance criteria for representing dependence. It appears that some movement is already taking place in this regard, given the use of spatial covariance in Ord and Getis (2001) in the development of the $O_i(d)$ local spatial autocorrelation statistic allowing for global dependence. In addition, the Getis filtering approach (Getis 1995; Getis and Griffith 2002) is distance based, and seems to admit prediction to new data locations using the distance criteria and filtering functions recorded in the fitted model. The Griffith eigenfunction decomposition approach discussed in Getis and Griffith (2002), and described in detail in Griffith (2000a, 2000b), does not, however, seem to open for prediction to new locations not contiguous with the locations on which the estimated model was fitted, because of its clear focus on the eigenvectors of the spatial weight matrix of the training data set. In addition, the selection of the eigenvectors to use for filtering may not transfer between geographical settings. For a detailed discussion on spatial filtering see Chapter B.5.

Finally, focusing on prediction using spatial econometric models does concentrate attention on assumptions about spatial homogeneity, including stationarity, support, multi-scale issues, and edge effects. Approaching modern statistical techniques as it were from the other side, we find work on geographically weighted regression (Brunsdon et al. 1996) and geographically weighted summary statistics (Brunsdon et al. 2002), in which many of these assumptions are addressed directly. In this context, it would be worthwhile to be able to test a geographically weighted regression fit against say a spatial error model fit, for instance by implementing a model comparison function like `anova(gwr.fit,sarlm.fit)`. But it is the flexibility of a language environment such as R, and the fruitfulness of class and method formalisms, that give rise to such projects for future research and implementation.

References

Anselin L (1988) Spatial econometrics: methods and models. Kluwer, Dordrecht
Anselin L (1995) SpaceStat version 1.80 user's guide. Regional Research Institute, West Virginia University, Morgantown [WV]
Becker RA, Chambers JM, Wilks AR (1988) The new S language. CRC Press (Taylor and Francis Group), Boca Raton [FL], London and New York
Bennett RJ, Haining RP, Griffith DA (1984) The problem of missing data on spatial surfaces. Ann Assoc Am Geogr 74(1):138-156

Bivand RS (2006) Implementing spatial data analysis software tools in R. Geogr Anal 38(1):23-40
Bivand RS, Gebhardt A (2000) Implementing functions for spatial statistical analysis using the R language. J Geogr Syst 2(3):307-317
Bivand RS, Portnov BA (2004) Exploring spatial data analysis techniques using R: the case of observations with no neighbours. In Anselin L, Florax R, Rey S (eds) Advances in spatial econometrics. Springer, Berlin, Heidelberg and New York, pp.121-142
Bivand RS, Pebesma EJ, Gómez-Rubio V (2008) Applied spatial data analysis with R. Springer, Berlin, Heidelberg and New York
Brunsdon C, Fotheringham AS, Charlton M (1996) Geographically weighted regression: a method for exploring spatial nonstationarity. Geogr Anal 28(4):281-289
Brunsdon C, Fotheringham AS, Charlton M (2002) Geographically weighted summary statistics: a framework for localised exploratory data analysis. Comput Environ Urban Syst 26(6):501-524
Chambers JM (1998) Programming with data. Springer, Berlin, Heidelberg and New York
Chambers JM, Hastie TJ (1992) Statistical models in S. CRC Press (Taylor and Francis Group), Boca Raton [FL], London and New York
Cliff, AD, Ord JK (1973) Spatial autocorrelation. Pion, London
Cox NJ, Jones K (1981) Exploratory data analysis. In Wrigley N, Bennett RJ (eds) Quantitative geography: a British view. Routledge, London, pp.135-143
Cressie NAC (1993) Statistics for spatial data (revised edition). Wiley, New York, Chichester, Toronto and Brisbane
Dalgaard P (2002) Introductory statistics with R. Springer, Berlin, Heidelberg and New York.
Diggle PJ, Tawn JA, Moyeed RA (1998) Model-based geostatistics. J App Stat 47(3):299-350
Getis A (1995) Spatial filtering in a regression framework: examples using data on urban crime, regional inequality, and government expenditure. In Anselin L, Florax RJGM (eds) New directions in spatial econometrics. Springer, Berlin, Heidelberg and New York, pp.172-185
Getis A, Griffith DA (2002) Comparative spatial filtering in regression analysis. Geogr Anal 34(2):130-140
Gilley OW, Pace RK (1996) On the Harrison and Rubinfeld data. J Environ Econ Manag 31(3):403-405
Gotway CA, Young LJ (2002) Combining incompatible spatial data. J Am Stat Assoc 97:632-648
Gould P (1970) Is *Statistix Inferens* the geographical name for a wild goose? Econ Geogr 46(Supp):439-448
Griffith DA (1989) Spatial regression analysis on the PC: spatial statistics using MINITAB. Discussion Paper, Institute of Mathematical Geography, Ann Arbor [MI]
Griffith DA (2000a) A linear regression solution to the spatial autocorrelation problem. J Geogr Syst 2(2):141-156
Griffith DA (2000b) Eigenfunction properties and approximations of selected incidence matrices employed in spatial analyses. Lin Algebra Appl 321(1-3):95-112
Griffith DA (2009) Spatial filtering. In Fischer MM, Getis A (eds) Handbook of applied spatial analysis. Springer, Berlin, Heidelberg and New York, pp.301-318
Griffith DA, Layne LJ (1999) A casebook for spatial statistical data analysis: a compilation of analyses of different thematic data sets. Oxford University Press, Oxford and New York

Griffith DA, Bennett RJ, Haining RP (1989) Statistical analysis of spatial data in the presence of missing observations: a methodological guide and an application to urban census data. Environ Plann A 21(11):1511-1523

Haining RP (1981) Spatial and temporal analysis: spatial modelling. In Wrigley N, Bennett RJ (eds) Quantitative geography: a British view, Routledge and Kegan Paul, London, pp.86-91

Haining RP (1990) Spatial data analysis in the social and environmental sciences. Cambridge University Press, Cambridge

Haining RP (1994) Diagnostics for regression modeling in spatial econometrics. J Reg Sci 34(3): 325-341

Haining RP, Griffith DA, Bennett RJ (1989) Maximum-likelihood estimation with missing spatial data and with an application to remotely sensed data. Comm Stat Theor Meth 18(5):1875-1894

Hartwig F, Dearing BE (1979) Exploratory data analysis. Sage, Beverly Hills [CA]

Ihaka R, Gentleman R (1996) R: a language for data analysis and graphics. J Comput Graph Stat 5(3):299-314

Johnson J, NiDardo J (1997) Econometric methods. Mc Graw Hill, New York

Kaluzny SP, Vega SC, Cardoso TP, Shelly AA (1996) S+SPATIALSTATS users manual version 1.0. MathSoft Inc., Seattle [WA]

Kelsall JE, Diggle PJ (1998) Spatial variation in risk of disease: a nonparametric binary regression approach. J Roy Stat Soc C Appl Stat 47(4):449-473

Martin RJ (1984) Exact maximum likelihood for incomplete data from a correlated Gaussian process. Comm Stat Theor Meth 13(10):1275-1288

Martin RJ (1990) The role of spatial statistical processes in geographical modelling. In Griffith DA (ed) Spatial statistics: past, present, and future. Institute of Mathematical Geography, Ann Arbor, Michigan, pp.109-127

Olsson, G (1968) Complementary models: a study of colonization maps. Geogr Ann B Hum Geogr 50(2):115-132

Olsson, G (1970) Explanation, prediction, and meaning variance: an assessment of distance interaction models. Econ Geogr 46(Supp):223-233

Ord JK, Getis A (2001) Testing for local spatial autocorrelation in the presence of global autocorrelation. J Reg Sci 41(3):411-432

Pinheiro JC, Bates DM (2000) Mixed-effects models in S and S-PLUS. Springer, Berlin, Heidelberg and New York

Ripley BD (1988) Statistical inference for spatial processes. Cambridge University Press, Cambridge

Ripley BD (1992) Applications of Monte-Carlo methods in spatial and image analysis. In Jöckel KJ, Rothe G, Sendler W (eds) Bootstrapping and related techniques. Springer, Berlin, Heidelberg and New York, pp.47-53

Stewart MB, Wallis KF (1981) Introductory econometrics. Blackwell, Oxford

Venables WN, Ripley BD (2000) S programming. Springer, Berlin, Heidelberg and New York

Venables WN, Ripley BD (2002) Modern applied statistics with S-PLUS (4th edition). Springer, Berlin, Heidelberg and New York

A.4 GeoDa: An Introduction to Spatial Data Analysis

Luc Anselin, Ibnu Syabri and *Youngihn Kho*

A.4.1 Introduction

The development of specialized software for spatial data analysis has seen rapid growth since the lack of such tools was lamented in the late 1980s by Haining (1989) and cited as a major impediment to the adoption and use of spatial statistics by GIS researchers. Initially, attention tended to focus on conceptual issues, such as how to integrate spatial statistical methods and a GIS environment (loosely vs. tightly coupled, embedded vs. modular, etc.), and which techniques would be most fruitfully included in such a framework. Familiar reviews of these issues are represented in, among others, Anselin and Getis (1992), Goodchild et al. (1992), Fischer and Nijkamp (1993), Fotheringham and Rogerson (1993, 1994), Fischer et al. (1996), and Fischer and Getis (1997). Today, the situation is quite different, and a fairly substantial collection of spatial data analysis software is readily available, ranging from niche programs, customized scripts and extensions for commercial statistical and GIS packages, to a burgeoning open source effort using software environments such as R, Java and Python. This is exemplified by the growing contents of the software tools clearing house maintained by the U.S.-based Center for Spatially Integrated Social Science [CSISS] (see http://www.csiss.org/clearinghouse/).

CSISS was established in 1999 as a research infrastructure project funded by the U.S. National Science Foundation in order to promote a spatial analytical perspective in the social sciences (Goodchild et al. 2000). It was readily recognized that a major instrument in disseminating and facilitating spatial data analysis would be an easy to use, visual and interactive software package, aimed at the non-GIS user and requiring as little as possible in terms of other software (such as GIS or statistical packages). *GeoDa* is the outcome of this effort. It is envisaged as an 'introduction to spatial data analysis' where the latter is taken to consist of visualization, exploration and explanation of *interesting* patterns in geographic data.

The main objective of the software is to provide the user with a natural path through an empirical spatial data analysis exercise, starting with simple mapping and geovisualization, moving on to exploration, spatial autocorrelation analysis, and ending up with spatial regression. In many respects, *GeoDa* is a reinvention of the original *SpaceStat* package (Anselin 1992), which by now has become quite dated, with only a rudimentary user interface, an antiquated architecture and performance constraints for medium and large data sets. The software was redesigned and rewritten from scratch, around the central concept of dynamically linked graphics. This means that different 'views' of the data are represented as graphs, maps or tables with selected observations in one highlighted in all. In that respect, *GeoDa* is similar to a number of other modern spatial data analysis software tools, although it is quite distinct in its combination of user friendliness with an extensive range of incorporated methods. A few illustrative comparisons will help clarify its position in the current spatial analysis software landscape.

In terms of the range of spatial statistical techniques included, *GeoDa* is most alike to the collection of functions developed in the open source R environment. For example, descriptive spatial autocorrelation measures, rate smoothing and spatial regression are included in the *spdep* package, as described by Bivand and Gebhardt (2000), Bivand (2002a, b), and Bivand and Portnov (2004). In contrast to R, *GeoDa* is completely driven by a point and click interface and does not require any programming. It also has more extensive mapping capability (still somewhat experimental in R) and full linking and brushing in dynamic graphics, which is currently not possible in R due to limitations in its architecture. On the other hand, *GeoDa* is not (yet) customizable or extensible by the user, which is one of the strengths of the R environment. In that sense, the two are seen as highly complementary, ideally with more sophisticated users 'graduating' to R after being introduced to the techniques in *GeoDa*.[1]

The use of dynamic linking and brushing as a central organizing technique for data visualization has a strong tradition in exploratory data analysis (EDA), going back to the notion of linked scatterplot brushing (Stuetzle 1987), and various methods for dynamic graphics outlined in Cleveland and McGill (1988). In geographical analysis, the concept of 'geographic brushing' was introduced by Monmonier (1989) and made operational in the *Spider/Regard* toolboxes of Haslett, Unwin and associates (Haslett et al. 1990; Unwin 1994). Several modern toolkits for exploratory spatial data analysis (ESDA) also incorporate dynamic linking, and, to a lesser extent, brushing. Some of these rely on interaction with a GIS for the map component, such as the linked frameworks combining XGobi or XploRe with ArcView (Cook et al. 1996, 1997; Symanzik et al. 2000), the SAGE

---

[1] Note that the CSISS spatial tools project is an active participant in the development of spatial data analysis methods in R, see, for example, http://sal.agecon.uiuc.edu/csiss/Rgeo/

toolbox, which uses ArcInfo (Wise et al. 2001), and the DynESDA extension for ArcView (Anselin 2000), *GeoDa*'s immediate predecessor. Linking in these implementations is constrained by the architecture of the GIS, which limits the linking process to a single map (in *GeoDa*, there is no limit on the number of linked maps). In this respect, *GeoDa* is similar to other freestanding modern implementations of ESDA, such as the cartographic data visualizer, or *cdv* (Dykes 1997), GeoVISTA Studio (Takatsuka and Gahegan 2002) and STARS (Rey and Janikas 2006). These all include functionality for dynamic linking, and to a lesser extent, brushing. They are built in open source programming environments, such as Tkl/Tk (cdv), Java (GeoVISTA Studio) or Python (STARS) and thus easily extensible and customizable. In contrast, *GeoDa* is (still) a closed box, but of these packages it provides the most extensive and flexible form of dynamic linking and brushing for both graphs and maps.

Common spatial autocorrelation statistics, such as Moran's I and even the local Moran are increasingly part of spatial analysis software, ranging from CrimeStat (Levine 2006), to the *spdep* and *DCluster* packages available on the open source Comprehensive R Archive Network (CRAN),[2] as well as commercial packages, such as the spatial statistics toolbox of the forthcoming release of ArcGIS 9.0 (ESRI 2004). However, at this point in time, none of these include the range and ease of construction of spatial weights, or the capacity to carry out sensitivity analysis and visualization of these statistics contained in *GeoDa*. Apart from the R *spdep* package, *GeoDa* is the only one to contain functionality for spatial regression modeling among the software mentioned here.

A prototype version of the software (known as *DynESDA*) has been in limited circulation since early 2001 (Anselin et al. 2002a, b), but the first official release of a beta version of *GeoDa* occurred on February 5, 2002. The program is available for free and can be downloaded from the CSISS software tools Web site (http://sal.agecon.uiuc.edu/geoda_main.php). The most recent version, 0.9.5-i, was released in January 2003. The software has been well received for both teaching and research use and has a rapidly growing body of users. For example, after slightly more than a year since the initial release (i.e., as of the end of April 2004), the number of registered users exceeds 1,800, while increasing at a rate of about 150 new users per month.

In the remainder of the chapter, we first outline the design and briefly review the overall functionality of *GeoDa*. This is followed by a series of illustrative examples, highlighting features of the mapping and geovisualization capabilities, exploration in multivariate EDA, spatial autocorrelation analysis, and spatial regression. The chapter closes with some comments regarding future directions in the development of the software.

---

[2] http://cran.r-project.org/

A.4.2 Design and functionality

The design of *GeoDa* consists of an interactive environment that combines maps with statistical graphs, using the technology of dynamically linked windows. It is geared to the analysis of *discrete* geospatial data, i.e., objects characterized by their location in space either as points (point coordinates) or polygons (polygon boundary coordinates). The current version adheres to ESRI's shape file as the standard for storing spatial information. It contains functionality to read and write such files, as well as to convert ascii text input files for point coordinates or boundary file coordinates to the shape file format. It uses ESRI's MapObjects LT2 technology for spatial data access, mapping and querying. The analytical functionality is implemented in a modular fashion, as a collection of C++ classes with associated methods. In broad terms, the functionality can be classified into six categories:

- *spatial data manipulation and utilities*: data input, output, and conversion
- *data transformation*: variable transformations and creation of new variables
- *mapping*: choropleth maps, cartogram and map animation
- *EDA*: statistical graphics
- *spatial autocorrelation*: global and local spatial autocorrelation statistics, with inference and visualization
- *spatial regression*: diagnostics and maximum likelihood estimation of linear spatial regression models.

The full set of functions is listed in Table A.4.1 and is documented in detail in the *GeoDa* User's Guides (Anselin 2003, 2004).[3]

The software implementation consists of two important components: the user interface and graphics windows on the one hand, and the computational engine on the other hand. In the current version, all graphic windows are based on Microsoft Foundation Classes (MFC) and thus are limited to MS Windows platforms.[4] In contrast, the computational engine (including statistical operations, randomization, and spatial regression) is pure C++ code and largely cross platform.

The bulk of the graphical interface implements five basic classes of windows: histogram, box plot, scatter plot (including the Moran scatter plot), map and grid (for the table selection and calculations). The choropleth maps, including the significance and cluster maps for the local indicators of spatial autocorrelation (LISA) are derived from MapObjects classes. Three additional types of maps were developed from scratch and do not use MapObjects: the map movie (map animation), the cartogram, and the conditional maps. The three dimensional scatter plot is implemented with the OpenGL library.

---

[3] A Quicktime movie with a demonstration of the main features can be found at http://sal.agecon.uiuc.edu/movies/GeoDaDemo.mov.

[4] Ongoing development concerns the porting of all MFC based classes to a cross-platform architecture, using wxWidgets. See also Section A.4.7.

Table A.4.1. GeoDa functionality overview

Category	Functions
Spatial data	Data input from shape file (point, polygon)
	Data input from text (to point or polygon shape)
	Data output to text (data or shape file)
	Create grid polygon shape file from text input
	Centroid computation
	Thiessen polygons
Data transformation	Variable transformation (log, exp, etc.)
	Queries, dummy variables (regime variables)
	Variable algebra (addition, multiplication, etc.)
	Spatial lag variable construction
	Rate calculation and rate smoothing
	Data table join
Mapping	Generic quantile choropleth map
	Standard deviational map
	Percentile map
	Outlier map (box map)
	Circular cartogram
	Map movie
	Conditional maps
	Smoothed rate map (EB, spatial smoother)
	Excess rate map (standardized mortality rate, SMR)
EDA	Histogram
	Box plot
	Scatter plot
	Parallel coordinate plot
	Three-dimensional scatter plot
	Conditional plot (histogram, box plot, scatter plot)
Spatial autocorrelation	Spatial weights creation (Rook, Queen, distance, k-nearest)
	Higher order spatial weights
	Spatial weights characteristics (connectedness histogram)
	Moran scatterplot with inference
	Bivariate Moran scatterplot with inference
	Moran scatterplot for rates (EB standardization)
	Local Moran significance map
	Local Moran cluster map
	Bivariate local Moran
	Local Moran for rates (EB standardization)
Spatial regression	OLS with diagnostics (e.g., LM test, Moran's I)
	Maximum likelihood spatial lag model
	Maximum likelihood spatial error model
	Predicted value map
	Residual map

The functionality of *GeoDa* is invoked either through menu items or directly by clicking toolbar buttons, as illustrated in Fig. A.4.1. A number of specific applications are highlighted in the following sections, focusing on some distinctive features of the software.

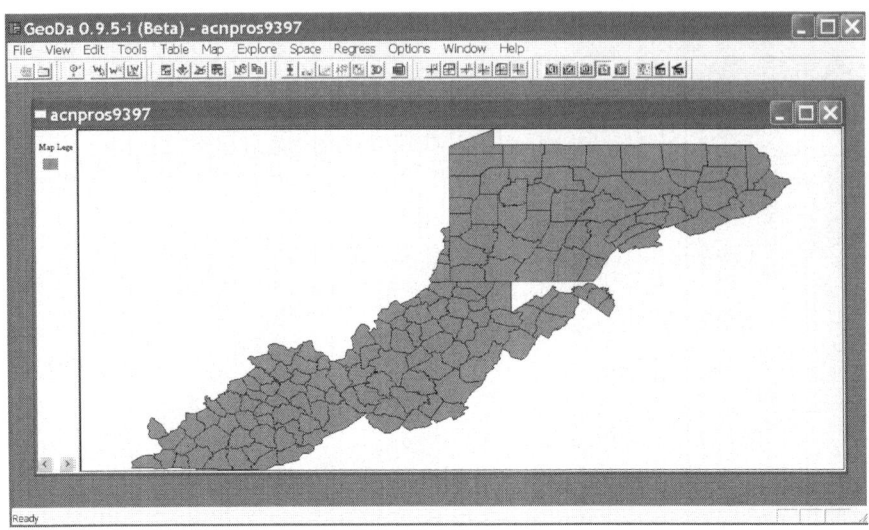

Fig. A.4.1. The opening screen with menu items and toolbar buttons

A.4.3 Mapping and geovisualization

The bulk of the mapping and geovisualization functionality consists of a collection of specialized choropleth maps, focused on highlighting outliers in the data, so-called *box maps* (Anselin 1999). In addition, considerable capability is included to deal with the intrinsic variance instability of rates, in the form of empirical Bayes (EB) or spatial smoothers.[5] As mentioned in Section A.2.2, the mapping operations use the classes contained in ESRI's MapObjects, extended with the capability for linking and brushing. *GeoDa* also includes a circular cartogram,[6] map animation in the form of a map movie, and conditional maps. The latter are nine micro choropleth maps constructed by conditioning on three intervals for two conditioning variables, using the principles outlined in Becker et al. (1996), and Carr et al. (2002).[7] In contrast to the traditional choropleth maps, the cartogram, map movie and conditional maps do not use MapObjects classes, and were developed from scratch.

---

[5] The EB procedure is due to Clayton and Kaldor (1987), see also Marshall (1991) and Bailey and Gatrell (1995, pp. 303-308). For an alternative recent software implementation, see Anselin et al. (2004). Spatial smoothing is discussed at length in Kafadar (1996).

[6] The cartogram is constructed using the non-linear cellular automata algorithm due to Dorling (1996).

[7] The conditional maps are part of a larger set of conditional plots, which includes histograms, box plots and scatter plots.

We illustrate the rate smoothing procedure, outlier maps and linking operations. The objective in this analysis is to identify locations that have elevated mortality rates and to assess the sensitivity of the designation as outlier to the effect of rate smoothing. Using data on prostate cancer mortality in 156 counties contained in the Appalachian Cancer Network (ACN), for the period 1993-97, we construct a box map by specifying the number of deaths as the numerator and the population as the denominator.[8] The resulting map for the crude rates (that is, without any adjustments for differing age distributions or other relevant factors) is shown as the upper-left panel in Fig. A.4.2. Three counties are identified as *outliers* and shown in dark grey.[9] These match the outliers *selected* in the box plot in the lower-left panel of the figure. The *linking* of all maps and graphs results in those counties also being cross-hatched on the maps.

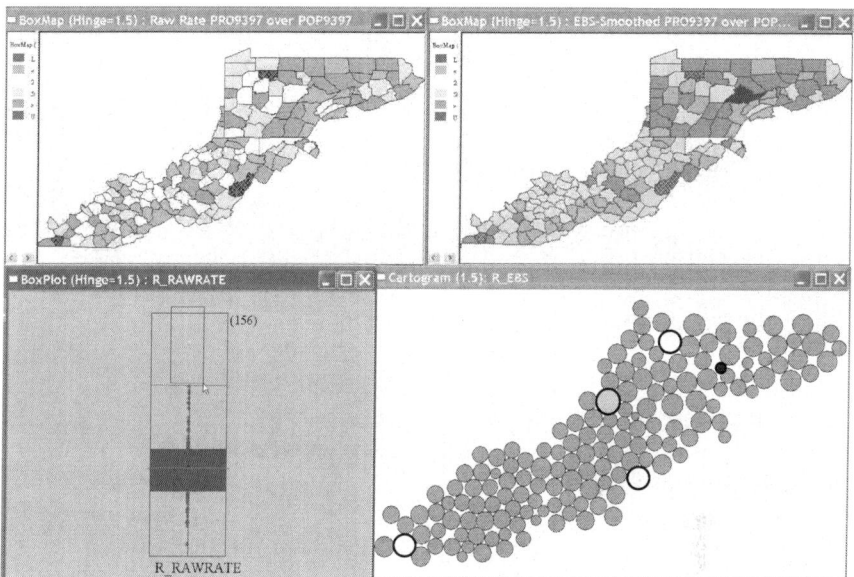

Fig. A.4.2. Linked box maps, box plot and cartogram, raw and smoothed prostate cancer mortality rates

The upper-right panel in the figure represents a smoothed rate map, where the rates were transformed by means of an Empirical Bayes procedure to remove the effect of the varying population at risk. As a result, the original outliers are no longer, but a different county is identified as having elevated risk. Also, a lower

---

[8] Data obtained from the National Cancer Institute SEER site (Surveillance, Epidemiology and End Results), http://seer.cancer.gov/seerstat/.

[9] The respective counties are Cumberland [KY], Pocahontas [WV], and Forest [PA].

outlier is found as well, shown as black in the box map.¹⁰ Note that the upper outlier is barely distinguishable, due to the small area of the county in question. This is a common problem when working with admininistrative units. In order to remove the potentially misleading effect of area on the perception of interesting patterns, a circular cartogram is shown in the lower-right panel of Fig. A.4.2, where the area of the circles is proportional to the value of the EB smoothed rate. The upper outlier is shown as a light grey circle, the lower outlier as a black circle. The white circles are the counties that were outliers in the crude rate map, highlighted here as a result of linking with the other maps and graphs.¹¹

A.4.4 Multivariate EDA

Multivariate exploratory data analysis is implemented in *GeoDa* through linking and brushing between a collection of statistical graphs. These include the usual histogram, box plot and scatter plot, but also a parallel coordinate plot (PCP) and three-dimensional scatter plot, as well as conditional plots (conditional histogram, box plot and scatter plot).

We illustrate some of this functionality with an exploration of the relationships between economic growth and initial development, typical of the recent "spatial" regional convergence literature (for an overview, see Rey 2004). We use economic data over the period 1980-1999 for 145 European regions, most of them at the NUTS-2 level of spatial aggregation, except for a few at the NUTS-1 level (for Luxembourg and the United Kingdom).¹²

Figure A.4.3 illustrates the various linked plots and map. The left-hand panel contains a simple percentile map (GDP per capital in 1989), and a three-dimensional scatter plot (for the percent agricultural and manufacturing employment in 1989 as well as the GDP growth rate over the period 1980-99). In the top right-hand panel is a PCP for the growth rates in the two periods of interest (1980-89 and 1989-99) and the GDP per capita in the base year, the typical components of a convergence regression. In the bottom of the right-hand panel is a

---

¹⁰ The new upper outlier is Ohio county [WV], the lower outlier is Centre county [PA].

¹¹ Note that the outliers identified may be misleading since the rate analyzed is not adjusted for differences in age distribution. In other words, the outliers shown may simply be counties with a larger proportion of older males. A much more detailed analysis is necessary before any policy conclusions may be drawn.

¹² The data are from the most recent version of the NewCronos Regio database by Eurostat. NUTS stands for 'Nomenclature of Territorial Units for Statistics' and contains the definition of administrative regions in the EU member states. NUTS-2 level regions are roughly comparable to counties in the U.S. context and are available for all but two countries. Luxembourg constitutes only a single region. For the United Kingdom, data is not available at the NUTS-2 level, since these regions do not correspond to local governmental units.

simple scatter plot of the growth rate in the full period (1980-99) on the base year GDP.

Both plots on the right hand side illustrate the typical empirical phenomenon that higher GDP at the start of the period is associated with a lower growth rate. However, as demonstrated in the PCP (some of the lines suggest a positive relation between GDP and growth rate), the pattern is not uniform and there is a suggestion of heterogeneity. A further exploration of this heterogeneity can be carried out by brushing any one of these graphs. For example, in Fig. A.4.3, a selection box in the three-dimensional scatter plot is moved around (*brushing*) which highlights the selected observations in the map (cross-hatched) and in the PCP, clearly showing opposite patterns in subsets of the selection. Furthermore, in the scatter plot, the slope of the regression line can be recalculated for a subset of the data without the selected locations, to assess the sensitivity of the slope to those observations. In the example shown here, the effect on convergence over the whole period is minimal (–0.147 vs. –0.144), but other selections show a more pronounced effect. Further exploration of these patterns does suggest a degree of spatial heterogeneity in the convergence results (for a detailed investigation, see LeGallo and Dall'erba 2003).

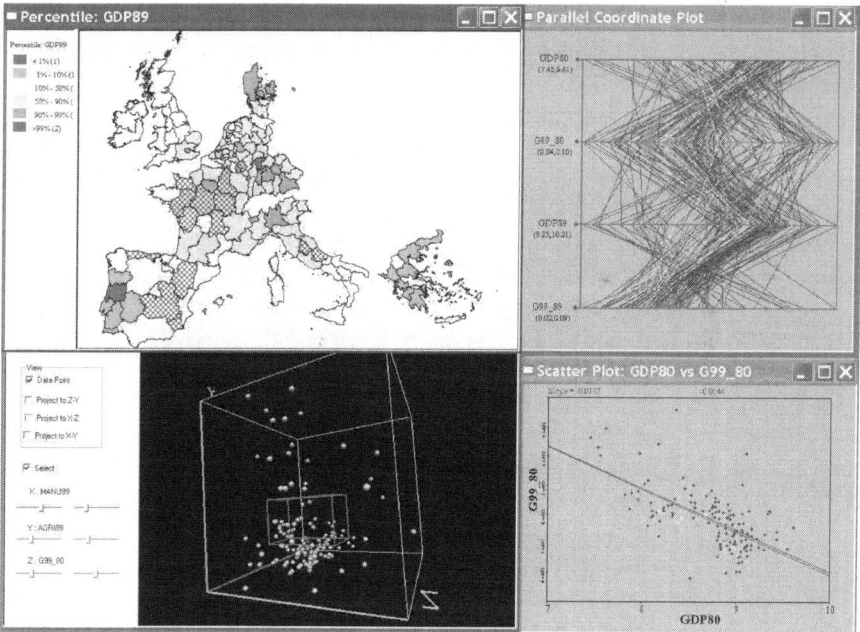

Fig. A.4.3. Multivariate exploratory data analysis with linking and brushing

A.4.5 Spatial autocorrelation analysis

Spatial autocorrelation analysis includes tests and visualization of both global (test for *clustering*) and local (test for *clusters*) Moran's I statistic. The global test is visualized by means of a Moran scatterplot (Anselin 1996), in which the slope of the regression line corresponds to Moran's I. Significance is based on a permutation test. The traditional univariate Moran scatterplot has been extended to depict bivariate spatial autocorrelation as well, that is, the correlation between one variable at a location, and a different variable at the neighboring locations (Anselin et al. 2002a). In addition, there also is an option to standardize rates for the potentially biasing effect of variance instability (see Assunção and Reis 1999).

Local analysis is based on the local Moran statistic (Anselin 1995), visualized in the form of significance and cluster maps. It also includes several options for sensitivity analysis, such as changing the number of permutations (to as many as 9,999), re-running the permutations several times, and changing the significance cut off value. This provides an ad hoc approach to assess the sensitivity of the results to problems due to multiple comparisons (that is, how stable is the indication of clusters or outliers when the significance barrier is lowered).

The maps depict the locations with significant local Moran statistics (LISA significance maps) and classify those locations by type of association (LISA cluster maps). Both types of maps are available for brushing and linking. In addition to these two maps, the standard output of a LISA analysis includes a Moran scatter plot and a box plot depicting the distribution of the local statistic. Similar to the Moran scatter plot, the LISA concept has also been extended to a bivariate setup and includes an option to standardize for variance instability of rates.

The functionality for spatial autocorrelation analysis is rounded out by a range of operations to construct spatial weights, using either boundary files (contiguity based) or point locations (distance based). A connectivity histogram helps in identifying potential problems with the neighbor structure, such as 'islands' (locations without neighbors).

We illustrate spatial autocorrelation analysis with a study of the spatial distribution of 692 house sales prices for 1997 in Seattle, WA. This is part of a broader investigation into the effect of subsidized housing on the real estate market.[13] For the purposes of this example, we only focus on the univariate spatial distribution, and the location of any significant clusters or spatial outliers in the data.

The original house sales data are for point locations, which, for the purposes of this analysis are converted to Thiessen polygons. This allows a definition of 'neighbor' based on common boundaries between the Thiessen polygons. On the left hand panel of Fig. A.4.4, two LISA cluster maps are shown, depicting the locations of significant local Moran's I statistics, classified by type of spatial

---

[13] The data are from the King County (Washington State) Department of Assessments.

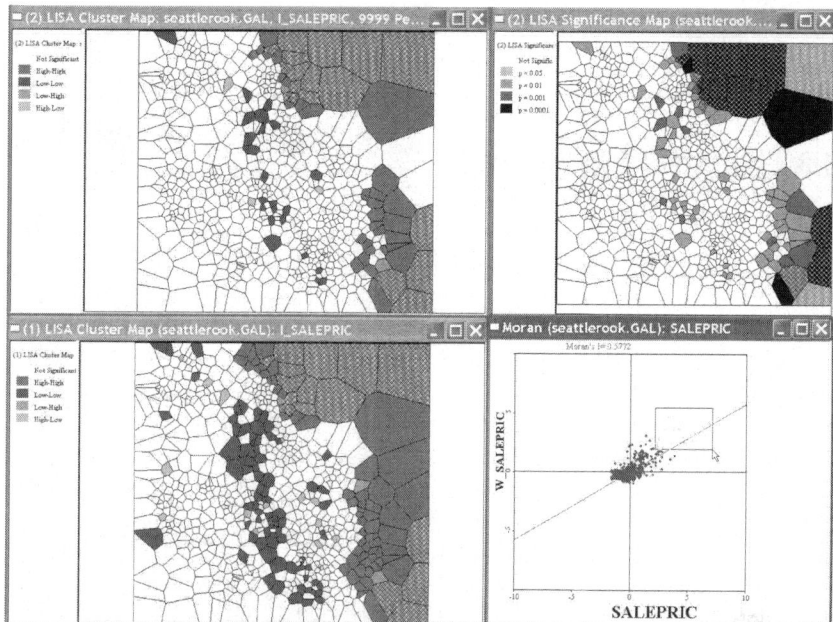

Fig. A.4.4. LISA cluster maps and significance maps

association. The dark grey locations are indications of spatial *clusters* (respectively, high surrounded by high, and low surrounded by low).[14] In contrast, the light grey are indications of *spatial outliers* (respectively, high surrounded by low, and low surrounded by high). The bottom map uses the default significance of $p = 0.05$, whereas the top map is based on $p = 0.01$ (after carrying out 9,999 permutations). The matching *significance map* is in the top right hand panel of Fig. A.4.4. Significance is indicated by darker shades of grey, with the darkest corresponding to $p = 0.0001$. Note how the tighter significance criterion eliminates some (but not that many) locations from the map. In the bottom right hand panel of the figure, the corresponding Moran scatterplot is shown, with the most extreme 'high-high' locations selected. These are shown as cross-hatched polygons in the maps, and almost all obtain highly significant (at $p = 0.0001$) local Moran's I statistics.

The overall pattern depicts a cluster of high priced houses on the East side, with a cluster of low priced houses following an axis through the center. Put in context, this is not surprising, since the East side represents houses with a lake view, while the center cluster follows a highway axis and generally corresponds with a lower income neighborhood. Interestingly, the pattern is not uniform, and

---

[14] More precisely, the locations highlighted show the 'core' of a cluster. The cluster itself can be thought of as consisting of the core as well as the neighbors. Clearly some of these clusters are overlapping.

several spatial outliers can be distinguished. Further investigation of these patterns would require a full hedonic regression analysis.

A.4.6 Spatial regression

As of version 0.9.5-i, *GeoDa* also includes a limited degree of spatial regression functionality. The basic diagnostics for spatial autocorrelation, heteroskedasticity and non-normality are implemented for the standard ordinary least squares regression. Estimation of the spatial lag and spatial error models is supported by means of the Maximum Likelihood (ML) method (see Anselin and Bera 1998, for a review of the technical issues). In addition to the estimation itself, predicted values and residuals are calculated and made available for mapping.

The ML estimation in *GeoDa* distinguishes itself by the use of extremely efficient algorithms, that allow the estimation of models for very large data sets. The standard eigenvalue simplification is used (Ord 1975) for data sets up to 1,000 observations. Beyond that, the sparse algorithm of Smirnov and Anselin (2001) is used, which exploits the characteristic polynomial associated with the spatial weights matrix. This algorithm allows estimation of very large data sets in reasonable time. In addition, *GeoDa* implements the recent algorithm of Smirnov (2003) to compute the asymptotic variance matrix for all the model coefficients (that is, including both the spatial and non-spatial coefficients). This involves the inversion of a matrix of the dimensions of the data sets. To date, *GeoDa* is the only software that provides such estimates for large data sets.

All estimation methods employ sparse spatial weights, but they are currently constrained to weights that are intrinsically symmetric (e.g., excluding k-nearest neighbor weights). The regression routines have been successfully applied to real data sets of more than 300,000 observations (with estimation and inference completed in a few minutes). By comparison, a spatial regression for the 3000+ U.S. counties takes a few seconds.

We illustrate the spatial regression capabilities with a partial replication and extension of the homicide model used in Baller et al. (2001) and Messner and Anselin (2004). These studies assessed the extent to which a classic regression specification, well-known in the criminology literature, is robust to the explicit consideration of spatial effects. The model relates county homicide rates to a number of socio-economic explanatory variables. In the original study, a full ML analysis of all U.S. continental counties was precluded by the constraints on the eigenvalue-based SpaceStat routines. Instead, attention focused on two subsets of the data containing 1,412 counties in the U.S. South and 1,673 counties in the non-South.

In Fig. A.4.5, we show the result of the ML estimation of a spatial error model of county homicide rates for the complete set of 3,085 continental U.S. counties in 1980. The explanatory variables are the same as before: a Southern dummy variable, a resource deprivation index, a population structure indicator, unemployment rate, divorce rate and median age.[15]

The results confirm a strong positive and significant spatial autoregressive coefficient ($\hat{\lambda} = 0.29$). Relative to the OLS results (for example, Messner and Anselin 2004, Table 7.1, p.137), the coefficient for unemployment has become insignificant, illustrating the misleading effect spatial error autocorrelation may have on inference using OLS estimates. The model diagnostics also suggest a continued presence of problems with heteroskedasticity. However, *GeoDa* currently does not include functionality to deal with this.

```
REGRESSION
SUMMARY OF OUTPUT: SPATIAL ERROR MODEL - MAXIMUM LIKELIHOOD ESTIMATION
Data set             :
Spatial Weight       : natrook.GAL
Dependent Variable   :         HR80   Number of Observations:    3085
Mean dependent var   :      6.927616  Number of Variables   :       7
S.D. dependent var   :      6.825088  Degree of Freedom     :    3078
Lag coeff. (Lambda)  :      0.293641

R-squared            :      0.462988  R-squared (BUSE)      :     -
Sq. Correlation      :         -      Log likelihood        :-9369.500716
Sigma-square         :     25.015000  Akaike info criterion :    18753
S.E of regression    :      5.0015    Schwarz criterion     :18795.241581

-----------------------------------------------------------------------
    Variable    Coefficient    Std.Error    z-value    Probability
-----------------------------------------------------------------------
    CONSTANT      8.463455     0.9765372    8.666803    0.0000000
       SOUTH     1.951838     0.2916178    6.693136    0.0000000
        RD80     3.461736     0.1350625   25.63063     0.0000000
        PS80     0.6745796    0.1185432    5.690582    0.0000000
        UE80    -0.04367847   0.03578631  -1.220535    0.2222621
        DV80     1.151325     0.07579833  15.18932     0.0000000
        MA80    -0.2395876    0.02761771  -8.675144    0.0000000
      LAMBDA     0.2936407    0.02562901  11.45736     0.0000000
-----------------------------------------------------------------------

REGRESSION DIAGNOSTICS
DIAGNOSTICS FOR HETEROSKEDASTICITY
RANDOM COEFFICIENTS
TEST                                    DF      VALUE         PROB
Breusch-Pagan test                       6     1187.417     0.0000000

DIAGNOSTICS FOR SPATIAL DEPENDENCE
SPATIAL ERROR DEPENDENCE FOR WEIGHT MATRIX : natrook.GAL
TEST                                    DF      VALUE         PROB
Likelihood Ratio Test                    1      120.599     0.0000000
============================ END OF REPORT ============================
```

Fig. A.4.5. Maximum Likelihood estimation of the spatial error model

---

[15] See the original papers for technical details and data sources. In Baller et al. (2001), a different set of spatial weights was used than in this example, but the conclusions of the specification tests are the same. Specifically, using the county contiguity, the robust Lagrange multiplier tests are 1.24 for the Lag alternative, and 24.88 for the Error alternative, strongly suggesting the latter as the proper alternative.

A.4.7 Future directions

GeoDa is a work in progress and still under active development. This development proceeds along three fronts. First and foremost is an effort to make the code cross-platform and open source. This requires considerable change in the graphical interface, moving from the Microsoft Foundation Classes (MFC) that are standard in the various MS Windows flavors, to a cross-platform alternative. The current efforts use wxWidgets,[16] which operates on the same code base with a native GUI flavor in Windows, MacOS X and Linux/Unix. Making the code open source is currently precluded by the reliance on proprietary code in ESRI's MapObjects. Moreover, this involves more than simply making the source code available, but entails considerable reorganization and streamlining of code (refactoring), to make it possible for the community to effectively participate in the development process.

A second strand of development concerns the spatial regression functionality. While currently still fairly rudimentary, the inclusion of estimators other than ML and the extension to models for spatial panel data are in progress. Finally, the functionality for ESDA itself is being extended to data models other than the discrete locations in the 'lattice' case. Specifically, exploratory variography is being added, as well as the exploration of patterns in flow data.

Given its initial rate of adoption, there is a strong indication that *GeoDa* is indeed providing the 'introduction to spatial data analysis' that makes it possible for growing numbers of social scientists to be exposed to an explicit spatial perspective. Future development of the software should enhance this capability and it is hoped that the move to an open source environment will involve an international community of like minded developers in this venture.

Acknowledgements. This research was supported in part by U.S. National Science Foundation Grant BCS-9978058, to the Center for Spatially Integrated Social Science (CSISS) and by grant RO1 CA 95949-01 from the National Cancer Institute. In addition, this research was made possible in part through a Cooperative Agreement between the Center for Disease Control and Prevention (CDC) and the Association of Teachers of Preventive Medicine (ATPM), award number TS-1125. The contents of the chapter are the responsibility of the authors and do not necessarily reflect the official views of NSF, NCI, the CDC or ATPM. Special thanks go to Oleg Smirnov for his assistance with the implementation of the spatial regression routines, and to Julie LeGallo and Julia Koschinsky for preparing, respectively, the data set for the European convergence study and for the Seattle house prices. GeoDa is a trademark of Luc Anselin.

---

[16] http://www.wxwidgets.org

References

Anselin L (1992) SpaceStat: a software program for the analysis of spatial data. National Center for Geographic Information and Analysis (NCGIA), University of California, Santa Barbara [CA]

Anselin L (1995) Local indicators of spatial association - LISA. Geogr Anal 27(2):93-115

Anselin L (1996) The Moran scatterplot as an ESDA tool to assess local instability in spatial association. In Fischer MM, Scholten H, Unwin D (eds) Spatial analytical perspectives on GIS in environmental and socio-economoc sciences. Taylor and Francis, London, pp.111-125

Anselin L (1999) Interactive techniques and exploratory spatial data analysis. In Longley PA, Goodchild MF, Maguire DJ, Rhind DW (eds) Geographical Information Systems: principles, techniques, management and applications. Wiley, New York, Chichestre, Toronto and Brisbane, pp.251-264

Anselin L (2000) Computing environments for spatial data analysis. J Geogr Syst 2(3):201

Anselin L (2003) GeoDa 0.9 User's Guide. Spatial Analysis Laboratory (SAL) Department of Agricultural and Consumer Economics, University of Illinois, Urbana-Champaign [IL]

Anselin L (2004) GeoDa 0.95i Release Notes. Spatial Analysis Laboratory (SAL) Department of Agricultural and Consumer Economics, University of Illinois, Urbana-Champaign [IL]

Anselin L, Bera AK (1998) Spatial dependence in linear regression models with an introduction to spatial econometrics. In Ullah A, Giles DEA (eds) Handbook of applied economic statistics. Marcel Dekker, New York. pp.237-289

Anselin L, Getis A (1992) Spatial statistical analysis and geographic information systems. Ann Reg Sci 26(1):19-33

Anselin L, Kim YW, Syabri I (2004) Web-based analytical tools for the exploration of spatial data. J Geogr Syst 6(2):197-218

Anselin L, Syabri I, Kho Y (2009) GeoDa: An introduction to spatial data analysis. In Fischer MM, Getis A (eds) Handbook of spatial data analysis. Springer, Berlin, Heidelberg and New York, pp.73-89

Anselin L, Syabri I, Smirnov O (2002a) Visualizing multivariate spatial correlation with dynamically linked windows. In Anselin L, Rey S (eds) New tools for spatial data analysis: proceedings of the specialist meeting. Center for Spatially Integrated Social Science (CSISS), University of California, Santa Barbara [CA] CD-ROM.

Anselin L, Syabri I, Smirnov O, Ren Y (2002b) Visualizing spatial autocorrelation with dynamically linked windows. Comput Sci Stat 33. CD-ROM.

Assunção R, Reis EA (1999) A new proposal to adjust Moran's I for population density. Stat Med 18(16):2147-2161

Bailey TC, Gatrell AC (1995) Interactive spatial data analysis. Longman, Harlow

Baller R, Anselin L, Messner S, Deane G, Hawkins D (2001) Structural covariates of U.S. county homicide rates: incorporating spatial effects. Criminol 39(3):561-590

Becker RA, Cleveland W, and Shyu M-J (1996) The visual design and control of Trellis displays. J Comput Graph Stat 5(2):123-155

Bivand RS (2002a) Implementing spatial data analysis software tools in R. In Anselin L and Rey S (eds) New tools for spatial data analysis. Proceedings of the specialist meeting. Center for Spatially Integrated Social Science (CSISS), University of California, Santa Barbara [CA] CD-ROM

Bivand RS (2002b) Spatial econometrics functions in R: classes and methods. J Geogr Syst 4(4):405-421

Bivand RS, Gebhardt A (2000) Implementing functions for spatial statistical analysis using the R language. J Geogr Syst 2(3):307-317

Bivand RS, Portnov BA (2004) Exploring spatial data analysis techniques using R: the case of observations with no neighbors. In Anselin L, Florax RJ, Rey SJ (eds) Advances in spatial econometrics: methodology tools and applications. Springer, Berlin, Heidelberg and New York, pp.121-142

Carr DB, Chen J, Bell S, Pickle L, Zhang Y (2002) Interactive linked micromap plots and dynamically conditioned choropleth maps. In Anselin L, Rey S (eds) New tools for spatial data analysis: proceedings of the specialist meeting. Center for Spatially Integrated Social Science (CSISS), University of California, Santa Barbara. [CA] CD-ROM

Clayton D, Kaldor J (1987) Empirical Bayes estimates of age standardized relative risks for use in disease mapping. Biometrics 43(3):671-681

Cleveland WS, McGill M (1988) Dynamic graphics for statistics. Wadsworth, Pacific Grove [CA]

Cook D, Majure J, Symanzik J, Cressie NAC (1996) Dynamic graphics in a GIS: a platform for analyzing and exploring multivariate spatial data. Comput Stat Data Anal 11:467-480

Cook D, Symanzik J, Majure JJ, Cressie NAC (1997) Dynamic graphics in a GIS: more examples using linked software. Comput Geosci 23:371-385

Dorling D (1996) Area cartograms: their use and creation CATMOG 59, Institute of British Geographers

Dykes JA (1997) Exploring spatial data representation with dynamic graphics. Comput Geosci 23:345-370

ESRI (2004) An overview of the spatial statistics toolbox. ArcGIS 9.0 Online Help System (ArcGIS 9.0 Desktop, Release 9.0, June 2004) Environmental Systems Research Institute, Redlands, CA

Fischer MM, Getis A (1997) Recent development in spatial analysis. Springer, Berlin, Heidelberg and New York

Fischer MM, Nijkamp P (1993) Geographic information systems, spatial modelling and policy evaluation. Springer, Berlin, Heidelberg and New York

Fischer MM, Scholten HJ, Unwin D (1996) Spatial analytical perspectives on GIS. Taylor and Francis, London

Fotheringham AS, Rogerson P (1993) GIS and spatial analytical problems. Int J Geogr Inform Syst 7(1):3-19

Fotheringham AS, Rogerson P (1994) Spatial analysis and GIS. Taylor and Francis, London

Goodchild MF, Anselin L, Appelbaum R, Harthorn B (2000) Toward spatially integrated social science. Int Reg Sci Rev 23(2):139-159

Goodchild MF, Haining RP, Wise S, and 12 others (1992) Integrating GIS and spatial analysis: problems and possibilities. Int J Geogr Inform Syst 6(5):407-423

Haining R (1989) Geography and spatial statistics: current positions, future developments In Macmillan B (ed) Remodelling Geography. Basil Blackwell, Oxford, pp.191-203

Haslett J, Wills G, Unwin A (1990) SPIDER: an interactive statistical tool for the analysis of spatially distributed data. Int J Geogr Inform Syst 4(3):285-296

Kafadar K (1996) Smoothing geographical data, particularly rates of disease. Stat Med 15(23):2539-2560

LeGallo J, Dall'erba S (2003) Evaluating the temporal and spatial heterogeneity of the European convergence process, 1980-1999. Technical report, Université Montesquieu-Bordeaux IV, Pessac Cedex, France

Levine N (2006) The CrimeStat program: characteristics, use and audience. Geogr Anal 38(1):41-56

Marshall RJ (1991) Mapping disease and mortality rates using empirical Bayes estimators. App Stat 40(2):283-294

Messner SF, Anselin L (2004) Spatial analyses of homicide with areal data. In Goodchild MF, Janelle D (eds) Spatially Integrated Social Science. Oxford University Press, New York, pp.127-144

Monmonier MS (1989) Geographic brushing: enhancing exploratory analysis of the scatterplot matrix. Geogr Anal 21(1):81-84

Ord JK (1975) Estimation methods for models of spatial interaction. J Am Stat Assoc 70(349):120-126

Rey SJ (2004) Spatial analysis of regional income inequality. In Goodchild MF, Janelle D (eds) Spatially integrated social science. Oxford University Press, Oxford, pp.280-299

Rey SJ, Janikas MV (2006) STARS: space-time analysis of regional systems. Geogr Anal 38(1):67-86

Smirnov O (2003) Computation of the information matrix for models of spatial interaction technical report, Regional Economics Applications Laboratory (REAL), University of Illinois, Urbana-Champaign [IL]

Smirnov O, Anselin L (2001) Fast maximum likelihood estimation of very large spatial autoregressive models: a characteristic polynomial approach. Comput Stat Data Anal 35(3):301-319

Stuetzle W (1987) Plot windows. J Amer Stat Assoc 82:466-475

Symanzik J, Cook D, Lewin-Koh N, Majure JJ, Megretskaia I (2000) Linking ArcView and XGobi: insight behind the front end. J Comput Graph Stat 9(3):470-490

Takatsuka M, Gahegan M (2002) GeoVISTA Studio: a codeless visual programming environment for geoscientific data analysis and visualization. Comput Geosci 28(10):1131-1141

Unwin A (1994) REGARDing geographic data. In Dirschedl P, Osterman R (eds) Computational Statistics. Physica Verlag, Heidelberg, pp.345-354

Wise S, Haining R, Ma J (2001) Providing spatial statistical data analysis functionality for the GIS user: the SAGE project. Int J Geogr Inform Sci 15(3):239-254

A.5 STARS: Space-Time Analysis of Regional Systems

Sergio J. Rey and *Mark V. Janikas*

A.5.1 Introduction

One of the active areas in the field of Geographic Information Sciences (GIS) is the development of new methods of exploratory spatial data analysis. A number of impressive efforts have recently appeared to provide researchers with powerful tools for both geospatial statistical analysis, data mining, as well as geovisualization. Well known efforts include the GeoDa environment (Anselin 2003), the GeoVista Studio (Takatsuka and Gahegan 2002), Cartographic Data Visualizer (Dykes 1995), SAGE (Wise et al. 2001) and the ArcView-XGobi project (Symanzik et al. 1998).

A new addition to this field is the package STARS: Space-Time Analysis of Regional Systems. STARS is an open source environment written in Python that supports exploratory *dynamic* spatial data analysis. Dynamic takes on two meanings in STARS. The first reflects a strong emphasis on the incorporation of time into the exploratory analysis of space-time data. To do so, STARS combines two sets of modules, *visualization* and *computation*. The visualization module consists of a family of geographical, temporal and statistical views that are interactive and interdependent. That is, they allow the user to explore patterns through various interfaces and the views are dynamically integrated with one another, giving rise to the second meaning of dynamic spatial data analysis. On the computational front, STARS contains a set of exploratory spatial data analysis modules, together with several newly developed measures for space-time analysis.

This chapter provides a detailed introduction to STARS and is organized as follows. The motivation giving rise to the creation of STARS is discussed in the following section. A detailed overview of the analytical components of the package are presented in Section A.5.3. The capabilities of these components are then illustrated in a series of examples drawing from the study of regional income dynamics in Section A.5.4. The chapter closes with an outline of future plans for the continued development of STARS.

A.5.2 Motivation

As is common with many open source packages, STARS was born out of a need to scratch an itch. In this instance the itch was the lack of an integrated statistical toolkit that supported the analysis of both the spatial and temporal dimensions of regional income growth and convergence. Regional convergence or divergence has both temporal and spatial dimensions, and in studying these processes researchers have relied on either spatial analysis (Rey and Montouri 1999) or time series methods (Carlino and Mills 1993).[1]

To consider both dimensions jointly requires the use of two different sets of methods, yet with the existing software this meant having to switch between software packages. This turns out to be a rather awkward way to do exploratory data analysis. It is clear that new tools are needed for an EDA toolkit that truly integrates *space and time*. While the question of time in GIS has attracted much conceptual attention (Peuquet 2002; Egenhofer and Golledge 1997), operational systems implementing both geocomputational and geovisualization components that also incorporate time are few in number.[2] STARS is an attempt to fill this niche. Although the initial motivation for STARS was the study of regional income dynamics, the methods and tools it contains can be applied to a wide set of socioeconomic or physical processes with data measured for areal units over multiple time periods.

A.5.3 Components and design

It was decided in the genesis of the STARS project that the exploratory geocomputational methods and the visualization techniques used to express them be developed separately. This facilitated the development of the STARS package in a modular fashion which has enabled users to interact with the program in a number of ways. First, the geocomputational and visualization modules can be linked together in a user friendly interactive graphical interface. Second, the individual modules can be used as a library and combined with scripts written in Python (or other scripting languages). The modularity also permits easy extension of STARS through the development of specialized modules. We return to this issue later on. Next we discuss the two core modules of STARS, geocomputation and visualization.

---

[1] For a recent overview of the empirical literature on spatial convergence see Rey and Janikas (2005).

[2] For an example of such a system focusing on geophysical data see Christakos et al. (2001).

Geocomputation. The methods used to explore the dynamics of space-time data have been broken into distinct categories, which are outlined in Table A.5.1. While STARS has many of the standard summary statistic capabilities that one would find in any number of data analysis packages, it is its inherent ability to identify and analyze the space-time characteristics of the data that makes it a unique environment.

Table A.5.1. Geocomputational methods contained in STARS

Category	Description
Descriptive statistics	Distribution and summary measures for variables by cross-section, time period, or pooled
Exploratory spatial data analysis	Various methods specifically designed to analyze spatial dependence. Global and local versions of Moran's I, Geary's c and the G statistic are provided
Inequality	Techniques that quantify and decompose inequality over time and space. Includes classic and spatial Gini coefficients as well as Theil decomposition
Mobility	Recent advances in internal mobility dynamics are presented through the τ and θ statistics
Markov analysis	Transitional dynamics of distributional attributes are examined through the use of classic Markov and spatial Markov techniques

STARS has focused on incorporating recent advances in the analysis of spatial dependence. Global measures of spatial autocorrelation are included for the analysis of dependence over a region. The program also contains Local Indicators of Spatial Autocorrelation (LISA's) which give a more disaggregated view at the nature of dependence (Anselin 1995). These have been extended to a dynamic context in a number of new empirical measures such as spatial Markov matrices, LISA Markov matrices, and indicators of spatial cohesion and flux introduced by Rey (2001).

A series of alternative computational categories that deal with inter/intra distributional dynamics are also contained in STARS. Measures such as Theil's T (Theil 1996) can be used to evaluate and decompose inequality over time and space (see Rey 2004a for an illustration). STARS also incorporates enhanced methods that identify various aspects of mobility within a distribution. These include spatially explicit rank correlation measures and regime based mobility decompositions introduced by Rey (2004b), as well as spatialized Gini coefficients. All these new measures provide insights as to the role of spatial context in the evolution of variable distributions over time and space.

STARS also provides a host of data and matrix utility functions. These can be used to create new or transform existing variables as well as to construct alternative forms of spatial weight matrices, network representations of spatial structure and temporal covariance matrices. The latter allow for detailed investigation and comparison of the implied relationships between spatial observations as reflected in various spatial weight matrices and those revealed from the temporal co-movement of variables for different cross-sectional units.

Vizualization. A list of the visualization capabilities of the STARS module is presented in Table A.5.2. STARS contains some views that are standard to an exploratory data package, however, the dynamic linking mechanisms enhance the users ability to analyze data over various dimensions (see Section A.5.4 for examples). Some of the views are multidimensional by nature. The conditional scatter plot can provide an additional facet to its traditional counterpart through a color weighting scheme based on a requisite variable. This supports the use of categorical variables for regime based analysis and a simple time variable which can identify hidden evolutions.

Table A.5.2. Visualization capabilities in STARS

Category	Description
Map	A variety of sequential, categorical and user-defined choropleth maps
Scatter plot	A basic two-dimensional view, the scatter plot can be used to analyze cross-sectional, time period or bivariate correspondence in X-Y space
Conditional scatter plot	Extends the traditional scatter plot to three dimensions by conditioning the color of the data points by the level of a third variable
Parallel coordinate plot	Allows the user to view multivariate relationships over space and time
Time series plot	Plots the evolution of a variable for a given spatial unit
Time path plot	Demonstrates the co-movement of a variable for two spatial units over time
Histogram	Creates a basic partitioning of a variable into respective bins.
Density	Contains empirical kernel density estimation for the analysis of dispersion, modality, and skewness
Box plots	Another distributional view with an added focus on quantiles and outliers

The time path plot illustrates the pair-wise movement of two variables and/or observations over time. This view is helpful in identifying levels of stability across a given structural process. Individual aspects of the co-movement progression can be dissected by interval gaps and distinct directional movements.

STARS also contains a series of maps which can be created and altered through the use of various commands. One example involves the visualization of covariance matrices over space. The covariance structure of a variable is portrayed as a series of links between the centroids of each polygon. Positive correlations are coloured differently than negative ones to more distinctly identify cross-sectional relationships. Threshold capabilities assures that the user can map covariance links based on specified criterion. These are illustrated later in the chapter.

Design. As mentioned previously, STARS is written entirely in the Python language. Python is an object-oriented scripting language gaining widespread acceptance as a language for scientific computing (Langtangen 2004; Saenz et al. 2002; Hinsen 2000; Schliep et al. 2001). As Python is open source and cross platform, researchers interested in using STARS are not limited in their choice of operating system or hardware platform. Moreover, Python has a clean and simple syntax which facilitates collaboration by researchers wanting to add extensions to STARS.

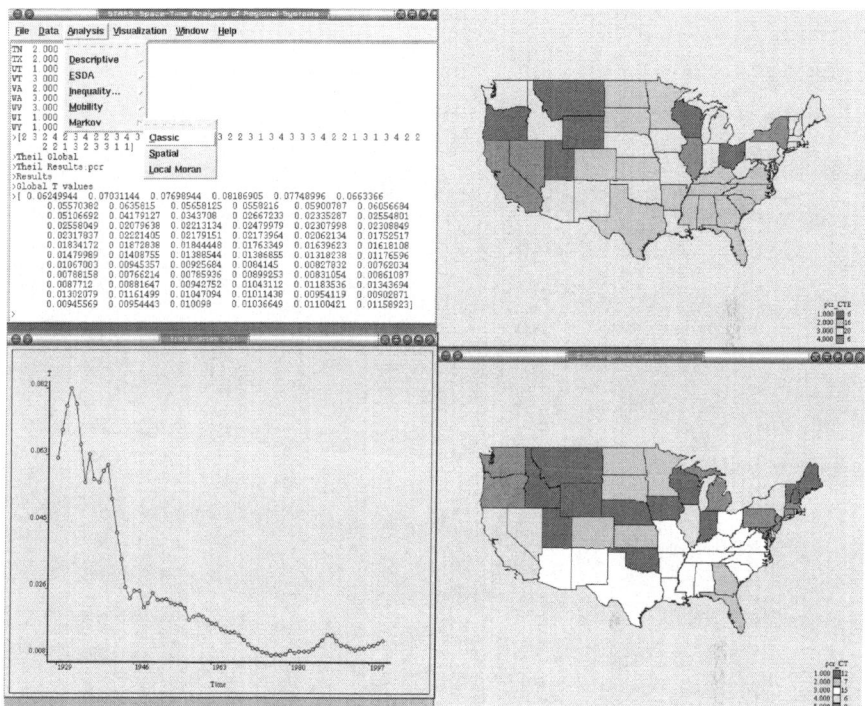

Fig. A.5.1. STARS in GUI mode

STARS is designed from the ground up as an object oriented system. This has a number of advantages. First, the internal architecture is accessible at a high-level, supporting the relatively easy enhancement of STARS via new specialized modules. Second, from an end-user's perspective, models, variables, matrices and other core elements of the system are all objects (for example, instances of classes in Python parlance), and thus are closer to the user's problem domain than is the case in a system designed around procedural programming.

In addition to being object oriented in design, STARS is also highly modularized. The geocomputational and visualization modules are orthogonal, that is, they can be used independently of one another, or they can be combined depending on the requirements of a particular project. This modularity permits the use of STARS in three different modes. The first is the GUI mode, where the two sets of modules are tightly integrated. Here the user accesses the analytical capabilities from a series of menu items as displayed in Fig. A.5.1. This mode is well suited to researchers wanting to apply exploratory space-time data analysis to a substantive problem.

```
For help on classes: from ModuleName import *
>>> shelp(Tau)
Classic Tau rank correlation

    Arguments:
        variable: STARS variable
        interval: integer for length of time interval (default=1)
        w:  STARS sparse weight matrix (optional)

    Attributes:
        concord: number of concordant pairs for each period
        discord: number of disconcordant pairs for each period
        T: number of unique pairs
        tau:    tau statistic (concord/T) for each period

        (if w is specified)
        nContiguities: number of unique contiguous pairs
        contConcordCount: number of contiguous concordant pairs
        contTau:    tau statistic for contiguous pairs
        nonContConcordCount: number of noncontiguous concordant pairs
        nonContTau:    tau statistic for noncontiguous pairs

    Example Useage:
        >>> from stars import Project
        >>> s=Project("s")
        >>> s.ReadData("csiss")
        >>> income=s.getVariable("pcincome")
        >>> r=s.getVariable("bea")
        >>> w=spRegionMatrix(r)
        >>> taus=Tau(income,w=w)
        >>> taus.tau[0]
        0.6313405797101449
        >>> taus.contTau[0]
        0.6875
        >>> taus.nonContTau[0]
        0.62291666666666667
        >>> tau=Tau(income,interval=10)
        >>> tau.tau
        array([ 0.55978261,  0.5923913 ,  0.60869565,  0.54076087,  0.55253623,  0.57880435,
                0.56612319,  0.58967391,  0.50724638,  0.59692029,  0.56431159,
                0.52083333,  0.52173913,  0.51449275,  0.42844203,  0.4692029 ,
                0.55615942,  0.47735507,  0.58152174,  0.5317029 ,  0.52626812,
                0.53713768,  0.48822464,  0.58423913,  0.57427536,  0.47101449,
                0.61865942,  0.5615942 ,  0.51539855,  0.5       ,  0.6195652 ,
                0.51539855,  0.53623188,  0.5923913 ,  0.5951087 ,  0.47373188,
                0.51539855,  0.51358696,  0.51811594,  0.39945652,  0.50996377,
                0.43387681,  0.49456522,  0.63677536,  0.49184783,  0.54981884,
                0.44384058,  0.53623188,  0.4701087 ,  0.52173913,  0.58423913,
                0.50362319,  0.44112319,  0.55706522,  0.48460145,  0.58695652,
                0.45742754,  0.48731884,  0.62228261,  0.48460145,  0.55253623,
                0.5       ])

For help on classes: from ModuleName import *
>>> ▮
```

Fig. A.5.2. STARS in command line interface mode

The second mode uses a command line interface (CLI) in which the computational module can be called directly from the Python interpreter. An example of such use is seen in Fig. A.5.2. This supports very efficient interactive computation, similar to that found in other data analysis environments such as R (R Development Core Team 2004). This mode also supports the wrapping of STARS modules inside larger Python scripts to implement simulation programs through batch processing.[3]

STARS can also be used in a combined CLI+GUI mode as shown in Fig. A.5.3. In this mode the user has access to the Python interpreter via the terminal window (upper left) and can create views either from that interpreter, or from the GUI (upper right). Results of interactive commands entered in the shell are reported in the text area of the GUI.

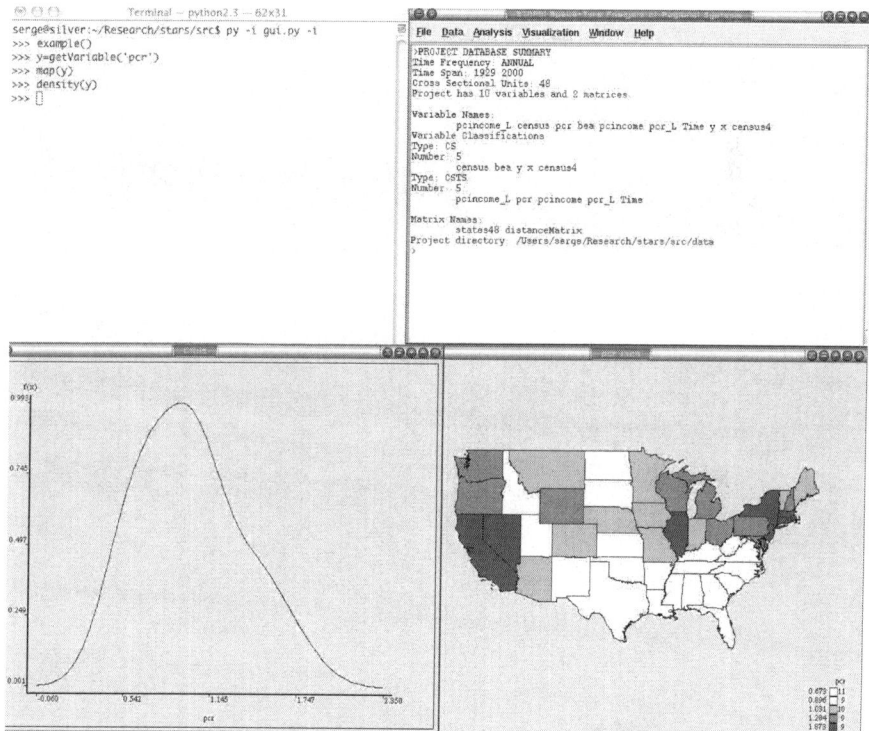

Fig. A.5.3. STARS in CLI+GUI mode

---

[3] An example of such an application is reported in Rey (2004a).

A.5.4 Illustrations

In this section a subset of the graphical and analytical capabilities of STARS are highlighted drawing on examples from regional income convergence studies. STARS stresses the need to study multiple dimensions underlying the data used in exploratory analysis. An illustration of this is provided in Fig. A.5.4 which contains four different views of data on U.S. regional incomes for the lower 48 states. The upper left view is a quintile map for incomes in 1929. Next to this is the Moran scatter plot (Anselin 1995), indicating strong positive spatial autocorrelation. Below the scatter plot, a histogram provides an a-spatial view of the income distribution, while the view to the left of the histogram portrays the time series for the global Moran statistic for the years 1929-2000. The latter figure reveals that the level of spatial clustering fluctuates substantially over time.

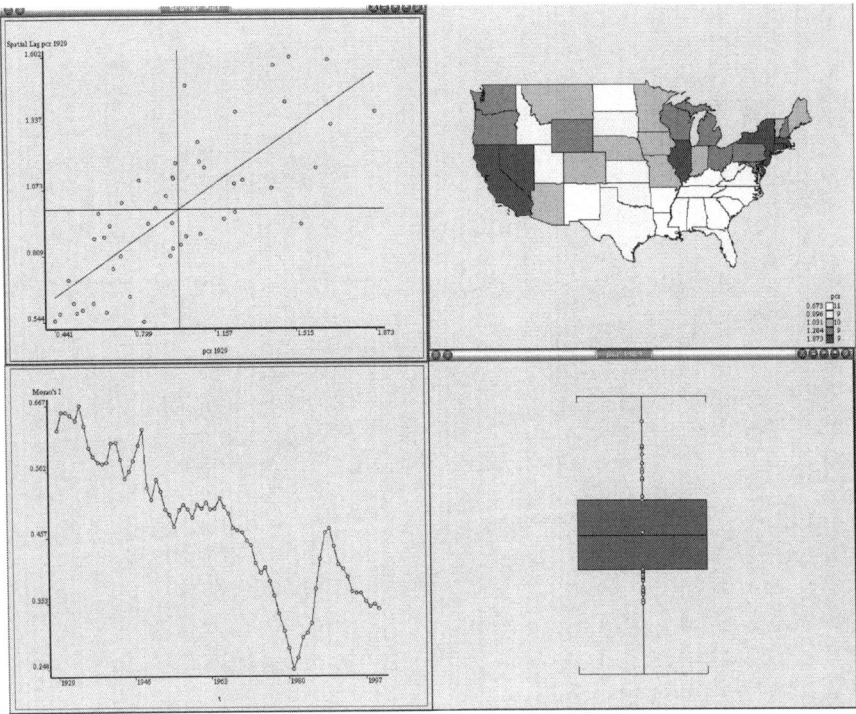

Fig. A.5.4. Multiple views of U.S. per capita income data

Linking and brushing views. In addition to providing views of the different dimensions (time, space, distribution), the views in STARS are also interactive. Interactivity can take on multiple forms. The first is *linking* in which the selection of observations in an origin view leads to the highlighting of associated observations in other destination views. An example of this can be seen in Fig. A.5.5, where the selection occurs on the origin view (map) using a rectangle created and sized with the mouse. When the user releases the mouse button, the polygons underneath the selection rectangle are selected and observations associated with these selected polygons are then highlighted in the three destination views.[4]

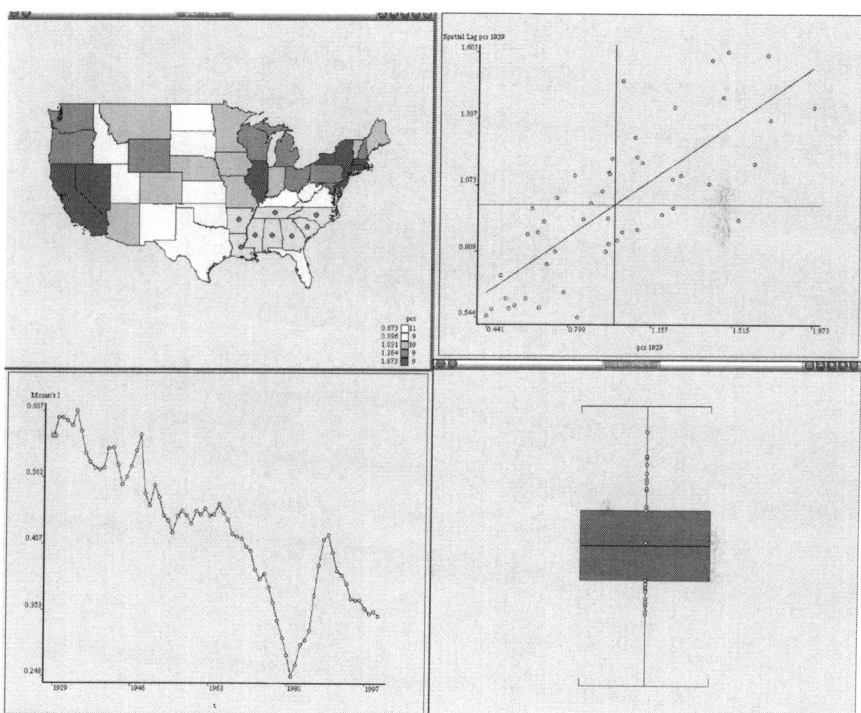

Fig. A.5.5. Linking multiple views

---

[4] The selection rectangle is not seen in Fig. A.5.5 as it is erased upon completion of the selection.

The second form of interaction is *brushing* which is illustrated in Fig. A.5.6. Here observations are selected in the same fashion as with linking, however the impact of the selected set is different, and results in a re-fitting of the global autocorrelation trend in the scatter plot to omit the states selected on the map. This provides insights as to the leverage of the selected states on the level of spatial clustering for that time period.

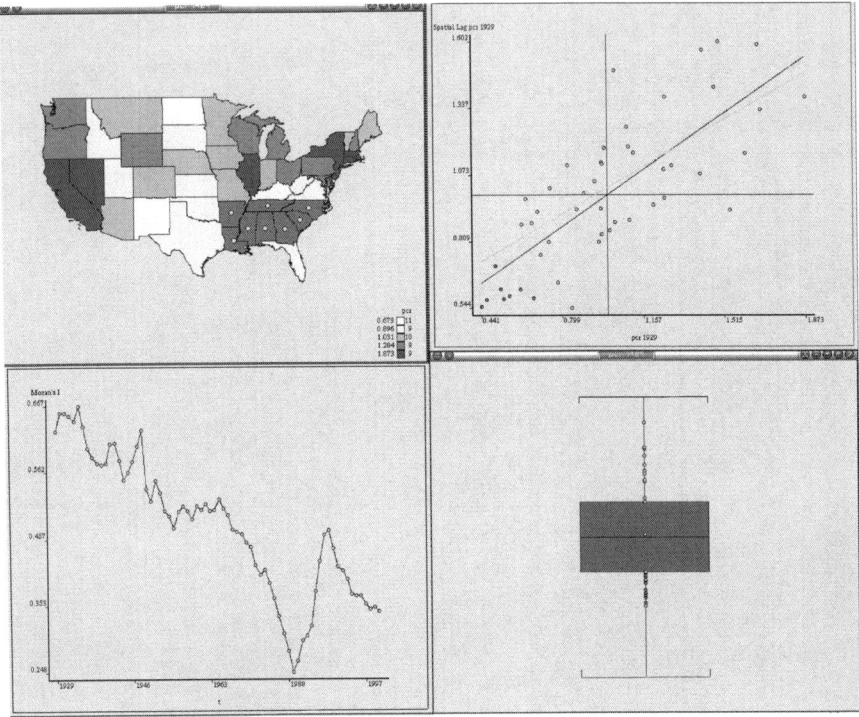

Fig. A.5.6. Brushing multiple views

Space-time traveling and roaming. Linking and brushing can also be combined with a third form of interaction referred to as *roaming*. When roaming, the selection rectangle remains on the screen and the user can move it around the origin view, as is reflected in Fig. A.5.7. Movement of the selection rectangle creates a new selection set of observations on the origin view to trigger the corresponding interaction signal (brushing or linking) on the destination views.

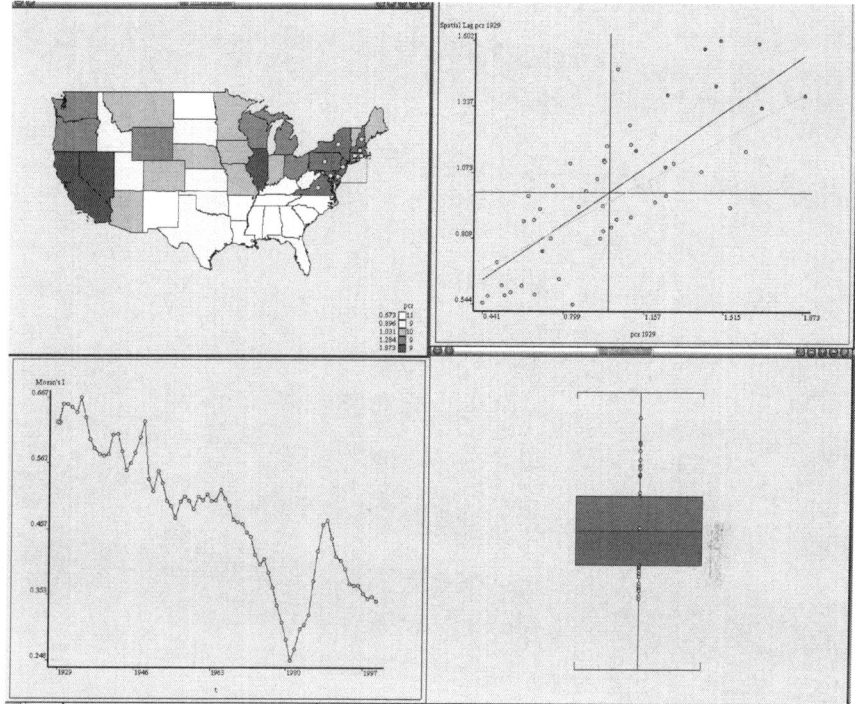

Fig. A.5.7. Roaming a map with brushing

Similar to roaming, linking and brushing can also be combined with *traveling*. Traveling on an origin view selects observations in a sorted order and triggers linking or brushing on the destination views. The traveling is done automatically over the entire set of observations on the origin view, giving the user a full depiction of the particular type of interaction (linking or brushing). An example of this is shown in Fig. A.5.8, which combines cumulative brushing on the scatter plot and box plot resulting from spatial traveling on the map.

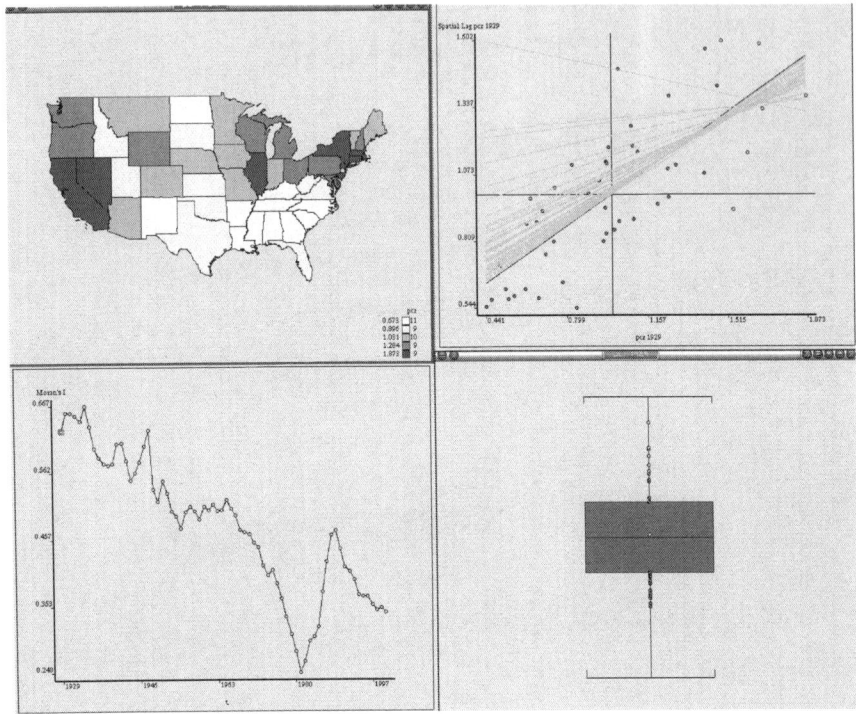

Fig. A.5.8. Spatial traveling with brushing

Traveling can also be done on a time series view to trigger temporal updating of destination views. The traveling proceeds from earliest period to the latest period given the user views of all destination views for each time period in the sample. Alternatively, the user can control the temporal updating by switching to roaming on a time series view. This is illustrated in Fig. A.5.9, where the vertical selector has been moved over the year 1990. Again the three destination views (scatter plot, map and box plot) are updated to this year, which reveals an outlier in the box plot. The user then selects that outlier observation on the box plot to trigger linking on the destination views (map, time series, scatter plot) to reveal that the outlier observation is Connecticut.

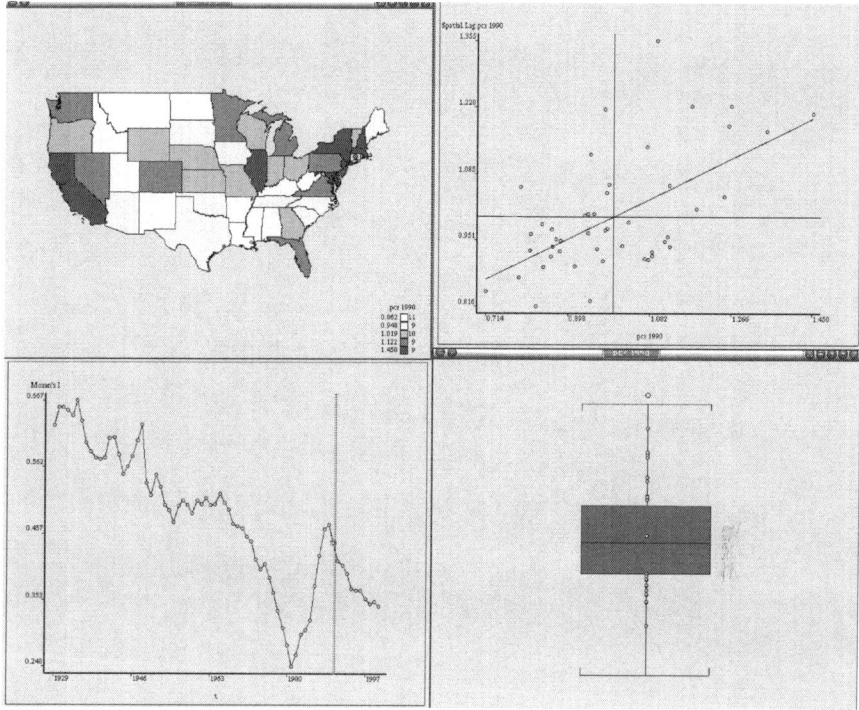

Fig. A.5.9. Time roaming

The combination of linking and brushing with either space-time roaming or traveling provides a powerful approach to exploratory visualization that can reveal patterns that otherwise would be very difficult to detect. An example of this can be seen in Fig. A.5.10 where a conditional scatter plot in the lower right corner is used to combine the Moran scatter plots from each year in a single view. The observations on each state's income and that of its spatial lag are then conditioned on a third variable, in this case Time, and the conditioning uses color depth to indicate early (light color) versus more recent (dark color) observations. The conditioning reveals that the dispersion in state incomes has declined substantially over time. The figure also reflects the result of the user selecting Illinois on the map to trigger linking in the destination views. The own-lag pairs for all time periods for Illinois are then highlighted in the conditional scatter plot to reveal that the spatial dynamics between Illinois and its neighbors have been qualitatively and quantitatively different from the overall space-time dynamics in the U.S. space economy.

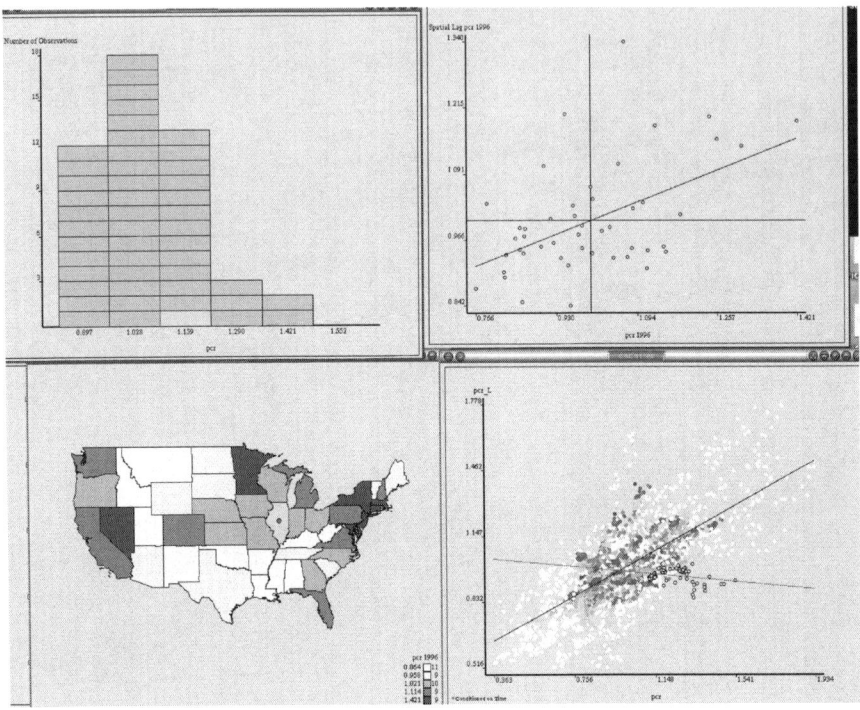

Fig. A.5.10. Space-time instabilities

View-generated-views. The view interactivity can be exploited to more fully explore these space-time instabilities depicted in the conditional scatter plot. While the latter shows that Illinois and its geographical neighbors have income dynamics moving in different directions, additional insights on these dynamics can be obtained by the user combining a key press (control) with a mouse-click on the Illinois specific observation in the Moran scatter plot which generates a new view called a time path as shown in the upper left of Fig. A.5.11. The time path shows the co-movement of Illinois per capita income and its spatial lag of per capita income for all time periods with subsequent time periods linked together.

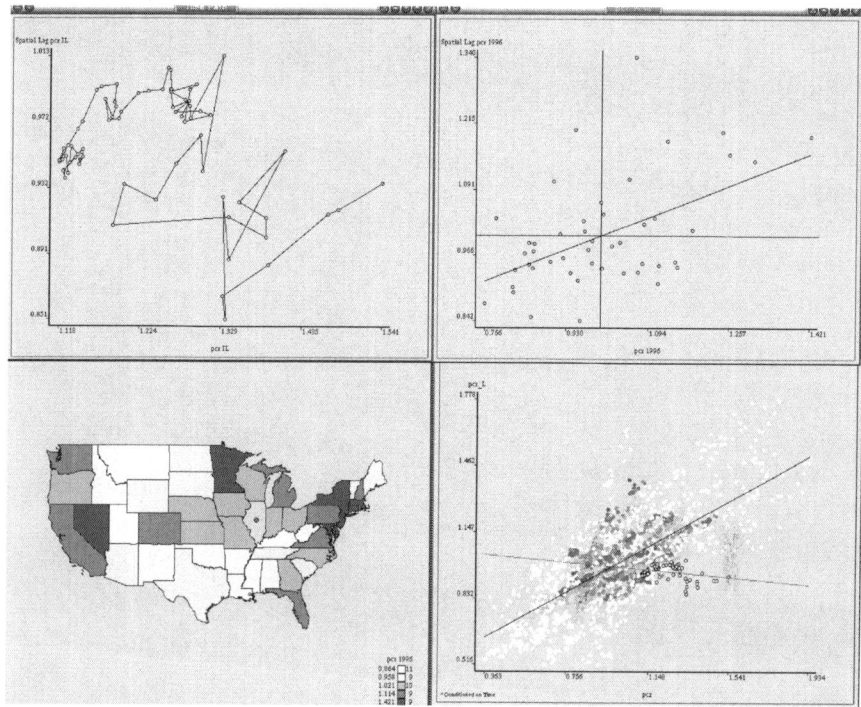

Fig. A.5.11. Scatter plot generated time path

The ability to generate new views through user actions on existing views offers a powerful exploratory device. View-generated-views can also be obtained from a map origin view as seen in Fig. A.5.12, where the user has issued the same selection event on Illinois in the map to generate the time series view of relative income for Illinois. This isolates the dynamics of Illinois income from the co-movement dynamics in the time path, in a similar manner to the way the co-movement dynamics for Illinois were isolated in the time path from the full set of state-lag co-movement dynamics depicted in the conditional scatter plot.

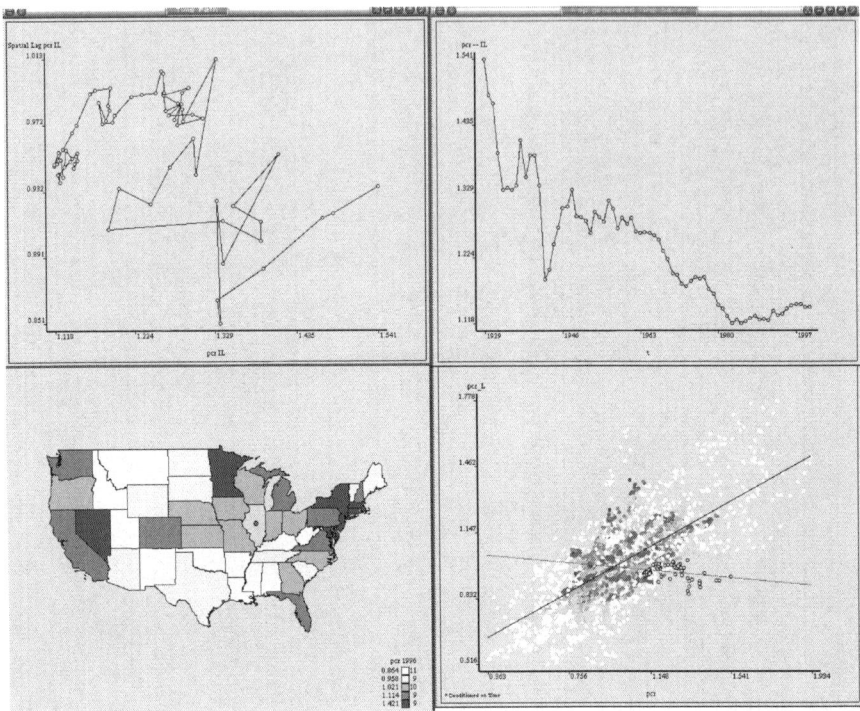

Fig. A.5.12. Map generated time series

Distribution dynamics. In addition to exploring spatial and temporal dimensions via view interactivity, the distributional dynamics can also be examined. One approach is displayed in Fig. A.5.13 in which two densities for state relative per capita incomes are displayed, one for the beginning of the period (1929) one for the last year of the sample (2000). To explore the movement of individual economies within the income distribution the user can trigger spatial traveling on the map serving as the origin view. This then highlights each state (from lowest income to highest income) on the map and identifies the positions of that state in the initial and terminal income densities. As the traveling is done automatically for the entire set of spatial units, the user sees the full extent of distributional dynamics. Following the automated traveling, the user can then select individual states on the map to isolate on their mobility characteristics. This is shown for Virginia which initially was a relatively poor economy but has shown substantial upward movement in the income distribution.

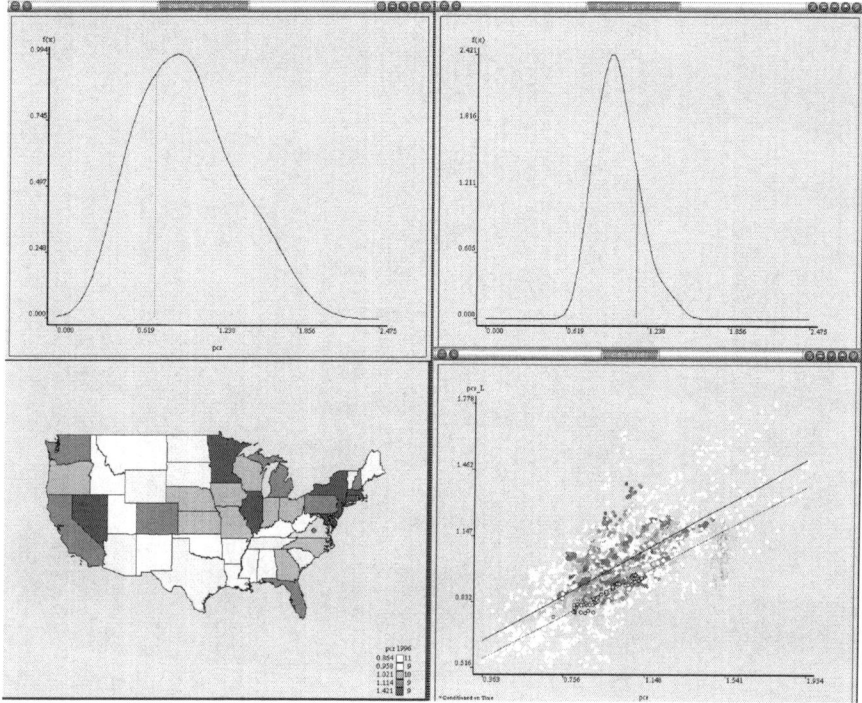

Fig. A.5.13. Distributional mixing

Spatial and temporal dependencies. In addition to providing dimension specific views, such as a time path, or box plot or quintile map, STARS enables the depiction of multiple dimensions on a single view. This is illustrated in Fig. A.5.14 which contrasts two forms of covariance in a graph representation. The linkages reflected in a spatial weight matrix based on contiguity are recorded as edges between polygon centroids for each state. These linkages are then conditioned on the strength of the temporal covariance between each pair of contiguous states, with light grey lines indicating strong temporal linkages.

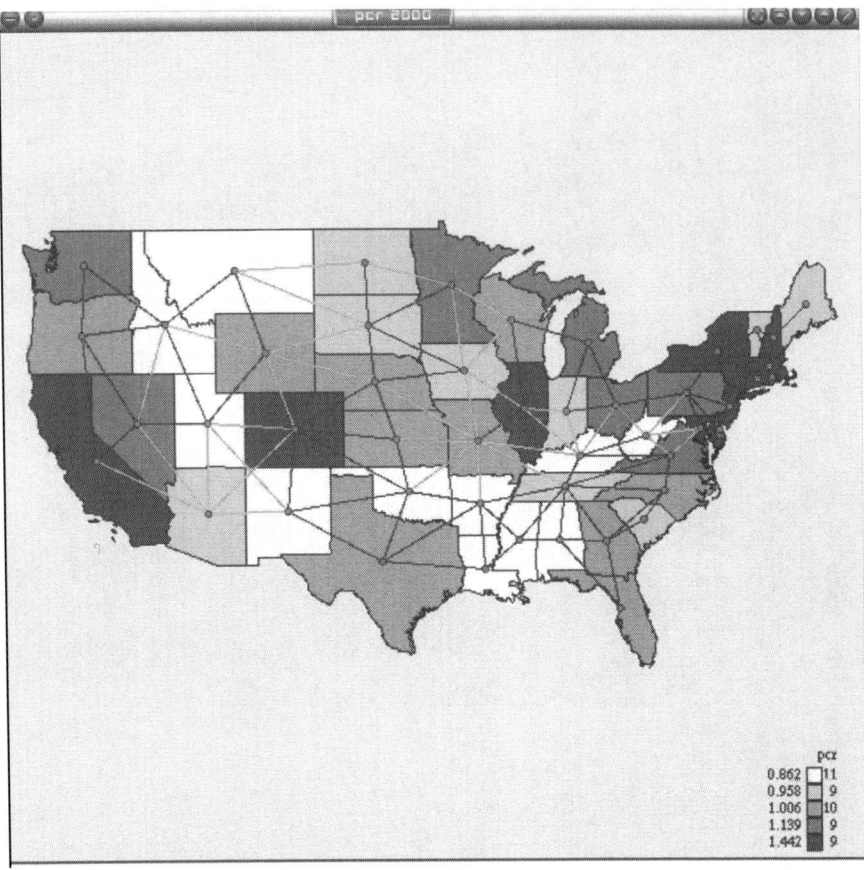

Fig. A.5.14. Spatial and temporal covariance networks

The nature of the specific temporal covariances between a state and the rest of the system can then be explored using the spider graph depicted in Fig. A.5.15. Here the user can step through each state to determine which other states it has the strongest temporal co-movements with. In this case the spider graph reveals that California income dynamics have not only been similar to some of its geographical neighbors, but also in sync with the northeast states. This type of interaction is useful for uncovering covariance relations that may not be obvious with traditional ESDA techniques.

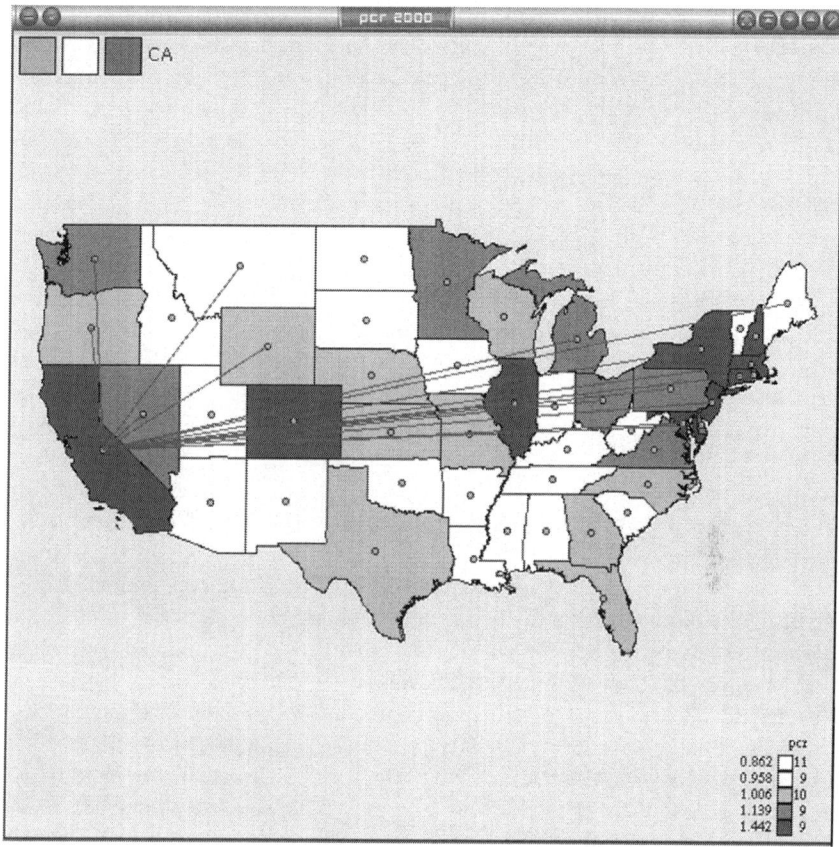

Fig. A.5.15. Spider graph of temporal networks

A.5.5 Concluding remarks

STARS has evolved quickly from its origins as specialized program to support research on regional income dynamics to now being used by researchers, outside of the development team, to examine such issues as spatial dynamics of fertility, land use cover change, segregation dynamics, migration, commodity flow patterns and housing market dynamics, among others. Each new application raises new demands for increased functionality and enhancement of STARS. Currently there are a number of such enhancement that are major priorities for the development team.

The first enhancement is the creation of a new type of map view to visualize substantive flows between cross-sectional observations.[5] There has been a growing interest in the extension of flow maps to include temporal-spatial dynamics which we believe STARS is quite posed to introduce. In short, the goal of this extension is to demonstrate how flows between cross-sectional units evolve over time. Although often used to study migration, the notion of flows is by no means confined to the movement of people. Flows of commodities, for example, could be considered a driver for many socioeconomic processes, and the inclusion of which could present some interesting research avenues; such as the covariation between these flows and economic growth, and the construction of hybrid weight matrices based on spatial constructs coupled with a-spatial flow linkages.

Another analytical front for the STARS module is cluster analysis. Although some basic forms of spatial clustering are identifiable by a number of graphs and maps produced in the current version of STARS, more analytical features on a-spatial cluster analysis seem a fruitful avenue for future work. The research team has an extensive body of code implementing agglomerative, partitive and medoid clustering methods written in variety of languages (R, Octave, Python) in support of on-going research on industrial cluster analysis (Rey and Mattheis 2000; Rey 2000a, b, c, d, e, 2002). The integration of these methods in STARS is currently underway.

We are also exploring new approaches towards recasting conventional measures of distributional dynamics, such as the so called σ-convergence measure, to incorporate spatially explicit dimensions (Rey and Dev 2004). Coupled with this is work on developing inferential methods for new space-time empirics based on both analytical distributions as well as computationally based approaches.

STARS is a powerful environment for exploring data that has both temporal and spatial dimensions. The interactivity of the various views helps to identify dependencies across various dimensions that may otherwise go unnoticed. These views are also tied to a suite of recently developed advanced methods for ESDA and ESTDA. Moreover, STARS has been designed for users with a wide range of demands and skill-sets. Researchers looking for a user-friendly GUI environment for exploratory space-time analysis should feel at home with STARS. Others who are developing new methods for exploratory analysis can easily integrate these into the modular framework underlying STARS. In between these two groups are researchers comfortable with writing simple macro type scripts (in Python) to use STARS for simulation experiments as well as for linkages with other model systems and statistical packages. We hope this design together with the commitment to the open source development model will attract researchers to collaborate on the enhancement and future development path of STARS.

---

[5] See Tobler's Flow Mapper at http://csiss.ncgia.ucsb.edu/clearinghouse/FlowMapper/ for a program designed for the sole purpose of studying flows.

Acknowledgements. This research was supported in part by U.S. National Science Foundation grant BCS-0433132.

References

Anselin L (1995) Local indicators of spatial association-LISA. Geogr Anal 27(2)::93-115
Anselin L (2003) An introduction to EDA with GeoDa. Technical report, Spatial Analysis Laboratory, University of Illinois
Carlino GA, Mills LO (1993) Are U.S. regional incomes converging? A time series analysis. J Monet Econ 32(2):335-346
Christakos G, Bogaert P, Serre M (2001) Temporal GIS. Springer, Berlin, Heidelberg and New York
Dykes JA (1995) Pushing maps past their established limits: a unified approach to cartographic visualization. In Innovations in GIS. Taylor and Francis, London, pp.177-187
Egenhofer MJ, Golledge RG (1997) Spatial and temporal reasoning in geographic information systems. Oxford University Press, Oxford and New York
Hinsen K (2000) The molecular modeling toolkit: a new approach to molecular simulations. J Comput Chem 21:79-85
Langtangen HP (2004) Python scripting for computational science. Springer, Berlin, Heidelberg and New York
Peuquet DJ (2002) Representations of space and time. Guilford, New York
R Development Core Team (2004) R: a language and environment for statistical computing. R Foundation for Statistical Computing, Vienna, Austria.
Rey SJ (2000a) Identifying regional industrial clusters in California, volume II: methods handbook. Technical Report. California Employment Development Department, Sacramento [CA]
Rey SJ (2000b) Identifying regional industrial clusters in California, volume III: technical documentation of the state's candidate industry clusters. Technical Report. California Employment Development Department, Sacramento [CA]
Rey SJ (2000c) Identifying regional industrial clusters in California, volume IV: the role of industrial clusters in California's economic recent economic expansion. Technical Report. California Employment Development Department, Sacramento [CA]
Rey SJ (2000d) A structural economic analysis of the biotechnology cluster in the San Diego Economy. Working Paper, San Diego State University, San Diego [CA]
Rey SJ (2000e) A structural economic analysis of the visitors industry cluster in the San Diego Economy. Working Paper, San Diego State University, San Diego [CA]
Rey SJ (2001) Spatial empirics for regional economic growth and convergence. Geogr Anal 33(3):195-214
Rey SJ (2002) Identifying regional industrial clusters in Imperial County California. Technical Report. California Center for Border and Regional Economic Studies, Diego State University, San Diego [CA]
Rey SJ (2004a) Spatial analysis of regional income inequality. In Goodchild M and Janelle D (eds) Spatially integrated social science: examples in best practice. Oxford University Press, Oxford and New York, pp.280-299
Rey SJ (2004b) Spatial dependence in the evolution of regional income distributions. In Getis A, Múr J, Zoeller H (eds) Spatial econometrics and spatial statistics. Palgrave, Hampshire, pp.194-214

Rey SJ, Dev B (2004) σ-convergence in the presence of spatial effects. Paper presented at the Western Regional Science Association Meetings. Maui [HI]

Rey SJ, Janikas MV (2005) Regional convergence, inequality, and space. Econ Geogr 5(2):155-176

Rey SJ, Mattheis DJ (2000) Identifying regional industrial clusters in California, volume I: conceptual design Technical Report. California Employment Development Department, Sacramento [CA]

Rey SJ, Montouri BD (1999) U.S. regional income convergence: a spatial econometric perspective. Reg Stud 33(2):143-156

Saenz J, Zubillaga J, Fernandez J (2002) Geophysical data analysis using Python. Comput Geosci 28(4):475-465

Schliep A, Hochstättler W, Pattberg T (2001) Rule-based animation of algorithms using animated data structures in gato. Technical report, Zentrum für Angewandte Informatik Köln, Arbeitsgruppe Faigle/Schrader

Symanzik J, Kötter T, Schmelzer S, Klinke S, Cook D, Swayne DF (1998) Spatial data analysis in the dynamically linked ArcView/XGobi/Xplore environment. Comput Sc Stat 29:561-569

Takatsuka M, Gahegan M (2002) GeoVista Studio: a codeless visual programming environment for geoscientific data analysis and visualization. Comput Geosci 28(10):1131-1144

Theil H (1996) Studies in global econometrics. Kluwer, Dordrecht

Wise S, Haining R, Ma J (2001) Providing spatial statistical data analysis functionality for the GIS user: the SAGE project. Int J of Geogr Inform Sci 15(3):239-254

A.6 Space-Time Intelligence System Software for the Analysis of Complex Systems

Geoffrey M. Jacquez

A.6.1 Introduction

The representation of geographies (e.g. census units), demographics and populations as unchanging rather than dynamic is due in part to the static world-view of GIS software, which has been criticized as not fully capable of representing temporal change and better suited to 'snapshots' of static systems (Goodchild 2000; Hornsby and Egenhofer 2002; Jacquez et al. 2005). This static view hinders the mapping, representation, and analysis of dynamic health, socioeconomic, and environmental information for populations that are dispersed and mobile – a key characteristic of the human condition (Schaerstrom 2003).

Several approaches to modifying GIS to better handle the temporal dimension have been proposed. Yearsley and Worboys (1995) proposed a space time object model that integrates abstract spatial data types with a geometric layer to construct a higher-level topological data model, Raper and Livingstone (1993) used an object oriented approach to represent dynamic spatial processes as spatio-temporal aggregations of point objects, and Peuquet and Duan (1993) formulated an event-based spatio-temporal data model (ESTDM) that maintains spatio-temporal data as a sequence of temporal events associated with a spatial object. See Miller (2005b) for a review of alternative data models.

Hägerstrand's (1970) seminal work in time geography has led to geometric and mathematical constructs for quantifying human mobility including geospatial lifelines, space-time prisms, and techniques for propagating location uncertainty through time (Miller 1991; Kwan 2003; Han et al. 2005; Miller 2005a). These, in turn, have provided a quantitative basis for the development of statistics and modeling approaches suited to the analysis of temporally dynamic systems. For example, Sinha and Mark (2005) proposed a Minkowski metric to quantify dissimilarity between geospatial lifelines; Han et al. (2005) present a K function calculated from the spatial pattern of place of residence at specific time slices; Q-statistics assess case-control clustering (Jacquez and Meliker 2008) and space-

Table A.6.1. Summary of STIS functionality

Category	Functionality	Characteristics
Data types	Points	Both static (e.g. space only) and temporally dynamic when points move through time
	Lines	Both static and temporally dynamic
	Polygons	Both static and dynamic such that polygons change shape (e.g. morph) and location
	Mobility histories	For representing geospatial lifelines and activity spaces
	Rasters	Both static and temporally dynamic for representing space-time fields
Visualization	Linked windows	Cartographic and statistical brushing with time-enabled spatial objects; time synchronization of maps and graphs
Tables	Tables	Attribute values can change through time
Maps	Maps	Of point, vector, polygons, mobility histories, raster data, spatial weights
	Cluster maps	Display locations of spatial outliers and clusters of low and high values and how they change through time
	Change maps	Also called difference maps, show absolute and relative change between time periods
	Disparity maps	Show where and when a target population differs significantly from a reference population (e.g. health disparities)
	Animations	Display movies of time-dynamic spatial data
Statistical graphics	Box plots	All statistical graphics except the time series plot are time enabled, displaying data relationships through time
	Histograms	
	Scattergrams	
	Principal coordinate plots	
	Time series plots	Show how attribute values for space-time objects change through time
	Variogram clouds	Visualize spatial variance at different spatial lags and through time
Weight sets	Spatial weights	Nearest neighbor, adjacency, distance, and inverse distance weights are time-dynamic when geography (for example, locations of points) changes through time
Smoothing	Empirical Bayesian	Bayesian smoother using user-specified spatial weights
	Poisson	Poisson smoother for point/rare event data
Pattern recognition methods	Local Moran	Univariate and bivariate, with or without temporal lags, for points or polygons
	Global Moran	Provided automatically with local Moran
	Local G and G^*	For points or polygons
	Besag and Newell	For case and population at risk, using points or polygons
	Turnbull	For case and population at risk data, points or polygons.
	Disparity statistics	For reference and target populations using rates and population sizes
	Variogram analysis	Isotropic, anisotropic, automated fitting, time dynamic
Modeling (all time-dynamic)	A-spatial regression	Linear, logistic and Poisson regression, using full model, best subset, forward or backward stepwise selection. Model is fitted through time.
	Geographically weighted regression	Linear, logistic and Poisson regression, using user-specified or automatically optimized bandwidth
	Kriging	Using traditional, standardized, residuals, weighted and Poisson estimators; includes simple, ordinary, kriging with a trend and Poisson kriging

time interaction in case data (Jacquez et al. 2007) that account for human mobility; and kernel functions weighted by duration at specific locations have been used to estimate risk functions in temporally dynamic systems (Sabel et al. 2000; Sabel et al. 2003).

Recent technological advances have resulted in Space Time Intelligence Systems (STIS) that implement constructs for representing temporal change (AvRuskin et al. 2004; Greiling et al. 2005; Jacquez et al. 2005; Meliker et al. 2005). The STIS technology has several advantages. First, it is founded on space-time data structures, enabling complex space-time queries not possible in conventional 'spatial only' GIS. Second, it incorporates statistical tests for space-time pattern such as univariate and bivariate local indicators of spatial autocorrelation and clustering that are automatically calculated through time, resulting in cluster animations that capture space-time change. Third, it employs dynamic linked windows that enable both cartographic and statistical brushing through time. Fourth, it calculates weight matrices for dynamic systems where points can move through time, and where polygons can morph, merge and divide such that pattern recognition and modeling readily account for dynamic and complex time geographies. Fifth, it constructs spatio-temporal statistical models including linear, Poisson and logistic regression, geographically weighted regression, variogram models, and kriging. Finally, it displays animated 'movies' for exploring how variables (for example, health outcomes such as maps of incidence, mortality, case counts and expectations, and clusters themselves) change through space and time. Development of the STIS software was funded by grants from the National Institutes of Environmental Health Sciences and the National Cancer Institute. This technology is well suited to the representation, visualization, modeling and simulation of dynamic patterns and processes, and its functionality (see Table A.6.1) is the topic of the balance of this Chapter.

A.6.2 An approach to the analysis of complex systems

Geographic systems typically are large, dynamic and complex. Our approach to analyzing complex systems in STIS consists of three stages: development of cognitive models, exploratory space-time data analysis, and modeling; with each stage informing the others.

Cognitive and ontological models have to do with the mental representation of the underlying causal mechanisms that drive the relationships observed in a complex system. These usually are based on speculation, an understanding of prior research findings, and by one's experience with similar systems. They guide exploratory data analysis, and form the basis on which more detailed data-based and process-based models are constructed. They are developed by visualizing and interacting with the data, and are continually refined through data analysis and modeling.

Exploratory Data Analysis (EDA) is founded on exploratory methods for quickly producing and visualizing simple summaries of data sets to reveal relationships and insights that often cause one to refine the cognitive model (Tukey 1977). *Exploratory Space-Time Data Analysis* (ESTDA) is made possible by software systems that incorporate spatial and temporal data, dynamic linked windows, statistical and cartographic brushing, and can generate hypotheses to be evaluated using clustering, inferential statistics and models. The objective of exploratory techniques is to illuminate and quantify relationships in order to increase the analyst's knowledge of the complex system, giving rise to testable hypotheses and to relationships that can be modeled.

Models of data include statistical tools such as ANOVA, regression and correlation, and are used to quantify relationships among variables, to test statistical hypotheses, and to identify factors that drive variability in the experimental system. These models require data of sufficient quality to estimate model coefficients (for example, regression intercept), and that the researcher has sufficient knowledge to be able to identify dependent and independent variables, and their relevant parameters. Models of data are often used for interpolation and for prediction but do not necessarily convey information regarding underlying causal mechanisms.

Models of process require a detailed understanding of the mechanics of the system being studied, and incorporate this understanding directly into the model itself. Examples of process models include infection transmission systems in which the population is structured into susceptible, infectious, and immune subgroups, and in which the model parameters describe mechanistic processes such as infection transmission to susceptible individuals (Koopman et al. 2001).

STIS provides a platform for analyzing complex space-time systems, from visualization, the quantification of geographic relationships using weight matrices that change through time, the identification of space-time pattern to generate hypotheses, to models that may be used for estimation and prediction, as described below.

A.6.3 Visualization

The first step is to enter data into STIS and to then create maps, animations, and statistical graphics to explore relationships in the data. Supported data types include points, mobility histories, lines and polygons. STIS reads ESRI shape files, excel, dbf and text files, using time series or time slice formats. A time slice means all objects in the geography change attribute values simultaneously, so that one may assign a time stamp defining an interval that applies to all objects in a data set. An example would be lung cancer mortality rates for white males in U.S. counties from 1950 to 1955. Time series data arise when the values of the attributes change asynchronously among different spatial objects. For example, one location may be sampled at hourly intervals, while another is sampled daily.

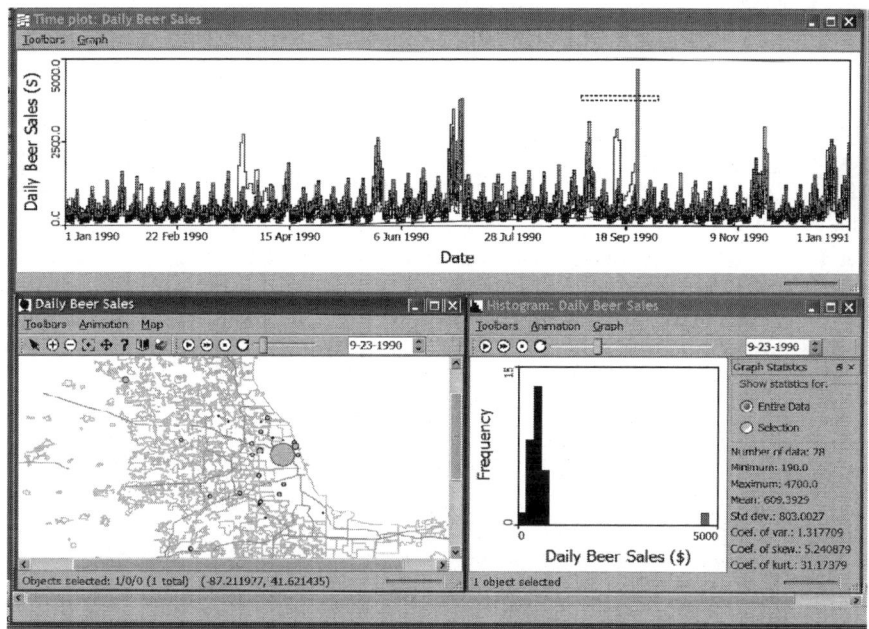

Fig. A.6.1. Visualization and exploration of space-time patterns in daily beer sales at Dominick's stores in the greater Chicago area in 1990. The user is brushing on the time plot (top) to identify the spike in sales that occurred at one store in central Chicago (map, lower left) on Sept 23, 1990. Notice the strong periodicity caused by increased beer sales on weekends

After the data are entered one next creates maps, and then animates them to obtain an initial impression of space-time patterns. Time series plots are used to explore how variable values change through time. Linked brushing on the maps, statistical graphics and tables, along with time animation, supports rapid identification of relevant space-time patterns (see Fig. A.6.1).

A.6.4 Exploratory space-time analysis

Dynamic spatial weights: Cluster analysis, autocorrelation analysis, spatial regression, geostatistics and other techniques in STIS rely on weights to model geographic relationships among the objects. STIS automatically calculates spatial weight matrices needed for cluster analyses, and prompts the user when more detailed weights or kernels are required for methods such as geographically weighted regression and geostatistics. In Fig. A.6.2 the user is exploring the spatial weight connections in counties in the Northeastern United States using centroids with five nearest neighbors (left) and polygon adjacencies (right). The use of centroid locations to represent geographic relationships among area-based data such as counties can produce misleading results since the spatial support (for example, area and configuration of the counties) is ignored (Jacquez and Greiling 2003).

The spatial weights in STIS are dynamic, so that changing geographies are modeled in a realistic fashion. Examples include census geography; zip-code geographies; area-codes; land parcel data; and land use maps, all of which change through time. Dynamic spatial weights are used by cluster analysis and modeling techniques (including geographically weighted regression, variogram models, and kriging) so that temporal change in both geographic relationships and attribute values are fully accounted for.

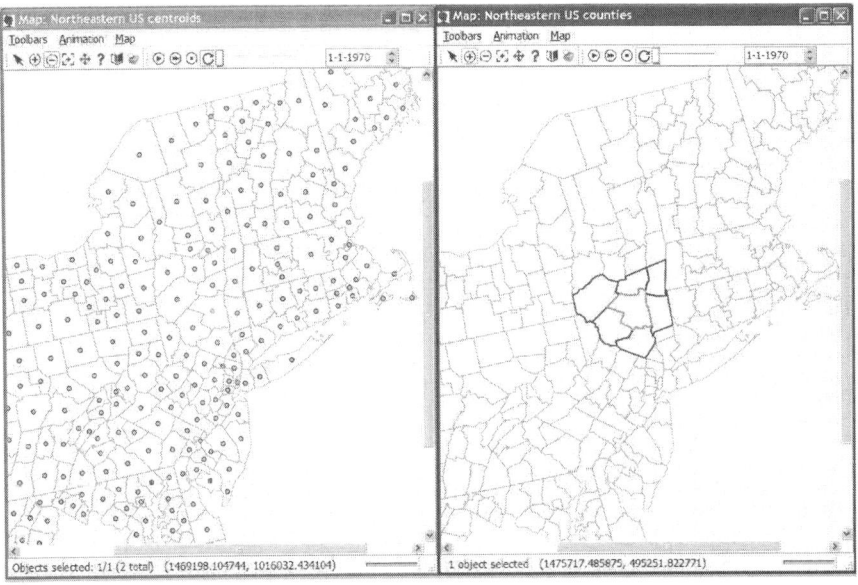

Fig. A.6.2. STIS visualizes spatial weights by outlining the selected location (centroid or polygon) in gold, and the localities to which it is connected in blue. The five nearest neighbors using centroids (left) differ from border adjacencies (right). The spatial weights for queried locations are written to the log view (not shown) for validation

Pattern recognition: STIS provides cluster tests for both point data and polygon data, including the local Moran (Anselin 1995), G statistics (Getis and Ord 1992; Ord and Getis 1995), Besag and Newell (1991) and Turnbull's (Turnbull et al. 1990) tests. Both absolute and relative disparity statistics identify significant differences in outcomes (for example, disease incidence and mortality, tumor staging, health screening utilization) through space and time (Goovaerts 2005). Spatial pattern recognition may also be accomplished using variogram analysis, the point of departure for which is the variogram cloud. STIS provides automatic variogram fitting and both the variogram cloud and variogram models are time-dynamic. Basic variogram models include spherical, exponential, cubic, Gaussian and power models (Fig. A.6.3). Automatic variogram fitting selects from among these models to find that model which provides the best fit, along with the corresponding parameter estimates.

Outlier detection: An important step in exploratory space-time data analysis is the identification of outliers – observations whose values are unusual when considered in the context of the sample. Outlier analysis methods in STIS include the box plot, anomaly detection using local indicators of spatial autocorrelation, detection of geostatistical outliers via statistical brushing on the variogram cloud, and the exploration of deviations from model predictions using these techniques applied to model residuals.

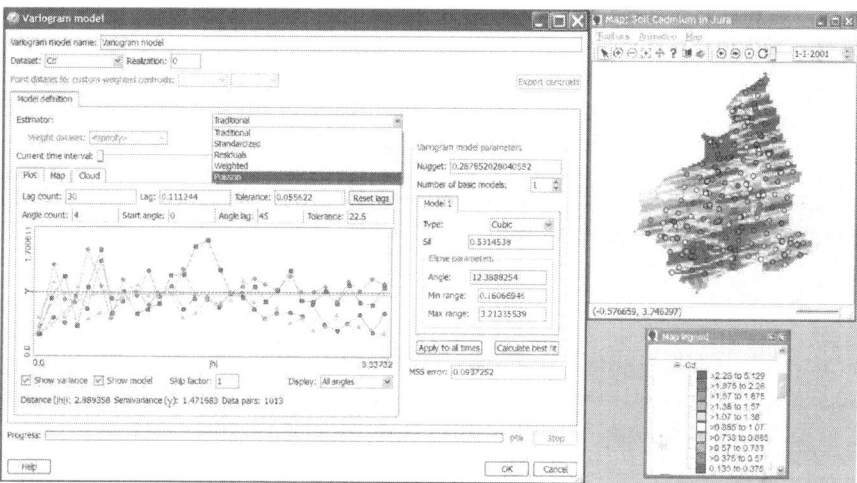

Fig. A.6.3. Automatic variogram model fitting of soil Cadmium concentrations in the Jura mountains, France. Notice the 'Calculate best fit' button in the variogram model window (left). Variogram estimators in STIS include traditional, standardized, residuals, weighted and Poisson. Here, an isotropic variogram model modeled a directional spatial pattern, and was then used to predict soil cadmium concentrations using kriging (raster map, right center). Data courtesy Pierre Goovaerts

A.6.5 Analysis and modeling

STIS provides advanced modeling techniques including a-spatial regression, geographically weighted regression, and geostatistics, as summarized below. All of these are time-enabled and automatically model changes in geography (for example, morphing polygons and moving points) as well as attributes (for example, how the value associated with a spatial object changes) through time.

A-spatial regression: Exploratory space-time data analysis using visualization and pattern recognition methods often generates hypotheses regarding dependencies and associations among the variables. Before invoking spatial modeling approaches a researcher may first choose to employ a-spatial models, and then evaluate pattern in the model residuals to determine whether more detailed space-time models are warranted. The rationale is one of parsimony – if an a-spatial model adequately explains the observed variability then a more complex spatial model may not be warranted.

STIS provides linear, logistic and Poisson regression, and for complex models with several variables evaluates the fit of subsets of the independent variables using the full model (all variables), forward stepwise, backward stepwise, and best subset. The criterion for finding the best subset – that combination of independent variables that does the best job of explaining variability in the dependent variable – for linear models include R-squared, adjusted R-squared, $C(p)$, and AIC. R-squared selects the model with the largest reduction in residual sum of squares, and thus favors complex model with the largest number of terms. The adjusted R-squared criterion punishes models with too many terms. The smallest AIC (Akaike information criterion) trades off model fit and model complexity using, for linear regression, the residual sum of squares (RSS) penalized by two times the number of regression term degrees of freedom (k = the number of regression parameters). Finally, the smallest Mallows $C(p)$ is another way of penalizing models with many independent variables. It is the residual sum of squares for the subset model being considered, divided by the error variance for the full model plus twice the number of regression degrees of freedom minus the total number of observations. Similar $C(p)$ values similar to the one for the full model are considered an indication of good candidate models. Appropriate model selection criteria are also provided for Poisson and logistic regression.

Geographically weighted regression (*GWR*): Most of the functionality and modeling approaches for a-spatial regression are available as well in GWR (see Chapter C.5 for more details). Whereas a-spatial regression makes strong assumptions regarding stationarity of the regression coefficients, GWR allows the regression coefficients to vary through geographic space and through time, and fits spatially and temporally local regression, with local estimates of model fit (for example, R^2, the regression coefficients and model residuals, correlations and other statistics). GWR has been pioneered by A. Stewart Fotheringham and Martin Charlton (currently at the National Center for Geocomputation, National University of Ireland), and Chris Brunsdon (University of Glamorgen, UK). Our implementation of this tool is based primarily upon their book on the topic (Fotheringham et al. 2002), but we have made some changes, which follow from the way in which many in the public health and environmental science fields are likely to use these tools. Our approach to GWR uses an unified framework for including both

A.6 Space-time intelligence system software

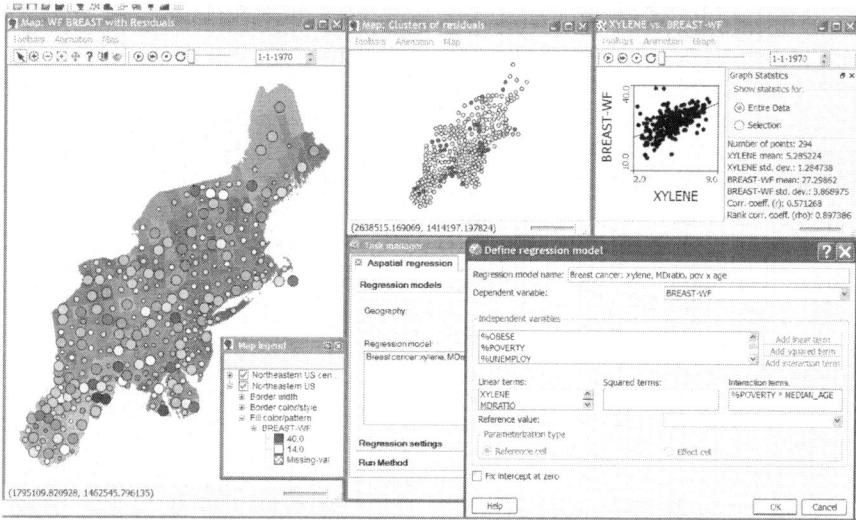

Fig. A.6.4. A-spatial regression analysis of breast cancer in the northeastern United States. The user has conducted a linear regression modeling breast cancer mortality in white females as a function of xylene and availability of physicians (MD ratio), with poverty and median age as interaction terms. The regression residuals have been mapped (circles) with the breast cancer mortality in white females (left). A local Moran analysis found significant clusters of high and low residuals (map top center) and a global Moran's I of 0.18 ($p < 0.001$). The presence of significant spatial autocorrelation in the residuals suggests an important predictor is missing and/or that a more detailed spatial model is needed

geographical weighting, and an extra non-geographical weight dataset that allows for user-supplied knowledge of the ratio variances at each source point. One example of this type of weight is the use of population data as a weight set for mortality rates, which has the effect of assigning higher 'confidence' to mortality rates derived from areas with higher populations. Our goal is to treat this type of weighting together with geographic weighting within a unified framework. As a result, STIS uses a maximum weighted likelihood approach to calculate the regression parameters, parameter variances, parameter R-square, expected y-values, residuals and y-standard errors as well as the 'local model' R-square. This approach boils down to treating geographically weighted regression as a local extension of weighted a-spatial regression. As a consequence GWR can be straightforwardly extended to non-linear regression procedures such as logistic and Poisson regression with parameter values and parameter variances calculated from a weighted log-likelihood formulation.

A key question when using GWR is the construction of the local kernel used to identify those observations to use when fitting a local regression. For kernels of fixed size STIS uses either a number of nearest neighbors or a range (distance) from the central observation, and allows weights to be assigned to the observations based on proximity to the center (for example, using Gaussian and bi-square decay functions). Researchers may also choose to use adaptive kernels that determine the kernel bandwidth by an iterative estimation procedure that minimizes the sum of the differences between the observed value of the dependent variable and the model's estimate of that value. This effectively results in a bandwidth specification that results in the best model 'fit' over the range of bandwidths specified by the researcher. We have found GWR to be particularly useful when concerned with prediction, since it typically results in mean local R^2 values that exceed the R^2 from the corresponding a-spatial regression. In the course of an analysis it is important to first derive a reasonable regression model using a-spatial techniques before proceeding to GWR.

Geostatistics: Geostatistics provides powerful techniques for prediction, interpolation and simulation (see Chapter A.7 for information on geostatistical software). As noted earlier, STIS provides automated variogram estimation methods for modeling spatial relationships through time (see Chapter B.6 for more details on the variogram and kriging). The variogram may then be used in kriging to develop models of how variable values change through space and time. The current release of STIS provides kriging of continuous attributes with or without secondary information. It supports simple kriging, ordinary kriging, kriging with a trend, factorial kriging and Poisson kriging. Underlying variogram models may account for directional components, and the search strategies for fitting the local kriging equations may be anisotropic as well. When the data are time-dynamic one can estimate the variogram model through time, or alternatively can specify one variogram model and then apply it over the entire time interval.

A.6.6 Concluding remarks

This chapter has provided a quick overview of some of the methods and functionality that are now available in the Space-Time Intelligence System software. The development of this software has been motivated by a desire to break the bonds of what has been called 'technological determinism'. This arises when tools and methods dictate the approaches that are used to solve problems, as summarized in the aphorism 'When one has a hammer everything starts to look like a nail'. In spatial analysis two factors lead to technological determinism. First, there still is a strong tradition of using static data models as the basis for developing statistical approaches for spatial data. One still often sees observations subscripted to identify their location, but we less often see a subscript denoting time – when that observation was observed. This is an oversimplification when data in reality are time-dynamic, and results in the application of statistical methods that assume static data to systems that are in fact highly time-dynamic. Second, many of the software tools, such as Geographical Information Systems, were originally founded on a 'static world view' that may be appropriate for geology and other fields where system change is slow, but is less appropriate in economic geography, medical geography and other fields where the systems under scrutiny are highly dynamic. It is unusual, for example, to find software in which the underlying assumption is that the location, extent and attributes associated with an object may change through time. The STIS software is a solution to this problem, and the assumption of dynamic objects leads naturally to time-enabled data views, tables, maps, statistical graphics, and analysis methods, including clustering, regression, geographically weighted regression, variogram analysis and kriging. In the near future we expect to include Q-statistics – methods for the analysis of case-control data that account for residential mobility, covariates and risk factors (Jacquez et al. 2006; Jacquez and Meliker 2009). STIS was created by BioMedware, and is being distributed by TerraSeer, Inc. Details on the methods are available on the TerraSeer website, www.Terraseer.com.

References

Anselin L (1995) Local indicators of spatial association-LISA. Geogr Anal 27(2):93-115
AvRuskin GA, Jacquez GM, Meliker JR, Slotnick MJ, Kaufmann AM, Nriagu JO (2004) Visualization and exploratory analysis of epidemiologic data using a novel space time information system. Int J Health Geographics 3(1):26
Besag J, Newell J (1991) The detection of clusters in rare diseases. J Roy Stat Soc Stat Soc 154(1):143-155
Fotheringham AS, Brundson C, Charlton M (2002) Geographically weighted regression: the analysis of spatially varying relationships. Wiley, New York, Chichester, Toronto and Brisbane
Getis A, Ord JK (1992) The analysis of spatial association by use of distance statistics. Geogr Anal 24(3):189-206
Goodchild MF (2000) GIS and transportation: status and challenges. GeoInformatica 4(2):127-139
Goovaerts P (2005) Analysis and detection of health disparities using geostatistics and a space-time information system. The case of prostate cancer mortality in the United States, 1970-1994 GIS Planet 2005, Estoril, Portugal
Goovaerts P (2009) Geostatistical software. In Fischer MM, Getis A (eds) Handbook of applied spatial analysis. Springer, Berlin, Heidelberg and New York, pp.125-134
Greiling DA, Jacquez GM, Kaufmann AM, Rommel RG(2005) Space time visualization and analysis in the cancer atlas viewer. J Geogr Syst 7(1):67-84
Hägerstrand T (1970) What about people in regional science? Papers in Reg Sci Assoc 24(1):7-21
Han D, Rogerson PA, Bonner MR, Nie J, Vena JE, Muti P, Trevisan M, Freudenheim JL (2005) Assessing spatio-temporal variability of risk surfaces using residential history data in a case control study of breast cancer. Int J Health Geographics 4(1):9
Hornsby K, Egenhofer MJ (2002) Modeling moving objects over multiple granularities. Ann Math Artif Intell 36(1-2):177-194
Jacquez GM, Greiling DA (2003) Local clustering in breast, lung and colorectal cancer in Long Island, New York. Int J Health Geogr 2(1):3
Jacquez GM, Meliker JR (2009) Case-control clustering for mobile populations. Chapter 19 In Fotheringham S, Rogerson P (eds) Sage handbook of spatial analysis. Sage Publications, Los Angeles, London, New Delhi and Singapore, pp.355-374

Jacquez GM, Greiling DA, Kaufmann AM (2005) Design and implementation of space-time information Systems. J Geogr Syst 7(1):7-31

Jacquez GM, Meliker JR, Kaufmann AM. (2007) In search of induction and latency periods: space-time interaction accounting for residential mobility, risk factors and covariates. Int J Health Geographics 6(1):35

Jacquez GM, Meliker JR, AvRuskin GA, Goovaerts P, Kaufmann AM, Wilson ML, Nriagu JO (2006) Case-control geographic clustering for residential histories accounting for risk factors and covariates. Int J Health Geographics 5(32), siehe auch http://www.ij-healthgeographics.com/articles/browse.asp?volume=5&page=2

Koopman JS, Jacquez GM, Chick SE (2001) New data and tools for integrating discrete and continuous population modeling strategies. Ann N Y Acad Sci 954: 268-294

Kwan MP (2003) Accessibility in space and time: a theme in spatially integrated social science. J Geogr Syst 5(1):1-3

Meliker JR, Slotnic MJ, AvRuskin GA, Kaufmann GM, Jacquez GM, Nriagu JO (2005) Improving exposure assessment for environmental epidemiology: applications of a space-time information system. J Geogr Syst 7(1):49-66

Miller HJ (1991) Modeling accessibility using space-time prism concepts within geographical information systems. Int J Geogr Inform Syst 5(3):287-301

Miller HJ (2005a) A measurement theory for time geography. Geogr Anal 37(1):17-45

Miller HJ (2005b) What about people in geographic information science? In Fisher P, Unwin D (eds) Re-presenting geographical information systems. Wiley, NewYork, Chichester, Toronto and Brisbane, pp. 215-242

Oliver MA (2009) The variogram and kriging. In Fischer MM, Getis A (eds) Handbook of applied spatial analysis. Springer, Berlin, Heidelberg and New York, pp.319-352

Ord J, Getis A (1995) Local spatial autocorrelation statistics: distributional issues and an application. Geogr Anal 27(4):286-306

Peuquet D, Duan N (1995) An event-based spatio-temporal data model for geographic information systems. Int J Geogr Inform Syst 9(1):7-24

Raper J, Livingstone D (1993) Development of geomorphological spatial model using object-oriented design. Int J Geogr Inform Syst 9(4):359-393

Sabel CE, Gatrell AC, Löytönen L, Maasilta P, Jokelainen M (2000) Modelling exposure opportunities: estimating relative risk for motor neurone disease in Finland. Soc Sci Med 50(7-8): 1121-1137

Sabel CE, Boyle PJ, Löytönen M, Gatrell AC, Jokelainen M, Flowerdew R, Maasilta P (2003) Spatial clustering of amyotrophic lateral sclerosis in Finland at place of birth and place of death. Am J Epidemiol 157(10):898-905

Schaerstrom A (2003) The potential for time geography in medical geography. In Toubiana L, Viboud C, Flahault A, Valleron A-J (eds) Geography and health. Inserm, Paris, pp.195-207

Sinha G, Mark D (2005) Measuring similarity between geospatial lifelines in studies of environmental health. J Geogr Syst 7(1):115-136

Tukey JW (1977) Exploratory data analysis. Addison-Wesley, Reading [MA]

Turnbull BW, Iwano EJ, Burnett WS, Howe HL, Clark LC (1990) Monitoring for clusters of disease: application to leukemia incidence in upstate New York. Am J Epidemiol 132(1 Suppl): pp.136-143

Wheeler D, Paéz A (2009) Geographically Weighted Regression. In Fischer MM, Getis A (eds) Handbook of applied spatial analysis. Springer, Berlin, Heidelberg and New York, pp.461-486

Yearsley C, Worboys M (1995) A deductive model of planar spatio-temporal objects. In Fisher P (ed) Innovations in GIS 2. Taylor and Francis, London, pp.43-51

A.7 Geostatistical Software

Pierre Goovaerts

A.7.1 Introduction

Geostatistical spatio-temporal models provide a probabilistic framework for data analysis and predictions that build on the joint spatial and temporal dependence between observations. Since its original development in the mining industry in the late 1950s and early 1960s, the geostatistical approach has been adopted in many disciplines, such as environmental sciences (remote sensing, characterization of contaminated sediments, estimation of fish abundance), meteorology (space-time distribution of temperature and rainfall), hydrology (modeling of subsurface hydraulic conductivity), ecology (characterization of population dynamics), agriculture (maps of soil properties and crop yields), and health (patterns of diseases and exposure to pollutants). Following the increasing popularity of geostatistics, the software market has expanded substantially since the late 1980s when it was restricted more or less to two public-domain applications running under DOS: Geo-EAS (Geostatistical Environmental Assessment Software, Englund and Sparks 1988) and the Geostatistical Toolbox (Froidevaux 1990). Nowadays geostatistical software encompasses a wide range of products in terms of price, operating systems, user-friendliness, functionalities, graphical and visualization capabilities. Several organizations, such as AI-GEOSTATS (www.aigeostats.org) or the Pedometrics commission of the International Union of Soil Sciences (www.pedometrics.org), provide a fairly complete list of geostatistical freeware and commercial packages on their website; the long list could intimidate any newcomer to the field and it is summarized in Table A.7.1. The following considerations should be taken into account when choosing a geostatistical package:

(i) Does the user need to have access to the source code (i.e. graduate student who plans to implement a new approach that is a variant of existing algorithms) or is (s)he content with a black-box product?
(ii) What are the characteristics of the data? Are the observations collected in 2D or 3D? Does the sampling domain span both space and time? Are the obser-

vations available at a limited number of discrete locations or over a large raster, such as DEM or satellite imagery?
(iii) What type of analysis is envisioned? A simple description of the major spatial pattern? Straightforward prediction (i.e. univariate kriging) at unsampled locations or more complex incorporation of secondary information? A modeling of local or spatial uncertainty?
(iv) What is the level of geostatistical expertise of the user? Does user-friendliness prevail over flexibility? Is the analysis restricted to geostatistics or does it involve several steps (for example sampling design) that require additional pieces of software? Would the user favor a completely automated approach where variogram modeling is done behind the scene?

Table A.7.1. List of main geostatistical software with the corresponding reference

Name	Code	Cost[a]	Reference
Agromet	C++	F	Bogaert et al. (1995)
AUTO-IK	Fortran	F	Goovaerts (2009)
BMELib	Matlab	F	Christakos et al. (2002)
COSIM	Fortran	F	ai-geostats website
EVS (C-Tech)		H	C Tech Development Corporation
GCOSIM3D/ISIM3D	C	F	Gomez-Hernandez and Srivastava (1990)
Genstat		F,L	Payne et al. (2008)
GEO-EAS	Fortran	F	Englund and Sparks (1988)
GeoR	R	F	Ribeiro and Diggle (2001)
Geostat Analyst		H	Extension for ArcGIS
Geostatistical Toolbox		F	Froidevaux (1990)
Geostokos Toolkit		H	ai-geostats website
GS+		M	Robertson (2008)
GSLIB	Fortran	F	Deutsch and Journel (1998)
Gstat	C,R	F	Pebesma and Wesseling (1998)
ISATIS (Geovariances)		H	www.geovariances.com
MGstat	Matlab	F	ai-geostats website
SADA (UT Knoxville)		F	Spatial analysis and decision assistance
SAGE 2001		M	Isaaks (1999)
SAS/STAT		H	SAS Institute Inc. (1989)
S-GeMS	C++	F	Remy et al. (2008)
SPRING		F	Camara et al. (1996)
Space-time routines	Fortran	F	De Cesare et al. (2002)
STIS (TerraSeer)		M	AvRuskin et al. (2004)
Surfer		M	Golden Software, Inc.
Uncert	C	F	Wingle et al. (1999)
Variowin		F	Pannatier (1996)
VESPER		F	Minasny et al. (2005)
WinGslib	Fortran	L	www.statios.com

Notes: [a] Cost: *H* high, *M* moderate, *L* low, *F* free

Each issue is discussed briefly in this chapter and appropriate software, among the ones the author is familiar with, are suggested for the main types of situation.

A.7.2 Open source code versus black-box software

As the geostatistical community increases, more researchers, particularly in academia, start sharing source code that is either posted online or published in journals such as Computers and Geosciences. Table A.7.1 (column 2) lists the programming language, such as Fortran or C++, whenever the source code is provided. While some programs require only the availability of a compiler, other routines necessitate more expensive packages, such as Matlab. Some software (for example, STIS, S-GeMS), also supports a plug-in mechanism to augment their functionalities, allowing for the addition of new geostatistical algorithms or adding supports for new types of grids on which geostatistics could be performed (Remy et al. 2008).

The Stanford Center for Reservoir Forecasting (SCRF) has been instrumental in the last 20 years in making source code for common, as well as advanced, geostatistical algorithms available to the academic community. The first attempt was the publication in 1992 of the Geostatistical Software LIBrary (GSLIB), a collection of Fortran 77 codes and executable files that cover variogram analysis, spatial interpolation and stochastic simulation (Deutsch and Journel 1998). The programs are well documented and the user manual provides both theoretical background and useful application tips. User-friendliness was greatly improved in the subsequent C++ product S-GeMS (Stanford Geostatistical Modeling Software) which offers a graphical user interface that enables interactive variogram modeling and facilitates the visualization of data and results in up to three dimensions.

Users who are statistically and computer-literate can take advantage of the rich collection of classical and modern spatial techniques implemented in the open source statistical program R (Ripley 2001). In particular, Gstat offers a robust and flexible suite of univariate and multivariate geostatistical methods for estimation and simulation. Simulation comprises conditional or unconditional (multi-) Gaussian sequential simulation of point values or block averages, or (multi-) indicator sequential simulation. The GeoR package implements model-based geostatistical methods but is limited to small (500 to 1,000 observations) univariate 2D datasets (Ribeiro et al. 2003).

Although space-time geostatistical routines are rather limited, most of these programs are public-domain. The BMElib library is a Matlab numerical toolbox that implements space/time variography and estimation using the Bayesian Maximum Entropy (BME) theory. This library is fairly complete, but it requires a strong statistical background and the Matlab package. On the other hand, Cesare et al. (2002) modified some of the GSLIB Fortran 77 routines to estimate and model space-time variograms, as well as to accommodate the use of such models in traditional kriging interpolation. Two general families of models are incorporated in the programs: the product model and the product-sum model, both based on the decomposition of the space-time covariance in terms of a spatial covariance and a temporal covariance. The commercial software STIS (Space-Time Information System) is one of the rare examples of GIS software where a time stamp is assigned to each piece of information, allowing the incorporation of time in the spatial data analysis. The geostatistical treatment of space-time data in STIS is currently limited, however, to the repetition of a purely spatial analysis for each time step, prohibiting any prediction at unmonitored times.

A.7.3 Main functionalities

As a consequence of the wide variety of geostatistical applications and the continuous development of new algorithms, finding all the functionalities required by a specific application within a single product might become increasingly difficult. Most geostatistical studies, however, share a similar sequence of steps: exploratory data analysis to get familiar with the data, characterization and modeling of the pattern of spatial variation, interpolation to the nodes of a grid or over blocks (upscaling), and modeling of local and spatial uncertainty.

Exploratory spatial data analysis. Except for a few products focusing on specific tasks, such as variography (for example, Variowin, SAGE 2001), estimation (for example, AUTO-IK, Vesper) or stochastic simulation (for example, COSIM, GCOSIM3D), most software in Table A.7.1 provides basic data mapping and exploratory tools, such as the histogram and scatterplots. These programs, however, differ in their ability to handle 3-dimensional and space-time databases, as well as dynamic exploration and visualization of the data. The S-GeMS and Uncert software offer public-domain visualization tools for three-dimensional datasets, but they lack basic GIS capabilities, such as data queries or linked windows. Such features are incorporated in the C-tech product EVS which is designed to integrate seamlessly with ESRI's ArcView® GIS and ArcGIS® or to operate in a stand-alone mode. Licenses for this high-end software can be expensive, however. TerraSeer STIS is less expensive and has excellent browsing and linking capability for exploratory analysis of space-time datasets in two dimensions.

Variogram estimation and modeling. Quantifying and modeling the pattern of variation in the data is the cornerstone of any geostatistical analysis. A wide range of options is available at present: from fully automated computation and modeling of variograms to highly interactive programs that allow the detection and elimination of spatial outliers (for example, variogram cloud cleaning), the exploration of spatial anisotropy through variogram maps or surfaces, and the manual fits of variograms. One of the first interactive programs for variography was Variowin (Pannatier 1996) that is public-domain. It provides several variogram estimators, and computes both variogram map and variogram cloud in addition to the traditional variogram plot. This program is limited to small 2D datasets, however, and does not include any interpolation or simulation procedure.

GIS software, such as ArcView® Geostat Analyst or TerraSeer STIS, offer similar options with better visualization capabilities than Variowin and a series of kriging and simulation algorithms that can use the variogram model in subsequent analysis. In particular, the variogram cloud in STIS is linked with the location map, which facilitates greatly the detection of data pairs with undue influence on the computation of the variogram. Other unique features of this program are the flexibility in variogram modeling (for example model parameters can be estimated automatically under constraints on the nugget effect and type of basic models), the ability to compute variograms from areal data (for example counties) and to derive the point-support model accounting for the shape and size of these geographical units (deconvolution).

Other programs, such as ArcView® Geostat Analyst or Surfer, also offer an automatic variogram modeling procedure but they either lack transparency, lead to unsatisfactory fits or do not allow anisotropic modeling. The SADA variogram module allows automatic variogram modeling as well and its exploration of anisotropy through the rose diagram is very appealing. The general-purpose statistical package Genstat (Payne et al. 2008) offers a wide variety of variogram models and allows automatic modeling, but its command language and procedure library are challenging for all users who are not statistically and computer-literate.

The SAGE 2001 software can be viewed as the 3D counterpart of the 2D stand-alone Variowin software. It is not free, but it has the capability of fitting 3D models automatically. Other commercial products, such as C-tech EVS and ISATIS, also provide an automatic 3D modeling procedure that is part of their kriging module. ISATIS is certainly the most flexible software since it allows identification of directions and scales of continuity through the unique 3D interactive variogram map. Public-domain software S-GEMS and UNCERT can compute variograms in three directions but only visual fitting is implemented. To the author's knowledge, there is currently no commercial software for the geostatistical treatment of space-time data, including the interpolation at unmonitored times and locations. Current public-domain software involves a lot of data manipulation and require expert knowledge in either the modeling of the variograms (De Cesare et al. 2002) or the use of the software itself (for example BMELib).

Spatial interpolation. Basic univariate kriging variants (simple, ordinary and universal kriging) are typically covered by geostatistical software. Products differ in their ability to handle irregular interpolation grids or uneven prediction supports (i.e. change of support through block kriging), their flexibility to set up a search strategy (for example stratified search windows), or the possibility of comparing various implementation schemes by cross-validation or jack-knifing. S-GeMS is an improvement over GSLIB and GEO-EAS because it allows the specification of user-defined interpolation grids instead of the traditional regular grids in mining applications. Point measurement supports and rectangular prediction supports only are implemented, which is not adequate for applications, such as those in epidemiology or the social sciences, where the units of measurement are irregular polygons. Such levels of complexity are handled in TerraSeer STIS where both measurement and prediction supports can be either points, polygons or raster cells. In addition, it is the only commercial software that implements Poisson kriging, an interpolation procedure that is tailored to the analysis of rate data, such as crime or mortality rates.

One of the key advantages of geostatistics over other spatial interpolation procedures is its ability to incorporate secondary information, which can be available at all locations where a prediction is sought (i.e. simple kriging with a local mean or external drift) or known at a limited number of locations (cokriging). All these algorithms are implemented in the public-domain GSLIB and in the commercial software ISATIS. Kriging with an external drift is lacking from S-GeMS, whereas cokriging is not implemented in STIS or C-tech EVS.

Probability mapping. An important contribution of geostatistics is the assessment of uncertainty about unsampled values, which usually takes the form of a map of the probability of exceeding critical values, such as regulatory thresholds. Such probabilities can be estimated using parametric (i.e. multi-Gaussian kriging) or non-parametric (i.e. indicator kriging) methods. Both sets of algorithms are available in S-GeMS as well as ISATIS. Indicator kriging is also implemented in SADA and the stand-alone AUTO-IK program (Goovaerts 2009).

Stochastic simulation. Stochastic simulation has certainly been one of the most active areas of research in geostatistics for the last decade. The basic idea is to generate a set of equiprobable representations (realizations) of the spatial distribution of attribute values and to use differences among simulated maps as a measure of uncertainty. Each simulated map looks more 'realistic' than the map of smooth kriging estimates because it reproduces the spatial variation modeled from the sample information. Simulation can be done using a growing variety of techniques that differ in the underlying random function model (multi-Gaussian or non-parametric), the amount and type of information that can be accounted for and the computer requirements.

S-GeMS implements the most common algorithms (i.e. sequential indicator and Gaussian simulations), as well as recent methods based on multiple point statistics. The most complete palette of simulation methods, covering both continuous and categorical variables, is in ISATIS. These two software packages also have modules to post-process the set of realizations, creating maps of averaged simulated values, the probability of exceeding critical thresholds or measures of differences among realizations. Table A.7.2 lists other products that include stochastic simulation, either as a stand-alone algorithm (COSIM, GCOSIM3D) or as part of the geostatistical module (STIS, Uncert).

Table A.7.2. List of functionalities for main geostatistical software (modified from the list on http://www.ai-geostats.org/)

Name	Data	V	K	CK	IK	MG	S	G
Agromet	2D	X	X	X				
AUTO-IK	2D	X			X			
BMELib	3D, ST	X	X	X			X	
COSIM	2D						X	
EVS (C-Tech)	3D	X	X		X			X
GCOSIM3D/ISIM3D	3D						X	
Genstat	3D	X	X	X				
GEO-EAS	2D	X	X					
GeoR	2D	X	X				X	
Geostat Analyst	2D	X	X	X	X	X		X
Geostatistical Toolbox	3D	X	X	X				
Geostokos Toolkit	3D	X	X	X	X		X	
GS+	2D	X	X	X			X	
GSLIB	3D	X	X	X	X	X	X	
Gstat	3D	X	X	X			X	
ISATIS	3D	X	X	X	X	X	X	X
MGstat	3D, ST	X	X					
SADA	3D	X	X		X			X
SAGE2001	3D	X						
SAS/STAT	2D	X	X					
S-GeMS	3D	X	X	X	X	X	X	
SPRING	2D	X	X		X		X	X
Space-time routines	2D, ST	X	X					
STIS (TerraSeer)	2D, ST	X	X			X	X	X
Surfer	2D	X	X					
Uncert	3D	X	X				X	
Variowin	2D	X						
VESPER	2D	X	X					
WinGslib	3D	X	X	X	X	X	X	

Notes: V variography, K kriging, CK cokriging, IK indicator kriging, MG multi-Gaussian kriging, S simulation, G GIS interface

A.7.4 Affordability and user-friendliness

A package can offer all geostatistical methods developed in the last 20 years, but it can scare away potential users by its price or design. In particular for academics, price and transparency typically drive the choice of geostatistical software. Consulting companies and federal agencies are likely to favor products that do not require advanced statistical background and provide all necessary functionalities within a single package. To appeal to users that are more task-oriented than method-oriented, several products such as SADA or STIS have task managers to guide the geostatistician through the sequence of steps required to accomplish the task at hand. For example, in SADA the task 'Interpolate my data' consists of eleven steps, starting with 'See the data' and ending at 'Add to results gallery'. This public-domain software also offers integrated modules for using the results of the geostatistical analysis in human health risk assessment, ecological risk assessment, cost/benefit analysis, sampling design, and decision analysis. On the other hand, STIS includes a complete regression module that is useful for calibrating the trend model used in multivariate kriging procedures.

Another approach to improve user-friendliness is to automate some of the steps, in particular the variogram modeling procedure which is typically the stumbling block for the adoption of kriging over more traditional methods, such as inverse distance weighting. The key is to provide transparency and use reasonable default options; for example, the user should have access to the variogram model computed behind the scene and it is puzzling that the unrealistic linear model is still used as the default variogram in Surfer. For example, C-tech MVS/EVS uses expert systems to analyze the input data, construct a multidimensional variogram which is a best fit to the dataset being analyzed, and then perform kriging in the domain to be considered in the visualization. The user is provided with the option to specify values for parameters that control the variogram and kriging procedures, and the subsequent display and analysis of the data. A public-domain alternative for 2D interpolation is VESPER that allows the automatic computation and modeling of local variograms, followed by spatial interpolation. Such a procedure capitalizes on high sampling density to adapt the process spatially to distinct local differences in the level of variation in the field. For non-parametric geostatistics, AUTO-IK is a free computer code that performs the following tasks automatically: selection of thresholds for binary coding of continuous data, computation and modeling of indicator variograms, modeling of probability distributions at unmonitored locations (regular or irregular grids), and estimation of the mean and variance of these distributions.

A.7.5 Concluding remarks

Summarizing the pros and cons of the geostatistical software currently available on the market in a few pages is a daunting task given the large number and diversity of these products. This brief chapter by no means pretends to provide a complete overview of all software, but rather offers a few pointers to guide the choice of a suitable product based on the task at hand, the user's expertise and financial resources. The main conclusion is that there is no such thing as a 'best all-purpose software'. Creating a geostatistical model is rarely a goal per se, but rather a preliminary step towards decision-making, such as design of a sampling or remediation scheme. The current trend is to have software that is tailored to the characteristics of the data of interest (for example areal health data, 3D pollution data, space-time climatic data) as well as the type of decision-making envisioned (for example detection of cancer clusters, estimation of volume of contaminated sediments, location of new monitoring stations). This customization of the products should improve their user-friendliness and expand their use while reducing common mistakes in the application of the geostatistical methodology.

Acknowledgments. This research was funded by grant R44-CA132347-01 from the National Cancer Institute. The views stated in this publication are those of the author and do not necessarily represent the official views of the NCI.

References

AvRuskin GA, Jacquez GM, Meliker JR, Slotnick MJ, Kaufmann AM, Nriagu JO (2004) Visualization and exploratory analysis of epidemiologic data using a novel space time information system. Int. J. Health Geog 3(1):26

Bogaert P, Mahau P, Beckers F (1995) The spatial interpolation of agro-climatic data. Cokriging Software and Source Code. Agrometeorology Series Working Paper 12, FAO Rome, Italy

Camara G, Souza RCM, Freitas UM, Garrido J (1996) SPRING: integrating remote sensing and GIS by object-oriented data modeling. Comput Graph 20(3):395-403

Christakos G, Bogaert P, Serre ML (2002) Temporal GIS: advanced functions for field-based applications. Springer, Berlin, Heidelberg and New York

De Cesare L, Myers DE, Posa D (2002) FORTRAN programs for space-time modeling. Comput Geosci 28(22):205-212

Deutsch CV, Journel AG (1998) GSLIB: geostatistical software library and user's guide (2nd edition). Oxford University Press, New York

Englund E, Sparks A (1988) Geo-EAS 1.2.1 user's guide. EPA Report 60018-91/008. EPA-EMSL, Las Vegas [NV]

Froidevaux R (1990) Geostatistical toolbox primer, version 1.30. FSS International, Troinex, Switzerland

Gomez-Hernandez JJ, Srivastava RM (1990) ISIM3D: an ANSI-C three dimensional multiple indicator conditional simulation program. Comput Geosci 16(4):395-440

Goovaerts P (2009) AUTO-IK: a 2D indicator kriging program for the automated non-parametric modeling of local uncertainty in earth sciences. Comput Geosci 35(6):1255-1270

Isaaks E (1999) SAGE 2001: a spatial and geostatistical environment for variography. Isaaks, San Mateo [CA]

Minasny B, McBratney AB, Whelan BM (2005) VESPER version 1.62. Australian Centre for Precision Agriculture, The University of Sydney, NSW

Pannatier Y (1996) VARIOWIN: Software for spatial data analysis in 2D. Springer, Berlin, Heidelberg and New York

Payne RW, Murray DA, Harding SA, Baird DB, Soutar DM (2008) GenStat for Windows (11th edition). VSN International, Hemel Hempstead

Pebesma EJ, Wesseling CG (1998) Gstat: a program for geostatistical modelling, prediction and simulation. Comput Geosci 24(1):17-31

Remy N, Boucher A, Wu J (2008) Applied geostatistics with SGeMS: a user's guide. Cambridge University Press, Cambridge

Ribeiro PJR, Diggle PJ (2001) GeoR: a package for geostatistical analysis. R NEWS 1(2):15-18

Ribeiro PJR, Christensen OF, Diggle PJ (2003) GeoR and geoRglm: software for model-based geostatistics. R-NEWS 1(2):15-18

Ripley BD (2001) Spatial statistics in R. R News 1:14–15

Robertson GP (2008) GS^+: Geostatistics for the environmental sciences. Gamma Design Software, Plainwell, Michigan

SAS Institute Inc. (1989) SAS/STAT user's guide 6(2) (4th edition). SAS Institute Inc., Cary [NC]

Wingle WL, Poeter EP, McKenna SA (1999) UNCERT: geostatistics, uncertainty analysis, and visualization software applied to groundwater flow and contaminant transport modeling. Comput Geosci 25(4):365-376

A.8 GeoSurveillance: GIS-based Exploratory Spatial Analysis Tools for Monitoring Spatial Patterns and Clusters

Gyoungju Lee, Ikuho Yamada and *Peter Rogerson*

A.8.1 Introduction

Spatial clusters are often formed by underlying non-random geographic processes generated from various factors (for example, a disease outbreak around a pollutant source). Spatial randomness is a theoretical baseline in comparison of which spatial clustering is assessed in statistical frameworks dealing with spatial uncertainty. Spatial statistical methods for investigating spatial clustering have been developed to reveal the locations of probable sources (for example, environmental factors) that may cause unusual concentrations of geographic events. Clustering tests assess the overall tendency for geographic events to concentrate in space, as well as measure the associated statistical significance, while clusters point to where geographic events are densely located in close proximity (Waller and Gotway 2004).

Three types of spatial statistical tests are categorized by Besag and Newell (1991). The categories are: (i) general tests, (ii) focused tests, and (iii) tests for the detection of clustering. General tests focus on identifying overall spatial pattern across an entire study region. These tests summarize the global spatial pattern using a single summary statistic (such as Moran's I), while omitting details associated with local variation. Focused tests focus on one or more prespecified geographic locations to examine whether there are spatial clusters around those foci. Tests for the detection of clustering are used to explore the local concentration of geographic phenomena when no *a priori* location information is given, unlike focused tests.

Global statistics fall into the category of general tests and are often used to test whether an overall spatial clustering propensity exists in a study region. Local statistics such as local Moran's I and local G statistics, when employed for many locations throughout the study region, are considered as tests for the detection of clustering, to detect the locations and sizes of geographic clusters which deviate from the null hypothesis of no clustering (Kim and O'Kelly 2008). The Geographical Analysis Machine (Openshaw et al. 1987), the Cluster Evaluation Permutation Procedure (Turnbull et al. 1990), and the Spatial Scan Statistic (Kulldorff 1997) also belong to this category. While tests for the detection of clustering search the entire study region for geographic clusters, focused tests use prior information on the locations of factors that may cause clustering. Tests in this category include the score statistic and Stone's test (Lawson 1993; Waller and Lawson 1995).

Methods for spatial cluster detection can also be classified with respect to whether they are retrospective and prospective (Rogerson 1997). Retrospective analysis is concerned with spatial data analysis that is carried out at a particular point in time, using data from the past, while prospective analysis, on the other hand, is designed for repeated statistical analysis on time-series data that are updated periodically. Most spatial statistical tests developed to date are retrospective in nature. Although they are effective in detecting static spatial patterns observed at a given time, they are insensitive to changes in spatial patterns, even when successively applied to time-series datasets, due to the temporal autocorrelation between tests (Rogerson and Sun 2001). Recently, considerable effort has been devoted to devising prospective tests that both take into account their dynamic nature and attempt to quickly find significant changes in spatial patterns of disease, crime, etc. (Rogerson 1997, 2001a; Kulldorff 2001).

Based on the powerful capability of GIS in dealing with spatial data, significant progress has been made in spatial analysis software development, and this has promoted applications of spatial statistical methodologies in various fields (for example, spatial epidemiology, crime analysis). This chapter introduces a GIS-based spatial analysis tool, *GeoSurveillance*[1] that can make some contribution in this regard. *GeoSurveillance* is stand-alone software designed to explore spatial patterns in both retrospective and prospective manners; the software and associated documentations are available from http://www.acsu.buffalo.edu/~rogerson/geosurv.htm. Other software, such as SaTScan and GeoDa, provides spatial statistical routines with other specific objectives or methodological foci for exploring spatial regimes in geographic phenomena.

In *GeoSurveillance*, a set of retrospective and prospective statistical tests is implemented in the framework of GIS, where basic GIS functions are provided for data exploration, including mapping analysis results and linking them to tables, charts, etc. in real time. A useful property of *GeoSurveillance* is the capability of simultaneously linking diverse analysis tools (maps, tables, and charts) in a single window, so that the user can perform exploratory spatial analysis in an integrated platform. Additionally, *GeoSurveillance* provides functionality for conducting

---

[1] This program was developed in Visual Basic 6.0 IDE. A simple GIS engine was developed for the task of mapping, zooming, panning, etc. Other third party GIS servers (for example, ESRI MapObjects) were not used.

cumulative sum (*cusum*) analysis, which is a type of prospective statistical procedure that can avoid the multiple testing problems associated with temporally autocorrelated tests. Although SaTScan allows prospective application of the spatial scan statistic to detect space-time clusters, the multiple testing problems are not accounted for explicitly. Details of the *cusum* statistic are discussed later in this chapter.

This chapter consists of five sections including this introductory one. Section A.8.2 describes the structure of *GeoSurveillance*, and Section A.8.3 provides a theoretical overview of retrospective and prospective tests implemented in *GeoSurveillance*. Section A.8.4 demonstrates spatial statistical analysis in *GeoSurveillance* using sample datasets included in the *GeoSurveillance* setup package. Section A.8.5 provides concluding remarks.

A.8.2 Structure of GeoSurveillance

In *GeoSurveillance*, two retrospective tests and one prospective test are implemented. The score statistic and the maximum local statistic (the M test) are available as retrospective tests, while the *cumulative sum* statistic for normal univariate variables is implemented as a prospective test. Also provided are some auxiliary tools that help users produce additional information relevant to these tests. Figure A.8.1 shows the overall software structure and Fig. A.8.2 illustrates the general procedure for performing statistical analysis in *GeoSurveillance*. Functional details of *GeoSurveillance* can be found in the user's manual from the website mentioned previously.

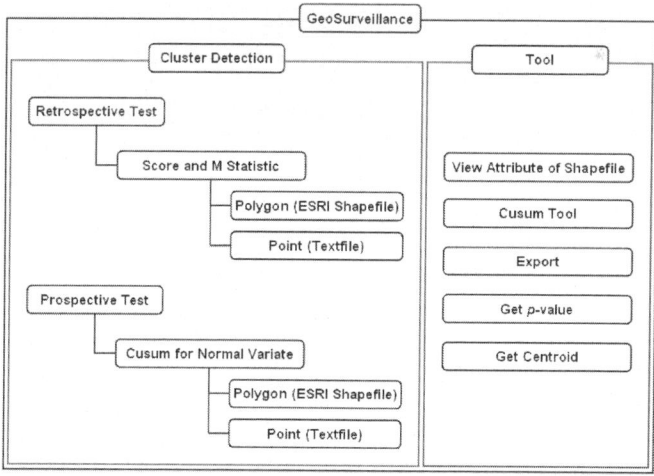

Fig. A.8.1. Structure of GeoSurveillance

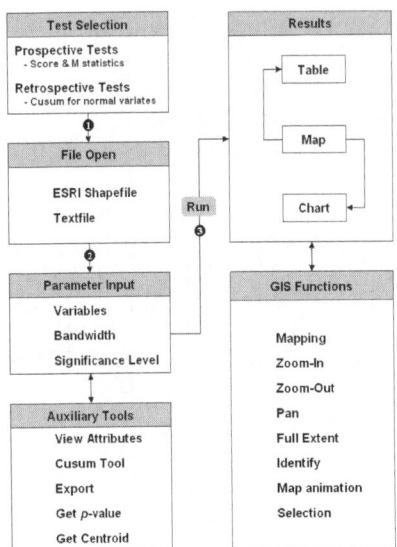

Fig. A.8.2. Statistical analysis procedures in GeoSurveillance

A.8.3 Methodological overview

This section provides a brief overview of the three statistical tests implemented in *GeoSurveillance*, namely, the score test and the maximum local statistic (the *M* statistic) for retrospective testing and the *cusum* statistic for prospective testing.

Consider a study region consisting of n subregions, and assume that observed and expected numbers of disease cases in subregion j ($j=1, \ldots, n$), are denoted by O_j and E_j, respectively, are available. The local score statistic is then defined as

$$U_i = \sum_{j=1}^{n} W_{ij}(O_j - E_j) \tag{A.8.1}$$

where W_{ij} represents a spatial weight that is usually specified as a function of the distance between subregions i and j. E_j can for example be estimated by multiplying the size of the at-risk population in subregion j by the overall disease rate for the entire study region in the simplest case, but other covariates such as age and gender structures of the population can also be taken into account. According to Waller and Lawson (1995), the local score statistic under the null hypothesis of no elevated risk in subregion j approximately follows a normal distribution with mean zero and variance

$$V[U_i] = \sum_{j=1}^{n} W_{ij}^2 \hat{\lambda} n_j - \left(\sum_{j=1}^{n} O_j\right) \frac{\left(\sum_{j=1}^{n} W_{ij}^2 n_j\right)^2}{\sum_{j=1}^{n} n_j} \quad (A.8.2)$$

so that an approximated z-value of the local score statistic is obtainable.

Rogerson (2005) defined a global score statistic as the sum of the squared local statistic in Eq. (A.8.1), that is,

$$U^2 = \sum_{i=1}^{n} U_i^2 = \sum_{i=1}^{n} \left(\sum_{j=1}^{n} W_{ij}(O_j - E_j)\right)^2. \quad (A.8.3)$$

To define the spatial weight w_{ij} between subregions i and j, GeoSurveillance uses a Gaussian function formulated as

$$W_{ij} = \frac{1}{\pi^{1/2}\sigma} \exp\left(-\frac{d_{ij}^2}{2\sigma^2(nA^{-1})}\right) \quad i,j = 1, \ldots, n \quad (A.8.4)$$

where σ is a bandwidth determining the level of spatial smoothness (or, equivalently, the size of the hypothesized cluster), d_{ij} is the distance between subregions i and j, and A is the area of the study region. A larger bandwidth gives more weight to distant subregions and more smoothing effects are induced. Because of irregular areal units and subregions near the edge of the study region, a scaled spatial weight

$$\widetilde{W}_{ij} = \frac{W_{ij}}{\left(\sum_{j=1}^{n} W_{ij}^2\right)^{1/2}} \quad (A.8.5)$$

needs to be used so that the sum of the squared W_{ij} in Eq.(A.8.4) will be equal to be one (Rogerson 2001b, 2005).

The scaled weight can further be adjusted for the expected value E_j as

$$\hat{W}_{ij} = \frac{\widetilde{W}_{ij}}{(E_j)^{1/2}}. \quad (A.8.6)$$

Using the readjusted spatial weight, Eqs. (A.8.1) and (A.8.3) can be rewritten respectively as follows:

$$U_i = \sum_{j=1}^{n} \hat{W}_{ij}(O_j - E_j) = \sum_{j=1}^{n} \left(\frac{\tilde{W}_{ij}}{(E_j)^{1/2}} \right)(O_j - E_j) \qquad (A.8.7)$$

$$U^2 = \sum_{i=1}^{n}\sum_{k=1}^{n} \tilde{W}_{ki}^2 \frac{(O_i - E_i)^2}{E_i} + \sum_{i=1}^{n}\sum_{j=1}^{n}\sum_{k=1}^{n} \tilde{W}_{ki}\tilde{W}_{kj} \frac{(O_i - E_i)(O_j - E_j)}{(E_i E_j)^{1/2}}. \qquad (A.8.8)$$

Rogerson (2005) shows that, since U^2 statistic takes the form of the general test by Tango (1995), its statistical significance can be assessed based on the null distribution of Tango's statistic. Focused tests for a particular region can be conducted based on Eq. (A.8.7), whereas a general test for assessing the overall clustering tendency can be based on Eq. (A.8.8).

The M statistic is defined as the maximum of the local statistics U_i given in Eq. (A.8.7) for a set of subregions (Rogerson 2001b). The critical value of the M statistic is given by

$$M^* = \left[-\pi^{1/2} \ln\left(\frac{4\sigma(1+0.81\sigma^2)}{n} \right) \right]^{1/2} \qquad (A.8.9)$$

where σ denotes a chosen probability of Type I errors.

The score statistic defined as above can be considered to be transforming O_j and E_j into z-values assuming that the counts follow the Poisson distribution. *Geo-Surveillance* provides two additional types of z-value transformation:

$$z_i = \frac{O_i - E_i}{(E_i)^{1/2}} \qquad \text{(Poisson method)} \qquad (A.8.10a)$$

$$z_i = (O_i)^{1/2} + (O_i + 1)^{1/2} - (4E_i + 1)^{1/2} \qquad \text{(Freeman-Tukey method)} \qquad (A.8.10b)$$

$$z_i = \frac{O_i - 3E_i + 2(O_i E_i)^{1/2}}{2(E_i)^{1/2}} \qquad \text{(Rossi method)} \qquad (A.8.10c)$$

Prospective test. The cumulative sum (cusum) statistic is used primarily in *statistical process control* in manufacturing environments to pinpoint persistent changes in the mean of a monitored variable to assess whether the process is in control or encountering abnormal deviations from what is expected (Hawkins and Olwell 1998). Rogerson (1997, 2001a) described how to use the *cusum* method in disease surveillance to monitor disease outbreaks, pointing out the increasing recognition of the need for detecting clusters quickly in a prospective manner. Rogerson and Sun (2001) also applied the method to detecting clusters of crime shifting in space and time. Rogerson and Yamada (2004) further extended the univariate basis of the method to a multivariate one. Quick detection of emerging clusters is made possible by continuously updating the *cusum* statistic in near real time as new data become available.

The basic form of the *cusum* statistic is defined as

$$S_t = \max\left(0, S_{t-1} + z_t - k\right) \tag{A.8.11}$$

where S_t represents the cumulative sum at time t, and z_t is the standardized value of a variable of interest with mean 0 and variance 1 at time t. Further, k is a parameter which is often set equal to ½, and therefore z_t exceeding k contributes positively to the accumulation of S_t. A value of S_t that exceeds a threshold parameter, h, indicates a significant shift or change in the mean value of the monitored variable z_t. An appropriate value of threshold h is determined according to a desired false alarm rate, characterized by the in-control average run length (ARL_0), defined as the mean time between false alarms under the null hypothesis of no change. Low values of h lead to too frequent false alarms, but a higher probability of detecting a real change. In contrast, higher values of h lead to a low probability of a false alarm, at the cost of a higher probability of not detecting a real change. Siegmund (1985) provided the following approximation for the in-control ARL_0 under the null hypothesis:

$$ARL_0 = \frac{\exp\left[2k(h+1.166)\right] - 2k(h+1.166) - 1}{2k^2}. \tag{A.8.12}$$

The user may first specify a desired ARL_0 and k, and then solve Eq. (A.8.12) for h. For the standardized variable z, k is often chosen to be 1/2 since it minimizes the time to detect an actual increase of $2k$ in the mean. Rogerson (2006) derived an approximating formula to compute h directly from given values of k and ARL_0

$$b \approx \left(\frac{2k^2 ARL_0 + 2}{2k^2 ARL_0 + 1} \right) \frac{\ln(1 + 2k^2 ARL_0)}{2k} \qquad (A.8.13a)$$

$$h = b - 1.166 \qquad (A.8.13b)$$

This formula provides generally accurate approximations in the range of $k^2 ARL_0 > 1$.

The relationship between h and ARL_0 in Eq. (A.8.13a) should be adjusted when the interest is in monitoring the *cusum* charts for all individual regions simultaneously. When each *cusum* chart is associated with observations in each of n subregions, a *Bonferroni* adjustment may be applied to adjust for the fact that n independent tests are carried out simultaneously. More specifically, to maintain the false alarm rate for the entire system, the threshold value h for individual subregions is obtained by replacing ARL_0 in Eq. (A.8.13a) by $n\,ARL_0$. When a *cusum* chart is for a local statistic for each subregion, one should take into account correlation between local statistics for nearby subregions, which may decrease the probability of detecting real change as well as the false alarm rate. A less conservative adjustment for multiple testing is therefore needed, and one possibility is to substitute $ARL_0\, n\, /\, (1+0.81\sigma)$ for ARL_0 in Eq. (A.8.13b) when the local statistics are calculated with the Gaussian weight defined above. Note that the former, Bonferroni adjustment is a special case of the latter where $\sigma = 0$. GeoSurveillance includes an auxiliary tool that returns an appropriate h value for given k, ARL_0, n, and σ.

It should also be pointed out that, if a variable of interest can be standardized to a z-value, the cusum scheme in Eq. (A.8.11) can be applied to detect significant change in the variable over time. Therefore, as Lee and Rogerson (2007) demonstrate using Moran's I and Getis's G statistics, potentially, any of the spatial statistics for detecting spatial clustering can be fruitfully utilized in the *cusum* scheme to monitor changes in spatial pattern over time (Rogerson and Sun 2001).

A.8.4 Illustration of GeoSurveillance

In this section, we illustrate how to conduct the statistical tests explained in the previous section in *GeoSurveillance* using sample datasets. For illustration of retrospective tests, the Sudden Infant Death Syndrome (SIDS) in North Carolina, 1974 data is used. It consists of the numbers of births and SIDS cases in 100 counties in the state. For the prospective test, the breast cancer mortality data in 217 northeastern counties of the U.S. is used; this contains annual, standard normal z-values computed based on observed and expected breast cancer counts for the period 1968-1988. They are both polygon datasets, but the same procedures may be applied to point datasets; see the user's manual for illustrations using point datasets.

In Figs. A.8.3 and A.8.4, the analysis form and map window linked to it shows the results from running the score and M statistics. The map in Fig. A.8.4 shows a distinctive spatial clustering pattern. The tables in Fig. A.8.3 provide additional information such as associated z-values, expected and observed counts. The detailed procedures for running the 'retrospective' tests are documented in the user's manual.

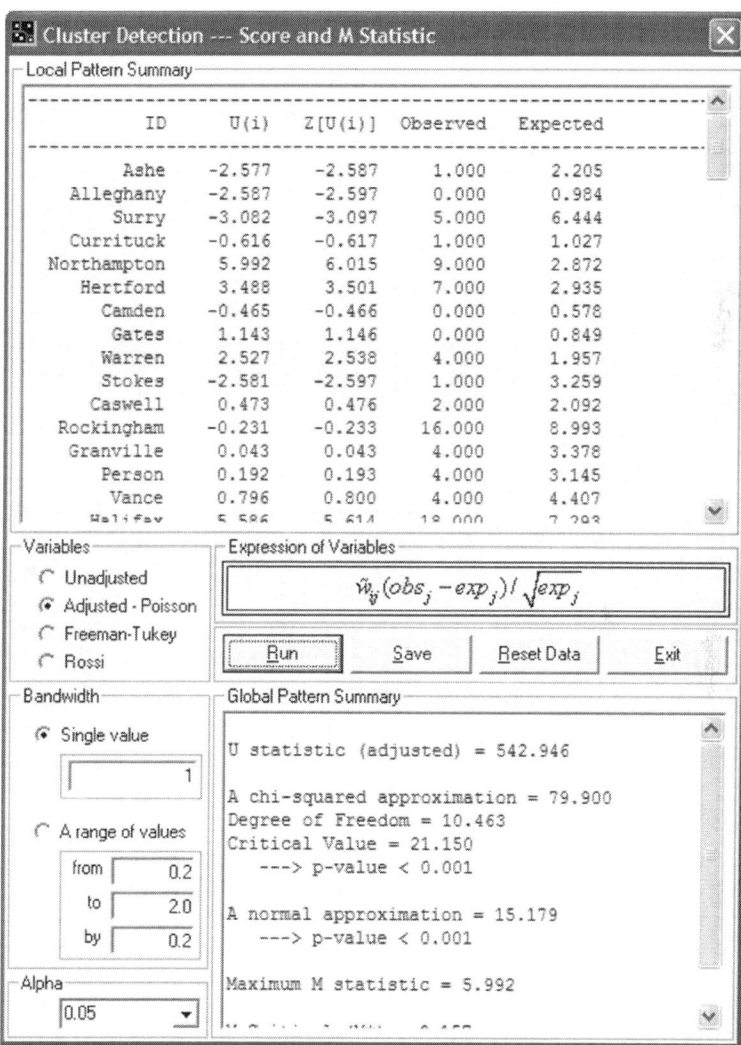

Fig. A.8.3. Result tables linked to the map in Fig. A.8.4

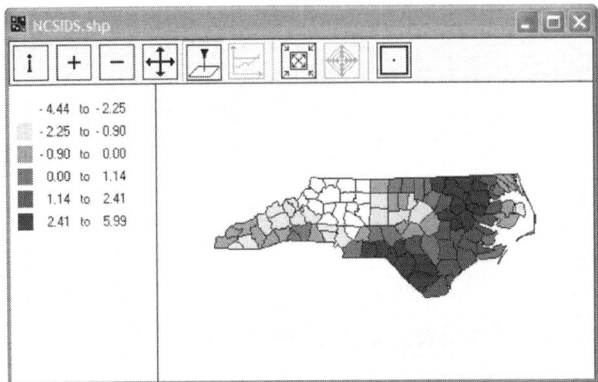

Fig. A.8.4. Map of the local score statistic for North Carolina SIDS data

The user can freely explore the spatial patterns by trying different combinations of options available in the analysis form. Figure A.8.5 illustrates the results of the score and M statistics for the four variable options with bandwidth $\sigma = 1$ and significance level, $\alpha = 0.05$. The hotspots (northeastern and southern parts) and cold spot (northwestern part) of SIDS cases show clear spatial separation and the overall spatial patterns look similar in all panels of the table.

Figure A.8.6 shows that as the bandwidth extends outward (0.2, 1.0, and 1.8), the separate spatial clustering tendency of cold and hot spots gets clearer, and more smooth results are produced. The spatial pattern where low and high local score statistics are spatially mixed for a small bandwidth ($\sigma = 0.2$) approaches the spatial patterns where three distinctive clusters emerge for larger bandwidths. The formation of the cold spot looks more conspicuous as σ increases. This tendency continues as the bandwidth grows; the differences of the local statistics among subregions eventually would become negligibly small with no meaningful spatial patterns yielded. For other variable types and bandwidth range options, similar results are expected.

Figures A.8.7 and A.8.8 present the results of carrying out a 'prospective test', namely the *cusum* test. Figure A.8.8 depicts an enlarged version of the lower right part of Fig. A.8.7, which contains tables and parameter boxes to set values of k, h, and σ. Figure A.8.9 presents the results when no spatial association among nearby subregions is assumed ($\sigma = 0$). Figure A.8.10 illustrates the cases for $\sigma = 1.5$ with h values adjusted for the induced spatial association. The charts in Figs. A.8.9 and A.8.10 represent maximum cusum values. In contrast to the map in Fig. A.8.9, clusters of signaled regions emerge in the mid-eastern and southern extremity of the study area in Fig. A.8.10. As the bandwidth increases to 1.5, the threshold h gets smaller based on the adjustment of Eq. (A.8.13) for the spatially correlated observations and the maximum *cusum* values get smaller – a consequence of the local maximum value being smoothed by nearby z-values.

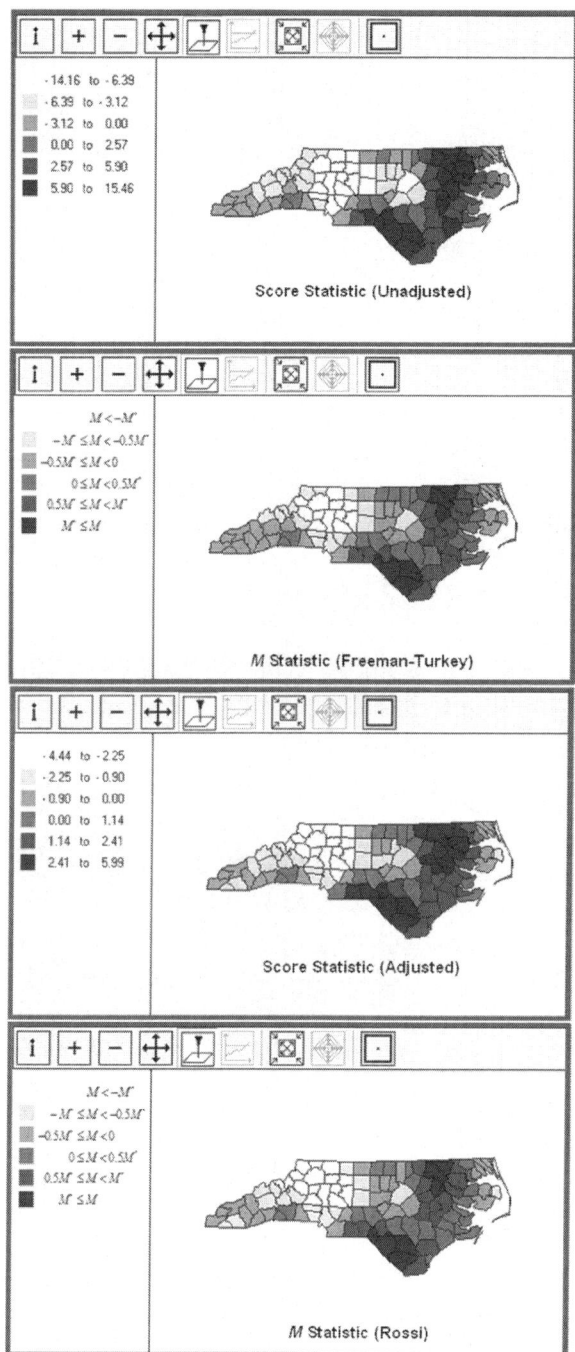

Fig. A.8.5. Results for the local score (upper) and *M* (lower) statistics

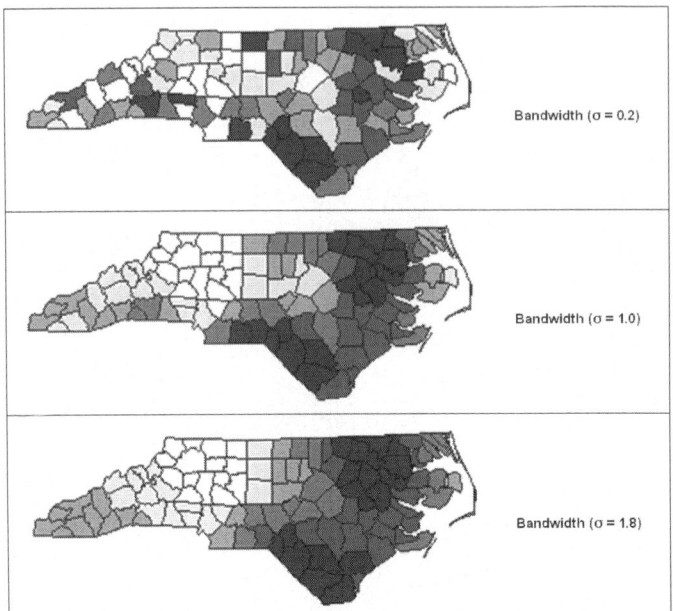

Fig. A.8.6. Maps of adjusted local score statistic for different bandwidths

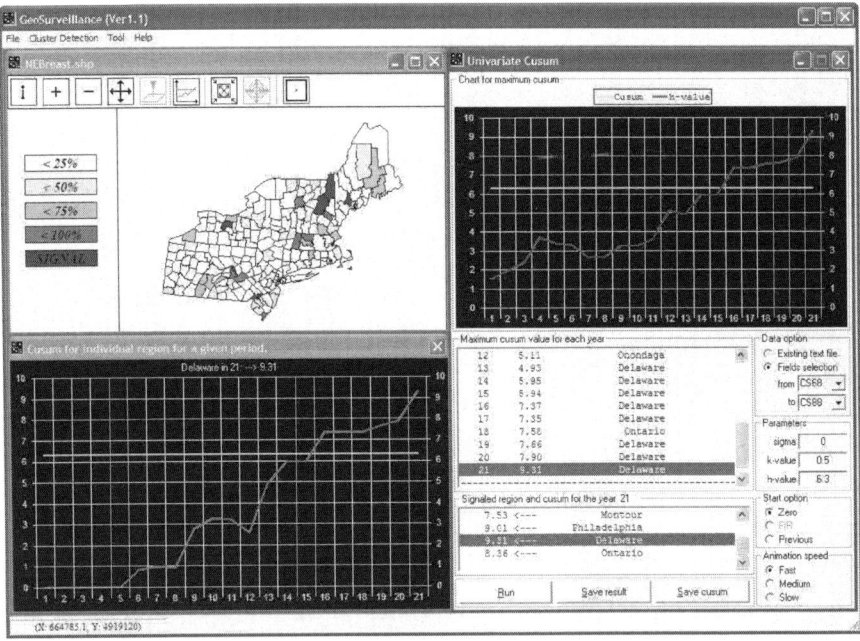

Fig. A.8.7. Linked windows of cusum map, tables and charts

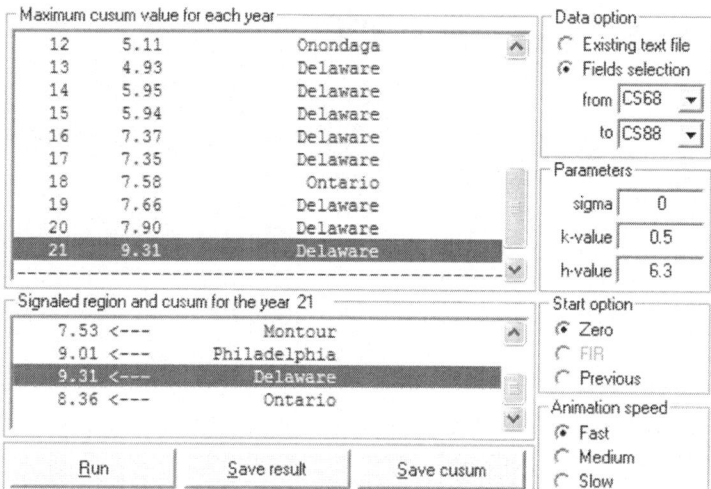

Fig. A.8.8. Enlarged image (tables and parameter control panels)

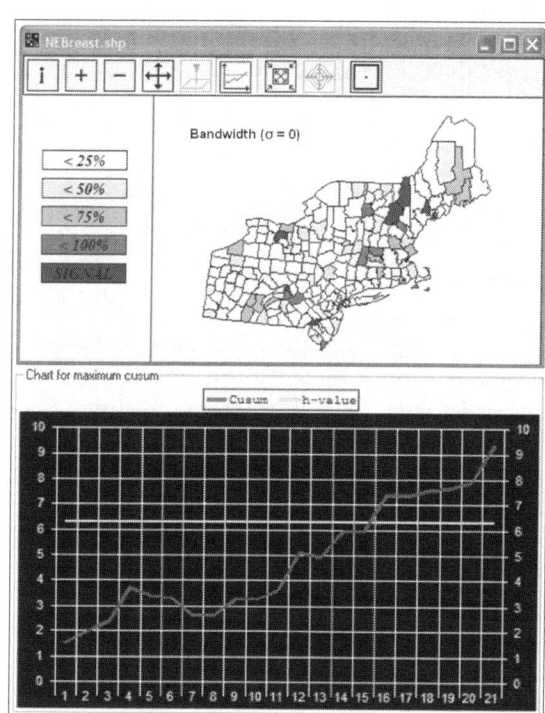

Fig. A.8.9. Maximum cusum charts and 1998 map when $\sigma = 0$

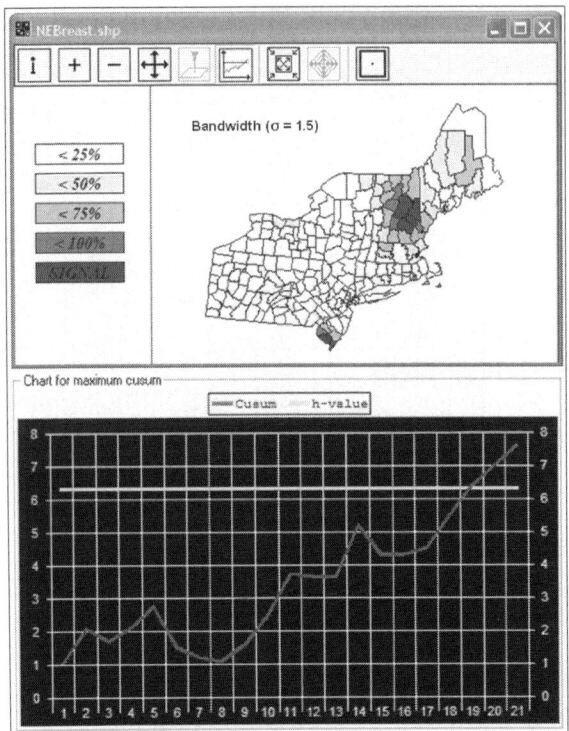

Fig. A.8.10. Maximum cusum charts and 1998 maps when $\sigma = 1.5$

A.8.5 Concluding remarks

In this chapter, *GeoSurveillance* was introduced as stand-alone software equipped with exploratory spatial analysis and monitoring tools. In *GeoSurveillance*, a set of associated spatial tests for cluster detection is implemented in both retrospective and prospective frameworks. The local score statistic and the M statistic can be used as a focused test and a test for detection of clusters in the retrospective framework. The associated global score statistic can be considered as a general test. The *cusum* statistic is used to monitor and detect spatial pattern changes in the prospective framework.

Analysis results are all interlinked in a map, tables, and charts. Various auxiliary tools are available in the program so that the user can transform various types of z-values based on observed and expected counts data, calculate p-values, determine the threshold values for *cusum* charts, etc. Based on the scheme of interlinked visual tools (map, table, chart), exploratory approaches are made possible to detect spatial clusters and spatial pattern changes. Although there are limitations, *GeoSurveillance* provides a useful analysis platform where some basic functions required for spatial statistical analysis and various exploratory tools are tightly integrated together.

We plan to upgrade *GeoSurveillance* by making the matrix calculation in the score statistic faster for relatively large datasets (over 217 observations). In addition, some GIS functions will also be improved.

References

Besag J, Newell J (1991) The detection of clusters in rare diseases. J Roy Stat Soc A 154(1):143-155

Hawkins DM, Olwell DH (1998) Cumulative sum charts and charting for quality improvement. Springer, Berlin, Heidelberg and New York

Kim YW, O'Kelly ME (2008) A bootstrap based space–time surveillance model with an application to crime occurrences. J Geogr Syst 10(2):141-165

Kulldorff M (1997) A spatial scan statistic. Comm Stat Theor Meth 26(6):1481-1496

Kulldorff M (2001) Prospective time periodic geographical disease surveillance using a scan statistic. J Roy Stat Soc A 164(1):61-72

Lawson A (1993) On the analysis of mortality events associated with a prespecified fixed point. J Roy Stat Soc A 156(3):363-377

Lee G, Rogerson PA (2007) Monitoring global spatial statistics. Stoch Environ Res Risk Assess 21(5):545-553

Openshaw SM, Charlton CW, Craft A (1987) A mark 1 geographical analysis machine for the automated analysis of point data set. Int J Geogr Inform Syst 1(4):335-358

Rogerson PA (1997) Surveillance systems for monitoring the development of spatial patterns. Stat Med 16:2081-2093

Rogerson PA (2001a) Monitoring point patterns for the development of space-time clusters. J Roy Stat Soc A 164(1): 87-96

Rogerson PA (2001b) A statistical method for the detection of geographic clustering. Geogr Anal 33(3):215-227

Rogerson PA (2005) A set of associated statistical tests for spatial clustering. Environ Ecol Stat 12(3):275-288

Rogerson PA (2006) Formulas for the design of cusum quality control charts. Comm Stat Theory Meth 35(2):373-383

Rogerson PA, Sun Y (2001) Spatial monitoring of geographical patterns: an application to crime analysis. Comput Environ Urban Syst 25:539-556

Rogerson PA, Yamada I (2004) Monitoring change in spatial patterns of disease: comparing univariate and multivariate cumulative sum approaches. Stat Med 23:2195-2214

Siegmund D (1985) Sequential analysis: test and confidence intervals. Springer, Berlin, Heidelberg and New York

Tango T (1995) A class of tests for detecting 'general' and 'focused' clustering of rare diseases. Stat Med 14(21):2323-2334

Turnbull BW, Iwano EJ, Burnett WS, Howe HL, Clark LC (1990) Monitoring for clusters of disease: application to leukemia incidence in upstate New York. Am J Epidemiol 132(1):136-143

Waller L, Gotway C (2004) Applied spatial statistics for public health data. Wiley, New York, Chichester, Torono and Brisbane

Waller L, Lawson A (1995) The power of focused tests to detect disease clustering. Stat Med 14(21-22):2291-2308

A.9 Web-based Analytical Tools for the Exploration of Spatial Data

Luc Anselin, Yong Wook Kim and *Ibnu Syabri*

A.9.1 Introduction

For close to twenty years now, there have been substantial efforts to extend Geographic Information Systems with functionality to carry out spatial analysis in general, and spatial statistical analysis in particular. Early work tended to emphasize objectives for the integration of GIS and spatial analysis, outline required functionality and describe overall frameworks, as exemplified in, among others, Goodchild (1987), Anselin and Getis (1992), Goodchild et al. (1992), Fotheringham and Rogerson (1993), and Fischer and Nijkamp (1993). More recently, this has translated into a range of software implementations of linked, embedded and otherwise integrated modules extending 'traditional' GIS functions with data exploration, visualization and analysis tools. For some recent reviews of the relevant literature, see, among others, Anselin (2000), Anselin et al. (2002), Symanzik et al. (2000), Zhang and Griffith (2000), Haining et al. (2000), and Gahegan et al. (2002).

The phenomenal growth of the world wide web has resulted in the development of so-called internet GIS, ranging from the delivery of static maps to interactive distributed computing frameworks. Most of the emphasis in internet GIS to date has arguably been on map delivery, cartographic presentation and providing access to a variety of distributed geographic information (see for example, Plewe 1997; Peng 1999; Kähkonen et al. 1999; Jankowski et al. 2001; Kraak and Brown 2001; Tsou and Buttenfield 2002).

Increasingly, more specialized spatial analytical capabilities are becoming implemented in an internet GIS environment as well. Some examples are virtual reality modeling (Huang and Lin 1999, 2002), hydrological modeling (Huang and Worboys 2001), as well as exploratory data analysis (Herzog 1998; Andrienko et al. 1999; Takatsuka and Gahegan 2001, 2002).

This chapter deals with efforts to incorporate methods for exploratory *spatial* data analysis in an internet GIS. The original motivation stemmed from the need to develop an interactive front end to the Atlas of U.S. Homicides of the National Consortium on Violence Research (Messner et al. 2000), which would include user-friendly ways to carry out a limited set of spatial data manipulations. The objective was to provide this functionality through a standard Web browser, so that the user would not need to have access to a GIS or specialized spatial data analysis software. Our focus is therefore on techniques to detect and visualize outliers in rate maps, to smooth these maps to correct for potential spurious inference, and to analyze and visualize patterns of spatial autocorrelation. Such methods are still largely absent in mainstream statistical and GIS software. A much more ambitious effort to provide ESDA and other spatial data analysis methods on the desktop is reflected in CSISS' *GeoDa* software project (Anselin 2003).[1]

In this chapter first a brief review of the methods included in our approach is provided, followed by an outline of the architecture of the software implementation. We illustrate the analytical tools with an application to the study of spatial pattens in county homicide rates around St. Louis, MO, and of colon cancer diagnoses in Appalachia. We close this chapter with some concluding remarks.

A.9.2 Methods

The techniques included in our analytical toolkit are aimed at the *exploration* of outliers in maps depicting rates or proportions, such as homicide rates, cancer incidence rates, mortality rates, etc. Three broad classes of methods are considered: outlier maps, smoothing procedures and spatial autocorrelation analysis. These methods are not new, and more extensive reviews and background can be found in, among others, Anselin (1994, 1998, 1999), Bailey and Gatrell (1995), Fotheringham et al. (2000) and Lawson et al. (1999). While familiar in the spatial analysis literature, they are typically not part of the standard functionality of a commercial statistical package or GIS, let alone included in an internet GIS.

The most basic set of techniques includes simple enhancements to standard choropleth maps in order to highlight extreme values. The maps are obtained by classifying the data in a particular way or by comparing the data to a reference value, as implemented in *percentile maps*, *box maps* and *excess rate maps*. A second set of methods encompasses smoothing procedures, in order to obtain 'more accurate' estimates of the underlying risk than produced by the raw rate maps. It is well known that when rates are estimated from unequal populations (such as widely varying county populations), the results are inherently unstable. Smoothing techniques address this issue by correcting ('shrinking') the raw rates

---

[1] *GeoDa* can be downloaded from http://sal.agecon.uiuc.edu/csiss/geoda.html.

while taking into account additional information (such as the indication provided by a reference rate). Two specific techniques are implemented here, the *Empirical Bayes* (EB) smoother and a *spatial rate* smoother. A final set of methods addresses the visualization of spatial autocorrelation by means of a *Moran scatterplot*. A brief review of some technical issues is provided next, for a more in-depth discussion we refer to the literature.

Outlier maps. Underlying any choropleth map is a sorting of the observed values into bins, similar to the classification used to construct a histogram. Each bin then corresponds to a color and all observations (locations) in the same bin are colored identically on the map.

In order to highlight extreme values in a distribution, and downplay the values around the median, a *percentile* map uses six categories for the classification of ranked observations: 0-1 percent, 1-10 percent, 10-50 percent, 50-90 percent, 90-99 percent and 99-100 percent. The lowest and highest percentile are extreme values, although this is only a simple ranking and does not imply that these observations are necessarily extreme relative to the rest of the distribution. In other words, they are candidates to be classified as outliers, but may not be outliers in a strict sense.

A more rigorous assessment of the characteristics of the complete distribution of the attributes is obtained in a *box map* (see, for example, Anselin 1998, 1999), a specialized form of a quartile map. Again, there are six categories. In addition to four categories corresponding to the four quartiles, an extra category is reserved at both the high and low end for those observations that can be classified as *outliers*, following the same definition as applied in the familiar *box plot*, also known as a box and whisker plot.[2] Consequently, when there are such outliers, the first and last quartile no longer contain exactly one fourth of the observations. The map shows the *location* of the outliers in the value distribution.

These first two types of maps are generic, in the sense that they apply to any kind of data. The *excess rate* (or, relative risk, standardized risk) maps are specific to rate or proportion data. Proportions are ratios of events (such as homicides, disease incidence or deaths) over a population at risk (the population in an areal unit, or, the population in a specific age/sex group in an areal unit). With E_i as the count of events, and P_i as the population at risk in area i, the 'raw rate' p_i is the simple proportion

$$p_i = E_i / P_i. \qquad (A.9.1)$$

---

[2] A box plot shows the ranking of observations by value and classified into four quartiles. Observations with values that are larger than (less than) the value corresponding to the 75th percentile (25th percentile) +(−)1.5 times the interquartile range are labeled outliers. See also Cleveland (1993) for an extensive discussion of data visualization issues. For an application of Tukey box plots see Chapter E.2.

Often, the result is scaled to yield a more meaningful number, such as homicides or deaths per ten thousand, per hundred thousand, etc. (typically, different disciplines have their own conventions about what is a 'standard' base value).

A measure of relative risk is obtained by comparing the rate at each location to the overall mean, computed as the ratio of all the events in the study region over the total population of the study region, or

$$\hat{\theta} = \frac{\sum_{i=1}^{N} E_i}{\sum_{i=1}^{N} P_i} \quad (A.9.2)$$

where N is the number of areal units in the study region. Note that this is not the same as the average of the individual p_i. Using the average risk and the population for each areal unit, an estimate of the *expected* number of events can be computed as

$$\hat{E}_i = \hat{\theta} P_i. \quad (A.9.3)$$

The ratio of actual to expected counts of events (or, their difference) is a commonly used indicator of the extent to which a location exceeds (or is below) what would be observed if the average risk applied to that location.[3] In an *excess rate* map, this is symbolized as a choropleth map. The map as such is purely for visualization and does not indicate whether of not the observed excess is 'significant' in a statistical sense.

Rate smoothing. Rate *smoothing* or *shrinkage* is the procedure used to statistically adjust the estimate for the underlying risk in a given spatial unit, by *borrowing strength* from the information provided by the other spatial units. The motivation for this approach comes from Bayesian statistics, where the estimate obtained from the data (the likelihood) is combined with *prior* information to derive a *posterior* distribution. This process is commonly referred to as borrowing strength, since it strengthens the original estimate. In practice, a wide range of approaches has been suggested that differ in the way additional information is incorporated into the estimation process. It is important to recognize that no method is *best*, and each will tend to result in (slightly) different adjustments to the raw rate estimate. The motivation for considering different smoothing techniques is to assess the degree of stability of the results. When two methods yield very different observations as 'outliers', additional investigation may be

---

[3] See the collection of papers in Lawson et al. (1999) for further discussion and several examples.

warranted. This contrasts with the situation where the same observation is consistently identified as an outlier across several methods.

An *Empirical Bayes* smoother uses Bayesian principles to guide the adjustment of the raw rate estimate by taking into account information in the rest of the sample. The principle is referred to as *shrinkage*, in the sense that the raw rate is moved (shrunk) towards an overall mean, as an inverse function of the inherent variance.[4]

In other words, if a raw rate estimate has a small variance (that is, is based on a large population at risk), then it will remain essentially unchanged. In contrast, if a raw rate has a large variance (that is, is based on a small population at risk, as in small area estimation), then it will be 'shrunk' towards the overall mean. From a Bayesian perspective, the overall mean is a *prior*, which is conceptualized as a random variable with its own ('prior') distribution.

Assume this prior distribution is characterized by a mean θ and variance ϕ. The Bayesian estimate for the underlying risk at i then becomes a weighted average of the raw rate p_i, given in Eq. (A.9.1), and the 'prior', with weights inversely related to their variance. This can be shown to yield

$$\hat{\pi}_i = w_i p_i + (1 - w_i)\theta \qquad (A.9.4)$$

with

$$w_i = \frac{\phi}{\phi + (\theta / P_i)}. \qquad (A.9.5)$$

Note that when the population at risk is large, the second term in the denominator of Eq. (A.9.5) becomes near zero, and $w_i \to 1$, giving all the weight in Eq. (A.9.4) to the raw rate estimate. As P_i gets smaller, more and more weight is given to the second term in Eq. (A.9.4). The *Empirical* Bayes approach (EB) consists of estimating the moments of the prior distribution from the data, rather than taking them as a 'prior' in a pure sense (for technical details, see, for example, Marshall 1991).

An important practical issue is the choice of the reference set from which the estimate for θ is computed. For example, one could argue that in a study of homicides in rural Minnesota counties (characterized by very low homicide counts, but also by small populations, such that a single homicide may cause an elevated rate), the proper prior would not necessarily be the national homicide rate, but rather an average calculated for the Great Plains 'region'. In any application of smoothing, it is important to consider the sensitivity of the results (in terms of how locations are classified as being outliers) to the choice of this

---

[4] The original reference is Clayton and Kaldor (1987), details are also given in Bailey and Gatrell (1995, pp 303-308).

reference region. One of the characteristics of the tools we implement is to make this straightforward for the user. Again, it is important to realize that there is no *best* reference region. Rather, in an exploratory exercise, an assessment of sensitivity of the identified 'patterns' to the choice of technique is an important consideration.

A spatial rate smoother (for example, Kafadar 1996) is based on the notion of a spatial moving average or *window average*. Instead of computing an estimate as the raw rate for each individual spatial unit, it is computed for that unit *together* with a set of 'reference' neighbors, S_i.[5] This contrasts with the EB technique, where the smoothed rate is an average of the raw rate and some *separately* computed reference estimate.

An important practical consideration in the implementation of a spatial smoother is the size of the 'window', or, the selection of the relevant neighbors. As with the EB method, there is no *best* solution, but rather, interest focuses on the sensitivity of the conclusions to the choice of the window. As a general rule, the larger the window (the more neighbors), the more of the original variability will be removed. In the extreme, if the spatial window includes all the observations in the data set, the smoothed rate will be the same everywhere. In practice, neighbors can be defined in similar fashion to the specification of spatial weights in spatial autocorrelation analysis. In our implementation, we use simple contiguity (common borders) to define the neighbors. The smoothed rate becomes

$$\hat{\pi}_i = \frac{E_i + \sum_{j=1}^{J_i} E_j}{P_i + \sum_{j=1}^{J_i} P_j} \qquad (A.9.6)$$

where $j \in S_i$ are the neighbors for i.[6] The spatially smoothed rate map is then a choropleth map based on the ranking of the smoothed rate values. It emphasizes broader regional trends and removes some of the spatial detail from the original map.

Visualizing Spatial Autocorrelation. The final component in our analytical framework is the visualization of spatial autocorrelation by means of a *Moran Scatterplot* (Anselin 1995, 1996). This is a specialized scatterplot with the spatially lagged transformation of a variable on the *y*-axis and the original variable on the *x*-axis, after standardizing the variable such that the mean is zero and variance one. With such a standardized variable as z_i, the spatial lag becomes

---

[5] A slightly different notion of spatial rate smoother is based on the median rate in the moving window, as used by Wall and Devine (2000).

[6] The total number of neighbors for each unit, J_i is not necessarily constant and depends on the contiguity structure.

$$[Wz]_i = \sum_j W_{ij} z_j \qquad (A.9.7)$$

where W_{ij} are elements of a row-standardized spatial weights matrix.[7] For the z_i and with a row-standardized spatial weights matrix, Moran's I coefficient of spatial autocorrelation is:

$$I = \frac{\sum_i \sum_j z_i W_{ij} z_j}{\sum_i z_i^2} \qquad (A.9.8)$$

or, the slope of the regression line of the spatially lagged variate $[Wz]_i$ on the original variate z_i (see Anselin 1996).

Since the variable z_i is standardized, the units on the axes of the scatterplot correspond to one standard deviation. Hence, points further than two standard deviations from the center (the mean) can be informally characterized as 'outliers'. However, the main contribution of the Moran scatterplot is the classification of the type of spatial autocorrelation into two categories, referred to as *spatial clusters* and *spatial outliers*. As explained in more detail in Anselin (1996), each quadrant of the Moran scatterplot corresponds to a different type of spatial correlation. The lower-left and upper-right quadrants indicate *positive* spatial autocorrelation, respectively of low values surrounded by neighboring low values, or high values surrounded by neighboring high values. Consequently, these are referred to as clusters. In contrast, the upper-left and lower-right quadrants suggest *negative* spatial autocorrelation, respectively of low values surrounded by neighboring high values, or high values surrounded by neighboring low values. These are therefore referred to as spatial outliers. It is important to note that the scatterplot provides the classification, but does not indicate 'significance'. The latter is obtained by applying a local Moran (LISA) test, as shown in Anselin (1995).

The scatterplot also provides a visual indication of the sign and strength of spatial autocorrelation in the form of the slope of the regression line. Finally, the scatterplot allows for an informal investigation of the leverage (influence) of specific observations (locations) on the autocorrelation measure.[8]

---

[7] The square spatial weights matrix has a row/column corresponding to each observation. For each row (observation) it indicates by a non-zero value those columns (observations) that are 'neighbors'. In our implementation, we only consider neighbors defined by simple contiguity. The weights matrix is row-standardized such that the elements of each row sum to one.

[8] In the latest incarnation of our tool, developed after the first version of this chapter was completed, a variance stabilization method due to Assunção and Reis (1999) is included as an option. This corrects the Moran's I statistic for potentially spurious inference due to the intrinsic variance instability of rates, similar to the EB smoother discussed above.

A.9.3 Architecture

Our point of departure for enabling an internet GIS with spatial analytical capability is the collection of Java classes contained in the *Geotools* open source mapping toolkit, originally developed at the University of Leeds.[9] *Geotools* implements choropleth mapping, cartograms, linking, zooming, panning and other standard functions of an internet GIS through a Java applet embedded in a standard html web page. The applet executes on the client's machine in the browser (provided the browser is Java-enabled). The toolkit is open source, which allows for easy customization and complete access to all the code.[10]

Basic Geotools architecture. In order to put our extensions into proper perspective, Fig. A.9.1 illustrates the basic logic of the standard *Geotools* internet mapping implementation. The main input is a file in ESRI's shape file format, from which an attribute (variable) is extracted for mapping. The attribute values are stored in *Geotools'* so-called *GeoData* object (data structure), which is essentially a two column matrix, with each row containing the value of a key (matching the ID of a corresponding feature in the shape file) and the attribute value (either numeric or character). Both the file name of the shape file as well as the name of the variable to be mapped are passed as parameters to the Java applet, but once the main applet is set up, they can no longer be changed.

Once the *GeoData* object is constructed, it is passed to the *ClassificationShader* class, which can be thought of as a central data dispatch center. The *ClassificationShader* moves the original data to the appropriate classification classes, such as *Quantile.class*, or *EqualInterval.class*. These classes implement the sorting and classification necessary to group the original data into bins for use in a thematic map. The result of the classification is passed back to the *ClassificationShader*, which transfers it to the main applet for mapping. This is both directly, for the map itself, and indirectly, via the specialized classes required to construct the legend (e.g., the *Key.class* and the *DiscreteShader.class*). The *ClassificationShader* also manages a rudimentary user interface (Popup dialog) to select the type of classification for the choropleth map, the number of intervals, start and end colors for a color ramp, etc. (see Fig. A.9.2).

---

[9] http://www.geotools.org. Our implementation is based on Geotools Version 0.8.0. More recently, Version 2.0 of Geotools has been released in alpha testing stage. The architecture of this new version is completely different and our framework cannot be ported 'as is' to the new architecture.

[10] An up to date source tree for the *Geotools* project is maintained in Sourceforge, at http://www.sourceforge.net/projects/geotools

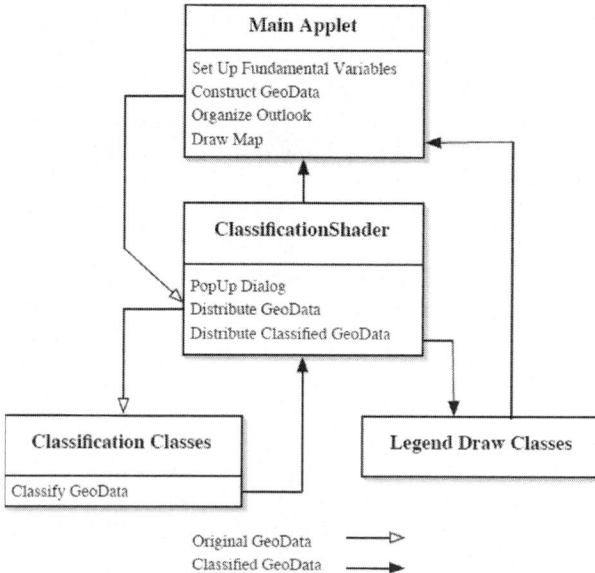

Fig. A.9.1. Basic Geotools architecture (original)

For our purposes, there were several limitations to the standard *Geotools* architecture. Foremost among these was the constraint that only a single variable could be handled. All manipulations within the *Geotools* classes (mapping, classification, linking) are limited to this single variable, i.e., the values contained in the *GeoData* object. In our application, the smoothing functions require at least two variables, i.e., an event count (numerator) and population at risk (denominator), and also need to allow for the computation of a new variable (the rate). Similarly, spatial correlation statistics necessitate that a new variable be calculated (the spatial lag) to provide the input to the statistic. This was not possible in the 'out of the box' *Geotools* release we used to implement our web analysis.

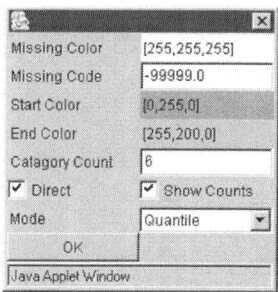

Fig. A.9.2. Geotools interface

The original architecture also makes it difficult to implement true subsetting, as opposed to zooming. In true subsetting, the classification of the selected subset of locations is recomputed each time the subset changes, whereas in zooming, the classification is unaffected. Again, the basic *GeoData* structure does not lend itself to subsetting and recomputation. Finally, there is limited user interaction. For example, it is not possible to specify a different shape file as input, or to select a different variable from what is hard coded in the original applet.

The need for flexible data manipulation, variable selection and subset computations required us to customize the basic toolkit. This took the form of several extensions to the standard collection of *Geotools* classes as well as the development of a number of new classes.

Geotools class extensions. An overview of the architecture of the extensions required to implement the smoothing and correlation computations is given in Fig. A.9.3. The main difference with Fig. A.9.1 is that the *GeoData* object is no longer constructed in the main applet, but instead only the Shape File Reader (SFR) is passed to the *ClassificationShader*. This input is obtained from the user, by extracting the name of the shape file through an html form embedded in the opening web page. The *ClassificationShader* remains the central data dispatch and handles a slightly more elaborate user interface through which the variable names and type of classification are selected (see Fig. A.9.4). This is implemented in a new class (Alert.class).

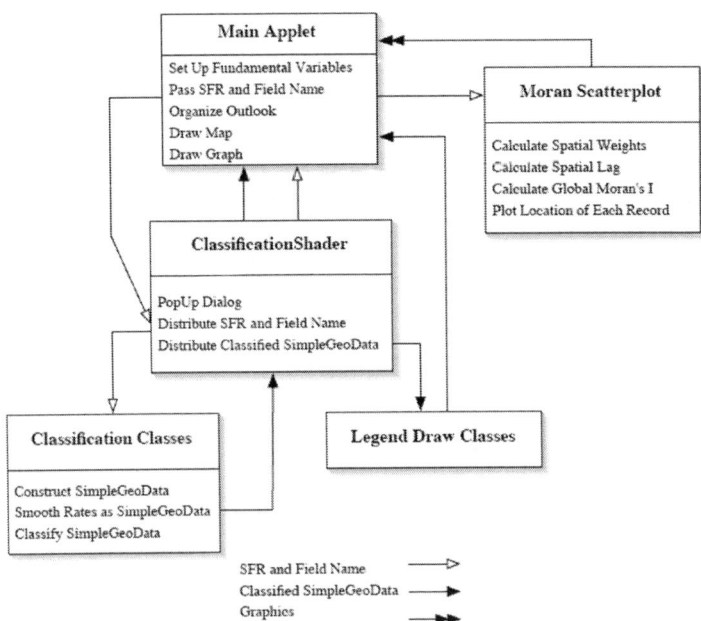

Fig. A.9.3. Extended Geotools architecture

In contrast to the original *Geotools*, where the hard coded variable does not require any additional computations, the construction of rates and the smoothing operations must be carried out internally. The main computational work to accomplish this is included in a number of extensions and new classes.

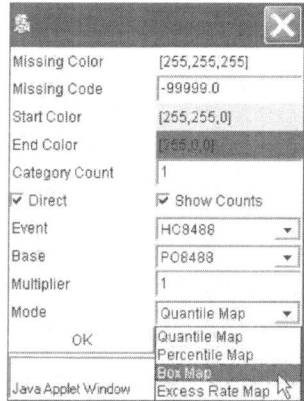

Fig. A.9.4. Customized interface

In our implementation, the Classification Classes handle both the construction of the data to be mapped as well as the customized classifications needed for the special outlier maps. The original Quantile class is extended to incorporate the computation of rates, based on the field names for the numerator (Event) and denominator (Base) passed by the user interface (Fig. A.9.4). This creates a *Geotools SimpleGeoData* object, which is somewhat more flexible than the basic *GeoData* object and can be used to handle most computed results (smoothed rates, spatial lags) as well as subsets. New classification classes were developed to handle each of the specialized outlier maps, the Percentile Map, Box Map and Excess Rate Map.[11] These are essentially specialized forms of the basic Quantile map, but using different criteria to construct the classification.

In addition to the specialized classifications, new classes were also needed to handle the computations required for the Empirical Bayes and spatial smoothing operations. These are included among the Classification Classes as well.

Moran scatterplot and spatial weights. The other main change from the original *Geotools* toolkit is the incorporation of spatial correlation analysis, implemented by the addition of the Moran Scatterplot class (the box included on the upper right side of Fig. A.9.2). At first sight, this might have been accomplished by customizing the available *Geotools* class for a scatterplot. However, the *ScatterPlot.class* included in the *Geotools* toolkit cannot properly

---

[11] Specifically, the Percentile.class, Box.class and Excess.class for, respectively, a percentile map, a box map and an excess rate map

accommodate subsetting, that is, where the slope of the Moran scatterplot is recalculated for a contiguous subset of locations. Also, linking does not function properly for subsets. The new class takes the shape field information from the main applet and constructs all the necessary auxiliary variables internally, that is, the contiguity based spatial weights, the spatial lag, and Moran's I. These internal computations yield the coordinates of the points in the plot (z_i on the x-axis and $[Wz]_i$ on the y-axis), and the slope and intercept of the regression line. This is recomputed and redrawn whenever a subset is selected.

It may be worthwhile to elaborate upon the way in which the spatial weigths are obtained. The *Geotools* toolkit includes a 'contiguity matrix', implemented as a *HashSet*, an internal data structure. However, this data structure includes considerable additional information (such as all point coordinates for each polygon). The spatial lag construction (for the spatial smoother and for the Moran scatterplot) only requires a subset of this, that is, the IDs of the neighbors for each location. Instead of using the built-in contiguity matrix, we derive our own data structure from the *HashSet* and store this information in a *SimpleGeoData* structure. This contains only the ID information and is kept in memory until a new data set is specified. Subsetting is applied directly to this structure as well.

User interaction. User interaction in a web-based spatial analysis is two-fold, one aspect dealing with the server, the other operating in the browser, on the client side. The latter is managed by the Java applet. The main choices (variable, smoothing procedure, etc.) are invoked by clicking on the legend box that appears when the map is first drawn. Initially, this is a single button, but after clicking, an interface appears as in Fig. A.9.4. Additionally, selected buttons appear in the web page to invoke specific methods (see the illustrations in Section A.9.4).

The interaction on the server side ensures that the initialization parameters are obtained to set the proper configuration for the Java applet. In a standard html page, a 'form' is used to record the selections, as illustrated in Fig. A.9.5. The form invokes a PHP script (on the server) that generates a web page corresponding to the selected options. This web page includes one of three Java applets, depending on the option selected. After this page is rendered on the client (and the applet downloaded) all further interaction is through the Java applet on the client. There are three basic options, as illustrated in Fig. A.9.5.[12] First, the screen resolution can be customized in order to make sure the maps and graphs fit on the user's screen (assuming the browser window is maximized). Second, a selection can be made from a series of maps/data sets included in a drop down list. These data sets must be present on the server in a directory specified by *Geotools*.

---

[12] This particular view is for a Safari web browser on a Mac G4 workstation, with the pages served using the Apache server on a Linux workstation.

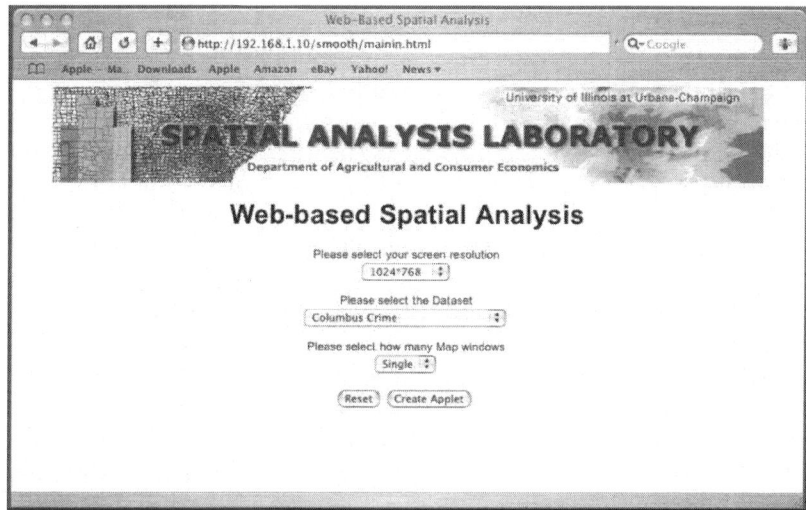

Fig. A.9.5. Welcome screen and general options

At this point it is not possible for the user to upload shape files to this directory without proper write permissions. The final option pertains to the type of analysis to be carried out. The *single* map option is primarily for visualization and smoothing, but only one map is rendered in the browser. This is the fastest option, with the shortest time required to download the applet. In contrast, the *two* map option renders both the smoothed map as well as the original (unsmoothed) map, to allow direct comparison of outliers and other features of the data. The *three* map option also provides space to draw the Moran Scatterplot for the selected variable. These two options take longer to download the applet.

Finally, the user can interact directly with the graphics, since all maps and graphs are linked, such that clicking on a location in one of them highlights the matching locations in the others. Also, all three graphics support zooming, panning and subsetting.

A.9.4 Illustrations

We provide a brief illustration of the functionality of the spatial analysis tools using two sample data sets. One is a subset of the NCOVR U.S. Homicide Atlas, limited to counties surrounding St. Louis, MO (Messner et al. 1999, 2000). The other contains data on colon cancer diagnoses in Appalachian counties.[13] Both

---

[13] Data compiled from individual cancer registry records and aggregated to the county level by Eugene J. Lengerich, Pennsylvania State Cancer Institute, Pennsylvania State University.

data sets are for rates, respectively homicide counts over population (for 1979-84) and colon cancer diagnosis counts over population (1994-98). Using standard practice, the counts are aggregated over a small number of years to avoid extreme heterogeneity.

We start with an Excess Rate map (or relative risk map) for the St. Louis region homicide rates (see Fig. A.9.6). The map is invoked by selecting the county homicide count in the period 1979-84 (HC7984) as the 'Event,' and the county population in the same period (PO7984) as the 'Base.' Also, the proper map type must be clicked in the Legend Interface (see Fig. A.9.4). The buttons at the top of the map allow zooming, panning and subsetting. For this particular map type, the legend is hard coded, showing six intervals for the relative risk.[14] Moving the mouse over each county triggers a pop up 'tooltip' with the ID value for that county (for example, St. Clair county in Fig. A.9.6).

The map illustrates how both St. Louis city and St. Clair county have homicide rates that far exceed the region-wide average. By contrast, outlying rural counties have relative risks well below the region-wide average. This highlights the dominance of the St. Louis-East St. Louis core when it comes to homicides in the period under consideration.

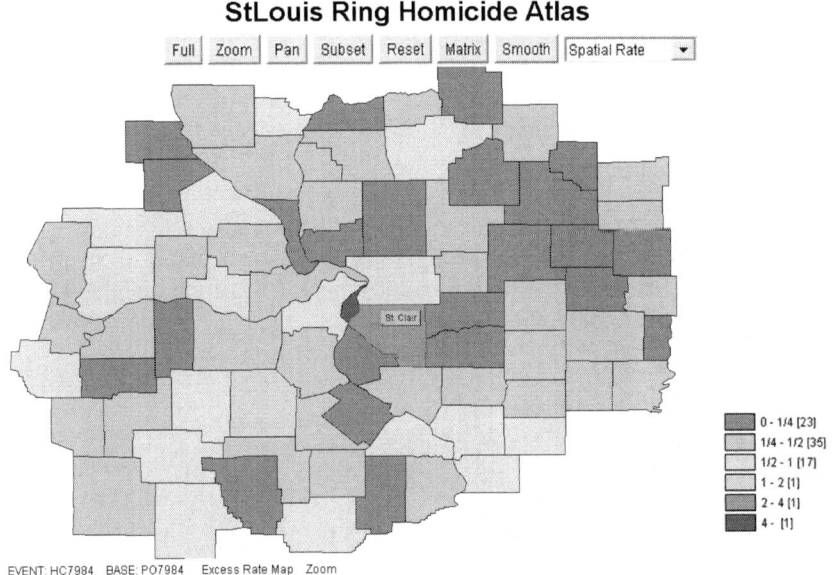

Fig. A.9.6. Excess Rate map, St. Louis region homicides (1979-84)

---

[14] The colors in the legend can be adjusted individually, but the default is based on recommendations from *ColorBrewer*, http://www.colorbrewer.org. The same approach is taken in all other thematic maps.

The second example highlights the use of two maps to compare 'raw' rates (the simple ratio of events over base) to their smoothed counterparts. The top map in Fig. A.9.7 shows an example for colon cancer rates that have been transformed using the Empirical Bayes approach, shrinking the raw rates towards the overall average for the Appalachian region. In this example, two box maps are shown in the browser, the top map with the smoothed rates, and the bottom map with the original raw rates. Note how Cameron county, identified as a high outlier in the raw rate map (shown as a tooltip), does not maintain that position in the smoothed smoothed map. The smoothing is invoked by clicking on the 'Smooth' button in the map window and selecting the specific smoothing method in the drop down

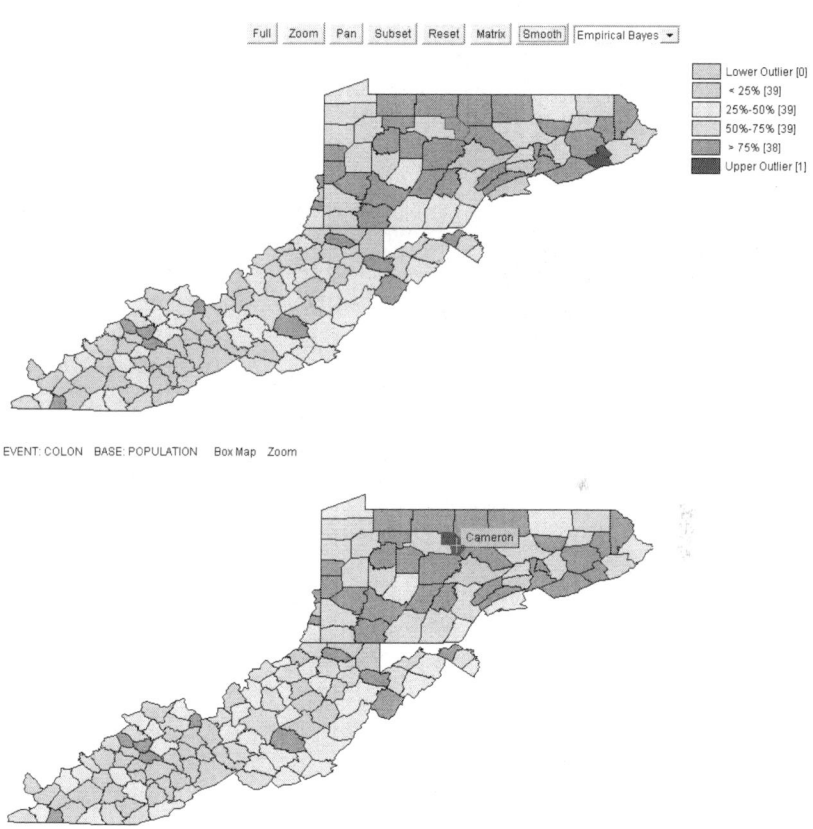

Fig. A.9.7. Empirical Bayes smoothing, colon cancer, Appalachia (1994-98). Two box maps with smoothed map on top and original raw rate on bottom

list. Counties that lose their outlier status after smoothing are so-called *spurious* outliers, where the extreme rate is likely due to a small population at risk.

In the Empirical Bayes smoothing method, a central role is played by the regional average to which the raw rates are shrunk. When the region is highly heterogeneous, the choice of the overall regional average as the reference rate may not be appropriate. More precisely, the choice of different subregions will yield varying subregional averages which affects the smoothing and the resulting indication of outliers. We provide a way to assess the sensitivity of the results to this choice by means of the *subset* command. Clicking on the corresponding button turns the cursor into a selection rectangle. The classification underlying the box map is recalculated for the selected counties, and, as a result, the indication of outlier may change. For example, in Fig. A.9.8, a county appears as a low end outlier, when the subset is reclassified for Pennsylvania counties only. In contrast,

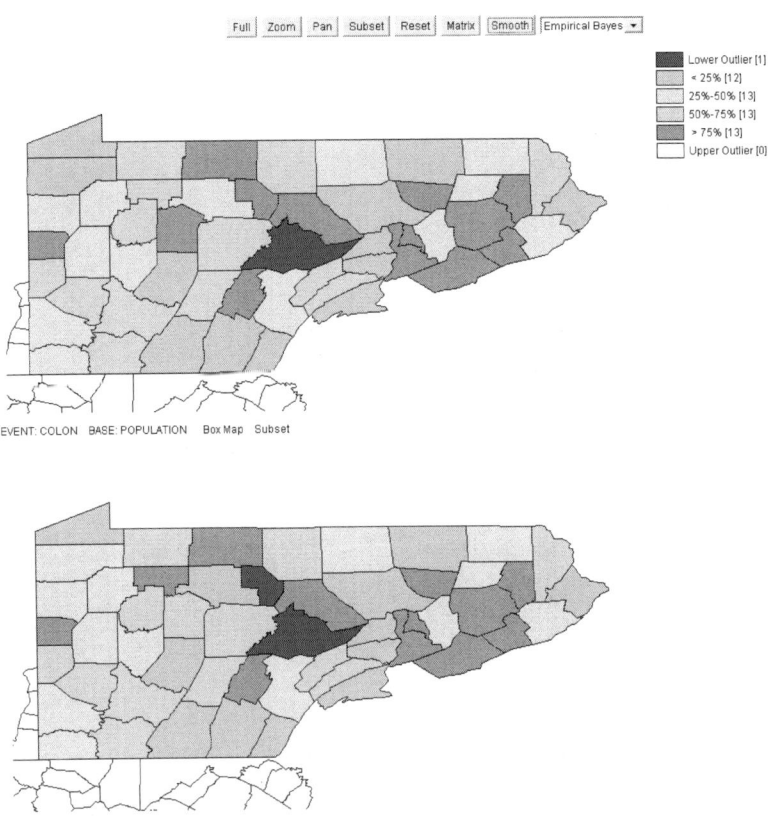

Fig. A.9.8. Empirical Bayes subset smoothing, colon cancer, Appalachia (1994-98). Two box maps with smoothed map on top and original raw rate on bottom

the overall map (see Fig. A.9.7) does not classify this county as a low end outlier. Again, note how an upper outlier in the raw rate map disappears in the EB smoothed map. Other changes are minor in this map, likely due to the smoothing of counts over time (the four year average used to compute the county rates).

Spatial smoothing, shown in Fig. A.9.9, tends to emphasize broad subregional trends. Note how the patterns are much stronger in the upper map than in the lower map. The smoothed map highlights a North-South divide in the region, suggesting spatial heterogeneity (and, possibly, spatial regimes). Again, the indication of outlier changes between the raw rate map and the smoothed map, supporting the importance of this type of sensitivity analysis before locations are classified as 'extreme.'

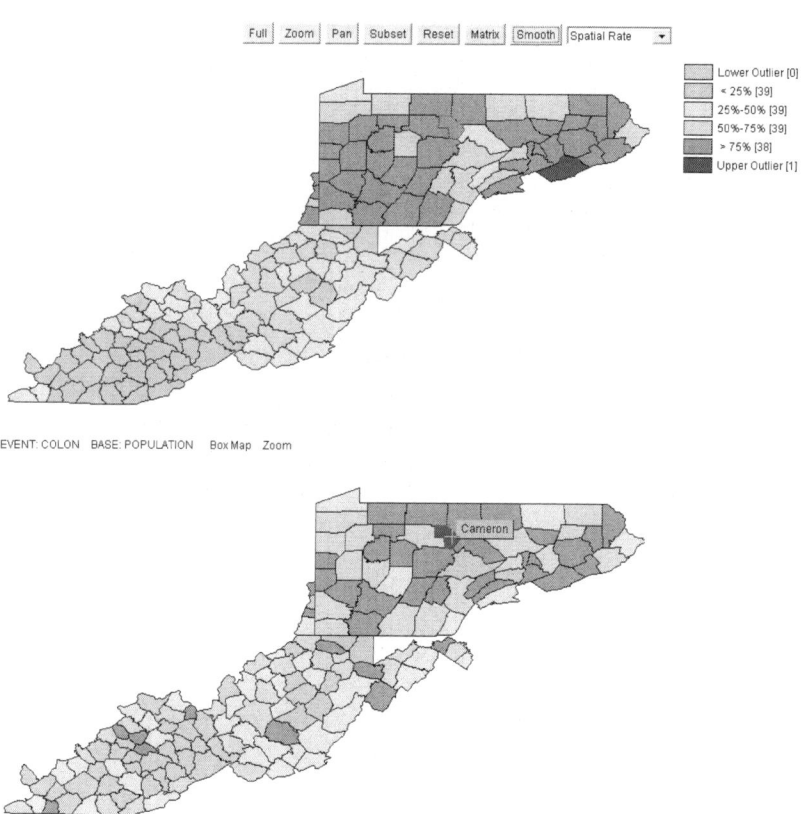

Fig. A.9.9. Spatial smoothing, colon cancer, Appalachia (1994-98)

The final element in our analytical toolbox pertains to the visualization of spatial autocorrelation by means of a Moran scatterplot. Fig. A.9.10 shows the bottom two graphs in the three graph plot generated by the Java applet.[15] The illustration is for the same homicide rate in the St. Louis region as used in Fig. A.9.6. The value of 0.196 is the slope of the regression line and suggests strong positive spatial autocorrelation in the homicide rates.[16]

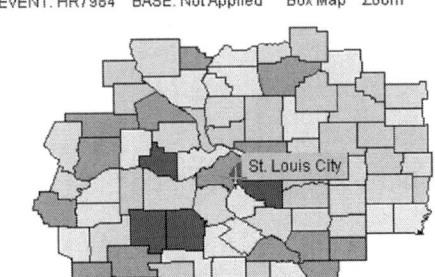

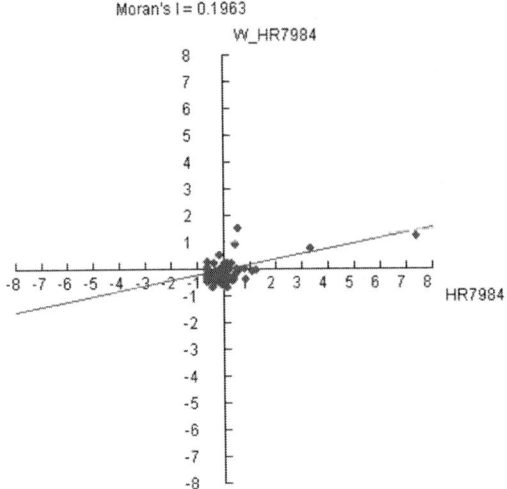

Fig. A.9.10. Moran scatterplot, St. Louis region homicide rate (1979-84)

---

[15] Since no smoothing is applied in the univariate Moran scatterplot, the smoothed and original map are identical.

[16] It is important to note that this does not indicate 'significance' of the spatial autocorrelation statistic, but only shows its magnitude. A formal hypothesis test is not currently included, but would be required before the value of 0.196 can be characterized as indicating significant spatial autocorrelation.

The highlighted point in the scatterplot corresponds to St. Louis City, as indicated by the linked graphs. Its position in the upper-right quadrant suggests that it is part of a 'cluster' of high homicide rates. The position of the point might also indicate potentially high leverage on the value of the statistic. To assess this, we select a subset of the counties to the East of St. Louis, but not including the city. The spatial pattern of the homicide rates, with a recalculated classification for the Box Map is shown in the top half of Fig. A.9.11. Note how in addition to St. Clair county (East St. Louis), an additional county in the Southern part of the map is now classified as an upper outlier (relative to the other values within the selected region). Also note how the recalculated Moran's I no longer suggests any spatial autocorrelation (the line is

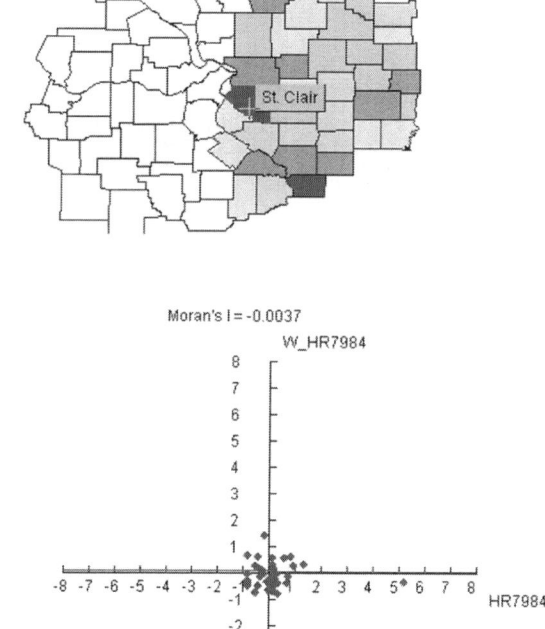

Fig. A.9.11. Moran scatterplot, East subregion homicide rate (1979-84)

essentially horizontal), illustrating the heavy leverage exerted by the single St.Louis observation.[17] In other words, once St. Louis city is removed from the sample, and the focus is on the more rural counties surrounding the city, the indication of strong spatial patterning disappears, and, instead, spatial randomness seems to be the appropriate conclusion. A complete analysis would assess this for other potential high leverage points as well.

Finally, note how the point to the utmost right in the Moran scatterplot of Fig. A.9.11 is more than five standard deviations from the mean. This qualifies it as an outlier in the traditional sense of descriptive statistics, as confirmed by its classification in the box map. Moreover, since it is in the lower-right quadrant of the scatterplot, it also corresponds to a *spatial* outlier, a location with a much higher homicide rate than its surrounding neighbors.

A.9.5 Concluding remarks

In this contribution, we outlined an initial framework to implement spatial data analysis functions in an internet GIS. Our efforts are a 'work in progress' and part of a much larger and more comprehensive endeavor to develop spatial analytical software tools as part of the program of the Center for Spatially Integrated Social Science (CSISS).[18] While the current tools serve their purpose, several important issues warrant further scrutiny.

The range of spatial analytical methods included in the framework is clearly limited. In part this is by design, given the specific objective to provide an interactive front end to an atlas. However, part of the limitation also has to do with performance issues encountered for medium size and larger data sets. The download time for the applet increases considerably when more functions are included, so it is easy to envisage a point where this approach becomes impractical.

In addition, Java as a language is not optimal as a platform for highly intensive numerical operations. While this is not a constraint for the currently included methods, techniques that require more computation (such as randomization tests for spatial autocorrelation) may need to be implemented in a different language and/or warrant the development of more optimal data structures in order to be completed within a time frame required for real time interaction with the data. This calls for a more careful consideration of the division of labor between the server and client. As many others have argued, the more computationally intensive operations should probably be carried out on the server,

---

[17] See Messner et al. (1999) for a more in-depth analysis of outliers in this data set. The overall findings of regional heterogeneity were similar to what is illustrated here.

[18] See http://sal.agecon.uiuc.edu/csiss/index.html

with user interaction and simple calculations allocated to the client. The exact nature of the tradeoffs associated with this balancing act merit further attention, and are the subject of ongoing research.

Finally, even given these limitations, the current framework provides some insight into the complexities of the characterization of spatial outliers and the sensitivity of the 'map' to various assumptions made in the process. This pedagogical objective is reached without requiring the user to have access to advanced statistical or GIS software, a main advantage of the web-based approach. It is hoped that continued work along these lines will further advance the dissemination of spatial analytical techniques to a broader audience.[19]

Acknowledgements. This research was supported in part by a number of grants from the U.S. National Science Foundation: NSF Grant SBR-9410612, BCS-9978058, to the Center for Spatially Integrated Social Science (CSISS), and a grant from the National Consortium on Violence Research (NCOVR) is supported under grant SBR-9513040 from the National Science Foundation). In addition, support was provided by grant RO1 CA 95949-01 from the National Cancer Institute. Special thanks to Dr. Eugene J. Lengerich of the Pennsylvania State Cancer Institute for providing the data on colon cancer diagnoses.

References

Andrienko GL, Andrienko NV, Voss H, Carter J (1999) Internet mapping for dissemination of statistical information. Comput Environ Urban Syst 23(6):425-441
Anselin L (1994) Exploratory spatial data analysis and geographic information systems. In Painho M (ed) New tools for spatial analysis, Eurostat, Luxembourg, pp.45-54
Anselin L (1995) Local indicators of spatial association - LISA. Geogr Anal 27(2):93-115
Anselin L (1996) The Moran scatterplot as an ESDA tool to assess local instability in spatial association. In Fischer MM, Scholten H, Unwin D (eds) Spatial analytical perspectives on GIS in environmental and socio-economic sciences, Taylor and Francis, London, pp.111-125
Anselin L (1998) Exploratory spatial data analysis in a geocomputational environment. In Longley PA, Brooks S, Macmillan B, McDonnell R (eds) Geocomputation: a primer. Wiley, New York, pp.77-94
Anselin L (1999) Interactive techniques and exploratory spatial data analysis. In Longley PA, Goodchild MF, Maguire DJ, Rhind DW (eds) Geographical information systems: principles, techniques, management and applications, Wiley, New York. pp.251-264
Anselin L (2000) Computing environments for spatial data analysis. J Geogr Syst 2(3):201-220

---

[19] The web tools described in this chapter are available for a sample of six data sets at http://sal.agecon.uiuc.edu/webtools/index.html

Anselin L (2003) GeoDa 0.9 user's guide. Spatial Analysis Laboratory (SAL). Department of Agricultural and Consumer Economics, University of Illinois, Urbana-Champaign [IL]

Anselin L, Getis A (1992) Spatial statistical analysis and geographic information systems. Ann Reg Sci 26(1):19-33

Anselin L, Syabri I, Smirnov O (2002) Visualizing multivariate spatial correlation with dynamically linked windows. In Anselin L, Rey S (eds) New tools for spatial data analysis: proceedings of the specialist meeting. Center for Spatially Integrated Social Science (CSISS), University of California, Santa Barbara [CA], CD-ROM

Assunção R, Reis EA (1999) A new proposal to adjust Moran's I for population density. Stat Med 18(16):2147-2161

Bailey TC, Gatrell AC (1995) Interactive spatial data analysis. Longman, Harlow

Clayton D, Kaldor J (1987) Empirical Bayes estimates of age-standardized relative risks for use in disease mapping. Biometrics 43(3):671-681

Cleveland WS (1993) Visualizing data. Hobart Press, Summit [NJ]

Fischer MM, Nijkamp P (1993) Geographic information systems, spatial modelling and policy evaluation. Springer, Berlin, Heidelberg and New York

Fischer MM, Stumpner P (2009) Income distribution dynamics and cross-region convergence in Europe. In Fischer MM, Getis A (eds) Handbook of applied spatial analysis. Springer, Berlin, Heidelberg and New York, pp.599-628

Fotheringham AS, Rogerson P (1993) GIS and spatial analytical problems. Int J Geogr Inform Syst 7(1):3-19

Fotheringham AS, Brunsdon C, Charlton M (2000) Quantitative geography: perspectives on spatial data analysis. Sage, London

Gahegan M, Takatsuka M, Wheeler M, Hardisty F (2002) Introducing GeoVISTA studio: an integrated suite of visualization and computational methods for exploration and knowledge construction in geography. Comput Environ Urban Syst 26(4):267-292

Goodchild MF (1987) A spatial analytical perspective on Geographical Information Systems. Int J Geogr Inform Syst, 1(4):327-334

Goodchild MF, Haining RP, Wise S, and 12 others (1992) Integrating GIS and spatial analysis – problems and possibilities. Int J Geogr Inform Syst 6(5):407-423

Haining RF, Wise S, Ma J (2000) Designing and implementing software for spatial statistical analysis in a GIS environment. J Geogr Syst 2(3):257-286

Herzog A (1998) Dorling cartogram. http://www.zh.ch/statistik/map/ dorling/dorling.html

Huang B, Lin H (1999) GeoVR: a web-based tool for virtual reality presentation from 2D GIS data. Comput Geosci 25(10):1167-1175

Huang B, Lin H (2002) A Java/CGI approach to developing a geographic virtual reality toolkit on the internet. Comput Geosci 28(1):13-19

Huang B, Worboys MF (2001) Dynamic modelling and visualization on the internet. Trans GIS 5(2):131-139

Jankowski P, Stasik M, Jankowska MA (2001) A map browser for an internet-based GIS data repository. Trans GIS 5(1):5-18

Kafadar K (1996) Smoothing geographical data, particularly rates of disease. Stat Med 15:2539-2560

Kähkonen J, Lehto L, Kiolpeläinen T, Sarjakoski T (1999) Interactive visualization of geographical objects on the internet. Int J Geogr Inform Sci 13(4):429-438

Kraak MJ, Brown A (2001) Web cartography. Taylor and Francis, London

Lawson A, Biggeri A, Böhning D, Lesaffre E, Viel J-F, Bertollini R (1999) Disease mapping and risk assessment for public health. Wiley, Chichester

Marshall RJ (1991) Mapping disease and mortality rates using Empirical Bayes estimators. J App Stat 40:283-294

Messner SF, Anselin L, Baller R, Hawkins D, Deane G, Tolnay S (1999) The spatial patterning of county homicide rates: an application of exploratory spatial data analysis. J Quant Criminol 15(4):423-450

Messner SF, Anselin L, Hawkins D, Deane G, Tolnay S, Baller R (2000) An atlas of the spatial patterning of county-level homicide, 1960-1990. National Consortium on Violence Research, Carnegie-Mellon University, Pittsburgh [PA], CD-ROM

Peng Z (1999) An assessment framework for the development of internet GIS. Environ Plann B 26(1):117-132

Plewe B (1997) GIS online. information retrieval, mapping and the internet. OnWorld Press Santa Fe [NM]

Symanzik J, Cook D, Lewin-Koh N, Majure JJ, Megretskaia I (2000) Linking ArcView and XGobi: insight behind the front end. J Comput Graph Stat 9(3):470-490

Takatsuka M, Gahegan M (2001) Sharing exploratory geospatial analysis and decision making using GeoVISTA studio: From a desktop to the web. J Geogr Inform Decis Anal 5(2):129-139

Takatsuka M, Gahegan M (2002) GeoVISTA Studio: A codeless visual programming environment for geoscientific data analysis and visualization. Comput Geosci 28(10):1131-1141

Tsou MH, Buttenfield B (2002) A dynamic architecture for distributing geographic information services. Trans GIS 6(4):355-381

Wall P, Devine O (2000) Interactive analysis of the spatial distribution of disease using a geographic information system. J Geogr Syst 2(3):243-256

Zhang Z, Griffith D (2000) Integrating GIS components and spatial statistical analysis in DBMSs. Int J Geogr Inform Sci 14(6):543-566

A.10 PySAL: A Python Library of Spatial Analytical Methods

Sergio J. Rey and *Luc Anselin*

A.10.1 Introduction

This chapter describes PySAL, an open source library for spatial analysis written in the object oriented language Python. PySAL grew out of the software development activities that were part of the Center for Spatially Integrated Social Sciences Tools Project (Goodchild et al. 2000). This National Science Foundation infrastructure project had as its goals to facilitate dissemination of spatial analysis software to social sciences, to develop a library of spatial data analysis modules, to develop prototypes implementing state of the art methods, and to initiate and nurture a community of open source developers.

PySAL is a collaborative effort between Luc Anselin's research group at UIUC and Sergio Rey's research group at SDSU to develop a cross-platform library of spatial analysis functions written in Python. This combines the development activities of GeoDA/PySpace (Anselin et al. 2006) and STARS – Space Time Analysis of Regional Systems (Rey and Janikas 2006). Both will continue to exist and exploit a common library of functions.

One particular subcomponent of PySAL is referred to as PySpace, an open source software development effort focused on the implementation of spatial statistical methods in general and spatial regression analysis in particular using Python and Numerical Python. Current activities deal with a set of classes and methods to carry out diagnostics for spatial correlation in linear regression models and to estimate spatial lag and spatial error specifications.

The goal of PySAL is to leverage existing software tools development underlying GeoDA/PySpace and STARS to yield a core library and application programming interface (API) that will serve three needs. First, to avoid duplication of effort in the development of core spatial data analysis functions, the teams are collaborating on key modules that can be shared across the different projects. As a result of this reorganization, the two projects will be able to focus on increased specialization and modularization of related functionality. For example, PySpace development can focus on advanced spatial econometric methods, while STARS development can continue implementing new space-time methods, yet both will draw on jointly developed spatial weights classes. This avoids the need for separate but largely parallel efforts and also increases standardization of core classes and methods.[1] By pooling developer time on the shared weights classes, we have freed up resources that are being used for advances along specialized interests of the two projects.

The third need that PySAL seeks to address is a current void in the Python community where advanced spatial analytic modules are largely absent. While much work is being done on cartographic and GIS libraries in Python (Coles et al. 2004; Butler and Gillies 2005; Gillies and Lautaportii 2006), functionality dealing with state-of-the-art spatial statistical and spatial econometric analysis is largely absent. Filling this void is important, given the rapidly growing scientific community that has adopted Python as the language of choice.[2]

The existing Python related cartographic and GIS efforts are part of a much larger movement in Open Source Geographic Information Systems. A recent inventory of open source packages that are designed to deal with spatial data identified over 237 such efforts (Lewis 2007). However, a close examination of the objectives of the projects listed reveals that the vast majority focus on spatial data manipulation and presentation. There is still a dearth of functionality that implements spatial statistical, econometric and modeling techniques. This lack of software tools for geospatial analysis in the open source GIS movement mimics the early days of commercial GIS development. This then prompted many scholars to identify the lack of software support as an impediment for the dissemination of spatial analysis methods in empirical research (for example, Haining 1989) and led to considerable efforts to remedy the situation (for a review, see Fischer and Getis 1997, Anselin 2005). The advantage of the current open source GIS efforts is that the very open source nature of the different projects facilitates their extension and integration with other software tools. Specifically, this provides opportunities to develop geospatial analysis tools that can be readily integrated with a wide range of mapping and other GIS functionality.

PySAL is intended to fill a particular niche in the growing field of spatial data analysis software.[3] Currently there are two broad classes of implementations of spatial analysis packages. The first are those that are self-contained and implement a subset of analytical methods in user friendly graphical interfaces. Chief among

---

[1] We provide illustrations in Section A.10.3.

[2] For example, see Langtangen (2006). Also, an overview of scientific computing projects using Python is given in http://wiki.python.org/moin/NumericAndScientific.

[3] For a recent overview of the field of spatial analysis software for the social sciences see Rey and Anselin (2006).

these are GeoDa, GeoVista Studio (Takatsuka and Gahegan 2002), CommonGIS (Andrienko and Andrienko 2005; Andrienko et al. 2003) among others. At the other extreme are efforts at implementing spatial analysis methods in packages for particular programming and data analysis environments. Prominent examples here include the R-Geo project (Bivand and Gebhardt 2000) and the econometrics toolbox for MATLAB (LeSage 1999). PySAL is envisaged as supporting both types of efforts, since the Python environment lends itself to command line execution through its interpreter as well as the bundling of code in user-friendly executables with a graphical user interface.

In the remainder of the chapter we first briefly outline the overall design and main components of the library. We next provide several illustrations of how the modules in the library can be combined and delivered in a number of different ways to address various spatial analytical questions, including computational geometry, the study of spatial dynamics, smoothing of rates, regionalization, spatial econometrics and spatial analytical web services. We close with some concluding remarks.

A.10.2 Design and components

PySAL is not intended to reinvent a complete Geographic Information System. Rather, it is designed as a library that would enable sophisticated spatial analysis through various delivery formats. This ranges from simple command line interactive scripts, to self-contained packages with a graphical user interface and add-on modules to commercial off the shelf programs (for example, to augment the spatial statistical toolbox of the ArcGIS software). The functionality of the library is geared to facilitate spatial statistical exploration and spatial econometric modeling and to avoid duplication of basic GIS functionality. The modular structure of the Python language effectively allows us to build upon other efforts in geovisualization and spatial data manipulation of the open source GIS movement.

We designed the modules in PySAL to be agnostic of the delivery mechanism, so that they can flexibly be integrated with alternative GUIs (for example, Tkinter or wxPython), combined as external libraries with other software (for example, ArcGIS), or mixed and matched with existing modules developed by others. The set of components in PySAL is designed to cover all steps of a spatial data analysis process, starting with reading various data formats and carrying out basic computational geometry, and moving on to a collection of specialized methods useful in spatial exploratory analysis and modeling. Intentionally, a key feature of PySAL is that it is self-contained and does not have any tight dependencies on external libraries beyond those available within Python. At the same time, because it is a library, components of PySAL can be combined with functionality from a different GIS or analytical package to carry out specialized analyses. Moreover,

PySAL gains the high degree of portability across different platforms and operating systems inherent in the Python language.

A graphical overview of the key components of the current incarnation of PySAL is presented in Fig. A.10.1. It is organized into six main categories of functionality, dealing with basic data operations, such as the construction and manipulation of spatial weights and essential computational geometry functions, data exploration, such as clustering methods and exploratory spatial data analysis, and spatial modeling, such as spatial dynamics and spatial econometrics. Table A.10.1 provides a complementary classification of the functionality included in PySAL. Here, a distinction is made between data analytic functions, intended to ease the reading, manipulation, and writing of common spatial data formats, and ESDA and modeling functions.

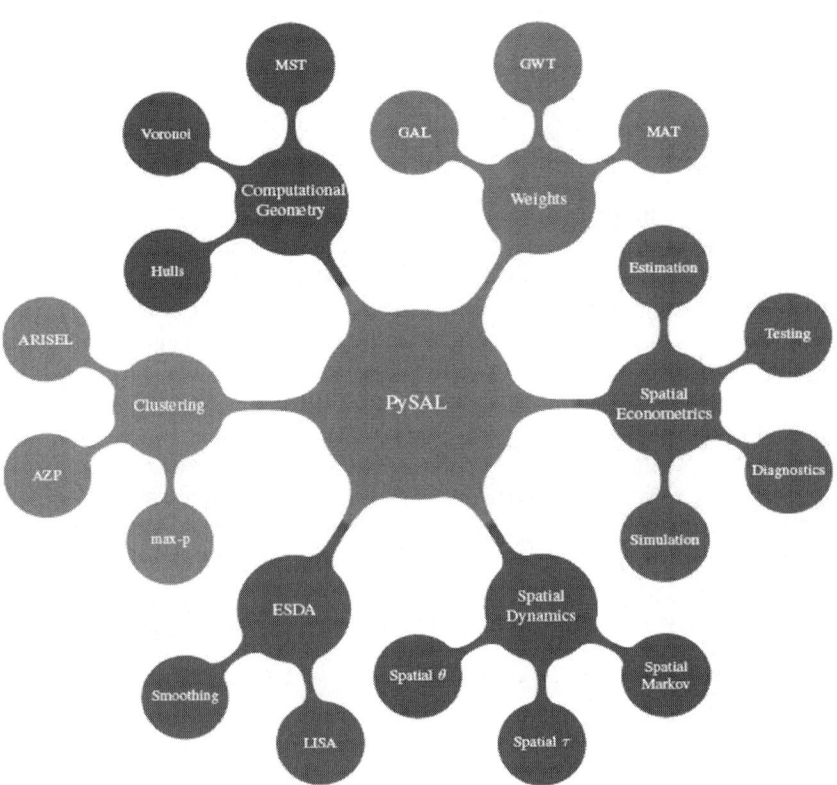

Fig. A.10.1. PySAL components

The weights module includes functionality to construct spatial weights from a range of input formats (including the standard ESRI shape files), and store the information efficiently in an internal data structure. This can then be exported to different file formats, such as the GAL and GWT formats used by GeoDa and R, and the MAT format used by the Matlab spatial econometrics libraries. The computational geometry module supports various other modules in providing basic manipulations of spatial data, such as the construction of Voronoi diagrams (Thiessen polygons), convex hulls and minimum spanning trees. These underlie the derivation of network based spatial weights as well as various computations in the clustering module.

Table A.10.1. PySAL functionality by component

Component	Capabilities
	Data analytic functions
File input-output	Read and write common spatial data formats
Map calculations	Map algebra
Computational geometry	Geometric summaries of spatial patterns
Spatial weights	Efficient construction/manipulation of spatial weights matrices
Rate Smoothing	Spatial and non-spatial smoothing of rate data
	ESDA and modeling functions
Spatial autocorrelation	Local and global spatial autocorrelation
Space-time correlation	Spatial and temporal correlation measures
Markov and mobility	Spatial Markov and distributional dynamics
Regionalization	Spatially constrained clustering
Spatial regression	Classic spatial econometric methods
Spatial panel regression	Spatial methods for panel data

Data exploration is supported by the clustering and ESDA modules. The clustering module implements a range of regionalization methods which can be used to simplify the data and provide alternatives to rate smoothing operations (in the ESDA module). They also form the basis for the construction of alternative spatial weights structures. The ESDA module contains different methods to implement the smoothing of rates as well as standard LISA functionality, such as the Moran scatter plot, local Moran and G_i statistics.

Spatial modeling is implemented in the spatial dynamics and spatial econometrics modules. The former contains a number of tools to track the change over time of spatial structure, developed with an eye towards applications in studies of regional economic convergence. These include spatial Markov analysis, as well as spatial θ and spatial τ measures of convergence. The spatial econometrics module contains a collection of diagnostics for spatial effects, specification tests and estimation methods, as well as simulation tools to embed various forms of spatial dependence in artificial data sets. Detailed illustrations of selected functionality are provided in the next section.

A.10.3 Empirical illustrations

We present a selection of applications of modules within PySAL and illustrate how they can be exposed through various delivery mechanisms, including alternative GUIs. The examples are intended to be suggestive, not exhaustive, and highlight how particular core modules, jointly developed in PySAL have been integrated into the two ongoing projects, GeoDA/PySpace and STARS.

Computational geometry and spatial weights. Figure A.10.2 contains the nearest neighbor graph for a point distribution. Here we have implemented efficient nearest neighbor algorithms for general k-nearest neighbor determination in large point sets. Combining these methods together with classes in the spatial weights module, we can generate alternative spatial weights matrices based on nearest neighbor relations for both point data sets, as well as areal/polygon data sets where representative points are used in developing the topological relationships.

The spatial weights module also supports additional graph based definitions of weights using point data. These include Gabriel, sphere of influence, and relative neighbor criteria. For polygon based shape files, the module also contains efficient classes for derivation of Queen and Rook based contiguity matrices on the fly. These classes free the user from the tedious and error-prone task of constructing weight matrices by hand. For all of these spatial weights, the associated classes implement manipulation and summarization methods that are commonly needed in spatial analysis, including measures of sparseness, connectivity, and various eigenvalue-based metrics, among many others. The weights module also supports the reading and writing of common spatial weights matrices formats including GAL, GWT and full matrices.

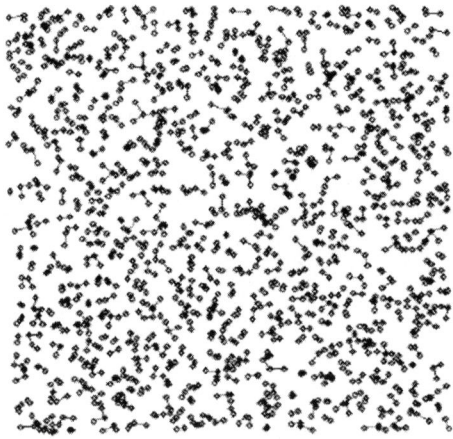

Fig. A.10.2. Nearest neighbor graphs

Spatial dynamics. With the increasing availability of spatial longitudinal data sets there is an growing demand for exploratory methods that integrate both the spatial and temporal dimensions of the data. The spatial dynamics component of PySAL implements a number of new exploratory space-time data analysis measures.

These new measures approach the issue of space-time analysis in two different ways. The first introduces a spatial dimension into what are classic measures of mobility or dynamics. For example, in the study of regional income distributions popular approaches to measure economic mobility include rank concordance statistics, rank correlation statistics, and Markov models. All of these generate indicators that summarize the amount of movement within the variate distribution over time. However, like many classic statistics they are silent about the role of geography in the dynamics. In PySAL, the spatial dynamic module implements spatialized versions of these three mobility indicators, including a spatial-τ statistic, spatial-θ (Rey 2004) and spatial Markov model (Rey 2001). Each of these methods speaks to the role of spatial clustering and context in the evolution of the distribution of interest. That is, they investigate the extent to which the dynamics of the process are spatially dependent.

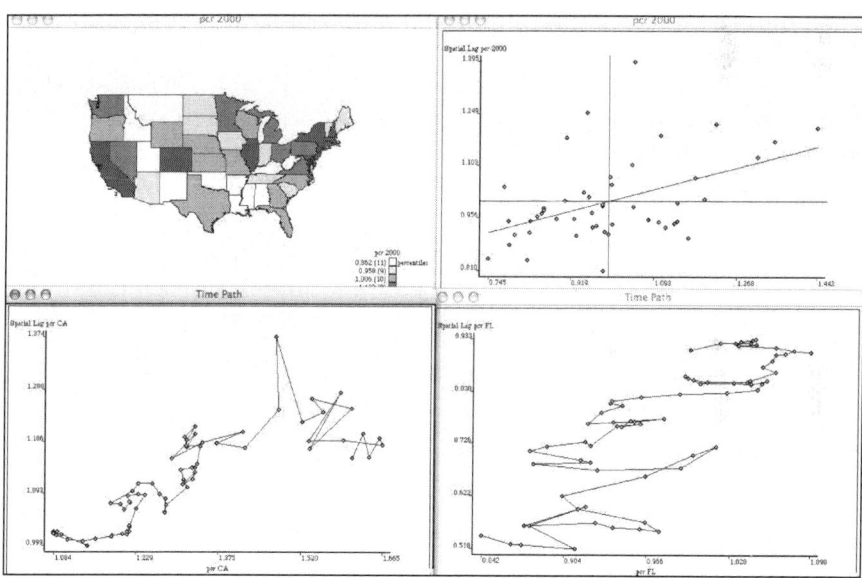

Fig. A.10.3. Spatial time paths

The second approach to spatial dynamics in PySAL starts with exploratory spatial data analysis methods and extends these measures to integrate the time dimension. One example of this is the *spatial time path*, two examples of which are shown in Fig. A.10.3. The time path can be viewed as a dynamic extension of a LISA statistic (Anselin 1995) in that the Y-axis of the graph corresponds to the value of the spatial lag of the variable while the X-axis is the original value for a particular spatial unit. In contrast to a Moran scatter plot (upper-right panel of Fig. A.10.3), which displays the (y_i, Wy_i) values for all locations at one point in time, the time path focuses on a single location i, but displays the ($y_{i,t}$, $Wy_{i,t}$) over all time periods.

These measures look at spatial dynamics from a slightly different perspective from the first in that they focus on the spatial dimension and explore its evolution over time. They can be used for comparative analyses, such as in Fig. A.10.3 where the paths for per capita incomes for California (bottom left) and Florida (bottom right) are contrasted. The spatial dynamics for Florida are more erratic than is the case for California. At the same time, a casual glance suggests the relationships are similar in that there is positive correlation between each state's income and that of its regional neighbors over time. However, by exploiting the interactive capabilities of the software, temporal animation reveals that the directionality of the dynamics is different in the two cases with Florida and its neighbors moving upward towards the center of the distribution, while California and its neighbors are moving downwards towards the mean.

In addition to the time paths, the spatial dynamics module includes a number of other new measures that are extensions of ESDA methods to incorporate time. These include a bi-variate LISA which allows for consideration of space-time lags between two different variables as well as space-time principal components which is a multivariate extension of the bi-variate LISA.

As with most of the modules in PySAL, the spatial dynamics classes can be combined with other modules to accomplish a complex analytical task. An example of this is seen in Fig. A.10.4 where a new type of spatial weights matrix is obtained through a consideration of the time series covariance of per capita incomes for each pair of states over a 72 year period. The join structure for the original simple contiguity matrix is presented as a simple network, yet each join is now colored to signify if that pair of states displays strong (dark grey) or weak (light grey) temporal co-movement. A hybrid contiguity matrix could be defined by only using the strong links. Also included on the figure is the spider graph for Colorado. These dark grey links show which states Colorado has its strongest temporal correlation with. This suggests a second type of hybrid contiguity matrix based on the intersection of the simple contiguity and the spider contiguity joins.

Smoothing of rates. An important aspect of exploratory spatial analysis of rates or proportions is to correct for the inherent variance instability of the rates. Ignoring this aspect may lead to spurious indications of outliers and clusters due to higher variance when the population at risk is small. Several techniques for smoothing rates have been incorporated into PySAL modules. They consist of a porting of the rate smoothing functionality in GeoDa (implemented in C++) to Python (for a more extensive discussion, see also Anselin et al. 2004, 2006).

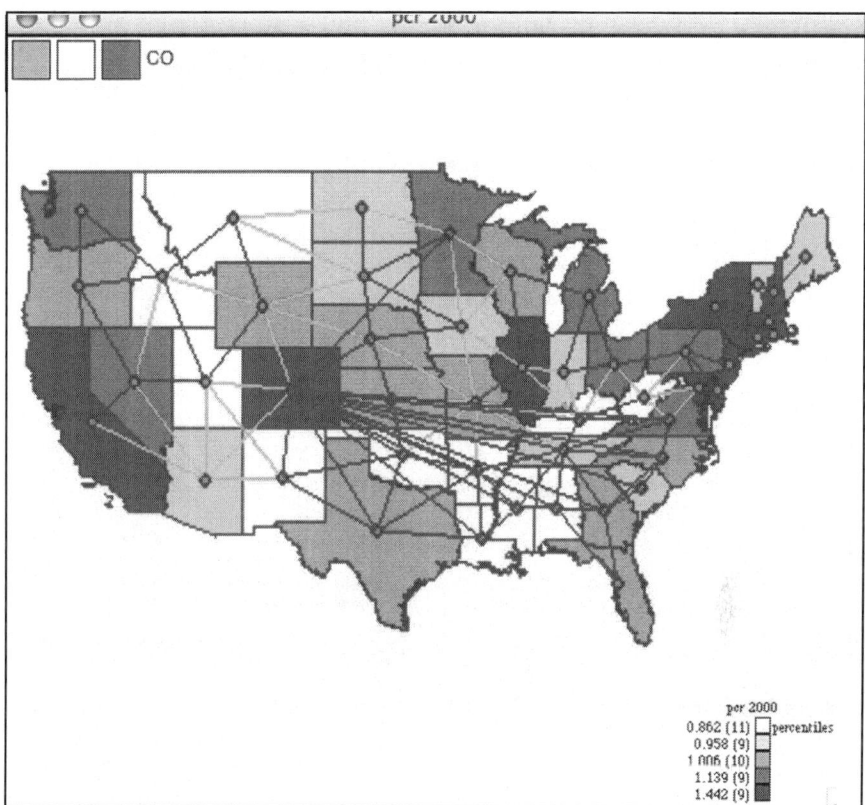

Fig. A.10.4. Spider and temporal contiguity graphs

Functionality of the rate smoothing modules can be classified into three major categories: data input, rate computation, and smoothing. The first includes the capacity to read in data on counts of events (e.g., number of diseased persons) and population at risk from various file formats, including SEER, either as aggregates or by age group. Rate computation takes the data and computes rates for individual spatial units (e.g., counties) as well as for aggregates (e.g., all the counties in a state) and implements both direct and indirect age standardization. Rate smoothing implements a number of common methods, including Empirical Bayes and spatial rate smoothing. The latter is an interesting instance where the modular nature of PySAL is exploited, since it requires functionality from the spatial weights module to implement the spatial averaging of rates.

Figure A.10.5 illustrates an application of spatial rate smoothing to age-standardized prostate cancer rates in counties covered by the Appalachian Cancer Network. This application utilizes the core rate manipulation and smoothing functionality of the library coupled to a graphical front end implemented in wxPython. This is an example of delivery of the functionality where the user is completely shielded from the Python programming environment, even though it is readily accessible if desired.

The wxPython graphical user interface is cross-platform and provides a local look and feel on each platform. It consists of a Python wrapper around the well known C++ wxWidgets library. In Fig. A.10.5, the particular look and feel is that of the Mac OS X operating system. Using simple menus, the user can select the data, spatial weights (for spatial rate smoothing) and smoothing technique and the result is presented on a map, as shown in the figure. Functionality such as this can also be readily delivered in compiled form, in which case the user no longer would have access to the original source code.

The same smoothing modules can also be used in conjunction with a different graphical user interface. For example, rate smoothing is included in STARS, which uses the Tkinter Python GUI. In addition, using the command line in with the Python interpreter, specific smoothing functions can be used individually in an interactive computing environment.

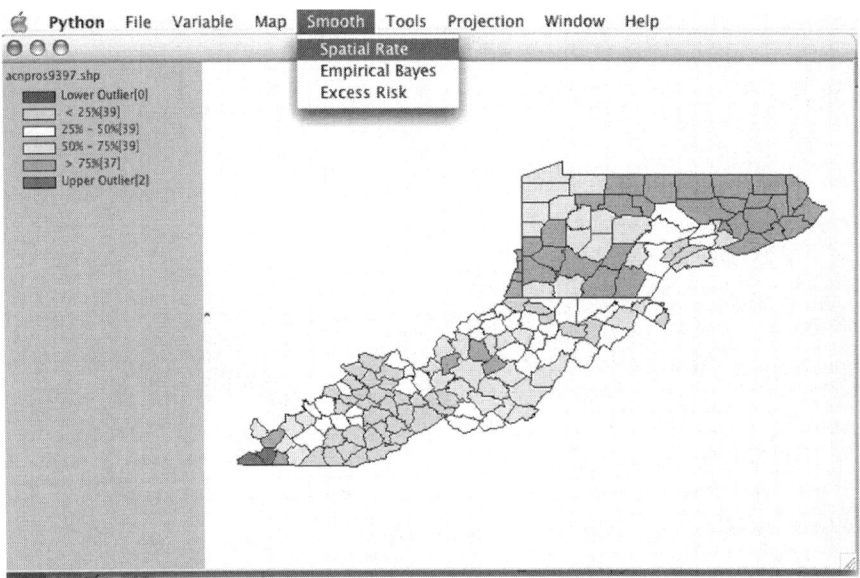

Fig. A.10.5. Spatial smoothing of ACN county prostate rates

Regionalization. The regionalization and clustering module of PySAL implements a number of new and existing methods that can be used to define groupings of fundamental units according to a variety of constraints. These methods include contiguity constrained clustering, Automatic Zoning Procedure (AZP) and the max-p region algorithm (Duque et al. 2007). Figure A.10.6 demonstrates the application of the AZP method to U.S. income dynamics.

The regionalization module can also be used together with other modules in PySAL to develop new approaches to spatial analytical problems. One example is the integration of the spatially constrained clustering algorithms together with the spatial smoothing module to develop new approaches towards spatial rate estimation (Rey et al. 2007). This work explored alternative ways in which the variance instability problem could be addressed by defining the neighborhood smoothing regions using the constrained clustering algorithms.

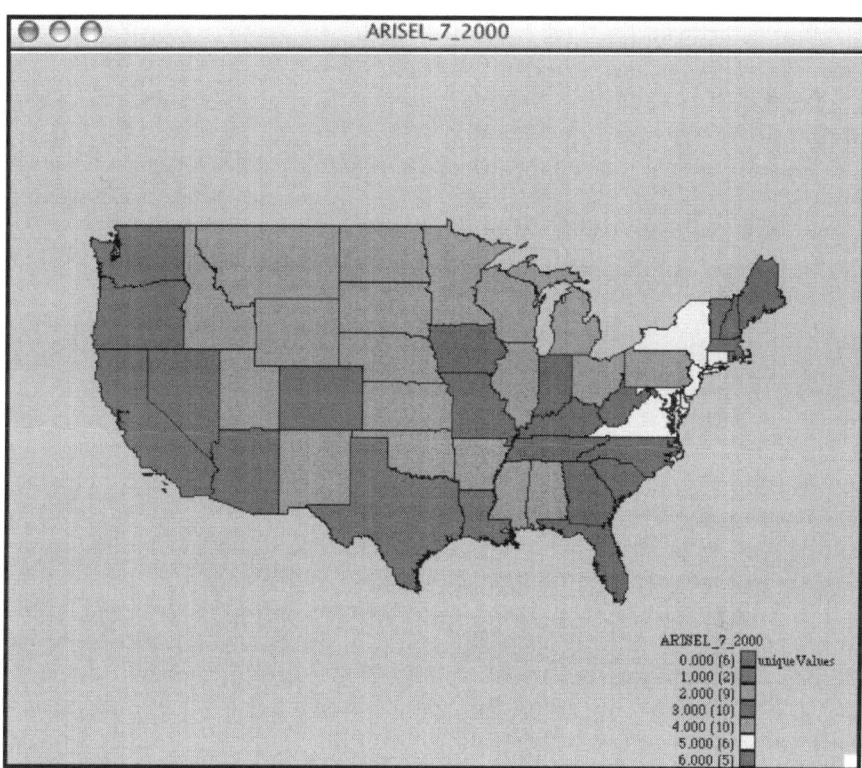

Fig. A.10.6. Regionalization of State incomes using AZP

Spatial econometrics. The spatial econometric modules in PySAL are primarily intended to provide support for two types of activities: (i) to allow rapid prototyping of newly suggested techniques; and (ii) to put together customized combinations of tests and estimation methods. The development efforts are focused on general method of moments estimators, semi-parametric approaches, spatial panel data models and specifications with discrete dependent variables. In this sense, these modules complement the spatial econometric functionality of GeoDa which is aimed at providing a user-friendly environment for more established spatial econometric techniques, such as Maximum Likelihood estimation.

For example, PySAL implements code to estimate regression models containing a spatially lagged dependent variable (a spatial lag model) by means of the spatial two stage least squares method (Anselin 1988; Kelejian and Prucha 1998). In addition to the traditional estimates of standard errors and a heteroskedastic robust form (White 1980; Anselin 1988), this also implements the recently suggested heteroskedastic and spatial autocorrelation robust form, or HAC estimator (Kelejian and Prucha 2007). The latter takes a non-parametric approach to allow for remaining spatial error autocorrelation of unspecified form, using a kernel estimation method.

The PySAL code for the HAC estimator was recently applied in Anselin and Lozano-Gracia (2008) to estimate a spatial hedonic model with over 100,000 observations, using a spatial lag model that included other endogenous variables as well. In addition to allowing for remaining spatial error autocorrelation in a spatial lag model, the spatial two stage least squares approach in PySAL is also not constrained to intrinsically symmetric spatial weights, as is the case for the ML estimators in GeoDa.

Figures A.10.7 to A.10.9 illustrate an application of the spatial econometric module to a replication of the analysis of U.S. county homicides in Baller et al. (2001). The implementation uses the command line only, taking the model specification information from a separate module that contains all the information on the data set, variables and spatial weights. For example, in Fig. A.10.7 the contents of such a model are shown, including a dictionary for the model variables and for the data (respectively, *spec* and *data*), as well as two lists of dictionaries with spatial weights needed for the spatial lag (*mweights*) and for the kernel estimation (*kweights*). Each of these dictionaries contains several attributes of the data and weights needed by the modules that implement data input and spatial weights construction. The module can be edited by means of a text editor and *imported* into the current session to be used by the spatial regression module. In the current example, an asymmetric spatial weights matrix for five nearest neighbors is used to construct the spatial lag.

The central element in the spatial econometric functionality is the *spmodel* class, similar in concept to the object-oriented design of model classes in the R language. Figure A.10.8 illustrates the construction of an object *model* of the *spmodel* class in the *spreg* module. Some of the arguments that are passed to the constructor include a data object (*spreg.db*), a model specification object (*spreg.spec*) as well as weights objects and some model options, e.g., the specification of a *lag* spatial model, using *gmm* as the estimation method and hac as the option for the variance-covariance estimator. Once the model object is created, its attributes can be accessed using the familiar dot notation. For example, in Fig. A.10.8, the name of the input data set, number of observations, number of variables, the model specification and the spatial weights are illustrated. Note how the spatial weights are themselves instances of the *weights* class constructed in the spatial weights modules.

The estimation results are obtained by invoking one of the methods in the *spmodel* class. In Fig A.10.9 this is illustrated for the *twosls* method. It is invoked on the command line by means of the dot notation, applied to the *model* instance of the *spmodel* class. This yields the output of the estimates, standard errors and measures of fit, in the familiar GeoDa format. Three tables are listed, for the traditional standard errors, the heteroskedastic robust form and the HAC. The latter is implemented using an Epanechnikov kernel function with an adaptive bandwidth for the 20 nearest neighbors. The standard errors increase slightly relative to the classic estimate.

```
# spec: model specification: y dep var, X exogenous, yend endogenous
#              H instruments
spec = {}
spec['y'] = 'HR90'
spec['X'] = ['RD90', 'PS90', 'MA90', 'DV90', 'UE90', 'SOUTH']

# data: data source
data = {}
data['fname']='natn.csv'
data['idvar']='FIPSNO'
data['dType']='listvars'
data['formatheader']= 0
data['numonly']=-0

# mweights: spatial weights for use in model lag 0 error 1
# if different
mweights = []
mw = {}
mw['wtfile'E-'natkS gwt'
mw['wtType']-'binary'
mw['headline']=0
mw['sep']='.'
mw['rowstand']=1
mw[' power]=1
mw['dmax]=0
mweights.append(mw)

# kweights: kernel weights
# if none specified, no kernel
kweights = []
kw = {}
kw['wtFile']= 'natk20.gwt'
kw['wtType']='epanech'
kw['headline']=0
kw['sep']='.'
kw['rowstand']=0
kw['power'J=1
kw['dmax']=0
kweights.append(kw)
```

Fig. A.10.7. Spatial regression model specification

```
>>> model = spreg.spmodel(spreg.db,spreg.spec,mweights=spreg.mw1,
... kweights=spreg.kw1,space='lag',method='gmm',option='hac')
>>> model.fname
'natn.csvV
>>> model.nobs
3085
>>> model.k
8
>>> model.spec
{'y': 'HR90', 'X' : ['RD90', 'PS90', 'MA90', 'DV90', 'UE90', 'SOUTH']
>>> model.mw1
<weights.spweight instance at 0x1641670>
>>> model.kw1
<weights.spweight instance at 0x14ce8f0>
```

Fig. A.10.8. Spatial regression model object attributes

```
>>> model.twosls()
Data: natn.csv N: 3085 df: 3077
Dependent Variable: HR90
Instruments: W_RD90 W_PS90 W_MA90 W_DV90 W_UE90 W-SOUTH
Spatial Weights: natk5.gwt Type: binary
Kernel Weights: natk20.gwt Type: epanech
RZ (var): 0.44097474      R2 (corr): 0.44015616
2SLS Results
CONSTANT       5.27970466     1.05421367      5.0081.9218     5,8040651e-07
RD90           3.70854698     0.14648314     25.31722759     1.2849701e-128
PS90           1.37504128     0.10015256     13.72946666     1.11636e-41
MA90          -0.08427221     0.02755448     -3.05838455     0.0022444905
DV90           0.54414517     0.05500481      9.89268364     9,7467151e-23
UE90          -0.28049426     0.04118603     -6.81042228     1.1655819e-11
SOUTH          1.31132254     0.28790335      4.55473172     5.4490157e-06
W-HR90         0.18870532     0.03971433      4.75156802     2.1109551e-06
Data: natn.csv N: 3085 df: 3077
Dependent Variable: HR90
Instruments: W_RD90 W_PS90 W_MA90 W-DV90 W_UE90 W_SOUTH
Spatial Weights: natk5.gwt Type: binary
Kernel Weights: natk20.gwt Type: epanech
R2 (var): 0.44097474      RZ (corr): 0.44015616
2SLS Results, White Variance
CONSTANT       5.27970466     1.04716434      5.04190649     4.8759527e-07
RD90           3.70854698     0.22598487     16.41059827     4.3949434e-58
PS90           1.37504128     0.16795804      8.18681414     3.8851091e-16
MA90          -0.08427221     0.02794873     -3.01524329     0.0025887092
DV90           0.54414517     0.08031388      6.77523167     1.4823565e-12
UE90          -0.28049426     0.05110650     -5.48842574     4.384525e-08
SOUTH          1.31132254     0.29345646      4.46854213     8.1594953e-06
W_HR90         0.18870532     0.04286644      4.40216873     1.1082167e-05
Data: natn.csv    N: 3085   df: 3077
Dependent Variable: HR90
Instruments: W_RD90 W_PS90 W_MA90 W_DV90 W_UE90 W_SOUTH
Spatial Weights: natk5.gwt Type: binary
Kernel Weights: natk20.gwt Type: epanech
R2 (var): 0.44097474      R2 (corr). 0.44015616
2SLS Results, HAC Variance with kernel epanech
CONSTANT       5.27970466     1.08618007      4.86080053     1.2276991e-06
RD90           3.70854698     0.24481531     15.14834572     4.7483522e-50
PS90           1.37504128     0.17801139      7.72445672     1.5089075e-14
MA90          -0.08427221     0.02796515     -3.01347276     0.002603817
DV90           0.54414517     0.08076990      6.73697932     1.9225394e-11
UE90          -0.28049426     0.05241424     -5.35148941     9.363273e-08
SOUTH          1.31132254     0.31097070      4.21686840     2.5485898e-05
W_HR90         0.18870532     0.04587635      4.11334626     4.0018136e-05
Anselin-Kelejian Test for Residual Spatial Autocorrelation
Moran's 1: -0.0552 LM: 11.02 p: 0,000903468
```

Fig. A.10.9. Spatial two stage least squares with HAC error variance

In the example, one diagnostic is included by default (it can also be invoked separately as a method of the *spmodel* class), the Anselin and Kelejian (1997) generalized Moran's *I* test for residuals in a spatial lag model. As shown in Fig. A.10.9, the null hypothesis is strongly rejected, providing a solid motivation for the use of the HAC standard errors.

Spatial analytical Web services. The core libraries are designed in such a way as to enable a variety of front ends through which users can interface with the functionality in PySAL. In previous examples, we have illustrated the use of two different GUIs and the shell/command line. A third form of user interface is the web browser, where the PySAL functionality is delivered in the form of a spatial analytical web service.

A straightforward way to accomplish this is to include components of the library as cgi (common gateway interface) scripts on a web server. The user interacts with this through a web page, which sends a form to the server that includes all the parameters needed to carry out the analysis. The results are then delivered as a new web page. To the user, the experience is similar to an interactive GUI on the desktop.

A more elaborate form of a web interface can be developed by exploiting the HTTP and SOAP (simple object access protocol) Web service functionality built into the Python language and extension modules. Figure A.10.10 illustrates the architecture of a prototype spatial analytical Web service to construct spatial weights from ESRI shape files, using standards supported by the Open GIS Consortium (OGC). This combines three components, that each can operate on a different physical server, allowing for a distributed system.

The information on the data source and weights type is then passed to the Analysis Server, using the SOAP protocol. This back end operation consists of a set of Python scripts to handle the interaction between the different services and to interface with the PySAL library for the actual computation of the weights. The

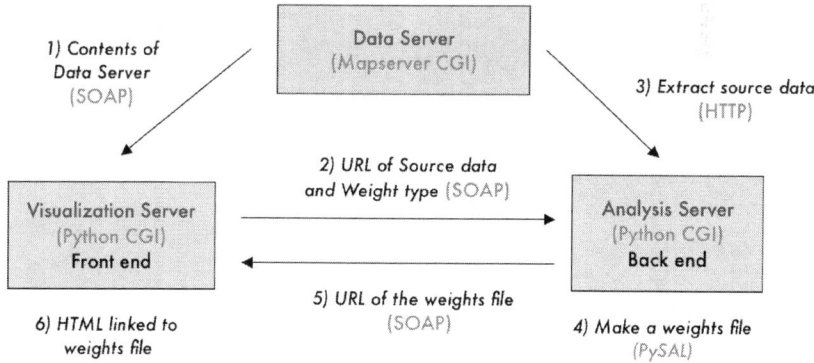

Fig. A.10.10. Architecture of spatial weights Web service

Data

Enter the url where your shp file resides with http:// or https://:
(ex: http://geog36.geog.uiuc.edu/~myunghwa/PyWebSpace/out_data/ohlung.shp or https://netfiles.uiuc.edu/mhwang4/www/ohlung.shp)

Or select one of the shp files from our data server.

[ohlung.shp]

Weight Type

[Rook]

[Make]

Fig. A.10.11. Weights Web service user interface

data are extracted from the data server, the weights are computed and stored on the analytical server and the URL of this location is passed back to the user interface. The weights information can also be transferred in other ways, using a standard XML format, as illustrated in Fig A.10.12.

```
- <ExecuteResponse version="0.4.0" xsi:schemaLocation="http://www.opengeospatial.net/wps
   http://www.bnhelp.cz/schema/wps/0.4.0/wpsExecute.xsd">
     <ows:Identifier> makeWeight </ows:Identifier>
   - <Status>
       <ProcessSucceeded/>
     </Status>
   - <ProcessOutputs>
     - <Output>
         <ows:Identifier> outputWeight </ows:Identifier>
         <ows:Title> Resulting weight file </ows:Title>
         <!--Element Abstract not set-->
       - <ComplexValue format="text/xml">
         - <Value>
           - SAL:weightfile inputfile="inputfile" numRec= "88" type="GAL" wtype="Rook Contiguity">
             - <SAL:record id = "1" numNeighbors="4">
                 <SAL:neighbors> 2,6,7,11 </SAL:neighbors>
               </SAL:record>
             - <SAL:record id = "2" numNeighbors="3">
                 <SAL:neighbors> 1,4,11 </SAL:neighbors>
               </SAL:record>
             - <SAL:record id = "3" numNeighbors="5">
                 <SAL:neighbors> 5,10,15,87,88 </SAL:neighbors>
               </SAL:record>
             - <SAL:record id = "4" numNeighbors="3">
                 <SAL:neighbors> 2,11,13 </SAL:neighbors>
```

Fig. A.10.12. Weights in XML format

The front end is the web interface (shown in Fig. A.10.11) through which the user interacts with the system by means of a set of Python cgi scripts that manage information flows between the front end and the two other components, the Data Server and the Analysis Server. Through the interface a web feature service (Data Server, using the Mapserver cgi) is queried for a list of available data sources, which then become available in a drop down list on the web interface, transparent to the user. This could easily be generalized to query a collection of web feature services for available data sets. Alternatively, users can specify the URL for the data source explicitly, which can be anywhere on the internet, including other compliant web feature services. In addition, the type of weights matrix (Rook or Queen) can be selected.

A.10.4 Concluding remarks

The main efforts thus far have been on the development of the core analytical functionality and coupling these modules with the graphical toolkits used in the two source projects: Tkinter for STARS and wxPython for OpenGeoDa/PySpace. Future work will explore use of PySAL with alternative front-ends including jython (Pedroni and Rappin 2002), RPy (Moriera and Warnes 2004), and ArcGIS. Additionally, we are investigating alternative shell/command line environments beyond the basic Python interpreter, such as iPython (Pérez 2006). At the same time we will regularly be integrating new developments in spatial analysis into the computational classes within PySAL.

Our plans are to continue refining the core components of the library and the associated application programming interface (API). We are also evaluating alternative licensing schemes with an eye towards leveraging the strengths of the open source and spatial analysis communities. We envisage a formal release of PySAL in the near future.

References

Andrienko GL, Andrienko NV (2005) Exploratory analysis of spatial and temporal data: a systematic approach. Springer, Berlin, Heidelberg and New York
Andrienko GL, Andrienko NV, Voss H (2003) GIS for everyone: the CommonGIS project and beyond. In Peterson MP (ed) Maps and the internet, Elsevier, Amsterdam, pp.131-146
Anselin L (1988) Spatial econometrics: methods and models. Kluwer, Dordrecht
Anselin L (1995) Local indicators of spatial association (LISA). Geogr Anal 27(2):93-116.
Anselin L (2005) Spatial statistical modeling in a GIS environment. In Maguire DJ, Batty M, Goodchild MF (eds) GIS, spatial analysis and modeling. ESRI Press, Redlands [CA], pp.93-111
Anselin L, Kelejian HH (1997) Testing for spatial error autocorrelation in the presence of endogenous regressors. Int Reg Sci Rev 20(1-2):153-182

Anselin L, Kim YW, Syabri I (2004) Web-based analytical tools for the exploration of spatial data. J Geogr Syst 6(2):197-218

Anselin L, Lozano-Gracia N (2008) Errors in variables and spatial effects in hedonic house price models of ambient air quality. Empirical Economics 34(1):5-34

Anselin L, Syabri I, Kho Y (2006) GeoDa: an introduction to spatial data analysis. Geogr Anal 38(1):5-22

Baller R, Anselin L, Messner S, Deane G, Hawkins D (2001) Structural covariates of U.S. county homicide rates: incorporating spatial effects. Criminology 39(3):561-590

Bivand RS, Gebhardt A (2000) Implementing functions for spatial statistical analysis using the R language. J Geogr Syst 2(3):307-317

Butler H, Gillies S (2005) Open source Python GIS hacks. In Open Source Geospatial '05, Minneapolis

Coles J, Wagner J-O, Koormann F (2004) User's manual for thuban 1.0. Technical report, Intevation GmbH

Duque JC, Anselin L, Rey SJ (2007) The max-p region problem. Regional Analysis Laboratory Working Paper 20070301

Fischer MM, Getis A (eds) (1997) Recent developments in spatial analysis. Springer, Berlin, Heidelberg and New York

Gillies S, Lautaportti K (2006) Python Cartographic Library (PCL). http://trac.gispython.org/projects/PCL/wiki

Goodchild MF, Anselin L, Appelbaum RP, Harthorn BH (2000) Toward spatially integrated social science. Int Reg Sci Rev 23(2):139-159

Haining R (1989) Geography and spatial statistics: current positions, future developments. In Macmillan B (ed) Remodelling geography. Blackwell, Oxford, pp.191-203

Kelejian HH, Prucha IR (1998) A generalized spatial two stage least squares procedures for estimating a spatial autoregressive model with autoregressive disturbances. J Real Estate Fin Econ 17(1):99-121

Kelejian HH, Prucha IR (2007) HAC estimation in a spatial framework. J Econometrics 140(1):131-154

Langtangen HP (2006) Python scripting for computational science. Springer, Berlin, Heidelberg and New York

LeSage JP (1999) Econometrics toolbox. Technical report, Department of Economics, University of Toledo

Lewis B (2007) Open source GIS. Technical report, OpenSourceGIS.org.

Moriera W, Warnes G (2004) 'rpy': a robust Python interface to the R programming language. http://ryp.sf.net.

Pedroni S, Rappin N (2002) Jython essentials. O'Reilly, Sebastopol [CA]

Pérez F (2006) IPython: an enhanced interactive Python. Department of Applied Mathematics, University of Colorado at Boulder

Rey SJ (2001) Spatial empirics for economic growth and convergence. Geogr Anal 33(3):195-214

Rey SJ (2004) Spatial dependence in the evolution of regional income distributions In Getis A, Múr J, Zoeller H (eds) Spatial econometrics and spatial statistics. Palgrave Macmillan, Hampshire and New York, pp.194-213

Rey SJ, Anselin L (2006) Recent advances in software for spatial analysis in the social sciences. Geogr Anal 38(1):1-4

Rey SJ, Janikas MV (2006) STARS: Space-time analysis of regional systems. Geogr Anal 38(1):67-86

Rey S, Anselin L, Duque J, Li X (2007) Max-p region based estimation of disease rates. Regional Analysis Laboratory Working Paper 20070420

Rey S, Duque J, Smirnov O, Kim Y, Stephens P (2005) Identifying value-added industry clusters in San Diego County. Technical report, Regional Analysis Laboratory Technical Report: REGAL 20050616

Takatsuka M, Gahegan M (2002) GeoVista Studio: a codeless visual programming environment for geoscientific data analysis and visualization. J Comp Geosci 28(10):1131-1144

White H (1980) A heteroskedastic-consistent covariance matrix estimator and a direct test for heteroskedasticity. Econometrica 48(4):817-838

Part B

Spatial Statistics and Geostatistics

B.1 The Nature of Georeferenced Data

Robert P. Haining

B.1.1 Introduction

Georeferenced data or spatial data (we use the terms interchangeably here) come in many forms. Geometrically speaking, such data refer either to points, lines or areas – spatial objects or features. Spatial interaction data record flows between the nodes (intersection points) of a network. These data are captured in an origin-destination matrix where the number of rows and columns of the matrix correspond to the nodes of the network and the entry on row i and column j records the total flow from node i to node j (Fischer 2000). Spatial tracking data records the movement of individuals (or groups) over time between areas or the nodes of a network (Goodchild 1998; Frank et al. 2001). The rows of the tracking matrix are the individuals, the columns are time periods and the entry on row i and column j records the location of individual i in time period j. These data can be used to estimate transition matrices where the entry on row i and column j of the transition matrix records the probability of any individual going from area i to area j in an interval of time (Wilson and Bennett 1985, pp.107-109 and pp.250-280). In these two cases the spatial objects (nodes, network links, areas) remain fixed – and motion takes place over this static spatial backdrop – but over time the point, line and area features themselves can for example move, grow, shrink, split and change form (Frank 2001).

It is another type of spatial data, which records attributes associated with spatial features (points or areas), that will be the focus here. It has the following generic form

$$\{z_j(s_i, t): j = 1, ..., k; \ i = 1, ..., n; \ t = 1, ..., T\} \equiv \{z_j(s_i, t)\}_{j,i,t} \tag{B.1.1}$$

where Z_j denotes the jth attribute (of which there are k) and the use of the lower case, z_j, denotes the measured value of the jth attribute. The terms s_i and t denote the ith point/area (of which there are n) and the tth time period (of which there are T) and these define the locations and time periods to which each attribute value refers. The georeferencing is associated with s_i. In the case of an area there may be further information contained in Eq. (B.1.1) which provides data, for each i, on those areas that are adjacent to or are the 'neighbors' of i, thus writing: $\{z_j(s_i, t), N(i)\}_{j,i,t}$. There would be for each i then, a listing of the neighbors of i which are denoted here as $N(i)$. These neighbors might, for example, be all the areas that share a common border with i, or those which have direct transport connections with i. So the neighbor data may reflect geometric properties of the set of areas but they could also capture interaction flows between the areas, or hierarchical relationships (Haining 1978). $\{N(i)\}$ is used in constructing the 'weights' matrix (usually denoted W) that appears in the specification of many spatial statistical techniques (for example, spatial autocorrelation statistics) and models (spatial regression models). The correspondence between a map and two types of weights matrix is shown in Fig. B.1.1.

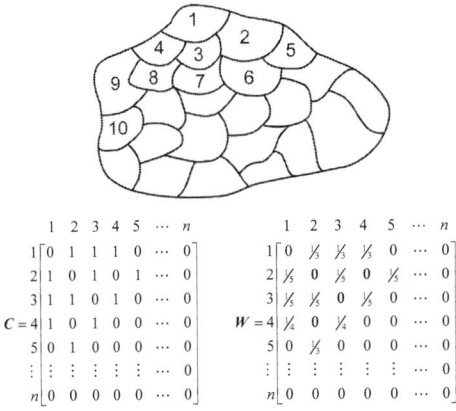

Fig. B.1.1. Binary and standardized connectivity matrices based on area adjacencies

The expression given by Eq. (B.1.1) has often been referred to in the quantitative geography literature as the *space-time data cube* (with or without data on $N(i)$). When t is fixed then in the geographic information science literature $\{z_j(s_i)\}_{j,i}$ or $\{z_j(s_i), N(i)\}_{j,i}$ is sometimes referred to as the *spatial data matrix* (SDM) with n rows corresponding to the set of locations and k columns referring to the measured attributes. As Goodchild and Haining (2004, p.365) remark, '*GIS and spatial data analysis come into contact ... at the spatial data matrix*'. A shapefile is a digital vector storage format for storing geometric location and associated attribute information for GIS. It has clear links with the SDM. The .shp file stores the feature geometry and the .dbf file the data on the attributes associated with each spatial object. Adjacencies and hence 'neighbors' can be calculated in GIS even though the geometric information contained in the shapefiles is not explicitly topological.

SDM attribute data may refer to properties of discrete entities such as people, businesses or houses assigned to point locations. Attribute data may also be in the form of counts of individuals located in a set of areas that possess a certain property (for example, total number of people; the number employed in the tertiary sector). Here, data values are only true of entire areas and the variable is called 'spatially extensive' (Goodchild and Lam 1980). A quantity expressed as a ratio, proportion or density (for example, population density; proportion of the workforce employed in the tertiary sector) is called 'spatially intensive' (Goodchild and Lam 1980). Spatially intensive data values could be true for every part of an area (if the variable was distributed uniformly). Attributes that are distributed continuously over an area are also defined as 'spatially intensive' – such as the average level of air pollution or rainfall. Often these attributes are calculated for arbitrarily constructed areas but there are other spatial attributes that can only be defined at the ecological or group level and only for areas or places that are in some sense well-defined – an example of such a variable is the level of social capital in an urban neighborhood or community. This quantity cannot be reduced to the level of individuals nor is it continuously distributed, rather it is an attribute of an area that has some functional meaning or significance (in this case a 'community').

Another type of spatial data refers to the properties of areas as they relate to each other, for example quantities such as the 'distance' or 'direction' from one place to another. It may be possible to extract such data (including adjacency data) directly from the geo-reference associated with $\{s_i\}_i$. There are also relational properties that combine attribute values and the spatial relationships between those values such as data on gradients (for example, the difference in material deprivation between two adjacent areas) and data on local area averages (for example, spatial averages based on sliding windows of different sizes over a map).

B.1.2 From geographical reality to the spatial data matrix

In this section we describe the transformational processes that turn 'real, continuous and complex geographic variation' (Goodchild 1989, p.108) into a finite number of discrete 'bits' of data that can be stored in a computer [see Eq. (B.1.1)]. Figure B.1.2 shows the sequence of stages associated with the construction of the SDM for the three elements (attributes, space, time) although we are only interested here in the stages associated with the locational data.

The object and field views are the two fundamentally distinct conceptualizations of geographical reality. The object view of the world conceptualizes space as populated by well-defined, indivisible and homogeneous entities (points, lines and polygons) set in an otherwise 'empty' space. The field view of the world conceptualizes space as covered by continuous surfaces. The object view is often used to conceptualize social, economic and demographic data (houses, people, factories, roads, towns) whilst the field view is often used to conceptualize environmental and physical data (rainfall, pollution, elevation) – although there is always some choice involved (Burrough and McDonnell 1998, p.20).

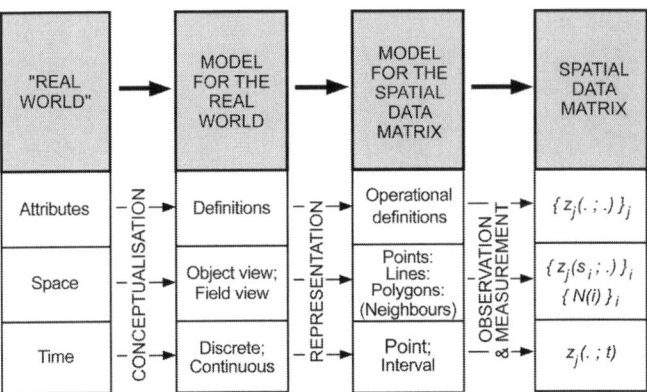

Fig. B.1.2. Stages in the construction of the spatial data matrix

The process of representation is the process of discretizing space that is, reducing these conceptualizations to a finite number of geometric entities that can then be stored in the computer usually as points, lines and polygons. A field view can be discretized using sample points, contour lines, regular polygons (for example, pixels obtained from remote sensing) or irregular polygons (for example, vegetation patches). An object view is captured using the same geometric objects but these objects are not always the same as the basic spatial entities that make up the object view. In a national Census, individual households (points) are aggregated into census tracts (polygons, often irregular in shape) for confidentiality reasons. These processes which discretize space inevitably involve a loss of information on spatial variability due to smoothing of attribute values and simplification of objects, with 'fuzzy' boundaries often becoming sharper and smoother than in reality they are. In any particular application this raises the question as to the quality of the model as a representation of the underlying geographical reality.

The entities that are created by this process and which discretize geographic space should ideally be *well-defined*. If there is individual level data then although an individual person is well defined, providing a georeference may raise problems because people move about daily and may live in different places over the course of their lives. Georeferencing people by their current residence might be satisfactory in the context of delivering health services to a population, but less satisfactory in the context of assessing population exposure to an environmental risk factor associated with a chronic disease.

The entities created by the process of discretizing geographic space should ideally be *internally homogeneous* in terms of their attributes. This will rarely be the case and the larger the scale, or resolution, at which polygons are constructed the more geographic variation will be smoothed. The choice of where to draw polygonal boundaries also have important implications for how attribute variation will appear on a map. If, subsequently, polygonal units are further aggregated into regional clusters through some process of region building then intra-area heterogeneity is likely to become still more marked. The usual benchmark, in assessing model quality, is whether the model is 'fit-for-purpose' and this can only be assessed against the particular application. These examples are designed to underline the problems inherent in creating a model for storing data about the world but they constitute only part of the larger process defined in Fig. B.1.2. Model quality assessment needs also to take into account the processes of conceptualization and representation applied to the definition of the attributes and the handling of the temporal dimension of Eq. (B.1.1). These will not be discussed here but see for example Haining (2003, pp.57-61).

The final step in the creation of the SDM is the process of measurement by which attributes at particular locations in space and time are assigned values (see Fig. B.1.2). Data values may be obtained by a sampling process or by a complete enumeration (for example a national Census of population). One way of thinking about the relationship between what is being measured and the recorded data value, is that any data value is thought of as an approximation (subject only to measurement error) to some 'true' value of the attribute at the particular space-time location. A different view of this relationship is that any attribute value is only one possible value from a distribution of possible data values (the so-called 'superpopulation' reading of spatial data). Underlying this latter view of data values is the assumption that the underlying data generating model is stochastic – to which may be added additional variation due to measurement error. This view of spatial data is common in many areas of statistics (see Cressie 1993) and geostatistics (Matheron 1963). Even Census data is sometimes analyzed with reference to a superpopulation. Godambe and Thompson (1971) noted that analysis of UK Census data is rarely concerned with the finite *de facto* population of the UK at a given point in time but rather with a conceptual superpopulation of people like those living in the UK on the date of the Census.

Data is classified by the level of measurement achieved: nominal, ordinal, interval or ratio. The level of measurement determines what logical and arithmetic operations can be performed on the data and hence, for example, what statistical procedures can be used. Nominal data allow data values to be compared using the operations: equal and not equal; ordinal data also allow ranking (greater than and less than); interval data also allow the operations of addition and subtraction; ratio data also allow the operations of multiplication and division. This provides a basis for a two way classification of data types: by level of measurement and the nature of the discretizing object to which they refer (see Fig. B.1.3). However in the case of map operations it is also necessary to distinguish between spatially intensive and spatially extensive variables. Both count data and rate data are examples of ratio level data. However count data are spatially extensive – when two areas are merged (as for example in the operation of areal interpolation or region building) the corresponding counts can be summed to give the count for the newly created map object. Rate data (for example, number of babies born with birth defects divided by the number of live births for an area) by contrast are spatially intensive and to arrive at the correct value for the newly created map object the numerator and the denominator must be aggregated separately.

The quality of the data in the SDM is assessed given the chosen model for the SDM. The combination of model quality and data quality is sometimes referred to as defining the 'uncertainty' of the relationship between the real world and what is held in the SDM. Data quality, for all three elements in Eq. (B.1.1), attributes, location and time, is assessed in terms of four criteria: accuracy, resolution, completeness and consistency (Guptill and Morrison 1995; Veregin and Hargitai 1995). There are a number of complications when considering spatial data quality. Data quality might vary across the map being linked to interaction between the process of measurement and the underlying geography, such as between topography and the quality of imagery for classifying land use (Haining 2003, p.62). Also, there can be interaction between location errors for example and attribute errors – an error in georeferencing a burglary event will introduce error into burglary counts by area if the location error is large enough to transfer the event from one polygon to another.

Accuracy is defined by Taylor (1982) as the inverse of error which is the difference between the value of an attribute as it appears in a database and its true value. Error is an inevitable consequence of taking measurements in the real world, reflecting for example imprecision associated with the measuring device.

		Discrete (object) space			Continuous (field) space
		Point	Line	Polygon	
Level of measurement	Nominal $(=)$	House: burgled/not	Road: under repair/not	Census tract, classified by lifestyle	Land use type
	Ordinal $(=, >, <)$	Quality of life preference rankings of towns	Road classification	Census tracts assigned to income classes	Soil texture (coarse, medium, fine)
	Interval $(=, >, <, \pm)$	Household deprivation score	Length using Greenwich meridian as reference	Index of multiple deprivation for a Census tract	Ground temperature (°C)
	Ratio $(=, >, <, \pm, x, /)$	Annual output from a factory	Annual freight tonnage carried	Regional per capita income	Rainfall (mm)

Fig. B.1.3. Examples of attributes by levels of measurement and types of space

Improvements in the quality of instrumentation and the skills of the people taking the measurements can be expected to reduce this error and thus improve accuracy. That said, the concept of a 'true' value is an idealization and may not even be useful in certain applications where it may be difficult to conceive of a 'true' value (for example, deprivation, social cohesion).

Resolution refers to the size of the spatial units (polygons) that partition the study area but can also refer to the density of sample points for capturing a continuous surface. The more spatial variability that is present, the higher the level of resolution that will be needed to capture that variation in detail. In geostatistics the term 'support' is used to refer to the geometrical size, shape and spatial orientation of the areal units used to take measurements (Olea 1991). The size of an area could be specified in terms of physical extent in the calculation of a quantity such as the proportion of contaminated land. When a quantity such as the proportion of unemployed or the proportion of burgled houses in an area is calculated support is best defined by the number of entities present in each area (population, houses, factories). Resolution issues become particularly problematic when data originating from different spatial frameworks, maybe not even at the same scale, have to be merged for the purpose of constructing the database (Gotway and Young 2002). Pre-2001 population Census data for the UK is available at the enumeration district level, health data is available by postcode but the two spatial frameworks do not correspond and postcodes do not nest within enumeration districts. We shall discuss the implications for spatial data analysis of spatial data resolution and the problems raised by data incompatibility (where different data sets needed for an analysis are collected on different frameworks) in Section B.1.3.

The other two data quality criteria also have significance in the context of the SDM and how it can be analyzed. Consistency is defined as the absence of contradictions in a database. Contradictions are most likely to arise when two different databases have been merged such as health outcome data and population data which may have been collected at different times. You cannot, by definition, have recorded deaths in an age cohort in a Census tract where there is supposedly zero population in that cohort. Georeferencing errors can give rise to inconsistencies – as for example when a traffic accident is mapped as occurring in the North Sea.

Finally completeness refers to the absence of missing data but may be extended to include the situation where there is no over or undercounting of an attribute. Haining (2003, pp.71-74) cites numerous examples. Census data suffer from these problems and agencies responsible for taking a Census often devote considerable resources to trying to plug the gaps created by non-response. This is important to researchers because national census data often provide essential denominator data for calculating rates of different social and economic attributes. A distinction is drawn between data being 'missing at random' and other forms of missing data because methods for imputing values to missing cells in the database (using the data that have been collected) depend on why the cells are empty. A weather station might be temporarily out of action because of equipment failure and this would be considered as a case of data being 'missing at random' if the reason for the equipment failure is not linked to the attribute being measured. Land use data obscured by cloud cover would probably not be considered 'missing at random' because the reason for the cloud cover may be linked to a topographic factor (height above sea level) that in turn might influence the type of land cover. Figure B.1.4 provides a summary of the quality issues described in this section.

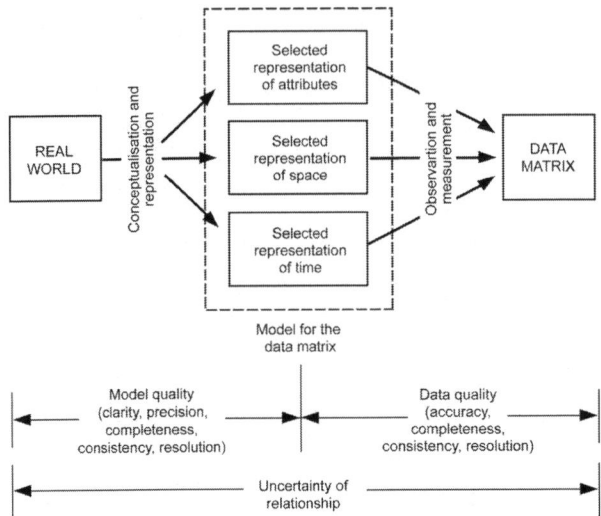

Fig. B.1.4. From geographical reality to the spatial data matrix

B.1.3 Properties of spatial data in the spatial data matrix

Methodology for analyzing data makes various statistical and mathematical assumptions about that data. It is important that the methodology for spatial data analysis makes assumptions that are consistent with data properties. It is these issues that we consider in this section. Methodology is also adapted to the types of questions asked of the data and the models used but the implications of this for spatial data will be considered in the next section.

In the previous section, it was shown how the spatial data matrix is populated with data as a result of three operations: conceptualization, representation and measurement. In order to put structure on the discussion in this section we shall distinguish between fundamental properties of spatial data, properties acquired due to the chosen representation and finally properties acquired as a consequence of the measurement process by which data are collected for storage in the SDM.

Fundamental properties. Fundamental spatial data properties are those properties that are in some sense inherent in attributes that are distributed across the earth's surface. Continuity is a fundamental property of attributes occurring in time. It would be a strange world if attributes changed suddenly and randomly from one second (indeed, one infinitesimal fraction of a second) to the next whether it was the weather or prices for commodities. Events occurring in time have an inherent continuity or structure so that knowing what the level of an attribute is at one point in time allows us to make some informed estimate of what it will be (or the bounds within which it is likely to lie) at some future point in time.

This is the conceptual basis for making predictions, but the further into the future we wish to predict the less certain we will be about an attribute's values.

Continuity, though of a different type, is a fundamental property of attributes located in space. In the GIScience literature it is often referred to as 'Tobler's First Law of Geography' although the observation that attribute values in space are not random has a statistical pedigree that in the published literature can be traced back to at least the early 20th century and a paper by Student (1914) and the observation that 'near things are more related than distant things' to observations by Fisher (1935, p.66) where in the context of agricultural field trials he remarked that 'patches in close proximity are commonly more alike, as judged by yield of crops, than those which are further apart'. Of course assessing spatial continuity is complicated, relative to the time series case, by the second dimension (dependency might not be the same in all directions) and by the lack of directionality (time has a natural uni-directional flow from past to present).

The idea of 'continuity' translates in the spatial context as follows. Measurements of the same attribute taken at several locations in geographical space will show evidence of correlation – or autocorrelation ('self-correlation'). If the value for an attribute at one location is high (low) it is probable that when the same attribute is measured at a nearby location, that value too will be high (low). This denotes positive spatial autocorrelation. As the distance separation of observations increases such autocorrelation will tend to weaken until at some distance observations on the same attribute will appear to be independent. At certain scales of separation there may be negative spatial autocorrelation.

A distinction should be drawn between autocorrelation in the case of entities in an object view of the world and in the case of a field view of the world. In the case of distinct entities it is possible to conceive of complete spatial randomness in their attribute values and indeed negative spatial autocorrelation between adjacent neighbors (for example, plants competing for nutrients in the soil which produce strong plants next to weaker plants). In the case of a surface of values it is not possible to imagine complete spatial randomness (for the same reason as in the temporal case) and certainly not possible to imagine negative spatial autocorrelation unless the surface has been partitioned – in the latter case, the scale of the partition coinciding with contrasting landscapes (for example, a ridge and valley topography partitioned into areas that coincide with the ridges and the valleys).

Quantifying and testing for the significance of spatial structure was an important research agenda during geography's 'quantitative revolution'. Statisticians including Moran (1950), Geary (1954) and Krishna-Iyer (1949) had developed tests for the presence of spatial autocorrelation on regular lattice systems, and a number of geographers including Dacey (1968) at Northwestern University and Cliff and Ord (1973) at Bristol University (Cliff a geographer, Ord a statistician) adapted these statistics to the situation usually encountered in the social sciences where data are collected from an irregular spatial partition or irregular distribution of locations.

Statistically, spatial autocorrelation is a second order property of an attribute distributed in geographical space signifying local variation about a mean value (see Chapter B.3 for more details). But in addition there may be a mean or first order component of variation represented by a linear, or quadratic or higher order trend. Statistical tests and models can be used to quantify the spatial structure present in spatial data but they are of course particular structures for representing or describing this fundamental property and should not be confused with the fundamental property itself. These early statistics focused on *global* descriptors of spatial structure (assuming the same structure is present irrespective of where the analyst is on the map – spatial stationarity) but there is no a priori reason to believe that spatial structure is homogeneous across the map or indeed that it can be defined using a global specification.

Structure might be quite localized with different structural forms present in different parts of the map. Indeed in a geographical context it has been argued that it is unlikely that structure will be location invariant (see for example Granger 1969 for an early statement of this point) and that as a consequence *spatial heterogeneity* is another fundamental property of spatial, particularly geographical, data. However it is arguable whether spatial heterogeneity should be considered a fundamental property of spatial data in the same sense that spatial dependence is considered a fundamental property. One reason for this is that the existence of spatial heterogeneity seems to be dependent on the extent of the area considered, another reason is that in statistical terms spatial heterogeneity is not distinguishable from clustering (Bartlett 1964). Notwithstanding this, many argue that heterogeneity should be expected in certain types of spatial data and that the concept of heterogeneity extends to the relationships between different variables (Brunsdon et al. 1998).

Properties due to the chosen representation. How the fundamental properties of a spatially distributed attribute are retained in the SDM will depend on the chosen representation (Jelinski and Wu 1996). If a surface is captured using point samples then the extent to which the underlying structure is captured will depend on the density of the sample points in relation to surface variability. If polygons are used, it will depend on the size of the polygons in relation to surface variability and whether the polygons represent an artificial, imposed grid (an intrinsic partition such as remote sensing pixels) or follow natural boundaries (a non-intrinsic partition such as vegetation patches, or neighborhoods). There is a loss of information which follows as a consequence of sampling and discretizing and if for example sample points are widely separated then important information about the extent to which nearby values are similar will be lost. In general non-intrinsic partitions produce less smoothing and the construction of such partitions can be carried out using region-building algorithms (see for example Haining 2003, pp.200-206).

In the case of the object view, the use of areas (polygons) for representing geography involves aggregation of the entities lying within them. This too results in a loss of information and a smoothing of spatial variability. The effect of this representation on data properties depends on the scale of the polygons and the positioning of the polygon boundaries. Intra-polygon heterogeneity will depend on the scale of the polygons in relation to the variability of entities within the polygon; inter-area heteroskedasticity (or the variation in the variance of attributes across the set of polygons) will depend on such factors as the variation in the number of entities within each polygon. If the set of polygons that partition the space vary greatly in say population size (with some polygons having large population and others small population counts) then calculated rates will have less precision in the case of those polygons with small populations relative to those calculated for polygons with larger population counts. This is referred to as the small number problem (see for example Haining 2003, pp.196-197).

Extreme rates tend to be associated with polygons with small populations but statistically significant rates (for example, disease rates significantly above or below a benchmark figure such as the area-wide mean) tend to be associated with polygons with large populations (Haining 2003, p.199). This is one of the conflicts at the heart of some applications of the new digital technology applied to spatial data. Spatial precision and the ability to store geographical data at fine spatial scale does not equate with statistical precision. Data errors or small random fluctuations, say in the number of events, can have a big effect on the calculation of rates for small populations, which also do not change smoothly or continuously in response to increases or decreases in the value of the numerator. Smoothing may be applied to try to reduce these effects (Banerjee et al. 2004, pp.69-70).

The size of the polygon partition will affect the calculated measure of spatial autocorrelation for an attribute that is a rate, for the reasons outlined above. If areas vary substantially in spatial support (population sizes on which the rates are calculated) then any test for spatial autocorrelation that assumes constant variance across the set of areas (such as the tests in Cliff and Ord 1973) should be used with caution. Significant spatial autocorrelation could be a consequence of the way areas with large and small population supports are distributed geographically (Gelman and Price 1999). This has led to revisions to the standard autocorrelation tests (see for example Assunção and Reis 1999).

Properties introduced through the measurement process. We conclude this section by considering how the process of taking measurements might introduce false properties into a spatial database. We consider a few selected examples. Location error may induce attribute error with overcounting in one area matched with undercounting in another because a case has been attributed to the wrong area. Such errors are often relatively short range (perhaps due to a small coding error) so it is possible that the errors in counts are negatively autocorrelated with an undercount in one area matched with an overcount in an adjacent area. False clusters of cases of an event can be induced if geo-coding follows some convention – such as assigning motor vehicle theft to the nearest cross roads.

When data on different attributes have been collected on different spatial frameworks and then have to be put onto a common spatial framework through the aggregation of smaller units *'the uniqueness of each unit and the dissimilarity amongst units is ... reduced'* (Gotway and Young 2002, p.633). If there is positive spatial autocorrelation in the attribute this decrease in variance is less marked. If the aggregation process is based on a moving average manipulation of the data the fact that overlapping spatial units will be used will induce (additional) positive spatial autocorrelation in the data. A somewhat similar effect is observed with the pixel values recorded through remote sensing. These values are not in one-to-one correspondence with an area of land on the ground because of the effects of light scattering (Forster 1980). The point spread function quantifies the relationship between pixel value and area of land included in the calculation. The form of the error is similar to a weak spatial filter passed over the surface introducing an additional level of positive spatial autocorrelation into the database because the values for adjacent pixels cover overlapping segments of geographical space. Subsequent map operations carried out on such data may further complicate the error properties for attributes derived from such data through error propagation (Haining and Arbia 1993).

Wherever problems of accurate measurement have a geography this is potentially a significant problem for any research that involves making comparative statements across the area of study. There tends to be more or less full counting of burglaries in suburban areas (where householders have insurance but need the police to provide a reference number in order to make a claim); however inner city residents tend to underreport burglaries perhaps because they may fear reprisals, perhaps because they do not expect the police to recover the stolen goods and they themselves have no household insurance. Area differences in crime rates between suburban and inner city areas may therefore not be fully reflected in the database.

B.1.4 Implications of spatial data properties for data analysis

We discuss how the nature of spatial data influences the form and conduct of spatial data analysis. We do so by examining aspects of exploratory and confirmatory spatial data analysis. We also consider how the nature of spatial data may influence the interpretation of the results of any analysis.

Cressie (1993) organized his book *Statistics for Spatial Data* based on a fourfold typology of spatial data: geostatistical data, lattice or area (or, to use Ripley's 1981 term 'regional') data, point pattern data and objects (Cressie 1993, pp.8-9). Geostatistical data is data from a continuous surface occupying a fixed subset, D, of two dimensional space (R^2) and where the attribute $Z(s_i)$ is a random vector at location s_i in D. Lattice or regional data is where D is a fixed, regular or irregular finite collection of points in R^2 or areas that partition R^2. D may be a graph so that the neighbors ($N(i)$) of each point are defined and $Z(s_i)$ is again a random vector at location s_i in D. Geostatistical data are distinguished most clearly from regional data by the ability of the spatial index s_i to vary continuously over the subset of R^2.

Point pattern data are data where D is a point process in R^2. $Z(s_i)$ is a random vector at location s_i in D which can be used to distinguish between the usual spatial point process and a marked spatial point process where points either possess or do not possess some attribute. Object data are data where again D is a point process in R^2 but $Z(s_i)$ is a random set. Point pattern data and object data are distinguished from geostatistical and regional data by the fact that points in D are the outcome of a random process. Methods of spatial data analysis depend at least in part on the type of data and the types of problems associated with that type of data – although Cressie (1993, p.11) notes: *'That is not to say that methods from one class of problems cannot be borrowed from methods usually associated with another class'*. Methods also depend on the types of inference problems associated with each type of data.

To help illustrate ideas and to provide a link with the earlier part of this chapter, Fig. B.1.5 shows the relationship between Cressie's (1993) typology of spatial data for the purpose of classifying different branches of spatial statistics and the GIScience conceptualization of geographic reality and includes some examples that fall into the different categories.

Exploratory spatial data analysis (ESDA) comprises a collection of visual and numerically resistant techniques for summarizing spatial data properties, detecting patterns in spatial data (both relationships between variables and geographical patterns), identifying unusual or interesting features in data including possible data errors and formulating hypotheses. The location of data points on the map and the spatial relationships between them carry important information that will be relevant to their analysis (Haining et al. 1998). The map is an essential visualization tool and the linkage between map windows and other graphics windows is an essential part of computer software designed for ESDA (Andrienko and Andrienko 1999; Monmonier 1989). However visualizing spatial data raises particular problems. In the case of area data, the areas that partition the map may be of different size. Two consequences may follow. The eye is drawn to the areas of greatest physical extent. Areas may contain different base populations and, as noted, rates are more robust in areas with large populations. These may not be the areas of greatest physical extent – indeed the reverse may be the case. Dorling (1994) advocates the use of cartograms so that areas are transformed in physical extent to reflect some key attribute such as population size. Geostatistical methods have been developed, 'downscaling' choropleth maps using area to point Poisson kriging, in order to create smoother maps of spatial variation (see for example Goovaerts 2006, 2008). Banerjee et al. (2004) review a number of exploratory techniques for geostatistical (spatial process) and area data types.

Turning to confirmatory spatial data analysis, at the core of statistical theory developed for spatial data is the recognition that spatial data are not independent so that classical statistical theory (developed under the assumption of observations from independent random variables) no longer applies. It is a fundamental property of much spatial data, as we have seen, that they are positively autocorrelated with an autocorrelation that decreases with increasing distance separation. This has a number of important consequences for modeling, parameter estimation and hypothesis testing using likelihood methods which we can only sketch here (for a treatment of inference with spatial data from a Bayesian perspective see Banerjee et al. 2004 where spatial effects are treated as latent within the data, never observed and modeled using random effects rather than directly).

		Conceptualizations of geographical reality	
		Object view	Field view
	Geostatistical		Rainfall
			Ground temperature
	Regional	D = points: prices at a set of retail outlets of the same type in a city	Soil pH recorded by soil regions (non-intrinsic polygons)
Cressie's data types		D = polygons: counts or rates of a particular cancer by health region	Vegetation data recorded by pixel from remotely sensed imagery (intrinsic polygons)
	Point process	Distribution of individual plants across an area (point process).	
		Distribution of individual plants across an area classified as diseased or not (marked point process)	
	Object	Distribution of craters across a landscape. The location of a crater and its spatial extent are the outcome of two processes	Vegetation map with patches of varying size. Location and extent of a patch are the outcome of two processes

Fig. B.1.5. Cressie's (1993) typology of spatial data and the two conceptualizations of geographic reality

Consider the sample mean as the estimator of the mean of a normal distribution. Classical theory underestimates the confidence interval associated with the sample mean when data are positively autocorrelated. This is because a sample of correlated normal data carries less information about the mean than an equivalent sized sample of independent normal observations. The 'effective' sample size (Clifford and Richardson 1985) for the purposes of inference is smaller than the number of observations. The interested reader is referred to Cressie (1993, pp.13-15) and Haining (1988) for a fuller discussion of this example where spatial models are used to explicitly measure the 'effective' sample size. Classical statistical inference based on ordinary least squares (OLS) theory also breaks down in the case of regression modeling with spatial data when model errors are positively autocorrelated (see for example Anselin and Griffith 1988). The conventional F and t tests of hypothesis on regression parameters are not valid and may lead the analyst to conclude that a statistically significant relationship exists between the dependent and an independent variable when such a conclusion is not justified at the chosen level of significance. Typically a regression model that explicitly models the spatial dependency structure replaces the normal regression model (see for example, Anselin 1988). We have also noted that, in the case of area data, the OLS assumption of constant variance is likely to be violated. Errors may not be homoskedastic if the dependent variable is a rate calculated for areas with widely varying base populations.

There is, however, another side to this coin. Suppose now inference is directed at predicting an unknown observation $z(s_{n+1})$ given n observed data values $\{z(s_1), ..., z(s_n)\}$ all from the same normal distribution with unknown mean and variance and independent. The best linear unbiased estimator (BLUE) of $z(s_{n+1})$ is the mean. However when the independence assumption is replaced by data that are spatially correlated the relevant theory is provided by ordinary kriging in which the BLUE is a linear function of the n data values with the largest weights attached to those locations closest to the location to be predicted, with adjustment for the effects of any observed data clustering (see, for example, Haining 2003, pp.167-174). Such a result accords well with our intuition. In this case the spatial dependence in the data is being exploited in order to derive a predictor of the unknown observation with a smaller mean squared prediction error than that given by classical theory which ignores information on the location and relative positions of observations.

It was suggested in Section B.1.3 that spatial heterogeneity of structure is also a property of geographical data sets particularly when the data are collected over large areas. This has been the motivation for developing *local* tests and model specifications. In the case of local statistics, geographically defined subsets of the data are analyzed rather than taking all the data together as is the case with global spatial autocorrelation tests (Anselin 1995). Local statistics have been developed to test for the presence of localized clusters of cases (of a disease for example). Perhaps the best known is the Scan test developed by Kulldorff (1997) but owing much to ideas contained in Openshaw's Geographical Analysis Machine (Openshaw et al. 1987). Openshaw's GAM, by using multiple moving windows of varying size and carrying out tests on each one, addressed the problem of searching for clusters across the map which may exist anywhere at any scale (thus avoiding prespecification bias), but it was Kulldorff's test that addressed the problem of multiple testing – one of the major statistical problems that has to be addressed when developing local statistics.

The problem of testing for clusters of cases of some attribute also illustrates a number of the core issues that arise due to representational choices and data measurement issues. If the data are in the form of area counts then the size of the areas (the data support) will have an important influence on our ability to detect clusters. Clusters will probably be lost or diluted if they are much smaller than the areal units that partition the study area. The position of the boundaries of each area could be critical if they cut through a cluster thereby dispersing the raised count over several areas. Most crucially, as implemented in GAM and the SCAN statistic, areas cannot be analyzed independently as was done in the case of earlier tests (see, for example, Choynowski 1959) – the spatial relationships between the areal units need to be allowed for in the construction of the test because a cluster may extend over several contiguous areal units. As a final point, if data are in the form of area counts or a marked point dataset, if there are cases in a cluster that are missing, misclassified or wrongly located then this might seriously affect the power of any test to detect that cluster (see Kulldorff 1998).

Data support raises further issues that have implications for spatial data analysis. Where data have been collected on different supports ('incompatible data') but for the purposes of analysis need to be combined, how should this be done? If different process mechanisms are important at different scales, data are required at the relevant scales but what types of models should be specified? What are the implications of data support for model inference? We conclude with consideration of these questions.

Combining data involves changing the support of a variable thereby creating a new variable. Gotway and Young (2002) define the change of support problem (COSP) as the problem of how the spatial variation of one variable associated with a given support relates to that of the new variable created on a different support. The authors review various solutions to the COSP including geostatistical solutions based on different forms of kriging and map overlay solutions (point in polygon, areal weighting, spatial smoothing and regression methods). Geographers and those who use GIS will be most familiar with the latter methods, physical geographers and earth scientists with the former. These solutions produce new maps with the original data transferred to the new support. Map overlay solutions give no indication of the prediction error associated with the new map whilst geostatistical solutions '*often rely on contrived parametric models*' (Gotway and Young 2002, p.645). Gotway and Young also review multiscale modeling solutions (multiscale spatial tree models and Bayesian hierarchical models).

There is further discussion of Bayesian models for the COSP in Banerjee et al. (2004, pp.175-212). These statistical modeling solutions allow the analyst to use data collected at different resolutions to model complex systems (see for example Fieguth et al. 1995; Gabrosek et al. 1999) but depend on making assumptions that are unverifiable from the data. As Gotway and Young (2002, p.645) observe, the common solution strategy for COSP is to build a model from data with small support and use that to estimate parameters and make valid inferences. The focus then turns to the validity of the assumptions needed to obtain solutions. Bierkens et al. (2000) provide an accessible discussion of the COSP in the context of environmental research focusing on a wide range of techniques for 'upscaling' (aggregation) and 'downscaling' (disaggregation).

Many environmental problems (for example, climate change research) require the construction of models that are defined at different scales with interactions occurring across those scales often within a hierarchical (global to local) structure. Wikle (2003) provides an overview of hierarchical modeling approaches to the analysis of complex environmental systems. Marceau (1999) in reviewing these issues discusses examples that demonstrate the effects of scale on the explanatory power of different sets of variables and, in the context of landscape ecology, describes the hierarchical patch dynamics paradigm (HPDP) of Wu and Louks (1995). HPDP combines two elements. The patch is the fundamental structural and functional unit of a landscape and deals with heterogeneity and interaction in a horizontal way; hierarchy theory defines the vertical structure of the system in terms of a limited number of discrete hierarchical levels with effects operating at different scales.

As Marceau (1999) notes, this paradigm and the consequent models recognize how spatio-temporal heterogeneity, scale and hierarchical organization affect the structure and dynamics of ecological systems. Similar conceptual frameworks are found in the social sciences. In the case of adolescent development Brooks-Gunn et al. (1993, p.354) comment: *'individuals cannot be studied without consideration of the multiple ecological systems in which they (the adolescents) operate'*. The contextual effect of place can operate at a hierarchy of scales from the immediate neighborhood up to regional scales and above. Criminologists analyze the contribution of different scales of influence from the individual to neighborhood scales on offending and victimization (Wikström and Loeber 2000). Sampson et al. (1997) use multi-level modeling to identify different scale effects in the distribution of violent crime.

Finally we turn to specific instances of inference problems that arise linked to the problem of support. The ecological inference problem and the modifiable areal units problem (MAUP) familiar to geographers are both linked to the COSP and are special instances of it. Much spatial data, especially in human geography, is made available in the form of counts or rates for areal units, or zones. Typically these zones are arbitrary and modifiable and all too often have no intrinsic geographical meaning (Hipp 2007). It has long been known that the results of statistical techniques such as regression and correlation are dependent on the spatial framework in terms of which data are collected. Results depend on the scale of the areal units (the scale problem) and their configuration (the aggregation problem). For early discussions see Gehlke and Biehl (1934) and McCarthy et al. (1956). This concern has come to be known as the modifiable areal units problem following Openshaw and Taylor (1979) who demonstrated how widely correlation coefficients can vary for the same set of underlying data but observed under different partitions of the space. As King (1997, p.252) and others have noted however the MAUP is not an empirical problem. If an analyst wants statistics that are invariant to the level of aggregation then do not use regression or correlation, use statistics that are invariant (see also Tobler 1990). As King (1997, p.252) points out, *'deriving statistics invariant to aggregation for the relationship between variables, corresponding to correlation or regression coefficients, may be more difficult, but it is a theoretical difficulty'*.

It follows that making statistical inferences about individuals based on aggregate data is flawed. Robinson (1950) used the term the ecological fallacy to describe the error of doing so. The term aggregation bias is used to denote the difference between the parameter estimate obtained from group data and the estimate obtained from individual level data. Where some individual level data are available, some progress can be made in using aggregate data to make individual level inference and this is a project of interest in spatial epidemiology where there is a focus on acquiring estimates of risk, at the individual level, to different types of exposures (Jackson et al. 2008).

B.1.5 Concluding remarks

An important class of spatial data is compiled through a process that starts with a continuous and complex reality and concludes with the construction of the spatial data matrix. Spatial data have properties, some deriving from the fundamental nature of geographical reality, others acquired through the representational and measurement processes that follow. Those properties and the resulting quality of spatial data provide an essential context within which methods for analyzing spatial data are developed and results interpreted. Equally important are the types of questions we ask, the inference problems we face including the spatially-structured models we build through which we seek to understand the world. It is these models and how they are constructed that determine the model for the spatial data matrix – that determine in any particular scientific context what data are needed, at what scales and according to what configurations.

References

Andrienko GL, Andrienko NV (1999) Interactive maps for visual data exploration. Int J Geogr Inform Sci 13(4):355-374
Anselin L (1988) Spatial econometrics. Kluwer, Dordrecht
Anselin L (1995) Local indicators of spatial association – LISA. Geogr Anal 27(2):93-115
Anselin L, Griffith D (1988) Do spatial effects really matter in regression analysis? Papers of the Reg Sci Assoc 65(1):11-34
Assunção RM, Reis EA (1999) A new proposal to adjust Moran's I for population density. Stat Med 18(16):2147-2162
Banerjee S, Carlin BP, Gelfand AE (2004) Hierarchical modeling and analysis for spatial data. CRC Press (Taylor and Francis Group),Boca Raton [FL], London and New York
Bartlett MS (1964) The spectral analysis of two dimensional point processes. Biometrika 51(3-4):299-311
Bierkens MFP, Finke PA, de Willigen P (2000) Upscaling and downscaling methods for environmental research. Kluwer, Dordrecht
Brooks-Gunn J, Duncan GJ, Klebanov PK, Sealand N (1993) Do neighborhoods influence child and adolescent development? Am J Soc 99:353-395
Brunsdon C, Fotheringham AS, Charlton M (1998) Geographically weighted regression. The Statistician 47(3):431-443

Burrough PA, McDonnell RA (1998) Principles of geographical information systems. Oxford University Press, Oxford

Choynowski M (1959) Maps based on probabilities. J Am Stat Assoc 54:385-388

Cliff AD, Ord JK (1973) Spatial autocorrelation. Pion, London

Clifford P, Richardson S (1985) Testing the association between two spatial processes. Statistics and Decisions 2:155-160

Cressie NAC (1993) Statistics for spatial data (revised edition). Wiley, New York, Chichester, Toronto and Brisbane

Dacey MF (1968) A review of measures of contiguity for two and k-color maps. In Berry BJL, Marble D (eds) Spatial analysis. Prentice-Hall, Englewood Cliffs [NJ], pp.479-495

Dorling D (1994) Cartograms for visualizing human geography. In Earnshaw HM, Unwin D (eds) Visualization in Geographic Information Systems. Wiley, New York, pp.85-102

Fieguth PW, Karl WC, Willsky AS, Wunsch C (1995) Multiresolution optimal interpolation and statistical analysis of TOPEX/POSEIDON satellite altimetry. IEEE Transact Geosci Remote Sens 33(2):280-292

Fischer MM (2000) Spatial interaction models and the role of geographic information systems. In Fotheringham AS, Wegener M (eds) Spatial models and GIS. Taylor and Francis, London, pp.33-43

Fisher R (1935) The design of experiments. Oliver and Boyd, Edinburgh

Forster BC (1980) Urban residential ground cover using LANDSAT digital data. Photog Eng Remote Sens 46:447-458

Frank AU (2001) Socio-economic units: their life and motion. In Frank AU, Raper J, Cheylan JP (eds) Life and motion of socio-economic units. Taylor and Francis, London, pp.21-34

Frank AU, Raper J, Cheylan J-P (2001) Life and motion of socio-economic units. Taylor and Francis, London

Gabrosek J, Huang H-C, Cressie NAC (1999) Spatio-temporal prediction of level 3 data for NASA's earth observing system. In Lowell K, Jaton A (eds) Spatial accuracy assessment: land information uncertainty in natural resources. Ann Arbor Press, Chelsea [MI], pp.331-337

Geary RC (1954) The contiguity ratio and statistical mapping. The Incorp Stat 5(3):115-145

Gehlke CE, Biehl K (1934) Certain effects of grouping upon the size of the correlation coefficient in census tract material. J Am Stat Assoc 29:169-170

Gelman A, Price PN (1999) All maps of parameter estimates are misleading. Stat Med 18(23):3221-3234

Godambe VP, Thompson ME (1971) Bayes, fiducial and frequency aspects of statistical inference in regression analysis in survey sampling. J Roy Stat Soc, Series B33:361-390

Goodchild MF (1989) Modelling error in objects and fields. In Goodchild MF, Gopal S (eds) Accuracy of spatial databases. Taylor and Francis, London, pp.107-113

Goodchild MF (1998) Geographic information systems and disaggregate transportation modeling. J Geogr Syst 5(1-2):19-44

Goodchild MF, Haining RP (2004) GIS and spatial data analysis: converging perspectives. Papers in Reg Sci 83(1):363-385

Goodchild MF, Lam NSM (1980) Areal interpolation: a variant of the traditional spatial problem. Geo-Processing 1:297-312

Goovaerts P (2006) Geostatistical analysis of disease data: accounting for spatial support and population density in the isopleths mapping of cancer mortality risk using area to point kriging. Int J Health Geographics 5:52

Goovaerts P (2008) Kriging and semi-variogram deconvolution in the presence of irregular geographical units. Math Geosci 40(1):101-128
Gotway CA Young LJ (2002) Combining incompatible spatial data. J Am Stat Assoc 97:632-648
Granger CWJ (1969) Spatial data and time series analysis. In Scott AJ (ed), Studies in Regional Science. Pion, London, pp.1-24
Guptill SC, Morrison JL (1995) Elements of spatial data quality. Elsevier Science, Oxford
Haining RP (1978) Interaction modeling on central place lattices. J Reg Sci 18(2):217-228
Haining RP (1988) Estimating spatial means with an application to remotely sensed data. Comm Stat Theor Meth 17(2):573-597
Haining RP (2003) Spatial data analysis: theory and practice. Cambridge University Press, Cambridge
Haining RP, Arbia G (1993) Error propagation through map operations. Technometrics 35(3):293-305
Haining RP, Wise S, Ma J (1998) Exploratory spatial data analysis in a geographic information system environment. The Statistician 47(3):457-469
Hipp JR (2007) Block, tract, and levels of aggregation: neighborhood structure and crime and disorder as a case in point. Am Soc Rev 72(5):659-680
Jackson C, Best N, Richardson S (2008) Hierarchical related regression for combining aggregate and individual data in studies of socio-economic disease risk factors. J Roy Stat Soc A 171:159-178
Jelinski DE, Wu J (1996) The modifiable areal unit problem and implications for landscape ecology. Landscape Ecology 11(3):129-140
King G (1997) A solution to the ecological inference problem: reconstructing individual behaviour from aggregate data. Princeton University Press, Princeton [N.J.]
Krishna-Iyer PV (1949) The first and second moments of some probability distributions arising from points on a lattice and their applications. Biometrika 36(1-2):135-141
Kulldorff M (1997) A spatial scan statistic. Comm Stat Theor Meth 26(6):1481-1496
Kulldorff M (1998) Statistical methods for spatial epidemiology: tests for randomness. In Gatrell A, Löytönnen M (eds) GIS and health. Taylor and Francis, London, pp.49-62
McCarthy HH, Hook JC, Knos DS (1956) The measurement of association in industrial geography. Department of Geography, State University of Iowa, Iowa City
Marceau DJ (1999) The scale issue in social and natural sciences. Canad J Rem Sens 25(4):347-356
Matheron G (1963) Principles of geostatistics. Econ Geol 58(8):1246-1266
Monmonier MS (1989) Geographic brushing: enhancing exploratory analysis of the scatterplot matrix. Geogr Anal 21(1):81-84
Moran PAP (1950) Notes on continuous stochastic phenomena. Biometrika 37(12):17-23
Olea RA (1991) Geostatistical glossary and multilingual dictionary. Oxford University Press, New York
Openshaw S, Taylor PJ (1979) A million or so correlation coefficients: three experiments on the modifiable areal units problem. In Wrigley N (ed) Statistical applications in the spatial sciences. Pion, London, pp.127-144
Openshaw S, Charlton M, Wymer C, Craft A (1987) A mark 1 geographical analysis machine for the automated analysis of point data sets. Int J Geogr Inform Syst 1(4):335-358
Robinson WS, (1950) Ecological correlations and the behavior of individuals. Am Socl Rev 15(3):351-357
Sampson RJ, Raudenbush SW, Earls F (1997) Neighborhoods and violent crime: a multilevel study of collective efficacy. Science 277:918-924

Student (1914) The elimination of spurious correlation due to position in time or space. Biometrika 10(1):179-180
Taylor JR (1982) An introduction to error analysis. University Science Books, Mill Valley [CA]
Tobler W (1990) Frame independent spatial analysis. In Goodchild MMF, Gopal S (eds) Accuracy of spatial databases. Taylor and Francis, London, pp.115-122
Veregin H, Hargitai P (1995) An evaluation matrix for geographical data quality. In Guptill SC, Morrison JL (eds) Elements of spatial data quality. Elsevier Science, Oxford, pp.167-188
Wikle CK (2003) Hierarchical models in environmental science. Int Stat Rev 71(2):181-199
Wikström PO, Loeber R (2000) Do disadvantaged neighborhoods cause well-adjusted children to become adolescent delinquents? A study of male juvenile serious offending, individual risk and protective factors and neighborhood context. Criminology 38(4):1109-1142
Wilson AG, Bennett RJ (1985) Mathematical methods in human geography and planning. Wiley, New York, Chichester, Toronto and Brisbane
Wu J, Loucks OL (1995) From balance of nature to hierarchical patch dynamics: a paradigm shift in ecology. Quart Rev Biol 70(4):439-466

B.2 Exploratory Spatial Data Analysis

Roger S. Bivand

B.2.1 Introduction

Exploratory spatial data analysis (ESDA) as used in spatial statistics, spatial econometrics and geostatistics, developed from exploratory data analysis (EDA). In particular, two threads that are central to a-spatial EDA have carried over to ESDA – the importance of the data themselves, and the importance of analytical graphics in representing chosen characteristics of the data.

This chapter will present some of the underlying intentions of ESDA, and survey some of the outcomes. This will necessarily involve the use of software, since most EDA and ESDA techniques presuppose the use of computing resources in some form. Here, we will use R-2.8.0 (R Development Core Team 2008), because the integration of its output with the printed page is somewhat less problematic than that of systems with graphical user interfaces. The choice of R also touches nicely on the Bell Labs' inheritance of the S language, with its links to John Tukey and Bill Cleveland, described by Chambers (2008), himself a major contributor to applied statistics.

In his recent book, Chambers (2008, p.1) proposes the principle that: '*our Mission, as users and creators of software for data analysis, is to enable the best and most thorough exploration of the data possible.*' In this tradition, exploration is part of the process of formulating the question and organising the data so as to be able to answer that question. As Cox and Jones (1981) note, the tradition stands some way from the classical division between descriptive and inferential statistics. The substantive research problem is what matters: '*As John Tukey often remarked, better an approximate answer to the right question than an exact answer to the wrong question*' (Chambers 2008, p.3). This may involve exploring distributional assumptions in relation to variables of interest, perhaps including transformations or the removal of trends, but does presuppose that the analyst wants to find the 'right' question, a point to which we will return in conclusion.

Attentive reading of classics in spatial data analysis, such as Cressie (1993), and Bailey and Gatrell (1995), shows that both EDA and ESDA have long played an important part in finding the 'right question'. This heritage is continued in newer presentations like Waller and Gotway (2004), and Schabenberger and Gotway (2005). While many point to the increasing availability of spatial data as such, it seems typical that the data we need to attack a given research problem is often costly to gather, often collected for other purposes, and often not with the support best suited to the problem. Consequently, we need to try to make the best possible use of the available data, both in connection with the thrust of our research problem, and looking out for signals suggesting potentially richer questions.

Our research problem focuses our attention on components of variation in our response variables of interest, on variables or spatial locations that account for observed variability. In terms proposed by Tukey, the response variables constitute the data, and what we know about the data based on previous knowledge is the smooth, leaving residual variation in the rough. Exploratory data analysis opens up two complementary possibilities: that our prior knowledge – choice of variables in the smooth and their functional form – deserves revision, and that patterning in the rough can be shifted to the smooth. In particular, spatial patterning in the rough can be used as a 'spatial' smooth in some cases, especially when observations on omitted variables shown in the spatial patterning are not available for any reason. Exploratory spatial data analysis plays an important role in the examination of a-spatial residual variation, to try to see whether spatial patterning can be used to account for the variation in the data in a more satisfactory way.

In this chapter, we will work with examples to show some of the available methods that build on the EDA approach to data analysis. The examples use legacy data sets, and will not necessarily start from univariate EDA as perhaps they should, but rather illustrate fresh groups of methods in turn in each section. One example data set that will be used frequently is the French 'Moral Statistics' data set discussed in detail by Friendly (2007) and taken up in connection with geographical visualization by Dykes and Brunsdon (2007).

B.2.2 Plotting and exploratory data analysis

Cox and Jones (1981, p.135) describe one of the basic attitudes of exploratory data analysis as: '*plot both your data and the results of data analysis*' – pointing directly to statistical graphics. Plotting multiple versions of a display by hand is so time-consuming that actually using EDA visualization had to wait until computer graphics resources became available, despite the hopes expressed in Tukey (1977) that paper and pencil would be enough. Naturally, in the 1970's computer graphics were not very sophisticated, and portability across graphics devices other than line printers was very hard to achieve, so early Minitab EDA output was formatted for line printers (as was output from the subroutines provided in Velleman and Hoaglin 1981).

Output to interactive user terminals was hard, with the initial exception of the first Apple Macintosh computers, which provided both a monochrome graphics screen and a pointing device. This was used by DataDesk and other software to provide ways of exploring data visually; other software for PC systems did not have such a standardised graphics library until much later; Systat for example used a pen plotter for graphics output. Workstation systems, largely running Unix, did have mature graphics libraries, but with a plethora of different versions – Silicon Graphics™ machines were well-liked but very costly.

Since those early years, cross-platform software accommodating differences in graphics devices has become more common, in addition to cross-platform graphics libraries – Xgobi transitioned to use the Gnome graphics library as Ggobi[1], and may now be used on many platforms (Cook and Swayne 2007). Other data visualization software has chosen to use Java as a virtualized platform, as we will see in Section B.2.2 in the case of Mondrian[2] (Theus 2002). This is not dissimilar to the use of XLISP to underpin XLispStat on a cross-platform basis, used by Brunsdon (1998) for exploratory spatial data analysis. The use of Tcl/Tk by Dykes (1997, 1998) is a further example of a developer 'borrowing strength' from an underlying programming language, which provided cross-platform support for interactive graphics, for exploratory spatial data analysis.

The concise introduction to exploratory data analysis by Jacoby (1997) provides us with a first data set and details of the computing environment used – S was used for demonstrating many of the techniques presented, and they may be reproduced using R. The univariate EDA methods used are described by Jacoby (1997), and implementation details of the graphics functions used can be found in Murrell (2005). Sarkar (2007) shows how to use lattice graphics in R to display panels accommodating both the variable(s) of interest and conditioning variables – this builds on Trellis graphics introduced in Cleveland (1993) and Becker et al. (1996). The data set contains Medicaid program quality scores for 48 U.S. contiguous states for 1986, here stored externally in a shapefile, and read into a `SpatialPolygonsDataFrame` object.

Figure B.2.1 shows a number of graphical representations of the observed values of the program quality scores (PQS), ranging from the simple but informative stem and leaf tally on panel a), through a jittered stripchart on panel b), and a boxplot [see panel c)]; [see Chapter E.2 for a further discussion of the use of boxplots], to a composite of a histogram with default bin widths and starting point, overlaid by density plots for three bandwidths, and furnished with a rug plot along the bottom axis showing the data values in panel d). As in the remainder of this chapter, the code snippets illustrate how the displays may be made, sometimes in abbreviated form to simplify presentation. The PQS variable belongs to the `medicaid` object, here a `SpatialPolygonsDataFrame` object, and is accessed using the $ operator.

---

[1] http://www.ggobi.org/

[2] http://www.theusrus.de/Mondrian/index.html

```
> stem(medicaid$PQS, scale = 2)
> stripchart (medicaid$PQS, method = 'jitter', vertical = TRUE)
> boxplot(medicaid$PQS)
> hist(medicaid$PQS, col = 'grey90', freq = FALSE)
> lines(density(medicaid$PQS, bw = 15), lwd = 2)
> rug(medicaid$PQS)
```

It is helpful to contrast the smoother generalisation of the boxplot, the histogram, and the density plot with the larger bandwidth to the stem and leaf plot, the stripchart, the rug plot, and the density plot with smaller bandwidth. The first group of techniques shows the 'big picture', while the second group gives more detail, and may even suggest some clustering of the observed values.

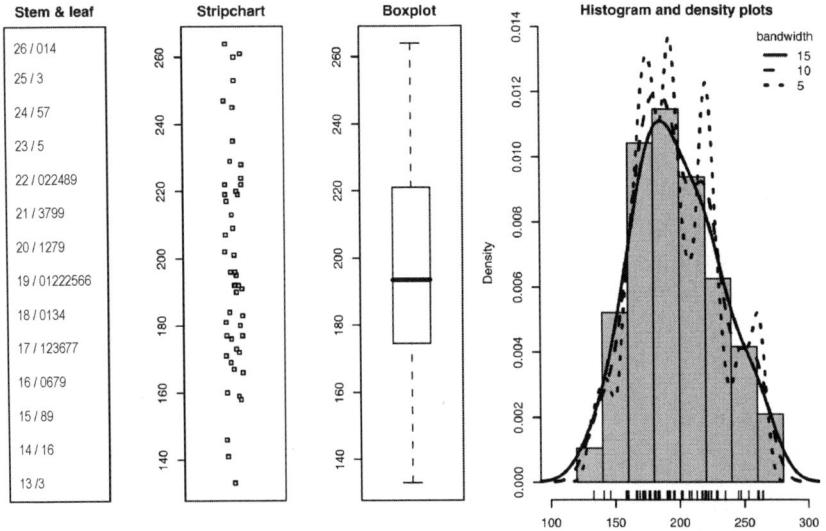

Fig. B.2.1. Displays of the reported Medicaid program quality score values 1986: a) stem and leaf display – here ordered with large values at the top to match the next two panels; b) stripchart with jittered points; c) boxplot with standard whiskers; d) histogram with overplotted density curves for selected bandwidths

All of these techniques use an ordering of the data, as do the two shown in Fig. B.2.2. The plot of the empirical cumulative distribution function of the observed values involves their ordering, and the tallying of ties, to be compared with their rank orders. A uniform distribution gives a more or less straight diagonal line, but the plot is perhaps most useful for exploring unusual breaks between values. The functions can be used in the following way.

```
> plot(ecdf(medicaid$PQS))
> o <- order(medicaid$PQS)
> dotchart(medicaid$PQS[o], labels = as.character (medicaid$STATE_ABBR) [o],
+ groups = medicaid$Division[o])
```

The accompanying dotchart displays all the observed values, with state labels and grouped by statistical division. It introduces the concept of conditioning, here on division, to permit the comparison of ordered values in relation to a structuring variable. With 48 observations, the dotchart is becoming illegible, and would probably benefit from aggregation: curiously, both stem and leaf displays and dotcharts may be viewed 'out of focus' to look at a 'big picture'. Zooming in, it does, however, permit the retrieval of values for identified observations, so that the analyst can see 'which are which'.

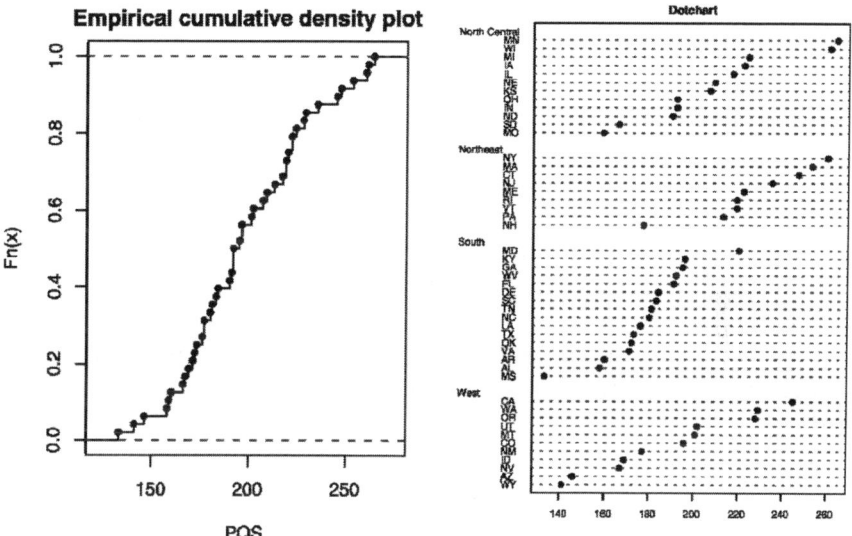

Fig. B.2.2. Medicaid program quality scores 1986: a) empirical cumulative distribution function, and b) dotchart

Dynamically linked graphics. The interactive identification of observations, and groups of observations with apparently shared characteristics, has emerged as an important exploratory tool in data analysis. A pointing device is used to select one or more observation on one graphics display, and the selection is dynamically displayed on all other data displays, both text and graphical. Naturally, this is hard to represent in print, but has generated a rich literature and many software implementations. The background for dynamically linked graphics is discussed in detail by Becker et al. (1987).

One implementation that has served as a research forum for exploring the possibilities offered by multivariate dynamically linked graphics is XGobi (Cook et al. 1996, 1997). From the beginning, XGobi developers were interested in linking to map displays (Symanzik et al. 2000), leaving geographical representation to a desktop GIS. Cook and Swayne (2007) show how dynamically linked graphics have developed and matured, and how dynamic data manipulation, such as 'flying through' clouds of multivariate data points, can be related to static but reproducible graphic displays. Theus (2002) describes the Mondrian software implementation of many multivariate dynamically linked graphics, including a map view. Naturally, showcasing dynamically linked graphics in print is not possible, but any `SpatialPolygonsDataFrame` object may be exported in the correct format for Mondrian in this way.

```
> library(maptools)
> sp2Mondrian(medicaid, 'medicaid.txt')
```

B.2.3 Geovisualization

While data visualization is perhaps more closely related to data analysis, the work of cartographers brings in scientific and information visualization. This crossfertilization has led to a range of innovative software tools, many of which are documented in the work of the Commission on GeoVisualization of the International Cartographic Association.[3] Work by cartographers is welcomed in statistical graphics; for example the results of studies into the use and abuse of colour in visualization have diffused widely. Geovisualization is not separate from exploratory spatial data analysis, but rather constitutes the backbone of ESDA, joining up the large range of techniques proposed for examining spatial data in a shared and easily comprehended visualization framework.

Monmonier (1989) introduced the concept of geographical brushing, borrowing from brushing in dynamically linked graphics, selecting observations for linked highlighting from a map representation, most often choosing observations within a map window. Many of these techniques for linked highlighting were implemented in software described by Haslett et al. (1991) and Haslett (1992), and followed up by Dykes (1997, 1998) in the 'cartographic data visualizer' crossplatform implementation. Progress made during the 1990s is summarised by Andrienko and Andrienko (1999) and Gahegan (1999).

Like Mondrian, GeoVISTA studio (Takatsuka and Gahegan 2002) uses Java as an integrating cross-platform framework linking the dynamic display of spatial data with its conceptual underpinnings. The treatment of ontologies as an integral part of geovisualization software is developed by MacEachren et al. (2004a, b). The approach taken by GeoDa (Anselin et al. 2006) is simpler, combining dynamically linked graphics, map views, and numerical exploratory techniques to be discussed in Section B.2.5.

Dykes and Mountain (2003) add the temporal dimension to interactive graphics with spatial data, while the data is smoothed by geographical weighting in the methods described by Dykes and Brunsdon (2007). Many of these proposals seem to address issues of importance for visualization research as such, rather than for

---

[3] http://geoanalytics.net/ica/

applied data analysis; by contrast, Wood et al. (2007) combine innovative geovisualization with 'mashups', permitting output graphics to be viewed using either browser-based mapping applications, or stand-alone software and geodata distribution systems like Google Earth™.

Thematic cartography. Just as graphical output may be described as lying on a continuum from analytical to presentation in terms of the requirements of its viewers, so may cartographic output (Slocum et al. 2005). Thematic cartography is an important part of exploratory data analysis with spatial data, as well as playing a vital role in presenting model results. It is also crucial in the communication of the intermediate and final results of research, both on screen in applications and documents, and in print. Bailey and Gatrell (1995, pp.48-61) describe the development of computer mapping for analytical purposes. We will not be considering the use of cartograms here, although arguments can be made for their importance in ESDA (Dorling 1993, 1995). There are issues concering the legibility of cartograms, and further difficulties in the algorithmic construction of legible polygons, which led Durham et al. (2006) to complete the construction of acceptable units for the British Census by hand. In this review, we will be using *R* graphics methods largely documented in Bivand et al. (2008, pp.57-80), in particular the spplot methods for suitable objects; the first argument here is the object, and the second, a vector of variables to display using the same class intervals, here a single variable.

```
> lbls <- as.character(medicaid$STATE_ABBR)
> spl <- list('sp.text', coordinates(medicaid), lbls, cex = 0.6)
> spplot(medicaid, 'PQS', col.regions = grey.colors(20,
+ 0.95, 0.4), sp.layout = spl, col = 'grey30')
```

The example (see Fig. B.2.3) shows a map view of the program quality score variable; the sp.layout argument allows additional graphics components to be added to the output. The spplot method can take an argument setting the class intervals, but where none is given, it uses a default of 'pretty' numbers encompassing the range of the data with 19 equally spaced internal intervals, so taking 20 colour values. The grey.colors function creates a ramp of grey shades from its second to third argument value for a default gamma of 2.2, which seems to match some computer displays, projectors, and printed output adequately.

The grey shades chosen are not the same as those proposed by Brewer et al. (1997); Brewer and Pickle (2002) in ColorBrewer, mostly in not using the lightest or darkest greys, and by using a larger gamma than the one proposed there.[4] Having good control of class intervals and colours used is an important part of thematic cartography, and is far from easy to achieve in print. Readers willing to try out the code underlying this review are invited to explore alternative palettes to see whether the 'message' of the presented thematic maps is affected.

---

[4] The gamma correction is a component of the colour space implementation intended to neutralise the effect of the display medium (the default value of 2.2).

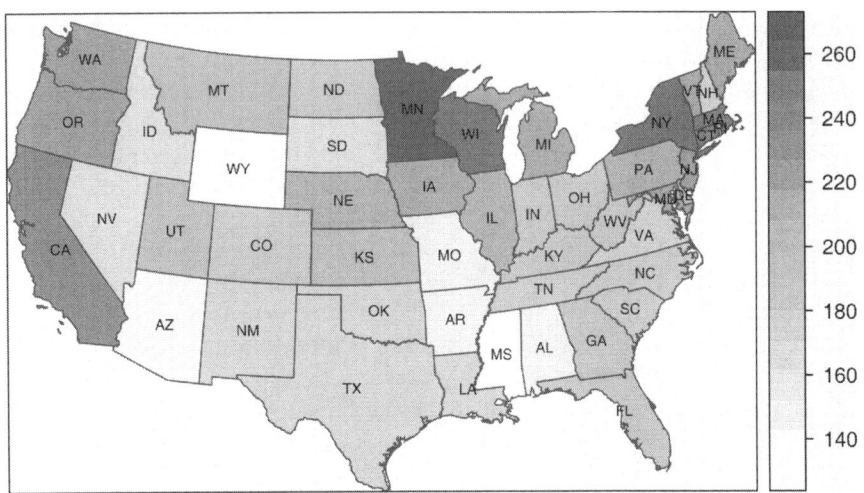

Fig. B.2.3. Medicaid program quality scores, 1986: thematic cartography as a method for statistical display

Conditioned choropleth maps. Trellis graphics displays are intended to permit the researcher to explore multivariate relationships by conditioning on potentially interesting variables (Becker et al. 1996). In an innovative paper building on modern statistical graphics, Carr et al. (2000) propose the use of linked micromaps, matching maps used to provide graphical indices for conditioned panels, and conditioned choropleth maps. They define CC maps in the following way: '*Similar to conditioning on sex and showing separate choropleth maps for males and females, CC maps provide for conditioning on the levels or values of variables and for the display of multiple choropleth maps. The basic difference in the examples here is that the conditioning does not distinguish separate populations within each unit of study but rather partitions the units of study*' (Carr et al. 2000, p.2530). More details and examples can be found in Carr et al. (2005).

In the classic North Carolina sudden infant death syndrome data set, a relationship is found between the Freeman-Tukey transformed SIDS rate for 1974-1978 by county and the Freeman-Tukey transformed nonwhite birth rate (Cressie 1993, pp.548-551).

We can use a lattice of conditioned choropleth maps to explore the spatial footprint of this relationship. We could convert the nonwhite birth rate into a categorical variable (factor) to partition the counties, but follow usual practice when conditioning panels on a numerical variable and use equal count overlapping shingles. The reasons for using overlapping shingles – to avoid the risk of giving the breaks in the conditioning variable too much influence in the display – are discussed by Becker et al. (1996, pp.142-147), and documented for R by Sarkar (2007, pp.177-187). With no overlap, `equal.count` would return members corresponding to quantiles for the number of conditioning levels required, but as can be seen from Fig. B.2.4, the shadings in the panel strips do overlap, reflecting the chosen degree of protection from interval choice artefacts. The `equal.count` function in lattice allows us to construct a shingle, and to use it in `CCmaps`.

```
> library(lattice)
> sh_nw4 <- equal.count(nc.sids$ft.NWBIR74, number = 4
+     overlap = 1/5)
> CCmaps(nc.sids, 'ft.SID74', list(Nonwhite_births = sh_nw4))
```

As we move from lower left to lower right, then upper left to upper right across the panels of Fig. B.2.4, we see that the counties in each level of the shingle seem to be clustered, and that the choropleth map values of the variable of interest increase. This corresponds to the positive relationship reported between the variables, but also suggests that including the conditioning variable may reduce residual autocorrelation in a model of Freeman-Tukey transformed SIDS rates.

```
> gfrance <- readOGR('.', 'gfrance1')
> gfrance$Pop_crime <- gfrance$Crime_prop/100
```

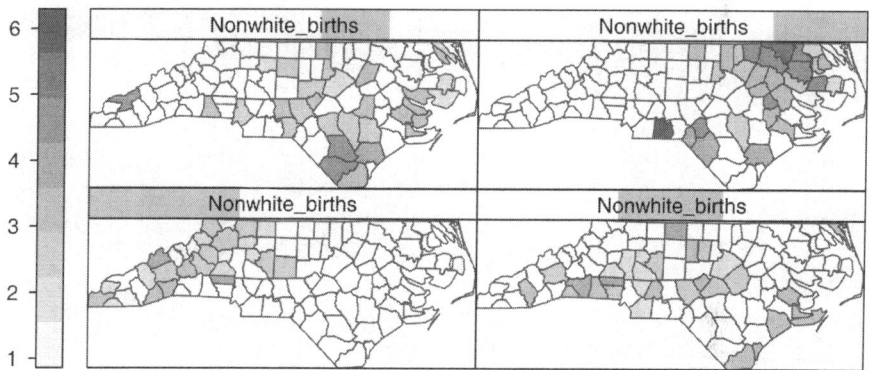

Fig. B.2.4. North Carolina Freeman-Tukey transformed SIDS rates by county for 1974-1978 conditioned on four shingles of the Freeman-Tukey transformed nonwhite live birth rates

Friendly (2007, p.395) includes a conditioned choropleth map of a variable from the Guerry French moral statistics data set: number of population per observation unit per crime against property, conditioned on wealth and literacy. The data set is available from the author's website[5] as a shapefile, which we read in as before. Figure B.2.5 shows the spatial distribution of the three variables being used here.

---

[5] http://www.math.yorku.ca/SCS/Gallery/guerry/maps.html#spatial

In order to plot a conditioned choropleth map, we need to construct two shingles, here, following Friendly (2007, p.395), with 10 percent overlap and two levels each.

```
> sh_wealth <- equal.count(gfrance$Wealth, number = 2,
+     overlap = 1/10)
> sh_literacy <- equal.count(gfrance$Literacy, number = 2,
+     overlap = 1/10)
> CCmaps(gfrance, 'Pop_crime', list(Wealth = sh_wealth,
+     Literacy = sh_literacy))
```

Figure B.2.6 differs from the original figure in a number of ways. The class intervals used for displaying the crime variable are not the same, and the legend is as provided by the underlying spplot and levelplot methods. The ordering of the panels also differs, but the spatial footprint is the same: wealthy and literate places experience higher rates of crime against property than poor and illiterate places. Note the inverted rate used – population per crime, rather than crime counts per inhabitant.

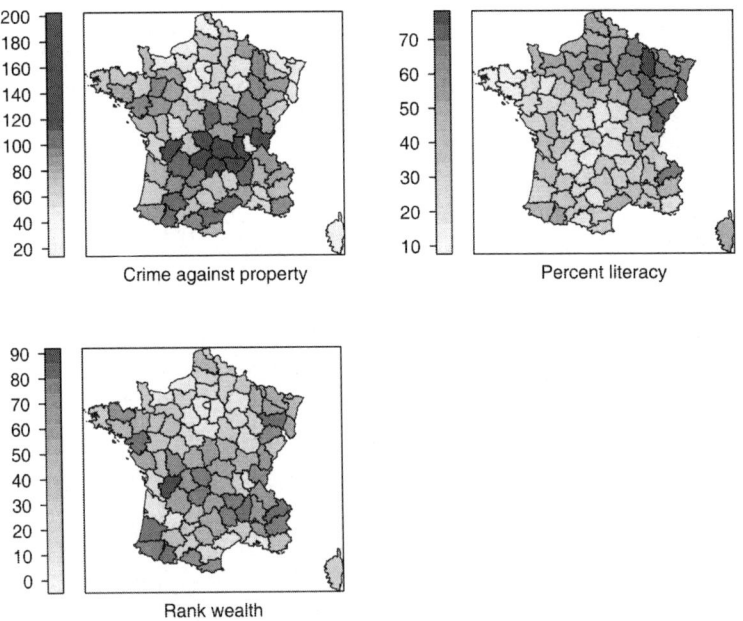

Fig. B.2.5. Choropleth maps of population per crime against property, rank wealth and percentage literacy, France (Friendly 2007)

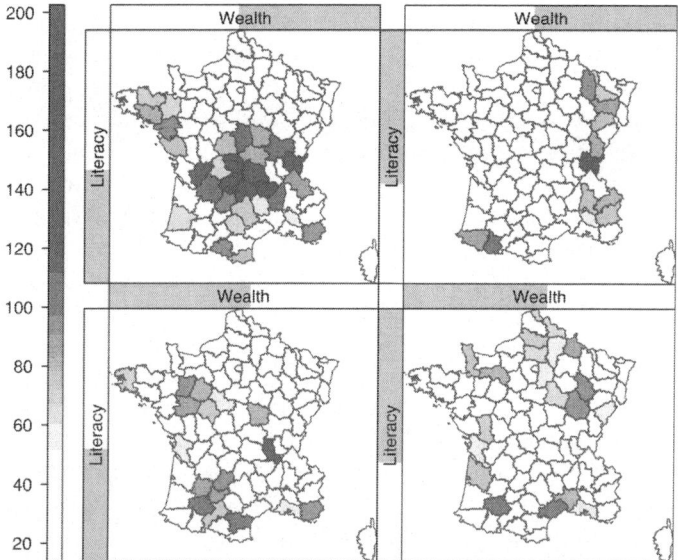

Fig. B.2.6. Choropleth maps of population per crime against property, conditioned on ranked wealth and percentage literacy, France (see Friendly 2007, p.395)

B.2.4 Exploring point patterns and geostatistics

Within the spatial analysis literature, ESDA has often been described as a subset of exploratory data analysis (Anselin 1998; Anselin et al. 2007). In a somewhat broader framework, however, it is perhaps difficult to distinguish ESDA as a subset of EDA, because many other strands feed into it, for example from information visualization and geographical information science, that are not present in EDA itself. It is tempting rather to see EDA as that part of ESDA of relevance to data where observations have no spatial location; such an over-arching view admits geovisualization as a part of ESDA, and places exchanges of knowledge and techniques between cartography and statistical graphics in a more natural context. Note that statisticians often use spatial data sets and objects as vehicles for their presentations (cf. Chambers 2008).

'*Analyzers of spatial data should ... be suspicious of observations when they are unusual with respect to their neighbours*' (Cressie 1993, p.33). This operational definition, buttressed by lively concern about data collection on the one hand and model specification on the other, is reflected in many of the examples presented in Cressie (1993), see also Unwin (1996), Kaluzny et al. (1998), Haining (2003), and Lloyd (2007). Often it is not sufficient to see ESDA as a toolbox of finished tools, because one frequently needs to 'get closer' to the data than the tools allow. This is one of the reasons for placing ESDA within an environment for statistical computing like R (Bivand et al. 2008), where users can engage the data as far as they might wish. Finally, it should be noted that there are topics not yet adequately covered, such as ESDA for categorical data, surveyed and advanced by Boots (2006).

Exploring point patterns. While ESDA is often seen as being applied to areal data, in fact approaches to data analysis derived from EDA are used throughout spatial data analysis. For example, the $\hat{G}$ nearest neighbour distance measure used in point pattern analysis is simply a binned empirical cumulative density function plot of the nearest neighbour distances. Levine (2006) describes how many exploratory tools are provided in CrimeStat in an accessible fashion, and with the possibility of using simulation to see whether the patterns detected by the user ought to be treated as significant. Diggle (2003) gives many examples of the ways in which care in data analysis – respecting the data – informs even the most technically advanced statistical procedures. Baddeley et al. (2005) show how residuals from modelling a point pattern may be explored diagnostically; the *spatstat* package for R provides many ways to explore point patterns (Baddeley and Turner 2005). We will not be considering scan tests in this chapter; their provision in R is reviewed in Gómez-Rubio et al. (2005), and Bivand et al. (2008).

One of the classic data sets provided with R shows the locations of earthquakes near Fiji since 1964; the points in geographical coordinates are accompanied by the depth detected, the magnitude of the event, and the number of stations reporting it. These mean that we can treat it as a marked point pattern, for example using non-overlapping shingles of depth. The `xyplot` function takes a `formula` object as its first argument – this is a symbolic expression of the model to be visualised, here with points to be plotted on longitude and latitude conditioned on a depth shingle.

```
> data(quakes)
> depthgroup <- equal.count(quakes$depth, number = 3, overlap = 0)
> xyplot(lat ~ long | depthgroup, data=quakes, main='Fiji earthquakes',
+        type = c('p', 'g'))
```

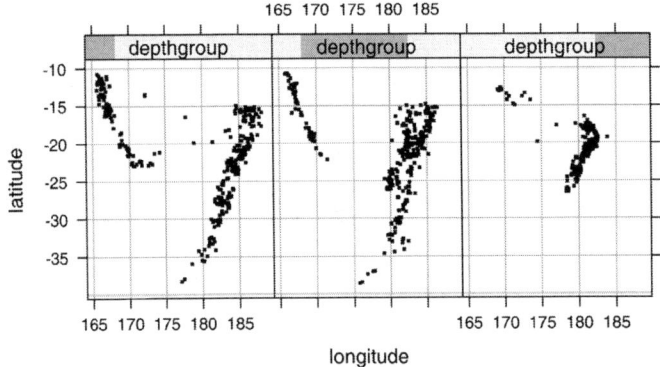

Fig. B.2.7. Seismic events near Fiji since 1964, conditioned on depth

Figure B.2.7 reproduces the conditioning of location on depth for the earthquake events discussed in detail by Murrell (2005, pp.126–141) and Sarkar (2007, pp.67–76). They also show how magnitude may also be visualized on conditioned scatterplots through a further shingle, or shaded symbols. Here we will consider how we might express the relative intensity of the point pattern using kernel smoothing. In order to do this we should project the geographical coordinates to the plane, using an appropriate set of parameters, here a Transverse Mercator projection used on Fiji. We use the default bisquare kernel with three chosen bandwidths, and set kernel values close to zero to NA.

```
> coordinates(quakes) <- c('long', 'lat')
> proj4string(quakes) <- CRS('+proj=longlat')
> quakes_tmerc <- spTransform(quakes, CRS('+init=epsg:3460'))
> library(splancs)
> pl <- bboxx(bbox(quakes_tmerc))
> h150k <- spkernel2d(as.points(coordinates(quakes_tmerc)),
+     poly = pl, h0 = 150000)
> is.na(h150k) <- h150k < .Machine$double.eps
```

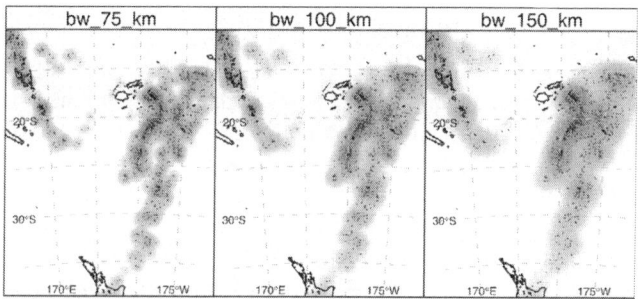

Fig. B.2.8. Kernel density plots of seismic events near Fiji; three increasing bandwidth settings

Figure B.2.8 shows density plots of the earthquake events for three increasing bandwidth values. The panels have also been furnished with shorelines and a graticule to aid in positioning the events. Had we additionally conditioned on depth or magnitude, or added tectonic boundaries, we might have come a little further. However, we can already see clearly that the observed pattern is not likely to be homogeneous. Exploration of point patterns is often helpful in drawing attention to the need to look for covariates that may account for inhomogeneity, or to possible use of a control point pattern to contrast with the observed cases.

Exploratory geostatistics. It is probable that more exploratory spatial data analysis is done in geostatistics than in the remaining domains of spatial data analysis; Cressie (1993) gives many examples. It is easy to grasp why interpolation is crucially dependent on identifying the 'right' model, in terms of the selection of observation locations, the fitting of models of spatial autocorrelation, detecting useful covariates, and checking the appropriateness of assumptions such as isotropy. If a seriously sub-optimal model is chosen, both the predictions themselves and estimates of uncertainty around those values will not be as satisfactory as might have been achieved with the same data. Lloyd (2007) and Müller (2007) provide further discussions of techniques for making good use of the data to hand, and of the design of patterns of sampling locations to improve prediction. Geostatistics is also discussed in Chapter B.6.

Here we will use a data set of precipitation values for Switzerland, discussed in Diggle and Ribeiro (2007, pp.118-121, pp.149-150, pp.169-172), and used in the 'Spatial Interpolation Comparison 97' contest[6]. The examples demonstrate that geostatistics software, here R packages, provides much support for exploratory spatial data analysis, discussed for example by Bivand et al. (2008, pp.192, pp.195-200). Other software adopts the same approach; the Geostatistical Analyst extension to ArcGIS™ is well furnished with ESDA tools.

```
> library(geoR)
> data(SIC)
> plot(sic.100, borders = sic.borders, lowess = TRUE)
```

In the geoR package, the plot method for a geodata object is to make an ESDA graphic display. Setting the lowess= argument permits a smoothed line to be drawn through scatterplots of the data against the x and y coordinates, so that the four-panel display, shown in Fig. B.2.9, conveys a lot of information. On screen, the map symbols are coloured, to draw more attention to the spatial patterning of the quartiles of the variable of interest. We could of course condition a scatterplot of the location coordinates on a shingle of the variable of interest, as presented above. The histogram overplotted with a density line and rug plot shows that the data deserves more exploration, especially if a trend is mixing distributions of precipitation values together. The trend is here taken as the mean of the data, but the smoothed lines suggest that a spatial trend is present, of course in addition to the effect of station elevation, which has not been included here.

Location diagnostics. Should we attempt to add in a spatial trend, or a covariate, we should pay attention of the warning given by Unwin and Wrigley (1987) to use the same diagnostic tools as in any other modelling exercise. It is, as Fig. B.2.10. shows, quite frequently the case that some observations exert a more than proportional influence on the fitted model. The circles are proportional to Cook's influence statistic, and indicate that the distinguished stations ought to be looked at carefully, to see why they differ so much from their near neighbours. Note that most of the distinguished stations are on the edge of the study area.

---

[6] http://www.ai-geostats.org/index.php?id=data

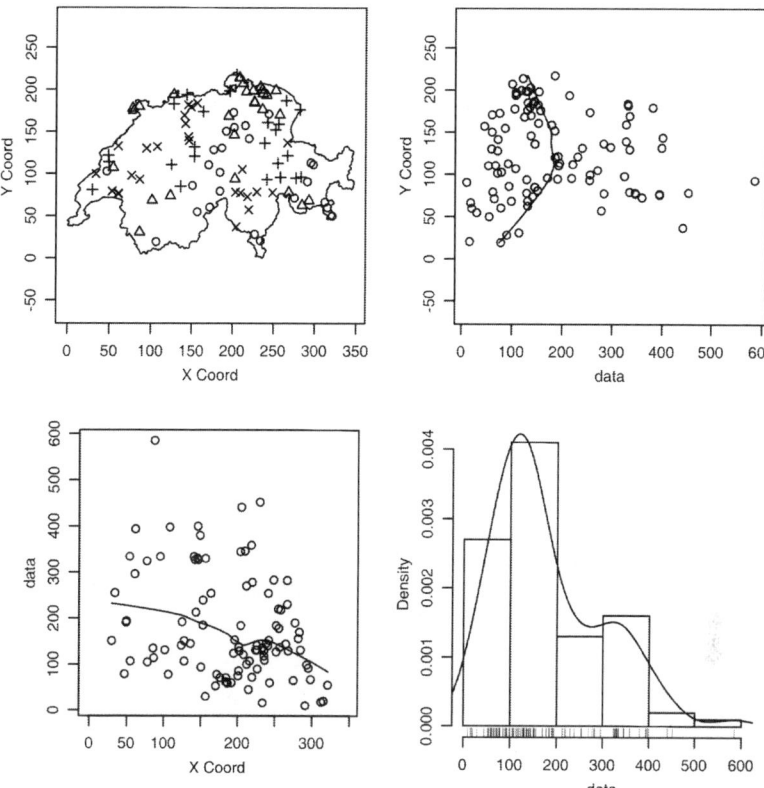

Fig. B.2.9. Exploratory geostatistical display of Swiss precipitation data from the 1997 Spatial Interpolation Comparison contest: a) precipitation quartiles; b) plot of precipitation by northings; c) plot of precipitation by eastings; d) histogram and density of precipitation

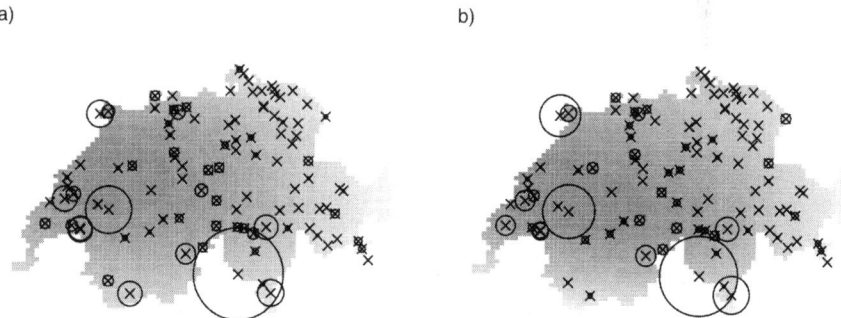

Fig. B.2.10. Influence plots for trend surfaces, Swiss precipitation data, circle radius proportional to Cook's influence statistic (Unwin and Wrigley 1987): a) quadratic trend surface; b) cubic trend surface

Variogram diagnostics. Variogram diagnostics are linked to other steps taken in exploring variables in geostatistics (Pebesma 2004). Using the spatial representations presented in Bivand et al. (2008), we can review some of the tools made available in the *gstat* package. First, we convert the Swiss precipitation data set to a suitable object form, and show a *h*-scatterplot of pairs of observed values conditioned on distance, expressed in the `breaks` argument to `hscat`. The formula interface used here places the variable of interest on the left hand side of the equation, and only the intercept term on the right hand side.

```
> library(gstat)
> sic.100SP <- SpatialPointsDataFrame(SpatialPoints(sic.100$coords),
+     data = data.frame(precip = sic.100$data))
> hscat(precip ~ 1, data = sic.100SP, breaks = seq(0, 120,
+     20))
```

The first diagnostic plot (Fig. B.2.11) is known as an *h*-scatterplot, and conditions a scatterplot of the values at pairs of locations on the binned distance h_{ij} between them; the diagonal lines represent perfect correlation. The sample correlations between the observed values at locations *i* and *j* are perhaps a little hard to read in a monochrome plot, so are repeated in text output, declining from 0.714 in the first 20km bin, to 0.344 between 20 and 40km, and going through zero in the third bin. It appears, then, that nearer observations are more like one another, and that the similarity declines with distance.

By defining a `gstat` object, we can easily create variograms of different kinds by passing this object and additional arguments to `variogram`.

```
> g <- gstat(id = 'precip', formula = precip ~ 1, data = sic.100SP)
> evgm <- variogram(g, cutoff = 100, width = 5)
> revgm <- variogram(g, cutoff = 100, width = 5, cressie = TRUE)
> cevgm <- variogram(g, cutoff = 100, width = 5, cloud = TRUE)
```

Figure B.2.12 shows a variogram cloud plot and a plot of empirical variogram values for twenty 5km wide bins, for classical and robust versions of the variogram. The bin borders are shown to highlight the way in which the empirical variogram is constructed as a measure of central tendency of squared differences in the variable of interest between pairs of points whose inter-point distance falls into the bin. Cressie (1993, pp.74-83) provides the development of a robust estimator, shown with a dashed line in Fig. B.2.12, that reduces the impact of unusually large differences in value between near neighbours. The *fields* package returns number summaries by bin in addition to the classical variogram estimator in output from the `vgram` function.

Figure B.2.13 shows a variogram map, and four empirical variograms for four axes at 0°, 45°, 90° and 135°; the variogram direction lines are coded in the same way on both panels. A variogram map is centred around (0, 0) and has map dimension and cell size similar to cutoff and interval width values; it is constructed by averaging pairs that have distance within a certain bin. In this case, we see that the structure aligned with the 45° direction corresponds to lower variogram values for nearer bins. Recall that here we taking the trend as the mean only, ignoring the impact of large scale spatial trends and covariates.

Directionality. Finally, we follow Bivand et al. (2008, pp.205-206) in examining possible anisotropy in the data set. Using the same bins as earlier, we add arguments to the `variogram` function to create objects for plotting.

```
> mevgm <- variogram(g, cutoff = 100, width = 5, map = TRUE)
> aevgm <- variogram(g, cutoff = 100, width = 5, alpha = c(0,
+ 45, 90, 135))
```

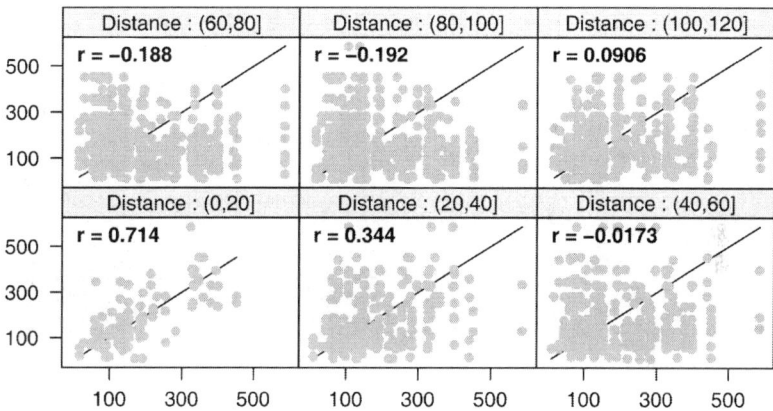

Fig. B.2.11. *h*-scatterplots: scatterplots of pairs of observed values conditioned on distance; sample correlations shown in panels

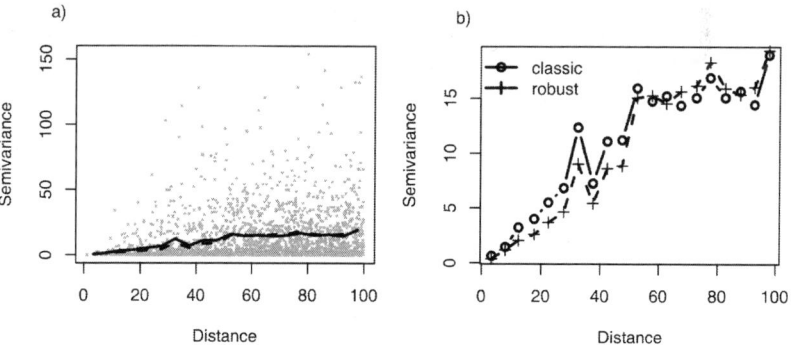

Fig. B.2.12. Swiss precipitation data – binned classic and robust variogram values: a) variogram cloud display; b) variogram values [note that the vertical axis is not in the same scale in a) and b)]

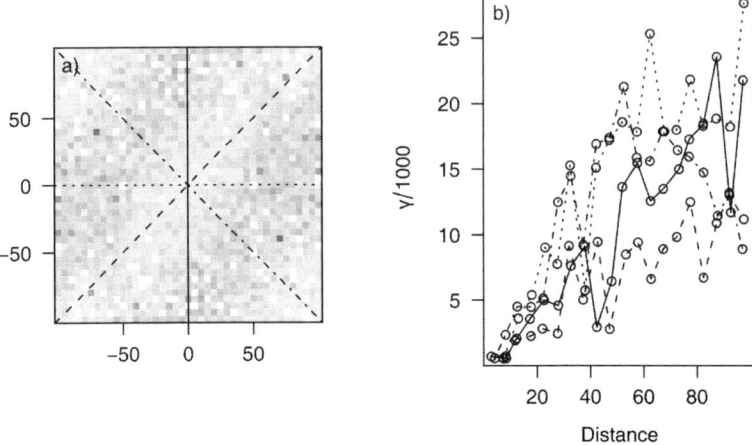

Fig. B.2.13. Detecting directionality in the variogram of Swiss precipitation data: a) variogram map showing binned semivariance values by direction and distance; b) classical variograms for four axes at 0°, 45°, 90° and 135°

B.2.5 Exploring areal data

Much of the literature on exploratory spatial data analysis has focussed on the exploration of areal data with respect to spatial association. In this section, we will look at local indicators of spatial association within this tradition, but will also consider how larger scale regularities may be revealed by using median polish smoothing and Moran eigenvector mapping. A topical area that has not been given enough attention is that of regression diagnostics for fitted spatial regression models (Haining 1994); while users appear to want heteroskedasticity-corrected standard errors, few seem to realise that the mis-specification could arguably be better handled if diagnostic methods had been used (see also Mur and Lauridsen 2007).

Median polish smoothing. Cressie (1993, pp.46-48, pp.393-400) discusses in some detail how smoothing may be used to partition the variation in the data into smooth and rough. Initial use of median polish smoothing is described by Cox and Jones (1981). In order to try it out on the North Carolina SIDS data set, we will use a coarse gridding into four columns and four rows given by Cressie (1993, pp.553-554), where four grid cells are empty; these are given by variables L_id and M_id in object nc.sids. Next we aggregate the number of live births and the number of SIDS cases 1974-1978 for the grid cells.

```
> L_id <- factor(nc.sids$L_id)
> M_id <- factor(nc.sids$M_id)
> both <- interaction(L_id, M_id)
> mBIR74 <- tapply(nc.sids$BIR74, both, sum)
> mSID74 <- tapply(nc.sids$SID74, both, sum)
```

Using the same Freeman-Tukey transformation as is used for the county data, we coerce the data into a correctly configured matrix, some of the cells of which are empty. The `medpolish` function is applied to the matrix, being told to remove empty cells; the function iterates over the rows and columns of the matrix using `median` to extract an overall effect, row and column effects, and residuals.

```
> mFT <- sqrt(1000) * (sqrt(mSID74/mBIR74) + sqrt((mSID74 +
+     1)/mBIR74))
> mFT1 <- t(matrix(mFT, 4, 4, byrow = TRUE))
> med <- medpolish(mFT1, na.rm = TRUE, trace.iter = FALSE)
> med

Median Polish Results (Dataset: 'mFT1')

Overall: 2.909650

Row Effects:
[1]   -0.05686791    -0.37236370     0.05686791     0.79541774

Column Effects:
[1]   -0.005484562   -0.446250551    0.003656375    0.726443256

Residuals:
          [,1]        [,2]       [,3]       [,4]
[1,]   NA         -0.45800    0.000000    0.37556
[2,]  -0.092554    0.00000    0.101695    0.00000
[3,]   0.092554    0.30464   -0.090726   -0.55364
[4,]   NA          NA         0.000000    NA
```

Returning to the factors linking rows and columns to counties, and generating matrices of dummy variables using `model.matrix`, we can calculate fitted values of the Freeman-Tukey adjusted rate for each county, and residuals by subtracting the fitted value from the observed rate. Naturally, the fitted value will be the same for counties in the same grid cell.

```
> mL_id <- model.matrix(~L_id - 1)
> mM_id <- model.matrix(~M_id - 1)
> nc.sids$pred <- c(med$overall + mL_id %*% med$row + mM_id %*%
+     med$col)
> nc.sids$mp_resid <- nc.sids$ft.SID74 - nc.sids$pred
> nc.sids$ft.SID74_c <- scale(nc.sids$ft.SID74, scale = FALSE)
> nc.sids$pred_c <- scale(nc.sids$pred, scale = FALSE)
```

Figure B.2.14 shows the median polish smoothing results as three maps, the observed Freeman-Tukey transformed SIDS rates, the fitted smoothed values, and the residuals.

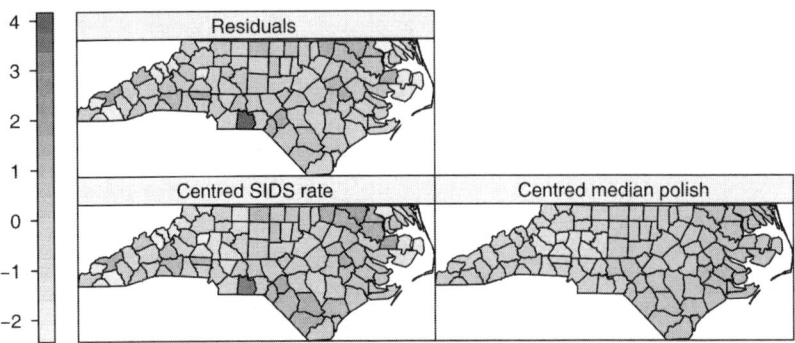

Fig. B.2.14. Median polish for North Carolina SIDS data – the Freeman-Tukey transformed SIDS rates and fitted smoothed values are mean-centred to use the same scale as the residuals

Local indicators of spatial association (LISA). While global measures permit us to test for spatial patterning over the whole study area, it may be the case that there is significant autocorrelation in only a smaller section, which is swamped in the context of the whole. Both distance statistics (Getis and Ord 1992, 1996; Ord and Getis 1995), and the local indicators of spatial association derived by Anselin (1995), resemble passing a moving window across the data, and examining dependence within the chosen region for the site on which the window is centred. The specifications for the window can vary, using perhaps contiguity or distance at some spatial lag from the considered zone or point.

There are clear connections here both to the study of point patterns – although methods for boundary correction have not been specifically added to weighting matrix definitions yet – and to geostatistics, since these statistics have application to the exploration of non-homogeneities in relationships between locations across the study area. They are however subject to a correlation problem when cast in a hypothesis testing framework, that estimated values of the local indicator for neighbouring zones or sites will be correlated with each other because they are necessarily calculated from many of the same values, recalling that neighbouring placements of the moving window will most likely overlap. Ord and Getis (1995) provide suitable adjustments to critical values of the G_i and G_i^* statistics. De Castro and Singer (2006) provide further developments for the appropriate handling of the false discovery rate.

The uses to which local statistics have been put are to identify 'hot-spots', to assess stationarity prior to the use of methods assuming that the data do conform to this assumption, and other checks for heterogeneity in the data series (Getis and Ord 1996). A thorny problem is that local indicators do pick up global patterns if they are present for whatever reason (Ord and Getis 2001). Measures of spatial autocorrelation are discussed in more detail in Chapter B.3.

Implementations of LISA techniques can be found in GeoDa (Anselin et al. 2006), in SAM (Rangel et al. 2006), and in the spatial statistics toolbox of Arc

GIS™, as well as the R versions discussed below (Bivand 2006; Bivand et al. 2008). The availability of software implementations has contributed to a wave of applications in many scientific domains. Scanning just the last two years, it appears that one key application area is in sociology and social policy, ranging from social medicine and fertility (Crighton et al. 2007; Schmertmann et al. 2008), through child care (Anselin et al. 2007; Freisthler et al. 2006; Lery 2008; Voss et al. 2006), to language neighbourhoods (Ishizawa and Stevens 2007), deprivation and mortality (Sridharan et al. 2007) and homicide (Ceccato et al. 2007). Another application area with many contributions is concerned with regional economic performance (Patacchini and Rice 2007; Patacchini and Zenou 2007; Yamamoto 2008), and regional and local development (Portnov 2006; Yu and Wei 2008). Penetration in other areas is also occurring, for example in local genetic structures (Sokal and Thomson 2006) and forestry (Räty and Kangas 2007).

Some but not all of the published cases using LISA techniques are exploratory. All of the papers introducing LISA techniques stress the need for caution in drawing conclusions, because apparent hotspots may rather reflect misspecification – for example the omission from the mean model of an important intermediate variable or the choice of an inappropriate functional form, because constructing tests for very small sets of neighbours even in the absence of misspecification is hard (Tiefelsdorf 2000, 2002; Bivand et al. 2009), and because of the multiple and dependent tests problem (de Castro and Singer 2006). Finally, as Waller and Gotway (2004, p.239) show, it may be necessary to create customised tests acknowledging the construction of the dependent variable, in their case using a constant risk hypothesis.

To present LISA techniques, we will return to the Guerry French moral statistics data set. To begin with, a list of contiguous neighbours is constructed, leaving Corsica with no neighbours (for details of the handling of no-neighbour observations, see Bivand and Portnov 2004).

```
> library(spdep)
> gf_cont <- poly2nb(gfrance)
```

Figure B.2.15 shows the G_i and G_i^* statistic values, scaled as standard deviates, for population per crime against property. The contiguity neighbours are converted into spatial weights using row-standardisation, after, in the G_i^* case, adding in the observations as their own neighbours.

```
> lwW <- nb2listw(gf_cont, zero.policy = TRUE)
> gfrance$local_G <- c(localG(gfrance$Pop_crime, lwW, zero.policy= TRUE))
> lwWs <- nb2listw(include.self(gf_cont))
> gfrance$local_G_star <- c(localG(gfrance$Pop_crime, lwWs))
```

Negative values show which observations are surrounded by observations with similar low values. while positive values show which observations are surrounded by observations with similar high values. Recall that high values show many inhabitants per crime, low values few inhabitants per crime. The value for Corsica, which has no neighbour, is missing for G_i and takes a value proportional to the difference between the global mean and its own inverse crime rate for G_i^*, because then Corsica is its own only neighbour.

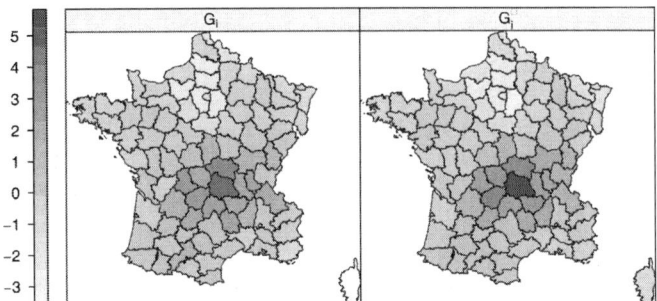

Fig. B.2.15. Local G_i and G_i^* statistics: population per crime against property, France

Since we are using G_i and G_i^* scaled as standard deviates, we will not apply them to residuals of models fitting global coefficients. The local Moran's I_i values are unscaled – they are not standard deviates, so the global Moran's I equals the mean of the local Moran's I_i values.

```
> gfrance$local_I <- localmoran(gfrance$Pop_crime, lwW,
+     zero.policy = TRUE)[, 1]
> mean(gfrance$local_I)

[1] 0.2606168

> moran.test(gfrance$Pop_crime, lwW, zero.policy = TRUE)$estimate[1]

Moran I statistic
        0.2606168
```

Since it may be the case that the local autocorrelation is driven by misspecification, we will try two variants on the null model of treating the mean of population per crime against property as all we know. In addition to the null model, we will fit a simultaneous autoregressive model with only an intercept; the autoregressive coefficient is significant, and the model fit improves from the null baseline.

```
> C_p_esar <- spautolm(Pop_crime ~ 1, gfrance, lwW, zero.policy = TRUE,
+     method = 'Matrix')
> coef(C_p_esar)

(Intercept)    lambda
 76.502332   0.470789

> gfrance$local_I_err <- localmoran(residuals(C_p_esar),
+     lwW, zero.policy = TRUE)[, 1]
```

The second variant is to fit a linear model using the percentage literacy and rank wealth variables as suggested in the conditioned choropleth map example. The coefficient for percentage literacy is negative, which – recalling that the crime rate is inverted – means that higher literacy is associated with more crime. The rank wealth coefficient is positive because lower rank means higher wealth, hence lower rank is linked to more crime.

```
> C_px_lm <- lm(Pop_crime ~ Literacy + Wealth, gfrance)
> coef(C_px_lm)

(Intercept)      Literacy       Wealth
 75.7783729    -0.4569233    0.4733127

> lm.morantest(C_px_lm, lwW, zero.policy = TRUE)$estimate[1]

Observed Moran's I
       0.06888486

> gfrance$local_I_xlm <- localmoran(residuals(C_px_lm),
+    lwW, zero.policy = TRUE)[, 1]
```

This model fits the data much better than the simultaneous autoregressive null model, and, as Friendly (2007, p.396) reports, accounts for somewhat over a quarter of the variation in the dependent variable. The residuals of this model show no global autocorrelation, and a simultaneous autoregressive model with these variables included does not improve the fit. As Fig. B.2.16 shows, there is much more 'action' in the left-hand panel, where we only model the data by the mean.

Both of the areas picked out in Fig. B.2.15: the Île-de-France in the north-central part of the country with low values of the statistic, and today's Auvergne region in the south-central part of the country with high values, corresponding to values of the inverted crime rate, have higher values of Moran's I_i. Observations with intermediate values of G_i have low values of I_i, because they represent places with neighbours with inverted crime rates unlike their own. Moving to the right in Fig. B.2.16, we see that the range of shading is compressed, as the effects of misspecification are removed. The very low value in Rhône (mid-southeast) in the map of I_i for the null model and the residuals of the simultaneous autoregressive null model is removed once the covariates are included (the large and relatively wealthy city of Lyon is atypical of its surroundings). In the map of I_i for the residuals of the linear model with covariates, Puy-de-Dôme in the Auvergne still has a large value of the statistic, suggesting that the inverse crime rate is even higher in the Auvergne than one would expect from the levels of wealth and literacy (or their absence) observed there.

We will make a LISA plot using a conditioned choropleth map, conditioning the observed Moran's I_i for the null model on factors capturing the Moran scatterplot quadrants in which the observations fall (Anselin 1996). The factors take values c ('L', 'H') depending on whether the observations are above or below the mean of the inverse crime rate, and above or below the mean of the spatial lag of the inverse crime rate.

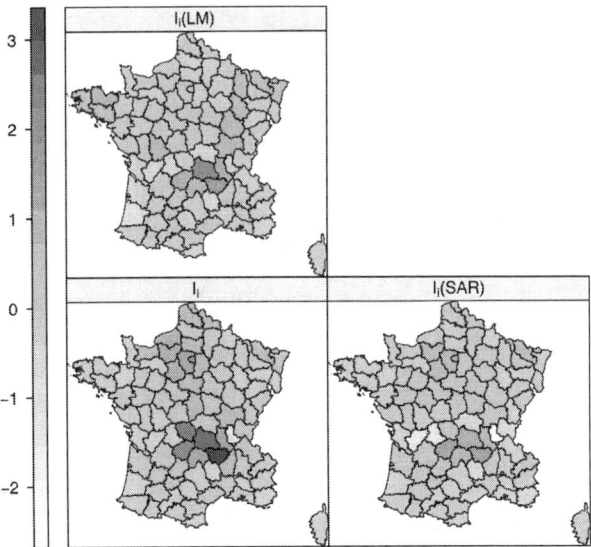

Fig. B.2.16. Local I_i statistics for the null model, the residuals of the simultaneous autoregressive model, and the residuals of the linear model including literacy and wealth: population per crime against property, France

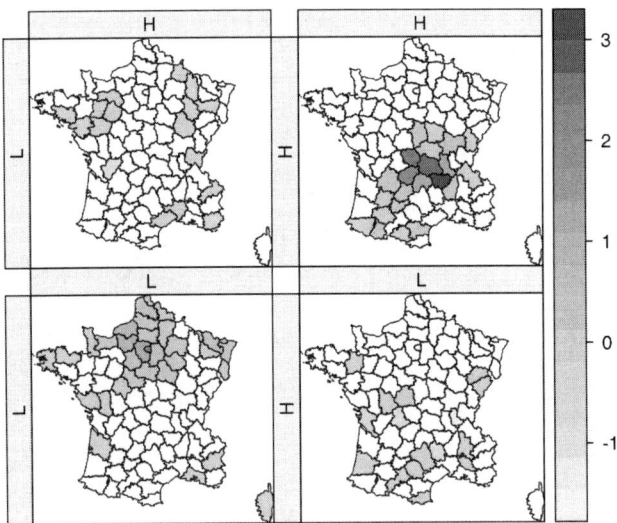

Fig. B.2.17. Conditioned choropleth LISA map: Moran's I_i for the null model conditioned on the LISA quadrant; first letter above, second letter left

Figure B.2.17 shows the split in the null model between the HH 'cluster' in the Auvergne, with high values of the inverse crime rate observed for the observations and their neighbours, and the LL 'cluster' in Île-de-France, with low values of the inverse crime rate observed for the observations and their neighbours. The HL and LH panels do not display patterns that are as clear.

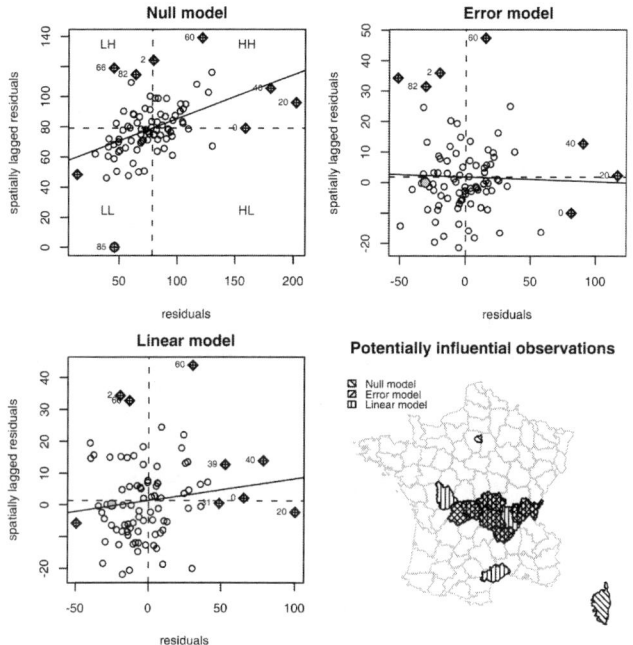

Fig. B.2.18. Moran scatterplots for a) null; b) simultaneous autoregressive; c) linear model with covariates; and d) influence map for the three models; the dashed lines divide the scatterplots into the LISA LL, HL, LH, and HH quadrants

Finally, Fig. B.2.18 shows Moran scatterplots for all three models, the null model, the simultaneous autoregressive null model, and the linear model with covariates (Anselin 1996). Interestingly, the observations found to extert influence on the linear relationship between the residuals from the models of the inverse crime rate and its spatial lag are largely the same ones across models, and form a belt stretching west and east from the Auvergne east to the Swiss border. These observations could be exerting such consistent influence because of measurement issues with the inverted crime rate, or because of remaining model mis-specification and pointing up such unusual observations is among the reasons for engaging in exploratory data analysis. Li et al. (2007) propose a approximate profile-likelihood estimator for spatial autocorrelation, which also has an ESDA extension, including a scatterplot and a local APLE measure.

```
> sPc <- scale(gfrance$Pop_crime, scale = FALSE)
> aple(sPc, lwW)
[1]    0.4810092
> aple_res <- aple.plot(sPc, lwW)
> crossprod(aple_res$Y, aple_res$X)/crossprod(aple_res$X)
          [,1]
[1,] 0.4810092
> gfrance$localAple <- localAple(sPc, lwW)
```

As Fig. B.2.19 shows, the new measure provides a view of the data that is not dissimilar to that of local Moran's I_i. The scatterplot shows that the same observations exert influence, and the map of values shows the same impact of higher positive local autocorrelation in the Auvergne and Île-de-France regions.

Two further avenues will be left unexplored here. First, it is possible that some of the problems in exploring the inverse crime rate come from the greater uncertainty of rate estimates for observations with small populations, and using an Empirical Bayes smoothing procedure may be appropriate. Second, the crime count with a log population offset term could be modelled using Poisson regression, and the deviance or Pearson residuals explored for spatial patterning.

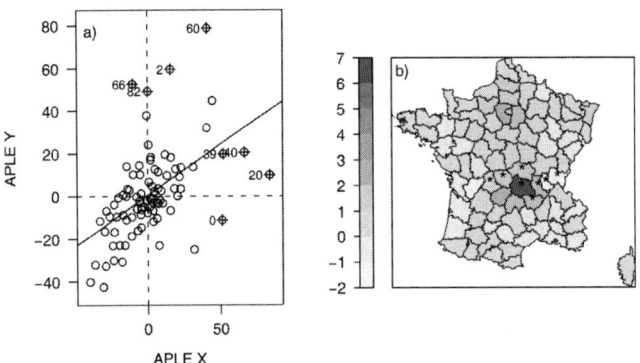

Fig. B.2.19. APLE plot and local APLE values for the population per crime rate:
a) approximate profile-likelihood estimator plot, showing observations with influence;
b) local APLE values, with observations with influence marked by asterisks

Scale. There are close relationships between the graph structure of spatial weights, and the structure exposed by examining the eigenfunctions of a centred weights matrix (Griffith 2003; Tiefelsdorf 2000), relationships underlying the understanding of Moran's I. It has been suggested by Griffith (2003) that maps of eigenvectors may be used to explore the effect of scale, because some eigenvectors will show large scale structures, others will capture regional differences, and others again will represent small scale patterns. Naturally, the choice of a different spatial weights matrix may give a different view on patterning at different spatial scales.

```
> SF1 <- SpatialFiltering(Pop_crime ~ 1, data = gfrance,
+ nb = gf_cont, style = 'W', zero.policy = TRUE, tol = 0.5,
+ verbose = FALSE)
> SF2 <- SpatialFiltering(Pop_crime ~ Literacy + Wealth,
+   data = gfrance, nb = gf_cont, style = 'W', zero.policy = TRUE,
+   tol = 0.5, verbose = FALSE)
```

Here we show the eigenvector maps for the eigenvectors chosen by semparametric spatial filtering for the null model and the linear model with covariates (Tiefelsdorf and Griffith 2007). Figure B.2.20 shows the six eigenvectors chosen to remove the residual spatial autocorrelation from the null model. The first eigenvector chosen is shown on the upper left, and displays a smooth, almost linear trend. The next two chosen on the upper row show regional patterns, something like quadratic and cubic trend surfaces. On the lower row, the chosen eigenvectors pick up smaller scale patterns.

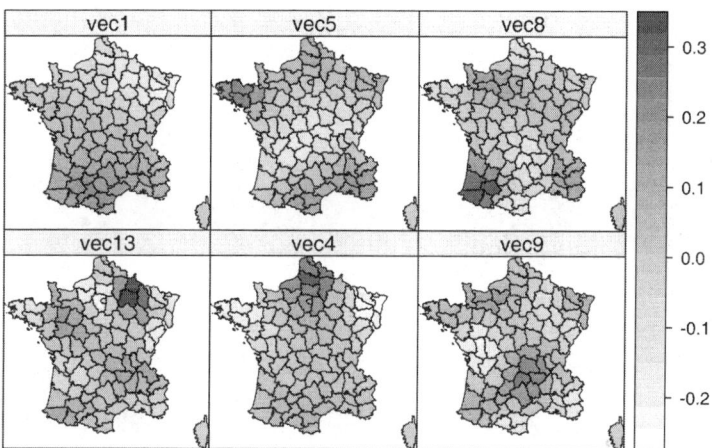

Fig. B.2.20. Six eigenvector maps for eigenvectors: null model

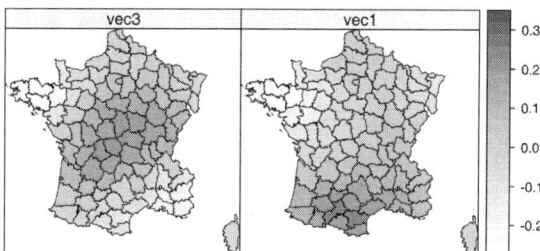

Fig. B.2.21. Two eigenvector maps for eigenvectors: linear model with covariates

Figure B.2.21 shows the two eigenvectors chosen to remove the residual spatial autocorrelation from the linear model with covariates. Because the same palette is used in Figs. B.2.20 and B.2.21, we can see how much of the residual autocorrelation has been removed by the covariates. Note that the eigenvectors differ because they are centred using the model projection matrices, so that their maps are not the same. Perhaps the patterning remaining in the linear model with covariates residuals signals that not all the mis-specification has been removed.

Geographically weighted approaches. Non-stationarity is a further source of misspecification, such as omitted variables or inappropriate functional forms. It may be approached through geographical weighting, passing a kernel with a given bandwidth over the map of data points in order to compute weighted regressions at fit points. The weights are proportional to the distances between the data points and the fit points (Brunsdon et al. 1998; Fotheringham et al. 2002). A change of support is involved, because the observation polygons are replaced by the polygon centroids, here both for the data points and the fit points.

```
> library(spgwr)
> GWfrance_bw100km1 <- gw.cov(gfrance, 'Pop_crime', bw = 1e+05,
+     cor = FALSE)
```

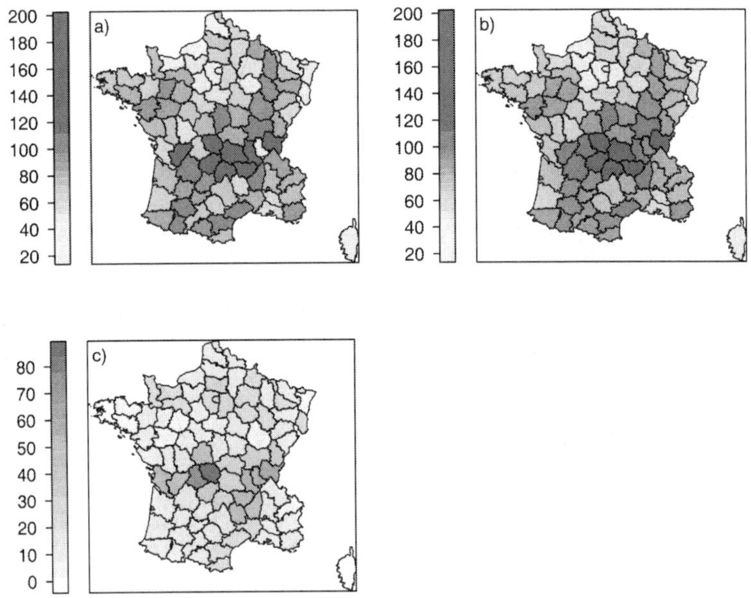

Fig. B.2.22. Population per crime against property: a) population per crime against property; b) geographically weighted means; and c) geographically weighted standard deviations

Taking a bandwidth of 100km and the default Gaussian kernel, we can calculate geographically weighted measures for the inverted crime rate (Dykes and Brunsdon 2007). Figure B.2.22 repeats the map of the inverted crime rate for reference, and shows the input variable and its geographically weighted mean using the same class intervals and palette. A smaller bandwidth would have yielded less smoothing, a larger bandwidth more, as Dykes and Brunsdon (2007) visualize.

Turning to the geographically weighted standard deviations, there seems to be some patterning, with observations apparently very unlike their neighbours being highlighted. However, recall that we are dealing with a rate variable, population per crime against property, where our confidence about the rate estimate should be related to population size. Figure B.2.23 shows a map of geographically weighted standard deviations for the chosen bandwidth conditioned on a shingle of the 1831 population. Although the picture is not very clear, it does seem that some of the observations with smaller populations have larger geographically weighted standard deviations. Obvious exceptions are the observations including the large cities of Lyon and Bordeaux, which were not like their rural neighbours in the first half of the Nineteenth century.

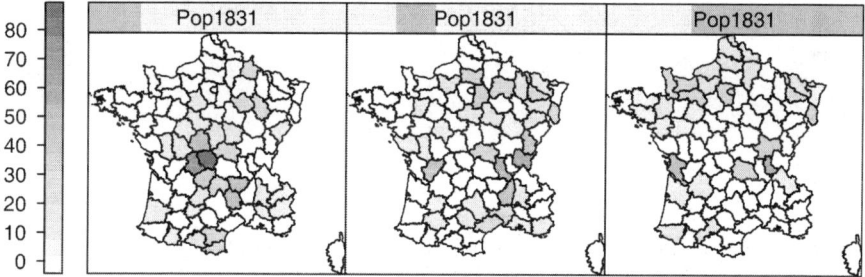

Fig. B.2.23. Conditioned choropleth map of the geographically weighted standard deviation on the inverted crime rate, conditioned on population size

Geographically weighted regression. Extending the geographically weighted approach to geographically weighted regression, we can fit our linear model with covariates using the same bandwidth and support.

```
> GWfrance_bw100km <- gwr(Pop_crime ~ Literacy + Wealth,
+     data = gfrance, bandwidth = 1e+05, hatmatrix = TRUE)
```

Figure B.2.24 shows maps of the geographically weighted regression coefficients and the coefficient of determination. As Wheeler and Tiefelsdorf (2005) point out, the GW coefficients may be highly negatively correlated with each other, as we see is the case between the intercept term and the percent literacy coefficient – the maps are almost mirror images of each other. It may be helpful to refer back to the maps of the variables shown in Fig. B.2.5; there are some similarities in spatial patterning between the covariates and the geographically weighted regression coefficients, given smoothing by the kernel employed. Since collinearity is present, it is hard to conclude unequivocally that the variation in the geographically weighted regression coefficients demonstrates non-stationarity, although it is very possible that the present linear model with covariates remains mis-specified.

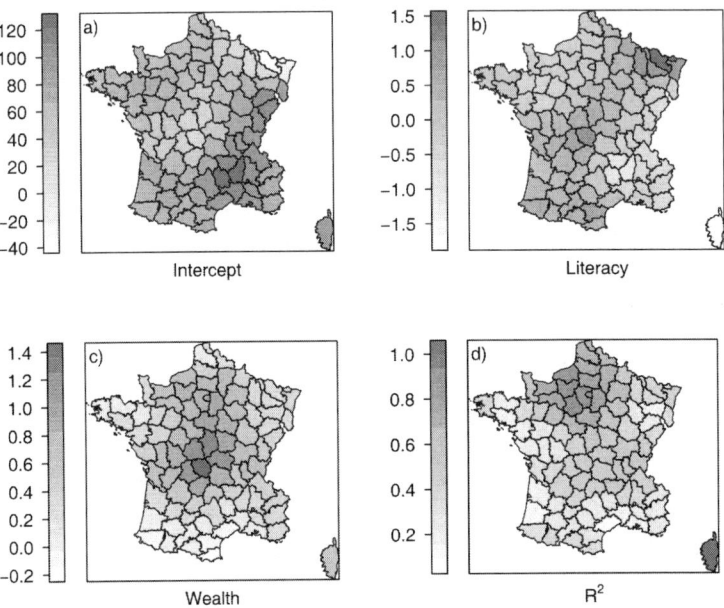

Fig. B.2.24. Maps of geographically weighted regression coefficients; a) intercept; b) percent literacy; c) rank wealth; and d) the coefficient of determination

Finally, as earlier, we also have a problem with Corsica, which had no contiguous spatial neighbours, and which here has almost no weight on any other observation for this bandwidth and kernel (sum.w).

```
> Corse <- which(gfrance$Department == 'Corse')
> as(GWfrance_bw100km$SDF, 'data.frame')[, c(1:5)][Corse,
+     ]
      sum.w X.Intercept. Literacy    Wealth       R2
85  1.032410    101.5770  -1.67253 0.7115072 0.9971373
> sapply(as(GWfrance_bw100km$SDF, 'data.frame')[, c(1:5)],
+     rank)[Corse, ]
     sum.w   X.Intercept.    Literacy    Wealth    R2
         1             77           1        64    86
```

It has extreme local coefficient values (shown by value and rank here) and a coefficient of determination of close to unity, which, although unimportant in themselves, do affect the visualization by stretching the range of values to be displayed. The use of an adaptive kernel perhaps have helped, but may make the interpretation of the output more complex.

B.2.6 Concluding remarks

This chapter should by now have shown that there are many EDA, geovisualization, and ESDA tools and techniques, and that many are implemented and available. There are however still two issues to be addressed: the tendency for *exploratory* analysis – looking for the 'right' question – to slide into inference, be it formalised or not, without considering the implications. In some cases, it can lead to the insertion of a kind of geographical particularism into our understanding of data generation processes. This is unfortunate, because it implies that our understanding of phenomena of interest is dominated by spatially structured (and/or unstructured) random effects, that the undocumented spatial autocorrelation is at the centre of our endeavours.

The second issue was taken up in the introduction: the assumption that the analyst does want to find the 'right' question. Krivoruchko and Bivand (2009, p.17) have discussed the wide range of user motivations encountered: '*In some cases, users are neither able to make nor interested in making an appropriate choice of method ... In other cases, users are more like developers, working much more closely with the software in writing scripts and macros, and in trying out new models.*'

This suggests that the problem may be addressed by making the methods easier to use, by documenting them better, and offering training. It may additionally mean drawing attention to the possible benefits of doing the analysis at hand responsibly, something which is far from simple in check-box organisations, or even when academic supervisors or referees impose their views on analyses rather than empower the analyst to move towards a better question. It is not a coincidence that many early publications on EDA appeared in newsletters concerned with the teaching of statistics and data analysis.

Perhaps it is the case that using EDA and ESDA may not get you tenure quickly, getting to right questions takes time, luck, experience, and often participation in a scientific community willing to share insights and advice. On the other hand, when the research questions actually do matter, improving the way that they are framed is not a trivial achievement, and it is this that is the purpose of exploratory data analysis.

Acknowledgements. The author would like to thank the editors, an anonymous referee, and participants at a spatial statistics session at the 55th North American Meetings of the Regional Science Association International, Brooklyn, November 2008, for helpful comments and suggestions for improvements.

References

Andrienko GL, Andrienko NV (1999) Interactive maps for visual data exploration. Int J Geogr Inform Sci 13(4):355-374
Anselin L (1995) Local indicators of spatial association – LISA. Geogr Anal 27(2):93-115
Anselin L (1996) The moran scatterplot as an esda tool to assess local instability in spatial association. In Fischer MM, Scholten HJ, Unwin D (eds) Spatial analytical perspectives on GIS. CRC Press (Taylor and Francis Group), Boca Raton [FL], London and New York, pp.111-125
Anselin L (1998) Exploratory spatial data analysis in a geocomputational environment. In Longley PA, Brooks SM, McDonnell R, MacMillan W (eds) Geocomputation: a primer.Wiley, New York, Chichester, Toronto and Brisbane, pp.77-94
Anselin L, Syabri I, Kho Y (2006) GeoDa: an introduction to spatial data analysis. Geogr Anal 38(1):5-22
Anselin L, Sridharan S, Gholston S (2007) Using exploratory spatial data analysis to leverage social indicator databases: the discovery of interesting patterns. Soc Ind Res 82(2):287-309
Baddeley A, Turner R (2005) spatstat: An R package for analyzing spatial point patterns. J Stat Software 12(6):1-42
Baddeley A, Turner R, Möller J, Hazelton M (2005) Residual analysis for spatial point processes (with discussion). J Roy Stat Soc B67(5):617-666
Bailey TC, Gatrell AC (1995) Interactive spatial data analysis. Longman, Harlow
Becker RA, Cleveland WS, Shyu MJ (1996) The visual design and control of trellis display. J Comput Graph Stat 5(2):123-155
Becker RA, Cleveland WS, Wilks AR (1987) Dynamic graphics for data analysis. Stat Sci 2(4):355-383
Bivand RS (2006) Implementing spatial data analysis software tools in R. Geogr Anal 38(1):23-40
Bivand RS, Portnov BA (2004) Exploring spatial data analysis techniques using R: the case of observations with no neighbours. In Anselin L, Florax RJGM, Rey SJ (eds) Advances in spatial econometrics: methodology, tools, applications. Springer, Berlin, Heidelberg and New York, pp.121-142
Bivand RS, Müller W, Reder M (2009) Power calculations for global and local Moran's I. Comput Stat Data Anal 53(8):2859-2872
Bivand RS, Pebesma EJ, Gómez-Rubio V (2008) Applied spatial data analysis with R. Springer, Berlin, Heidelberg and New York
Boots B (2006) Local configuration measures for categorical spatial data: binary regular lattices. J Geogr Syst 8(1):1-24
Brewer CA, Pickle L (2002) Comparison of methods for classifying epidemiological data on choropleth maps in series. Ann Assoc Am Geogr 92(4):662-681
Brewer CA, MacEachren AM, Pickle LW, Herrmann DJ (1997) Mapping mortality: evaluating color schemes for choropleth maps. Ann Assoc Am Geogr 87(3):411-438
Brunsdon C (1998) Exploratory spatial data analysis and local indicators of spatial association with XLISP-STAT. The Statistician 47(3):471-484

Brunsdon C, Fotheringham AS, Charlton M (1998) Geographically weighted regression – modelling spatial non-stationarity. The Statistician 47(3):431-443

Carr DB, Wallin J, Carr D (2000) Two new templates for epidemiology applications: linked micromap plots and conditioned choropleth maps. Stat Med 19(17/18):2521-2538

Carr DB, White D, MacEachren A (2005) Conditioned choropleth maps and hypothesis generation. Ann Assoc Am Geogr 95(1):32-53

de Castro MC, Singer BH (2006) Controlling the false discovery rate: a new application to account for multiple and dependent tests in local statistics of spatial association. Geogr Anal 38(2):180-208

Ceccato V, Haining R, Kahn T (2007) The geography of homicide in Sao Paulo, Brazil. Environm Plann A39(7):1632-1653

Chambers JM (2008) Software for data analysis: programming with R. Springer, New York

Cleveland WS (1993) Visualizing data. Hobart Press, Summit [NJ]

Cook D, Swayne DF (2007) Interactive and dynamic graphics for data analysis. Springer, Berlin, Heidelberg and New York

Cook D, Majure J, Symanzik J, Cressie NAC (1996) Dynamic graphics in a GIS: exploring and analyzing multivariate spatial data using linked software. Comput Stat 11(4):467-480

Cook D, Symanzik J, Majure J, Cressie NAC (1997) Dynamic graphics in a GIS: more examples using linked software. Comput Geosci 23(4):371-385

Cox NJ, Jones K (1981) Exploratory data analysis. In Wrigley N, Bennett RJ (eds) Quantitatve geography. Routledge and Kegan Paul, London, pp.135-143

Cressie NAC (1993) Statistics for spatial data (revised edition). Wiley, New York, Chichester, Toronto and Brisbane

Crighton EJ, Elliott SJ, Moineddin R, Kanaroglou P, Upshur REG (2007) An exploratory spatial analysis of pneumonia and influenza hospitalizations in Ontario by age and gender. Epidemi Infect 135(2):253-261

Diggle PJ (2003) Statistical analysis of spatial point patterns (2nd edition). Arnold, London

Diggle PJ, Ribeiro PJR (2007) Model-based geostatistics. Springer, Berlin, Heidelberg and New York

Dorling D (1993) Map design for Census mapping. Cartogr J 30(2):167-183

Dorling D (1995) Visualizing changing social-structure from a Census. Environm Plann A27(3):353-378

Durham H, Dorling D, Rees P (2006) An online Census atlas for everyone. Area 38(3):336-341

Dykes JA (1997) Exploring spatial data representation with dynamic graphics. Comput Geosci 23(4):345-370

Dykes JA (1998) Cartographic visualization: exploratory spatial data analysis with local indicators of spatial association using Tcl/Tk and cdv. The Statistician 47(3):485-497

Dykes JA, Brunsdon C (2007) Geographically weighted visualization: interactive graphics for scale-varying exploratory analysis. IEEE Transact Visual Comput Graph 13(6):1161-1168

Dykes JA, Mountain D (2003) Seeking structure in records of spatio-temporal behaviour: visualization issues, efforts and applications. Comput Stat Data Anal 43(4):581-603

Fischer MM, Stumpner P (2009) Income distribution dynamics and cross-region convergence in Europe. In Fischer MM, Getis A (eds) Handbook of applied spatial analysis. Springer, Berlin, Heidelberg and New York, pp.599-627

Fotheringham AS, Brunsdon C, Charlton M (2002) Geographically weighted regression: the analysis of spatially varying relationships. Wiley, New York, Chichester, Toronto and Brisbane

Freisthler B, Lery B, Gruenewald PJ, Chow J (2006) Methods and challenges of analyzing spatial data for social work problems: the case of examining child maltreatment geographically. Soct Work Rest 30(4):198-210

Friendly M (2007) A.-M. Guerry's moral statistics of France: challenges for multivariable spatial analysis. Stat Sci 22(3):368-399

Gahegan M (1999) Four barriers to the development of effective exploratory visualisation tools for the geosciences. Int J Geogr Inform Sci 13(4):289-309

Getis A (2009) Spatial Autocorrelation. In Fischer MM, Getis A (eds) Handbook of applied spatial analysis. Springer, Berlin, Heidelberg and New York, pp.255-279

Getis A, Ord JK (1992) The analysis of spatial association by the use of distance statistics. Geogr Anal 24(2):189-206

Getis A, Ord JK (1996) Local spatial statistics: an overview. In Longley P, Batty M (eds) Spatial analysis: modelling in a GIS environment. GeoInformation International, Cambridge, pp.261-277

Gómez-Rubio V, Ferrándiz-Ferragud J, López-Quílez A (2005) Detecting clusters of disease with R. J Geogr Syst 7(2):189-206

Griffith DA (2003) Spatial autocorrelation and spatial filtering: gaining understanding through theory and scientific visualization. Springer, Berlin, Heidelberg and New York

Haining R (1994) Diagnostics for regression modeling in spatial econometrics. J Reg Sci 34(3):325-341

Haining RP (2003) Spatial data analysis: theory and practice. Cambridge University Press, Cambridge

Haslett J (1992) Spatial data-analysis challenges. The Statistician 41(3):271-284

Haslett J, Bradley R, Craig P, Unwin A, Wills G (1991) Dynamic graphics for exploring spatial data with application to locating global and local anomalies. Am Stat 45(3):234-242

Ishizawa H, Stevens G (2007) Non-english language neighborhoods in Chicago, Illinois, 2000. Soc Sci Res 36(3):1042-1064

Jacoby WG (1997) Statistical graphics for univariate and bivariate data. Sage, Thousand Oaks [CA]

Kaluzny SP, Vega SC, Cardoso TP, Shelly AA (1998) S+SpatialStats, user manual for Windows and UNIX. Springer, Berlin, Heidelberg and New York

Krivoruchko K, Bivand R (2009) GIS, users, developers, and spatial statistics: on monarchs and their clothing. In Pilz J (ed) Interfacing geostatistics and GIS. Springer, Berlin, Heidelberg and New York, pp.203-222

Lery B (2008) A comparison of foster care entry risk at three spatial scales. Subst UseMisuse 43(2):223-237

Levine N (2006) Crime mapping and the CrimeStat program. Geogr Anal 38(1):41-56

Li H, Calder CA, Cressie NAC (2007) Beyond Moran's I: testing for spatial dependence based on the spatial autoregressive model. Geogr Anal 39(4):357-375

Lloyd CD (2007) Local models for spatial analysis. CRC Press (Taylor and Francis Group), Boca Raton [FL], London and New York

MacEachren A, Gahegan M, Pike W (2004a) Visualization for constructing and sharing geo-scientific concepts. Proceedings of the National Academy of Sciences of the United States of America 101 (Suppl. 1), pp.5279-5286

MacEachren A, Gahegan M, Pike W, Brewer I, Cai G, Lengerich E, Hardisty F (2004b) Geovisualization for knowledge construction and decision support. IEEE Comp Graph Appl 24(1):13-17

Monmonier MS (1989) Geographic brushing: enhancing exploratory analysis of the scatterplot matrix. Geogr Anal 21(1):81-84

Müller W (2007) Collecting spatial data. Springer, Berlin, Heidelberg and New York

Mur J, Lauridsen J (2007) Outliers and spatial dependence in cross-sectional regressions. Environ Plann A39(7):1752-1769

Murrell P (2005) R Graphics. CRC Press (Taylor and Francis Group), Boca Raton [FL], London and New York

Oliver M (2009) The variogram and kriging. In Fischer MM, Getis A (eds) Handbook of applied spatial analysis. Springer, Berlin, Heidelberg and New York, pp.319-352

Ord JK, Getis A (1995) Local spatial autocorrelation statistics: distributional issues and an application. Geogr Anal 27(3):286-306

Ord JK, Getis A (2001) Testing for local spatial autocorrelation in the presence of global autocorrelation. J Reg Sci 41(3):411-432

Patacchini E, Rice P (2007) Geography and economic performance: exploratory spatial data analysis for Great Britain. Reg Stud 41(4):489-508

Patacchini E, Zenou Y (2007) Spatial dependence in local unemployment rates. J Econ Geogr 7(2):169-191

Pebesma E (2004) Multivariable geostatistics in S: the gstat package. Comput Geosci 30(7):683-691

Portnov BA (2006) Urban clustering, development similarity, and local growth: a case study of Canada. Europ Plann Stud 14(9):1287-1314

R Development Core Team (2008) R: a language and environment for statistical computing. R Foundation for Statistical Computing, Vienna, Austria, URL http://www.R-project.org, ISBN 3-900051-07-0

Rangel TFLVB, Diniz-Filho JAF, Bini LM (2006) Towards an integrated computational tool for spatial analysis in macroecology and biogeography. Glob Ecol Biogeogr 15(4):321-327

Räty M, Kangas A (2007) Localizing general models based on local indices of spatial association. Europ J Forest Res 126(2):279-289

Sarkar D (2007) Lattice multivariate data visualization with R. Springer, Berlin, Heidelberg and New York

Schabenberger O, Gotway CA (2005) Statistical methods for spatial data analysis. CRC Press (Taylor and Francis Group), Boca Raton [FL], London and New York

Schmertmann CP, Potter JE, Cavenaghi SM (2008) Exploratory analysis of spatial patterns in Brazil's fertility transition. Popul Res Pol Rev 27(1):1-15

Slocum TA, McMaster RB, Kessler FC, Howard HH (2005) Thematic cartography and geographical visualization. Prentice-Hall, Upper Saddle River [NJ]

Sokal R, Thomson B (2006) Population structure inferred by local spatial autocorrelation: an example from an Amerindian tribal population. Am J Phys Anthr 129(1):121-131

Sridharan S, Tunstall H, Lawder R, Mitchell R (2007) An exploratory spatial data analysis approach to understanding the relationship between deprivation and mortality in Scotland. Soc Sci Med 65(9):1942-1952

Symanzik J, Cook D, Lewin-Koh N, Majure J, Megretskaia I (2000) Linking ArcView (TM) and XGobi: insight behind the front end. J Comput Graph Stat 9(3):470-490

Takatsuka M, Gahegan M (2002) GeoVISTA studio: a codeless visual programming environment for geoscientific data analysis and visualization. Comput Geosci 28(10):1131-1144

Theus M (2002) Interactive data visualization using mondrian. J Stat Software 7(11):1-9

Tiefelsdorf M (2000) Modelling spatial processes: the identification and analysis of spatial relationships in regression residuals by means of Moran's I. Springer, Berlin, Heidelberg and New York

Tiefelsdorf M (2002) The saddlepoint approximation of Moran's I and local Moran's I_i reference distributions and their numerical evaluation. GeogrAnal 34(3):187-206

Tiefelsdorf M, Griffith DA (2007) Semiparametric filtering of spatial autocorrelation: the eigenvector approach. Environ Plann A39(5):1193-1221

Tukey JW (1977) Exploratory data analysis. Addison-Wesley, Reading [MA]

Unwin A (1996) Exploratory spatial analysis and local statistics. Comput Stat 11(4):387-400

Unwin DJ, Wrigley N (1987) Towards a general-theory of control point distribution effects in trend surface models. Comput Geosci 13(4):351-355

Velleman P, Hoaglin D (1981) The ABC's of EDA: applications, basics, and computing of exploratory data analysis. Duxbury, Boston

Voss PR, Long DD, Hammer RB, Friedman S (2006) County child poverty rates in the US: a spatial regression approach. Popul Res Pol Rev 25(4):369-391

Waller LA, Gotway CA (2004) Applied spatial statistics for public health data. Wiley, New Jersey

Wheeler DC, Tiefelsdorf M (2005) Multicollinearity and correlation among local regression coefficients in geographically weighted regression. J Geogr Syst 7(2):161-187

Wood J, Dykes J, Slingsby A, Clarke K (2007) Interactive visual exploration of a large spatio-temporal dataset: reflections on a geovisualization mashup. IEEE Transact Visual Compu Graph 13(6):1176-1183

Yamamoto D (2008) Scales of regional income disparities in the USA, 1955- 2003. J Econ Geogr 8(1):79-103

Yu D, Wei YD (2008) Spatial data analysis of regional development in Greater Beijing, China, in a GIS environment. Papers in Reg Sci 87(1):97-117

B.3 Spatial Autocorrelation

Arthur Getis

B.3.1 Introduction

In this chapter we review the concept of spatial autocorrelation and its attributes. Our purpose is to outline the various formulations and measures of spatial autocorrelation and to point out how the concept helps assess the spatial nature of georeferenced data. For a fuller treatment of the subject, a number of texts, written at various junctures in the development of the concept and at differing levels of mathematical sophistication, spell out many of the details not discussed here (Cliff and Ord 1973, 1981; Miron 1984; Upton and Fingleton 1985; Goodchild 1986; Odland 1988; Anselin 1988; Haining 1990a; Legendre 1993; Dubin 1998; Griffith 1987, 1988, 2003). In addition, and as background to this chapter, Haining's contribution in this volume (see Chapter B.1) gives a clear view of the nature of georeferenced data. Our goal is to briefly describe the literature on this subject so that the spatial autocorrelation concept is accessible to those who (i) are new to dealing with georeferenced data in a research framework or (ii) have worked with georeferenced data previously but without explicit knowledge of how the concept can be beneficial to them in their research. We are constrained by space and, as a result, our plan is to be short on explanations but identify key literature where the reader will find further details.

After defining and briefly giving the background for the concept of spatial autocorrelation in this section, we explain the concept's attributes and uses in Section B.3.2. In the next section, we discuss the matrices that must be created in order to assess most measures of the spatial autocorrelation concept. We outline the various spatial autocorrelation formulations in Section B.3.4. This is followed in Section B.3.5 with a short discussion of the problems in applying the concept in research situations. Finally, Section B.3.6 provides a brief description of available spatial autocorrelation software. The reference list can serve as a guide to the literature in this area.

Definitions

The simplest definition of the spatial autocorrelation concept is that it represents the relationship between nearby spatial units, as seen on maps, where each unit is coded with a realization of a single variable. Adding more detail and conciseness, as Hubert et al. (1981, p.224) put it:

> '*Given a set S containing n geographical units, spatial autocorrelation refers to the relationship between some variable observed in each of the n localities and a measure of geographical proximity defined for all n(n–1) pairs chosen from S.*'

If a matrix Y represents all of the (n^2-n) associations between all realizations of the Y variable in region $\mathscr{R}$ and W represents all of the (w^2-w) associations of the spatial units to each other in region $\mathscr{R}$, irrespective of Y, then the degree to which the two matrices are positively (negatively) correlated is the degree of positive (negative) spatial autocorrelation. Thus, if it is assumed that neighboring spatial units are associated and so are represented in the W matrix as high positive numbers and low numbers or zero for all others, and the Y matrix has high values in spatial units neighboring other high values, then the two matrices are similar in structure with the result that positive spatial autocorrelation exists.

Development of the concept

The spatial autocorrelation concept was bred at the University of Washington in the late 1950s, principally by Michael F. Dacey, mainly in the presence of William L. Garrison and Edward Ullman, two geographers very much influenced by the central place work of the 1930s German economic geographer Walter Christaller. Earlier, an extensive literature had been developed on the principal of nearness, that is, the strong effect that nearby areas have on each other versus the relatively weak influence of areas further away (for example, Ravenstein 1885; von Thünen 1826; Zipf 1949) with the implication that near spatial units are similar to one another. This notion is best summarized by Tobler's First Law, '*Everything is related to everything else, but near things are more related than distant things*' (Tobler 1970, p.234). The roots of the idea go back to Galton, Pearson, Student, and Fisher. Until 1964, in the social science and statistics literature, spatial autocorrelation had been called 'spatial dependence,' 'spatial association,' 'spatial interaction,' 'spatial interdependence,' among other terms. In geography, the modern meaning of the term 'spatial autocorrelation' was first mentioned by Garrison in or before 1960 (Thomas 1960, in Berry and Marble 1968), and first developed in a statistical framework by Cliff and Ord (1969).

Three statisticians laid out the mathematical characteristics of spatial autocorrelation, although they used the term *contiguity ratio* to describe their work. Moran (1948), Krishna-Iyer (1949), and Geary (1954) developed join count statistics based on the probability that joined spatial units were of the same nominal type (black or white) more than chance would have it. Their work was extended to take interval data into account. Geary, in particular, made the point that the mapped residuals from an ordinary least squares regression analysis must display the characteristic of independence. Dacey further explicated join count statistics, extending the number of colors studied from two to k, and clearly showed the link between using nominal and interval data (Dacey 1965). Also, Dacey recognized the possible effect of the shapes, sizes, and boundaries of regions (topological invariance) on the results of analyses that used georeferenced data (Dacey 1965).

In the field of geostatistics, Matheron (1963) had already developed in considerable detail the mathematics that accompanies the assumption of *intrinsic stationarity*, the notion that inherently characteristic of spatial distributions is a distance effect. Without using the term spatial autocorrelation, the *correlogram* (the inverse of the semivariogram), was invented to represent intrinsic stationarity, the declining similarity of variable values assumed to exist among spatial units as distance increased from each other.

The monograph *Spatial Autocorrelation* by Cliff and Ord (1973) sheds light on the problem of model mis-specification owing to spatial autocorrelation and demonstrated statistically how one can test residuals of regression analysis for spatial randomness by using spatial autocorrelation statistics. Models that require traditional statistics for their evaluation are mis-specified if they do not take spatial autocorrelation into account. The moments of Moran's distribution, called Moran's I, were fully developed by Cliff and Ord (1973, 1981) under varying sampling assumptions.

B.3.2 Attributes and uses of the concept of spatial autocorrelation

The following list gives some idea of the range of uses for the concept and for the formulas created to measure the degree of spatial autocorrelation in modeling situations. The list should convince all of those who deal with georeferenced data that an explicit recognition of the concept is basic to any spatial analysis.

- *A test on model mis-specification.* Properly specified models that call for normally distributed residuals also require that residuals map onto the study region in such a way that one cannot detect any association between nearby spatial units. Proper specification requires that any spatial association is subsumed within the model proper. The most used, and statistically most powerful, test for detecting the spatial independence of residuals is that of the spatial autocorrelation statistic, Moran's I (Cliff and Ord 1972, 1981; Anselin 1988).
- *A measure of the strength of the spatial effects on any variable.* A thorough understanding of the effects of regressor variables on a dependent variable re-

quires that any spatial effects on both dependent and independent variables be quantified. Spatial autocorrelation coefficients in regression models help us to understand the strength of spatial effects (Haining 1990b; Anselin and Rey 1991).
- *A test on assumptions of spatial stationarity and spatial heterogeneity.* Before engaging in many types of spatial analysis, it is necessary to make the assumption that spatial stationarity exists. There are many definitions of spatial stationarity; most common is that the mean and variance of a variable under consideration do not vary appreciably from subregion to subregion in the study region. Spatial autocorrelation measures allow for tests on hypotheses of no spatial differences in distribution parameters such as the mean and variance (Haining 1977; Leung 2000).
- *A means of identifying spatial clusters.* Spatial clustering algorithms are dependent on the conjecture that there is spatial autocorrelation among some nearby values of one or more variables of interest. The basis of clustering computer routines such as ClusterSeer, StatScan, and AMOEBA is the concept of spatial autocorrelation (Aldstadt and Getis 2006).
- *A means of identifying the role that distance decay or spatial interaction might have on any spatial autoregressive model.* Measures of spatial autocorrelation can identify the parameters of spatial decay (for example, the parameters of a negative exponential model) or the parameters of spatial interaction models (Fotheringham 1981).
- *A way to understand the influence that the geometry of spatial units has on a variable.* Measures of spatial autocorrelation will change in certain known ways when the configuration of spatial units changes. These measures are ideal for understanding the role that spatial scale might have on relationships among georeferenced variables (Arbia 1989; Wong 1997). Also see Okabe et al. (2006) on network configurations and spatial autocorrelation.
- *A test on hypotheses about spatial relationships.* Spatial autocorrelation statistics are usually designed to test the null hypothesis that there is no relationship among realizations of a single variable, but the tests may be extended to consider spatial relations between variables (Wartenberg 1985).
- *A means of weighing the importance of temporal effects.* A series of measures of spatial autocorrelation taken over time sheds light on temporal effects (Rey and Janikas 2006).
- *A focus on a single spatial unit's effect on other units and vice versa.* The local view of spatial autocorrelation (see below) allows for focused tests where a particular spatial unit is the focus (Ord and Getis 1995; Anselin 1995; Sokal et al. 1998).
- *A means of identifying outliers, both spatial and non-spatial.* Certain statistical and graphical routines allow for the exact identification of units that unduly influence spatial effects (Anselin 1995).
- *A help in designing an appropriate spatial sample.* If the goal is to avoid, as much as possible, spatial autocorrelation in the sample, then a reasonable sam-

ple design would benefit from a study of spatial autocorrelation in the region where the sample is to be selected (Fortin et al. 1989; Legendre et al. 2002; Griffith 2005).

The list can be expanded, but suffice it to say here are many characteristics of spatial autocorrelation that add depth and understanding to any spatial analysis.

B.3.3 Representation of spatial autocorrelation

Since the types of studies in which the concept of spatial autocorrelation is used vary considerably, many methods and techniques of analysis have been created for special purposes. The following simple representation of spatial autocorrelation is the key to the proper choice of measure or test (Hubert and Golledge 1981; Getis 1991).

The cross-product statistic

$$\Gamma_{ij} = \sum_{i=1}^{n}\sum_{j=1}^{n} W_{ij} Y_{ij} \tag{B.3.1}$$

where Γ is a measure of spatial autocorrelation for n georeferenced observations. It is made up of W, a matrix of values that represents the spatial relationships of each location i to all other sites j. The Y matrix shows the non-spatial relationship of realizations of a variable Y at site i with all other realizations at all other sites j. When W, the spatial weights matrix, and Y, the variable matrix have similar structures [for example, both have high values in the same (i, j) cells in their respective matrices and low values in the same (i, j) cells] one can say that there is a high degree of spatial autocorrelation. The correlation can be positive or negative depending on whether respective cells are similarly matched or oppositely matched. If realizations of Y are randomly placed in the spatial units, no matter how the spatial weights matrix is structured, the result will be a Γ of zero, or no spatial autocorrelation. The same is true if the W matrix is based on random spatial associations and the Y happens to be spatially structured. Thus, it is clear that for any meaningful assessment of spatial autocorrelation the W matrix must be a careful representation of spatial structure and the Y matrix must represent a meaningful association between realizations of the Y variable. Equation (B.3.1), as it is presented, is not a test of spatial autocorrelation, but only a measure. Tests on the existence of spatial autocorrelation, however, take on the same cross-product structure. In the next section, the structure of W matrices is discussed.

The W matrix

The **W** matrix embodies our preconceived or derived understanding of spatial relationships. If we believe or if theory tells us that a particular spatial relationship is distance dependent, then the **W** matrix should reflect that supposition. For example, if it is assumed that a spatial relationship declines in strength as distance increases from any given site, then the **W** matrix will show that nearby areas are weighted more highly than sites that are far from one another. Various distance-decay formulations theorized or derived for such phenomena as travel behavior, economic interaction, or disease transmission would require the elements within the **W** matrix to reflect these effects. Thus, a typical **W** matrix might contain matrix elements (represented as lower case letters) of the form

$$W_{ij} = d_{ij}^{-\alpha} \quad \text{with } \alpha \geq 1. \tag{B.3.2}$$

Or, in words, the weight entered into cell (i,j) is the inverse of distance d between the two sites, i and j, reduced by the exponent α, where α is greater than one. The **W** matrix can represent distances other than those derived from Cartesian geometry. For example, friendship or cell phone networks may be distance related in sociological terms. A bevy of schemes have been created to attempt to fashion **W** (Getis and Aldstadt 2004). Some of the schemes are:

- Spatially contiguous neighbors (default for many studies),
- Inverse distances raised to some power (distance decline function),
- Lengths of shared borders divided by the perimeter (a geometric view),
- Bandwidth as the nth nearest neighbor distance (point density dependent),
- Ranked distances (non-Cartesian approach),
- All centroids within distance d (density dependent),
- n nearest neighbors (equal weighting of matrix entries),
- Bandwidth distance decay (required for geographically weighted regression),
- Gaussian distance decline (based on the square term),
- Derived spatial autocorrelation (based on observed spatial association).

Perhaps the most used **W** is the first in the list above. **W** is made up of ones for contiguous neighbors and zero for all others, whether the data are raster or vector. By convention, the ith observation is not considered a neighbor of itself. The contiguity **W** matrix is often row-standardized, that is, each row sum in the matrix is made to equal one, the individual W_{ij} values are proportionally represented. Row-standardization of **W** in contiguity schemes is desirable so that each neighbor of a spatial unit is given equal weight and the sum of all W_{ij} is equal to n. As we will later see, these characteristics enhance understanding of spatial autocorrelation measures and coefficients. Researchers should be aware, however, that row-standardization may give too much weight to observations with few spatial links and not enough weight to observations having many contiguous neighbors (Tiefelsdorf et al. 1998).

Those developing spatial models consider the spatial weights matrix to be one of the following three types of representations:

(i) a theoretical notion of spatial association, such as a distance decline function,
(ii) a geometric indicator of spatial nearness, such as the representation of contiguous spatial units,
(iii) some descriptive expression of the spatial association already existing within a set of data.

For viewpoint one, modelers argue that a W matrix is exogenous to any system and should be based on a pre-conceived matrix structure. A typical theoretical formulation for W would be based on a strict distance decline function such as shown in Eq. (B.3.2). Since little theory is available for the creation of these matrices, many researchers follow viewpoint two, that is, they resort to geometric W specifications, such as a contiguity matrix, reasoning that it is the nearest neighboring spatial units that bear most heavily on spatial association in a typical set of georeferenced data. Tiefelsdorf (2000) has created a system for coding these and other matrices based on geometric structure that goes well beyond simple contiguity matrices.

For viewpoint three, modelers allow study data to 'speak for themselves,' that is, they extract from the already existing data whatever spatial relationships appear to be the case and then create a W matrix from the observed spatial associations. As a result, models based on this type of endogenous specification have limited explanatory power, the limit being the reference region. Kooijman (1976) proposed to choose W in order to maximize Moran's coefficient (see next section). Reinforcing this view is Openshaw (1977), who selects that configuration of W which results in the optimal performance of the spatial model. Getis and Aldstadt (2004) construct W by using a local spatial autocorrelation statistic to generate the W_{ij} from the data.

The nature of the variables being studied for spatial effects is the key to an appropriate W. Variables that show a good deal of local spatial heterogeneity at the scale of analysis chosen would probably be more appropriately modeled by few links in W, while a homogeneous or spatial trending variable would better be modeled by a W with many links. This implies that the scale characteristics of data are crucial elements in the creation of W. As spatial units become large, spatial dependence between units tends to fall (Can 1996).

*The **Y** matrix*

The non-spatial matrix, **Y**, provides a view of how the realizations of a variable are associated with one another. **Y** represents the interaction of the elements y_{ij}. They may interact by an additive $(y_i + y_j)$, multiplicative $(y_i\, y_j)$, differencing $(y_i - y_j)$, or division $(y_i\, /\, y_j)$ process. A useful type of multiplicative matrix is the covariance matrix $(y_i - \bar{y})(y_j - \bar{y})$. All of these matrices can be scaled in order to serve a particular view of relationships within a variable. In the following section, we present the scaling of these processes for the creation of various views of spatial autocorrelation. In sum, the measures and tests for spatial autocorrelation differ by use and by the structure of their **W** and **Y** matrices.

B.3.4 Spatial autocorrelation measures and tests

Spatial autocorrelation measures can be differentiated from tests on spatial autocorrelation by purpose, but both allow for the assessment of spatial effects in any analysis of georeferenced data. Moran's *I*, discussed below, is both the leading measure of and leading test on spatial autocorrelation, while, for example, the Kelejian-Robinson test on spatial autocorrelation is not used as a measure. Also, measures of spatial autocorrelation of the correlogram are not used as tests on spatial autocorrelation.

Spatial autocorrelation measures and tests can be differentiated by the scope or scale of analysis. Traditionally, they are separated into 'global' and 'local' categories. Global implies that all elements in the **W** and **Y** matrices taken together are brought to bear on an assessment of spatial autocorrelation, that is, all associations of spatial units one with another are included in any calculation of spatial autocorrelation. This results in one value for spatial autocorrelation for any one **W** and **Y** matrix taken together. Local measures are focused, that is, they usually assess the spatial autocorrelation associated with one particular spatial unit. Thus, only one row of the **W** and the matching row of the **Y** matrix reflect on the measure of spatial autocorrelation although all elements' interactions may be used as a scalar.

Global measures and tests

Gamma (Γ). As discussed earlier, this measure was used in our discussion as the basis on which all spatial autocorrelation measures and tests are structured. A test on the statistical significance of Γ is made practical by randomizing *Y* values in a number of simulations. The observed Γ can then be compared to the envelope created by the results of the simulations. Statistical significance implies that spatial autocorrelation exists.

Join-count. The purpose here is to identify for an exhaustive nominal classification of spatial units, such as for land use types – residential (*A*), industrial (*B*), commercial (*C*) – whether there are statistically significant numbers of spatially associated *AA, AB, AC, BB, BC,* and/or *CC* occurrences. In a system of spatial units, the expected number of *AA*, for example is a function of the type of test that is selected for identifying statistical significance. Here we use the free sampling test (Cliff and Ord 1981).

Given the probability p_r that a spatial unit is a particular type of land use, and the number of units of that type is n_r, the expected number of joins of the same type is

$$E(J) = \tfrac{1}{2} \sum_{i=1}^{n} \sum_{j=1}^{n} W_{ij} \, p_r^2 . \tag{B.3.3}$$

For different types, the expectation is

$$E(J) = \sum_{i=1}^{n} \sum_{j=1}^{n} W_{ij} \, p_r \, p_s . \tag{B.3.4}$$

The p values are usually estimated from the data (n_r / n). The **W** matrix is made up of ones and zeros representing joined spatial units (one) and non-joined spatial units (zero). There is a series of **Y** matrices, one for each test, where each is made up of ones and zeros representing specified types of associated spatial units (for example, *AB* is one and not *AB* is zero) and summarized as the probabilities of occurrence of *A* and *B* (p_r and p_s). In order to perform tests on spatial autocorrelation, the variance must be known and the assumption invoked of an asymptotic normal distribution of the frequency of cells (see Cliff and Ord 1981 for details).

Moran's I. This statistic is structured as the Pearson product moment correlation coefficient. The crucial difference is that space is included by means of a **W** matrix and instead of finding the correlation between two variables, the goal is to find the correlation of one variable with itself vis-à-vis a spatial weights matrix. The **Y** is a covariance matrix, that is, Moran's *I* focuses on each observation as a difference from the mean of all observations. Set **W** to a preferred or required spatial weights matrix (any of those listed above), set **Y** equal to the auto-covariance $(y_i - \bar{y})(y_j - \bar{y})$, and scale the measure (invoking a Pearson limit structure) by multiplying by

$$\frac{n}{W} \left[\sum_{i=1}^{n} (y_i - \bar{y})^2 \right] \tag{B.3.5}$$

where

$$W = \sum_{i=1}^{n}\sum_{j=1}^{n} W_{ij} \tag{B.3.6}$$

and, by convention, i is not to equal j (no self association). We have then

$$I = \frac{n}{\sum_{i=1}^{n}\sum_{j=1}^{n} W_{ij}} \frac{\sum_{i=1}^{n}\sum_{j=1}^{n} W_{ij}(y_i - \bar{y})-(y_j - \bar{y})}{\sum_{i=1}^{n}(y_i - \bar{y})^2} \quad i \neq j. \tag{B.3.7}$$

The expected value is $E(I) = -1/(n-1)$ and the variance is calculated somewhat differently under an assumption of randomness versus an assumption of normality. These two assumptions represent the supposed theoretical way the Y values were produced under the hypothesis of randomly placed Y values. Thus, Moran's I is a test for spatial randomness; rejection of the null hypothesis implies with a certain degree of certainty (for example, 95 percent) that spatial autocorrelation exists. The randomness assumption (R) implies that the values of y are realizations of a single uniformly distributed Y variable (that is, a variable where all possible realizations are equally likely). The normal assumption means that each y value is a randomly selected realization of a different normal distribution, one representing each spatial unit. It should be pointed out that a variation exists for Moran's I when residuals of regression are being tested for spatial randomness. This is

$$I = \frac{n}{\sum_{i=1}^{n}\sum_{j=1}^{n} W_{ij}} \frac{\varepsilon^{\mathrm{T}} W \varepsilon}{\varepsilon^{\mathrm{T}} \varepsilon} \tag{B.3.8}$$

where ε is a vector of ordinary least squares residuals and ε^{T} is the matrix transpose. The expected value and variance are a function of the number of independent variables in the system (Cliff and Ord 1972).

Moran's I can be used in a wide variety of circumstances. As a global statistic, Moran's I quickly indicates not only the existence of spatial autocorrelation (positive or negative) but also the degree of spatial autocorrelation If the variable of interest is the error term in a regression model, the question of model misspecification can be evaluated by applying Moran's I. In spatial econometrics, the test has power for testing residuals for many types of spatial autoregressive models (Anselin 2006). Since Moran's *I* is distributed normally, its value may be assessed by the *z* values of the normal distribution. The statistic is flexible in that the **W** matrix may be of any form – it has no restrictions on the spatial system used. Of course, outliers in one or both of the **W** and **Y** matrices will yield meaningless results. The local version of Moran's *I*, discussed later, lends itself to spatial cluster identification and spatial filtering. A large literature has been developed to explore the properties of Moran's *I*. In addition to the basic references given in the first paragraph of this contribution, see Tiefelsdorf and Boots (1995, 1997); Hepple (1998).

Geary's c. The particular test employed for spatial autocorrelation is a function of the type of hypothesis required for the analysis. In the case of Moran's *I*, the null hypothesis was based on a covariance structure, that is, the expectation that related neighbors co-vary in no consistent way. For Geary's *c*, the null hypothesis is that related spatial units do not differ from one another. The implication of this hypothesis is that the expectation is that there is no consistency to the differences between neighbors; sometimes the differences are large and sometimes small. In this case, as for Moran's *I*, the **W** matrix is made up of any meaningful spatial relations between spatial units. The **Y** matrix is simply made up of the differences in the realizations of the variable *Y* among all observations: $(y_i - y_j)^2$. A scale is included so that the resulting structure is normal, thereby lending Geary's *c* to statistical tests. Thus, we have

$$c = \frac{(n-1) \sum_{i=1}^{n} \sum_{j=1}^{n} W_{ij} (y_i - y_j)^2}{2 W \sum_{i=1}^{n} (y_i - \bar{y})^2} \quad i \neq j. \quad (B.3.9)$$

Note that the scale results in an expected value of Geary's *c* as one. In tests, values less than one indicate positive spatial autocorrelation (small differences) and values greater than one imply negative spatial autocorrelation (consistently large differences). Geary's *c* is negatively related to Moran's *I*. Many of the references already given for Moran's spatial autocorrelation statistic contain references to Geary's measure.

The variogram. Central to the field of geostatistics is the semivariogram. Cressie (1993) provides a detailed treatment of the concept. Suffice is to say here that the semivariogram is a distribution of differences among spatially associated units and therefore is related to Geary's *c*. The major difference is that the semivariogram hypothesizes that the differences decline with distance from each other in a systematic way. Thus, the semivariogram describes a continuous view of differences, while Geary's statistic is relegated to one **W**. A typical semivariogram has the shape of a positive exponential distribution, where close distances display small differences and low variances, and far distances are not affected by distance effects in such a way that when all differences are taken together the value of the global variance obtains. The semivariogram has the form

$$\gamma(ad) = \tfrac{1}{2} \sum_{i=1}^{n} \sum_{j=1}^{n} W_{ij} (y_i - y_j)^2 \tag{B.3.10}$$

where there is a W for each constant distance d controlled by an integer multiplier a. Thus, a particular constant distance (say, one kilometer) has a W for each a. But, the W matrices are constrained to contain ones and zeros. In effect, the W matrix identifies spatial units that are related at a particular distance (ad) or a particular band from each observation. The display of the spatial autocorrelation is called the correlogram, a function that decreases with distance until the *range* is reached. The *range* represents a distance where the global variance is unaffected by distance effects. The scale of the semivariogram, 1/2, is a recognition that there is double counting, the differences between i and j are the same as between j and i. Cressie (1993) provides a comprehensive treatment of geostatistics, and Rosenberg et al. (1999) emphasizes the spatial autocorrelation aspects of the analysis.

Ripley's K function. As is true of the correlogram, Ripley's K function (Ripley 1976; Besag 1977) represents a continuous set of spatial autocorrelation indicators. The K function, unlike the measures discussed previously, emphasizes only location and not the other attributes of a random variable. So here we are restricted to point patterns based on the number of pairs of points found at a series of distances from each ith point. In this case, the object is to count all pairs of points at each distance. If there are more pairs of points than spatial random chance (spatial Poisson distribution) would have it, there is statistically significant clustering; fewer pairs of points implies a statistically significant dispersion of points, the opposite of clustering. The null hypothesis obtains when there are about as many pairs of points as one might find in a point distribution created by a random process. A random spatial process is called a homogeneous Poisson process over the study plane, that is, all sites within the area of study are equally likely to receive a point, and the siting of a point in no way bears on the siting of another point. The statistic is estimated in the following way

$$\hat{K}(d) = \frac{R}{n^2} \sum_{i=1}^{n} \sum_{j=1}^{n} \frac{W_{ij}}{e_{ij}} \quad i \neq j. \tag{B.3.11}$$

Within the study region of size R, for distance d we count all of the pairs of points that are not larger than d apart. Thus for a number of increasing distances the value of $K(d)$ will increase as more pairs are added to the total. The W matrix is made up of one for (i, j) pairs within d of one another and zero otherwise. As distance increases, the boundary of the region is more likely to be closer to a point i than to any j point. In that case, an edge correction e_{ij} is invoked which assumes that any point outside of the boundary is unobserved but that the point process continues for at least a short distance beyond the boundary. Center a circle of radius d_{ij} on i, and if the circle crosses the boundary, the proportion of the circumference of the circle that lies inside the study area replaces that particular pair count of one to a value greater than one, thus insuring consistency of the presumed point process. Of course, points close to the boundary but far from a neighbor distort any result. Further, by including the estimate of $\hat{K}(d)$ in the following formula, a significant improvement is made for recognizing spatial autocorrelation in a point process.

$$\hat{L}(d) = \sqrt{\frac{\hat{K}(d)}{\pi}}. \qquad (B.3.12)$$

When this formula is used, the expectation based on the hypothesis of Poisson randomness becomes a positive straight line where $\hat{L}(d) = d$. Typically a series of Poisson random distributions is simulated, helping to create an envelope containing, say, 95 percent of possible point patterns under the hypothesis of randomness. An observed pattern whose $L(d)$ value falls outside of the envelope indicates the existent of positive (clustering) or negative (dispersion) spatial autocorrelation. The value of this analysis is particularly great when it is assumed that some non-Poisson point process is responsible for the observed spatial pattern. Thus, a clustered pattern may itself be considered a null hypothesis that can be tested for further clustering. In addition, a number of patterns in the same area representing different variables may be compared. See Bailey and Gatrell (1995), and Getis and Franklin (1987).

Spatial autocorrelation coefficients. In regression models where estimation is based on georeferenced data, it is mandatory that any statistically significant spatial effect must be accounted for in the model. The spatial effects can be diagnosed by means of Moran's I tests on residuals or on variables that are to be included in the model. Also, regardless of diagnostics, spatial dependencies may be subsumed by creating spatial autoregressive models of one kind or another. Two popular autoregressive models are (i) the mixed regressive spatial autoregressive model, often called the spatial lag model,

$$y = \rho Wy + X\beta + \varepsilon \quad \text{(B.3.13)}$$

and (ii) the linear regression with a spatial autoregressive error, or simultaneous autoregressive model (SAR), often called the spatial error model

$$y = X\beta + (I - \lambda W)^{-1}\mu. \quad \text{(B.3.14)}$$

In both of these cases, parameters representing spatial effects, ρ and λ must be determined. Note that in each case they precede the W matrix, which takes any of the forms discussed above. In essence, the coefficients reveal the strength or influence of the W matrix. In so doing, they become spatial autocorrelation coefficients; high positive or negative values represent strong spatial effects and low values the opposite. When ρ, λ are zero, there are no spatial effects. This is true since the error terms ε and μ respectively are randomly distributed in space. If, in estimation of the models, errors are spatially correlated, the models are misspecified. In addition to Moran's I regression residual test, specialized tests such as the the Kelejian and Robinson (KR) test (1993), or the Wald, Likelihood Ratio, and Lagrange multiplier tests are used to identify spatial autocorrelation in spatial lag or spatial error type models (Anselin 2006). For example, for the KR test, normality of errors is not required, nor is it necessary to hypothesize a strictly linear model. In addition, KR studies only certain selected contiguity relationships (Kelejian and Robinson 1993). For details on spatial autocorrelation coefficients, see Anselin (1988). Anselin (2006) presents a comprehensive review of spatial econometrics.

Local measures and tests

Among spatial analysts, there has always been an interest in focused measures, that is, a desire to describe precisely the 'situation' or proximity characteristics of a particular site. But it was not until the invention of local statistics that it became possible to measure and test for certain situational characteristics. What better way is there to investigate situational characteristics of sites than to use spatial autocorrelation measures and tests? The basis for local tests for and measures of spatial autocorrelation comes from the cross-product statistic. This time the structural form is

$$\Gamma_i = \sum_{j=1}^{n} W_{ij} Y_{ij} \quad i \neq j. \quad \text{(B.3.15)}$$

Note that here we are finding the interaction between spatial weights in the *i*th vector only and the *y* values in *Y*'s *i*th vector. Γ_i allows for autocorrelative comparisons between the two vectors for a given site *i*.

Getis and Ord local statistics. These statistics are additive in that the focus is on the sum of the *j* values in the vicinity of *i*. The fact that there are two statistics, G_i and G_i^*, allows researchers to choose hypotheses based on proximity (G_i) or on clustering (G_i^*). G_i^* is written as

$$G_i^*(d) = \frac{\sum_{j=1}^{n} W_{ij}(d) y_j - W_i^* \bar{y}}{s\{[(nS_{1i}^*) - W_i^{*2}]/(n-1)\}^{1/2}} \quad \text{for all } j \quad (B.3.16a)$$

where

$$W_i^* = W_i + W_{ii} \quad \text{and} \quad S_{1i}^* = \sum_{j=1}^{n} W_{ij}^2 \quad \text{for all } j \quad (B.3.16b)$$

and $\bar{y}$ and *s* are the mean and standard deviation, respectively.

The mathematical distinction between the two statistics depends on the role of the *i*th observation. If our concern is with the effect of the influence of *i* on the *j* values, the focus is on the site *i* but not the *y*-value associated with it. Thus, the view is one of proximity (situation). The null hypothesis would be: there is no association between *i* and its neighbors *j* up to distance *d*. The G_i^* statistic, on the other hand, includes the value y_i in its calculations; it sums associations between *i* and *j* including *i* (the value for W_{ii} – usually one – is added to W_i). Thus, G_i^* lends itself to studies of clustering since a cluster usually contains its focus as a member of the cluster.

Both statistics are distributed normally. They are scaled in such a way that $G_i(d)$ and $G_i^*(d)$ are equivalent to standard deviations of the normal distribution. Thus, there is no need to convert the statistics. It is interesting to note that G_i^* is mathematically associated with global Moran's *I*(*d*) so that Moran's *I* may be interpreted as a weighted average of the local statistics (Getis and Ord 1992; Ord and Getis 1995). For these statistics as well as all other spatial autocorrelation statistics, boundary effects may lessen the number of associations between *i* and *j*. To avoid the resulting bias, boundary effects should be minimized by judicially selecting the area of study. Hot spots identified by these statistics can be interpreted as clusters or indications of spatial nonstationarity.

Local indicators of spatial association – LISA. LISA statistics were created by Anselin (1995), whose motivation was to decompose global statistics such as Moran's *I* and Geary's *c* into their local components for the purpose of identifying influential observations and outliers. The individual components of I_i are related to I. Just as the Γ_i sum to Γ, so too will all I_i sum to I, subject to a factor of pro-portionality. Local Moran's I_i is defined as

$$I_i = \frac{y_i - \bar{y}}{\frac{1}{n}\sum_{i=1}^{n}(y_i - \bar{y})^2} \sum_{j=1}^{n} W_{ij}(y_i - y_j) \qquad i \neq j, \text{ for } j \text{ within } d \text{ of } i \qquad \text{(B.3.17)}$$

and the factor of proportionality is

$$\gamma = \frac{1}{n}\sum_{i=1}^{n}\sum_{j=1}^{n} W_{ij} \sum_{i=1}^{n}(y_i - \bar{y})^2 . \qquad \text{(B.3.18)}$$

The expected value of

$$E(I_i) = -\frac{1}{n-1}\sum_{j=1}^{n} W_{ij} . \qquad \text{(B.3.19)}$$

Tests for spatial autocorrelation may be carried out either using the moments of the I_i distribution (see Anselin 1995) or by random permutations. The second technique, the strategy of conditional randomization, is preferred for LISA since the possible existence of global autocorrelation would otherwise affect the interpretation of I_i (Anselin 1995).

For Geary's c, the local version is

$$c_i = \frac{1}{\frac{1}{n}\sum_{i=1}^{n}(y_i - \bar{y})^2} \sum_{j=1}^{n} W_{ij}\left[(y_i - \bar{y}) - (y_j - \bar{y})\right]^2 . \qquad \text{(B.3.20)}$$

Here the factor of proportionality is

$$\gamma = \frac{2n}{(n-1)} \sum_{i=1}^{n}\sum_{j=1}^{n} W_{ij} . \qquad \text{(B.3.21)}$$

The LISA statistics are particularly useful for identifying spatial clusters. High spatial autocorrelation values indicate clusters of high or low values. Software provided in GeoDa (discussed below and in an earlier section of this book) provides graphics in which the ++, − −, + −, and − + types of spatial association are differentiated. Sokal et al. (1998) take a different view of local analysis, and Boots (2002) analyzes local measures of spatial autocorrelation.

Geographically weighted regression. A local version of an ordinary least squares regression analysis has been proposed by Fotheringham et al. (1995). The point of geographically weighted regression (GWR) is that regression parameters are not constant over space as characterized by traditional regression models and that the variation can be explicitly modeled. By using a W, usually a Gaussian or near-Gaussian spatial weights decline function for each i as elements in the matrix, a regression can be estimated for each ith location. Although each weight matrix need not be focused on data sites, the point of the analysis is to estimate the variation in parameters across space. The form of GWR can be written as

$$Y = (\beta \otimes X) \mathbf{1} + \varepsilon \qquad (B.3.22)$$

where the logical operator (Kronecker product) $\otimes$ requires that corresponding elements in each matrix are multiplied by each other. Since each matrix has n-by-$(k+1)$ dimensions, where the number of independent variables is k, the vector of ones with dimensions $(k+1)$-by-1 yields the required n-by-1 matrix for Y. This allows β to consist of n sets of local parameters. Each set contains a slope and intercept for each independent variable for each i. The *betas* are estimated by use of a W for each i. The d for all Ws is either selected in advance or estimated from the data. A typical W is based on a pre-selected outer distance bandwidth b

$$W_{ij} = \begin{cases} \left[1 - \left(\dfrac{d_{ij}}{b}\right)^2\right]^2 & \text{if } d_{ij} < b \\ 0 & \text{otherwise.} \end{cases} \qquad (B.3.23)$$

Often, in a single study, b is allowed to vary because standard errors might be particularly high when the b-radius includes only a few data points around i. Various systems are provided for selecting b (Fotheringham et al. 2002). A result of GWR is a map of what might be called 'parameter space.' Areas with high parameter values indicate particularly strong correlative relationships between regressor and response variables, but the parameters are not directly indicative of spatial autocorrelation. Since the beta values are a function of the spatial weighting scheme, to the extent that W captures the spatial autocorrelation effects in each of the variables, it is reasonable to say that high beta values reflect on the pattern of spatial autocorrelation in the system. It is possible, however, to specify autoregressive instead of OLS models, thus the GWR parameters can play the same role as in spatial autoregressive models. The implication is that one or more spatial autocorrelation maps can be produced for each equation in the system. Fotheringham, Brunsdon, and Charlton continue to write many articles on this subject. Reviews and analyses are found in Páez et al. (2002), Leung (2000) and Wheeler and Tiefelsdorf (2005) and in Chapter C.6 of this Handbook.

Local spatial autocorrelation in the presence of global spatial autocorrelation. As mentioned in our necessarily short discussion of local statistics, when global spatial autocorrelation exists, it becomes difficult to interpret the nature of local spatial autocorrelation. How much of a statistically significant result for a local test on i is due to pervasive global autocorrelation? It may be that local statistical significance is just an artifact of the larger scale effect due to global associations. Ord and Getis (2001) provide a test, called O, that includes separate information on the observations within d of i (regular or irregular areas) representing the hypothesized hot spot, and on observations immediately outside of the hot spot. The statistic is

$$O_i(d) = \overline{Y}_d - \overline{Y}_0 \qquad (B.3.24)$$

where $\overline{Y}_d$ is the mean of the $n(d)$ observations within d and $\overline{Y}_0$ is the mean of the $m = M - n(d)$ observations, the M being a regionally partitioned set of observations that displays 'relative homogeneity.' The M should be considerably larger than $n(d)$ [at least 10 times] but considerably smaller than all n observations in the study area. M can be selected to include all observations from i [(except the $n(d)$] up to the *range* (in the geostatistical sense) derived from all observations. The idea of the statistic is to compare characteristics of data at two spatial scales; $E[O_i(d)] = 0$. Testing procedures are given in Ord and Getis (2001). Boots and Tiefelsdorf (2000) consider the relationship of global to local measures of spatial autocorrelation.

B.3.5 Problems in dealing with spatial autocorrelation

It is clear that spatial autocorrelation can be defined precisely, but it is not always clear whether the various measures and tests just described can actually find spatial autocorrelation in georeferenced data. Each of them has its own shortcomings, but more important, they perform better or worse depending on the way in which W and Y are specified. For example, results depend on the nature of W, again emphasizing the importance of a meaningful specification of W. Much work remains to be done to better understand the effects of various W matrices on results. Similarly, Y will yield different results depending on the nature of the associations specified for the realizations of Y. The fundamental question for researchers in this area is: What is responsible for any spatial autocorrelation that exists in a particular data set? Is it the way the boundaries of the spatial units were drawn (the geometry and/or scale of the spatial units under study) or is it a function of the nature of the variables under study? When spatial autocorrelation is embedded within a variable, is it because of the geometry of the spatial units or something else? A number of commentators have discussed the problems in dealing with spatial autocorrelation including Bao and Henry (1996), Legendre (1993), and Pace and Barry (1997).

A particularly difficult area of research is the selection of tests that can withstand the simultaneity effects of multiple tests. Especially in local statistics, usually there are tests on spatial autocorrelation for each data site. This results in very large numbers of tests that are in fact dependent on one another. Thus we come to the ironic situation where in the search for spatial autocorrelation we are subject to the effects of spatial autocorrelation itself. For example, many of the observations used to find a local measure of spatial autocorrelation will be used again for a test focused on a neighboring observation. There have been several attempts to resolve this problem of simultaneous, dependent tests (Getis and Ord 2000; Castro and Singer 2006; Benjamini and Hochberg 1995). Also see Chapter B.4. Researchers must be conscious of Bonferroni-type confidence intervals when they select their diagnostic and testing devices.

Many traditional tests require the assumption of stationarity. Checking for stationarity in empirical work is a good practice. GWR, while attempting to get around this problem, falls prey to problems of sample size (necessarily small for estimates of Y_i) and to the overlapping test problem.

The problem of the effect of global spatial autocorrelation on local effects was alluded to above. Is that relationship fully understood? What about the effect of boundaries on levels of confidence? Sample size and thus the number of degrees of freedom are affected by the spatial extent of study regions. For example, does the distance d include suitable numbers of observations that allow for acceptable levels of confidence in results? How is d to be selected? Careful attention must be given to the effect of various d on results. A promising technique of analysis, spatial filtering, may be particularly useful in answering many of these questions (Getis 1990, 1995; Griffith 1996, 2003; Griffith 2002). See also Chapter B.5 in this volume. Many of these problems can be better understood in a framework of exploratory spatial data analysis. The software packages mentioned in the next session are designed to assist in exploration and model development and testing.

B.3.6 Spatial autocorrelation software

Tests and measures of spatial autocorrelation are available in a number of software packages. Most often in these packages, finding and testing for spatial autocorrelation is only one part of a large variety of spatial analytic procedures.

GeoDa. Perhaps the most comprehensive package is GeoDa (Anselin et al. 2006), which provides a number of exploratory procedures that elicit information about spatial patterns. In addition, tests and analysis of spatial autocorrelation are available in a number of different segments of the software, including the estimation and testing of a variety of spatial econometric models. Novel graphical and mapping procedures allow for detailed study of global and local spatial autocorrelation results. Non-stationarity and outliers can be assessed by means of maps of statistically significant clusters. See Chapter A.4 of this volume for further explanations.

R *Packages.* Two noteworthy packages are based on the R language environment. One is the *spdep*, a package with many spatial data exploratory functions, graphics, and hypothesis tests on spatial autocorrelation (Bivand 2006). A package specifically designed for the study of point pattern processes is *Spatstat* (Baddeley and Turner 2005). A special feature of this package is simulation routines for different types of point pattern processes. Tests and diagnostics are included. See Chapter A.3 for a fuller treatment of this package.

PPA (Point Pattern Analysis). This small package includes routines for global and local spatial autocorrelation statistics. Included are nearest neighbor and K function procedures and tests. Graphics are not included (see Aldstadt et al. 2002).

SANET is a toolbox that allows for the study of spatial autocorrelation on networks (Okabe et al. 2006).

STARS (Space-Time Analysis of Regional Systems) is an exploratory package that brings together a number of recently developed methods of space-time analysis into a graphical environment. Spatial autocorrelation can be studied on dynamically-viewed time-dependent maps. Many descriptive statistics are available, as in *GeoDa*, that are keyed directly to individual observations on maps (Rey and Janikas 2006). See Chapter A.5.

ArcGIS. This large system of spatial data management and analysis contains modules that allow for map study with *K* functions and autocorrelation statistics (ArcVIEW 9.3). More detail is available in Chapter A.1. Recent versions contain routines for GWR. One module, *Geostatistical Analyst*, provides a large number of descriptive and analytical routines for the study of semivariograms (ESRI 2001).

ClusterSeer 2. Developed primarily for health science spatial research, this package makes available a number of pattern analytic routines popular in disease and crime research. The routines identify statistically significant spatial clusters whether or not the focus is on a particular observation or a particular site. The concept of spatial autocorrelation is embedded in many of the routines (Jacquez et al. 2002).

Le Sage's Spatial Econometrics Toolbox. This package contains an extensive collection of MATLAB econometric functions, many of which were created for spatial data (LeSage 1999, 2004).

Spatial Statistics and SAS. By means of SAS procedures, Griffith (see Chapter A.2 of this volume) has created specialized routines that allow for the analysis of spatial econometric systems, in particular, spatial filtering (see Chapter B.5 of this volume).

References

Aldstadt J (2009) Spatial clustering. In Fischer MM, Getis A (eds) Handbook of applied spatial analysis. Springer, Berlin, Heidelberg and New York, pp.279-300
Aldstadt J, Getis A (2006) Using AMOEBA to create a spatial weights matrix and identify spatial clusters. Geographical Analysis 38(4):327-343
Aldstadt J, Chen D-M, Getis A (2002) PPA: point pattern analysis (version 1.0a). http://www.nku.edu/~longa/cgi-bin/cgi-tcl-examples/generic/ppa/ppa.cgi
Anselin L (1988) Spatial econometrics: methods and models. Kluwer, Dordrecht
Anselin L (1995) Local indicators of spatial association - LISA. Geogr Anal 27(2):93-115
Anselin L (2006) Spatial econometrics. In Mills TC, Patterson K (eds.) Palgrave handbook of econometrics, volume 1. Palgrave Macmillan, New York, pp.901-969
Anselin L, Rey S (1991) Properties of tests for spatial dependence in linear regression models. Geogr Anal 23(2):112-131
Anselin L, Syabri I, Kho Y (2009) GeoDa: an introduction to spatial data analysis. In Fischer MM, Getis A (eds) Handbook of applied spatial analysis. Springer, Berlin, Heidelberg and New York, pp.73-89
Arbia G (1989) Spatial data configuration in statistical analysis of regional economic and related problems. Kluwer, Dordrecht
Baddeley A, Turner R (2005) Spatstat: an R package for analyzing spatial point patterns. J Stat Software 12(6):1-42
Bailey TC, Gatrell AC (1995) Interactive spatial data analysis. Longman, Harlow
Bao S, Henry M (1996) Heterogeneity issues in local measurements of spatial association. J Geogr Syst 3(1):1-13
Benjamini Y, Hochberg Y (1995) Controlling the false discovery rate: a practical and powerful approach to multiple testing. J Roy Stat Soc B 57(1):289-300
Berry BJL, Marble DF (1968) Spatial analysis: a reader in statistical geography. Prentice-Hall, Englewood Cliffs [NJ]
Besag J (1977): Discussion following Ripley. J Roy Stat Soc B 39(2):193-195
Bivand R (2006) Implementing spatial data analysis software tools in R. Geogr Anal 38(1):23-40
Bivand R (2009) Spatial economic functions in R. In Fischer MM, Getis A (eds) Handbook of applied spatial analysis. Springer, Berlin, Heidelberg and New York, pp.53-71
Boots B (2002) Local measures of spatial association. Ecoscience 9:168-176
Boots B, Tiefelsdorf M (2000) Global and local spatial autocorrelation in bounded regular tesselations. J Geogr Syst 2(4):319-348
Can A (1996) Weight matrices and spatial autocorrelation statistics using a topological vector data model. Int J Geogr Inform Sys 10(8):1009-1017
Casetti E (2009) Expansion method, dependency, and multimodeling. In Fischer MM, Getis A (eds) Handbook of applied spatial analysis. Springer, Berlin, Heidelberg and New York, pp.487-505

de Castro MC, Singer BH (2006) Controlling the false discovery rate: a new application to account for multiple and dependent tests in local statistics of spatial association. Geogr Anal 38(2):180-208

Cliff AD, Ord JK (1969) The problem of spatial autocorrelation. In Scott AJ (ed) Studies in regional science. Pion, London, pp.25-55

Cliff AD, Ord JK (1972) Testing for spatial autocorrelation among regression residuals. Geogr Anal 4(3):267-284

Cliff AD, Ord JK (1973) Spatial autocorrelation. Pion, London

Cliff AD, Ord JK (1981) Spatial processes: models and applications. Pion, London

Cressie NAC. (1993) Statistics for spatial data (revised edition). Wiley, New York, Chichester, Toronto and Brisbanet

Dacey MF. (1965) A review of measures on contiguity for two and k-color maps. Technical Report no.2, Spatial Diffusion Study, Department of Geography, Northwestern University, Evanston [IL]

Dubin R (1998) Spatial autocorrelation: a primer. J Hous Econ 7(4):304-327

ESRI (2001) Using ArcGIS geostatstical analyst. ESRI, Redlands [CA]

Fortin MJ, Drapeau P, Legendre P (1989) Spatial autocorrelation and sampling design. Vegetatio 83:209-222

Fotheringham AS (1981) Spatial structure and distance-decay parameters. Ann Assoc Am Geogr 71(3):425-436

Fotheringham AS, Brunsdon C, Charlton M (2002) Geographically weighted regression: the analysis of spatially varying relationships. Wiley, New York, Chichester, Toronto and Brisbane

Geary RC. (1954) The contiguity ratio and statistical mapping. The Incorp Stat 5(3):115-145.

Getis A (1990) Screening for spatial dependence in regression analysis. Papers in Reg Sci Assoc 69:69-81

Getis A (1991) Spatial interaction and spatial autocorrelation: a cross-product approach. Environ Plann A 23(9):1269-1277

Getis A (1995) Spatial filtering in a regression framework: experiments on regional inequality, government expenditures, and urban crime. In Anselin L, Florax RJG (eds) New directions in spatial econometrics. Springer, Berlin, Heidelberg and New York, pp.172-188

Getis A, Aldstadt J (2004) Constructing the spatial weights matrix using a local statistic. Geogr Anal 36(2):90-104

Getis A, Franklin J (1987) Second-order neighborhood analysis of mapped point patterns. Ecology 68(3):473-477

Getis A, Griffith DA (2002) Comparative spatial filtering in regression analysis. Geogr Anal 34(2):130-140

Getis A, Ord JK (1992) The analysis of spatial association by distance statistics. Geogr Anal 24(3):189-206

Getis A, Ord JK (2000) Seemingly independent tests: addressing the problem of multiple simultaneous and dependent tests. Paper presented at the 39th Annual Meeting of the Western Regional Science Association, Kauai [HI]

Goodchild MF (1986) Spatial autocorrelation. Geo Books, Norwich

Griffith DA (1987) Spatial autocorrelation: a primer. Association of American Geographers, Washington [DC]

Griffith DA (1988) Advanced spatial statistics. Kluwer, Dordrecht

Griffith DA (1996) Spatial autocorrelation and eigenfunctions of the geographic weights matrix accompanying geo-referenced data. Canad Geogr 40(4):351-367

Griffith DA (2002) A spatial filtering specification for the auto-Poisson model. Stat & Prob Lett 58(2):254-251

Griffith DA (2003) Spatial autocorrelation and spatial Filtering. Springer, Berlin, Heidelberg and New York

Griffith DA (2005) Effective geographic sample size in the presence of spatial autocorrelation. Ann Assoc Am Geogr 95(4):740-760

Griffith DA (2009) Spatial filtering. In Fischer MM, Getis A (eds) Handbook of applied spatial analysis. Springer, Berlin, Heidelberg and New York, pp.301-318

Haining R (1977) Model specification in stationary random fields. Geographical Analysis 9:107-129

Haining R (1990a) Spatial data analysis in the social and environmental sciences. Cambridge University Press, Cambridge

Haining R (1990b) The use of variable plots in regression modeling with spatial data. Prof Geogr 42(3):336-344

Haining R (2009) The nature of georeferenced data. In Fischer MM, Getis A (eds) Handbook of applied spatial analysis. Springer, Berlin, Heidelberg and New York, pp.197-217

Hepple LW (1998) Exact testing for spatial autocorrelation among regression residuals. Environ Plann A 30(1):85-108

Hubert LJ, Golledge RG (1981) A heuristic method for the comparison of related structures. J Math Psych 23(3):214-226

Hubert LJ, Golledge RG, Costanzo CM (1981) Generalized procedures for evaluating spatial autocorrelation. Geogr Anal 13(3):224-233

Jacquez GM, Greiling DA, Durbeck H, Estberg L, Do E, Long A, Rommel B (2002) ClusterSeer: software for identifying disease clusters. TerraSeer Inc., Ann Arbor [MI]

Kelejian HH, Robinson DP (1993) A suggested method of estimation for spatial interdependent models with autocorrelated errors, and an application to a county expenditure model. Papers in Reg Sci 72(3):297-312

Kooijman S (1976) Some remarks on the statistical analysis of grids especially with respect to ecology. Ann Syst Res 5:113-132

Krishna-Iyer PVA (1949) The first and second moments of some probability distributions arising from points on a lattice, and their applications, Biometrics 36(1-2):135-141

Legendre P (1993) Spatial autocorrelation: trouble or a new paradigm. Ecology 74(6):1659-1673

Legendre P, Dale MRT, Fortin M-J, Gurevitch J, Hohn M, Myers D (2002) The consequences of spatial structure for the design and analysis of ecological field surveys. Ecography 25(5):601-616

LeSage JP (1999) Spatial econometrics, the Web book of regional science. Regional Research Institute, Morgantown [WV]

LeSage JP (2004) The MATLAB spatial econometrics toolbox. http://www.spatial-econometrics.com

Leung Y, Mei C-L, Zhang W-X (2000) Testing for spatial autocorrelation among the residuals of geographically weighted regression, Env Plann A 32(5):871-890

Matheron G (1963) Principles of geostatistics. Econ Geol 58(8):1246-1266

Miron JR (1984) Spatial autocorrelation in regression analysis: a beginners guide. Reidel, Toronto

Moran PAP (1948) The interpretation of statistical maps. J Roy Stat Soc B 10(2):243-251

Odland J (1988) Spatial autocorrelation. Sage, Newbury Park [CA]

Okabe A, Okunuki K-I, Shiode S (2006) SANET: a toolbox for spatial analysis on a network. Geogr Anal 38(1):57-66

Openshaw S (1977) Optimal zoning system for spatial interaction models. Environ Plann A 9(2):169-184

Ord JK, Getis A (1995) Local spatial autocorrelation statistics: distributional issues and an application. Geogr Anal 27(4):286-306

Ord JK, Getis A (2001) Testing for local spatial autocorrelation in the presence of global autocorrelation. J Reg Sci 41(3):411-432

Pace RK, Barry R (1997) Sparse spatial autoregressions. Stat Probab Lett 33(3):291-297

Páez A, Uchida T, Miyamoto K (2002) A general framework for estimation and inference of geographically weighted regression models: 1. location-specific kernel bandwidths and a test for locational heterogeneity. Environ Plann A34(4):733-754

Ravenstein EG (1885) The laws of migration. J Roy Stat Soc 48(2):167-235

Rey SJ, Janikas MV (2006) STARS: Space-time analysis of regional systems. Geogr Anal 38(1):67-86

Rey SJ, Janikas MV (2009) STARS: Space-time analysis of regional systems. In Fischer MM, Getis A (eds) Handbook of applied spatial analysis. Springer, Berlin, Heidelberg and New York, pp.91-112

Ripley BD (1977) Modeling spatial patterns. J Roy Stat Soc B 39(2):172-194

Rosenberg MS, Sokal RR, Oden NL, DiGiovanni D (1999) Spatial autocorrelation of cancer in Western Europe. Europ J Epidemiol 15(1):15-22

Rura MJ, Griffith DA (2009) Spatial statistics in SAS. In Fischer MM, Getis A (eds) Handbook of applied spatial analysis. Springer, Berlin, Heidelberg and New York, pp.43-52

Scott LM, Janikas MV (2009) Spatial statistics in ArcGIS. In Fischer MM, Getis A (eds) Handbook of applied spatial analysis. Springer, Berlin, Heidelberg and New York, pp. 27-41

Sokal RR, Oden NL Thomsen BA (1998) Local spatial autocorrelation in a biological model. Geogr Anal 30(4):331-356

Thomas E (1960) Maps of residuals from regression. In Berry BJL, Marble DF (eds) (1968) Spatial analysis: a reader in statistical geography. Prentice-Hall, Englewood Cliffs [NJ]

Tiefelsdorf M (2000) Modelling spatial processes. Springer, Berlin, Heidelberg and New York

Tiefelsdorf M, Boots B (1995) The exact distribution of Moran's I. Environ Plann A 27(6):985-999

Tiefelsdorf M, Boots B (1997) A note of the extremities of local Moran's I_i and their impact on global Moran's I. Geogr Anal 29(3):248-257

Tiefelsdorf M, Griffith DA, Boots B (1998) A variance stabilizing coding scheme for spatial link matrices. Environ Plann A 31(1):165-180

Tobler WR (1970) A computer movie simulating urban growth in the Detroit region. Econ Geogr 46(2):234-240

Upton GJ, Fingleton B (1985) Spatial statistics by example: point pattern and quantitative data. Wiley, New York, Chichester, Toronto and Brisbane

von Thünen JH (1826) Der isolierte Staat in Beziehung auf Landwirtschaft und Nationalökonomie. In Kapp KW, Kapp LL (eds) (1956) Readings in economics. Barnes and Noble [NY]

Wartenberg D (1985) Multivariate spatial correlation: a method for exploratory geographical analysis. Geogr Anal 17(4):263-283

Wheeler DC, Tiefelsdorf M (2005) Multicollinearity and correlation among local regression coefficients in geographically weighted regression. J Geogr Syst 7(2):161-188

Wong DWS (1997) Spatial dependency of segregation indices. Canad Geogr 41(2):128-136

Zipf, GK (1949) Human behavior and the principle of least effort: an introduction. Cambridge University Press, Cambridge

B.4 Spatial Clustering

Jared Aldstadt

B.4.1 Introduction

Spatial clustering analysis has become common in many fields of research, and is most commonly used in epidemiology and criminology applications. Knox (1989, p.17) defines a spatial cluster as, '*a geographically bounded group of occurrences of sufficient size and concentration to be unlikely to have occurred by chance.*' This is a useful operational definition, but there are very few situations when phenomena are expected to be distributed randomly in space. In most cases an implicit assumption in spatial cluster analysis is that the researcher has accounted for all the factors known to influence the variable of study. This would lead to an examination of residual spatial variation in a spatial modeling exercise. Spatial clustering analysis is carried out on raw variables or rates when there are no *a priori* hypotheses regarding the process.

There are an ever increasing number of methods available for the analysis of spatial clustering. These techniques can be divided into two categories: those that are used to determine if clustering is present in the study region, and those that attempt to identify the location of clusters. The first category of tests is called global clustering techniques and these methods provide a single statistic that summarizes the spatial pattern of the region. These will be discussed in the section that follows. The second type of methodology is called local clustering. Local methods examine specific sub-regions or neighborhoods within the study to determine if that area represents a cluster of high values (a hot spot) or low values (a cold spot). These methods can be further differentiated as either focused or non-focused tests. Focused tests examine one or a small set of pre-defined foci of interest. Non-focused tests are designed to find clusters that exist throughout the entire region of analysis. Local clustering methods will be discussed in Section B.4.3. Considerations for choosing a spatial clustering method and some concluding remarks are provided in Section B.4.4.

B.4.2 Global measures of spatial clustering

The methods developed to detect global clustering are also called general tests of clustering. In most cases, the null hypothesis is one of spatial randomness. These methods provide a single summary statistic which describes the degree of clustering present in the mapped pattern. The value of the statistic indicates whether the pattern is clustered, random, or dispersed. In contrast to a clustered pattern, a dispersed pattern is one where high values and low values are nearby each other more often than would be expected in a random pattern. Clustered and dispersed patterns may also be labeled positive and negative spatial autocorrelation respectively.

Areal data methods

The first set of methods deal with areal data, or the attributes of units that are mapped as polygons. These attributes are most often aggregate data such as a density or a rate per unit of population. It does not usually make sense to carry out spatial analysis with a raw count of events within a spatial unit. Much of the variation in the attribute is likely to be a function of the size of the unit or the population at risk within the unit. The use of rates may also confound cluster analysis when there is substantial variation in the size of the denominator to be used to calculate rates. Consequently, variants of general tests have been developed that account for this variation in population size and examine the spatial pattern of the excess or deficiency of events occurring in each spatial unit. These analyses are not limited to scale data, and a method that examines clustering in a map with two classes will also be discussed.

Global clustering statistics take a common form that compares the similarity of values at locations to the spatial proximity of the locations. This type of statistic is called a general cross-product statistic, and it was introduced by Mantel (1967) for computing the similarity between two matrices. The spatial proximity between each pair of locations i and j is denoted W_{ij} and entered into an n-by-n matrix called the spatial weights matrix. The spatial weights matrix is most often denoted as W, and is discussed further below. The similarity of two data values x_i and x_j is denoted S_{ij} and can be entered into an n-by-n matrix that is labeled S. Clustering is indicated when spatial proximity and similarity are positively related. In summation notation, the general form of the statistic is

$$\sum_{i=1}^{n}\sum_{j=1}^{n} W_{ij}\, S_{ij}. \qquad (B.4.1)$$

Each of the techniques presented in this section are a variation of this form, with the distinguishing variant being the measure of similarity between values. Often the indices are normalized by global measures of similarity and spatial connectivity.

The spatial weights matrix defines the structure of spatial relationships in the study region. It delimits the extent of clustering that the clustering technique is able to detect. The choice of **W**, therefore, should be considered carefully in clustering analysis. The simplest and perhaps most commonly used set of spatial weights is the binary contiguity matrix. Here, W_{ij} is equal to one if units i and j share a common boundary and zero otherwise. There are two variants of the binary contiguity matrix. The Rook case requires that neighbors share a common edge. A common vertex or point is all that is required for contiguity in the Queen case. Other binary weights matrices include a number of nearest neighbors and the complete set of neighbors with a given distance. Spatial relationships may also be defined as a function of the distance between units. Most commonly elements are defined as

$$W_{ij} = d_{ij}^{-\alpha} \quad (B.4.2)$$

where d_{ij} is the distance between units i and j *and* α is larger than zero. It should also be noted that the diagonal of the weights matrix, the values W_{ii}, are usually set to zero.

The weights matrix used in cluster analysis is often standardized so that the elements of each row sum to one (row standardization). This procedure serves to equalize the weight given each observation in the analysis with respect to its number of neighbors. The elements of this standardized matrix are calculated as

$$\widetilde{W}_{ij} = \frac{W_{ij}}{\sum_{j=1}^{N} W_{ij}}. \quad (B.4.3)$$

Standardization should not be carried out in cases when the weights have meaningful interpretation with regards to the analysis (Anselin 1988). For example, standardizing inverse distance matrices will distort the relative spatial relationships between units and cloud interpretation of the clustering index. The effects of standardization are examined and an alternative to row standardization is provided by Tiefelsdorf et al. (1999). A more complete examination of the spatial weights matrix with references to many alternative forms and several reviews is given by Getis and Aldstadt (2004).

Join-count statistic. The join count statistic is a measure of clustering for a binary classification of data. These values could be visualized as a two-category choropleth map. The two classes are usually referred to as black (B) and white (W). A join is another name for the contiguity relationship of two areas sharing a boundary. The statistic value is the number of joins of a given type. Each boundary may connect two black units (BB), two white units (WW) or one unit of each type (BW). Cliff and Ord (1973) define the number of BW joins as the general cross product statistic

$$BW = \tfrac{1}{2}\sum_{i=1}^{n}\sum_{j=1}^{n} W_{ij}(x_i - x_j)^2 \tag{B.4.4}$$

where x_i equal to one corresponds to B and x_i equal to zero corresponds to W. Following from the definition of join, the weights, W_{ij} are usually restricted to a binary contiguity representation. Under a free sampling assumption, the expected number of BW joins in a random spatial distribution is

$$E[BW] = 2Jpq \tag{B.4.5}$$

where J is the total number of joins. p is the probability that a unit is coded B and is often estimated as the proportion of units that are in the class B. q is the probability that a unit is coded W and is equal to *one minus p*. The number of joins may be calculated from the binary contiguity weights as

$$J = \tfrac{1}{2}\sum_{i=1}^{n}\sum_{j=1}^{n} W_{ij}. \tag{B.4.6}$$

If the classes are clustered together, there would be fewer observed BW joins than expected. Likewise, if the pattern is dispersed or similar to a checkerboard pattern, there would be more BW joins than expected in a spatially random pattern. The variance of the BW statistic under both free and non-free sampling are derived in Cliff and Ord (1973) along with an extension to the case when there are more than two classes.

Moran's I. Moran's I is a well known test for spatial autocorrelation (Moran 1950). The index is similar to covariance and correlation statistics. The measure of similarity between values at two locations i and j is the product of the deviation between the value at each location and the estimate of the global mean $\bar{x}$. This product is weighted by the spatial proximity of the two locations, and the sum of the resulting values for all pairs of locations is the spatial autocovariance. The standardized index is given as

$$I = \frac{N}{S_0} \frac{\sum_{i=1}^{n}\sum_{j=1}^{n} W_{ij}(x_i - \bar{x})(x_j - \bar{x})}{\sum_{i=1}^{n}(x_i - \bar{x})^2} \quad i \neq j \tag{B.4.7}$$

where

$$S_0 = \sum_{i=1}^{n}\sum_{j=1}^{n} W_{ij}. \tag{B.4.8}$$

The expected value for a spatially random distribution is minus one over $(n-1)$. This quantity tends towards zero as the sample size increases. Values greater than this indicate clustering of units with high and or low values. Values that are smaller than the expected value indicate negative association between proximate locations. Unlike the Pearson's correlation coefficient, Moran's I is not bounded between negative one and one, but usually falls within this interval (Bailey and Gatrell 1995). A correlogram displays the Moran's I value calculated for a number of increasing distances. The distances are most often mutually exclusive distance bands or orders of contiguity. The correlogram can be used to determine the extent of spatial autocorrelation and at what distance spatial autocorrelation is maximized.

Cliff and Ord (1973) derive the distribution of Moran's I under the null hypothesis for two different sampling assumptions. Under the randomization assumption the n observed values are fixed, but they are relocated randomly among the locations in a random fashion. The normality assumption assumes that the values at each location are drawn from independent and identical normal distributions. Underlying both of these assumptions is the additional assumption of stationarity. In the spatial context, stationarity implies that the mean and variance of the variable of interest is constant throughout the study region. Cliff and Ord (1973) prove that under both the randomization and normality assumptions Moran's I is asymptotically normally distributed. When n is large, a reliable significance value can be computed based on this distribution. Tiefelsdorf and Boots (1995) show that the rate of convergence to normality is a function of the spatial weights matrix and the distribution of the data values as well as sample size. A Monte Carlo approach, as outlined by Besag and Newell (1991), is often used to generate significance values under either the randomization or normality assumptions.

Adjusting for heterogeneous variance. When the spatial units vary significantly in size, the assumption of constant variance is violated. Specifically, units with large populations are less likely to deviate from the global mean with respect to units with small populations (Haining 2003). Walter (1992) demonstrates that variation in size of population at risk can result in incorrectly rejecting the null hypothesis. Several methods have been proposed to test the spatial randomness hypothesis when the background population is heterogeneous (Waller and Gotway 2004). Oden (1995) proposed a version of Moran's I, I_{pop}, that is based on individual level data. Inference is again based on the randomization assumption. However, the randomization refers to the status of individuals. This is most often applied in studies of disease clustering where cases are denoted as one and the remaining individuals are denoted zero. Tango (1995) proposed the excess events test (*EET*) that is defined as

$$EET = \sum_{i=1}^{n}\sum_{j=1}^{n} W_{ij}\left(c_i - n_i \tfrac{C}{n}\right)\left(c_j - n_j \tfrac{C}{n}\right) \qquad (B.4.9)$$

where c_i is the number of cases in unit i, n_i is the population of unit i, and C is the total number of cases in the study region. Like I_{pop} a large variation from the expected number of cases within a region contribute to large statistics, and I_{pop} is an affine transformation of *EET* (Oden et al. 1998; Tango 1998). Tango suggested an exponentially decreasing function of distance as the weight between units exp $(-d_{ij}/\lambda)$, where d_{ij} is the distance between locations i and j, and λ is a measure of the spatial scale of clustering. The maximized excess events test (*MEET*) searches over a plausible range of λ for the minimum *p-value* (Tango 2000). This methodology examines clustering at a number of scales while accounting for multiple testing. Assunção and Reis (1999) propose an Empirical Bayes method for standardizing rates when variances are not stable. In this approach x_i is

$$x_i^{adj} = \frac{x_i - E[x_i]}{\sqrt{Var(x_i)}}. \qquad (B.4.10)$$

In the accompanying simulation study, the authors determine that the standardized index is more powerful than the traditional Moran's I. Assunção and Reis (1999) also compare their method to Oden's I_{pop} which is powerful in detecting rate heterogeneity within units, but is not as useful for detecting spatial correlation of rates.

Geary's c. Geary's c is an alternative measure of spatial clustering that takes the familiar cross-product form (Geary 1954). The similarity of two locations is quantified as the difference between the values at each location squared. This leads to the statistic

$$c = \frac{(n-1)}{2S_0} \frac{\sum_{i=1}^{n}\sum_{j=1}^{n} W_{ij}(x_i - x_j)^2}{\sum_{i=1}^{n}(x_i - \bar{x})^2}. \tag{B.4.11}$$

Two values that are similar will have a small contribution to the global value, therefore low values of c are indicative of a clustered pattern. The expected value of a random pattern is one, and c ranges between zero and two. Cliff and Ord (1973) derived the variance under the randomization and normalization assumptions.

Getis-Ord G. The Getis-Ord G statistic quantifies the relationship between two locations as the product of the values at the locations (Getis and Ord 1992). The statistic is

$$G = \frac{\sum_{i=1}^{n}\sum_{j=1}^{n} W_{ij} x_i x_j}{\sum_{i=1}^{n}\sum_{j=1}^{n} x_i x_j}. \tag{B.4.12}$$

Use of the general G requires that the variable of analysis is positive valued with a natural origin. The expected value under a random pattern is

$$E[G] = \frac{\sum_{i=1}^{n}\sum_{j=1}^{n} W_{ij}}{n(n-1)}. \tag{B.4.13}$$

G values greater than the expected value result from a pattern that is dominated by concentrations of high values because the product of neighboring units is large. A low G value results from a pattern dominated by clusters of low values. Acceptance of the null does not necessarily imply a random pattern, but may result in the case that clusters of both high and low values exist in the study region. The G statistic differs from the other indexes discussed in this section in that it is not strictly a measure of clustering, but provides an indication of the type of clustering that is present in the study region.

Point data methods

A second set of methods is used to analyze phenomena that are mapped as points. These could be the location of a set of objects or the locations of a set of events. Complete spatial randomness (CSR) describes the pattern of points that would occur by chance in a completely undifferentiated environment. The process that generates this pattern is called the homogeneous planar Poisson point process. In this process points are generated in a study are under the conditions: (a) each location in the study area has an equal probability of receiving a point; and (b) the selection of a location for a point is independent of the location of existing points. As with areal data, patterns may deviate from CSR by being either clustered or dispersed. In a clustered pattern, points are on average closer than expected in CSR. In a dispersed pattern, points are uniformly distributed throughout the study area.

The CSR hypothesis is limiting and rejection of this null may not be meaningful. There are few instances when the homogeneous and independent probability of occurrence is plausible. To avoid this limiting assumption, comparative analysis of two or more point patterns is conducted. This allows for examination of clustering above and beyond what would be expected due to spatial variation in the probability of occurrence. The aim is often to determine whether some attribute is clustered in a population given its heterogeneous distribution. When analyzing one or more types of events or objects, the point patterns are often referred to as marked point patterns.

Quadrat analysis. Quadrat analysis is one of the first techniques used to test the CSR hypothesis. Quadrat analysis involves partitioning the study area into a number of scattered or contiguous equal sized quadrats and was originally developed in the plant ecology literature (Greig-Smith 1952). The number of events in each cell is tabulated and a frequency table of these cell counts is computed. A goodness-of-fit test is then performed to determine if the frequencies are significantly different from those expected under a Poisson process. An excess number of low and high cell counts indicate a clustered pattern. An excess number of cells with average density indicate a dispersed pattern. The results are dependent on the size of the quadrats, and often the analysis is repeated for a range of quadrat sizes (Boots and Getis 1988). The general clustering methods described above are also used to analyze the pattern of events aggregated into quadrats.

Nearest neighbor analysis. Nearest neighbor analysis also has it origins in the plant ecology literature. These methods are based on the distance between each point and its closest neighbor. Clark and Evans (1954) derived the expected value and variance of the average nearest neighbor distance in a CSR pattern. The use of the mean nearest neighbor distance provides an easy to interpret summary statistic, but is a crude representation of a point pattern. For instance, a few very large nearest neighbor distances associated with isolated points could obscure an otherwise clustered pattern. Refined nearest neighbor analysis overcomes this issue by examining the entire distribution of nearest neighbor distances. The test statistic is the maximum difference between the observed nearest neighbor distance frequency distribution and the distribution expected under the null hypothesis (Diggle 1990). A rigorous analysis of a point data set can also include the analysis of higher order neighbors.

Ripley's K function. One problem with quadrat analysis and nearest neighbor analysis is that they examine only one scale of interaction at a time (Bailey and Gatrell 1995). Most commonly these techniques detect clustering at short distances. Advances in computational capabilities have enabled the examination of all inter point distances. Ripley's K function can be computed over a range of distances and be used to identify the scales over which clustering occurs (Ripley 1976). The estimator is defined as

$$\hat{K}(d) = \frac{R}{n^2} \sum_{i=1}^{n} \sum_{j=1}^{n} W_{ij} \quad \text{for } i \neq j \tag{B.4.14}$$

where R is the size of the study area. The weights matrix is binary and equal to one when points i and j are within distance d, and zero otherwise. A standardized measure that simplifies interpretation is given as

$$\hat{L}(d) = \sqrt{\frac{\hat{K}(d)}{\pi}}. \tag{B.4.15}$$

The expected value of $\hat{L}(d)$ under CSR is d. A value greater than d indicates clustering and a value less than d indicates dispersion. The statistical significance of the results is determined through Monte Carlo simulations under an appropriate null hypothesis (Besag and Diggle 1977).

The points outside the study region are unobserved and cannot be included in the summation. In order to correct for this edge effect, points near the boundary may be given a larger weight in the analysis. Ripley (1976) provided one such correction for rectangular study areas (see Chapter B.3). The boundary problem is also overcome by transforming or duplicating the existing dataset to create points outside the boundary. A comparison of the various edge correction methods is provided by Yamada and Rogerson (2003).

Ripley's K function is a form of second order analysis because it is examining the interaction or dependence between points. This is in contrast to the intensity of points, which are termed first-order effects. There is an implicit assumption that the density of points is uniform within the study area (Diggle 2003). When the density of points is heterogeneous within the study area, this first-order effect may be captured in the K function. To avoid this ambiguity the distances of analysis should be limited so that they are small relative to the size of the study area. One rule of thumb is to limit the maximum distance of analysis to no longer than one-half the length of the shorter side of a rectangular study area.

Bivariate point patterns. The methods above have only considered points of a single type. Bivariate point pattern methods may be used to answer questions concerning the spatial dependence of two types of events. One set of points may also be used as a control group to correct for the variations in density within the study area. This type of analysis is especially relevant to epidemiological studies where inhomogeneous populations at risk are the norm.

The cross K function is a useful tool for examining the relationship between two sets of events (Bailey and Gatrell 1995). The estimator is given as

$$\hat{K}_{12}(d) = \frac{R}{n_1 n_2} \sum_{i=1}^{n_1} \sum_{j=1}^{n_2} W_{ij} \qquad (B.4.16)$$

where n_1 and n_2 are the number of each type of points. The result can be standardized in the same manner as above (see Eq. (B.4.15)). In this case, a value greater than d indicates that attraction between the two types of events and a value lower than d indicates repulsion between the two types of events. Significance is calculated through randomization. In this case, the patterns are preserved in their original form, but they are shifted relative to one another. These shifts may be performed using a toroidal transformation of the study area.

Spatial randomness may not always be an important hypothesis to test. Very often the potential locations of an event are limited within the study area. Examples include crimes which are geocoded to the nearest available street address or cases of disease which are distributed among the population at risk. This type of heterogeneity can be accounted for using bivariate point pattern analysis. Cuzick and Edwards (1990) presented a method based on the number of nearest neighbors of each type of point. The method depends on a scale parameter, k, that indicates the extent of analysis in terms of the number of nearest neighbors. The method was designed to detect clusters in epidemiological datasets, and the events of interest are usually cases of disease. The second set of events is called controls and is selected as being representative of the population at risk. The statistic is given as

$$T_k = \sum_{i=1}^{n_1} m_i(k) \qquad (B.4.17)$$

where n_1 is the number of cases, and $m_i(k)$ is the number of cases among the k nearest neighbors. When cases are clustered, the resulting statistic will be large. T_k will be small when the cases are dispersed and therefore, surrounded by controls. Jacquez (1994) developed a modification to the Cuzick and Edwards' test that can be used to evaluate aggregate data as well.

A form of the K function can be employed in the same situation (Diggle and Chetwynd 1991). The statistic becomes the difference between the two univariate K functions,

$$\mathit{Diff}(d) = \hat{K}_1(d) - \hat{K}_2(d) \qquad (B.4.18)$$

where $\hat{K}_1(d)$ and $\hat{K}_2(d)$ are the K functions for each set of points. If the events of type one are distributed randomly in relation to the remaining points, the difference will be approximately zero. A positive difference indicates that points of type one are more clustered than points of type two. A negative value indicates that points of type one are more dispersed than points of type two. The significance of both T_k and $\mathit{Diff}(d)$ can be examined under the random labeling null hypothesis. The designation of event type is randomly permuted or shuffled among the points for each realization in a Monte Carlo procedure.

B.4.3 Local measures of spatial clustering

When the null hypothesis of spatial randomness is rejected by a general test for spatial clustering two additional questions are raised: where are the clusters and what is their spatial extent. Local clustering statistics are used to answer these questions. It should be noted, however, that there may be significant local clustering even in the case that the general test results in acceptance of the null hypothesis. Local measures can be either tests of clustering or focused tests.

Areal data methods

As with global clustering statistics, the local tests take a general form. A local clustering statistic is the product of a spatial weights vector and a similarity vector. It is represented in summation notation as

$$\sum_{j=1}^{n} W_{ij} S_{ij} . \qquad (B.4.19)$$

Several of the global methods presented in Section B.4.2 have a local equivalent that is the ith unit's contribution to the global statistic.

Getis-Ord G_i and G_i^*. Getis and Ord (1992) present a local clustering test that is based on the concentration of values in the neighborhood of a unit. The original statistic was given as

$$G_i = \frac{\sum_{j=1}^{n} W_{ij} x_j}{\sum_{j=1}^{n} x_j} \quad \text{for } i \neq j. \quad (B.4.20)$$

The authors derive the expected value and variance of G_i when W_{ij} are elements of a binary spatial weights matrix. Most often the weights are based on proximity with the value at all units within a given distance being summed in the numerator. The G_i^* statistic includes the contribution of the ith unit in the calculation of local concentration. This amounts to adding the value x_i to both the numerator and denominator in Eq. (B.4.20). The G_i^* matches the usual definition of cluster as a contiguous and non-perforated set of units. In this original formulation, the statistics are intended for use with variables that possess a natural origin.

Modified versions of the G_i and G_i^* statistics are presented by Ord and Getis (1995). The newer formulation standardizes the statistic by subtracting the expected value and dividing the difference by the standard error. This eases interpretation as the result can be interpreted as approximately following a standard normal distribution. A positive value indicates clustering of high values and a negative value indicates a cluster of low values. This update also allows for the use of non-binary weights matrices and variables without a natural origin. The standardized G_i^* statistic is given in Chapter B.3.

The Moran scatter plot and local Moran's I. The Moran scatter plot was introduced by Anselin (1996) as an exploratory spatial data analysis (ESDA) tool for assessing local patterns of spatial association (see also Chapter B.1). This bivariate scatter plot places the unit values (x_i) on the horizontal axis and the spatial lag (lag_i) for the same variable on the vertical axis (see Fig. B.4.1). The spatial lag is the spatially weighted average of the values at neighbouring units, and is calculated as

$$lag_i = \frac{\sum_{j=1}^{n} W_{ij} x_j}{\sum_{j=1}^{n} W_{ij}}. \quad (B.4.21)$$

The axes of the plot are drawn so that they cross at the average value of x_i and lag_i, respectively. The four quadrants of the plot separate the spatial association into four components. The first letter in the quadrant labels indicates whether the value of x_i is higher (H) or lower (L) than the average of all values. Correspondingly, the second letter in the quadrant labels indicates whether the value of lag_i is higher (H) or lower (L) than the average of all the spatial lags. Units that fall into the quadrants labelled 'HH' and 'LL' represent clustering of high and low values respectively. The remaining quadrants contain units that have negative association with their neighbours and can be considered as spatial outliers. A spatial outlier may arise from a cluster consisting of just one unit. The Moran scatter plot is a useful visualization tool for assessing spatial pattern and spatial clustering.

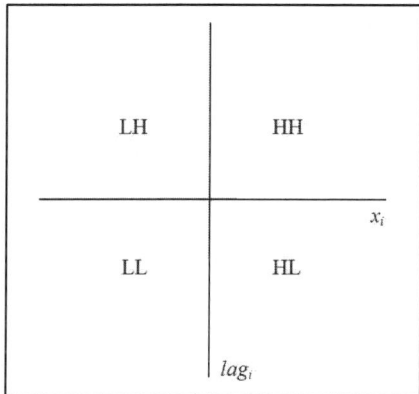

Fig. B.4.1. The Moran scatter plot

The significance of extreme points in the Moran scatter plot can be assessed using local Moran's I or I_i (Anselin 1995). For each region, I_i is calculated as

$$I_i = \frac{\sum_{j=1}^{n} W_{ij}(x_i - \bar{x})(x_j - \bar{x})}{\frac{1}{n}\sum_{j=1}^{n}(x_j - \bar{x})} \ . \tag{B.4.22}$$

As discussed in Chapter B.3, I_i represents a decomposition of the global Moran's I. This form of local method is called a Local Indicator of Spatial Association (LISA). Anselin (1995) also presents the formulation of the local Geary's c or c_i. Statistical significance can be determined through the provided expected value and variance or by Monte Carlo procedure. A positive I_i indicates clustering of high or low values. A negative I_i indicates a spatial outlier. Several results are, therefore, reported for each unit. These include the statistic value, the significance value, and the label of corresponding quadrant of the Moran scatter plot.

Local clustering of categorical data. In the case of global clustering statistics, the methods for categorical data preceded the methods for metric data. This was not the case for local methods of pattern analysis. Boots (2003, 2006) details the issues in this research area and presents ESDA methods for describing and understanding patterns of categorical data.

Accounting for multiple and dependent testing

Local spatial statistics are often used in an exploratory mode to test for clustering at each location in the study area simultaneously. In this case, the issue of multiple and dependent testing is a concern when assessing the significance of clustering. Multiple testing problems arise whenever more than one hypothesis test is carried out using the same dataset. The probability of rejecting the null hypothesis at least once when it is true in all cases is much higher than the nominal type I error rate, α. The dependence part of the problem is a result of nearby local tests relying on many of the same data values. The results of these tests are, therefore, correlated. Failure to account for these effects results in over identification of clusters by local spatial statistics (Anselin 1995; Ord and Getis 1995).

The Bonferroni correction is commonly used to account for multiple testing (Warner 2007). In this approach, a new critical value is calculated for the individual tests by dividing the overall level of type I error by the number of tests. For example, if an overall significance level of 0.05 is desired for 20 simultaneous tests, a significance level of 0.0025 is used in each separate test. Caldas de Castro and Singer (2006) demonstrate the usefulness of a less conservative approach called the false discovery rate (FDR). FDR controls for the rate of false positives among the nominally significant results and was introduced by Benjamini and Hochberg (1995). It is based on the distribution of significance values for a set of tests, and is therefore adaptive to the characteristics of each dataset. Simulation studies performed by Caldas de Castro and Singer (2006) compared uncorrected local statistics with the Bonferroni and FDR corrected versions. FDR was superior in properly identifying the location and extent of spatial clusters. Another common approach to account for multiple testing is to examine just the most extreme value of all the individual tests (Baker 1996; Tango 2000). This approach provides a satisfying solution in a general test of clustering, but it does not address each local test individually.

Local spatial tests are most often evaluated under the assumption that there is no global spatial autocorrelation. Some attempts have been made to relax this assumption and evaluate clustering in the presence of spatial autocorrelation. One technique, that of Ord and Getis (2001), is described in Chapter B.3 of this handbook. Goovaerts and Jacquez (2004) present a geostatistical technique for generating datasets under a realistic null hypothesis. These models include both spatial autocorrelation and heterogeneous populations in the examinations of clustering.

Cluster detection algorithms

A second set of local methods are the automated search procedures and their associated test statistics. These computational techniques involve testing a large number of regions within the study area for spatial clustering. These methods have primarily been applied to spatial analysis of epidemiological data. They are flexible in that they can, for the most part, be applied to both point and aggregate data. In the case of aggregate data, the location associated with spatial units is most often taken as the centroid of the unit. They differ from the methods presented above in that they are not limited to a fixed definition of neighborhood, and thus cluster size, but are designed to detect clusters of varying sizes. It should be noted, however, that the test statistics discussed in the previous section could be used in conjunction with the search procedures outlined below.

Geographical Analysis Machine (GAM). The Geographical Analysis Machine (GAM) was the first automated approach to finding cluster locations in spatial patterns (Openshaw et al. 1987). The original GAM involves searching a large number of circles across the study area. The circles are centered on a grid, and the radius of these circles is allowed to vary over a suitable range of values. The number of cases in each circle is counted and the significance of the count is evaluated. A Monte Carlo procedure is used, and the circles that fall within a given threshold are retained. The resulting set of circles is then mapped to show cluster centers. One weakness of the GAM is the lack of control for multiple testing (Besag and Newell 1991). The GAM did, however, show the utility of a geo-computational approach to cluster detection and has inspired several modifications and improvements. Each of the methods described below has built on the foundation of the GAM. There have also been several improvements to the GAM procedure itself. One example is the method of Conley et al. (2005). This technique uses genetic algorithms to speed search times and reduce over-reporting of cluster sizes.

Besag and Newell's method. One additional shortcoming of the original GAM is that the circles examined are based on a distance only approach. If the population at risk varies, then circles of the same size contain different size populations. This variation in population at risk must be included in the analysis. The Besag and Newell (1991) method overcomes this difficulty by requiring the expected cluster size, say k, as a user input. Each unit with at least one case of disease is examined as a potential center of clustering. The circle is expanded in order of nearest neighbor distance until at least k cases are included within the circle. The inference is then based on the number of units, L_i, containing k cases. The significance of each potential cluster is evaluated using the Poisson cumulative distribution function under the uniform risk null hypothesis

$$P(L_i \le l_i) = 1 - \sum_{j=1}^{k-1} \exp(-\mu) \frac{\mu^j}{j!} \tag{B.4.23}$$

where l_i is the observed number of units containing k cases, and μ is the expected number of cases within those units. μ is calculated as the product of the global risk and the population at risk within the set of units under examination. Fotheringham and Zhan (1996) compare GAM, Besag and Newell's method and their own modification of the GAM search algorithm. All methods are deemed successful at detecting clusters, but Besag and Newell's method is the least likely to result in false positives. Additionally, Fotheringham and Zhan (1996) provide a formulation of Besag and Newell's method for use with point data, as the original presentation was based on areal spatial units.

The SaTScan procedure. The SatScan procedure is another cluster finding procedure inspired by the GAM (Charlton 2006). Like the GAM, SaTScan searches a large number of circles and examines the number of cases in relation to the population at risk (Kulldorff 2004). Most analysts choose to examine clusters that are centered on cases or region centroids as in the Besag and Newell method, but any number of potential clusters could be examined. At each center, the size of the circle is increased until a user defined maximum cluster size is reached. The maximum cluster size could be given in terms of geographic area or population at risk. The minimum cluster size does not need to be specified. During the search procedure, the likelihood that each cluster has occurred by change is evaluated using the spatial scan statistic. Kulldorff (1997) derived the spatial scan statistic for count or marked point pattern data. Variants of the spatial scan statistic appropriate for other types of data have also been developed (Huang et al. 2007; Jung et al. 2007). The spatial scan statistic based on the Poisson distribution is employed for aggregate case data. A uniform risk null hypothesis is evaluated. $L(R)$ is the likelihood that there is a cluster in a region R, and L_0 is the likelihood under the null. A likelihood ratio test statistic is given by

$$\frac{L(R)}{L_o} = \left(\frac{c_R}{\mu_R}\right)^{c_R} \left(\frac{C - c_R}{C - \mu_R}\right)^{C - c_R} \tag{B.4.24}$$

if $c_R > \mu_R$, and one otherwise. Here C is the total number of cases for the population, c_R is the number of cases in region R, and μ_R is the expected number of cases in the region R. The most likely cluster or clusters are those with the largest likelihood ratio values. An exact p-value is calculated using a Monte Carlo procedure. A primary advantage of the spatial scan statistic is that it takes multiple testing into account. A version of the SaTScan procedure that examines elliptical regions as potential clusters is presented by Kulldorff et al. (2006).

Finding arbitrarily shaped clusters. To this point, each of the cluster detection methods discussed are limited to either a prespecified and fixed definition of neighborhood or the examination of a large number of circles or ellipses. In most cases there is little reason to expect that spatial clustering would take a regular shape. To overcome this limitation, a variety of tests have been developed to locate irregularly shaped clusters. Each of these approaches uses a definition of proximity equivalent to the binary contiguity matrix. Spatial units are treated as nodes on a connected graph. The resulting clusters are not limited to being regular shapes, but must be contiguous regions or connected sub-graphs.

Tango and Takahashi (2005) proposed an examination of all possible connected sub-graphs up to a pre-selected maximum cluster size. This approach works well for clusters containing a small number of units, but is not feasible for finding larger clusters. Two approaches use stochastic optimization techniques to overcome this shortcoming. Duczmal and Assunção (2004) employ simulated annealing, and Duczmal et al. (2007) a genetic algorithm. These techniques are not restricted to a maximum cluster size, but they require additional inputs, known as hyper parameters, that govern the search process.

Aldstadt and Getis (2006) proposed an iterative region growing approach to finding arbitrarily shaped clusters called AMOEBA. To begin this procedure a single unit is selected as the seed location. All possible combinations of contiguous units are examined and the set that maximizes the clustering statistic is retained. The algorithm then continues by examining the units at each order of contiguity until the addition of units no longer increases the test statistic. At this point a cluster based on the first seed location is delimited. The procedure can be repeated using every location as the seed location. The significance of each delimited cluster is evaluated using a Monte Carlo procedure. The iterative approach ensures that low value units will not be included in clusters of high values. This prohibits the linking of two or more disjoint clusters as one, which is possible in the other approaches.

Focused clustering methods

Focused clustering tests start with a predetermined set of foci, and examine the likelihood that each of these foci is the center of a cluster. Foci are most often represented as points, but they may also be linear or areal features. The most common application of these tests is the examination of disease clusters in proximity to a pollution source. The null hypothesis is that disease risk is not elevated in proximity to the foci. It bears repeating that the potential sources should be identified before the initiation of these focused tests. If potential foci are selected based on their proximity to areas of raised incidence identified through cluster detection procedures, the inference is biased toward rejection of the null hypothesis (Waller and Gotway 2004). This is known as the 'Texas sharpshooter fallacy.' The name comes from the Texan that shoots into the side of a barn and then paints a target centered on the hits so that it appears he is a sharpshooter.

The Lawson-Waller score test. Waller et al. (1992) and Lawson (1993) independently developed a score tests for focused clustering. The global risk can be estimated as the total number of cases, C, divided by the total population at risk, n. The resulting score statistic is a local version of Tango's EET statistic. The score statistic for a focus i is given as

$$T_i = \sum_{j=1}^{n} W_{ij}\left(c_j - n_j \frac{C}{n}\right) \quad \text{(B.4.25)}$$

where c_j is the number of cases in unit j, n_j is the population of unit j. Here, the spatial weight can take a variety of forms. A distance decay function depicts the setting where exposure decreases as distance to the foci increases. A binary weight may also be used to indicate that all units within a given distance are experiencing similar exposure. Under the constant risk null hypothesis, the expected value of the statistic is zero. The variance of T_i is

$$Var(T_i) = \frac{C}{n}\left(\sum_{j=1}^{n} W_{ij}^2 n_j\right) - \frac{C}{n^2}\left(\sum_{j=1}^{N} W_{ij} n_j\right)^2. \quad \text{(B.4.26)}$$

The standardized statistic

$$Z(T_i) = \frac{T_i}{\sqrt{Var(T_i)}} \quad \text{(B.4.27)}$$

can then be compared to the standard normal distribution. Monte Carlo tests may be more appropriate when there is a small number of regions or for a vary rare disease (Waller et al. 1992). A method of determining the exact distribution of T_i is provided by Waller and Lawson (1995). Rogerson (2005) defines both a global test and a local clustering statistic based on the score test.

Other focused clustering tests. Stone (1988) developed a group of tests based on the first isotonic regression estimator. This method assumes that the relationship between exposure and risk is monotonic, but the relationship does not have to take a parametric form. This flexibility is unique among focused clustering tests.

Bithell (1995) provided a set of tests that are called linear risk score tests. These tests are based on the notion of the relative risk function. Under this alternative hypothesis, relative risk of disease declines as distance to the focus increases. The test statistic is the sum of these estimated relative risk values. This test is commonly performed using the rank of distances to neighboring units. In this case the risk becomes a function of relative location as opposed to exact location. Tango (2002) provides an extended score test that allows for non-monotonic relative risk functions. The extended score test would be most useful in the situation where exposure is expected to peak at some distance form the putative source.

A focused clustering test for individual or point level data is provided by Diggle (1990) and refined by Diggle and Rowlingson (1994). This method can be applied to inhomogeneous point patterns when the locations of disease cases and a representative control group are known. The method is flexible in terms of the functional form of the spatial risk, but the type of model must be specified. The parameters of the kernel are estimated using non-linear binary regression. The regression framework allows for straightforward inclusion of covariates when they are available. If the kernel function is log-linear or a step function, the model reduces to logistic regression (Diggle and Rowlingson 1994).

B.4.4 Concluding remarks

The choice of clustering method depends on several factors. The first consideration is whether the method is appropriate for the available data type. Beyond this practical consideration it is of primary importance that the method evaluates an appropriate null and alternative hypotheses (Waller and Gotway 2004). Some null hypotheses that have been mentioned are spatial randomization, constant risk, and random labeling. Possible alternative hypotheses include variations of regional, local, or focused clustering. Beyond these criteria, an analyst might consider the power of the test in choosing between appropriate methods. In the case of spatial clustering, power refers to the probability of rejecting the null hypothesis given that the data have been generated under the alternative hypothesis. Monte Carlo methods are useful in this regard, and can be used to generate data under a variety of hypotheses. Kulldorff et al. (2003) developed a set of benchmark data, generated under a variety of alternative hypotheses, that can be used to evaluate and compare methods. A later paper compares a large set of methods using the benchmark data (Song and Kulldorff 2003). The power of a test can also be affected by the properties of the data and choice of parameters for clustering methods (Waller et al. 2006). For example, the power can vary widely based on the choice of spatial weights (Song and Kulldorff 2005). Takahashi and Tango (2006) provide a modified test for power that takes into account not only the ability to reject the null hypothesis but also whether the detected clusters are of the correct size and in the proper location. A discussion on method choice and statistical power can be found in Waller and Gotway (2004).

There was a time when, due to a lack of clustering methodologies, researchers could be excused for applying techniques without strict adherence to assumptions. For the most part, this is no longer the case. There are now tools available to handle most data types and a variety of hypotheses. The research in this field will progress by improving existing methods and developing new ones. These developments combined with the rapid innovation in software for spatial data analysis, as covered in Part A of this handbook, will increase the utility of spatial clustering analysis as a research tool.

References

Aldstadt J, Getis A (2006) Using AMOEBA to create a spatial weights matrix and identify spatial clusters. Geogr Anal 38 (4):327-343
Anselin L (1988) Spatial econometrics: methods and models. Kluwer, Dordrecht
Anselin L (1995) Local indicators of spatial association – LISA. Geogr Anal 27(2):93-115
Anselin L (1996) The Moran scatterplot as an ESDA tool to assess local instability in spatial association. In Fischer MM, Scholten HJ, Unwin D (eds) Spatial analytical perspectives on GIS. Taylor and Francis, London, pp.111-125
Assunção RM, Reis EA (1999) A new proposal to adjust Moran's I for population density. Stat Med 18(16):2147-2162
Bailey TC, Gatrell AC (1995) Interactive spatial data analysis. Longman, Harlow
Baker RD (1996) Testing for space-time clusters of unknown size. J Appl Stat 23(5):543-554
Benjamini Y, Hochberg Y (1995) Controlling the false discovery rate: a practical and powerful approach to multiple testing. J Roy Stat Soc B 57(1):289-289
Besag J, Diggle PJ (1977) Simple Monte Carlo tests for spatial pattern. J Appl Stat 26(3):327-333
Besag J, Newell J (1991) The detection of clusters in rare diseases. J Roy Stat Soc A 154(1):143-155
Bithell JF (1995) The choice of test for detecting raised disease risk near a point source. Stat Med 14:2309-2322
Boots BN (2003) Developing local measures of spatial association for categorical data. J Geogr Syst 5(2):139-160
Boots BN (2006) Local configuration measures for categorical spatial data: binary regular lattices. J Geogr Syst 8(1):1-24
Boots BN, Getis A (1988) Point pattern analysis. Sage, London
Caldas de Castro M, Singer BH (2006) Controlling the false discovery rate: a new application to account for multiple and dependent tests in local statistics of spatial association. Geogr Anal 38(2):180-208
Charlton ME (2006) A mark 1 geographical analysis machine for the automated analysis of point data sets: twenty years on. In Fisher PF (ed) Classics from IJGIS: twenty years of the International Journal of Geographical Information Science and Systems. CRC Press (Taylor and Francis Group), Boca Raton [FL], London and New York, pp.35-40
Clark PJ, Evans FC (1954) Distance to nearest neighbor as a measure of spatial relationships in populations. Ecology 35(4):445-453
Cliff AD, Ord JK (1973) Spatial autocorrelation. Pion, London
Conley J, Gahegan M, Macgill J (2005) A genetic approach to detecting clusters in point data sets. Geogr Anal 37(3):286-314
Cuzick J, Edwards R (1990) Spatial clustering for inhomogeneous populations. J Roy Stat Soc B 52(1):73-104
Diggle PJ (1990) A point process modelling approach to raised incidence of a rare phenomenon in the vicinity of a prespecified point. J Roy Stat Soc A 153(3):349-362

Diggle PJ (2003) Statistical analysis of spatial point patterns. Edward Arnold, New York
Diggle PJ, Chetwynd AG (1991) Second-order analysis of spatial clustering for inhomogeneous populations. Biometrics 47(3):1155-1163
Diggle PJ, Rowlingson BS (1994) A conditional approach to point process modelling of elevated risk. J Roy Stat Soc A157(3):433-440
Duczmal L, Assunção R (2004) A simulated annealing strategy for detection of arbitrarily shaped spatial clusters. Compu Stat Data Anal 45(2):269-286
Duczmal L, Cançado ALF, Takahashi RHC, Bessegato LF (2007) A genetic algorithm for irregularly shaped spatial scan statistics. Comp Stat Data Anal 52(1):43-52
Fotheringham AS, Zhan FB (1996) A comparison of three exploratory methods for cluster detection in spatial point patterns. Geogr Anal 28(3):200-218
Geary RC (1954) The contiguity ratio and statistical mapping. The Incorp Stat 5(3):115-145
Getis A (2009) Spatial autocorrelation. In Fischer MM, Getis A (eds) Handbook of applied spatial analysis. Springer, Berlin, Heidelberg and New York, pp.255-278
Getis A, Aldstadt J (2004) Constructing the spatial weights matrix using a local statistic. Geogr Anal 36(2):90-105
Getis A, Ord JK (1992) The analysis of spatial association by distance statistics. Geogr Anal 24(3):189-206
Goovaerts P, Jacquez GM (2004) Accounting for regional background and population size in the detection of spatial clusters and outliers using geostatistical filtering and spatial neutral models: the case of lung cancer in Long Island, New York. Int J Health Geogr 3(14), http://www.ij-healthgeographics.com/content/3/1/14
Greig-Smith P (1952) The use of random and contiguous quadrats in the study of the structure of plant communities. Ann Bot 16(2):293-316
Haining RP (2003) Spatial data analysis: theory and practice. Cambridge University Press, Cambridge
Haining RP (2009) The nature of geoferenced data. In Fischer MM, Getis A (eds) Handbook of applied spatial analysis. Springer, Berlin, Heidelberg and New York, pp.197-217
Huang L, Kulldorff M, Gregorio D (2007) A spatial scan statistic for survival data. Biometrics 63(1):109-118
Jacquez GM (1994) Cuzick and Edwards' test when exact locations are unknown. Am J Epidemiol 140(1):58-64
Jung I, Kulldorff M, Klassen AC (2007) A spatial scan statistic for ordinal data. Stat Med 26(7):1594
Knox EG (1989) Detection of clusters. In Elliott P (ed) Methodology of enquiries into disease clustering. Small Area Health Statistics Unit, London, pp.17-20
Kulldorff M (1997) A spatial scan statistic. Comm Stat Theor Meth 26(6):1481-1496
Kulldorff M (2004) SaTScan v4.0: Software for the spatial and space-time scan statistics. Information Management Services Inc.
Kulldorff M, Tango T, Park P (2003) Power comparisons for disease clustering tests. Comput Stat Data Anal 42(4):665-684
Kulldorff M, Huang L, Pickle L, Duczmal L (2006) An elliptic spatial scan statistic. Stat Med 25(22):3929-3943
Lawson AB (1993) On the analysis of mortality events associated with a prespecified fixed point. J Roy Stat Soc A156(3):363-377
Mantel N (1967) The detection of disease clustering and a generalized regression approach. Cancer Res 27:209-220
Moran PAP (1950) Notes on continuous stochastic phenomena. Biometrika 37(12):17-23
Oden N (1995) Adjusting Moran's I for population density. Stat Med 14(1):17-26

Oden N, Jacquez GM, Crimson R (1998) Authors reply. Stat Med 17:1058-1062

Openshaw S, Charlton ME, Wymer C, Craft A (1987) A mark 1 geographical analysis machine for the automated analysis of point data sets. Int J Geogr Inform Sci 1(4):335-358

Ord JK, Getis A (1995) Local spatial autocorrelation statistics: distributional issues and an application. Geogr Anal 27(4):286-306

Ord JK, Getis A (2001) Testing for local spatial autocorrelation in the presence of global autocorrelation. J Reg Sci 41(3):411-432

Ripley BD (1976) The second-order analysis of stationary point processes. J Appl Prob 13(2):255-266

Rogerson PA (2005) A set of associated statistical tests for spatial clustering. Environ Ecol Stat 12(3):275-288

Song C, Kulldorff M (2003) Power evaluation of disease clustering tests. Int J Health Geographics 2(1):9

Song C, Kulldorff M (2005) Tango's maximized excess events test with different weights. Int J Health Geographics 4(1):32

Stone R (1988) Investigations of excess environmental risks around putative sources: statistical problems and a proposed test. Stat Med 7(6):649-660

Takahashi K, Tango T (2006) An extended power of cluster detection tests. Stat Med 25(5):841

Tango T (1995) A class of tests for detecting 'general' and 'focused' clustering of rare diseases. Stat Med 14:2323-2334

Tango T (1998) Adjusting Moran's I for population density by N. Oden. Stat Med 17(9):1055-1058

Tango T (2000) A test for spatial disease clustering adjusted for multiple testing. Stat Med 19(2):191-204

Tango T (2002) Score tests for detecting excess risks around putative sources. Stat Med 21 (4):497-514

Tango T, Takahashi K (2005) A flexibly shaped spatial scan statistic for detecting clusters. Int J Health Geographics 4(11), http://www.ij-healthgeographics.com/content/4/1/11

Tiefelsdorf M, Boots B (1995) The exact distribution of Moran's I. Environ Plann A 27(6): 985-999

Tiefelsdorf M, Griffith DA, Boots B (1999) A variance-stabilizing coding scheme for spatial link matrices. Environ Plann A 31(1):165-180

Waller LA, Gotway CA (2004) Applied spatial statistics for public health data. Wiley-Interscience, Hoboken [NJ]

Waller LA, Lawson AB (1995) The power of focused tests to detect disease clustering. Stat Med14:2291-2308

Waller LA, Hill EG, Rudd RA (2006) The geography of power: statistical performance of tests of clusters and clustering in heterogeneous populations. Stat Medicine 25(5):853

Waller LA, Turnbull BW, Clark LC, Nasca P (1992) Chronic disease surveillance and testing of clustering of disease and exposure: application to leukemia incidence and TCE-contaminated dumpsites in upstate New York. Environmetrics 3(3):281-300

Walter SD (1992) The analysis of regional patterns in health data – I. Distributional considerations. Am J Epidemiol 136(6):730-741

Warner RM (2007) Applied statistics: from bivariate through multivariate techniques. Sage, Thousand Oaks [CA]

Yamada I, Rogerson PA (2003) An empirical comparison of edge effect correction methods applied to K function analysis. Geogr Anal 35(2):97-110

B.5 Spatial Filtering

Daniel A. Griffith

B.5.1 Introduction

In spatial statistics and spatial econometrics, spatial filtering is a general methodology supporting more robust findings in data analytic work, and is based upon a posited linkage structure that ties together georeferenced data observations. Constructed mathematical operators are applied to decompose geographically structured noise from both trend and random noise in georeferenced data, enhancing analysis results with clearer visualization possibilities and sounder statistical inference. In doing so, nearby/adjacent values are manipulated to help analyze attribute values at a given location. Spatial filtering mathematically manipulates data in order to correct for potential distortions introduced by such factors as arbitrary scale, resolution and/or zonation (i.e., surface partitioning).

The primary idea is that some spatial proxy variables extracted from a spatial relationship matrix are added as control variables to a model specification. The principal advantage of this methodology is that these control variables, which identify and isolate the stochastic spatial dependencies among georeferenced observations, allow model building to proceed as if these observations were independent.

Population counts data from the 2005 census of Peru, by district, for the 108 districts forming the Cusco Department are presented here to empirically illustrate the various spatial filtering approaches; an ArcGIS shapefile furnishes area measures for these districts. Population density, which ranges from 0.8 to 11,512.8 per unit area here, tends to be skewed, with a natural lower bound of zero, and few areal units with relatively sizeable concentrations. Accordingly, analyses based upon the normal probability model require application of a Box-Cox power transformation to better align the empirical population density frequency distribution with a bell-shaped curve; here the transformation is

$$\frac{10}{\left(\frac{\text{population}}{\text{area}}+13.7\right)^{0.56}}. \qquad (B.5.1)$$

This population density forms an elongated mound map pattern with a single peak. The highest density is in the city of Cusco, which has existed for more than 500 years, with the next-highest densities stretching along an economic corridor formed by the Vilcanota River valley; the lowest densities are in the most rural areas of this Department. This population density tends to covary specifically with elevation variability, $s_{\text{elevation}}$. Here the Box-Cox transformation is

$$\frac{1000}{s_{\text{elevation}}+407.6}. \qquad (B.5.2)$$

The bivariate correlation for these two transformed variables is –0.48345, which is statistically significant.

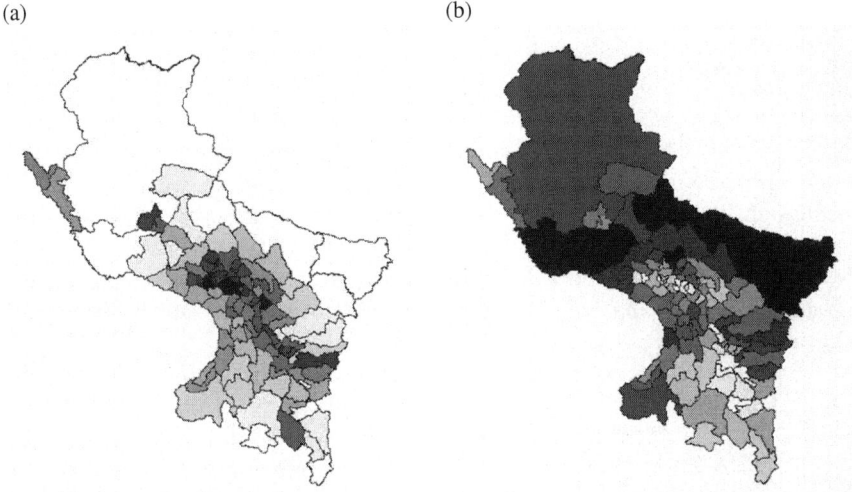

Fig. B.5.1. Geographic distributions across the Cusco Department of Peru; magnitude is directly related to gray tone darkness. (a): transformed population density. (b): transformed elevation standard deviation

The geographic distributions (see Table B.5.1 and Fig. B.5.1) of both transformed population density and elevation variability display moderate, positive, and statistically significant spatial autocorrelation.

Table B.5.1. Transformed population density and elevation variability: spatial autocorrelation in terms of *MC* and *GR*

Attribute	MC	z_{MC}	GR
Y: population density	0.51461	8.85	0.41358
X: elevation standard deviation	0.45545	7.98	0.46650

Notes: *MC* denotes the Moran Coefficient, and *GR* denotes the Geary Ratio

B.5.2 Types of spatial filtering

A limited number of implementations of this methodology currently exist for georeferenced data analysis purposes, and include autoregressive linear operators (*à la* Cochrane-Orcutt type of prewhitening), Getis's G_i-based specification (Getis 1990, 1995), linear combinations of eigenvectors extracted from distance-based principal coordinates of neighboring matrices (PCNM; Borcard et al. 2002, 2004; Dray et al. 2006), and topology-based spatial weights matrix eigenfunctions (Griffith 2000, 2002, 2003, 2004). The first of these is written in terms of a variance component, whereas the other three are written in terms of a mean response component, allowing especially the last two to be incorporated into generalized linear model (GLM) specifications.

One technical advantage of the latter three types of spatial filter is that probability density/mass function normalizing factors no longer are problematic. These constants ensure that the probability density/mass function integrates/sums to one. They are a function of the eigenvalues of matrix **C** for the normal probability model. They are intractable for the binomial and Poisson probability models, requiring Markov Chain Monte Carlo (MCMC) techniques to calculate parameter estimates for these models. Another advantage is that the basis for the control variables does not change unless the spatial relationship matrix is changed. In other words, any attribute variables geographically distributed across a landscape tagged to the same geocoding scheme can be treated with the same spatial filtering. One disadvantage is that, for example, eigenfunctions may need to be extracted numerically from perhaps very large *n*-by-*n* matrices. Fortunately, the asymptotic analytically eigenfunctions for a regular square tessellation forming a rectangular region (for example, a remotely sensed image) are known.

Various studies (for example, Getis and Griffith 2002; Griffith and Peres-Neto 2006) report that results obtained with these different spatial filter approaches essentially are equivalent.

Autoregressive linear operators

Impulse-response function filtering of time series data predates a parallel approach for spatial filtering, and motivated the development of spatial autoregressive linear operators (Tobler 1975), whose error term is correlated with some response variable, Y. Consider the simultaneous spatial autoregressive (SAR) model specification

$$Y = X\beta + (I - \rho C)^{-1} \varepsilon \qquad (B.5.3)$$

where X is a n-by-$(P+1)$ matrix of covariates, β is a $(P+1)$-by-1 vector of regression coefficients, ρ is a spatial autocorrelation parameter, n is the number of areal units, I is an n-by-n identity matrix, and C is a topology-based n-by-n geographic connectivity/weights matrix (for example, $c_{ij} = 1$ if areal units i and j are nearby/adjacent, and $c_{ij} = 0$ otherwise; $c_{ii} = 0$). Here these spatial filters take the matrix form $(I - \rho C)$. The parameter ρ is estimated for Y (denoted $\hat{\rho}$, and then used in the two multiplications $(I - \hat{\rho} C)Y$, for the n-by-1 vector of response values, and $(I - \hat{\rho} C)X$, for the n-by-$(p+1)$ vector of p covariates and intercept term.

This spatial filter is almost always coupled with the normal probability model, and if properly specified, renders independent and identically distributed random error terms. Smoothing occurs in that each dataset value is rewritten as the difference between the observed value and a linear combination of neighboring values.

The pure spatial autoregressive (SAR) maximum likelihood parameter estimates for the transformed population density (pd) and elevation standard deviation ($s_{elevation}$) attribute variables are, respectively, 0.79164 and 0.77455. According to their corresponding pseudo-R^2 calculations, positive spatial autocorrelation latent in the transformed population density variable accounts for roughly 60 percent, whereas that in the transformed $s_{elevation}$ accounts for roughly 55 percent, of its geographic variability. The bivariate correlation coefficient calculated for the spatially filtered variate pair, $(I - 0.79164\ W)Y$ and $(I - 0.77455\ W)X$, where matrix W is the row-standardized version of matrix C, and both of which continue to conform closely to a normal distribution, decreases in absolute value to -0.42070. Although both variables have roughly the same level of positive spatial autocorrelation, this decrease is rather modest because their map patterns are noticeably different (see Fig. B.5.1).

Getis's G_i specification

This specification involves a multistep procedure exploiting Ripley's second-order statistic or the range of a geostatistical semivariogram model coupled with the Getis-Ord (1992) G_i statistic, and converts each spatially autocorrelated variable into a pair of synthetic variates, one capturing spatial dependencies and one capturing non-spatial systematic and random effects. Regressing a response variable on the set of constructed spatial and a-spatial variates allows geographically structured noise to be separated from trend and random noise in georeferenced data. But it is restricted to non-negative random variables having a natural origin.

The primary pair of equations is given by

$$y_i^* = y_i \frac{\frac{\sum_{j=1}^{n} c_{ij}(d)}{n-1}}{\frac{\sum_{j=1}^{n} c_{ij}(d) y_i}{\sum_{j=1}^{n} y_j - y_i}} \qquad (B.5.4)$$

and

$$L_y = Y - Y^* \qquad (B.5.5)$$

where d denotes distance separating location j from location i, the denominator is $G_i(d)$, the numerator is $E[G_i(d)]$, Y^* is the a-spatial variable realization, and L_y is the spatial variable. Distance d is selected such that $G_i(d)$, which initially tends to increase with increasing distance, begins to decrease.

Figure B.5.2(a) displays the areal unit centroids for the Cusco region. Figure B.5.2(b) indicates that a 3-parameter gamma distribution (parameter estimates: shape = 0.7533, scale = 0.6984, and threshold = 0.0297) furnishes a good description of the set of distances. Figure B.5.2(c) illustrates the concavity of the Eq. (B.5.4) trajectories across the distance range of [0, 3.996]. Of note is that some trajectories encounter local peaks that are not global peaks. The number of geographic connections used for the transformed population density $G_i(d)$ is 3,262, whereas that for the transformed elevation standard deviation is 4,787; in contrast, the number of connections in matrix C is 570.

Figure B.5.3 portrays the maps of the synthetic spatial variates given by Eq. (B.5.5). The correlation between the two a-spatial synthetic variates is –0.20744, indicating that spatial autocorrelation dramatically inflates the observed coefficient. The regression equations may be written as follows:

$$Y = a + b_1 Ly + b_2 X^* + b_3 L_x + e. \qquad (B.5.6)$$

The variance in Y, the transformed population density, is accounted for as follows: 11.41 percent by L_y, the synthetic spatial variate; 15.39 percent by X^*, the synthetic a-spatial covariate; and, 6.46 percent by L_x, the synthetic spatial covariate. Moderate multicollinearity is present in this model specification, but with virtually no impact of the regression coefficient variance inflation factors (VIFs).

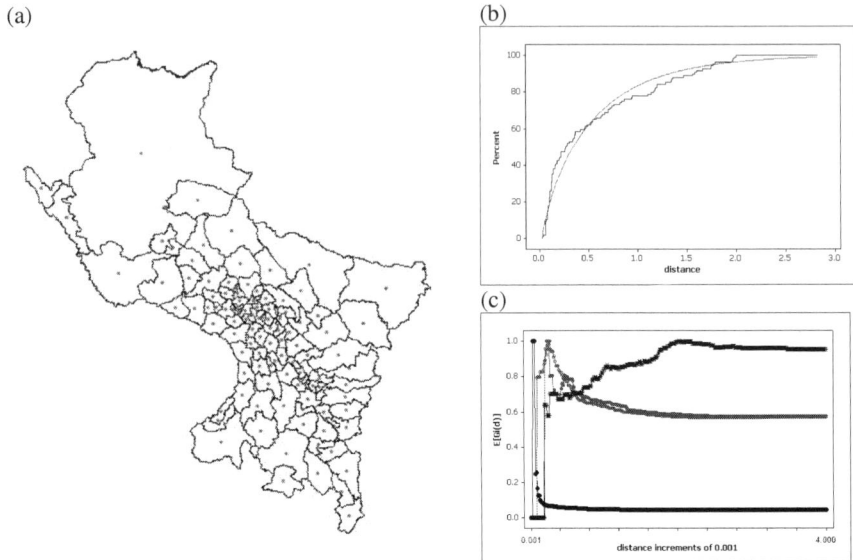

Fig. B.5.2. The Cusco Department of Peru:
(a) geographic distribution of areal unit centroids; (b) a three-parameter gamma distribution description of the d_i values for $G_i(d)$ – the black line denotes the empirical, and the gray line denotes the theoretical, cumulative distribution function (CDF); (c) four selected areal unit trajectories for identifying the d_i values for transformed population density – solid black circle denotes the smallest d_i, black asterisk denotes the largest d_i, and gray circles denote median d_is

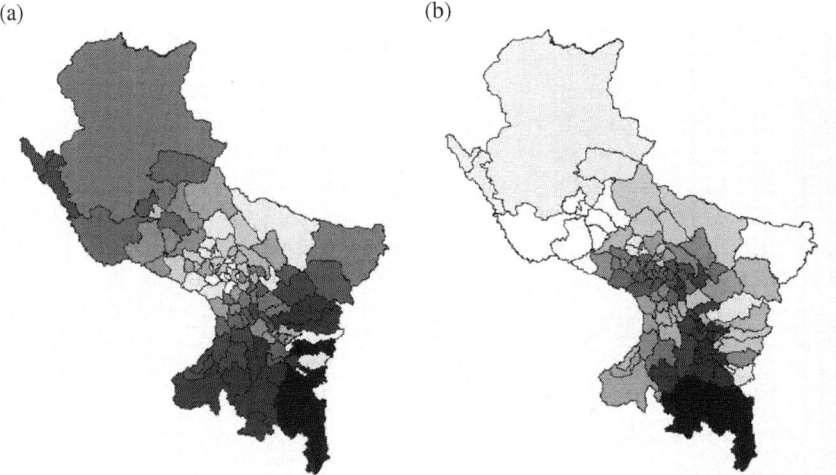

Fig. B.5.3. Geographic distributions across the Cusco Department of Peru of $G_i(d)$-based spatial variates; magnitude is directly related to gray tone darkness: (a) extracted from the transformed population density; (b) extracted from the transformed elevation standard deviation

Linear combinations of distance matrix-based eigenvectors

Dray et al. (2006) specify the PCNM transformation procedure that depends on mathematical expressions, known as eigenfunctions, of a truncated inter-location distance matrix, where the truncation value is the maximum distance that maintains all sampling units being connected using a minimum spanning tree. The PCNM specification relates to semivariogram modeling. Distance-based eigenvector maps with large eigenvalues (that is, strong positive spatial autocorrelation) tend to have only a few large clusters of values on a map and represent global trends [for example, Fig. B.5.4(b)]. Eigenvectors with intermediate size eigenvalues tend to have a number of moderate-sized clusters of values on a map and represent regional trends [for example, Fig. B.5.4(c) and Fig. B.5.4(d)]. And, eigenvectors with small eigenvalues tend to have numerous small clusters of values on a map and represent patchiness and hence more local trends across a landscape [for example, Fig. B.5.4(e)]. Moreover, distance-based eigenvector maps capture a range of geographic scales encapsulated in a given georeferenced dataset, portraying increasing fragmentation as the corresponding eigenvalues decrease in magnitude.

This specification utilizes eigenvectors extracted from the modified geographic weights matrix $(I - 11^T / n) W (I - 11^T / n)$ where 1 is an n-by-1 vector of ones, and ^{T} denotes the matrix transpose operation. The elements of the n-by-n geographic weights matrix W are defined as follows:

$$W_{ij} = \begin{cases} 0 & \text{if } i=j \\ 0 & \text{if } d_{ij} > t \\ 1-[d_{ij}/(4t)]^2 & \text{if } 0 < d_{ij} \leq t \end{cases} \quad (B.5.7)$$

where t is the maximum distance for a minimum spanning tree connecting all n locations (for example, Fig. B.5.4(a)). Here the great circle distance value for t is 16.022 km.

The eigenvalues associated with the PCNM eigenvectors do not have a simple relationship with their affiliated MCs (see Table B.5.1); some non-zero eigenvalues even represent weak negative spatial autocorrelation. Employing an adjusted value of $MC/MC_{max} > 0.25$, where MC_{max} denotes the maximum MC value, reduces the candidate set of eigenvectors for constructing PCNM spatial filters to 15 (that is, eigenvectors E_1 to E_{12}, E_{14}, E_{16} and E_{17}). The spatial autocorrelation contained in a response variable Y may be described with these eigenvectors as follows

$$Y = \mu_Y 1 + E_k \beta_k + \varepsilon_Y \quad (B.5.8)$$

where E_k is an n-by-K matrix of selected eigenvectors (using stepwise regression techniques), μ_Y is the mean of variable Y (because all of the eigenvectors have a mean of zero), β_k is a K-by-1 vector of regression coefficients, and ε_Y is a random error term that is iid $\mathcal{N}(0, \sigma_\varepsilon^2)$. For transformed population density in the Cusco Department, Eq. (B.5.8) contains seven eigenvectors that account for 52.42 percent of its geographic variation. The z_{MC} (z-score for the MC under a null hypothesis of zero spatial autocorrelation) value decreases from 8.79 to 2.83, and residuals continue to mimic a normal distribution, with $MC = 0.76944$ ($GR = 0.25051$) for the spatial filter.

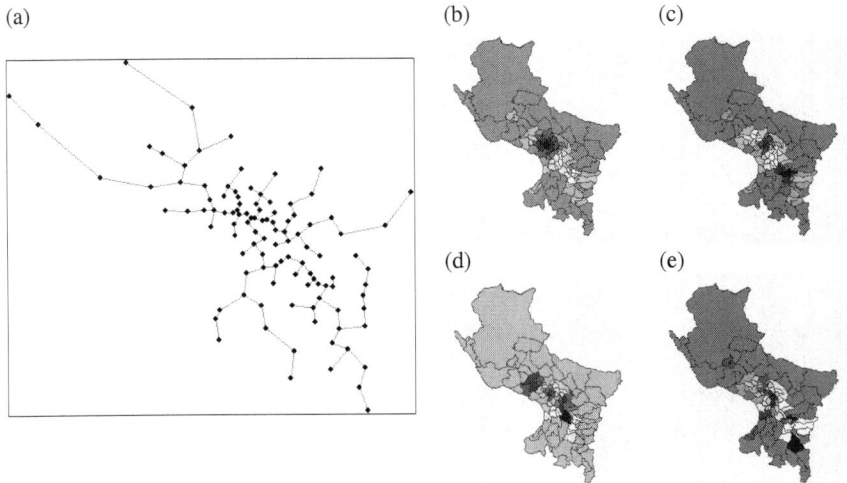

Fig. B.5.4. The Cusco Department of Peru; magnitude in the choropleth maps is directly related to gray tone darkness: (a) the minimum spanning tree connecting the areal unit centroids; (b) E_1, $MC/MC_{max} = 1$; (c) E_3, $MC/MC_{max} = 0.78$; (d) E_9, $MC/MC_{max} = 0.52$; (e): E_{14}, $MC/MC_{max} = 0.25$

The correlation between the two sets of residuals for Eq. (B.5.8), after the respective spatial filters have been subtracted from transformed population density and transformed $s_{\text{elevation}}$, is -0.39203, indicating that spatial autocorrelation dramatically inflates the observed bivariate correlation coefficient. This inflation primarily is attributable to the three common eigenvectors, whose correlation is -0.91928; but it is suppressed by the presence of two sets of unique eigenvectors, whose correlations are exactly zero.

Table B.5.2. Spatial autocorrelation contained in the 30 PCNM eigenvectors with non-zero eigenvalues

Eigenvalue	MC	GR	Eigenvalue	MC	GR
6.757698	0.879567	0.199937	1.115993	0.255954	0.644689
5.387761	0.830660	0.257939	0.986066	0.246825	0.870757
4.428140	0.683353	0.326650	0.968562	0.192463	0.462175
3.891251	0.687340	0.395729	0.890959	0.168454	0.909110
3.390504	0.586984	0.444859	0.779474	0.098354	1.126354
2.960842	0.611523	0.439971	0.714815	0.151115	0.818756
2.796129	0.523140	0.534097	0.664565	−0.034844	1.188295
2.389961	0.350517	0.691055	0.578630	−0.109281	1.295845
2.285282	0.458341	0.611829	0.540622	0.045193	1.032415
2.176693	0.495837	0.470299	0.386237	0.003223	0.917556
1.932853	0.407144	0.637691	0.291445	−0.127027	1.275769
1.467388	0.355678	0.725449	0.228748	−0.025737	1.200260
1.359360	0.196455	0.720122	0.213037	−0.036185	1.254641
1.345400	0.220722	0.719404	0.158005	0.043036	1.033882
1.164052	0.209691	0.768738	0.083016	−0.067713	1.261137

Linear combinations of topological matrix-based eigenvectors

This specification (see Tiefelsdorf and Griffith 2007) is a transformation procedure that also depends on eigenvectors extracted from the adjusted geographic weights matrix $(I - 11^T / n) \, C \, (I - 11^T / n)$, a term appearing in the numerator of the *MC* spatial autocorrelation index. This decomposition also could be based upon the *GR* index, and rests on the following property: the first eigenvector, say E_1, is the set of real values that has the largest *MC* achievable by any set for the spatial arrangement defined by the geographic connectivity matrix *C*; the second eigenvector is the set of real values that has the largest achievable *MC* by any set that is uncorrelated with E_1; the third eigenvector is the third such set of real values; and so on through E_n, the set of real values that has the largest negative *MC* achievable by any set that is uncorrelated with the preceding (*n*–1) eigenvectors. As such, these eigenvectors furnish distinct map pattern descriptions of latent spatial autocorrelation in georeferenced variables, because they are both orthogonal and uncorrelated. Their corresponding eigenvalues, which can be easily converted to *MC* values, index the nature and degree of spatial autocorrelation portrayed by each eigenvector.

As with PCNM, the resulting spatial filter is constructed from some linear combination of a subset of these eigenvectors. The candidate set can begin with all eigenvectors portraying the same nature (that is, positive or negative) of spatial autocorrelation as is measured in a response variable. Next, those eigenvectors representing inconsequential levels of spatial autocorrelation (that is, with very small eigenvalues) should be removed from this candidate set. Finally, a stepwise regression procedure can be used to select those eigenvectors that account for the spatial autocorrelation in the response variable. This stepwise selection can be based upon, say, the conventional R^2-maximiation criterion, or a residual MC minimization criterion.

In practice, this spatial filter specification replaces the autoregressive spatial filter with its eigenfunction counterpart, and its single autoregressive parameter with a set of parameter estimates, one for each eigenvector, removing those from the model whose estimates essentially are zero.

Table B.5.3. Eigenvector spatial filter regression results using a 10 percent level of significance selection criterion

Component	Population density (Y) and elevation standard deviation (X), for the Cusco Department, Peru ($n = 108$)	
	Transformed Y	Transformed X
Common eigenvectors	$R^2 = 0.4645$	$R^2 = 0.5189$
Unique eigenvectors	$R^2 = 0.1543$	$R^2 = 0.0565$
All selected eigenvectors	$R^2 = 0.6188$	$R^2 = 0.5753$
Residual MC	$z_{MC} \approx -0.23$	$z_{MC} \approx -0.19$
Shapiro-Wilk (S-W) statistic	0.987 ($prob = 0.393$)	0.986 ($prob = 0.313$)
MC for spatial filters	0.4719	0.4019

Spatial filters were constructed for the two Cusco transformed attribute variables, where the candidate eigenvector set was restricted to those 24 vectors portraying positive spatial autocorrelation and having a $MC/MC_{max} > 0.25$; the maximum possible MC value for Cusco's topological surface partitioning, MC_{max}, is 1.09315, the MC value for the principal eigenvector. The resulting spatial filters appear in Fig. B.5.5, each portraying strong positive spatial autocorrelation, and each closely reflecting its parent map (see Fig. B.5.1). Summary measures for them are reported in Table B.5.2. The bivariate correlation coefficient between $(X - F_X)$ and $(Y - F_Y)$, where F_j denotes the spatial filter for variable j, and both of which continue to conform closely to a normal distribution, decreases in absolute value to –0.42688. Here spatial autocorrelation roughly accounts for, respectively, 62 percent and 58 percent of the geographic variability in these transformed attribute variables. The filtered residuals contain negligible spatial autocorrelation. Although both variables have roughly the same level of positive spatial autocorrelation, the correlation coefficient decrease is rather modest because their map patterns are noticeably different: their spatial filters have nine eigenvectors in common, and seven that are specific to one or the other of them. The decompositions highlighted here may be written as

$$Y = \mu_Y \mathbf{1} + \mathbf{E}_c \boldsymbol{\beta}_{c_Y} + \mathbf{E}_{u_Y} \boldsymbol{\beta}_{u_Y} + \boldsymbol{\varepsilon}_Y \tag{B.5.9}$$

$$X = \mu_X \mathbf{1} + \mathbf{E}_c \boldsymbol{\beta}_{c_X} + \mathbf{E}_{u_X} \boldsymbol{\beta}_{u_X} + \boldsymbol{\varepsilon}_X \tag{B.5.10}$$

where E is an n-by-H matrix for X and an n-by-K matrix for Y (with H and K not necessarily equal) of selected eigenvectors, subscripts c and u respectively denote common and unique sets of eigenvectors, β is a vector of regression coefficients, and ε_Y and ε_X respectively are the iid $\mathcal{N}(0, \sigma_{\varepsilon_j}^2)$, $j = X$ or Y, a-spatial variates for variables X and Y. As with PCNM, the linear combinations of eigenvectors are the spatial filters.

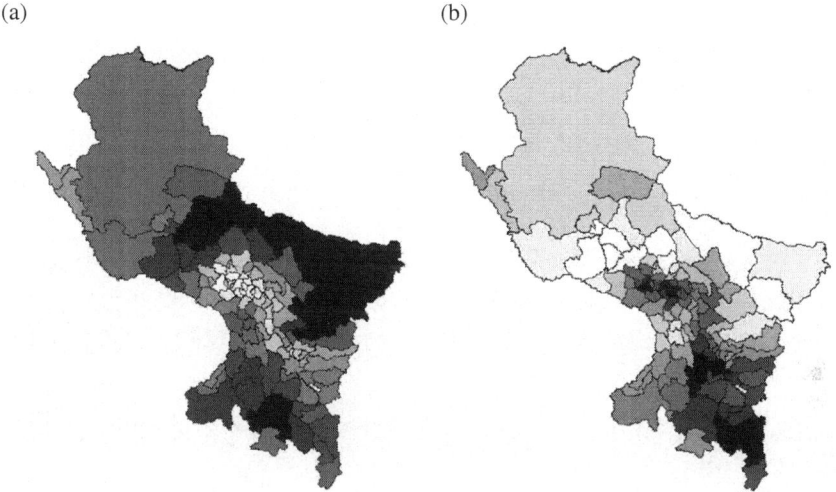

Fig. B.5.5. Typology-based spatial filters for the Cusco Department of Peru; eigenvector values are directly related to gray tone darkness: (a) for transformed population density; (b) for transformed elevation standard deviation

Now the bivariate correlation coefficient can be rewritten as the following weighted combination of different correlation coefficients, where the weights are the square roots of relative variance term products (see Table B.5.2)

$$r_{X,Y} = r_{resid_X, resid_Y} \sqrt{(1-R_X^2)(1-R_Y^2)} + r_{E_{c_X}, E_{c_Y}} \sqrt{R_{E_{c_X}}^2 R_{E_{c_Y}}^2} +$$

$$+ r_{E_{u_X}, resid_Y} \sqrt{R_{E_{u_X}}^2 (1-R_Y^2)} + r_{resid_X, E_{u_Y}} \sqrt{(1-R_X^2) R_{E_{u_Y}}^2} + 0 \sqrt{R_{E_{u_X}}^2 R_{E_{u_Y}}^2} \quad \text{(B.5.11)}$$

where *resid* denotes the residuals, R^2 is a linear regression multiple correlation coefficient, and the subscripts X and Y denote with which variable a term is associated. The zero correlation arises because the unique sets of eigenvectors are orthogonal and uncorrelated. Substituting the corresponding Cusco case study values into this equation (see Table B.5.1; some rounding error is present) yields

$$-0.48345 = -0.43904\sqrt{(1-0.6188)(1-0.5753)} - 0.60486\sqrt{(0.4645)(0.5189)}$$
$$-0.10384\sqrt{(0.1543)(1-0.5753)} + 0.11396\sqrt{(1-0.6188)(0.0565)}$$
$$+0\sqrt{(0.1543)(0.0565)}.$$

This decomposition equation like that for PCNM, emphasizes that common eigenvectors tend to increase the magnitude of a correlation coefficient, whereas unique eigenvectors tend to suppress it.

B.5.3 Eigenfunction spatial filtering and generalized linear models

A spatial filter can be constructed for GLM specifications again using a stepwise selection technique. By doing so, MCMC techniques can be avoided when estimating model parameters in the presence of spatial autocorrelation; rather, standard GLM procedures can be used.

Because population is a count variable, it can be treated as a Poisson random variable, and the area variable in the denominator of a population density can be converted to a GLM offset variable (that is, its coefficient is set to one and not estimated) by including its logarithm as a special covariate (that is, an offset) in a model specification. For the Cusco Departmental data, the GLM estimation, including $\log(s_{elevation})$ as a covariate, yields the spatial filter appearing in Fig. B.5.6, whose $MC = 0.86030$ ($z_{MC} = 14.94$) and $GR = 0.31022$. This spatial filter has nine eigenvectors, six of which are contained in the set of eleven for the corresponding normal-approximation spatial filter. Including the previously specified transformed $s_{elevation}$ as a covariate in the normal approximation specification increases its R^2 to 0.6821. Switching to the correct probability function here results in a more parsimonious model whose predicted values better align with actual population density across the entire range of density values [see Fig. B.5.6(b) and Fig. B.5.6(c)].

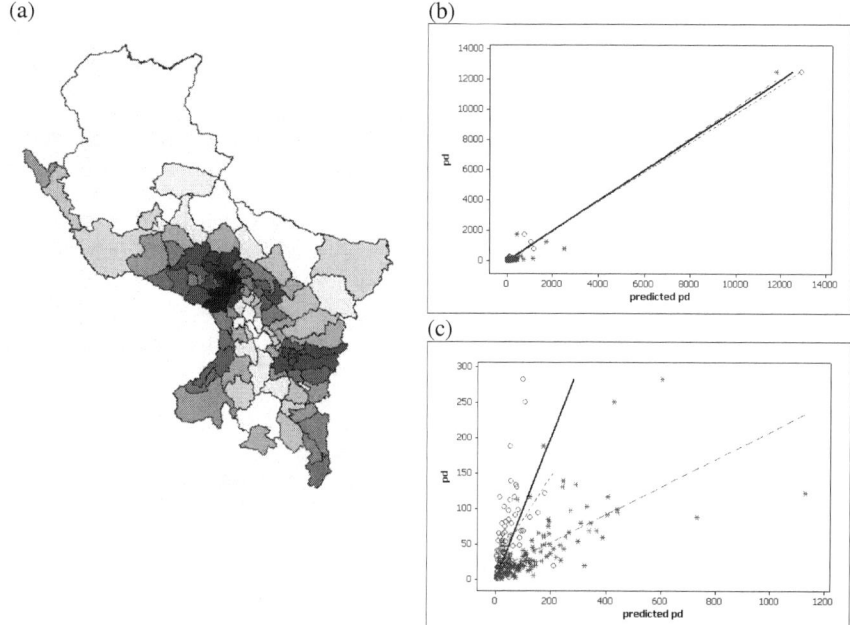

Fig. B.5.6. Generalized linear model (GLM) results: (a) the population density GLM spatial filter; eigenvector values are directly related to gray tone darkness; (b) scatterplot of the predicted versus the observed *pd*; (c) scatterplot of the predicted versus the observed *pd* with the four largest values set aside. The solid black line denotes observed *pd*, open circles denote GLM-predicted *pd*, and asterisks denote back-transformed normal approximation predicted *pd*

B.5.4 Eigenfunction spatial filtering and geographically weighted regression

Eigenfunction spatial filters allow geographically varying coefficient models to be specified, along the lines of geographically weighted regression (GWR). Interaction terms can be created by multiplying each variable in a set of covariates by each eigenvector in a candidate set. In other words, these interaction variates are cross-products of each synthetic spatial variate and each covariate. Again stepwise regression can be used to select the relevant variables. The stepwise procedures can be used to select from the candidate eigenvector set (which relates to the intercept term), the set of covariates, and the set of interaction terms. Once the subset has been identified, it can be grouped into sets having a common covariate so that this covariate can be factored from each set. What remains for each set is a linear combination of the synthetic spatial variates used to construct a cross-product, which when added together constitutes geographically varying coefficients. The affiliated equation may be written as follows:

$$Y = f(\mu_Y \mathbf{1} + \mathbf{E}_1 \boldsymbol{\beta}_1 + \mathbf{X} \odot \mathbf{E}_x \boldsymbol{\beta}_x)) \tag{B.5.12}$$

where f denotes some function (for example, the natural antilogarithm, e, for the Poisson probability model), the subscript 1 denotes the eigenvector and the regression coefficient associated with the intercept term, the subscript X denotes eigenvectors and their regression coefficients associated with the slope coefficient, and $\odot$ denotes the Hammard matrix product (that is, element-by-element matrix multiplication).

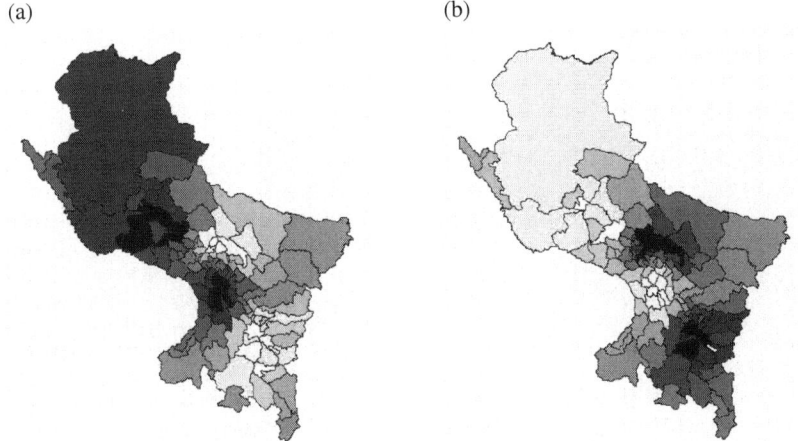

Fig. B.5.7. Geographically varying coefficients for the GLM population density model; coefficient magnitudes are directly related to gray tone darkness: (a) spatially varying intercept term; (b) spatially varying slope coefficient

Consider the preceding GLM model describing population density across the Cusco Department. The geographically varying intercept can be rewritten as

$8.8834 - 4.7838\, E_1 - 4.3226\, E_3 + 48.3641\, E_4 + 1.8258\, E_6 - 2.0448\, E_{12} +$
$2.1773\, E_{13} - 2.7006\, E_{14} - 1.6251\, E_{16} - 1.7334\, E_{19}$.

Meanwhile, the geographically varying slope coefficient can be rewritten as

$-0.9446 - 8.7899\, E_4 - 0.3106\, E_{10} - 0.4664\, E_{11} + 0.7628\, E_{15}$.

This is the term that is factored from the set of cross-product terms (i.e., each eigenvector multiplied by $s_{elevation}$); each element of this term is multiplied by its corresponding log($s_{elevation}$) value. The geographic distributions of the spatially varying coefficients appear in Fig. B.5.7. Because eigenvector E_4 is common to both coefficient expressions, and it dominates the intercept term, the correlation between these two geographically varying coefficients is very high (−0.98036). Because each of the eigenvectors has a mean of zero, these two geographically varying coefficients are centered on their respective global values [that is, the intercept constant, and the slope coefficient for log($s_{elevation}$), itself]. Furthermore, because the coefficient variability is a function of the eigenvectors, these geographically varying coefficients contain (as well as account for) spatial autocorrelation in the response variable Y.

Table B.5.4. Geographically varying coefficients: spatial autocorrelation in terms of MC and GR

Coefficient	MC	z_{MC}	GR
Intercept	0.92345	16.02	0.22664
Log($s_{elevation}$) slope	0.92090	15.98	0.23104

Notes: MC denotes the Moran Coefficient, and GR denotes the Geary Ratio

Each coefficient contains statistically significant, weak positive spatial autocorrelation.

B.5.5 Eigenfunction spatial filtering and geographical interpolation

Spatial interpolation is a problem frequently encountered in spatial analysis. Its solution exploits spatial autocorrelation in order to predict an unknown value at some location from known values at nearby locations. The redundant information interpretation of spatial autocorrelation, which relates to the amount of geographic variance it accounts for within an attribute variable, supports this interpolation.

The best imputation of a missing response value is its expected value given a set of available data. In other words, it equals the prediction equation estimated with a set of observed data. This value can be calculated by inserting a binary indicator variable into a regression equation, where this variable is assigned a value of minus one for the single observation with a missing response value, and a zero for all other observations. The regression coefficient calculated for this indicator variable is an imputation. For a Poisson model specification, this requires the missing response variable value to be replaced with a one

$$\exp(\alpha + \beta_X X_i + \sum_{k=1}^{K} E_{ki} \beta_k - \beta_m 1) = 1 \tag{B.5.13}$$

when

$$\beta_m = \alpha + \beta_X X_i + \sum_{k=1}^{K} E_{ki} \beta_k. \tag{B.5.14}$$

Imputed values for population density across the Cusco Department were calculated and are portrayed in Fig. B.5.8. The expected values were computed with the covariate log($s_{elevation}$) coupled with a spatial filter. Of note is that Fig. B.5.8(a) is very similar to Fig. B.5.6(b); more variability appears here because each density value is not used in the calculation of the GLM, increasing the uncertainty in its prediction. Nevertheless, given their alignment with the ideal line in Fig. B.5.8, the imputed values obtained here appear to be reasonable.

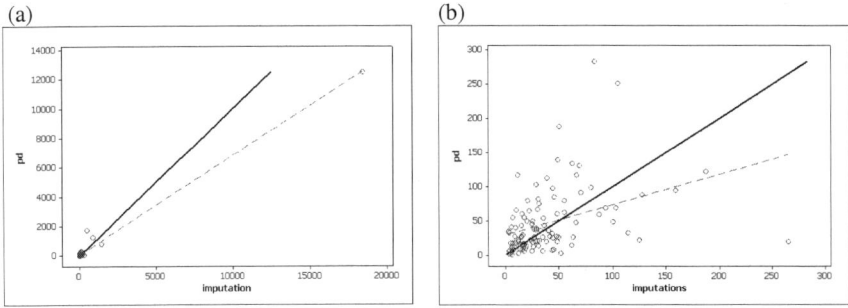

Fig. B.5.8. Generalized linear model (GLM) imputation results: (a) scatterplot of the imputed versus the observed population densities (*pd*); (b) scatterplot of the imputed versus the observed population densities (*pd*) with the four largest values set aside. The solid black line denotes observed *pd*, and the open circle denotes GLM-imputed *pd*

B.5.6 Eigenfunction spatial filtering and spatial interaction data

Recent work has returned attention to the role spatial autocorrelation plays in the estimation of model parameters describing spatial interaction data. LeSage and Pace (2008) propose a formulation that is autoregressive-based, and relates to the autoregressive linear operator spatial filter. Fischer and Griffith (2008) compare this autoregressive linear operator specification with an eigenfunction spatial filter specification. One finding is that the spatial autocorrelation involved transcends that latent in attribute variables representing characteristics of origins/destinations. Rather, the spatial autocorrelation relates to flows leaving nearby origins and arriving in nearby destinations. This conceptualization is reminiscent of the hierarchical component affiliated with geographic diffusion. This topic is at the research frontiers of spatial filtering work.

B.5.7 Concluding remarks

Spatial filtering methodology seeks to account for spatial autocorrelation in georeferenced data in a way that enables conventional statistical estimation techniques to be exploited. It also allows impacts of spatial autocorrelation to be uncovered in a more data analytic manner. Two geographically distributed attribute variables for the Cusco Department of Peru – 2005 population density and elevation variation – are used here to illustrate this contention, with special reference to their bivariate correlation coefficient. The naive correlation coefficient is -0.48345. Adjusting this value for the presence of positive spatial autocorrelation results in a decrease in its absolute value; in other words, positive spatial autocorrelation tends to inflate correlation coefficients. But this reduction is a function of the spatial filter specification employed. The autoregressive linear operator, PCNM, and eigenfunction spatial filtering results are very comparable. They are, respectively, -0.42070, -0.39203, and -0.43904. This finding is not surprising, because all three of these methodologies share a common mathematical foundation. In contrast, the $G_i(d)$-based spatial filtering yields a value of -0.20744. Part of its deviation from the other three results may well be attributable to its more restrictive assumptions.

Spatial filtering can be employed not only with the normal probability model, but also with the entire family of probability models affiliated with generalized linear models. It also supports spatial interpolation, and offers a vehicle for addressing spatial autocorrelation in geographic flows data.

Acknowledgements. We are indebted to Marco Millones, Clark University, for providing us with the Cusco Department GIS files, and its 2005 Peru Census data numbers.

References

Borcard D, Legendre P (2002) All-scale spatial analysis of ecological data by means of principal coordinates of neighbour matrices. Ecol Mod 153(1/2):51-68

Borcard D, Legendre P, Avois-Jacquet C, Tuomisto H (2004) Dissecting the spatial structure of ecological data at multiple scales. Ecology 85(7):1826-1832

Dray S, Legendre P, Peres-Neto P (2006) Spatial modeling: A comprehensive framework for principal coordinate analysis of neighbor matrices (PCNM). Ecol Mod 196(3/4):483-493

Fischer MM, Griffith D (2008) Modeling spatial autocorrelation in spatial interaction data: an application to patent citation data in the European Union. J Reg Sci 48(5):969-989

Getis A (1990) Screening for spatial dependence in regression analysis. Papers in Reg Sci Assoc 69(1):69-81

Getis A (1995) Spatial filtering in a regression framework: experiments on regional inequality, government expenditures, and urban crime. In Anselin A, Florax, R (eds) New directions in spatial econometrics. Springer, Berlin, Heidelberg and New York, pp.172-188

Getis A, Griffith D (2002) Comparative spatial filtering in regression analysis. Geogr Anal 34(2):130-140

Getis A, Ord JK (1992) The analysis of spatial association by use of distance statistics. Geogr Anal 24(3):189-206

Griffith D (2000) A linear regression solution to the spatial autocorrelation problem. J Geogr Syst 2(2):141-156

Griffith D (2002) A spatial filtering specification for the auto-Poisson model. Stat Prob Letters 58(3):245-251

Griffith D (2003) Spatial autocorrelation and spatial filtering: gaining understanding through theory and scientific visualization. Springer, Berlin, Heidelberg and New York

Griffith D (2004) A spatial filtering specification for the autologistic model. Environ Plann A 36(10):1791-1811

Griffith D, Peres-Neto P (2006) Spatial modeling in ecology: the flexibility of eigenfunction spatial analyses. Ecology 87(10):2603-2613

Haining R (1991) Bivariate correlation with spatial data. Geogr Anal 23(3):210-227

LeSage JP, Pace K (2008) Spatial econometric modeling of origin-destination flows. J Reg Sci 48(5):941-968

Tiefelsdorf M, Griffith D (2007) Semi-parametric filtering of spatial autocorrelation: the eigenvector approach. Environ Plann A 39(5):1193-1221

Tobler WR (1975) Linear operators applied to areal data. In Davis J, McCullagh M (eds) Display and analysis of spatial data. Wiley, London, pp.14-37

B.6 The Variogram and Kriging

Margaret A. Oliver

B.6.1 Introduction

Spatial statistics and geostatistics have developed to describe and analyze the variation in both natural and man-made phenomena on, above or below the land surface. Spatial statistics includes any of the formal techniques that study entities that have a spatial index (Cressie 1993). Geostatistics is embraced by this general umbrella term, but originally it was more specifically concerned with processes that vary continuously, i.e. have a continuous spatial index. The term geostatistics applies essentially to a specific set of models and techniques developed largely by Matheron (1963) in the 1960s to evaluate recoverable reserves for the mining industry. These ideas had arisen previously in other fields; they have a long history stretching back to Mercer and Hall (1911), Youden and Mehlich (1937), Kolmogorov (1941), Gandin (1965), Matérn (1960) and Krige (1966). Geostatistics has since been applied in many different fields, such as agriculture, fisheries, hydrology, geology, meteorology, petroleum, remote sensing, soil science and so on. In most of these fields the data are fragmentary and often sparse, therefore there is a need to predict from them as precisely as possible at places where they have not been measured. This chapter covers two of the principle techniques of geostatistics that solve this need for prediction; the variogram and kriging.

B.6.2 The theory of geostatistics

A brief summary only is given here of the theory that underpins geostatistics (for more detail see Journel and Huijbregts, 1978; Goovaerts, 1997; Webster and Oliver 2007). Most spatial properties vary in such a complex way that the variation cannot be defined deterministically. To deal with this spatial uncertainty a different approach from the traditional deterministic methods of spatial analysis was required that relies on a stochastic or probabilistic approach. The basis of modern geostatistics is to treat the variable of interest as a random variable. This implies that at each point x in space there is a series of values for a property, $Z(x)$, and the one observed, $z(x)$, is drawn at random according to some law, from some probability distribution. At x, a property $Z(x)$ is a random variable with a mean, μ and variance, σ^2. The set of random variables, $Z(x_1)$, $Z(x_2)$, ..., is a random process, and the actual value of Z observed is just one of potentially any number of realizations of the random process. In classical statistics this set of observed values, the realization, is the population.

To define the variation of the underlying random process, we can take into account the fact that the values of regionalized variables at places near to one another tend to be related. As well as estimating the mean and variance of the property, we can also estimate the spatial covariance to describe this relation between pairs of points. The covariance for the random variables is given by

$$C(x_1,x_2) = E[\{Z(x_1) - \mu(x_1)\} \{Z(x_2) - \mu(x_2)\}] \tag{B.6.1}$$

where $\mu(x_1)$ and $\mu(x_2)$ are the means of Z at x_1 and x_2, and E denotes the expected value. This solution is unavailable, however, because the means are unknown as there is only ever one realization of Z at each point. To proceed we have to invoke assumptions of stationarity.

Stationarity

Under the assumptions of stationarity certain attributes of the random process are the same everywhere. We assume that the mean, $\mu = E[Z(x)]$, is constant for all x, and so $\mu(x_1)$ and $\mu(x_2)$ can be replaced by μ, which can be estimated by repetitive sampling. When x_1 and x_2 coincide, Eq. (B.6.1) defines the variance (or the *a priori* variance of the process), $\sigma^2 = E[\{Z(x) - \mu\}^2]$, which is assumed to be finite and, as for the mean, the same everywhere. When x_1 and x_2 do not coincide, their covariance depends on their separation and not on their absolute positions, and this applies to any pair of points x_i, x_j separated by the lag $h = x_i - x_j$ (a vector in both distance and direction), so that

$$C(x_i, x_j) = E[\{Z(x_i) - \mu\} \{Z(x_j) - \mu\}] = E[\{Z(x)\} \{Z(x+h)\} - \mu^2] = C(h) \tag{B.6.2}$$

which is also constant for a given h. This constancy of the first and second moments of the process constitutes second-order or weak stationarity. Equation (B.6.2) indicates that the covariance is a function of the lag and it describes quantitatively the dependence between values of Z with changing separation or lag distance. The autocovariance depends on the scale on which Z is measured; therefore, it is often converted to the dimensionless autocorrelation by

$$\rho(h) = C(h)/C(0) \qquad (B.6.3)$$

where $C(0) = \sigma^2$ is the covariance at lag zero.

Intrinsic variation and the variogram

The mean often appears to change across a region and then the variance will appear to increase indefinitely as the extent of the area increases. The covariance cannot be defined because there is no value for μ to insert into Eq. (B.6.2). This is a departure from weak stationarity. Matheron's (1965) solution to this was the weaker intrinsic hypothesis of geostatistics. Although the general mean might not be constant, it would be for small lag distances and so the expected differences would be zero as follows:

$$E[Z(x) - Z(x+h)] = 0 \qquad (B.6.4)$$

and the expected squared differences for those lags define their variances

$$E[\{Z(x) - Z(x+h)\}^2] = \text{var}[Z(x) - Z(x+h)] = 2\gamma(h). \qquad (B.6.5)$$

The quantity $\gamma(h)$ is known as the semivariance at lag h, or the variance per point when points are considered in pairs. As for the covariance, the semivariance depends only on the lag and not on the absolute positions of the data. As a function of h, $\gamma(h)$ is the semivariogram or more usually the variogram.

If the process $Z(x)$ is second-order stationary, the semivariance and covariance are equivalent:

$$\gamma(h) = C(0) - C(h) = \sigma^2\{1 - \rho(h)\}. \qquad (B.6.6)$$

However, if the process is intrinsic only there is no equivalence because the covariance function does not exist. The variogram is valid, however, and therefore it can be applied more widely than the covariance function. This makes the variogram a valuable tool and as a consequence it has become the cornerstone of geostatistics.

B.6.3 Estimating the variogram

This section describes two methods for estimating the variogram from data, Matheron's method of moments and the residual maximum likelihood (REML) method, together with the main features that variograms are likely to have.

The method of moments estimator

The empirical semivariances can be estimated from data, $z(x_1)$, $z(x_2)$, ..., by

$$\hat{\gamma}(h) = \frac{1}{2m(h)} \sum_{i=1}^{m(h)} \{z(x_i) - z(x_i + h)\}^2 \tag{B.6.7}$$

where $z(x_i)$ and $z(x_i+h)$ are the actual values of Z at places (x_i) and (x_i+h), and $m(h)$ is the number of paired comparisons at lag h. By changing h, an ordered set of semivariances is obtained; these constitute the experimental or sample variogram. Equation (B.6.7) is the usual formula for computing semivariances; it is often referred to as Matheron's method of moments (MoM) estimator. The way that this equation is implemented as an algorithm depends on the configuration of the data. For a regular transect the lag becomes a scalar, $h = |h|$, for which the semivariances can be computed only at integral multiples of the sampling interval. The number of paired comparisons decreases one at a time as the lag interval is increased. The maximum lag should be set to no more than a third of the length of the transect. For a regular grid, semivariances can be calculated along the rows and columns of the grid and the lag increment is the grid interval. For irregularly sampled data in one or more dimensions, or to compute the omnidirectional variogram of data on a regular grid, the separations between pairs of points are placed into bins with limits in both separating distance and direction, Fig. B.6.1. In this figure, 0L is the nominal lag interval of length h, w is the width of the bin, $\alpha/2$ is the angular tolerance and θ is one of a set of directions. To calculate the variogram over all directions, the omnidirectional variogram, $\alpha/2$ is set to 180° and θ is set to zero.

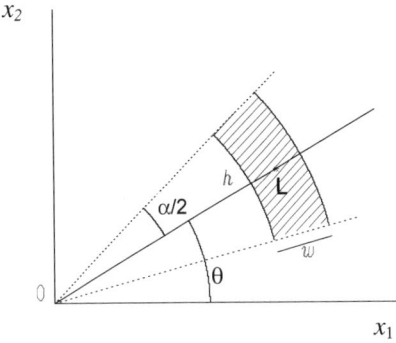

Fig. B.6.1. Discretization of the lag into bins for irregularly scattered data

The choice of narrow bins tends to give rise to erratic variograms, whereas wide bins tend to smooth and result in a loss of detail. You can see the effect of this in Fig. B.6.4. For a grid, it is usual to choose the grid interval as the nominal lag interval and for irregularly scattered data, the average distance between sampling points.

Webster and Oliver (1992) have shown that at least 100 sampling points are required to estimate the MoM variogram reliably. For many situations these are more data than can be afforded, for example where the costs of sampling and or sample analysis are considerable. In other situations this sample size might result in a closer sample spacing than is needed to resolve the variation adequately; this occurs where the property of interest has a large scale of spatial variation relative to the extent of the study area. This would result in over-sampling and a waste of resources. Pardo-Igúzquiza (1997) suggested the maximum likelihood (ML) approach as an alternative to Matheron's estimator. He also suggested that where the number of data is relatively small (a few dozen), the ML variogram estimator offers an alternative that gives an estimate of the variogram parameters and of their uncertainty (Pardo-Igúzquiza 1998, pp. 462-464).

The residual maximum likelihood (REML) variogram estimator

By contrast to the MoM approach, the ML methods are parametric and they also assume that the process, Z, is second-order stationary. Following the notation of Kerry and Oliver (2007), it is assumed that the data, $z(x_i)$, $i = 1, ..., n$, a realization of this process, follow a multivariate Gaussian distribution with the joint probability density function (pdf) of the measurements defined by

$$p(z|\beta, \theta) = (2\pi)^{-\frac{n}{2}} |V|^{-\frac{1}{2}} \exp\left\{-\tfrac{1}{2}(z-X\beta)^\mathrm{T} V^{-1}(z-X\beta)\right\} \qquad (B.6.8)$$

where z is a vector that contains the n data, θ contains the parameters of the covariance matrix, V is the n-by-n variance-covariance matrix, and $X\beta$ represents the trend. The matrix V can be factorized as

$$V = \sigma^2 A \qquad (B.6.9)$$

where σ^2 is the variance and A is the autocorrelation matrix. The pdf can then be rewritten as

$$p(z|\beta, \sigma^2, \theta) = (2\pi)^{-\frac{n}{2}} \sigma^{-n} |A|^{-\frac{1}{2}} \exp\left\{-\tfrac{1}{2\sigma^2}(z-X\beta)^\mathrm{T} A^{-1}(z-X\beta)\right\} \qquad (B.6.10)$$

where θ is the set of covariance parameters excluding the variance. The parameters, β, σ^2, θ, are estimated in such a way that they minimize the negative log-likelihood function given by

$$\ln L(\beta,\hat{\sigma}^2,\theta|z) = \tfrac{n}{2}\ln(2\pi)+n\ln(\sigma)+\tfrac{1}{2}\ln|A| + \tfrac{1}{2\sigma^2}(z-X\beta)^{\mathrm{T}} A^{-1}(z-X\beta). \qquad (B.6.11)$$

In the ML approach the drift parameter, β, is estimated at the same time as the set of covariance parameters.

Simultaneous estimation of the trend and covariance parameters in the ML approach results in biased covariance parameter estimates (Matheron 1971; Kitanidis and Lane 1985). Residual maximum likelihood (REML) developed by Patterson and Thompson (1971) avoids this problem because instead of working with the original data, it uses linear combinations of the data. These latter, known as generalized increments, filter out the trend. The generalized increments, g, can be represented as

$$g = \Lambda z \qquad (B.6.12)$$

where the matrix Λ is derived from the projection matrix

$$P = I - X(X^{\mathrm{T}}X)^{-1}X^{\mathrm{T}} \qquad (B.6.13)$$

by dropping p rows in Λ because there are p generalized increments that are linearly dependent on others (Kitanidis 1983). The matrix P has the property that

$$PX = 0 \qquad (B.6.14)$$

then

$$Pz = PX\beta + Pe = Pe \qquad (B.6.15)$$

which filters out the trend regardless of what the coefficients β are. The e are the residuals. Then

$$E(g) = 0 \qquad (B.6.16)$$

and

$$E(g\,g^{\mathrm{T}}) = \Lambda V \Lambda^{\mathrm{T}}. \qquad (B.6.17)$$

The increments, g, are assumed to be Gaussian and the covariance parameters are estimated by minimization of the negative log-likelihood function (NLLF), given by

$$\ln L^{\mathrm{T}}(\hat{\sigma}^2, \theta|g) = \tfrac{n-p}{2}\ln(2\pi) + \tfrac{n-p}{2} - \tfrac{n-p}{2}\ln(n-p) + \tfrac{1}{2}\ln|A A A^{\mathrm{T}}| + \tfrac{n-p}{2}\ln[g^{\mathrm{T}}(A A A^{\mathrm{T}})^{-1}g].$$

(B.6.18)

The covariance parameters, θ, can include the nugget variance (see below for the definition), long- and short-range distance components for isotropic and anisotropic situations, together with the anisotropy ratio for the latter. Pardo-Igúzquiza's (1997) MLREML program computes these parameters for three covariance models, the spherical, exponential and Gaussian.

For both the ML and REML approaches there is no experimental variogram, and as a consequence there is no smoothing of the spatial structure because there is no *ad hoc* definition of lag classes (bins). This is particularly advantageous for irregularly spaced data.

Features of the variogram

Continuity. Most environmental variables are continuous, therefore we should expect $\gamma(h)$ to pass through the origin at $h = 0$ [Fig. B.6.2(a)]. In practice, however, the variogram often appears to approach the ordinate at some positive value as h approaches zero, Fig. B.6.2(b), which suggests that the process is discontinuous. This discrepancy is known as the *nugget variance*. For properties that vary continuously the nugget variance usually includes some measurement error, but mostly comprises variation that occurs over distances less than the shortest sampling interval. Figure B.6.2(c) is a pure nugget variogram which usually indicates that the sampling interval is too large to resolve the variation present.

Monotonic increasing. Figure B.6.2(a) and (b) shows that the semivariance increases with increasing lag distance. This indicates that at short distances the values of the $Z(x)$ are similar, but as the lag distance increases they become increasingly dissimilar on average. The monotonic increasing slope indicates that the process is *spatially dependent*.

Sill and range. Figure B.6.2(b) shows a variogram that reaches an upper bound after the initial slope; this bound is known as the *sill variance*. It is the *a priori* variance, σ^2, of the process. A bounded variogram describes a process that is second-order stationary. The distance at which the variogram reaches its sill is the *range*, i.e. the range of spatial dependence. Places further apart than the range are spatially independent, Fig. B.6.2(b).

Hole effect and periodicity. The variogram may decrease from its maximum to a local minimum and then increase again. This maximum is equivalent to a minimum in the covariance function in which it appears as a 'hole'. It suggests fairly regular repetition in the process. A variogram that fluctuates in a periodic way with increasing lag distance indicates greater regularity of repetition.

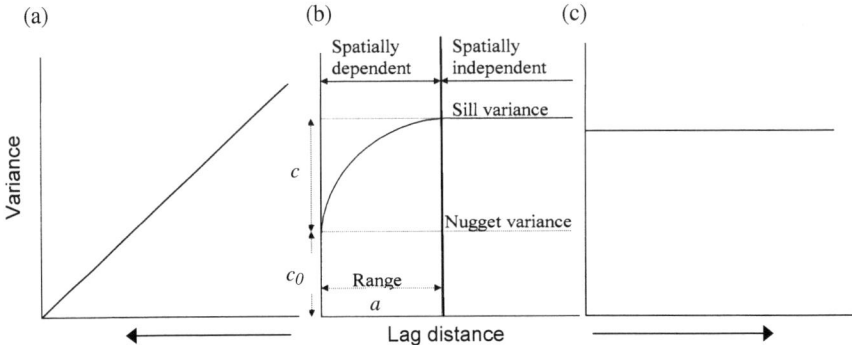

Fig. B.6.2. Three idealized variogram forms: (a) unbounded; (b) bounded; and (c) is the spatially correlated component [c_0 is nugget variance, a is the range of spatial dependence, $c + c_0$ is the sill variance, and c pure nugget]

Unbounded variogram. If the variogram increases indefinitely with increasing lag distance as in Fig. B.6.2(a), the process is intrinsic only.

Anisotropy. Spatial variation might not be the same in all directions. To explore data for any anisotropy, i.e. directional variation, the variogram must be computed in at least three directions. For a regular grid, it is usual to compute the variogram along the rows, columns and the principal diagonals. If there are four directions, start by setting the angular discretization to 22.5°, for example, and this angle can be decreased if there appears to be anisotropy. If the initial gradient or range of the variogram changes with direction and a simple transformation of the coordinates will remove it, then this is known as *geometric anisotropy*. An example of this is given in Fig. B.6.5 later in the case study; it shows the variogram of pH at Broom's Barn Farm computed in four directions from data on a regular grid. If the sill variance fluctuates with changes in direction, this might indicate the presence of preferentially orientated zones with different means. This is known as *zonal anisotropy*. It can sometimes be dealt with by stratifying the area of interest and then computing the variogram from the residuals of the class means. This is sometimes called the pooled within-class variogram.

Nested variation. Variation in the environment often occurs at several spatial scales simultaneously, and patterns in the variation can be nested within one another. This is usually evident when there are many data, for example from remote sensing etc. The experimental variogram will often appear more complex if more than one spatial scale is present; this can be seen in Fig. B.6.6. A combination of two or more simple models that are authorized can be used to model such a variogram. The simplest combined model is one with a nugget component. Spatial dependence may occur at two distinct scales and these can be represented in the variogram as two spatial components. Models describing more than one spatial structure are often known as nested functions; the nested or double spherical model has been the most commonly fitted, Fig. B.6.6(b).

B.6.4 Modeling the variogram

The experimental MoM variogram comprises a set of discrete estimates at particular lag intervals, which are subject to error that arises largely from sampling fluctuation. The underlying variogram, which represents the regional variation, is continuous. To obtain an approximation to this we can fit what are known as authorized functions that are conditional negative semi-definite (CNSD) to the experimental values. Functions that are CNSD will not give rise to negative variances when random variables are combined (see Webster and Oliver 2007 for more detail on this). There are a few principal features that the function must be able to represent:

(i) a monotonic increase with increasing lag distance from near the ordinate,
(ii) a constant maximum or asymptote (the sill),
(iii) a positive intercept on the ordinate (the nugget),
(iv) anisotropy.

There are a few simple functions only that encompass the above features and that are CNSD. They can be divided into those that are bounded, which represent processes that are second-order stationary, and those that are unbounded that are intrinsic only. There are several functions, but here we shall focus on those that are fitted most commonly in the environmental sciences. The formulae for the selected functions are given in their isotropic form, i.e. for $h = |h|$. A nugget variance, c_0, has been included because most experimental variograms if extended to the ordinate would have a positive intercept. The Gaussian model is included in many popular geostatistical packages, but it is excluded here. Its use can give rise to unstable kriging equations because the model approaches the origin with zero gradient (the limit for random variation), and this function will be replaced with the stable exponential model (Wackernagel 2003). Webster and Oliver (2007) describe a wide range of suitable variogram functions.

Circular model. The equation for the circular function is

$$\gamma(h) = \begin{cases} c_0 + c\left\{1 - \dfrac{2}{\pi}\cos^{-1}\left(\dfrac{h}{a}\right) + \dfrac{2h}{\pi a}\sqrt{1 - \dfrac{h^2}{a^2}}\right\} & \text{for } h \leq a \\ c_0 + c & \text{for } h > a \\ 0 & \text{for } h = 0 \end{cases} \quad (B.6.19)$$

where $\gamma(h)$ is the semivariance at lag h, c is the *a priori* variance of the autocorrelated process, c_0 is the nugget variance which represents the spatially uncorrelated variation at distances less than the sampling interval and measurement error, and a is the distance parameter, the range of spatial dependence or spatial autocorrelation. Values at places less than this apart are correlated, whereas those further apart are not. The combined $c_0 + c$ is the sill of the model. Theoretically the semivariance at lag zero is itself zero, but in practice there are usually too few estimates of $\gamma(h)$ near to the ordinate to fit a model through the origin. This function is CNSD in two dimensions. It curves tightly as it approaches the range (see Fig. B.6.4(i)).

Spherical function. This is one of the two most widely fitted models in the environmental sciences. Its equation is

$$\gamma(h) = \begin{cases} c_0 + c\left\{\dfrac{3h}{2a} + \dfrac{1}{2}\left(\dfrac{h}{a}\right)^3\right\} & \text{for } h \leq a \\ c_0 + c & \text{for } h > a \\ 0 & \text{for } h = 0. \end{cases} \quad (B.6.20)$$

The symbols have the same meaning as above. This model curves more gradually as the sill is reached than the circular one, see Fig. 6.4.4(c). This function is CNSD in three dimensions. It represents transition features that have a common extent that appear as patches, some with large values and other with small ones. The average diameter of the patches is represented by the range of the model.

Pentaspherical function. This model curves more gently as it approaches its sill than the preceding models, see Fig. B.6.3(b). It is CNSD in three dimensions. The pentaspherical function has the equation

$$\gamma(h) = \begin{cases} c_0 + c\left\{\dfrac{15h}{8a} - \dfrac{5}{4}\left(\dfrac{h}{a}\right)^3 + \dfrac{3}{8}\left(\dfrac{h}{a}\right)^5\right\} & \text{for } h \leq a \\ c_0 + c & \text{for } h > a \\ 0 & \text{for } h = 0 \end{cases} \quad (\text{B.6.21})$$

Exponential function. The exponential and spherical functions together account for a large proportion of the models fitted in the environmental sciences. Its equation is

$$\gamma(h) = c_0 + c\left\{1 - \exp\left(-\frac{h}{r}\right)\right\} \quad (\text{B.6.22})$$

where c_0 and c have the same meanings as above, but the distance parameter is now r. The exponential model approaches its sill even more gently than the preceding models and also asymptotically so that it does not have a finite range. In practice, an effective range is assigned at the distance at which the function has reached 95 percent of c. The effective range, a', is $3r$. It is CNSD in three dimensions. The exponential function also represents transition structures, but they now have random extents.

Stable exponential. This is a useful substitute for the Gaussian function for experimental variograms that appear to approach the origin with a reverse curvature; they can be represented by the general equation

$$\gamma(h) = c_0 + c\left\{1 - \exp\left(-\frac{h^\alpha}{r^\alpha}\right)\right\} \quad (\text{B.6.23})$$

in which $1 < \alpha < 2$. For the Gaussian function $\alpha = 2$, which is excluded because it represents differentiable variation in the process, which is not random. Webster and Oliver (2006) used the stable exponential function to describe topographic variation.

Unbounded models. Variograms that are intrinsic only increase without bound as the lag distance increases. These can usually be fitted by power functions, which have the general equation including a nugget variance of

$$\gamma(h) = c_0 + wh^\alpha \quad (\text{B.6.24})$$

where w describes the intensity of the process, and the exponent, α, describes the curvature. If $\alpha < 1$, the curve is convex upwards; if it is one it is a straight line and

w is the gradient; and if $\alpha > 1$ the curve is concave upwards. The exponent must lie strictly between zero and two.

Modeling anisotropy. If the experimental variogram is anisotropic, then the variation is a function of distance, h, and direction, θ. Geometric anisotropy can be made isotropic by a linear transformation of the coordinates. The transformation is defined by reference to an ellipse

$$\Omega(\theta) = \sqrt{A^2 \cos^2(\theta - \phi) + B^2 \sin^2(\theta - \phi)} \tag{B.6.25}$$

where A and B are the long and short diameters of the ellipse, respectively, and ϕ is its orientation, i.e. the direction of the long axis. For bounded models, Ω replaces the distance parameter of the isotropic variogram as follows for the exponential variogram (see Fig. B.6.5(b)).

$$\gamma(h, \theta) = c_0 + c\left[1 - \exp\left\{-\frac{|h|}{\Omega(\theta)}\right\}\right] \tag{B.6.26}$$

and for the power function it replaces the gradient

$$\gamma(h, \theta) = c_0 + [\Omega(\theta) h]^\alpha \tag{B.6.27}$$

Nested models. The nested spherical function is given by

$$\gamma(h) = \begin{cases} c_0 + c_1\left\{\frac{3h}{2a_1} - \frac{1}{2}\left(\frac{h}{a_1}\right)^3\right\} + c_2\left\{\frac{3h}{2a_2} - \frac{1}{2}\left(\frac{h}{a_2}\right)^3\right\} & \text{for } 0 < h \leq a_1 \\ c_0 + c_1 + c_2\left\{\frac{3h}{2a_2} - \frac{1}{2}\left(\frac{h}{a_2}\right)^3\right\} & \text{for } a_1 < h \leq a_2 \\ c_0 + c_1 + c_2 & \text{for } h = a_2 \end{cases} \tag{B.6.28}$$

where c_1 and a_1 are the sill and range of the short-range component of the variation, and c_2 and a_2 are the sill and range of the long-range component. A nugget component can also be added as above (see Fig. B.6.6(b)).

B.6.5 Case study: The variogram

We illustrate some of the principles of geostatistics with results from a recent study on precision farming for the British Home-Grown Cereals Authority (Oliver and Carroll 2004). The field (UK National Grid reference SU 458174) covers 23ha on the Yattendon Estate, Berkshire, England. It is on part of the Chalk downland of southern England and has the typical undulating topography of this region. From the extensive set of survey data obtained during 2002 we have selected topsoil (0–15 cm) available potassium. Data on yield of winter wheat were available for 2001 to illustrate nested variation. Table B.6.1 gives the summary statistics for these two variables.

Sampling for the soil survey was at the nodes of a 30m × 30m grid, with additional observations at 15m intervals along short transects from randomly selected grid nodes. The sampling intervals were based on scales of variation determined from several years of yield data with the aim of ensuring that the variation in the soil (of which there was no prior knowledge) would be represented adequately and efficiently. At each site ten cores of soil were bulked from a support of 5m × 2m to form the sample; this helps to reduce the locally erratic variation that contributes to the nugget variance. There were 230 data points, which enabled any anisotropy in the variation to be determined; this sample size is close to the 250 data recommended by Webster and Oliver (1992).

Table B.6.1. Summary statistics

Statistic	Topsoil K [mg l^{-1}]	Yield 2001 [t ha^{-1}]
Number	230	4060
Mean	142.5	6.838
Median	143.0	7.050
Minimum	48.1	1.000
Maximum	254.4	14.600
Variance	1367.5	3.909
Standard deviation	37.0	1.977
Skewness	0.1	–0.298

Experimental variograms were computed by Eq. (B.6.7) in four directions to reveal any anisotropy in the variation. The results for topsoil K are shown in Fig. B.6.3(a) for the directions 0°, 45°, 90° and 135°. There is little divergence among the different directions until lag 130m, after which the sills start to diverge. This suggests that there is zonal anisotropy in the variation of topsoil K in this field. Since the directional variograms are close together for the initial lags, the variation can be treated as isotropic for kriging, and the solid line shows the best fitting isotropic exponential function to the omnidirectional variogram.

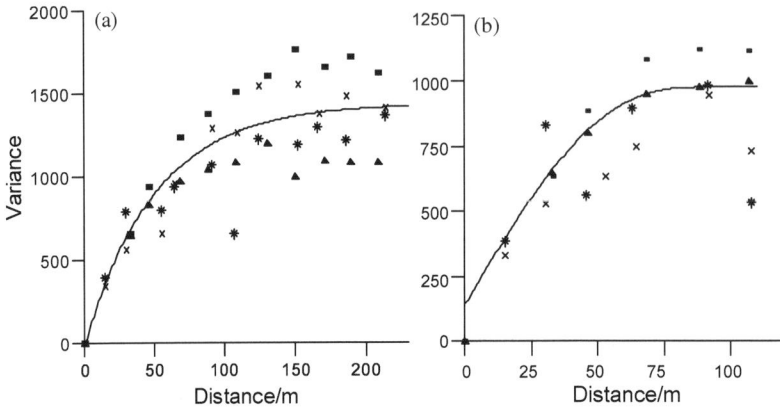

Fig. B.6.3. (a) Directional variogram computed on the raw data (230 points) from the Yattendon Estate, and (b) directional variogram computed on the residuals from the class means. The symbols represent: * denotes 0° (E–W), ■ denotes 45°, × denotes 90° (N–S), ▲ denotes 135°

To illustrate the effect of sample size on the variogram, we subsampled the complete set of data (230 sampling points) to give subsets of 94 and 47 data. Experimental omnidirectional variograms were computed from the total data and two subsets for topsoil K. To explore the effect of different bin widths, variograms were computed for lag intervals of 15m (the sampling interval for the transects), 20m (mid-way between the transect and overall grid interval) and 40m (for illustration). Models were fitted to the experimental values using GenStat (Payne 2008).

Figure B.6.4 shows the experimental values as symbols and the fitted models as solid lines. The experimental variograms suggest that the 20m lag interval is a good comprise between the rather erratic result for the 15m interval and the loss of detail with the 40m lag interval. The experimental variograms also show the effect of decreasing the number of data; the variograms becomes more erratic and that computed from 47 data also shows a serious loss of variance.

Table B.6.2 gives the models and their parameters fitted to the experimental variograms. These show how sensitive the model parameters are to changes in lag interval and number of data. For the 230 data, the main difference in the model parameters for the variograms computed with different lag intervals is in the nugget variance, which is zero for the 15m lag. This suggests that the data from the transect sampling have resolved the local variation in topsoil K well. This is an important consideration when designing a sampling scheme. For a grid survey, it is worthwhile having some additional sampling points at shorter distances than the grid interval as in this survey because it helps to reduce the nugget variance. There were 40 sampling points at the shorter interval which is only 17 percent of the total data.

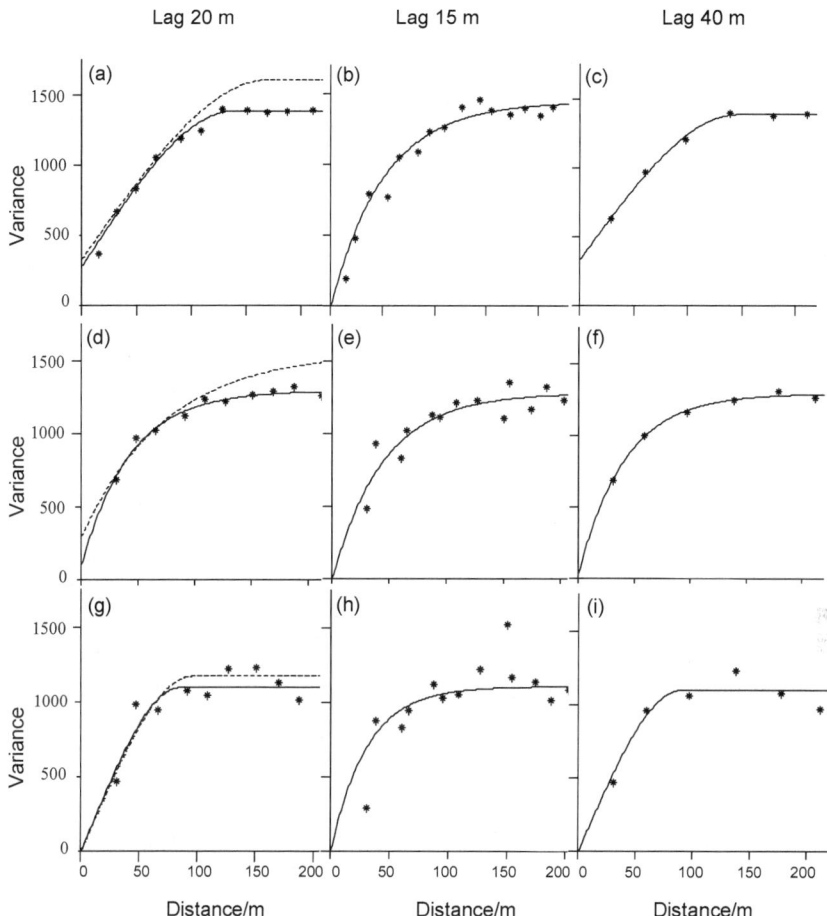

Fig. B.6.4. Experimental variograms (∗) computed by the method of moments (MoM) estimator for lag distances of 20m, 15m and 40m, and for the complete data set of 230 sites [(a), (b) and (c), respectively], subset of 94 data [(d), (e), (f)] and subset of 47 data [(g), (h), (i)] for topsoil K on the Yattendon Estate. The solid line is the model fitted to the MoM variogram and the dashed line is the variogram estimated by residual maximum likelihood (REML)

For the subsample of 94 data, the difference in model parameters from those for complete set of data is small; this indicates that Webster and Oliver's (1992) recommendation of a minimum of 100 data is adequate to obtain a reliable variogram. The model parameters for the smallest data set are considerably different from those of the complete set of data, suggesting that the variograms of the smallest data set are not an accurate reflection of the structure of the variation. For example, the sill variances are markedly less and the ranges of spatial dependence

Table B.6.2. Variogram model parameters

Topsoil Property	Model type	Parameter			
		Nugget variance	Correlated component	Range (m)	Sill variance
MoM estimator		c_0	c_1 c_2♦	a_1/m a_2/m+ or r/m*	
K (230 sites)					
Lag 20 m	Spherical	319.3	1070.0	142.9	1389.3
Lag 15 m	Exponential	0	1441.0	151.7	1441.0
Lag 40 m	Spherical	355.6	1035.7	148.4	1391.3
K residuals					
Lag 20 m	Pentaspherical	145.5	830.7	90.8	976.2
K (94 sites)					
Lag 20 m	Exponential	163.7	1138.0	44.6	1301.7
Lag 15 m	Exponential	0	1282.0	44.6	1109.0
Lag 40 m	Spherical	338.9	1051.0	146.4	1389.9
K (47 sites)					
Lag 20 m	Spherical	0	1098.0	85.3	1098.0
Lag 15 m	Exponential	0	1109.0	30.5	1109.0
Lag 40 m	Circular	0	1100.0	79.6	1100.0
pH Broom's Barn	Exponential	0	0.37	89.70	0.37
	Anisotropic Exponential	0	0.38	69.54 114.50	ϕ=1.09
Yield 1995	Double Spherical	1.76	1.04 1.16♦	44.19 277.50+	0.8882
REML estimator					
REML 230	Spherical	334.5	1273.5	170.6	1608.0
REML 94	Exponential	300.0	1262.6	74.0	1562.6
REML 47	Spherical	1.9	1171.1	95.7	1173.0

Notes: ♦ is the spatially correlated variance of the long-range spatial component, + is the range of the long-range spatial component, * is the distance parameter of the exponential function; to obtain a working range $a' = 3r$

are shorter. Table B.6.2 shows that the models are all bounded functions indicating that the variation has a patchy distribution.

Variograms were also computed by REML for the 20m grid interval, and are shown as the dashed line in Fig B.6.4(a), (d) and (g). The variograms estimated by REML for the two larger data sets are not as similar to those computed by MoM as one might expect. The sill variances are larger than the variance of the data. The range of the exponential model for the subset of 94 data is also much longer than that for the MoM variogram. The variograms estimated by REML and MoM are more similar to one another for the smallest data set, yet it is for these data that one would expect the greatest difference in model parameters. Although Kerry and Oliver (2007) showed a distinct advantage in computing variograms by REML for small sets of data, this is not particularly evident in the study described here.

The experimental variogram computed from the yield data of a crop of winter wheat (2001) shows a complex structure [see Fig. B.6.6(a)]. The best fitting model was a spherical function with two spatial components; one with a range of 44m and the other of 278m. Figure B.6.6(b) shows the experimental variogram with the fitted model; the nugget, short- and long-range components of the model are also shown separately.

Anisotropy. Figure B.6.3(a) shows the directional variogram for topsoil K. It is evident that the sill variances disperse after a lag of about 130m. Zonal anisotropy cannot be dealt with by a simple transformation of the coordinates. If the region can be stratified into zones, then this is one way in which zonal anisotropy can be resolved. The variogram models suggest that the variation is patchy, which could arise from zones that are preferentially orientated and with different means. A classification of these data had been done previously (see Frogbrook and Oliver 2007 for details), therefore the class means were subtracted from the values of K for the appropriate class.

The directional variogram was then computed on the residuals from the class means, Fig. B.6.3(b). The directional variogram is shown by the symbols for the four directions and the isotropic models fitted to the omnidirectional variograms by the solid black line for both the raw data and the residuals. Stratification has effectively removed the zonal anisotropy – some scatter remains in the different directions but this is to be expected from sampling fluctuations. The model parameters have also changed considerably; the best fitting model is now a pentaspherical function with a sill variance of less than 1000 and a range of 91m. The model now has a much shorter range of spatial dependence, Table B.6.2, and so the variogram has been plotted to a maximum lag of 150m to take into account this difference. There is no marked evidence of anisotropy over distances less than the range.

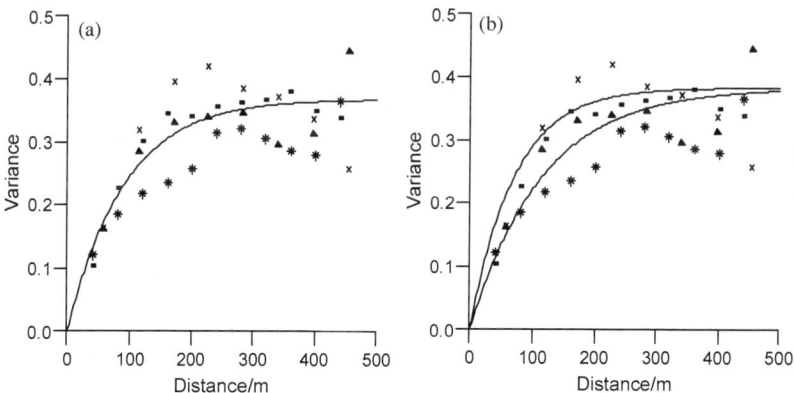

Fig. B.6.5. Directional variogram computed on the pH data from Broom's Barn Farm (433 sampling points): (a) with the best fitting isotropic model (solid line), and (b) with an isotropic exponential function (the solid lines show the envelope of this function). The symbols represent: ∗ denotes 0° (E–W), ■ denotes 45°, × denotes 90° (N–S), ▲ denotes 135°, and the solid lines are the isotropic models fitted to the omnidirectional variograms

To illustrate geometric anisotropy we have used the data for pH from Broom's Barn Farm. This is an experimental sugar beet farm near to Bury St. Edmunds, Cambridgeshire, UK (see Webster and Oliver 2007, for more detail on these data). Figure B.6.5(a) shows the directional variogram which illustrates how the semivariances in the different directions start to diverge after a lag of 80m. The solid line is the best fitting isotropic model, an exponential function (Table B.6.2). Figure B.6.5(b) shows the directional variogram with the fitted anisotropic exponential function. The two lines show the envelope of this function and Table B.6.2 gives the parameters of the fitted function. The direction of maximum variation and of the shorter range is about 60° (where 0° is E–W) and the direction of minimum variation is perpendicular to this.

Nested variation. Figure B.6.6(a) shows the experimental variogram for yield 2001 at the Yattendon Estate; it appears to have a complex structure. Several models were fitted and the one with the smallest residual sums of squares was a nested spherical function, which is shown as the solid line fitted to the experimental values in Fig. B.6.6(b). The model parameters for yield 2001 are given in Table B.6.2. To illustrate the individual components of this model, we have shown them separately in Fig. B.6.6(b) as lines with different ornament. The complex structure identified from the experimental variogram is evident as two markedly different ranges of spatial variation of 44m and 278m.

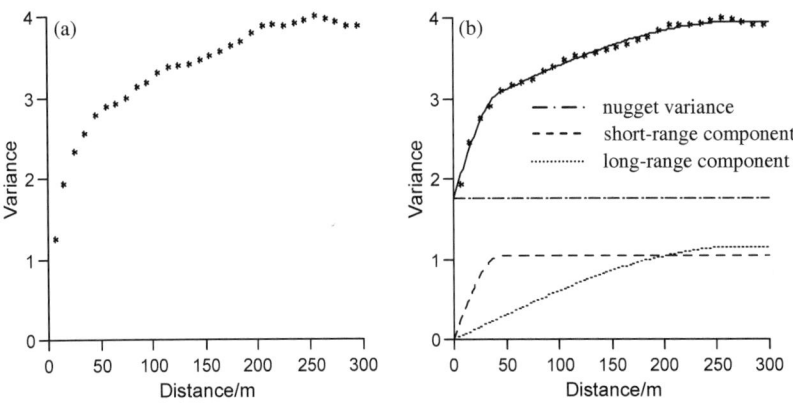

Fig. B.6.6. Variogram of yield 2001 for the Yattendon Estate: (a) experimental variogram (symbols), and (b) the experimental variogram with the fitted double spherical model (solid line); the ornamented lines represent the individual model components

B.6.6 Geostatistical prediction: Kriging

Kriging is a method of optimal prediction or estimation in geographical space, often known as a best linear unbiased predictor (BLUP). It is the geostatistical method of interpolation for random spatial processes. Matheron (1963) first used the term 'kriging' for the method in recognition of D. G. Krige's contribution to improving the precision of estimating concentrations of gold and other metals in ore bodies. Krige (1951) had observed that he could improve estimates of ore grades in mining blocks by taking into account the grades in neighbouring blocks. Matheron (1963) expanded Krige's empirical ideas and put them into the theoretical framework of geostatistics. However, Matheron's developments were not in isolation; the mathematics of simple kriging had been worked out by A. N. Kolmogorov in the 1930s (Kolmogorov 1939, 1941), by Wold (1938) for time series analysis and later by Wiener (1949). Cressie (1993) gives a brief history of the origins of kriging.

Kriging provides a solution to a fundamental problem faced by environmental scientists of predicting values from sparse sample data based on a stochastic model of spatial variation. Most properties of the environment (soil, vegetation, rocks, water, oceans and atmosphere) can be measured at any of an infinite number of places, but for economic reasons they are measured at relatively few. Several mathematical methods of interpolation are available, for example, Thiessen polygons, triangulation, natural neighbour interpolation, inverse functions of distance, least-squares polynomials (trend surfaces) and splines. Most of these methods take account of systematic or deterministic variation only and disregard the errors of prediction. Kriging, on the other hand, overcomes the weaknesses of these mathematical interpolators. It makes the best use of existing knowledge by taking account of the way a property varies in space through the variogram or covariance function. Kriging also provides not only predictions but also the kriging variances or errors. It can be regarded simply as a method of local weighted moving averaging of the observed values of a random variable, Z, within a neighbourhood, V. Kriging can be done for point (punctual kriging) or block supports of various size (block kriging), depending upon the aims of the prediction, even though the sample information is often for points.

Since its original formulation, kriging has been elaborated to tackle increasingly complex problems in disciplines that use spatial prediction and mapping. It is used in mining, petroleum engineering, meteorology, soil science, precision agriculture, pollution control, public health, monitoring fish stocks and other animal densities, remote sensing, ecology, geology, hydrology and other disciplines. As a consequence, kriging has become a generic term for a range of BLUP least-squares methods of spatial prediction in geostatistics. The original formulation of kriging, now known as ordinary kriging (Journel and Huijbregts 1978), is the most robust method and the one most often used.

Types of kriging

Ordinary kriging assumes that the mean is unknown and that the process is locally stationary. Simple kriging, which assumes that the mean is known, is used little because the mean is generally unknown. However, it is used in indicator and disjunctive kriging in which the data are transformed to have known means. Lognormal kriging is ordinary kriging of strongly positively skewed data transformed by logarithms to approximate a lognormal distribution. Kriging with trend enables data with a strong deterministic component (non-stationary process) to be analyzed;

Matheron (1969) originally introduced universal kriging for this purpose, but the state-of-the-art is empirical-BLUP (Stein 1999), which uses the REML variogram (Lark et al. 2006). Matheron (1982) developed factorial kriging or kriging analysis for variation that is nested. It estimates the long- and short-range components of the variation separately, but in a single analysis. Ordinary cokriging (Matheron 1965) is the extension of ordinary kriging to two or more variables that are spatially correlated. If some property that can be measured cheaply at many sites is spatially correlated or coregionalized with others that are expensive to measure and recorded at many fewer sites, the latter can be estimated more precisely by cokriging with the spatial information from the former.

Disjunctive kriging (Matheron 1973) is a non-linear parametric method of kriging. It is valuable for decision-making because the probabilities of exceeding (or not) a predefined threshold are determined in addition to the kriged estimates. Indicator kriging (Journel 1982) is a non-linear, non-parametric form of kriging in which continuous variables are converted to binary ones (indicators). It can handle distributions of almost any kind and can also accommodate 'soft' qualitative information to improve prediction. Probability kriging was proposed by Sullivan (1984) because indicator kriging does not take into account the proximity of a value to the threshold, but only its geographic position. Bayesian kriging was introduced by Omre (1987) for situations in which there is some prior knowledge about the drift or trend.

Ordinary kriging

Ordinary kriging is by far the most widely used type of kriging. It is based on the assumption that the mean is unknown. Consider that a random variable, Z, has been measured at sampling points, x_i, $i = 1, \ldots n$, and we want to use this information to estimate its value at a point x_0 (punctual kriging) with the same support as the data by

$$\hat{Z}(x_0) = \sum_{i=1}^{n} \lambda_i z(x_i) \tag{B.6.29}$$

where n usually represents the data points within the local neighbourhood, V, and is much less than the total number in the sample, N, and λ_i are the weights. To ensure that the estimate is unbiased the weights are made to sum to one

$$\sum_{i=1}^{n} \lambda_i = 1 \qquad (B.6.30)$$

and the expected error is $E[\hat{Z}(x_0) - Z(x_0)] = 0$. The prediction variance is

$$\mathrm{var}[\hat{Z}(x_0)] = E[\{\hat{Z}(x_0) - Z(x_0)\}^2] = 2\sum_{i=1}^{n} \lambda_i \gamma(x_i, x_0) - \sum_{i=1}^{n}\sum_{j=1}^{n} \lambda_i \lambda_j \gamma(x_i, x_j)$$

$$(B.6.31)$$

where $\gamma(x_i, x_j)$ is the semivariance of Z between points x_i and x_j, $\gamma(x_i, x_0)$ is the semivariance between the ith sampling point and the target x_0. The semivariances are derived from the variogram model because the experimental semivariances are discrete and at limited distances.

Kriged predictions are often required over areas (block kriging) that are larger than the sample support of the data. The estimate is a weighted average of the data, $z(x_1), z(x_2), \ldots, z(x_n)$, at the unknown block,

$$\hat{Z}(B) = \sum_{i=1}^{n} \lambda_i z(x_i). \qquad (B.6.32)$$

The estimation variance of $Z(B)$ is:

$$\mathrm{var}[\hat{Z}(B)] = E[\{\hat{Z}(B) - Z(B)\}^2] = 2\sum_{i=1}^{n} \lambda_i \bar{\gamma}(x_i, B) - \sum_{i=1}^{n}\sum_{j=1}^{n} \lambda_i \lambda_j \gamma(x_i, x_j) - \bar{\gamma}(B, B)$$

$$(B.6.33)$$

where $\bar{\gamma}(x_i, B)$ is the average semivariance between data point x_i and the target block B, and $\bar{\gamma}(B, B)$ is the average semivariance within B, the within block variance.

Equation (B.6.31) for a point leads to a set of $n + 1$ equations in the $n + 1$ unknowns

$$\sum_{i=1}^{n} \lambda_i \gamma(x_i, x_j) + \psi(x_0) = \gamma(x_j, x_0) \qquad \text{for all } j \qquad (B.6.34)$$

$$\sum_{i=1}^{n} \lambda_i = 1 \qquad (B6.35)$$

the Lagrange multiplier, ψ, is introduced to achieve minimization. The kriging equations in matrix form for punctual kriging are

$$A\lambda = b \qquad (B.6.36)$$

where A is the matrix of semivariances between data points, $\gamma(x_i, x_j)$, b is the vector of semivariances between data points and the target, $\gamma(x_i, x_0)$ and λ is the vector of weights and the Lagrange multiplier. The kriging weights are obtained as follows by inverting matrix A,

$$\lambda = A^{-1} b. \qquad (B.6.37)$$

The weights, λ_i, are inserted into Eq. (B.6.29) to give the prediction of Z at x_0. The kriging (prediction or estimation) variance is then

$$\sigma^2(x_0) = \sum_{i=1}^{n} \lambda_i \gamma(x_i, x_0) + \psi(x_0) \qquad (B.6.38)$$

and in matrix form

$$\sigma^2(x_0) = b^T \lambda. \qquad (B.6.39)$$

Punctual kriging is an exact interpolator – the kriged value at a sampling site is the observed value there and the estimation variance is zero. The equivalent kriging system for blocks is

$$\sum_{i=1}^{n} \lambda_i \gamma(x_i, x_j) + \psi(B) = \bar{\gamma}(x_j, B) \qquad \text{for all } j \qquad (B.6.40)$$

$$\sum_{i=1}^{n} \lambda_i = 1 \qquad (B6.41)$$

and the block kriging variance is obtained as

$$\sigma^2(B) = \sum_{i=1}^n \lambda_i \bar{\gamma}(x_i, B) + \psi(B) - \bar{\gamma}(B, B) \tag{B.6.42}$$

and in matrix form

$$\sigma^2(B) = b^T \lambda - \bar{\gamma}(B, B). \tag{B.6.43}$$

Block kriging results in smoother estimates and smaller estimation variances overall because the nugget variance is contained entirely in the within-block variance, $\bar{\gamma}(B, B)$, and it does not contribute to the block kriging variance.

For many environmental applications kriging is most likely to be used for interpolation and mapping. The values of the property are usually estimated at the nodes of a fine grid, and the variation can then be displayed by isarithms or by layer shading. The estimation variances or standard errors can also be mapped similarly: they are a guide to the reliability of the estimates, where sampling is irregular, such a map may indicate if there are parts of a region where sampling should be increased to improve the estimates.

Kriging weights

The kriging weights depend on the variogram and the configuration of the sampling. The way in which the data points within the search radius are weighted is one feature that makes kriging different from classical methods of prediction where the weights are applied arbitrarily. Webster and Oliver (2007) illustrate how the weights vary according to changes in the nugget: sill ratio, the range, type of model, sampling configuration and the effect of anisotropy. The weights are particularly sensitive to the nugget variance and anisotropy. Weights close to the point or block to be estimated carry more weight than those further away, which shows that kriging is a local predictor. As the nugget: sill ratio increases the weights near to the target decrease and those further away increase. For a pure nugget variogram, the kriging weights are all the same and the estimate is simply the mean of the values in the neighbourhood. The effect of the range is more complex than for the nugget: sill ratio because it is also affected by the type of variogram model. In general, however, as the range increases the weights increase close to the target. For data that are irregularly distributed, points that are clustered carry less weight individually than those that are isolated.

The fact that the points nearest to the target generally carry the most weight has practical implications. It means that the search neighbourhood need contain no more than 16–20 data points, which in turn means that matrix A in the kriging system need never be large.

Factorial kriging

If the variogram of $Z(x)$ is nested, it can be represented as a combination of S individual variograms

$$\lambda(h) = \gamma^1(h) + \gamma^2(h) + \cdots + \gamma^S(h) \tag{B.6.44}$$

where the superscripts refer to the component variograms. If we assume that the processes represented by these components are uncorrelated, then Eq. (B.6.44) can be written as

$$\lambda(h) = \sum_{k=1}^{S} b^k g^k(h) \tag{B.6.45}$$

where $g^k(h)$ is the kth basic variogram function and b^k is a coefficient that measures the relative contribution of the variance $g^k(h)$ to the sum.

The components on the right-hand side of Eq. (B.6.45) correspond to S random functions that in sum form $Z(x)$, which can be represented as

$$Z(x) = \sum_{k=1}^{S} Z^k(x) + \mu \tag{B.6.46}$$

in which μ is the mean of the process. Each $Z^k(x)$ has an expectation zero, and the squared differences are

$$\tfrac{1}{2} E\left[\{Z^k(x) - Z^k(x+h)\}\{Z^{k'}(x) - Z^{k'}(x+h)\}\right] = \begin{cases} b^k g^k(h) & \text{if } k = k' \\ 0 & \text{otherwise.} \end{cases} \tag{B.6.47}$$

The last component, $Z^S(x)$ could be intrinsic only, so that $g^S(h)$ in Eq. (B.6.45) is unbounded with gradient b^S. This equation expresses the mutual independence of the S random functions, and enables the values of the contributing processes to be estimated separately by factorial kriging. Each spatial component $Z^k(x)$ is estimated as a linear combination of the observations, $z(x_i)$, $i = 1, \ldots, n$

$$\hat{Z}^k(x_0) = \sum_{i=1}^{n} \lambda_i^k \, z(x_i). \tag{B.6.48}$$

The λ_i^k are weights assigned to the observations, but now they must sum to zero, not to one, to ensure that the estimate is unbiased and to accord with Eq. (B.6.46).

Subject to this condition, they are chosen to minimize the kriging variance. This leads to the kriging system

$$\sum_{j=1}^{n} \lambda_j^k \gamma(x_i, x_j) - \psi^k(x_0) = b^k g^k(x_i, x_0) \qquad \text{for all } i = 1, \ldots, n \qquad \text{(B.6.49)}$$

$$\sum_{j=1}^{n} \lambda_j^k = 0 \qquad \text{(B.6.50)}$$

where $\psi^k(x_0)$ is the Lagrange multiplier for the kth component. This system of equations is solved for each spatial component, k, to find the weights, λ_j^k, which are then inserted into Eq. (B.6.48) for that component. Estimates are made for each spatial scale, k, by solving Eq. (B.6.49).

Kriging is usually done in small moving neighbourhoods centred on x_0, as for ordinary kriging. Thus, from a theoretical point of view, it is necessary only that $Z(x)$ is locally stationary. Equation (B.6.46) can then be rewritten as

$$Z(x) = \sum_{k=1}^{S} Z^k(x) + \mu(x) \qquad \text{(B.6.51)}$$

where $\mu(x)$ is a local mean that can be considered as a long-range spatial component. We need to krige the local mean, which is again a linear combination of the data:

$$\hat{\mu}(x_0) = \sum_{j}^{n} \lambda_j^{\text{mean}} z(x_j). \qquad \text{(B.6.52)}$$

The weights are obtained by solving the kriging system:

$$\sum_{j=1}^{n} \lambda_j^{\text{mean}} \gamma(x_i, x_j) - \psi^{\text{mean}}(x_0) = b^k g^k(x_i, x_0) \quad \text{for all } i=1, \ldots, n \qquad \text{(B.6.53)}$$

$$\sum_{j=1}^{n} \lambda_j^k = 1. \qquad \text{(B.6.54)}$$

Estimating the long-range component can be affected by the size of the moving neighbourhood (Galli et al. 1984). To estimate a spatial component with a given range, the distance across the neighbourhood should be at least equal to that range. If the sampling is intensive and the range is large, there are so many data within the chosen neighbourhood that only a small proportion of them is retained for kriging, and those are all near to the target. Although modern computers can handle many data at a time, the inversion of such large matrices can be unstable. Further, only the nearest few data to the target contribute to the estimate because they screen the more distant data. Consequently, the neighbourhood used is smaller than the one specified, which means that the range of the component estimated is smaller than that determined from the variogram. Galli et al. (1984) suggested a way of overcoming this shortcoming by selecting only a proportion of the data within the specified neighbourhoods. Such a selection is arbitrary, and Jaquet (1989) proposed an alternative that involves adding the estimate of the local mean to the estimated long-range component. Following Oliver et al. (2000), this is the solution we have adopted for the case study below.

B.6.7 Case study: Kriging

The case study describes applications of ordinary kriging with an isotropic variogram model and with an anisotropic one where there are directional differences in the variation. Factorial kriging is applied to explore variation that is described best by a nested variogram function.

Ordinary kriging

The complete set of data and the two data subsets of topsoil potassium from Yattendon are used to illustrate ordinary kriging. Predictions were made at unsampled places at the nodes of a 5m × 5m grid by ordinary punctual and block kriging. A minimum of seven and a maximum of 20 points were the limits set for the number of data in the neighbourhood. For block kriging, estimates were made over blocks of 10m × 10m. The parameters of the variogram models fitted to the MoM experimental variograms of each data set for the 20m lag (Table B.6.2) were used with the respective data for kriging. The kriged predictions were mapped in Gsharp. Figure B.6.7 shows the maps of block kriged estimates; those from punctual kriging are not shown as they appear so similar. The map based on the 230 data, Fig. B.6.7(a), shows the detail in the variation of topsoil K from the intensive sampling. The areas of small concentrations are where the soil is more sandy and the largest concentrations are in a dry valley that extends from NW to SE across the field where the soil contains more clay and silt. The map based on the sample size of 94, Fig. B.6.7(b), which is close to Webster and Oliver's (1992) minimum recommended size for computing an accurate variogram, shows the main features of the variation in topsoil K, albeit with some loss of detail. From a management perspective this map would form a sound basis to manage applications of K in this field. This smaller sample size represents a saving of almost 60 percent in sampling effort. Figure B.6.7(c) is the block kriged map based on 47 data and the loss of detail is evident. It is clear that to reduce the sample size to this level would be unadvisable for managing K applications in this field.

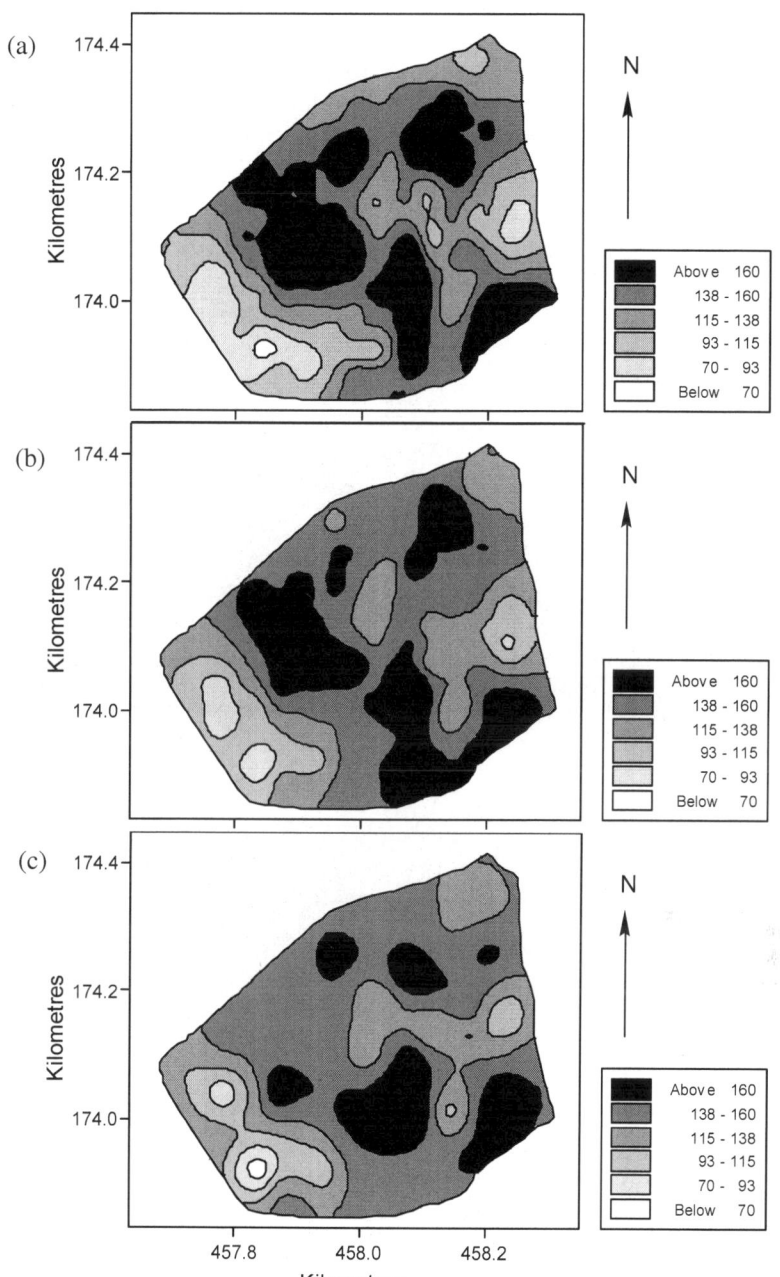

Fig. B.6.7. Maps of block kriged predictions of topsoil potassium at the Yattendon Estate for: (a) complete set of 230 data, (b) subset of 94 data, and (c) subset of 46 data

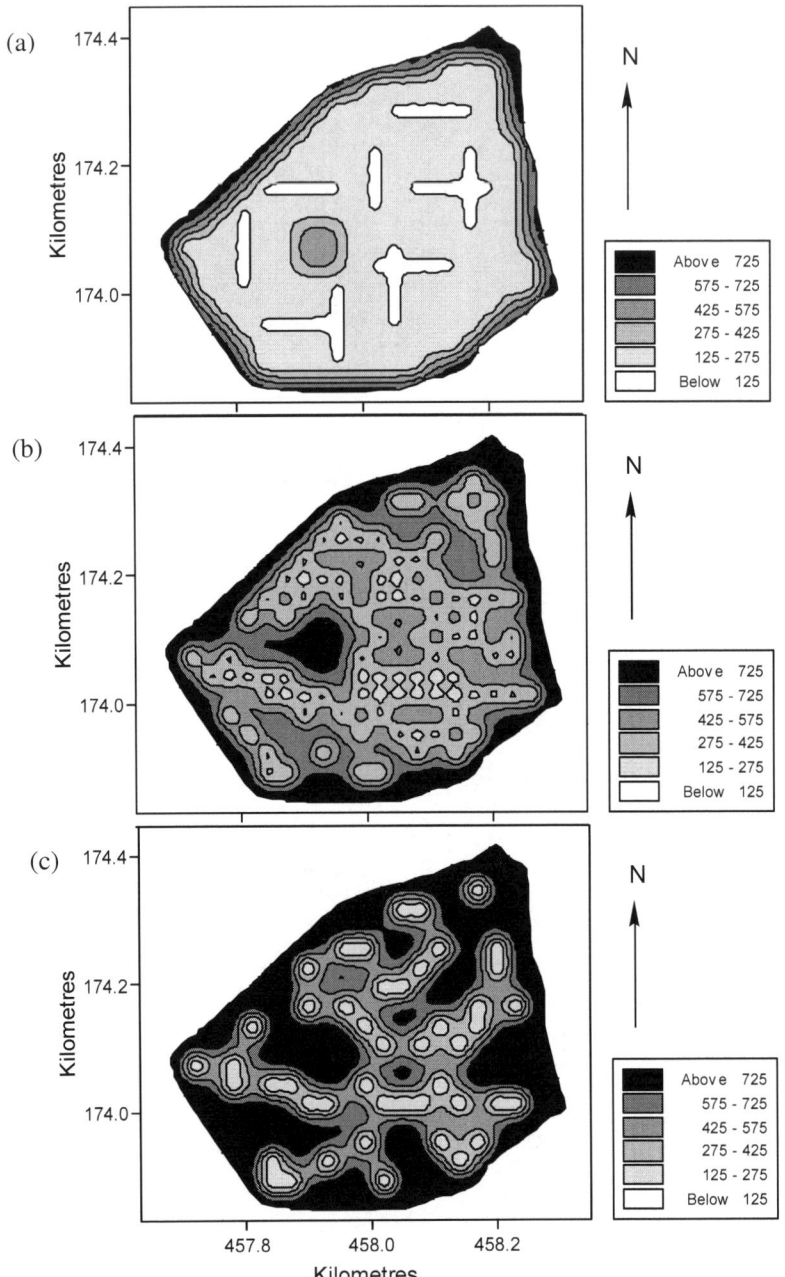

Fig. B.6.8. Maps of block kriged kriging variances for topsoil potassium at the Yattendon Estate for: (a) total of 230 data, (b) subset of 94 data, and (c) subset of 46 data

Figure B.6.8 (a), (b) and (c) shows the maps of block kriging variances for the three sizes of sample (230, 94 and 47, respectively); they show clearly how the variances of the predictions increase markedly with fewer data. The large kriging variances in the central part of the field in Fig. B.6.8(a) and (b) indicate an area with no sampling points where there is a copse. Figure B.6.8(a) shows that the smallest errors are along the short transects where the sampling was most intensive. Figure B.6.8(b) and (c) also shows that the kriging variances are smallest close to sampling points. The large variances around the field margins show the edge effects where there were fewer data from which to predict. These maps show that economizing on sampling to a sample size of 47 results in a loss of accuracy in the predictions that could have implications for subsequent management.

Figure B.6.9 shows the map of kriging variances from punctual kriging of the complete data set. Although the maps of estimates for punctual and block kriging were almost indistinguishable, the maps of kriging variance are quite different. The punctual kriging variances are much larger because the nugget variance sets a lower limit to the kriging variance. For block kriging the nugget variance disappears from the block kriging variance [see Eqs. (B.6.31) and (B.6.37)]. The larger is the proportion of nugget variance, the greater is the difference between the block and punctual kriging variances.

Kriging with an anisotropic model

The pH data from Broom's Barn Farm were used with the anisotropic exponential model for ordinary punctual kriging on a 10m × 10m grid. Figure B.6.10 shows the map of predictions. It is evident that there is more variation in pH from SSE to NNW than at right angles to this as the model in Table B.6.2 above describes.

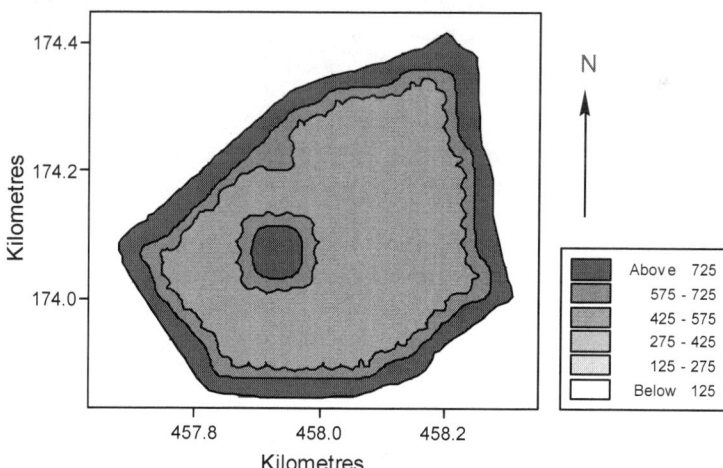

Fig. B.6.9. Map of punctually kriged kriging variances for topsoil potassium at the Yattendon Estate for the complete set of 230 data

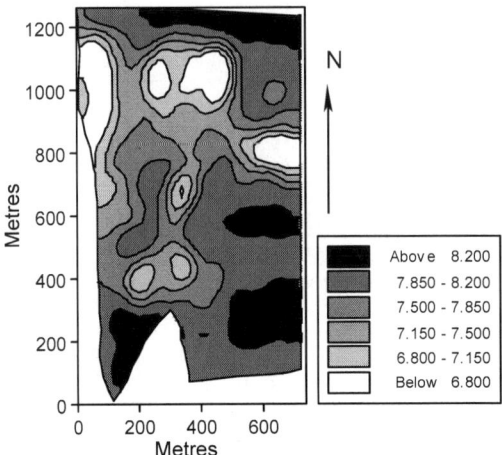

Fig. B.6.10. Map of punctually kriged predictions for topsoil pH at the Broom's Barn Farm

Nested variation: factorial kriging

The yield of winter wheat for 2001 from the Yattendon Estate is used to illustrate factorial kriging; its variogram (see Fig. B.6.6 and Table B.6.2) shows that there is more than one scale of variation present. Predictions were made at the nodes of a 5m × 5m grid as for topsoil K at Yattendon. The parameters of the double spherical model were used for ordinary kriging first; Fig. B.6.11(a) is the map of predictions. The pattern of variation appears complex because of the long- and short-range components of the variation. These components were then extracted separately and predicted by factorial kriging. Figure B.6.11(b) is the map of the long-range predictions. It is similar to that from ordinary kriging but it is less noisy because the short-range variation is no longer present. The regions of the field with large and small yields are clear in both maps. Many of the areas with large yield correspond to areas of large topsoil K concentrations [see Fig. B.6.11(a)]. The map of the short-range predictions, Fig. B.6.11(c) is quite different from the other two maps. It shows a much smaller scale of variation with a strong regular pattern.

This component of the variation appears to relate to the lines of management within the field in a NE–SW direction. The larger values are probably between the tramlines where the soil has suffered less compaction from machinery. There is some weak evidence of variation perpendicular to these lines that might reflect tramlines of previous operations. These management effects that have given rise to the short-range variation are not evident in the map of ordinary kriged predictions, Fig. B.6.11(a).

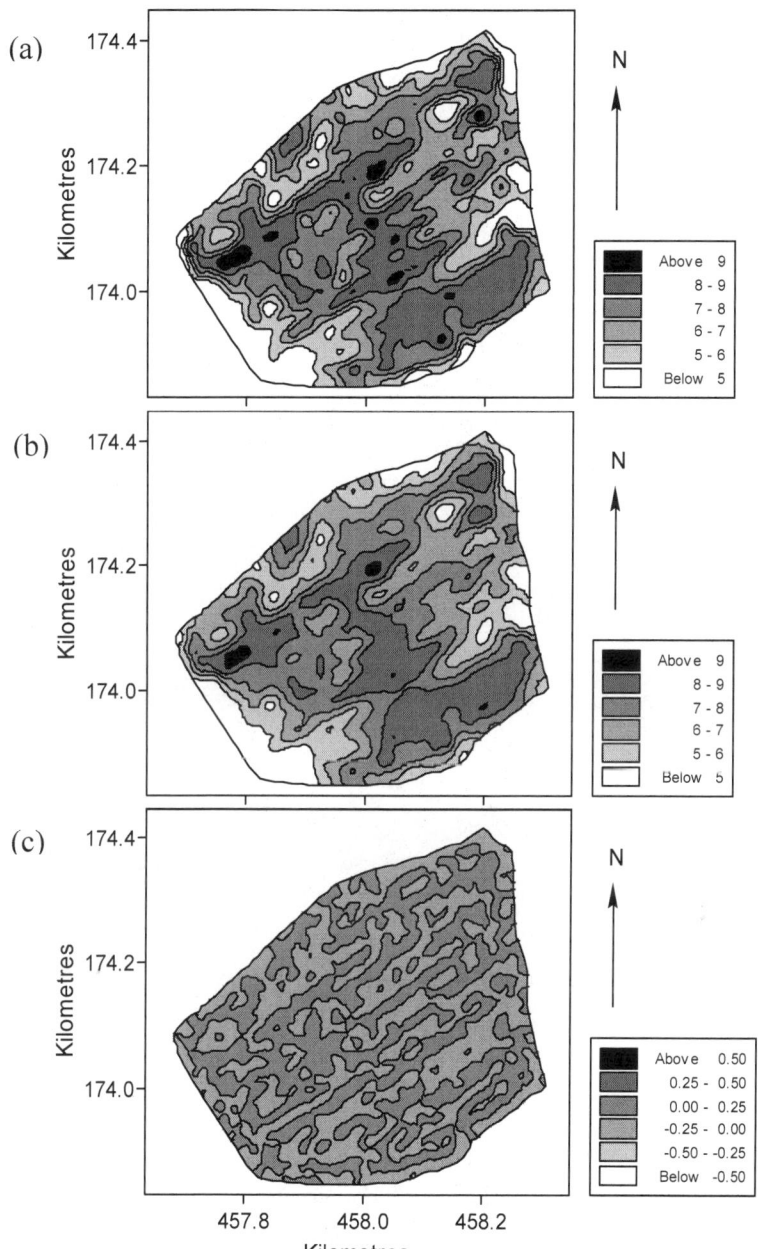

Fig. B.6.11. Maps of wheat yield for 2001 at the Yattendon Estate for: (a) ordinary kriged predictions, (b) predictions of the long-range component of the variation, and (c) predictions of the short-range component of the variation

For intensive data such as those from yield monitors, digital elevation models and satellites, factorial kriging is a valuable technique to explore the variation at different spatial scales. In this way it might be possible to gain some insight into the underlying processes that are responsible for variation at the different spatial scales.

Acknowledgements. The majority of the results in the case studies were from the author's project which was funded by the Home Grown Cereals Authority. We thank them for their support. We also thank Dr Z. L Frogbrook and Dr S. J Baxter for their work on this project. The data for Broom's Barn Farm were from an original survey of the Farm. We thank Dr J. D. Pidgeon for their use.

References

Cressie NAC (1993) Statistics for spatial data (revised edition). Wiley, New York, Chichester, Toronto and Brisbane

Frogbrook ZL, Oliver MA (2007) Identifying management zones in agricultural fields using spatially constrained classification of soil and ancillary data. Soil Use Mgmt 23(1):40-51

Galli A, Gerdil-Neuillet F, Dadou C (1984) Factorial kriging analysis: a substitute to spectral analysis of magnetic data. In Verly G, David M, Journel AG, Marechal A (eds) Geostatistics for natural resource characterization. Reidel, Dordrecht, pp.543-557

Gandin LS (1963) Objective analysis of meterological fields. Leningrad, Gidremeterologicheskoe Izdatel'stvo (GIMIZ) (translated by Israel Program for Scientific Translations, Jerusalem 1965)

Goovaerts P (1997) Geostatistics for natural resources evaluation. Oxford University Press, New York

Jaquet O (1989) Factorial kriging analysis applied to geological data from petroleum exploration. Math Geol 21(7):683-691

Journel AG (1983) Non-parametric estimation of spatial distributions. J Int Ass Math Geol 15(3):445-468

Journel AG, Huijbregts CJ (1978) Mining geostatistics. Academic Press, London

Kerry R, Oliver MA (2007) Comparing sampling needs for variograms of soil properties computed by the method of moments and residual maximum likelihood. Geoderma, Pedometrics 2005 140(4):383-396

Kitanidis PK (1983) Statistical estimation of polynomial generalized covariance functions and hydrological applications. Water Resources Research 19(4):909-921

Kitanidis PK, Lane RW (1985) Maximum likelihood parameter estimation of hydrologic spatial processes by the Gauss-Newton method. J Hydrol79(1-2):53-71

Kolmogorov AN (1939) Sur l'interpolation et l'extrapolation des suites stationnaires. C R Acad Sci 208:2043-2045

Kolmogorov AN (1941) The local structure of turbulence in an incompressible fluid at very large Reynolds numbers. Doklady Academii Nauk SSSR 30:301-305

Krige DG (1951) A statistical approach to some basic mine valuation problems on the Witwatersrand. J Chem Met Min Soc of South Africa 52(6):119-139

Krige DG (1966) Two-dimensional weighted moving average trend surfaces for ore-evaluation. J South African Inst Min Met 66(1):13-38

Lark RM, Cullis BR, Welham SJ (2006) On optimal prediction of soil properties in the presence of spatial trend: the empirical best linear unbiased predictor (E-BLUP) with REML. Europ J Soil Sci 57(6):787-799

Matérn B (1966) Spatial variation: Stochastic models and their applications to problems in forest surveys and other sampling investigations. Meddelanden från Statens Skogsforskningsinstitut 49(5):1-144

Matheron G (1963) Principles of geostatistics. Econ Geol 58:1246-1266

Matheron G (1965) Les variables régionalisées et leur estimation, une application de la theorie de fonctions aleatoires aux sciences de la nature. Masson et Cie, Paris

Matheron G (1969) Le krigeage universel. Cahiers du Centre de Morphologie Mathématique No.1, Ecole des Mines de Paris, Fontainebleau

Matheron G (1971) The theory of regionalized variables and its applications. Cahiers No 5 du Centre de Morphologie Mathematique, Fontainebleau, France

Matheron G (1973) The intrinsic random functions and their applications. Adv Applied Probab 5(3):439-468

Matheron G (1982) Pour une analyse krigreante de données regionalisées. Note N-732 du Centre de Géostatistique, Ecole des Mines de Paris, Fontainebleau, France

Mercer WB, Hall AD (1911) The experimental error of field trials. J Agricult Sci 4:107-132

Oliver MA, Carroll ZL (2004) Description of spatial variation in soil to optimize cereal management. Project Report 330 Home Grown Cereals Authority (HGCA), London

Oliver MA, Webster R, Slocum K (2000) Filtering SPOT imagery by kriging analysis. Int J Remote Sens 21(4):735-752

Omre H (1987) Bayesian kriging-merging observations and qualified guesses in kriging. Math Geol 19(1):25-39

Pardo-Igúzquiza E (1997) MLREML: a computer program for the inference of spatial covariance parameters by maximum likelihood and restricted maximum likelihood. Comp Geosci 23(2):153-162

Pardo-Igúzquiza E (1998) Inference of spatial indicator covariance parameters by maximum likelihood using MLREML. Comp Geosci.24(5):453-464

Payne R (ed) (2008) The guide to genstat release 10 - Part 2: statistics. VSN International, Hemel Hempstead

Patterson HD, Thompson R (1971) Recovery of interblock information when block sizes are unequal. Biometrika 58(3):545-554

Stein ML (1999) Interpolation of spatial data: some theory for kriging. Springer, Berlin, Heidelberg and New York

Sullivan J (1984) Conditional recovery estimation through probability kriging: theory and practice. In Verly G, David M, Journel AG, Marechal A (eds). Geostatistics for natural resource characterization. Reidel, Dordrecht, pp.365-384

Wackernagel H (2003) Multivariate geostatistics (3rd edition). Springer, Berlin, Heidelberg and New York

Webster R, Oliver MA (1992) Sample adequately to estimate variograms of soil properties. J Soil Sci 43(1):177- 192

Webster R, Oliver MA (2007) Geostatistics for environmental scientists (2nd edition). Wiley, New York, Chichester, Toronto and Brisbane

Wiener N (1949) Extrapolation, interpolation and smoothing of stationary time series. MIT Press, Cambridge [MA]

Wold H (1938) A study in the analysis of stationary time series. Almqvist and Wiksell, Uppsala

Youden WJ, Mehlich A (1937) Selection of efficient methods for soil sampling. Contributions of the Boyce Thompson Institute for Plant Research 9:59-70

Part C

Spatial Econometrics

C.1 Spatial Econometric Models

James P. LeSage and *R. Kelley Pace*

C.1.1 Introduction

Spatial regression models allow us to account for dependence among observations, which often arises when observations are collected from points or regions located in space. The spatial sample of observations being analyzed could come from a number of sources. Examples of point-level observations would be individual homes, firms, or schools. Regional observations could reflect average regional household income, total employment or population levels, tax rates, and so on. Regions often have widely varying spatial scales (for example, European Union regions, countries, or administrative regions such as postal zones or census tracts).

Each observation is linked to a location which in the case of point-level samples could be latitude-longitude coordinates. For region-level observations we can rely on latitude-longitude coordinates of a point located within the region, perhaps a centroid point.

It is commonly observed that sample data collected for regions or points in space are not *independent*, but rather *positively spatially dependent*, which means that observations from one location tend to exhibit values similar to those from nearby locations.

The data generating process (DGP) that produced the sample data determines the type of spatial dependence. Of course, we never truly know the DGP, so alternative approaches to applied modeling situations have been advocated. One approach is to rely on flexible model specifications that can accommodate a wide range of different possible data generating processes. For example, LeSage and Pace (2009) advocate use of the spatial Durbin model (SDM), since it nests a number of other models as special cases.

A second approach would be to rely on economic or other types of theory to motivate the DGP. For example, Ertur and Koch (2007) use a theoretical model that posits physical and human capital externalities as well as technological interdependence between regions. They show that this leads to a reduced form growth regression that should include an average of growth rates from neighboring regions as an explanatory variable in the model.

A third approach might be to rely on a purely econometric argument that favors use of particular models to protect against heterogeneity, omitted variables or other types of problems that arise in applied practice. For example, LeSage and Pace (2008) show that in the case of spatial interaction models of the type discussed in Chapter C.3, omitted variables or latent unobservable influences will lead to a model that includes a spatial lag of the dependent variable.

A fourth approach is to formally incorporate our uncertainty regarding the DGP into the estimation and inference procedure, which is illustrated in Chapter C.4. This involves drawing conclusions about the phenomena being modeled from a host of different model specifications, where each model is probabilistically weighted according to its consistency with the sample data evidence.

Conventional regression models commonly used to analyze cross-section and panel data assume that observations are independent of one another. In the case of spatial data samples where each observation represents a point or region located in space, this means that nearby regions are no more closely related than those more distant. A fundamental tenant of regional analysis is that regions located nearby tend to be more similar than those separated by great distances. This means that *positive spatial dependence* seems more plausible than *spatial independence* when analyzing regional data samples.

As an example, a conventional regression model that relates commuting times to work for region i to the number of persons in region i assumes that these commuting times are independent of those for persons located in a neighboring region j. Since it seems unlikely that regions i and j do not share parts of the road network, we would expect this assumption to be unrealistic. In addition to lack of realism, ignoring a violation of independence between observations can produce estimates that are biased and inconsistent. We pursue a demonstration of this in the sequel.

In our commuting time example, it may seem intuitively appealing to include an average of dependent variables observations from other nearby regions as a right-hand-side explanatory variable in the cross-sectional regression model. This could be formally implemented using a spatial indicator matrix that identifies neighboring observations in our sample. For example, in the case of regions located on a regular lattice we might specify that neighboring observations are the eight regions surrounding each region (ignoring the fact that regions on the edge have less than eight neighbors). This is sometimes referred to as Queen-based contiguity using an analogy to the board moves of the queen piece in the game of Chess. This would result in an extension of the regression model for observation i taking the form shown in Eq. (C.1.1), where the sample contains n observations.

$$y_i = \rho \sum_{j=1}^{n} W_{ij} y_j + \sum_{r=1}^{k} X_{ir} \beta_r + \varepsilon_i. \tag{C.1.1}$$

In Eq. (C.1.1), the dependent variable for observation i is y_i, the k explanatory variables are X_{ir}, $r = 1, \ldots, k$ with associated coefficients β_r, and the disturbance term is ε_i.[1] The n-by-n matrix W reflects the Queen's contiguity relations between the n regions and we use W_{ij} to denote the (i,j)th element. The matrix W is defined so that each element in row i of the matrix W contains values of zero for regions that are not neighbors to region i, and values of 1/8 for the eight contiguous neighbors to region i. By definition we do not allow region i to be a neighbor to itself, leading to the matrix W having zeros on the main diagonal. This leads to the product: $\Sigma_{j=1}^{n} W_{ij} y_i$ representing a scalar value equal to the average of values taken by the eight regions neighboring region i. The scalar ρ in model given by Eq. (C.1.1) is a parameter to be estimated that will determine the strength of the average (over all observations $i = 1, \ldots, n$) association between the dependent variable values for regions/observations and the average of those values for their neighbors.

There are of course numerous other ways to define the connectivity structure of the sample observations/regions embodied in the matrix W, details of which are beyond the scope of this chapter. In cases involving irregular lattices or point observations these become a consideration in specifying a spatial regression model. For example, one could use some fixed number of nearest neighbors for the case of irregular lattices, a number of neighbors selected using a distance cut-off or some other contiguity definition such as Rook-based contiguity in lieu of 'Queen-based' contiguity described above. There is also flexibility in the way that weights are assigned to neighboring regions/observations. For example, weighting schemes based on the length of shared borders separating regions have been proposed as well as weights exhibiting distance decay (LeSage and Pace 2009, Chapter 4). Conventional wisdom is that the specification of the matrix W exerts a great deal of influence on estimates and inferences regarding the parameters of these models. However, LeSage and Pace (2009) argue that this is an incorrect conclusion that has arisen from invalid interpretation of parameters from these models, a subject that we take up later.

It should be clear that if the parameter $\rho = 0$, we have a conventional regression model: $y_i = \Sigma_{r=1}^{k} X_{ir} \beta_r + \varepsilon_i$, so a point of interest would be the statistical significance of the coefficient estimate for ρ.

We can write the model in Eq. (C.1.1) using matrix/vector notation as shown in Eq. (C.1.2), where y is an n-by-1 vector containing the dependent variable observations, W is our n-by-n spatial weight matrix that identifies the connectivity or

---

[1] Without loss of generality, one of the variable vectors X_r could represent an intercept vector of ones.

neighbor structure of the sample observations, X is the n-by-k matrix of explanatory variables which may include an intercept term. The n-by-1 vector ε represents zero mean, constant variance, zero covariance, normally distributed disturbances, for example, $\varepsilon \sim \mathcal{N}(0, \sigma^2 I_n)$, where we use I_n to denote an n-by-n identity matrix. The scalar parameter ρ and the k-by-1 vector β along with the scalar variance parameter σ^2 represent model parameters to be estimated. The associated DGP for this model which we label SAR is shown in Eq. (C.1.3), and the expected value or prediction from this model is shown in Eq. (C.1.4).

$$y = \rho W y + X\beta + \varepsilon \qquad (C.1.2)$$

$$y = (I_n - \rho W)^{-1} X\beta + (I_n - \rho W)^{-1} \varepsilon \qquad (C.1.3)$$

$$E(y) = (I_n - \rho W)^{-1} X\beta \qquad (C.1.4)$$

$$\varepsilon \sim \mathcal{N}(0, \sigma^2 I_n). \qquad (C.1.5)$$

The expectation follows from the assumption that elements of the matrix W are fixed/non-stochastic as are observations in the matrix X. This results in $E\,[I_n - \rho W)^{-1}\,\varepsilon] = (I_n - \rho W)^{-1}\,E\,[\varepsilon] = 0$.

There are of course other ways we could envision spatial dependence arising as part of the DGP and these lead to other extensions of the conventional regression model. For example, it may be the case that dependence arises only in the disturbance process leading to the model in Eq. (C.1.6) (which we label SEM), associated DGP in Eq. (C.1.7), and expectation in Eq. (C.1.8).

$$y = X\beta + u \qquad (C.1.6a)$$

$$u = \rho W u + \varepsilon \qquad (C.1.6b)$$

$$y = X\beta + (I_n - \rho W)^{-1} \varepsilon \qquad (C.1.7)$$

$$E(y) = X\beta \tag{C.1.8}$$

$$\varepsilon \sim \mathcal{N}(0,\, \sigma^2 I_n). \tag{C.1.9}$$

Another elaboration of the basic model is one we label SDM shown in Eq. (C.1.10) with associated DGP in Eq. (C.1.11) and expectation in Eq. (C.1.12). In setting forth the SDM model we need to separate out the intercept term from the explanatory variables matrix X because $W\iota_n = \iota_n$, where the n-by-1 intercept vector of ones is denoted by ι_n. This model includes spatial lags of the dependent variable is denoted by the matrix Wy, and spatial lags of the explanatory variables denoted by the matrix product WX in addition to the conventional explanatory variables X. The matrix product WX creates an average of explanatory variable values from neighboring regions which are added to the set of explanatory variables.

$$y = \rho W y + \alpha \iota_n + X\beta + W X \theta + \varepsilon \tag{C.1.10}$$

$$y = (I_n - \rho W)^{-1} (\alpha \iota_n + X\beta + W X \theta + \varepsilon) \tag{C.1.11}$$

$$E(y) = (I_n - \rho W)^{-1} (\alpha \iota_n + X\beta + W X \theta) \tag{C.1.12}$$

$$\varepsilon \sim \mathcal{N}(\sigma^2 I_n). \tag{C.1.13}$$

There are also models based on moving average spatial error processes, $u = (I_n - \rho W)\varepsilon$ rather than the autoregressive spatial error process, $u = (I_n - \rho W)^{-1}\varepsilon$ which we have described here (see LeSage and Pace 2009).

An important point to note is that the SEM model has an expectation equal to that from a conventional regression model where independence between the dependent variable observations is part of the maintained hypothesis. In large samples, point estimates for the parameters β from the SEM model and conventional regression will be the same, but in small samples there may be an efficiency gain from correctly modeling spatial dependence in the disturbance process. In contrast, the SAR and SDM models which are sometimes referred to as spatial lag models (because they contain terms Wy on the right-hand-side) produce expectations that differ from those of the conventional regression model. Use of least-squares regression methods to estimate the parameters of these models will result in biased and inconsistent estimates for the parameters β as well as ρ.

C.1.2 Estimation of spatial lag models

From the DGP associated with the SAR model, it should be clear that there is a Jacobian term involved in the transformation from ε to **y**. The log-likelihood function for the SAR model takes the form in Eqs. (C.1.14) – (C.1.16) (see Ord 1975), where ω is an *n*-by-1 vector containing eigenvalues of the matrix **W**. If ω contains only real eigenvalues, a positive definite variance-covariance matrix is ensured by conditions relating to the minimum and maximum eigenvalues of the matrix **W**. LeSage and Pace (2009, Chapter 4) provide a discussion of situations involving complex eigenvalues that can arise for certain types of spatial weight matrices **W**. Lee (2004) shows that maximum likelihood estimates are consistent for these models.

$$\ln L = -\frac{n}{2}\ln(\pi\sigma^2) + \ln|I_n - \rho W| - \frac{e^T e}{2\sigma^2} \qquad (C.1.14)$$

$$e = y - \rho W\, y - X\beta \qquad (C.1.15)$$

$$\rho \in \left[\min(\omega)^{-1}, \max(\omega)^{-1}\right]. \qquad (C.1.16)$$

A simple manipulation of the SAR model shown in Eq. (C.1.2): $y - \rho W y = X\beta + \varepsilon$ suggests that the log-likelihood in Eq. (C.1.14) can be concentrated with respect to the parameters β and σ^2. This is accomplished using: $\beta = (X^T X)^{-1} X^T (I_n - \rho W) y$ to replace this parameter vector in the full likelihood function. We also replace the parameter σ^2 with $e^T e = (y - \rho W y - X\beta)^T (y - \rho W y - X\beta) n^{-1}$, where β is as defined above. Concentrating the full likelihood in this fashion results in a univariate optimization problem over the parameter ρ. Since the parameter ρ has a well-defined range based on the eigenvalues of the matrix **W**, this is a well-defined optimization problem. Given a maximum likelihood estimate for ρ, which we label ρ^*, we can use this estimate to recover maximum likelihood estimates for the parameters $\beta^* = (X^T X)^{-1} X^T (I_n - \rho^* W) y$, and $\hat{\sigma}^2 = e^T e = (y - \rho^* W y - X\beta^*)^T (y - \rho^* W y - X\beta^*) n^{-1}$.

Of course, similar likelihood functions exist for other spatial regression models such as the SEM, SDM and moving average processes. See LeSage and Pace (2009) for details regarding these and computationally efficient approaches to optimization. The most computationally challenging part of solving for maximum likelihood estimates using the concentrated log-likelihood function is evaluating the log-determinant for the *n*-by-*n* matrix: $\ln|I_n - \rho W|$, since the number of observations *n* can be large in spatial samples. There has been a great deal of research on computationally efficient ways to calculate this term. As a brief overview of the alternative approaches we note that Pace and Barry (1997) discuss use of sparse LU and Cholesky algorithms and set forth a vector expression for the concentrated log-likelihood as a gridded function of values taken by the parameter ρ involved in the univariate optimization problem. Barry and Pace (1999) describe an approach to producing a statistical estimate of this term along with confidence intervals for the estimate. There has been a great deal of literature on approximation approaches (see Pace and LeSage 2003, 2009b; Smirnov and Anselin 2009). In cases involving regular lattices and a repeating pattern of connectivity relations (a regular locational grid such as arises in satellite remote sensing) between the spatial units of observation, analytical formulas can be used to calculate the determinant (LeSage and Pace 2009).

An alternative to tackling what have been perceived as computational difficulties associated with maximum likelihood estimation is to rely on an estimation method that is not likelihood-based. Examples include the instrumental variables approach of Anselin (1988, pp.81-90), the instrumental variables/generalized moments estimator from Kelejian and Prucha (1998, 1999), or the maximum entropy method of Marsh and Mittelhammer (2004). These alternative methods suffer from a number of drawbacks. One is that they can produce dependence parameter estimates (ρ in our discussion) that fall outside the interval defined by the eigenvalue bounds arising from the matrix W. In addition, inferential procedures for these methods can be sensitive to implementation issues such as the interaction between the choice of instruments and model specification, which are not always obvious to the practitioner.

There are alternative model specifications such as the matrix exponential spatial specification introduced by LeSage and Pace (2007) which they label MESS that can be estimated using maximum likelihood or Bayesian methods. This spatial regression model specification can be used in situations where the model DGP is that of the SAR or SDM to produce equivalent estimates and inferences. The MESS model eliminates the troublesome determinant term from the likelihood function, allowing rapid maximum likelihood and Bayesian estimation of these models for large spatial samples. LeSage and Pace (2007) provide a closed-form solution for estimates of this model. It is also possible to produce a closed-form solution for maximum likelihood estimates of the SAR, SDM and SEM models discussed here, a recent innovation introduced by LeSage and Pace (2009). These approaches greatly reduce the motivation for reliance on non likelihood-based methods which have been traditionally advocated as a work-around for the perceived computational difficulties of maximum likelihood estimation. These difficulties have been largely resolved with the recent advances described in LeSage and Pace (2009).

The bias of least-squares

As noted, one focus of inference is the magnitude and significance of the parameter ρ, since this distinguishes the SAR model from conventional regression and provides information regarding the strength of spatial dependence between dependent variable observations.

To contrast the maximum likelihood estimate ρ^* to that from least-squares which we label $\hat{\rho}$, consider the matrix expressions in Eq. (C.1.17).

$$y = (Wy\ X)\begin{pmatrix} \rho \\ \beta \end{pmatrix} + \varepsilon \tag{C.1.17}$$

$$\begin{pmatrix} \hat{\rho} \\ \hat{\beta} \end{pmatrix} = \left[\begin{pmatrix} y^T W^T \\ X^T \end{pmatrix} (Wy\ X) \right]^{-1} \begin{pmatrix} y^T W^T \\ X^T \end{pmatrix} y = \begin{bmatrix} y W^T W y & y^T W^T X \\ X^T W y & X^T X \end{bmatrix}^{-1} \begin{pmatrix} y^T X^T y \\ X^T y \end{pmatrix}. \tag{C.1.18}$$

If we assume zero covariance (or orthogonality) between Wy and X, the inverse matrix in Eq. (C.1.18) becomes diagonal having a simple analytical inverse, leading to: $\hat{\rho} = (y^T W^T W y)^{-1} y^T W^T y$. Of course, for the case of non-zero covariance between Wy and X we could rely on a partitioned matrix inverse formulation to produce a similar, but more complicated result than the one we present here.

We can show that the least-squares estimate for the parameter ρ in this simple case of zero covariance is biased and inconsistent. This involved considering whether the definition of consistency: $\text{plim}(\hat{\rho}) = \rho$, holds true.

$$\hat{\rho} = (y^T W^T W y)^{-1} y^T W^T y = (y^T W^T W y)^{-1} y^T W^T (\rho W y + X \beta + \varepsilon)$$

$$= \rho + (y^T W^T W y)^{-1} y^T W^T X \beta + (y^T W^T W y)^{-1} y^T W^T \varepsilon$$

$$= \rho + (y^T W^T W y)^{-1} y^T W^T \varepsilon \tag{C.1.19}$$

where the last equation follows from zero covariance, $y^T W^T X = 0$. Now consider the probability limit (plim) of the expression: $\text{plim}\ (y^T W^T W y)^{-1} y^T W^T \varepsilon$.

The term: $Q = \text{plim}\ (1/n)(y^T W^T W y)^{-1}$ could obtain the status of a finite non-singular matrix with reasonable restrictions/assumptions made in typical applications. Specifically, we must view W as non-stochastic sample data information and assume that as the sample size increases the number of non-zero elements in each row of the matrix W has a finite limit. In addition, the parameter ρ must obey the eigenvalue bounds to ensure bounded y.

We turn attention to the term: $R = \text{plim}\,(1\,/\,n)\,\mathbf{y}^\text{T}\,\mathbf{W}^\text{T}\,\boldsymbol{\varepsilon}$. Using the model DGP: $\mathbf{y} = (\mathbf{I}_n - \rho\mathbf{W})^{-1}(\mathbf{X}\boldsymbol{\beta} + \boldsymbol{\varepsilon})$, we find

$$R = \text{plim}\,(1\,/\,n)\,\mathbf{y}^\text{T}\mathbf{W}^\text{T}\,\boldsymbol{\varepsilon} \qquad (\text{C.1.20})$$

$$R = \text{plim}(1/n)[(\mathbf{I}_n - \rho\mathbf{W})^{-1}(\mathbf{X}\boldsymbol{\beta} + \boldsymbol{\varepsilon})]^\text{T}\,\mathbf{W}^\text{T}\,\boldsymbol{\varepsilon} \qquad (\text{C.1.21})$$

$$R = \text{plim}\,(1\,/\,n)\,\boldsymbol{\varepsilon}^\text{T}\,(\mathbf{I}_n - \rho\,\mathbf{W}^\text{T})^{-1}\,\mathbf{W}^\text{T}\,\boldsymbol{\varepsilon} \qquad (\text{C.1.22})$$

$$\hat{\rho} = \rho + R. \qquad (\text{C.1.23})$$

It should be clear that the plim (the probability limit operator) of the quadratic form in the disturbances shown in Eq. (C.1.22), will not equal zero except in the trivial case where $\rho = 0$, or if the matrix W is strictly triangular. As noted, under the simplifying assumption that $\mathbf{Wy}$ and X are uncorrelated, the matrix inverse in Eq. (C.1.18) becomes diagonal having a simple analytical inverse, leading to: $\hat{\boldsymbol{\beta}} = (\mathbf{X}^\text{T}\mathbf{X})^{-1}\mathbf{X}^\text{T}\mathbf{y}$. It should be clear that a similar proof of inconsistency could be constructed for the least-squares estimate of this parameter vector. As already noted the maximum likelihood estimate should equal: $\boldsymbol{\beta}^* (\mathbf{X}^\text{T}\mathbf{X})^{-1}\mathbf{X}^\text{T} (\mathbf{I}_n - \rho^*\mathbf{W})\mathbf{y}$, which requires an unbiased estimate for ρ.

Pace and LeSage (2009a) discuss the biases of OLS when applied to spatially dependent data in more detail. In a richer setting spatial dependence in the explanatory variables as well as in the disturbances can further amplify the bias discussed here.

Bayesian estimation

An alternative to maximum likelihood estimation is Bayesian Markov Chain Monte Carlo (MCMC) estimation set forth in LeSage (1997) for the SAR model.[2] MCMC is based on the idea that a large sample from the Bayesian posterior distribution of our parameters can be used in place of an analytical Bayesian solution where this is difficult or impossible. We designate the posterior distribution using

---

[2] For an introduction to Bayesian methods in econometrics see Koop (2003).

$p(\boldsymbol{\theta} \mid D)$, where $\boldsymbol{\theta}$ represents the parameters ρ, $\boldsymbol{\beta}$, σ^2 and D the sample data. If the sample from $p(\boldsymbol{\theta}|D)$ were large enough, we could approximate the form of the posterior density using kernel density estimators or histograms, eliminating the need to know the precise analytical form of this complicated density. Simple statistics could also be used to construct means and variances based on the sampled values taken from the posterior.

The parameters $\boldsymbol{\beta}$ and σ^2 in the SAR model can be estimated by drawing sequentially from the conditional distributions of these two sets of parameters, a process known as Gibbs sampling because of its origins in image analysis, (Geman and Geman 1984). The conditional distributions for these sets of parameters take the form of a multivariate normal distribution (for $\boldsymbol{\beta}$) and inverse Gamma distribution (for σ^2). Gibbs sampling has also been labeled alternating conditional sampling, which seems a more accurate description of the procedure.

To illustrate how this works, assume for simplicity that we knew the true value for the parameter ρ. As already motivated in our discussion of concentrating the likelihood function, the parameter vector $\boldsymbol{\beta}$ can be expressed as: $\boldsymbol{\beta} = (X^T X)^{-1} X^T (I_n - \rho W) y$, which is the mean of the normal conditional posterior distribution $\boldsymbol{\beta} \sim \mathcal{N}[(X^T X)^{-1} X^T (I_n - \rho W) y, \sigma^2 (X^T X)^{-1}]$. We can use this mean expression in conjunction with the associated variance-covariance matrix: $\sigma^2 (X^T X)^{-1}$, to construct a multivariate normal draw for the k-by-1 parameter vector $\boldsymbol{\beta}$. We note that being able to condition on the parameter σ^2 (that is assume it is known) is what makes this calculation and multivariate normal draw simple. Similarly, the conditional posterior distribution for the parameter σ^2 takes the form of an inverse Gamma distribution that we denote $IG(a,b)$ with $a = n/2$, and $b = [(I_n - \rho W) y - X\boldsymbol{\beta}]^T [(I_n - \rho W) y - X\boldsymbol{\beta}] / 2$. Again, the fact that we can treat the parameter vector $\boldsymbol{\beta}$ as known makes the calculations required to produce this draw simple.

On each pass through the sequence of sampling from the two conditional distributions for $\boldsymbol{\beta}$, σ^2, we collect the parameter draws which are used to construct a joint posterior distribution for these model parameters. (We are ignoring the parameter ρ here, assuming it is known.) Gelfand and Smith (1990) demonstrate that sampling from the complete sequence of conditional distributions for all parameters in the model produces a set of estimates that converge in the limit to the true (joint) posterior distribution of the parameters. That is, despite the use of conditional distributions in our sampling scheme, a large sample of the draws can be used to produce valid posterior inferences regarding the joint posterior mean and moments of the parameters.

For the case of the SAR, SEM and SDM models, the conditional distribution for the spatial dependence parameter ρ does not take the form of a known distribution. However, LeSage (1997) describes an approach for sampling from the conditional distribution of this parameter using what has been labeled Metropolis-Hastings sampling, (Metropolis et al. 1953; Hastings 1970). This allows us to estimate spatial regression models using MCMC sampling which involves producing samples from the complete sequence of conditional distributions for the model parameters $\boldsymbol{\beta}$, σ^2 and ρ.

C.1.3 Estimates of parameter dispersion and inference

In addition to maximum likelihood or Bayesian estimates for the parameters ρ, β and σ^2, we are often interested in inference regarding these. Bayesian MCMC estimation leads to large samples of draws for the model parameters that can be used to construct measures of dispersion used in Bayesian inference. Maximum likelihood inference usually employs *likelihood ratio* (LR), *Lagrange multiplier* (LM), or *Wald* (W) tests. These are equivalent asymptotically, but can differ in small samples. The choice between these methods often comes down to computational convenience or personal preference.

Pace and Barry (1997) propose likelihood ratio tests for hypotheses such as the deletion of a single explanatory variable that exploit the computational advantages of being able to rapidly evaluate the likelihood. Pace and LeSage (2003) discuss use of *signed root deviance* statistics which can be used to transform likelihood ratio tests for single variable deletion to a form similar to *t*-tests.[3] The signed root deviance is the square root of the deviance statistic with a sign matching the sign of the coefficient estimates β (Chen and Jennrich 1996). These statistics behave similar to *t*-ratios for large samples, and can be used like a *t*-statistics for hypothesis testing.

Wald inference employs either an analytical or numerical version of the Hessian or the related information matrix to produce a variance-covariance matrix for the estimated parameters. This can be used to construct conventional regression *t*-statistics. An implementation issue is that constructing the analytical Hessian (or information matrix) involves computing the trace of a dense n-by-n matrix inverse $(I_n - \rho W)^{-1}$. LeSage and Pace (2009) provide a number of alternative ways to rapidly approximate elements of the Hessian.

From a computational speed perspective the vector expressions from Pace and Barry (1997) for rapidly evaluating the log-likelihood function makes a purely numerical Hessian feasible for these models. However, there are some drawbacks to implementing this approach in software for general use, since practitioners often work with poorly scaled and multicollinear sample data. Such data can greatly degrade the accuracy of numerical estimates of the derivatives populating the Hessian. A second point is that univariate optimization takes place using the likelihood concentrated with respect to the parameters β and σ^2, so a numerical approximation to the full Hessian from the maximum likelihood estimation procedure requires additional work. LeSage and Pace (2009) show how a single computationally difficult term within the analytical Hessian can be replaced with a numerical approximation. This allows the remaining analytical terms to be employed, increasing the accuracy and overcoming scaling problems.

---

[3] Deviance is minus twice the log-likelihood ratio.

C.1.4 Interpreting parameter estimates

Simultaneous feedback is a feature of the spatial regression model that comes from dependence relations embodied in spatial lag terms such as Wy. These lead to feedback effects from changes in explanatory variables in a region that neighbors i, say region j, that will impact the dependent variable for observation/region i. This can of course be a valuable feature of these models if we are interested in quantifying spatial spillover effects associated with the phenomena we are attempting to model.

To see how these feedback effects work, consider the data generating process associated with the SAR model, shown in Eq. (C.1.24), to which we have applied the well-known infinite series expansion in Eq. (C.1.25) to express the inverse.

$$y = (I_n - \rho W)^{-1} X\beta + (I_n - \rho W)^{-1} \varepsilon \qquad (C.1.24)$$

$$(I_n - \rho W)^{-1} = I_n + \rho W + \rho^2 W^2 + \rho^3 W^3 + \ldots \qquad (C.1.25)$$

$$y = X\beta + \rho WX\beta + \rho^2 W^2 X\beta + \ldots + \varepsilon + \rho W\varepsilon + \rho^2 W^2 \varepsilon + \rho^3 W^3 \varepsilon + \ldots \qquad (C.1.26)$$

The model statement in Eq. (C.1.26) can be interpreted as indicating that the expected value of each observation y_i will depend on the mean value plus a linear combination of values taken by neighboring observations scaled by the dependence parameter $\rho, \rho^2, \rho^3, \ldots$

Consider powers of the row-stochastic spatial weight matrices $W^2, W^3, \ldots$ that appear in Eq. (C.1.26), where we assume that rows of the weight matrix W are constructed to represent *first-order* contiguous neighbors. The matrix W^2 will reflect *second-order* contiguous neighbors, those that are neighbors to the first-order neighbors. Since the neighbor of the neighbor (second-order neighbor) to an observation i includes observation i itself, W^2 has positive elements on the diagonal. That is, higher-order spatial lags can lead to a connectivity relation for an observation i such that $W^2 X\beta$ and $W^2 \varepsilon$ will extract observations from the vectors $X\beta$ and ε that point back to the observation i itself. This is in stark contrast with the conventional independence relation in ordinary least-squares regression where the Gauss-Markov assumptions rule out dependence of ε_i on other observations j, by assuming zero covariance between observations i and j in the data generating process.

Steady-state equilibrium interpretation

One might suppose that feedback effects would take time, but there is no explicit role for passage of time in our cross-sectional model. Instead, we view the cross-sectional sample of regions as the result of an equilibrium outcome or steady state of the regional process we are modeling. To elaborate on this point, consider a relationship where *y* represents regional income at time *t*, denoted by y_t, and this depends on current period own-region characteristics X_t such as labor, human and physical capital and associated parameters β, plus observed income levels of neighboring regions from the past period, $t-1$. This type of *space-time dependence* could be represented by a *space-time lag* variable Wy_{t-1}, leading to the model in Eq. (C.1.27). It seems reasonable to assume that regional characteristics such as labor, human and physical capital change slowly over time, so we make the simplifying assumption that these do not change over time, that is we set $X_t = X$ in Eq. (C.1.27).[4]

$$y_t = \rho W y_{t-1} + X\beta + \varepsilon_t. \qquad (C.1.27)$$

Note that we can replace y_{t-1} on the right-hand-side of Eq. (C.1.27) with $y_{t-1} = \rho W y_{t-2} + X\beta + \varepsilon_{t-1}$, and continue this type of recursive substitution and in the limit with large *t* and *q* produce (LeSage and Pace 2009):

$$\lim_{q \to t} E(y_t) = \lim_{q \to t} E\left[\left(I_n + \rho W + \rho^2 W^2 + \ldots + \rho^{q-1} W^{q-1}\right) X\beta + \rho^q W^q y_{t-q} + u\right]$$

$$= (I_n - \rho W)^{-1} X\beta. \qquad (C.1.28)$$

We conclude from this that the long-run expectation of the model in Eq. (C.1.27), can be interpreted as having a steady-state equilibrium that takes a form consistent with the data generating process for our cross-sectional SAR model. In other words, simultaneous feedback is a feature of the equilibrium steady-state for spatial regression models that include spatial lags of the dependent variable. In the context of our static cross-sectional SAR model where we treat the observed sample as reflecting a steady state equilibrium outcome, these feedback effects appear as instantaneous, but they should be interpreted as showing a movement to the next steady state.

---

[4] LeSage and Pace (2009) show that one can produce a similar result to that presented here if the explanatory variables X_t evolve over time in a number of ways.

Interpreting the parameters β

LeSage and Pace (2009) point out that interpretation of the parameter vector β in the SAR model is different from a conventional least squares interpretation. In least-squares the rth parameter, β_r, from the vector β, is interpreted as representing the partial derivative of y with respect to a change in the rth explanatory variable from the matrix X, which we write as X_r. In standard least-squares regression where the dependent variable vector contains *independent* observations, changes in observation i of the rth variable which we denote X_{ir} only influence observation y_i, whereas the SAR model allows this type of change to influence y_i as well as other observations y_j, where $j \neq i$. This type of impact arises due to the interdependence or connectivity between observations in the SAR model.

To see how this works, consider the SAR model expressed as shown in Eq. (C.1.29).

$$(I_n - \rho W)y = X\beta + \varepsilon \tag{C.1.29}$$

$$y = \sum_{r=1}^{k} S_r(W) X_r + V(W) \varepsilon \tag{C.1.30}$$

$$S_r(W) = V(W)(I_n \beta_r) \tag{C.1.31}$$

$$V(W) = (I_n - \rho W)^{-1} = I_n + \rho W + \rho^2 W^2 + \rho^3 W^3 + \ldots \tag{C.1.32}$$

To illustrate the role of $S_r(W)$, consider the expansion of the data generating process in Eq. (C.1.30) as shown in Eq. (C.1.33).

$$\begin{pmatrix} y_1 \\ y_2 \\ \vdots \\ y_n \end{pmatrix} = \sum_{r=1}^{k} \begin{pmatrix} S_r(W)_{11} & S_r(W)_{12} & \cdots & S_r(W)_{1n} \\ S_r(W)_{21} & S_r(W)_{22} & & \\ \vdots & \vdots & \ddots & \\ S_r(W)_{n1} & S_r(W)_{n2} & \cdots & S_r(W)_{nn} \end{pmatrix} \begin{pmatrix} X_{1r} \\ X_{2r} \\ \vdots \\ X_{nr} \end{pmatrix} + V(W) \iota_n \alpha + V(W)\varepsilon. \tag{C.1.33}$$

To make the role of $S_r(W)$ clear, consider the determination of a single dependent variable observation y_i shown in Eq. (C.1.34).

$$y_i = \sum_{r=1}^{k}\left[(S_r(W))_{i1} X_{1r} + S_r(W)_{i2} X_{2r} + \ldots + S_r(W)_{in} X_{nr}\right] + V(W)_i \ln\alpha + V(W)_i \varepsilon.$$
(C.1.34)

It follows from Eq. (C.1.34) that the derivative of y_i with respect to X_{jr} takes the form shown in Eq. (C.1.35), where we use $S_r(W)_{ij}$ to represent the (i,j)th element from the matrix $S_r(W)$.

$$\frac{\partial y_i}{\partial X_{jr}} = S_r(W)_{ij}.$$
(C.1.35)

In contrast to the least-squares case, the derivative of y_i with respect to X_{ir} usually does not equal β_r, and the derivative of y_i with respect to X_{jr} for $j \neq i$ usually does not equal zero. Therefore, any change to an explanatory variable in a single region (observation) can affect the dependent variable in other regions (observations). This is of course a logical consequence of our simultaneous spatial dependence model. A change in the characteristics of neighboring regions can set in motion changes in the dependent variable that will impact the dependent variable in neighboring regions. These impacts will continue to diffuse through the system of regions.

Since the partial derivative impacts now take the form of a matrix, LeSage and Pace (2009) propose scalar summary measures for these impacts. These cumulate the impacts across all observations that arise from changes in all observations of the explanatory variables and then construct an average impact to simplify interpretation.

The scalar summary measures of impact are based on the idea that the *own derivative* for the *i*th region takes the form in Eq. (C.1.36), representing the *i*th diagonal element of the matrix $S_r(W)$, which we denote $S_r(W)_{ii}$.

$$\frac{\partial y_i}{\partial X_{ir}} = S_r(W)_{ii}$$
(C.1.36)

Of course, the cross-derivative would take the form shown in Eq. (C.1.35) for $i \neq j$, so we can construct scalars by averaging over elements of the matrix $S_r(W)$. Averaging over the main diagonal elements of the matrix produces a scalar summary that reflects own-derivatives while averaging over off-diagonal elements reflect cross-derivatives. The total impact arising from a change in explanatory variable X_r is reflected by all elements of the matrix $S_r(W)$. This can be decomposed into direct and indirect or spatial spillover impacts that sum to a total impact arising from a change (on average across all observations) in the variable X_r.

Formally, the LeSage and Pace (2009) definitions for the scalar summary measures of impact are

(a) *Average Direct Impact.* The impact of changes in the ith observation of X_r – which we denote X_{ir} – on y_i could be summarized by measuring the average of main diagonal elements $S_r(W)_{ii}$, from the matrix $S_r(W)$.
(b) *Average Total Impact.* The sum across the ith row of $S_r(W)$ represents the total impact on individual observation y_i resulting from changing the rth explanatory variable by the same amount across all n observations (for example, $X_r + \delta \iota_n$ where δ is the scalar change). On the other hand, the sum across the ith column reflects the total impact on all y_i arising from changing the rth explanatory variable by an amount in the jth observation (for example, $X_{jr} + \delta$). Averaging either the sum of the row or column sums will produce the same number, which represents the total impact.
(c) *Average Indirect Impact.* This is by definition the difference between the total and direct impacts. This summary impact measure reflects what are commonly thought of as spatial spillovers, or impacts falling on regions other than the own-region.

LeSage and Pace (2009) point to an interpretative distinction between the average total impact summary measure that arises from averaging row-sums versus that from averaging columns-sums. Despite the equality of these two scalar summaries, the average of row-sums could be viewed as reflecting the (average) *Total Impact to an Observation*, whereas the average column-sums are more appropriately interpreted as the (average) *Total Impact from an Observation*.

To elaborate on the distinction between these two interpretative viewpoints, consider a modeling situation where interest centers on how a financial crisis in a single country/observation spills over to produce contagion in financial markets of other countries (Kelejian et al. 2006). This situation can be viewed as a change in the jth observation/country (for example, $X_{jr} + \delta$) impact on all countries y_i, $i = 1$, ..., n, or the (average) *Total Impact from an Observation*.

In contrast, if interest centers on how a rise in human capital levels across all regions by some amount will (on average) influence a single region's growth rate, then we are working with the (average) *Total Impact to an Observation* interpretative viewpoint (Dall'erba and LeGallo 2007).

It is easy to see that the numerical values of the summary measures for the two forms of average total impacts set forth above are equal, since the average of the column of row-sums $c_r = S_r(W)\iota_n$, equal to $n^{-1} \iota_n^T c_r = n^{-1} \iota_n^T S_r(W) \iota_n$. On the other hand, the average of the row of column-sums $r_r = \boldsymbol{\iota}_n^T S_r(W)$, equals $n^{-1} r_r \boldsymbol{\iota}_n$ which is also equal to $n^{-1} \boldsymbol{\iota}_n^T S_r(W) \boldsymbol{\iota}_n$.

The summary measure of total impacts, $n^{-1} \boldsymbol{\iota}_n^T S_r(W)\boldsymbol{\iota}_n$, for the SAR model take the simple form in Eq. (C.1.37) for a model that relies on a row-stochastic W matrix (where the row-sums equal one).

$$n^{-1} \boldsymbol{\iota}_n^T S_r(W) \boldsymbol{\iota}_n = n^{-1} \boldsymbol{\iota}_n^T (I_n - \rho W)^{-1} \beta_r \boldsymbol{\iota}_n = (1-\rho)^{-1} \beta_r. \qquad (C.1.37)$$

One point to note is that even the average direct impact for this model does not equal the coefficient β_r as in the case of a conventional regression model. The difference between the coefficient estimate β_r and the scalar summary measure of average direct impact arises from the feedback loop reflecting how initial changes in y_i give rise to impacts on neighboring regions y_j which in turn pass through neighboring regions and feedback to region i. Of course, the magnitude of this type of feedback will depend on aspects of the spatial regression model used and the resulting parameter estimates. For example, the nature of the connectivity structure W used in the model and the magnitude of the parameter estimates for ρ and β both play a role in determining the impacts.

Finally, we should bear in mind the discussion in Section C.1.4, indicating that we should interpret these scalar summary measures of impact as reflecting how changes in the explanatory variables work through the simultaneous dependence system over time to culminate in a new steady state equilibrium. For example, if we find that a ten percent increase in regional levels of human capital give rise to a five percent direct impact on regional income growth and a ten percent indirect impact, we would conclude that these changes would be associated with regional income levels in the new steady-state equilibrium. In the context of our static cross-sectional model we cannot make informative statements about the time that will be required to reach this new equilibrium. Another point is that the indirect impacts will often exceed the direct impacts because the scalar summary measures cumulate impacts over all regions in the model. LeSage and Pace (2009) provide ways to decompose these cumulative impacts into those falling on first-order, second-order and higher-order neighboring regions. These decompositions result in the more intuitive situation where direct impacts exceed indirect impacts falling on first-order, second-order and higher-order neighbors. However, the cumulative impact scalar summary measures add up impacts falling on neighbors of all orders, which often results in indirect or spatial spillover impacts that exceed the direct impacts.

One applied illustration that uses these scalar summary impact estimates can be found in Chapter E.1. The application considers the direct, indirect and total impacts of changes in human capital on labor productivity levels in European Union regions. A number of other applications can be found in LeSage and Pace (2009) in a wide variety of applied contexts.

Inference regarding the impacts

For inference regarding the significance of these impacts, we need to determine their empirical or theoretical distribution. Since the impacts reflect a non-linear combination of the parameters ρ and β in the case of the SAR model, working with the theoretical distribution is not particularly convenient. Given the model estimates as well as associated variance-covariance matrix along with the knowledge that maximum likelihood estimates are (asymptotically) normally distributed, we can simulate the parameters ρ and β. These empirically simulated magnitudes can be used in expressions for the scalar summary measures to produce an empirical distribution of the scalar impact measures.

For the case of Bayesian MCMC estimates we already have a sample of parameter draws for ρ and β which can be used in conjunction with the expressions for the scalar summary measures to produce a posterior distribution of the total, direct and indirect impact measures. Gelfand et al. (1990) show that this is a valid approach to derive the posterior distribution for non-linear combinations of model parameters.

For the case of the SAR model, this is relatively straightforward requiring that we need only evaluate the expression: $(1-\rho)^{-1}\beta_r$ to find the total impacts. Calculating the direct impacts requires that we work with the main diagonal of the matrix $(I_n - \rho W)^{-1}$ for which LeSage and Pace (2009) provide computationally efficient methods. Recall that we would need to carry out these calculations thousands of times using the simulated parameter values or MCMC draws to determine the empirical measures of dispersion. These measures are used to determine the statistical significance of direct, indirect and total impacts associated with the various explanatory variables in the model, in a fashion similar to use of t-statistics in conventional regression models. In more complicated models such as the SDM, the scalar summary measures of impact take more complicated forms, but LeSage and Pace (2009) provide computationally efficient approaches for evaluating these expressions.

An applied illustration of a simulation approach to determining measures of dispersion for these scalar summary impact estimates can be found in Chapter E.1. Another illustration is given in LeSage and Fischer (2008) in the context of model averaging methods discussed in Chapter C.4.

Spatial heterogeneity, spatial dependence, and impacts

Many authors draw a distinction between models of spatial dependence and those of spatial heterogeneity. Typically, spatial dependence models estimate a parameter for each variable while spatial heterogeneity models effectively estimate an n-by-n matrix of parameters. The Casetti expansion method (Casetti 1997, see also Chapter C.6) and GWR (Fotheringham et al. 2002; see also Chapter C.5) exemplify this approach.

However, the distinction between models of spatial dependence and those of spatial heterogeneity is not as clear as it might initially appear. To motivate this discussion, consider the usual linear model (C.1.38) with the parameters written in matrix form in Eqs. (C.1.39) to (C.1.41).

$$E(y) = X_1 \beta_1 + X_2 \beta_2 + \ldots + X_k \beta_k \qquad (C.1.38)$$

$$E(y) = \Theta^{(1)} X_1 + \Theta^{(2)} X_2 + \ldots + \Theta^{(k)} X_k \qquad (C.1.39)$$

$$B_{ii}^{(r)} = \beta_r \quad r = 1, \ldots, k, \quad i = 1, \ldots, n \qquad (C.1.40)$$

$$\Theta^{(r)} = B^{(r)}. \qquad (C.1.41)$$

Obviously, in the usual linear model the impact of changing the explanatory variable is the same across observations and a change in the explanatory variable for one observation does not affect the others.

What if we gave geometrically declining weights to the values of the parameters at the neighbors, including parameters at the neighbors of neighbors, and so forth as shown in Eq. (C.1.42). Given the formula for the infinite series expansion, this leads to Eq. (C.1.43). Interestingly, the matrix of parameters implied by this process equals the matrix of impacts ($S_r(W)$) discussed previously. As before, we can view the expected value of the dependent variable as a sum of the impacts from all the explanatory variables as in Eq. (C.1.44).

$$\Theta^{(r)} = I_n B^{(r)} + \rho W B^{(r)} + \rho^2 W^2 B^{(r)} + \ldots \qquad (C.1.42)$$

$$\Theta^{(r)} = (I_n - \rho W)^{-1} B^{(r)} = S_r(W) \qquad (C.1.43)$$

$$E(y) = S_1(W) X_1 + S_2(W) X_2 + \ldots + S_k(W) X_k. \qquad (C.1.44)$$

To summarize, spatial dependence involving a spatial lag of the dependent variable implies a form of spatial heterogeneity where the impacts measure the heterogeneity across observations. Error models, however, do not result in heterogenous impacts over space. Therefore, the traditional distinction between spatial heterogeneity and spatial dependence is meaningful in the case of error models but misleading in the case of spatial autoregressive models.

C.1.5 Concluding remarks

Spatial autoregressive processes represent a parsimonious way to model spatial dependence between observations that often arises in regional economic research. We have shown how basic regression models can be augmented with spatial autoregressive processes to produce models that incorporate simultaneous feedback between regions located in space. It was also shown that conventional regression model estimates that ignore this feedback are biased and inconsistent.

Interpretation of estimates and inferences regarding the spatial connectivity relationships modeled require interpretation based on a steady-state equilibrium view. These models produce a situation where changes in the explanatory variables lead to a series of simultaneous feedbacks that ultimately result in a new steady-state equilibrium. Because we are working with cross-sectional sample data, these model adjustments appear as if they are simultaneous, but we argued that these models can be viewed as containing an implicit time dimension.

The availability of public domain software to implement estimation and inference for the models described here should make these methods widely accessible (Anselin 2006; Bivand 2002; LeSage 1999; Pace 2003).

References

Anselin L (1988) Spatial econometrics: Methods and models. Kluwer, Dordrecht
Anselin L (2006) GeoDa™ 0.9 User's guide, at geoda.uiuc.edu
Barry R, Pace RK (1999) A Monte Carlo estimator of the log determinant of large sparse matrices. Lin Algebra Appl 289(1-3):41-54
Bivand R (2002) Spatial econometrics functions in R: classes and methods. J Geogr Syst 4(4):405-421
Casetti E (1997) The expansion method, mathematical modeling, and spatial econometrics. Int Reg Sci Rev 20(1-2):9-33
Casetti E (2009) Expansion method, dependency, and modeling. In Fischer MM, Getis A (eds) Handbook of applied spatial analysis. Springer, Berlin, Heidelberg and New York, pp.487-505
Chen J, Jennrich R (1996) The signed root deviance profile and confidence intervals in maximum likelihood analysis. J Am Stat Assoc 91:993-998
Dall'erba S, LeGallo J (2007) Regional convergence and the impact of European structural funds over 1989-1999: a spatial econometric analysis. Papers in Reg Sci 87(2):219-244
Ertur C, Koch W (2007) Convergence, human capital and international spillovers. J Appl Econ 22(6):1033-1062
Fischer MM, Bartkowska M, Riedl A, Sardadvar S, Kunnert A (2009) The impact of human capital on regional labor productivity growth in Europe. In Fischer MM, Getis A

(eds) Handbook of applied spatial analysis. Springer, Berlin, Heidelberg and New York, pp.585-597
Fotheringham AS, Brunsdon C, Charlton M (2002) Geographically weighted regression: the analysis of spatially varying relationships. Wiley, New York, Chichester, Toronto and Brisbane
Gelfand AE, Smith AFM (1990) Sampling-based approaches to calculating marginal densities. J Am Stat Assoc 85:398-409
Gelfand AE, Hills SE, Racine-Poon A, Smith AFM (1990) Illustration of Bayesian inference in normal data models using Gibbs sampling. J Am Stat Assoc 85:972-985
Geman S, Geman D (1984) Stochastic relaxation, Gibbs distributions, and the Bayesian restoration of images. IEEE Transact Patt Anal Machine Intell 6(6):721-741
Hastings WK (1970) Monte Carlo sampling methods using Markov Chains and their applications. Biometrika 57(1):97-109
Kelejian H, Prucha IR (1998) A Generalized spatial two-stage least squares procedure for estimating a spatial autoregressive model with autoregressive disturbances. J Real Est Fin Econ 17(1):99-121.
Kelejian H, Prucha IR (1999) A generalized moments estimator for the autoregressive parameter in a spatial model. Int Econ Rev 40(2):509-533
Kelejian H, Tavlas HGS, Hondronyiannis G (2006) A spatial modeling approach to contagion among emerging economies. Open Econ5 Rev5 17(4/5):423-442
Koop G (2003) Bayesian econometrics. Wiley, West Sussex [UK]
Lee LF (2004) Asymptotic distributions of quasi-maximum likelihood estimators for spatial econometric models. Econometrica 72(6):1899-1926
LeSage JP (1997) Bayesian estimation of spatial autoregressive models. Int Reg Sci Rev 20(1-2):113-129
LeSage JP (1999) Spatial econometrics using MATLAB a manual for the spatial econometrics toolbox functions, available at www.spatial-econometrics.com
LeSage JP, Fischer MM (2008) Spatial growth regressions: model specification, estimation and interpretation. Spat Econ Anal 3(3):275-304
LeSage JP, Fischer MM (2009) Spatial econometric methods for modeling origin-destination flows. In Fischer MM, Getis A (eds) Handbook of applied spatial analysis. Springer, Berlin, Heidelberg and New York, pp.409-433
LeSage JP, Pace RK (2007) A matrix exponential spatial specification. J Econometrics 140(1):190-214
LeSage JP, Pace RK (2008) Spatial econometric modeling of origin-destination flows. J Reg Sci 48(5):941-967
LeSage JP, Pace RK (2009) Introduction to spatial econometrics. CRC Press (Taylor and Francis Group), Boca Raton [FL], London and New York
Marsh TL, Mittelhammer RC (2004) Generalized maximum entropy estimation of a first order spatial autoregressive model. In LeSage JP, Pace RK (eds) Advances in econometrics, volume 18. Elsevier, Oxford, pp.203-238
Metropolis N, Rosenbluth AW, Rosenbluth MN, Teller AH, Teller E (1953) Equation of state calculations by fast computing machines. J Chem Physics 21(6):1087-1092
Ord JK (1975) Estimation methods for models of spatial interaction. J Am Stat Assoc 70(349):120-126
Pace RK (2003) Matlab spatial statistics toolbox 2.0, available at www.spatial-statistics.com.
Pace RK, Barry B (1997) Quick computation of spatial autoregressive estimators. Geogr Anal 29(3):232-246
Pace RK, LeSage JP (2003) Likelihood dominance spatial inference. Geogr Anal 35(2):133-147

Pace RK, LeSage JP (2009a) Omitted variables biases of OLS and spatial lag models. In Páez A, LeGallo J, Buliung R, Dall'Erba S (eds) Progress in spatial analysis: theory and methods, and thematic applications. Springer, Berlin, Heidelberg and New York (forthcoming)

Pace K, LeSage JP (2009b) A sampling approach to estimating the log determinant used in spatial likelihood problems. J Geogr Syst (forthcoming)

Parent O, LeSage JP (2009) Spatial econometric model averaging. In Fischer MM, Getis A (eds) Handbook of applied spatial analysis. Springer, Berlin, Heidelberg and New York, pp.435-460

Smirnov O, Anselin L (2009) An O(N) parallel method of computing the Log-Jacobian of the variable transformation for models with spatial interaction on a lattice. Comput Stat Data Anal 53(8):2980-2988

Wheeler DC, Páez A (2009) Geographically weighted regression. 1er MM, Getis A (eds) Handbook of applied spatial analysis. Springer, Berlin, Heidelberg and New York, pp.465-486

C.2 Spatial Panel Data Models

J. Paul Elhorst

C.2.1 Introduction

In recent years, the spatial econometrics literature has exhibited a growing interest in the specification and estimation of econometric relationships based on spatial panels. Spatial panels typically refer to data containing time series observations of a number of spatial units (zip codes, municipalities, regions, states, jurisdictions, countries, etc.). This interest can be explained by the fact that panel data offer researchers extended modeling possibilities as compared to the single equation cross-sectional setting, which was the primary focus of the spatial econometrics literature for a long time. Panel data are generally more informative, and they contain more variation and less collinearity among the variables. The use of panel data results in a greater availability of degrees of freedom, and hence increases efficiency in the estimation. Panel data also allow for the specification of more complicated behavioral hypotheses, including effects that cannot be addressed using pure cross-sectional data (see Hsiao 2005 for more details).

Elhorst (2003) has provided a review of issues arising in the estimation of four panel data models commonly used in applied research extended to include spatial error autocorrelation or a spatially lagged dependent variable: fixed effects, random effects, fixed coefficients, and random coefficients models. In addition, Matlab routines to estimate the fixed effects and random effects models have been provided at his website, see <www.regroningen.nl/elhorst> or <http://www.rug.nl/staff/j.p.elhorst/projects>. Many studies have applied these routines by now to estimate regional labor market models, economic growth models, public expenditures or tax setting models, and agricultural models. These applications have led to new insights, developments and extensions, but also to new questions and misunderstandings. This chapter reviews and organizes these recent methodologies. It deals with the possibility to test for spatial interaction effects in standard panel data models, the estimation of fixed effects and the determination of their significance levels, the possibility to test the fixed effects specification against the random effects specification of panel data models extended to include spatial error autocorrelation or a spatially lagged dependent variable using Hausman's specification test, the determination of the variance-covariance matrix of the parameter estimates of these extended models, the determination of goodness-of-fit measures and the best linear unbiased predictor when using these models for prediction purposes. For reasons of space, attention is limited to models with spatial fixed effects or spatial random effects. The concluding section also briefly discusses the possibility to test for endogeneity of one or more of the explanatory variables and the possibility to include dynamic effects.

C.2.2 Standard models for spatial panels

First, a simple pooled linear regression model with spatial specific effects is considered, but without spatial interaction effects

$$y_{it} = X_{it}\beta + \mu_i + \varepsilon_{it} \tag{C.2.1}$$

where i is an index for the cross-sectional dimension (spatial units), with $i = 1, ..., N$, and t is an index for the time dimension (time periods), with $t = 1, ..., T$. y_{it} is an observation on the dependent variable at i and t, X_{it} an 1-by-K row vector of observations on the independent variables, and β a matching K-by-1 vector of fixed but unknown parameters. ε_{it} is an independently and identically distributed error term for i and t with zero mean and variance σ^2, while μ_i denotes a spatial specific effect. The standard reasoning behind spatial specific effects is that they control for all space-specific time-invariant variables whose omission could bias the estimates in a typical cross-sectional study.

When specifying interaction between spatial units, the model may contain a spatially lagged dependent variable or a spatial autoregressive process in the error term, known as the spatial lag and the spatial error model, respectively. The spatial lag model posits that the dependent variable depends on the dependent variable observed in neighboring units and on a set of observed local characteristics

$$y_{it} = \delta \sum_{j=1}^{N} W_{ij} y_{jt} + X_{it}\beta + \mu_i + \varepsilon_{it} \tag{C.2.2}$$

where δ is called the spatial autoregressive coefficient and W_{ij} is an element of a spatial weights matrix W describing the spatial arrangement of the units in the sample. It is assumed that W is a pre-specified non-negative matrix of order N.[1] According to Anselin et al. (2006, p.6), the spatial lag model is typically considered as the formal specification for the equilibrium outcome of a spatial or social interaction process, in which the value of the dependent variable for one agent is jointly determined with that of the neighboring agents. In the empirical literature on strategic interaction among local governments, for example, the spatial lag model is theoretically consistent with the situation where taxation and expenditures on public services interact with taxation and expenditures on public services in nearby jurisdictions (Brueckner 2003).

The spatial error model, on the other hand, posits that the dependent variable depends on a set of observed local characteristics and that the error terms are correlated across space

$$y_{it} = X_{it}\beta + \mu_i + \phi_{it} \qquad (\text{C.2.3a})$$

$$\phi_{it} = \rho \sum_{j=1}^{N} W_{ij}\phi_{jt} + \varepsilon_{it} \qquad (\text{C.2.3b})$$

where ϕ_{it} reflects the spatially autocorrelated error term and ρ is called the spatial autocorrelation coefficient. According to Anselin et al. (2006, p.7), a spatial error specification does not require a theoretical model for a spatial or social interaction process, but, instead, is a special case of a non-spherical error covariance matrix. In the empirical literature on strategic interaction among local governments, the spatial error model is consistent with a situation where determinants of taxation or expenditures on public services omitted from the model are spatially autocorrelated, and with a situation where unobserved shocks follow a spatial pattern. A spatially autocorrelated error term may also be interpreted to reflect a mechanism to correct rent-seeking politicians for unanticipated fiscal policy changes (Allers and Elhorst 2005).

In both the spatial lag and the spatial error model, stationarity requires that $1/\omega_{min} < \delta < 1/\omega_{max}$ and $1/\omega_{min} < \rho < 1/\omega_{max}$, where ω_{min} and ω_{max} denote the smallest (i.e., most negative) and largest characteristic roots of the matrix W. While it is often suggested in the literature to constrain δ or ρ to the interval $(-1, +1)$, this may be unnecessarily restrictive. For row-normalized spatial weights, the largest characteristic root is indeed +1, but no general result holds for the smallest characteristic root, and the lower bound is typically less than −1. See also the lively dis-

---

[1] The regularity conditions W should satisfy in a cross-sectional setting have been derived by Lee (2004), but some of these regularity conditions may change in a panel data setting (Yu et al. 2007).

cussion at GeoDa's Openspace mailing list about the bounds on the spatial lag coefficient.

As an alternative to row-normalization, W might be normalized such that the elements of each column sum to one. This type of normalization is sometimes used in the social economics literature (Leenders 2002). Note that the row elements of a spatial weights matrix display the impact *on* a particular unit *by* all other units, while the column elements of a spatial weights matrix display the impact *of* a particular unit *on* all other units (see Chapter C.1 for a more detailed discussion of this issue). Consequently, row normalization has the effect that the impact on each unit by all other units is equalized, while column normalization has the effect that the impact of each unit on all other units is equalized.

If W_0 denotes the spatial weights matrix before normalization, one may also divide the elements of W_0 by its largest characteristic root $\omega_{0,max}$ to get $W = (1/\omega_{0,max})W_0$, or normalize W_0 by $W = D^{-1/2}W_0 D^{-1/2}$, where D is a diagonal matrix containing the row sums of the matrix W_0. The first operation may be labeled matrix normalization, since it has the effect that the characteristic roots of W_0 are also divided by $\omega_{0,max}$, as a result of which $\omega_{max}=1$, just like the largest characteristic root of a row- or column-normalized matrix. The second operation has been proposed by Ord (1975) and has the effect that the characteristic roots of W are identical to the characteristic roots of a row-normalized W_0. Importantly, the mutual proportions between the elements of W remain unchanged as a result of these two alternative normalizations. This is an important property when W represents an inverse distance matrix, since scaling the rows or columns of an inverse distance matrix so that the weights sum to one would cause this matrix to lose its economic interpretation for this decay (Anselin 1988, pp.23-24). One concomitant advantage of spatial weights matrices that do not lose their property of symmetry as a result of normalization is that notation, in some cases, is considerably simplified and that computation time will speed up (Elhorst 2001, 2005a).

Two main approaches have been suggested in the literature to estimate models that include spatial interaction effects. One is based on the maximum likelihood (ML) principle and the other on instrumental variables or generalized method of moments (IV/GMM) techniques. Although IV/GMM estimators are different from ML estimators in that they do not rely on the assumption of normality of the errors, both estimators assume that the disturbance terms ε_{it} are independently and identically distributed for all i and t with zero mean and variance σ^2. The Jarque-Bera (1980) test may be used to investigate the normality assumption when applying ML estimators.[2] One disadvantage of IV/GMM estimators is the possibility of

---

[2] This test has a chi-squared distribution with one degree of freedom. In addition, the Jarque-Bera test may be used to test for serial independence and homoskedasticity of the regression residuals. These tests have a chi-squared distribution with p degrees of freedom when testing for p-order serial autocorrelation, and q degrees of freedom when testing for homoskedasticity, one degree for every variable that might explain heteroskedasticity. Although informative, it should be noted that these tests were not developed in the context of a model with spatial interaction effects.

ending up with a coefficient estimate for δ or for ρ outside its parameter space $(1/\omega_{min}, 1/\omega_{max})$. Whereas this coefficient is restricted to its parameter space by the Jacobian term in the log-likelihood function of ML estimators, it is unrestricted using IV/GMM since these estimators ignore the Jacobian term.

Franzese and Hays (2007) compare the performance of the IV estimator and the ML estimator of panel data models with a spatially lagged dependent variable in terms of unbiasedness and efficiency, but unfortunately without considering spatial fixed or random effects. They find that the ML estimator offers weakly dominant efficiency and generally solid performance in unbiasedness, although it sometimes falls a little short of IV on unbiasedness grounds at lower values of δ.

The main focus in this chapter will be on ML estimation, because the number of studies considering IV/GMM estimators of spatial panel data models is still relatively sparse. One exception is Kelejian et al. (2006), who considered IV estimation of a spatial lag model with time period fixed effects. They point out that this model cannot be combined with a spatial weights matrix whose non-diagonal elements are all equal to $1/(N-1)$. In this situation, the spatially lagged dependent variable can be written in vector form as

$$\left\{ \tfrac{1}{N-1}\sum_j y_{j1} - \tfrac{y_{11}}{N-1}, \ldots, \tfrac{1}{N-1}\sum_j y_{j1} - \tfrac{y_{N1}}{N-1}, \ldots, \tfrac{1}{N-1}\sum_j y_{jT} - \tfrac{y_{1T}}{N-1}, \ldots, \tfrac{1}{N-1}\sum_j y_{jT} - \tfrac{y_{NT}}{N-1} \right\}^T \quad (C.2.4)$$

which is asymptotically proportional and thus collinear with the time period fixed effects as N goes to infinity. Another exception is Kapoor et al. (2007), who considered the GMM estimator of a spatial error model and time period random effects. However, neither of these studies considered spatial fixed or random effects, while just these effects often appear to be important in panel data studies.

One shortcoming of the spatial lag model and the spatial error model is that spatial patterns in the data may be explained not only by either endogenous interaction effects or correlated error terms, but also by endogenous interaction effects, exogenous interaction effects and correlated error terms at the same time (Manski 1993). The best strategy would, therefore, seem to be to include the spatially lagged dependent variable, the K spatially lagged independent variables, and the spatially autocorrelated error term simultaneously.[3] However, Manski (1993) has also pointed out that at least one of these $2+K$ spatial interaction effects must be excluded, because otherwise their interaction parameters are not identified. In ad-

---

[3] In his keynote speech at the First World Conference of the Spatial Econometrics Association 2007, Harry Kelejian advocated models that include both a spatially lagged dependent variable and a spatially autocorrelated error term, while James LeSage in his Presidential Address at the 54th North American Meeting of the Regional Science Association International 2007 advocated models that include both a spatially lagged dependent variable and spatially lagged independent variables.

dition to this, the spatial weights matrix of the spatially lagged dependent variable must be different from the spatial weights matrix of the spatially autocorrelated error term, an additional requirement for identification when applying ML estimators (Anselin and Bera 1998). One ostensible advantage of IV/GMM estimators is that the same spatial weights matrix can be used to estimate a model extended to include a spatially lagged dependent variable and a spatially autocorrelated error term (Kelejian and Prucha 1998; Lee 2003). However, these estimators on their turn are unable to estimate models with spatially lagged independent variables, since they use these variables as instruments.

Alternatively, one may first test whether spatially lagged independent variables must be included and then whether the model should be extended to include a spatially lagged dependent variable or a spatially autocorrelated error term (Florax and Folmer 1992; Elhorst and Freret 2009) or adopt an unconstrained spatial Durbin model and then test whether this model can be simplified (Elhorst et al. 2006; Ertur and Koch 2007). An unconstrained spatial Durbin model with spatial fixed effects takes the form

$$y_{it} = \delta \sum_{j=1}^{N} W_{ij} y_{jt} + X_{it} \beta + \sum_{j=1}^{N} W_{ij} X_{jt} \gamma + \mu_i + \varepsilon_{it} \qquad (C.2.5)$$

where γ, just as β, is a K-by-1 vector of fixed but unknown parameters. The hypothesis $H_0: \gamma = 0$ can be tested to investigate whether this model can be simplified to the spatial lag model and the hypothesis $H_0: \gamma + \delta\beta = 0$ whether it can be simplified to the spatial error model. A simulation study by Florax et al. (2003) showed that the specific-to-general approach outperforms the general-to-specific approach when using cross-sectional data. However, one objection to this study is that the comparison between the two approaches is invalid because the null rejection frequencies have not been standardized (Hendry 2006). Another objection is that the model that has been used as point of departure did not include spatially lagged independent variables. Hence, a more careful elaboration of the relative merits of both approaches when using spatial panel data remains a topic of further research.

C.2.3 Estimation of panel data models

The spatial specific effects may be treated as fixed effects or as random effects. In the fixed effects model, a dummy variable is introduced for each spatial unit, while in the random effects model, μ_i is treated as a random variable that is independently and identically distributed with zero mean and variance σ_μ^2. Furthermore, it is assumed that the random variables μ_i and ε_{it} are independent of each other.

Throughout this chapter it is assumed that the data are sorted first by time and then by spatial units, whereas the classic panel data literature tends to sort the data first by spatial units and then by time. When y_{it} and X_{it} of these T successive cross-sections of N observations are stacked, we obtain an NT-by-1 vector for y and an NT-by-K matrix for X.

Fixed effects model

If the spatial specific effects are treated as fixed effects, the model in Eq. (C.2.1) can be estimated in three steps. First, the spatial fixed effects μ_i are eliminated from the regression equation by demeaning the dependent and independent variables. This transformation takes the form

$$y_{it}^* = y_{it} - \tfrac{1}{T}\sum_{t=1}^{T} y_{it} \quad \text{and} \quad X_{it}^* = X_{it} - \tfrac{1}{T}\sum_{t=1}^{T} X_{it}. \tag{C.2.6}$$

Second, the transformed regression equation $y_{it}^* = X_{it}^* \boldsymbol{\beta} + \varepsilon_{it}^*$ is estimated by OLS: $\boldsymbol{\beta} = (X^{*T} X^*)^{-1} X^{*T} y^*$ and $\sigma^2 = (y^* - X^* \boldsymbol{\beta})^T (y^* - X^* \boldsymbol{\beta})/(NT{-}N{-}K)$. This estimator is known as the least squares dummy variables (LSDV) estimator. The main advantage of the demeaning procedure is that the computation of $\boldsymbol{\beta}$ involves the inversion of a K-by-K matrix rather than $(K{+}N)$-by-$(K{+}N)$ as in Eq. (C.2.1). This would slow down the computation and worsen the accuracy of the estimates considerably for large N.

Instead of estimating the demeaned equation by OLS, it can also be estimated by ML. Since the log-likelihood function of the demeaned equation is

$$\ln L = -\tfrac{NT}{2}\ln(2\pi\sigma^2) - \tfrac{1}{2\sigma^2} \sum_{i=1}^{N}\sum_{t=1}^{T}(y_{it}^* - X_{it}^* \boldsymbol{\beta})^2 \tag{C.2.7}$$

the ML estimators of $\boldsymbol{\beta}$ and σ^2 are $\boldsymbol{\beta} = (X^{*T} X^*)^{-1} X^{*T} y^*$ and $\sigma^2 = (y^* - X^*\boldsymbol{\beta})^T (y^* - X^*\boldsymbol{\beta})/NT$, respectively. In other words, the ML estimator of σ^2 is slightly different from the LSDV estimator in that it does not correct for degrees of freedom. The asymptotic variance matrix of the parameters is (see Greene 2008, p.519)

$$\mathrm{AsyVar}(\boldsymbol{\beta},\sigma^2) = \begin{bmatrix} \tfrac{1}{\sigma^2} X^{*T} X^* & 0 \\ 0 & \tfrac{NT}{2\sigma^4} \end{bmatrix}^{-1}. \tag{C.2.8}$$

Finally, the spatial fixed effects may be recovered by

$$\mu_i = \tfrac{1}{T}\sum_{t=1}^{T}(y_{it} - X_{it}\boldsymbol{\beta}) \qquad i = 1, \ldots, N. \tag{C.2.9}$$

It should be stressed that the spatial fixed effects can only be estimated consistently when T is sufficiently large, because the number of observations available for the estimation of each μ_i is T. Also note that sampling more observations in the cross-sectional domain is no solution for insufficient observations in the time domain, since the number of unknown parameters increases as N increases, a situation known as the incidental parameters problem. Fortunately, the inconsistency of μ_i is not transmitted to the estimator of the slope coefficients $\boldsymbol{\beta}$ in the demeaned equation, since this estimator is not a function of the estimated μ_i. Consequently, the incidental parameters problem does not matter when $\boldsymbol{\beta}$ are the coefficients of interest and the spatial fixed effects μ_i are not, which is the case in many empirical studies. Finally, note that the incidental parameters problem is independent of the extension of the model with spatial interaction effects.

In case the spatial fixed effects μ_i do happen to be of interest, their standard errors may be computed as the square roots of their asymptotic variances (see Greene 2008, p.196)

$$AsyVar(\hat{\mu}_i) = \frac{\hat{\sigma}^2}{T} + \hat{\sigma}^2 (\tfrac{1}{T}\sum_{t=1}^{T}X_{it})(X^{*T}X^*)^{-1}(\tfrac{1}{T}\sum_{t=1}^{T}X_{it})^{T}. \tag{C.2.10}$$

An alternative and equivalent formulation of Eq. (C.2.1) is to introduce a mean intercept α, provided that $\Sigma_i \mu_i = 0$. Then the spatial fixed effect μ_i represents the deviation of the ith spatial unit from the individual mean (see Hsaio 2003, p.33).

To test for spatial interaction effects in a cross-sectional setting, Anselin et al. (1996) developed Lagrange multiplier (LM) tests for a spatially lagged dependent variable, for spatial error correlation, and their counterparts robustified against the alternative of the other form.[4] These tests have become very popular in empirical research. Recently, Anselin et al. (2006) also specified the first two LM tests for a spatial panel

---

[4] Software programs, such as SpaceStat and GeoDa, have built-in routines that automatically report the results of these tests. Matlab routines have been made available by Donald Lacombe at ≤ http://oak.cats.ohiou.edu/~lacombe/research.html ≥.

$$\mathrm{LM}_\delta = \frac{[e^\mathrm{T}(I_T \otimes W)y\,\hat{\sigma}^{-2}]^2}{J} \quad \text{and} \quad \mathrm{LM}_p = \frac{[e^\mathrm{T}(I_T \otimes W)e\,\hat{\sigma}^{-2}]^2}{T\,T_W} \qquad (C.2.11)$$

where the symbol $\otimes$ denotes the Kronecker product, I_T denotes the identity matrix and its subscript the order of this matrix, and e denotes the residual vector of a pooled regression model without any spatial or time-specific effects or of a panel data model with spatial and/or time period fixed effects. Finally, J and T_W are defined by

$$J = \frac{1}{\hat{\sigma}^2}\left\{\left((I_T \otimes W)X\hat{\beta}\right)^\mathrm{T}[I_{NT} - X(X^\mathrm{T}X)^{-1}X^\mathrm{T}](I_T \otimes W)X\hat{\beta} + TT_W\,\hat{\sigma}^2\right\} \qquad (C.2.12)$$

$$T_W = trace(WW + W^\mathrm{T}W) \qquad (C.2.13)$$

where *trace* denotes the trace of a matrix. In view of these formulas, the robust counterparts of these LM tests for a spatial panel will take the form

$$\text{Robust}\,\mathrm{LM}_\rho = \frac{[e^\mathrm{T}(I_T \otimes W)y\,\hat{\sigma}^{-2} - e^\mathrm{T}(I_T \otimes W)e\,\hat{\sigma}^{-2}]^2}{J - TT_W} \qquad (C.2.14)$$

$$\text{Robust}\,\mathrm{LM}_\rho = \frac{[e^\mathrm{T}(I_T \otimes W)e\,\hat{\sigma}^{-2} - [TT_W/J]\,e^\mathrm{T}(I_T \otimes W)y\,\hat{\sigma}^{-2}]^2}{TT_W\,[1 - TT_W/J]^{-1}}. \qquad (C.2.15)$$

Note that the performance of these tests when having panel data instead of cross-sectional data and when having a model extended to include spatially lagged independent variables must still be investigated.

Applied researchers often find weak evidence in favor of spatial interaction effects when time period fixed effects are also accounted for. The explanation is that most variables tend to increase and decrease together in different spatial units along the national evolution of these variables over time. The labor force participation rate and its evolution over the business cycle is one of the best examples (Elhorst 2008a). In the long term, after the effects of shocks have been settled, variables return to their equilibrium values. In equilibrium, neighboring values tend to be more similar than those further apart, but this interaction effect is often weaker than its counterpart over time. The mathematical explanation is that time period fixed effects are identical to a spatially autocorrelated error term with a spatial weights matrix whose elements are all equal to $1/N$, including the diagonal elements. When this spatial weights matrix would be adopted, one obtains

$$y_{it} - \sum_{j=1}^{N} W_{ij} y_{jt} = y_{it} - \tfrac{1}{N}\sum_{j=1}^{N} y_{jt} \text{ and } X_{it} - \sum_{j=1}^{N} W_{ij} X_{jt} = X_{it} - \tfrac{1}{N}\sum_{j=1}^{N} X_{jt} \qquad (C.2.16)$$

which is equivalent to the demeaning procedure of Eq. (C.2.6) but then for fixed effects in time. Even though spatial weights matrices with non-zero diagonal elements are unusual in spatial econometrics, these expressions show that accounting for time period fixed effects is one way to correct for spatial interaction effects among the error terms. If, in addition to time period fixed effects, a spatial error term is considered with a spatial weights matrix with zero diagonal elements, the magnitude of this spatial interaction effect will automatically fall as a result.

Applied researchers also often find significant differences among the coefficient estimates from models with and without spatial fixed effects. These models are different in that they utilize different parts of the variation between observations. Models with controls for spatial fixed effects utilize the time-series component of the data, whereas models without controls for spatial fixed effects utilize the cross-sectional component of the data. As a result, some studies argue that models with controls for spatial fixed effects tend to give short-term estimates and models without controls for spatial fixed effects tend to give long-term estimates (Baltagi 2005, pp.200-201; Partridge 2005). A related problem of controlling for spatial fixed effects is that any variable that does not change over time or only varies a little cannot be estimated, because it is wiped out by the demeaning transformation. This is the main reason for many studies not controlling for spatial fixed effects.

On the other hand, if one or more relevant explanatory variables are omitted from the regression equation, when they should be included, the estimator of the coefficients of the remaining variables is biased and inconsistent (Greene 2008, pp.133-134). This also holds true for spatial fixed effects and is known as the omitted regressor bias. One can test whether the spatial fixed effects are jointly significant by performing a Likelihood Ratio (LR) test of the hypothesis H_0: $\mu_1 = \ldots = \mu_N = \alpha$, where α is the mean intercept. The corresponding test statistic is $-2s$, where s measures the difference between the log-likelihood of the restricted model and that of the unrestricted model. The LR test has a chi-squared distribution with degrees of freedom equal to the number of restrictions that must be imposed on the unrestricted model to obtain the restricted model, which in this particular case is $N-1$. Thanks to the availability of the log-likelihood of the restricted as well as of the unrestricted model, the LR test can be carried out instead of, or in addition to, the classical *F*-test spelled out in Baltagi (2005, p.13). It is another advantage of estimating models by ML.

Random effects model

A compromise solution to the all or nothing way of utilizing the cross-sectional component of the data is the random effects model. This model avoids the loss of degrees of freedom incurred in the fixed effects model associated with a relatively large *N* and the problem that the coefficients of time-invariant variables cannot be estimated. However, whether the random effects model is an appropriate specification in spatial research remains controversial. When the random effects model is implemented, the units of observation should be representative of a larger population, and the number of units should potentially be able to go to infinity. There are two types of asymptotics that are commonly used in the context of spatial observations: (a) the 'infill' asymptotic structure, where the sampling region remains bounded as $N \to \infty$. In this case more units of information come from observations taken from between those already observed; and (b) the 'increasing domain' asymptotic structure, where the sampling region grows as $N \to \infty$. In this case there is a minimum distance separating any two spatial units for all *N*.

According to Lahiri (2003), there are also two types of sampling designs: (a) the stochastic design where the spatial units are randomly drawn; and (b) the fixed design where the spatial units lie on a nonrandom field, possibly irregularly spaced. The spatial econometric literature mainly focuses on increasing domain asymptotics under the fixed sample design (Cressie 1993, p.100; Griffith and Lagona 1998; Lahiri 2003). Although the number of spatial units under the fixed sample design can potentially go to infinity, it is questionable whether they are representative of a larger population. For a given set of regions, such as all counties of a state or all regions in a country, the population may be said '*to be sampled exhaustively*' (Nerlove and Balestra 1996, p.4), and '*the individual spatial units have characteristics that actually set them apart from a larger population*' (Anselin 1988, p.51). According to Beck (2001, p.272), '*the critical issue is that the spatial units be fixed and not sampled, and that inference be conditional on the observed units*'. In addition, the traditional assumption of zero correlation between μ_i in the random effects model and the explanatory variables, which also needs to be made, is particularly restrictive.

An iterative two-stage estimation procedure may be used to obtain the ML estimates of the random effects model (Breusch 1987). Note that the random effects model also includes a constant term, as a result of which the number of independent variables is *K*+1. The log-likelihood of the random effects model in Eq. (C.2.1) is

$$\ln L = -\tfrac{NT}{2}\ln(2\pi\sigma^2) + \tfrac{N}{2}\ln\theta^2 - \tfrac{1}{2\sigma^2}\sum_{i=1}^{N}\sum_{t=1}^{T}(y_{it}^{\bullet} - X_{it}^{\bullet}\beta)^2 \qquad \text{(C.2.17)}$$

where θ denotes the weight attached to the cross-sectional component of the data, with $0 \leq \theta^2 = \sigma^2 / (T\sigma_\mu^2 + \sigma^2) \leq 1$, and the symbol $\bullet$ denotes a transformation of the variables dependent on θ

$$y_{it}^{\bullet} = y_{it} - (1-\theta)\tfrac{1}{T}\sum_{t=1}^{T}y_{it} \quad \text{and} \quad X_{it}^{\bullet} = X_{it} - (1-\theta)\tfrac{1}{T}\sum_{t=1}^{T}X_{it}. \qquad \text{(C.2.18)}$$

If $\theta = 0$, this transformation simplifies to the demeaning procedure of Eq. (C.2.6) and hence the random effects model to the fixed effects model.

Given θ, β and σ^2 can be solved from their first-order maximizing conditions: $\beta = (X^{\bullet T}X^{\bullet})^{-1}X^{\bullet T}y^{\bullet}$ and $\sigma^2 = (Y^{\bullet} - X^{\bullet}\beta)^{\mathrm{T}}(Y^{\bullet} - X^{\bullet}\beta) / NT$. Conversely, θ may be estimated by maximizing the concentrated log-likelihood function with respect to θ, given β and σ^2,

$$\ln L = -\tfrac{NT}{2}\ln\left\{\sum_{i=1}^{N}\sum_{t=1}^{T}[(y_{it} - (1-\theta)\tfrac{1}{T}\sum_{t'=1}^{T}y_{it'}] - [X_{it} - (1-\theta)\tfrac{1}{T}\sum_{t'=1}^{T}X_{it'}]\beta)^2\right\} + \tfrac{N}{2}\ln\theta^2.$$

(C.2.19)

The use of θ^2 instead of θ ensures that both the argument of $\ln(\theta^2)$ and of $\sqrt{(\theta^2)}$ are positive (see Magnus 1982 for details). The asymptotic variance matrix of the parameters is

$$\text{AsyVar}(\beta, \theta, \sigma^2) = \begin{bmatrix} \tfrac{1}{\sigma^2}X^{\bullet \mathrm{T}}X^{\bullet} & 0 & 0 \\ 0 & N\left(1+\tfrac{1}{\theta^2}\right) & -\tfrac{N}{\sigma^2} \\ 0 & -\tfrac{N}{\sigma^2} & \tfrac{NT}{2\sigma^4} \end{bmatrix}^{-1}. \qquad \text{(C.2.20)}$$

One can test whether the spatial random effects are significant by performing a LR test of the hypothesis H_0: $\theta = 1$.[5] This test statistic has a chi-squared distribution with one degree of freedom. If the hypothesis is rejected, the spatial random effects are significant.

C.2.4 Estimation of spatial panel data models

This section outlines the modifications that are needed to estimate the fixed effects model and the random effects model extended to include a spatially lagged dependent variable or a spatially autocorrelated error. It is assumed that W is constant over time and that the panel is balanced. Although the estimators can be modified for a spatial weights matrix that changes over time, as well as for an unbalanced panel, their asymptotic properties, in the event of an unbalanced panel, may become problematic if the reason why data are missing is not known.

Fixed effects spatial lag model

According to Anselin et al. (2006), the extension of the fixed effects model with a spatially lagged dependent variable raises two complications. First, the endogeneity of $\Sigma_j W_{ij} y_{jt}$ violates the assumption of the standard regression model that $E\left[(\Sigma_j W_{ij} y_{jt}) \varepsilon_{it}\right] = 0$. In model estimation, this simultaneity must be accounted for. Second, the spatial dependence among the observations at each point in time may affect the estimation of the fixed effects.

In this section, we derive the ML estimator to account for the endogeneity of $\Sigma_j W_{ij} y_{jt}$. The log-likelihood function of the model in Eq. (C.2.2) if the spatial specific effects are assumed to be fixed is

$$\ln L = -\frac{NT}{2}\ln(2\pi\sigma^2) + T\ln|I_n - \delta W| - \frac{1}{2\sigma^2}\sum_{i=1}^{N}\sum_{t=1}^{T}\left[y_{it} - \delta\sum_{j=1}^{N}W_{ij}y_{jt} - X_{it}\beta - \mu_i\right]^2$$

(C.2.21)

where the second term on the right-hand side represents the Jacobian term of the transformation from ε to y taking into account the endogeneity of $\Sigma_j W_{ij} y_{jt}$ (Anselin 1988, p.63).

The partial derivatives of the log-likelihood with respect to μ_i are

---

[5] $\theta = 1$ implies $\sigma_\mu^2 = 0$, since σ_μ^2 may be calculated from θ by $[(1-\theta^2)/\theta^2][\sigma^2/T]$.

$$\frac{\partial \ln L}{\partial \mu_i} = \frac{1}{\sigma^2}\sum_{t=1}^{T}\left[y_{it}-\delta\sum_{j=1}^{N}W_{ij}\,y_{jt}-X_{it}\,\beta-\mu_i\right]=0 \qquad i=1,\dots,N. \qquad \text{(C.2.22)}$$

When solving μ_i from Eq. (C.2.22), one obtains

$$\mu_i = \frac{1}{T}\sum_{t=1}^{T}\left[y_{it}-\delta\sum_{j=1}^{N}W_{ij}\,y_{jt}-X_{it}\,\beta)\right] \qquad i=1,\dots,N. \qquad \text{(C.2.23)}$$

This equation shows that the standard formula for calculating the spatial fixed effects, Eq. (C.2.9), applies to the fixed effects spatial lag model in a straightforward manner. Corrections for the spatial dependence among the observations at each point in time, other than the addition of the spatially lagged dependent variable to these formulas, are not necessary.[6]

Substituting the solution for μ_i into the log-likelihood function, and after rearranging terms, the concentrated log-likelihood function with respect to β, δ and σ^2 is obtained

$$\ln L = -\frac{NT}{2}\ln(2\pi\sigma^2) + T\ln|I_N - \delta W| - \frac{1}{2\sigma^2}\sum_{i=1}^{N}\sum_{t=1}^{T}\left[y_{it}^{*} - \delta\left(\sum_{j=1}^{N}W_{ij}\,y_{jt}\right)^{*} - X_{it}^{*}\beta\right]^2$$
$$\text{(C.2.24)}$$

where the asterisk denotes the demeaning procedure introduced in Eq. (C.2.6).

Anselin and Hudak (1992) have spelled out how the parameters β, δ and σ^2 of a spatial lag model can be estimated by ML starting with cross-sectional data. This estimation procedure can also be used to maximize the log-likelihood function in Eq. (C.2.24) with respect to β, δ and σ^2. The only difference is that the data are extended from a cross-section of N observations to a panel of NT observations. This estimation procedure consists of the following steps.

First, stack the observations as successive cross-sections for $t = 1, \dots, T$ to obtain NT-by-1 vectors for y^* and $(I_T \otimes W)y^*$, and an NT-by-K matrix for X^* of the demeaned variables. Note that these calculations have to be performed only once and that the NT-by-NT diagonal matrix $(I_T \otimes W)$ does not have to be stored. This would slow down the computation of the ML estimator considerably for large data sets. Second, let b_0 and b_1 denote the OLS estimators of successively regressing y^*

---

[6] Anselin et al. (2006) asked for a more careful elaboration of this.

and $(I_T \otimes W)y^*$ on X^*, and e_0^* and e_1^* the corresponding residuals. Then the ML estimator of δ is obtained by maximizing the concentrated log-likelihood function

$$\ln L = C - \tfrac{NT}{2} \ln\left[(e_0^* - \delta e_1^*)^T (e_0^* - \delta e_1^*)\right] + T\ln|I_N - \delta W| \qquad \text{(C.2.25)}$$

where C is a constant not depending on δ. Unfortunately, this maximization problem can only be solved numerically, since a closed-form solution for δ does not exist. However, since the concentrated log-likelihood function is concave in δ, the numerical solution is unique (Anselin and Hudak 1992). To speed up computation time and to overcome numerical difficulties one might face in evaluating ln | $I_N - \delta W$|, Pace and Barry (1997) propose to compute this determinant once over a grid of values for the parameter δ ranging from $1/\omega_{min}$ to one prior to estimation, provided that W is normalized. This only requires the determination of the smallest characteristic root of W. They suggest a grid based on 0.001 increments for δ over the feasible range. Given these predetermined values for the log determinant of $(I_N - \delta W)$, they point out that one can quickly evaluate the concentrated log-likelihood function for all values of δ in the grid and determine the optimal value of δ as that which maximizes the concentrated log-likelihood function over this grid.[7]

Third, the estimators of β and σ^2 are computed, given the numerical estimate of δ,

$$\beta = b_0 - \delta b_1 = (X^{*T} X^*)^{-1} X^{*T} [y^* - \delta(I_T \otimes W) y^*] \qquad \text{(C.2.26a)}$$

$$\sigma^2 = \tfrac{1}{NT} (e_0^* - \delta e_1^*)^T (e_0^* - \delta e_1^*). \qquad \text{(C.2.26b)}$$

Instead of the demeaned variables, one may also use the original variables y and X, since $y^* = Qy$, $(I_T \otimes W) y^* = Q (I_T \otimes W) y$, and $X^* = QX$, where Q denotes the demeaning operator in matrix form

$$Q = I_{NT} - \tfrac{1}{T} \iota_T \iota_T^T \otimes I_N \qquad \text{(C.2.27)}$$

---

[7] The computation of the log determinant may be carried out using the Matlab routine 'lndet' from LeSage's website <www.spatial-econometrics.com> (LeSage 1999).

and ι_T is a vector of ones whose subscript denotes the length of this vector. Since Q is a symmetric idempotent matrix, the estimator of β starting with the original variables may also be written as

$$\beta = (X^T Q^T Q X)^{-1} X^T Q^T Q [y - \delta(I_T \otimes W) y] =$$

$$(X^T Q X)^{-1} X^T Q [y - \delta(I_T \otimes W) y]. \quad (C.2.28)$$

Anselin et al. (2006) have pointed out that this estimator may also be seen as the GLS estimator of a linear regression model with disturbance covariance matrix $\sigma^2 Q$, but the difficulty of this interpretation is that Q is singular. Their conclusion that the singularity of Q also limits the practicality of this model has been contradicted by Hsaio (2003, p.320), Magnus and Neudecker (1988, pp.271-273) and Baltagi (1989) in that Q may be replaced by its general inverse,[8] which again produces (C.2.28).

Finally, the asymptotic variance matrix of the parameters is computed for inference (standard errors, t-values). This matrix has been derived by Elhorst and Freret (2009) and takes the form (since this matrix is symmetric the upper diagonal elements are left aside)

$$AsyVar(\beta, \delta, \sigma^2) =$$

$$\begin{bmatrix} \frac{1}{\sigma^2} X^{*T} X^* & & \\ \frac{1}{\sigma^2} X^{*T}(I_T \otimes \widetilde{W}) X^* \beta & T\, trace(\widetilde{W}\widetilde{W} + \widetilde{W}^T \widetilde{W}) + \frac{1}{\sigma^2} \beta^T X^{*T}(I_T \otimes \widetilde{W}^T \widetilde{W}) X^* \beta & \\ 0 & \frac{T}{\sigma^2} trace(\widetilde{W}) & \frac{NT}{2\sigma^4} \end{bmatrix}^{-1}$$

(C.2.29)

where $\widetilde{W} = W(I_N - \delta W)^{-1}$. The differences with the asymptotic variance matrix of a spatial lag model in a *cross-sectional* setting (see Anselin and Bera 1998; Lee 2004) are the change in dimension of the matrix X^* from N to NT observations and the summation over T cross-sections involving manipulations of the N-by-N spa-

---

[8] Q^+ is called the generalized (Moore-Penrose) inverse of Q if it satisfies the conditions: $Q Q^+ Q = Q$, $Q^+ Q Q^+ = Q^+$, $(Q^+ Q)^T = Q^+ Q$ and $(Q Q^+)^T = Q Q^+$ (Magnus and Neudecker 1988, p.32).

tial weights matrix **W**. For large values of N the determination of the elements of the variance matrix may become computationally impossible. In that case the information may be approached by the numerical Hessian matrix using the maximum likelihood estimates of β, δ and σ^2.

Fixed effects spatial error model

Anselin and Hudak (1992) have also spelled out how the parameters β, ρ and σ^2 of a linear regression model extended to include a spatially autocorrelated error term can be estimated by ML starting with cross-sectional data. Just as for the spatial lag model, this estimation procedure can be extended to include spatial fixed effects and from a cross-section of N observations to a panel of NT observations. The log-likelihood function of model in Eq. (C.2.3) if the spatial specific effects are assumed to be fixed is

$$\ln L = -\tfrac{NT}{2}\ln(2\pi\sigma^2) + T\ln|I_N - \rho W|$$

$$-\frac{1}{2\sigma^2}\sum_{i=1}^{N}\sum_{t=1}^{T}\left\{ y_{it}^* - \rho\left[\sum_{j=1}^{N} W_{ij} y_{jt}\right]^* - \left[X_{it}^* - \rho\left(\sum_{j=1}^{N} W_{ij} X_{jt}\right)^*\right]\beta \right\}^2. \quad (C.2.30)$$

Given ρ, the ML estimators of β and σ^2 can be solved from their first-order maximizing conditions, to get

$$\beta = \{[X^* - \rho(I_T \otimes W)X^*]^T [X^* - \rho(I_T \otimes W)X^*]\}^{-1}$$

$$[X^* - \rho(I_T \otimes W)X^*]^T [y^* - \rho(I_T \otimes W)y^*] \quad (C.2.31a)$$

$$\sigma^2 = \frac{e(\rho)^T e(\rho)}{NT} \quad (C.2.31b)$$

where $e(\rho) = y^* - \rho(I_T \otimes W)y^* - [X^* - \rho(I_T \otimes W)X^*]\beta$. The concentrated log-likelihood function of ρ takes the form

$$\ln L = -\tfrac{NT}{2}\ln[e(\rho)^T e(\rho)] + T\ln|I_N - \rho W|. \quad (C.2.32)$$

Maximizing this function with respect to ρ yields the ML estimator of ρ, given β and σ^2. An iterative procedure may be used in which the set of parameters β and σ^2 and the parameter ρ are alternately estimated until convergence occurs. The asymptotic variance matrix of the parameters takes the form

$$AsyVar(\beta,\rho,\sigma^2) = \begin{bmatrix} \frac{1}{\sigma^2}X^{*T}X & & \\ 0 & T\,trace(\widetilde{\widetilde{W}}\widetilde{\widetilde{W}}+\widetilde{\widetilde{W}}^T\widetilde{\widetilde{W}}) & \\ 0 & \frac{T}{\sigma^2}trace(\widetilde{\widetilde{W}}) & \frac{NT}{2\sigma^4} \end{bmatrix}^{-1} \quad (C.2.33)$$

where $\widetilde{\widetilde{W}} = W(I_N - \rho W)^{-1}$. The spatial fixed effects can finally be estimated by

$$\mu_i = \frac{1}{T}\sum_{t=1}^{T}(y_{it} - X_{it}\beta) \quad i=1,...,N. \quad (C.2.34)$$

Random effects spatial lag model

The log-likelihood of model in Eq. (C.2.2) if the spatial effects are assumed to be random is

$$\ln L = -\frac{NT}{2}\ln(2\pi\sigma^2) + T\ln|I_N - \delta W| - \frac{1}{2\sigma^2}\sum_{i=1}^{N}\sum_{t=1}^{T}\left[y_{it}^{\bullet} - \delta\left(\sum_{j=1}^{N}W_{ij}y_{jt}\right)^{\bullet} - X_{it}^{\bullet}\beta\right]^2$$

(C.2.35)

where the symbol $^{\bullet}$ denotes the transformation introduced in Eq. (C.2.18) dependent on θ. Given θ, this log-likelihood function is identical to the log-likelihood function of the fixed effects spatial lag model in Eq. (C.2.24). This implies that the same procedure can be used to estimate β, δ and σ^2 as described above [Eqs. (C.2.25), (C.2.26a) and (C.2.26b)], but that the superscript * must be replaced by $^{\bullet}$. Given β, δ and σ^2, θ can be estimated by maximizing the concentrated log-likelihood function with respect to θ

$$\ln L = -\tfrac{NT}{2}\ln[e(\theta)^T e(\theta)] + \tfrac{N}{2}\ln\theta^2 \qquad (C.2.36)$$

where the typical element of $e(\theta)$ is

$$e(\theta)_{it} = y_{it} - (1-\theta)\tfrac{1}{T}\sum_{t=1}^{T} y_{it} - \delta\left[\sum_{j=1}^{N} W_{ij} y_{jt} - (1-\theta)\tfrac{1}{T}\sum_{t=1}^{T} W_{ij} y_{jt}\right]$$

$$- \left[X_{it} - (1-\theta)\tfrac{1}{T}\sum_{t=1}^{T} X_{it}\right]\beta. \qquad (C.2.37)$$

Again an iterative procedure may be used where the set of parameters β, δ and σ^2 and the parameter θ are alternately estimated until convergence occurs. This procedure is a mix of the estimation procedures used to estimate the parameters of the fixed effects spatial lag model and those of the non-spatial random effects model.

The asymptotic variance matrix of the parameters takes the form

$$AsyVar(\beta, \delta, \theta, \sigma^2) =$$

$$\begin{bmatrix} \tfrac{1}{\sigma^2}X^{\bullet T}X^{\bullet} & & & \\ \tfrac{1}{\sigma^2}X^{\bullet T}(I_T \otimes \widetilde{W})X^{\bullet}\beta & T\,trace(\widetilde{W}^T\widetilde{W}+\widetilde{W}^T\widetilde{W}) + \tfrac{1}{\sigma^2}\beta^T X^{\bullet T}(I_T \otimes \widetilde{W}^T\widetilde{W})X^{\bullet}\beta & & \\ 0 & -\tfrac{1}{\sigma^2}trace(\widetilde{W}) & N(1+\tfrac{1}{\theta^2}) & \\ 0 & \tfrac{T}{\sigma^2}trace(\widetilde{W}) & -\tfrac{N}{\sigma^2} & \tfrac{NT}{2\sigma^4} \end{bmatrix}^{-1}$$

$$(C.2.38)$$

Random effects spatial error model

The log-likelihood of model in Eq. (C.2.3) if the spatial effects are assumed to be random is (Anselin 1988; Elhorst 2003; Baltagi 2005)

$$\ln L = -\tfrac{NT}{2}\ln(2\pi\sigma^2) - \tfrac{1}{2}\ln|V| + (T-1)\sum_{i=1}^{N}\ln|B|$$

$$-\tfrac{1}{2\sigma^2}e^{\mathrm{T}}(\tfrac{1}{T}\iota_T\iota_T^{\mathrm{T}}\otimes V^{-1})e - \tfrac{1}{2\sigma^2}e^{\mathrm{T}}(I_T - \tfrac{1}{T}\iota_T\iota_T^{\mathrm{T}})\otimes(B^{\mathrm{T}}B)e \qquad (C.2.39)$$

where $V = T\varphi I_N + (B^{\mathrm{T}}B)^{-1}$ with $\varphi = \sigma_\mu^2/\sigma^2$,[9] $B = I_N - \rho W$ and $e = y - X\beta$. It is the matrix V that complicates the estimation of this model considerably. First, the Pace and Barry (1997) procedure to overcome numerical difficulties one might face in evaluating $\ln|B| = \ln|I_N - \rho W|$ cannot be used to calculate $\ln|V| = \ln|T\varphi I_N + (B^{\mathrm{T}}B)^{-1}|$. Second, there is no simple mathematical expression for the inverse of V. Baltagi (2006) solves these problems by considering a random effects spatial error model with equal weights, i.e., a spatial weights matrix W whose non-diagonal elements are all equal to $1/(N-1)$. Due to this setup, the inverse of V and a feasible GLS estimator of β can be determined mathematically. Furthermore, by considering a GLS estimator the term $\ln|V|$ in the log-likelihood function does not have to be calculated.

Elhorst (2003) suggests to express $\ln|V|$ as a function of the characteristic roots of W based on Griffith (1988, Table 3.1)

$$\ln|V| = \ln|T\varphi I_N + (B^{\mathrm{T}}B)^{-1}| = \sum_{i=1}^{N}\ln\left[T\varphi + \frac{1}{(1-\rho\omega_i)^2}\right]. \qquad (C.2.40)$$

Furthermore, he suggests to adopt the transformation

$$y_{it}^\circ = y_{it} - \rho\sum_{j=1}^{N}W_{ij}y_{jt} + \sum_{j=1}^{N}\left\{[p_{ij} - (1-\rho W_{ij})]\frac{1}{T}\sum_{t=1}^{T}y_{jt}\right\} \qquad (C.2.41)$$

and the same for the variables X_{it}, where p_{ij} is an element of an N-by-N matrix P such that $P^{\mathrm{T}}P = V^{-1}$. P can be the spectral decomposition of V^{-1}, $P = \Lambda^{-1/2}R$, where R is an N-by-N matrix of which the ith column is the characteristic vector r_i of V, which is the same as the characteristic vector of the spatial weights matrix W (see Griffith 1988, Table 3.1), $R = (r_1, \ldots, r_N)$, and Λ an N-by-N diagonal matrix with the ith diagonal element the corresponding characteristic root, $c_i = T\varphi + (1 - \rho\omega_i)^{-2}$.

---

[9] Note that $\varphi = \sigma_\mu^2/\sigma^2$ is different from θ^2 in the random effects model and in the random effects spatial lag model.

A similar procedure has been adopted by Yang et al. (2006). It is clear that for large N the numerical determination of P can be problematic. However, Hunneman et al. (2007) find that if W is kept symmetric by using one of the alternative normalizations discussed in Section C.2.2, this procedure works well within a reasonable amount of time for values of N up to 4,000.

As a result of Eqs. (C.2.40) and (C.2.41), the log-likelihood function simplifies to

$$\ln L = -\tfrac{NT}{2}\ln(2\pi\sigma^2) - \tfrac{1}{2}\sum_{i=1}^{N}\ln\!\left[1+T\phi(1-\rho\omega_i)^2\right] + T\sum_{i=1}^{N}\ln(1-\rho\omega_i) - \tfrac{1}{2\sigma_2}e^{\mathrm{oT}}e^{\mathrm{o}}$$

(C.2.42)

where $e^{\mathrm{o}} = y^{\mathrm{o}} - X^{\mathrm{o}}\beta$. β and σ^2 can be solved from their first-order maximizing conditions: $\beta = (X^{\mathrm{oT}}X^{\mathrm{o}})^{-1}X^{\mathrm{oT}}y^{\mathrm{o}}$ and $\sigma^2 = (y^{\mathrm{o}} - X^{\mathrm{o}}\beta)^{\mathrm{T}}(y^{\mathrm{o}} - X^{\mathrm{o}}\beta)/NT$. Upon substituting β and σ^2 in the log-likelihood function, the concentrated log-likelihood function of ρ and φ is obtained

$$\ln L = C - \tfrac{NT}{2}\ln\!\left[e(\rho,\phi)^{\mathrm{T}}e(\rho,\phi)\right] - \tfrac{1}{2}\sum_{i=1}^{N}\ln\!\left[1+T\phi(1-\rho\omega_i)^2\right] + T\sum_{i=1}^{N}\ln(1-\rho\omega_i)$$

(C.2.43)

where C is a constant not depending on ρ and φ and the typical element of e (ρ, φ) is

$$e(\rho,\theta)_{it} = y_{it} - \rho\sum_{j=1}^{N}W_{ij}y_{jt} + \sum_{j=1}^{N}\left\{[p(\rho,\varphi)_{ij} - (1-\rho W_{ij})]\tfrac{1}{T}\sum_{t=1}^{T}y_{jt}\right\}$$

$$-\left\{X_{it} - \rho\sum_{j=1}^{N}W_{ij}X_{jt} + \sum_{j=1}^{N}\left[p(\rho,\varphi)_{ij} - (1-\rho W_{ij})\right]\tfrac{1}{T}\sum_{t=1}^{T}X_{jt}\right\}\beta. \qquad \text{(C.2.44)}$$

The notation $p_{ij} = p\,(\rho,\varphi)_{ij}$ is used to indicate that the elements of the matrix P depend on ρ and φ. One can iterate between β and σ^2 on the one hand, and ρ and φ on the other, until convergence. The estimators of β and σ^2, given ρ and φ, can be obtained by OLS regression of the transformed variable y^{o} on the transformed variables X^o. However, the estimators of ρ and φ, given $\boldsymbol{\beta}$ and σ^2, must be attained by numerical methods because the equations cannot be solved analytically.

The asymptotic variance matrix of this model has been derived by Baltagi et al. (2007). They develop diagnostics to test for serial error correlation, spatial error correlation and/or spatial random effects. They also derive asymptotic variance matrices provided that one or more of the corresponding coefficients are zero. One objection to this study is that serial and spatial error correlation are modeled sequentially instead of jointly. Elhorst (2008b) demonstrates that jointly modeling serial and spatial error correlation results in a trade-off between the serial and spatial autocorrelation coefficients and that ignoring this trade-off causes inefficiency and may lead to non-stationarity. However, if the serial autocorrelation coefficient is set to zero, this problem disappears. Consequently, the asymptotic variance matrix that is obtained if the serial autocorrelation coefficient is set to zero exactly happens to be the variance matrix of the random effects spatial error model.

One difference is that Baltagi et al. (2007) do not derive the asymptotic variance matrix of $\boldsymbol{\beta}, \rho, \varphi$ and σ^2, but of $\boldsymbol{\beta}, \rho, \sigma_\mu^2$ and σ^2. This matrix takes the following form[10]

$$AsyVar(\boldsymbol{\beta}, \rho, \sigma_\mu^2, \sigma^2) =$$

$$\begin{bmatrix} \frac{1}{\sigma^2} X^{oT} X^o & & & \\ 0 & \frac{T-1}{2} trace(\boldsymbol{\Gamma})^2 + \frac{1}{2} trace(\boldsymbol{\Sigma\Gamma})^2 & & \\ 0 & \frac{T}{2\sigma^2} trace(\boldsymbol{\Sigma\Gamma V}^{-1}) & \frac{T^2}{2\sigma^4} trace(\boldsymbol{V}^{-1})^2 & \\ 0 & \frac{T-1}{2\sigma^2} trace(\boldsymbol{\Gamma}) + \frac{1}{2\sigma^2} trace(\boldsymbol{\Sigma\Gamma\Sigma}) & \frac{T}{2\sigma^4} trace(\boldsymbol{\Sigma V}^{-1}) & \frac{1}{2\sigma^4}[(T-1)N + trace(\boldsymbol{\Sigma})^2] \end{bmatrix}^{-1}$$

(C.2.45)

where $\boldsymbol{\Gamma} = (\boldsymbol{W}^T \boldsymbol{B} + \boldsymbol{B}^T \boldsymbol{W})(\boldsymbol{B}^T \boldsymbol{B})^{-1}$ and $\boldsymbol{\Sigma} = \boldsymbol{V}^{-1}(\boldsymbol{B}^T \boldsymbol{B})^{-1}$. Since $\varphi = \sigma_\mu^2/\sigma^2$, the asymptotic variance of φ can be obtained using the formula (Mood et al. 1974, p.181)

$$var(\varphi) = \varphi^2 \left[\frac{var(\sigma_\mu^2)}{(\varphi\sigma^2)^2} + \frac{var(\sigma^2)}{(\sigma^2)^2} - 2\frac{var(\sigma_\mu^2, \sigma^2)}{(\varphi\sigma^2)\sigma^2} \right].$$

(C.2.46)

---

[10] Note that the matrix Z_0 in Baltagi et al. (2007, pp.39-40) has been replaced by $Z_0 = [T\sigma_\mu^2 \boldsymbol{I}_N + \sigma^2 (\boldsymbol{B}^T \boldsymbol{B})^{-1}]^{-1} = \frac{1}{\sigma^2}[T\varphi \boldsymbol{I}_N + (\boldsymbol{B}^T \boldsymbol{B})^{-1}]^{-1} = \frac{1}{\sigma^2}\boldsymbol{V}^{-1}$.

In conclusion, we can say that the estimation of the random effects spatial error model is far more complicated than that of the other spatial panel data models. Since a spatial error specification also does not require a theoretical model for a spatial or social interaction process, but is a special case of a non-spherical error covariance matrix, and the random effects models in spatial research is controversial, the random effects spatial error model will probably be of limited value in empirical research.

C.2.5 Model comparison and prediction

This section sets forth Hausman's specification test for statistically significant differences between random effects models and fixed effects models, two goodness-of-fit measures, one that includes the impact of spatial fixed or random effects and the impact of a spatial lag and one that does not, and the best linear unbiased predictor of the different models.

Random effects versus fixed effects

The random effects model can be tested against the fixed effects model using Hausman's specification test (Baltagi 2005, pp.66-68). The hypothesis being tested is $H_0: h = 0$, where

$$h = d^{\mathrm{T}}[var(d)]^{-1}d, \ d = \hat{\beta}_{FE} - \hat{\beta}_{RE} \ \text{ and } \ var(d) = \hat{\sigma}_{RE}^2(X^{\bullet\mathrm{T}}X^{\bullet})^{-1} - \hat{\sigma}_{FE}^2(X^{*\mathrm{T}}X^{*})^{-1}.$$

(C.2.47)

Note the reversed sequence with which d and $var(d)$ are calculated. This test statistic has a chi-squared distribution with K degrees of freedom (the number of explanatory variables in the model, excluding the constant term). Hausman's specification test can also be used when the model is extended to include spatial error autocorrelation or a spatially lagged dependent variable. Since the spatial lag model has one additional explanatory variable, one might calculate d by $d = [\hat{\beta}^{\mathrm{T}} \ \hat{\delta}]_{FE}^{\mathrm{T}} - [\hat{\beta}^{\mathrm{T}} \ \hat{\delta}]_{RE}^{\mathrm{T}}$ to obtain a test statistic that has a chi-squared distribution with $K+1$ degrees of freedom. To calculate $var(d)$ in this particular case, one should extract the first $K+1$ rows and columns of the variance matrices in Eqs. (C.2.29) and (C.2.38). If the hypothesis is rejected, the random effects models must be rejected in favor of the fixed effects model.

Goodness-of-fit

The computation of a goodness-of-fit measure in spatial panel data models is difficult because there is no precise counterpart of the R^2 of an OLS regression model with disturbance covariance $\sigma^2 I$ to a generalized regression model with disturbance covariance matrix $\sigma^2 \Omega$ ($\Omega \neq I$). Most people use

$$R^2(e,\Omega) = 1 - \frac{e^T \Omega e}{(y-\bar{y})^T (y-\bar{y})} \quad \text{or} \quad R^2(\tilde{e}) = 1 - \frac{\tilde{e}^T \tilde{e}}{(y-\bar{y})^T (y-\bar{y})} \quad (C.2.48)$$

where $\bar{y}$ denotes the overall mean of the dependent variable in the sample and e is the residual vector of the model. Alternatively, $e^T \Omega e$ can be replaced by the residual sum of squares of transformed residuals $\tilde{e}^T \tilde{e}$.

One objection to the measures in Eq. (C.2.48) is that there is no assurance that adding (eliminating) a variable to (from) the model will result in an increase (decrease) of R^2. This problem is at issue in the fixed effects spatial error model, the random effects spatial lag model and the random effects spatial error model, because the coefficients ρ, θ or φ may change when changing the set of independent variables. The problem is not at issue in the fixed effects spatial lag model, even though it may be seen as a linear regression model with disturbance covariance matrix $\sigma^2 Q$. This is because the demeaning procedure was only meant to speed up computation time and to improve the accuracy of the estimates of β. If the R^2 is calculated after the spatial fixed effects have been added back to the model, it will have the same properties as the R^2 of the OLS model.

An alternative goodness-of-fit measure that meets the above objection is the squared correlation coefficient between actual and fitted values (Verbeek 2000, p.21)

$$corr^2(y,\hat{y}) = \frac{[(y-\bar{y})^T (\hat{y}-\bar{y})]^2}{[(y-\bar{y})^T (y-\bar{y})][(\hat{y}-\bar{y})^T (\hat{y}-\bar{y})]} \quad (C.2.49)$$

where $\hat{y}$ is an NT-by-1 vector of fitted values. Unlike the R^2, this goodness-of-fit measure ignores the variation explained by the spatial fixed effects. The argumentation is that the estimator of β in the fixed effects model is chosen to explain the time-series rather than the cross-sectional component of the data, as well as that the spatial fixed effects capture rather than explain the variation between the spatial units (Verbeek 2000, p.320). This is also the reason why the spatial fixed effects are often not computed, let alone reported. The difference between R^2 and $corr^2$ indicates how much of the variation is explained by the fixed effects, which in many cases is quite substantial. A similar type of argument applies to spatial random effects.

Another difficulty is how to cope with a spatially lagged dependent variable. If the spatial lag is seen as a variable that helps to explain the variation in the dependent variable, the first measure (R^2) should be used. By contrast, if the spatial lag is not seen as variable that helps to explain the variation in the dependent variable, simply because it is a left-hand side variable in principle, the second measure ($corr^2$) should be used. The latter measure is adopted by LeSage (1999) to calculate the goodness-of-fit of the spatial lag model in a cross-sectional setting.[11] In vector notation, the reduced form of the spatial lag model in Eq. (C.2.2) is

$$y=[I_{NT}-\delta(I_T \otimes W)]^{-1}[X\beta + (\tau_T \otimes I_N)\mu + \varepsilon] \qquad (C.2.50)$$

where μ is an N-by-1 vector of the spatial specific effects, $\mu = (\mu_1, \ldots, \mu_N)^T$. From this equation it can be seen that the squared correlation coefficient between actual and fitted values in spatial lag models, no matter whether μ is fixed or random, should also account for the spatial multiplier matrix $[I_{NT} - \delta(I_T \otimes W)]^{-1}$.

Table C.2.1. Two goodness-of-fit measures of the four spatial panel data models

	Fixed effects spatial lag model
$R^2(e, I_n)$	$e = y - \hat{\delta}(I_T \otimes W)y - X\hat{\beta} - (\tau_T \otimes I_N)\hat{\mu}$
$Corr^2$	$corr^2\left\{y^*, [I_{NT} - \hat{\delta}(I_T \otimes W)]^{-1} X^* \hat{\beta}\right\}$
	Fixed effects spatial error model
$R^2(\tilde{e})$	$\tilde{e} = y - \hat{\rho}(I_T \otimes W)y - [X - \hat{\rho}(I_T \otimes W)X]\hat{\beta} - (\tau_T \otimes I_N)\hat{\mu}$
$Corr^2$	$corr^2(y^*, X^* \hat{\beta})$
	Random effects spatial lag model
$R^2(\tilde{e})$	$\tilde{e} = y^\bullet - \hat{\delta}(I_T \otimes W)y^\bullet - X^\bullet \beta$
$Corr^2$	$corr^2\left\{y, [I_{NT} - \hat{\delta}(I_T \otimes W)]^{-1} X\hat{\beta}\right\}$
	Random effects spatial error model
$R^2(\tilde{e})$	$\tilde{e} = y^\circ - X^\circ \hat{\beta}$
$Corr^2$	$corr^2(y, X\hat{\beta})$

Notes: $R^2(e, I_N)$ and $R^2(\tilde{e})$ are defined by Eq. (C.2.48), $corr^2$ is defined by Eq. (C.2.49)

---

[11] See the routine 'sar' posted at LeSage's website <www.spatial-econometrics.com>

The two measures for the different spatial panel data models are listed in Table C.2.1. It shows that in the fixed and random effects spatial lag model not only the spatially lagged dependent variable, but also the spatial fixed or random effects are ignored when calculating the squared correlation coefficient between actual and fitted values.

Prediction

Finally, prediction formulas are presented for fixed effects and random effects models with spatial interaction effects. Goldberger (1962) shows that the best linear unbiased predictor (BLUP) for the cross-sectional units in a linear regression model with disturbance covariance matrix Ω at a future period $T+C$ is given by

$$\hat{y}_{T+C} = X_{T+C}\hat{\beta} + \psi^T \Omega^{-1} e \tag{C.2.51}$$

where $\psi = E(\varepsilon_{T+C}\,\varepsilon)$ is the covariance between the future disturbance ε_{T+C} and the sample disturbances ε, X covers the independent variables of the model, $\hat{\beta}$ is the estimator of β, and e denotes the residual vector of the model. Baltagi and Li (2004) derive the prediction formulas for the fixed effects and random effects model with spatial autocorrelation. Here, we also present these formulas for the fixed effects and random effects model extended to include a spatially lagged dependent variable based on own derivations. The prediction formulas are listed in Table C.2.2.

Baltagi and Li (2004) point out that $\psi = 0$ in the fixed effects model, provided that error terms are not serially correlated over time. Unlike the fixed effects model, the correction term $\psi^T \Omega e$ in the random effects model is not zero. In the random effects spatial lag model, the correction term $\psi^T \Omega\, e$ is identically equal to its counterpart in a standard random effects model, which has been reported in Baltagi and Li (2004). To calculate this correction term (see Table C.2.2), the residuals of each spatial unit are first averaged over the sample period and then multiplied with $(1-\theta^2)$, a factor that can take values between zero and one.[12] However, in addition to the standard random effects model, both $X_{T+C}\hat{\beta}$ and the correction term should also be premultiplied with the N-by-N spatial multiplier matrix $(I_{NT} - \delta W)^{-1}$.

---

[12] Note that $(1 - \theta^2) = T\sigma_\mu^2 / (T\sigma_\mu^2 + \sigma^2)$ (see Baltagi 2005, p. 20, for the second part of this formula).

Table C.2.2. Prediction formula of the four spatial panel data models

Fixed effects spatial lag model

$$\hat{y}_{T+C} = (I_N - \hat{\delta}W)^{-1} X_{T+C} \hat{\beta} + (I_N - \hat{\delta}W)^{-1} \hat{u}$$

Fixed effects spatial error model

$$\hat{y}_{T+C} = X_{T+C} \hat{\beta} + \hat{u}$$

Random effects spatial lag model

$$\hat{y}_{T+C} = (I_N - \hat{\delta}W)^{-1} X_{T+C} \hat{\beta} + (I_N - \hat{\delta}W)^{-1} (1-\hat{\theta}^2) \left\{ \frac{1}{T} \sum_{t=1}^{T} \begin{pmatrix} y_{1t} - X_{1t} \hat{\beta} \\ \vdots \\ y_{Nt} - X_{Nt} \hat{\beta} \end{pmatrix} \right\}$$

Random effects spatial error model

$$\hat{y}_{T+C} = X_{T+C} \hat{\beta} + \hat{\varphi} \hat{V}^{-1} \sum_{t=1}^{T} \begin{pmatrix} y_{1t} - X_{1t} \hat{\beta} \\ \vdots \\ y_{Nt} - X_{Nt} \hat{\beta} \end{pmatrix}$$

Just as in the random effects spatial lag model, the residuals in the random effects spatial error model are first averaged over the sample period (see Table C.2.2). However, the sum of the residuals is not just divided by T, but premultiplied by $V^{-1} = [T\varphi I_N + (B^T B)^{-1}]^{-1}$, a matrix that also accounts for the interaction effects among the residuals. Finally, the 'average' residuals are multiplied by φ, which measures the ratio between σ_μ^2 and σ^2.

One problem of predictors based on fixed or random effects models is that one has no information on the spatial fixed effects or the averaged residuals of spatial units outside the sample. For this reason, some researchers abandon fixed or random effects models. However, they better stick to the fixed effects or random effects models, provided that these effects appear to be (jointly) significant, and set the spatial fixed effects or the averaged residuals of spatial units outside the sampling region to zero or, alternatively, try to approach them from proximate spatial units within the sample region.

C.2.6 Concluding remarks

The spatial econometrics literature has exhibited a growing interest in the specification and estimation of econometric relationships based on spatial panels. Many empirical studies have found their way to the Matlab routines of the fixed effects and random effects models the author of this chapter has provided at his website.

Updated versions have been made available and include the (robust) LM tests, the estimation of fixed effects and the determination of their significance level, the determination of the variance-covariance matrix of the parameters estimates, the determination of good-of-fit measures, Hausman's specification test and the formulas for the best linear unbiased predictor, as discussed in this chapter.

Two other areas where more insight has been gained into the extension of spatial panel data models with spatial interaction effects is the possibility to test for endogeneity of one or more of the explanatory variables and the possibility to include dynamic effects. However, this literature has not yet been crystallized.

Fingleton and LeGallo (2007) consider models including an endogenous spatial lag, additional endogenous variables due to a system feedback and an autoregressive or a moving average error process, and suggest an IV/GMM estimator based on Kelejian and Prucha (1998) and Fingleton (2008). Elhorst et al. (2007) present a framework to determine the best of three estimators (2SLS, fixed effects 2SLS and first-difference 2SLS) in the presence of potential endogeneity using two Hausman type test-statistics. Using this framework, they conclude that the first-difference 2SLS is the preferred estimator of the East German wage curve, since the regional unemployment rate, the main explanatory variable of the wage rate, is not strictly exogenous and the spatial specific effects are not uncorrelated to the explanatory variables. To investigate the possible endogeneity of the regional unemployment rate in combination with time-specific effects, a similar framework is used, except for the first-difference 2SLS estimator. This is because first differencing does not assist in eliminating time specific effects. For this reason, they develop a spatial first-difference 2SLS estimator where the values of y and X in every spatial unit are taken in deviation of y and X in one reference spatial unit.

Finally, Elhorst (2008a) adopts the use of matrix exponentials, a transformation recently introduced by LeSage and Pace (2007). This transformation is different from the spatial lag model in Eq. (C.2.2) or the spatial error model in Eq. (C.2.3) in that its Jacobian term is zero. This zero Jacobian term opens the opportunity to use an estimation method partly based on IV and partly based on ML to control for endogeneity of one or more of the explanatory variables.

There has also been a growing interest in the estimation of dynamic panel data models. Elhorst (2005a) derives the ML estimator and Su and Yang (2007) the corresponding regularity conditions of a dynamic panel data model extended to include spatial error autocorrelation. Elhorst (2005b), Korniotis (2005), Yu et al. (2007) and Vrijburg et al. (2007) consider a dynamic panel data model extended to include a spatially lagged dependent variable. Up to now, the first of these six studies has also been applied successfully in the empirical work of other researchers (Kholodilin et al. 2008).

Acknowledgements. The author of this chapter is grateful to Maarten Allers, Jan Jacobs, Thomas Seyffertitz and the editors of this Handbook for useful suggestions and comments on an earlier draft.

References

Allers MA, Elhorst JP (2005) Tax mimicking and yardstick competition among governments in the Netherlands. Int Tax Publ Fin 12(4):493-513
Anselin L (1988) Spatial econometrics: methods and models. Kluwer, Dordrecht
Anselin L, Bera AK (1998) Spatial dependence in linear regression models with an introduction to spatial econometrics. In Ullah A, Giles DEA (eds) Handbook of applied economic statistics. Marcel Dekker, New York, pp.237-289
Anselin L, Hudak S (1992) Spatial econometrics in practice: a review of software options. Reg Sci Urban Econ 22(3):509-536
Anselin L, Le Gallo J, Jayet H (2006) Spatial panel econometrics. In Matyas L, Sevestre P. (eds) The econometrics of panel data, fundamentals and recent developments in theory and practice (3rd edition). Kluwer, Dordrecht, pp.901-969
Anselin L, Bera AK, Florax R, Yoon MJ (1996) Simple diagnostic tests for spatial dependence. Reg Sci Urban Econ 26(1):77-104
Baltagi BH (1989) Applications of a necessary and sufficient condition for OLS to be BLUE. Stat Prob Letters 8(5):457-461
Baltagi BH (2005) Econometric analysis of panel data (3rd edition). Wiley, New York, Chichester, Toronto and Brisbane
Baltagi BH (2006) Random effects and spatial autocorrelation with equal weights. Econ Theory 22(5):973-984
Baltagi BH, Li D (2004) Prediction in the panel data model with spatial autocorrelation. In Anselin L, Florax RJGM, Rey SJ (eds) Advances in spatial econometrics: methodology, tools, and applications. Springer, Berlin, Heidelberg and New York, pp.283-295
Baltagi BH, Song SH, Jung BC, Koh W (2007) Testing for serial correlation, spatial autocorrelation and random effects using panel data. J Econometrics 140(1):5-51
Beck N (2001) Time-series-cross-section data: what have we learned in the past few years? Ann Rev Pol Sci 4(1):271-293
Breusch TS (1987) Maximum likelihood estimation of random effects models. J Econometrics 36(3):383-389
Brueckner JK (2003) Strategic interaction among local governments: an overview of empirical studies. Int Reg Sci Rev 26(2):175-188
Cressie NAC (1993) Statistics for spatial data (revised edition). Wiley, New York, Chichester, Toronto and Brisbane
Elhorst JP (2001) Dynamic models in space and time. Geogr Anal 33(2):119-140
Elhorst JP (2003) Specification and estimation of spatial panel data models. Int Reg Sci Rev 26(3):244-268
Elhorst JP (2005a) Unconditional maximum likelihood estimation of linear and log-linear dynamic models for spatial panels. Geogr Anal 37(1):62-83
Elhorst J.P. (2005b) Models for dynamic panels in space and time; an application to regional unemployment in the EU. Paper presented at the Spatial Econometrics Workshop, April 8-9, 2005, Kiel

Elhorst JP (2008a) A spatiotemporal analysis of aggregate labour force behaviour by sex and age across the European Union. J Geogr Syst 10(2):167-190

Elhorst JP (2008b) Serial and spatial autocorrelation. Econ Letters 100(3):422-424

Elhorst JP, Freret S (2009) Evidence of political yardstick competition in France using a two-regime spatial Dublin model with fixed effects. J Reg Sci. DOI: 10.1111/j.1467-9787.2009.00613.x [forthcoming]

Elhorst JP, Blien U, Wolf K (2007) New evidence on the wage curve: a spatial panel approach. Int Reg Sci Rev 30(2):173-191

Elhorst JP, Piras G, Arbia G (2006) Growth and convergence in a multi-regional model with space-time dynamics. Paper presented at the Spatial Econometric Workshop, May 25-27, 2006, Rome

Ertur C, Koch W (2007) Growth, technological interdependence and spatial externalities: theory and evidence. J Appl Econ 22(6):1033-1062

Fingleton B (2008) A generalized method of moments estimator for a spatial panel data model with endogenous spatial lag and spatial moving average errors. Spat Econ Anal 3(1):27-44

Fingleton B, Le Gallo J (2007) Estimating spatial models with endogenous variables, a spatial lag en spatially dependent disturbances: finite sample properties. Paper presented at the First World Conference of the Spatial Econometrics Association, July 11-14, 2007, Cambridge

Florax RJGM, Folmer H (1992) Specification and estimation of spatial linear regression models. Reg Sci Urban Econ 22(3):405-432

Florax RJGM, Folmer H, Rey SJ (2003) Specification searches in spatial econometrics: the relevance of Hendry's methodology. Reg Sci Urban Econ 33(5):557-579

Franzese Jr RJ, Hays JC (2007) Spatial econometric models of cross-sectional interdependence in political science panel and time-series-cross-section data. Pol Anal 15(2):140-164

Goldberger AS (1962) Best linear unbiased prediction in the generalized linear regression model. J Am Stat Assoc 57:369-375

Greene WH (2008) Econometric analysis (6th edition). Pearson, Upper Saddle River [NJ]

Griffith DA (1988) Advanced spatial statistics. Kluwer, Dordrecht

Griffith DA, Lagona F (1998) On the quality of likelihood-based estimators in spatial autoregressive models when the data dependence structure is mis-specified. J Stat Plann Inference 69(1):153-174

Hendry DF (2006) A comment on 'Specification searches in spatial econometrics: The relevance of Hendry's methodology'. Reg Sci Urban Econ 36(2):309-312

Hsiao C (2003) Analysis of Panel Data (2nd edition). Cambridge University Press, Cambridge

Hsiao C (2005) Why panel data? University of Southern California, IEPR Working Paper 05.33

Hunneman A, Bijmolt T, Elhorst JP (2007) Store location evaluation based on geographical consumer information. Paper presented at the Marketing Science Conference, June 28-30, 2007, Singapore

Jarque CM, Bera AK (1980) Efficient tests for normality, homoskedasticity and serial independence of regression residuals. Econ Letters 6(3):255-259

Kapoor M, Kelejian HH Prucha IR (2007) Panel data models with spatially correlated error components. J Econometrics 140(1):97-130

Kelejian HH, Prucha IR (1998) A generalized spatial two stage least squares procedure for estimating a spatial autoregressive model with autoregressive disturbances. J Real Est Fin Econ 17(1):99-121

Kelejian HH, Prucha IR, Yuzefovich Y (2006) Estimation problems in models with spatial weighting matrices which have blocks of equal elements. J Reg Sci 46(3):507-515

Kholodilin KA, Siliverstovs B, Kooths S (2008) A dynamic panel data approach to the forecasting of the GDP of German Länder. Spat Econ Anal 3(2):195-207

Korniotis GM (2005) A dynamic panel estimator with both fixed and spatial effects. Paper presented at the Spatial Econometrics Workshop, April 8-9, 2005, Kiel

Lahiri SN (2003) Central limit theorems for weighted sums of a spatial process under a class of stochastic and fixed designs. Sankhya 65(2): 356-388

Lee LF (2003) Best spatial two-stage least squares estimators for a spatial autoregressive model with autoregressive disturbances. Econ Rev 22(4):307-335

Lee LF (2004) Asymptotic distribution of quasi-maximum likelihood estimators for spatial autoregressive models. Econometrica 72(6):1899-1925

Leenders RTAJ (2002) Modeling social influence through network autocorrelation: Constructing the weight matrix. Soc Netw 24(1):21-47

LeSage JP (1999) Spatial econometrics. www.spatial-econometrics.com/html/sbook.pdf

LeSage JP, Pace RK (2007) A matrix exponential spatial specification. J Econometrics 140(1):190-214

Magnus JR (1982) Multivariate error components analysis of linear and non-linear regression models by maximum likelihood. J Econometrics 19(2):239-285

Magnus JR, Neudecker H (1988) Matrix differential calculus with applications in statistics and econometrics. Wiley, New York, Chichester, Toronto and Brisbane

Manski CF (1993) Identification of endogenous social effects:the reflection problem. Rev Econ Stud 60:531-542

Mood AM, Graybill F, Boes DC (1974) Introduction to the theory of statistics (3rd edition). McGraw-Hill, Tokyo

Nerlove M, Balestra P (1996) Formulation and estimation of econometric models for panel data. In Mátyás L, Sevestre P (eds) The econometrics of panel data (2nd edition). Kluwer, Dordrecht, pp.3-22

Ord JK (1975) Estimation methods for models of spatial interaction. J Am Stat Assoc 70:120-126

Pace RK, Barry R (1997) Quick computation of spatial autoregressive estimators. Geogr Anal 29(3):232-246

Partridge MD (2005) Does income distribution affect U.S. state economic growth. J Reg Sci 45(2):363-394

Su L, Yang Z (2007) QML Estimation of dynamic panel fata models with spatial errors. Paper presented at the First World Conference of the Spatial Econometrics Association, July 11-14, 2007, Cambridge

Verbeek M (2000) A guide to modern econometrics. Wiley, New York, Chichester, Toronto and Brisbane

Vrijburg H, Jacobs JPAM, Ligthart JE (2007) A spatial econometric approach to commodity tax competition. Paper presented at the NAKE Research Day, October 24, 2007, Utrecht

Yang Z, Li C, Tse YK (2006) Functional form and spatial dependence in spatial panels. Econ Letters 91(1):138-145

Yu J, Jong R de, Lee L (2007) Quasi-maximum likelihood estimators for spatial dynamic panel data with fixed effects when both n and T are large. J Econometrics 146(1):118-134

C.3 Spatial Econometric Methods for Modeling Origin-Destination Flows

James P. LeSage and *Manfred M. Fischer*

C.3.1 Introduction

Spatial econometric theory and practice have been dominated by a focus on object data. In economic analysis these objects correspond to economic agents with discrete locations in geographic space, such as addresses, census tracts and regions. In contrast spatial interaction or flow data pertain to measurements each of which is associated with a link or pair of origin-destination locations that represent points or areas in space. While there is a voluminous literature on the specification and estimation of models for cross-sectional object data (see, Chapter C.1 in this volume), less attention has been paid to sample data consisting of origin-destination pairs that form the basic units of analysis in spatial interaction models.

Spatial interaction models represent a class of methods which are used for modeling origin-destination flow data. The interest in such models is motivated by the need to understand and explain flows of tangible entities such as persons and commodities or intangible ones such as capital, information or knowledge across geographic space. By adopting a spatial interaction modeling perspective attention is focused on interaction patterns at the aggregate rather than the individual level.

The basis of modeling is the use of a discrete zone system. Discrete zone systems can obviously take many different forms, both in relation to the level of resolution and the shape of zones. The subdivision of the geography into zones introduces spatial aggregation problems. Such problems come from the fact that substantially different conclusions can be obtained from the same dataset and the same spatial interaction model, but at another spatial aggregation level (see, for example, Batty and Sidkar 1982). Spatial aggregation problems involve both a scale issue and a zoning issue. The tidiest, and often most convenient system to use would be a square grid. But quite often one is forced to use administratively defined regions, such as NUTS-2 regions in Europe, counties in a country or the wards of a city.

The subject of spatial interaction modeling has a long and distinguished history that has led to the emergence of three major schools of analytical thought: the macroscopic school based upon a statistical equilibrium approach (see Wilson 1967; Roy 2004), the microscopic school based on a choice-theoretic approach (see Smith 1975; Sen and Smith 1995), and the geocomputational school based upon the neural network approach that processes spatial interaction models as universal function approximators (see Fischer 2002; Fischer and Reismann 2002). In these schools there is a deep-seated view that spatial interaction implies movement of entities, and that this has little to do with spatial association (Getis 1991).

Spatial interaction models typically rely on three types of factors to explain mean interaction frequencies between origins and destinations of interaction: (i) origin-specific attributes that characterize the ability of the origins to produce or generate flows, (ii) destination-specific attributes that represent the attractiveness of destinations, and (iii) origin-destination variables that characterize the way spatial separation of origins from destinations constrains or impedes the interaction. They implicitly assume that using spatial separation variables such as distance will eradicate the spatial dependence among the sample of spatial flows.

However, research dating back to the 1970s, noted that spatial dependence or autocorrelation might be intermingled in spatial interaction model specifications. This idea was first put forth in a theoretical context by Curry (1972), with some subsequent debate in Curry et al. (1975). Griffith and Jones (1980) documented the presence of spatial dependence in conventional spatial interaction models. Despite this, most practitioners assume independence among observations and few have used spatial lags of the dependent variable or disturbances in spatial interaction models. Exceptions are Bolduc et al. (1992), and Fischer and Griffith (2008) who rely on spatial lags of the disturbances, and LeSage and Pace (2008) who use lags of the dependent variable.

The focus of this chapter is on problems that plague empirical implementation of conventional regression-based spatial interaction models and econometric extensions that have recently appeared in the literature. These new models replace the conventional assumption of independence between origin-destination flows with formal approaches that allow for spatial dependence in flow magnitudes. We follow LeSage and Pace (2008) and extend the generic version of the spatial interaction model to include spatial lags of the dependent variable.

C.3.2 The analytical framework

Spatial interaction data represent phenomena that may be described in their most general terms as interactions between populations of actors and opportunities distributed over some relevant geographic space. Such interactions may involve movements of individuals from one location to another, such as daily traffic flows in which case the relevant actors are individual travellers (commuters, shoppers, etc.) and the relevant opportunities are their destinations (jobs, stores, etc.). Similarly, one may consider annual migration flows, where the relevant actors are migrants (individuals, family units, firms, etc.) and the relevant opportunities are their possible new locations. Interactions may also involve flows of information such as telephone calls or electronic messages. Here the callers or message senders may be the relevant actors, and the possible receivers of calls or electronic messages may be considered as the relevant opportunities (Sen and Smith 1995). With this range of examples in mind, the purpose of this section is to outline a framework in which all such spatial interaction behaviour can be studied.

The classical spatial interaction model

Suppose we have a spatial system consisting of n discrete zone (locations, regions) where i ($i = 1, ..., n$) denotes the origin and j ($j = 1, ..., n$) the destination of interaction. Let $m(i, j)$ denote observations on random variables, say $M(i, j)$, each of which corresponds to a movement of tangible or intangible entities from i to j. The $M(i, j)$ are assumed to be independent random variables. They are sampled from a specified probability distribution that is dependent upon some mean, say $\mu(i, j)$. Let us assume that no a priori information is given about the origin and destination totals of the observed flow matrix. Then the mean interaction frequencies between origin i and destination j may be modeled by

$$\mu(i, j) = C \ A(i) \ B(j) \ S(i, j) \tag{C.3.1}$$

where $\mu(i, j) = E[M(i, j)]$ is the expected flow, C denotes a constant term, the quantities $A(i)$ and $B(j)$ are called origin and destination factors or variables respectively, and $S(.)$ is some unspecified distance deterrence function (see Fischer and Griffith 2008). Note if the outflow totals for each origin zone and/or the inflow totals into each destination zone are known a priori, then model (C.3.1) would need to be modified to incorporate the explicitly required constraints to match exact totals. Imposing origin and/or destination constraints leads to socalled *production-constrained*, *attraction-constrained* and *production-attraction-constrained* spatial interaction models that may be convincingly justified using entropy maximizing methods (see Fotheringham and O'Kelly 1989; Bailey and Gatrell 1995 for a discussion).

Equation (C.3.1) is a very general version of the classical (unconstrained) spatial interaction model. The exact functional form of the three terms $A(.)$, $B(.)$ and $S(.)$ on the right hand side of Eq. (C.3.1) is subject to varying degrees of conjecture. There is wide agreement that the origin and destination factors are generally best given by power functions

$$A(i) = (A_i)^\beta \tag{C.3.2a}$$

$$B(j) = (B_j)^\gamma \tag{C.3.2b}$$

where A_i represents some appropriate variable measuring the propulsiveness of origin i, and B_j some appropriate variable measuring the attractiveness of destination j in a specific spatial interaction context. The product $A(i)B(j)$ can be interpreted simply as the number of distinct (i, j)-interactions that are possible. Thus, for origin-destination pairs (i, j) with the same level of separation, it follows from Eq. (C.3.1) that mean interaction levels are proportional to the number of possible interactions between such (i, j)-pairs. The exponents, β and γ, indicate the origin and destination effects respectively, and are treated as statistical parameters to be estimated.

If more than one origin and one destination variable are relevant in a specific context the above specification may be extended to

$$A(i) = \prod_{q \in Q} (A_{iq})^{\beta_q} \tag{C.3.3a}$$

$$B(j) = \prod_{r \in R} (B_{jr})^{\gamma_r} \tag{C.3.3b}$$

where A_{iq} $(q \in Q)$ and B_{jr} $(r \in R)$ represent sets of relevant (positive) origin-specific and destination-specific variables, respectively. The exponents $(\beta_q : q \in Q)$ and $(\gamma_r : r \in R)$ are parameters to be estimated. See Fotheringham and O'Kelly (1989) for a range of explicit variable specifications.

The distance deterrence function $S(i, j)$ constitutes the very core of spatial interaction models. Hence, a number of alternative specifications have been proposed in the literature (for a discussion see Sen and Smith 1995). One prominent example is the following power function specification given by

$$S(i, j) = [D(i, j)]^\theta \tag{C.3.4}$$

for any positive scalar distance measure, $D(i, j)$, and negative distance sensitivity parameter θ that has to be estimated. Another popular specification is the exponential function $S(i, j) = \exp[-\theta D(i, j)]$, where θ has to be an univariate parameter with specific value depending on the choice of units for distance (see Sen and Smith 1995).

The deterrence function reflects the way in which spatial separation or distance constrains or impedes movement across space. In general we will refer to this as distance between an origin i and a destination j, and denote it as $D(i, j)$. At relatively large scales of geographical inquiry this might be simply the great circle distance separating an origin from a destination zone measured in terms of the distance between their respective centroids. In other cases, it might be transportation or travel time, cost of transportation, perceived travel time or any other sensible measure such as political distance, language distance or cultural distance measured in terms of nominal or categorical attributes. To allow for the possibility of multiple measures of spatial separation, the power function specification in Eq. (C.3.4) can be extended to the following class of multivariate power deterrence functions

$$S(i, j) = \prod_{k \in K} [^k D(i, j)]^{\theta_k} \tag{C.3.5}$$

with corresponding distance sensitivity vector $\theta = (\theta_k : k \in K)$.

From the positivity of the functions $A(.)$, $B(.)$ and $S(.)$, it follows that the spatial interaction model (C.3.1) with the specifications (C.3.3) and (C.3.4) can be expressed equivalently as a log-additive model of the form

$$y(i, j) = c + \sum_{q \in Q} \beta_q \, a_q(i) + \sum_{r \in R} \gamma_r \, b_r(j) + \theta \, d(i, j) \tag{C.3.6}$$

where $y(i, j) = \log \mu(i, j)$, $c = \log C$, $a_q(i) = \log A_{iq}$, $b_r(j) = \log B_{jr}$, and $d(i, j) = \log D(i, j)$. In the sequel we will illustrate how these $n^2 (= N)$ equations can be written more compactly using vector and matrix notation.

The spatial interaction model in matrix notation

Let Y denote an n-by-n square matrix of origin-destination flows from each of the n origin zones to each of the n destination zones as shown in Eq. (C.3.7) where the n columns represent different origins and the n rows different destinations. The elements on the main diagonal of the matrix represent intrazonal flows, and we use $N = n^2$ for notational simplicity.

$$Y = \begin{bmatrix} y(1,1) & \cdots & y(1,i) & \cdots & y(1,n) \\ \vdots & & \vdots & & \vdots \\ y(j,1) & \cdots & y(j,i) & \cdots & y(j,n) \\ \vdots & & \vdots & & \vdots \\ y(n,1) & \cdots & y(n,i) & \cdots & y(n,n) \end{bmatrix} \quad \text{(C.3.7)}$$

LeSage and Pace's (2008) introduction of notational conventions allow use of *origin-centric* or *destination-centric* flow matrices. An *origin-centric* ordering of the flow matrix Y is shown in Table C.3.1, where the dyad label denotes the overall index from 1, ..., N for the ordering. The first n elements in the stacked vector y reflect flows from origin zone $i = 1$ to all n destinations and the last n elements flows from origin zone $i = n$ to destinations 1, ..., n. This case often arises in practice when intraregional flows cannot be measured or are difficult to measure.

Table C.3.1. Data organization convention

Dyad label	ID origin	ID destination	Flows	Origin variables	Destination variables	Distance variable
1	1	1	$y(1,1)$	$a_1(1)...a_Q(1)$	$b_1(1)...b_R(1)$	$d(1,1)$
$\vdots$	$\vdots$	$\vdots$	$\vdots$	$\vdots$	$\vdots$	$\vdots$
n	1	n	$y(1,n)$	$a_1(1)...a_Q(1)$	$b_1(n)...b_R(n)$	$d(1,n)$
$n+1$	2	1	$y(2,1)$	$a_1(2)...a_Q(1)$	$b_1(1)...b_R(1)$	$d(2,1)$
$\vdots$	$\vdots$	$\vdots$	$\vdots$	$\vdots$	$\vdots$	$\vdots$
$2n$	2	n	$y(2,n)$	$a_1(2)...a_Q(2)$	$b_1(n)...b_R(n)$	$d(2,n)$
$\vdots$	$\vdots$	$\vdots$	$\vdots$	$\vdots$	$\vdots$	$\vdots$
$N-n+1$	n	1	$y(n,1)$	$a_1(n)...a_Q(n)$	$b_1(1)...b_R(1)$	$d(n,1)$
$\vdots$	$\vdots$	$\vdots$	$\vdots$	$\vdots$	$\vdots$	$\vdots$
N	n	n	$y(n,n)$	$a_1(n)...a_Q(n)$	$b_1(n)...b_R(n)$	$d(n,n)$

The least-squares regression approach widely used in practice to explain variation in origin-destination flows relies on two sets of explanatory variable matrices. One is an N-by-Q matrix of Q origin-specific variables for the n regions that we label X_o. This matrix reflects an n-by-q matrix of explanatory variables X_q ($q = 1, ..., Q$) that is *repeated n* times using $X_o = X \otimes \iota_n$, where ι_n is an n-by-1 vector of ones. The matrix Kronecker product ($\otimes$) works to multiply the right-hand argument ι_n times each element in the matrix X, which strategically repeats the explanatory variables so they are associated with observations treated as origins. Specifically, the matrix product would repeat the origin characteristics of the first zone to form the first n rows, the origin characteristics of the second zone n times for the next n rows and so on (see Table C.3.1), resulting in the N-by-Q matrix X_o. LeSage and Pace (2008) point out that if we organized the matrix of flows Y

using a *destination-centric* ordering based on Y^T, then the matrix of origin-specific explanatory variables would consist of $X_o = \iota_n \otimes X$.

The second matrix is an N-by-R matrix $X_d = \iota_n \otimes X_r$ ($r = 1, ..., R$) that represents the R destination characteristics of the n regions. The Kronecker product works to repeat the matrix X_r n times to produce an N-by-R matrix representing destination characteristics (see Table C.3.1) that we label X_d.

In addition to explanatory variables consisting of origin and destination characteristics, a vector of distances between each origin-destination dyad is included in the regression model. This vector is formed using the n-by-n distance matrix D containing distances between each origin and destination zone. The N-by-1 vector of distances is formed using $d = \text{vec}(D)$, where vec is an operator that converts a matrix to a vector by stacking the columns of the matrix, as shown in Table C.3.1.

This results in a regression model of the type shown in Eq. (C.3.8) that represents the *log-additive power deterrence function spatial interaction model* in matrix notation

$$y = \alpha \iota_n + X_o \beta + X_d \gamma + \theta d + \varepsilon \qquad (C.3.8)$$

where

y N-by-*1* vector of origin-destination flows,
X_o N-by-Q matrix of Q origin-specific variables that characterize the ability of the origin zones to produce flows,
β the associated Q-by-1 parameter vector that reflects the origin effects,
X_d N-by-R matrix of R destination-specific variables that represent the attractiveness of the destination zones,
γ the associated R-by-1 parameter vector that reflects the destination effects,
d N-by-1 vector of distances between origin and destination zones,
θ scalar distance sensitivity parameter that comes from the power deterrence function and reflects the distance effects,
ι_n N-by-1 vector of ones,
α constant term parameter on ι_n,
ε N-by-1 vector of disturbances with $\varepsilon \sim \mathcal{N}(0, \sigma^2 I_N)$.

This spatial interaction model is based on the independence assumption for the case of a square matrix where each origin zone is also a destination zone and where no a priori information is given on the row and/or column totals of the interaction data matrix. In the sequel we will refer to this model as the *independence* (log-normal) *model*.

C.3.3 Problems that plague empirical use of conventional spatial interaction models

There are several problems that arise in applied practice when estimating the conventional spatial interaction model given by Eq. (C.3.8). We enumerate each of these problems in the following section and discuss solutions that have been proposed in the literature. These solutions often rely on elaborations of the basic model specification given in Eq. (C.3.8).

Efficient computation

One problem that can arise in cases where the sample of regions n is large involves computational memory. For the case of the U.S. counties, for example, we have $n > 3,000$ leading to N-by-Q and N-by-R matrices for the explanatory variables involving $N = n^2 > 9,000,000$. LeSage and Pace (2008) propose a solution for the case where $Q = R = k$ and we rely on the same n-by-k explanatory variables matrix X for both origin and destination characteristics. They point out that repeating the same sample of n-by-k explanatory variable information is not necessary if we take a moment matrix approach to the estimation problem.

If we let $Z = (\iota_N \ X_d \ X_o \ d)$, we can form the moment matrix $Z^T Z$ shown in Eq. (C.3.9), with the symbol 0_k denoting a 1-by-k vector of zeros, and tr representing the *trace* operator

$$Z^T Z = \begin{pmatrix} N & 0_k & 0_k & 0 \\ 0_k^T & n X^T X & 0_k^T 0_k & X^T D \iota_n \\ 0_k^T & 0_k^T 0_k & n X^T X & X^T D \iota_n \\ 0 & \iota_n D^T X & \iota_n D^T X & tr(D^2) \end{pmatrix} \quad (C.3.9)$$

where we assume that the matrix X and vector d are in deviation from means form. This leads to many of the entries in Eq. (C.3.9) taking values of zero.

For the case of the $Z^T y$ required to produce least-squares estimates for the parameters, $\delta = (Z^T Z)^{-1} Z^T y$, we have

$$Z^T y = \begin{pmatrix} \iota_n Y \iota_n \\ X^T Y \iota_n \\ X^T Y^T \iota_n \\ tr(DY) \end{pmatrix}. \quad (C.3.10)$$

Kronecker products prove extremely useful in working with origin-destination flows, as we will see. However, there are limitations associated with this approach that were not fully elaborated by LeSage and Pace (2008). One limitation is that the system of flows is a *closed system* with the same number of origins (n) as destinations (n). This will be required when we discuss modeling spatial dependence by constructing spatial lags of the dependent variable or disturbance terms. For example, if we were modeling shopping trips from various residential locations to a *single* store, this limitation would come into play.

Another limitation pertains to moment-based expressions in Eqs. (C.3.9) and (C.3.10) for working with large problems. These require that the same matrix X is used to form both the origin and destination characteristics matrices so that $X_d = \iota_n \otimes X$ and $X_o = X \otimes \iota_n$. This is equivalent to imposing the restriction that $Q = R$ in Table C.3.1. The moment-based expressions in Eqs. (C.3.9) to (C.3.10) also assume the matrix X is in deviation from means form, but LeSage and Pace (2009a) provide moment expressions that relax this requirement.

If these limitations are consistent with the problem at hand, the moment-based approach to estimation of the model parameters saves a great deal of computer memory. This is accomplished by working with n-by-n matrices rather than n^2-by-$(2k + 2)$, where we have k explanatory variables for regions treated as origins, k for the destination regions in addition to the intercept term and distance vector.

Spatial dependence in origin-destination flows

As already indicated, numerous applied work has pointed to the presence of spatial dependence in the least-squares disturbances from models involving origin-destination data samples (Porojan 2001; Lee and Pace 2005; Fischer and Griffith 2008).

One way to incorporate spatial dependence into a log-normal spatial interaction model of the form (C.3.8) is to specify a spatial process that governs the spatial interaction variable y. This approach leads to a family of models depending on restrictions imposed on the spatial origin-destination filter specification set forth in LeSage and Pace (2009a). Specifically, this type of model specification takes the form

$$y = \rho_o W_o\, y + \rho_d W_d\, y + \rho_w W_w\, y + \alpha \iota_n + X_o\, \beta + X_d\, \gamma + \theta d + \varepsilon \qquad \text{(C.3.11a)}$$

$$\varepsilon \sim \mathcal{N}(0,\ \sigma^2 I_N) \qquad \text{(C.3.11b)}$$

where the spatial weight matrix $W_o = W \otimes I_n$ is used to form a spatial lag vector $W_o\, y$ that captures *origin-based dependence* arising from flows (observation dyads) that neighbor the origins. The n-by-n spatial weight matrix W is a non-negative sparse matrix with diagonal elements set to zero to prevent an observation from being defined as a neighbor to itself. Non-zero values for element pairs (i,j) denote that zone i is *a neighbor* to zone j. Neighbors could be defined using contiguity or other measures of spatial proximity such as cardinal distance (for example, kilometers) and ordinal distance (for example, the five closest neighbors). The spatial weight matrix is typically standardized to have row sums of unity, and this is required to produce linear combinations of flows from neighboring regions in the model given by Eq. (C.3.11).

Given an origin-centric organization of the sample data, the spatial weight matrix $W_o = W \otimes I_n$ will form an N-by-1 vector containing a linear combination of flows from regions neighboring each observation (dyad) treated as an origin. In the case where neighbors are weighted equally, we would have an average of the neighboring region flows. Similarly, a spatial lag of the dependent variable formed using the weight matrix $W_d = I_n \otimes W$ to produce an N-by-1 vector $W_d y$ captures *destination-based dependence* using an average (or linear combination) of flows associated with observations (dyads) that neighbor the destination regions. Finally, a spatial weight matrix, $W_w = W \otimes W$ can be used to form a spatial lag vector that captures *origin-to-destination based dependence* using a linear combination of neighbors to both the origin and destination regions.

This model specification can also be written as

$$(I_n - \rho_o W_o)(I_n - \rho_d W_d) y = Z \delta + \varepsilon \qquad (C.3.12a)$$

$$(I_n - \rho_o W_o - \rho_d W_d + \rho_o \rho_d W_o W_d) y = Z \delta + \varepsilon \qquad (C.3.12b)$$

$$\{I_n - \rho_o [W \otimes I_n] - \rho_d [I_n \otimes W] + \rho_o \rho_d [W \otimes W]\} = Z \delta + \varepsilon \qquad (C.3.12c)$$

where the matrix cross-product term, $\rho_o \rho_d W_o W_d \equiv \rho_w W_w$ motivates the term reflecting *origin-to-destination based dependence*. LeSage and Pace (2008) note that this specification implies that $\rho_w = -\rho_o \rho_d$, but these restrictions need to be applied during estimation. There is a need to impose restrictions on the values of the scalar dependence parameters ρ_d, ρ_o, ρ_w to ensure stationarity in the case where ρ_w is free of the restriction. LeSage and Pace (2008) discuss maximum likelihood estimation of this specification, and LeSage and Pace (2009a) set forth a Bayesian heteroscedastic variant of the model along with Markov Chain Monte Carlo (MCMC) estimation methods.

This variant allows for non-constant variance in the disturbances by introducing a set of N scalar variance parameters. Specifically, $\varepsilon \sim \mathcal{N}(\mathbf{0}_N, \Sigma)$, where the N-by-N *diagonal matrix* Σ contains variance scalar parameters to be estimated on the diagonal and zeros elsewhere.

A virtue of the model in Eq. (C.3.11) is that changes in the value of an explanatory variable associated with a single region will potentially impact flows to all other regions. For example, a *ceteris paribus* change in observation i of the explanatory variables matrix X for variable X_r implies that region i will be viewed differently as both an origin and destination. Given the structure of the matrices X_o, X_d changes in observation i imply changes in $2n$ observations from the explanatory variables matrices. This is true for the *independence model* as well as the spatial model. In the case of the *independence model* such a ceteris paribus change will lead to changes in the flows associated with the same $2n$ observations and no others. Intuitively, if, for example, the labor market opportunities in a single region i decrease, this region will look less attractive as a destination when considered by workers residing in the own and other $n-1$ regions in a migration application context, for example. This should lead to a decrease in migration *pull* from within and outside region i, the impact of changing the n-elements in X_d and associated parameter. Region i will exert more *push* leading to an increase in out-migration to the other $n-1$ regions (as well as a decrease in within-region migration). This impact is reflected by the n-elements in X_o and associated parameter. In the independence model, changes in the explanatory variables associated with the $2n$ observations can only impact changes in flows in the *same* $2n$ observations (by definition).

Turning to the spatial model that includes spatial lags of the dependent variable, these $2n$ changes will lead to changes in flows involving more than the $2n$ observations whose explanatory variables have changed. The additional impacts arising from changes in a single region's characteristics represent *spatial spillover effects*. Intuitively, a decrease in labor market opportunities for region i will indirectly impact the attractiveness of a region that neighbors i, say region j. Region j will become less attractive as a destination for migrants given the decrease in labor market opportunities in neighboring region i. Residents of region j who work in region i and suffer from the labor market downturn in this neighboring region might also find out-migration more attractive. In-migrants to region j may consider labor market opportunities not only in region j but also in neighboring regions such as i. The partial derivative impacts on observations y_i arising from changes in the explanatory variables associated with observations j are zero (by definition) in the independence model, but not in the spatial model containing lags of the dependent variable (see LeSage and Pace 2009a for a discussion of this). Correct calculation and interpretation of the partial derivative impacts associated with the spatial lag model allow one to quantify the spatial spillover impacts.

LeSage and Polasek (2008) provide a minor modification to the model that can be used in the case of commodity flows. In an application involving truck and train commodity flows between 40 Austrian regions, they provide a procedure that adjusts the spatial weight matrix to account for the presence or absence of interregional transport connectivity. Since the mountainous terrain of Austria precludes the presence of major rail and highway infrastructure in all regions, they use this priori non-sample knowledge regarding the transportation network structure connecting regions to produce a modified spatial weight structure. Bayesian model comparison methods indicate that these adjustments to the spatial weight matrix result in an improved model.

Another approach to dealing with spatial dependence in origin-destination flows is to specify a spatial process for the disturbance terms, structured to follow a (first-order) spatial autoregressive process (see Fischer and Griffith 2008). This specification could be estimated using maximum likelihood methods. In this framework, the spatial dependence resides in the disturbance process ε, as in the case of serial correlation in time series regression models. Griffith (2007) also takes this specification approach that focuses on dependence in the disturbances but relies on a *spatial filtering* estimation methodology.

Specifically, the most general variant of this type of model specification takes the form

$$y = \alpha \iota_n + X_o \beta + X_d \gamma + \theta d + u \tag{C.3.13a}$$

$$u = \rho_o W_o u + \rho_d W_d u + \rho_w W_w u + \varepsilon \tag{C.3.13b}$$

$$\varepsilon \sim \mathcal{N}(0, \sigma^2 I_N) \tag{C.3.13c}$$

where the definitions for the spatial lags involving the disturbance terms in Eq. (C.3.13), $W_0 u$, $W_d u$ and $W_w u$, are analogous to those for the spatial lags of the dependent variable in Eq. (C.3.12).

Simpler models can be constructed by imposing restrictions on the general specification in Eq. (C.3.13). For example, we could specify the disturbances using

$$u = \rho \widetilde{W} u + \varepsilon \tag{C.3.14a}$$

$$\varepsilon \sim \mathcal{N}(0, \sigma^2 I_N) \tag{C.3.14b}$$

which merges origin- and destination-based dependence to produce a single (row-normalized) spatial weight matrix $\widetilde{W}$ consisting of the sum of W_o and W_d which is row-normalized to produce a single vector $\widetilde{W}u$ reflecting a spatial lag of the disturbances. This specification also restricts the origin-to-destination based dependence in the disturbances to be zero, since ρ_w is implicitly set to zero.

The virtue of a simpler model such as this is that conventional software for estimating spatial error models could be used to produce an estimate for the parameter ρ along with the remaining model parameters α, β, γ and θ. It may or may not be apparent that estimating the more general models that involve more than a single spatial dependence parameter requires customized algorithms of the type set forth in LeSage and Pace (2008). These are needed to maximize a log-likelihood that is concentrated with respect to the parameters α, β, γ, θ and σ^2 resulting in an optimization problem involving the three dependence parameters ρ_d, ρ_o, ρ_w. Of note is the fact that an extended version of the moment-based expressions involving the matrix Z from Eq. (C.3.9) and Eq. (C.3.10) can be used for both maximum likelihood and Bayesian MCMC estimation (see LeSage and Pace 2009a for details).

One point to note regarding modeling spatial dependence in the model disturbances is that the coefficient estimates α, β, γ, θ will be asymptotically equal to those from least-squares estimation. However, there may be an efficiency gain that arises from modeling dependence in the disturbances. Another point is that the partial derivative impacts associated with this model are the same as those from the *independence model*. That is, no spatial spillover impacts arise in this type of model so that ceteris paribus changes in region i's explanatory variable only result in changes in the $2n$ regions associated with the n^2 dyad relationships involving region i.

A third approach to modeling spatial dependence is motivated by the use of *fixed effects* parameters for origin and destination regions in non-spatial versions of the gravity model in the empirical trade literature (Feenstra 2002). Assuming the origin-centric data organization set forth in Table C.3.1, a fixed effects model would take the form in Eq. (C.3.15). The N-by-n matrix Δ_o contains elements that equal one if region I is the origin region and zero otherwise, and θ_o is an n-by-1 vector of associated fixed effects estimates for regions treated as origins. Similarly, the N-by-n matrix Δ_d contains elements that equal one if region j is the destination region and zero otherwise leading to an n-by-1 vector θ_d of fixed effects estimates for regions treated as destinations

$$y = \alpha + \beta_o X_o + \beta_d X_d + \gamma d + \Delta_o \theta_o + \Delta_d \theta_d + \varepsilon. \tag{C.3.15}$$

LeSage and Llano (2007) extend this model to the case of spatially structured random effects. This involves introduction of latent effects parameters that are structured to follow a spatial autoregressive process. This is accomplished using a Bayesian prior that the origin and destination effects parameters are similar for neighboring regions.

In the context of commodity flows between Spanish regions, the model takes the form

$$y = Z\delta + \Delta_d \theta_d + \Delta_o \theta_o + \varepsilon \quad \text{(C.3.16a)}$$

$$\theta_d = \rho_d W \theta_d + u_d \quad \text{(C.3.16b)}$$

$$\theta_o = \rho_o W \theta_o + u_o \quad \text{(C.3.16c)}$$

$$u_d \sim \mathcal{N}(0, \sigma_d^2 I_n) \quad \text{(C.3.16d)}$$

$$u_o \sim \mathcal{N}(0, \sigma_o^2 I_n). \quad \text{(C.3.16e)}$$

Given our origin-centric orientation of the flow matrix (columns as origins and rows as destinations), the matrices $\Delta_d = I_n \otimes \iota_n$ and $\Delta_o = \iota_n \otimes I_n$ produce N-by-n matrices. It should be noted that estimates for these two sets of random effects parameters are identified, since a set of n sample data observations are aggregated through the matrices Δ_d and Δ_o to produce each estimate in θ_d and θ_o.

The spatial autoregressive prior structure placed on the destination effects parameters θ_d (conditional on the parameters ρ_d and σ_d^2) is shown in Eq. (C.3.17) and that for the spatially structured origin effects parameters θ_o in Eq. (C.3.18), where we use the symbol $\pi(.)$ to denote a prior distribution:

$$\pi(\theta_d | \rho_d, \sigma_d^2) \sim (\sigma_d^2)^{n/2} |B_d| \exp\left(-\frac{1}{2\sigma_d^2} \theta_d^\mathrm{T} B_d^\mathrm{T} B_d \theta_d\right) \quad \text{(C.3.17)}$$

$$\pi(\theta_o | \rho_o, \sigma_o^2) \sim (\sigma_o^2)^{n/2} |B_o| \exp\left(-\frac{1}{2\sigma_0^2} \theta_o^\mathrm{T} B_o^\mathrm{T} B_o \theta_o\right) \quad \text{(C.3.18)}$$

$$B_d = (I_n - \rho_d W) \quad \text{(C.3.19)}$$

$$B_o = (I_n - \rho_o W). \quad \text{(C.3.20)}$$

Estimation of the spatially structured effects parameters requires that we estimate the dependence parameters ρ_d, ρ_o and associated variances σ_d^2, σ_o^2. LeSage and Llano (2007) provide details regarding using of Markov Chain Monte Carlo methods for estimation of this model.

This model does not allow directly for spatial spillover effects. It does, however, provide a *spatially structured effect* adjustment for each origin and destination region. These act in the same fashion as non-spatial effects parameters producing an intercept shift adjustment that would be added to the parameters β and γ when considering the partial derivative impacts arising from ceteris paribus changes in region i's explanatory variable. Another point about the spatially structured prior is that if the scalar spatial dependence parameters (ρ_o, ρ_d) are not significantly different from zero, the spatial structure of the effects vectors disappears, leaving us with normally distributed random effects parameters for the origins and destinations similar to the conventional effects models described in Feenstra (2002).

Large diagonal flow matrix elements

Another problem that arises in empirical work is the fact that the diagonal elements of the flow matrix Y representing intraregional flows are often quite large relative to the off-diagonal elements reflecting interregional flows. Since the objective of spatial interaction modeling is typically a model that attempts to explain variation in interregional rather than intraregional flows, practitioners often view intraregional flows as a nuisance, and introduce dummy variables for these observations (see, for example, Koch et al. 2007). For the case of the independence model this approach is fine, but it can have deleterious impacts on models involving spatial lags of the dependent variable. To see this, consider the case of a simple model involving

$$y = \rho \tilde{W} y + Z \delta + \varepsilon \qquad (C.3.21a)$$

$$\varepsilon \sim \mathcal{N}(0, \sigma^2 I_N) \qquad (C.3.21b)$$

where $\tilde{W}$ is a row-normalized version of the sum of the spatial weight matrices W_o, W_d, W_w. The n zero elements associated with the diagonal of the vectorized flow matrix $y = \text{vec}(Y)$ in the N-by-1 vector of flows will have the impact of producing outliers in the spatial lags when these observations are involved in the linear combination used to form $\tilde{W} y$.

To avoid this problem, LeSage and Pace (2008) suggest a procedure that embeds a separate model for the intraregional flows into the spatial interaction model. This is accomplished by adjusting the explanatory variables matrices X_o, X_d and the intercept vector ι_n to have zero values for the n observations associated with the main diagonal elements (intraregional flows) of the flow matrix Y. We use $\tilde{X}_o, \tilde{X}_d$ to denote these adjusted matrices. A new matrix that we label X_i is introduced containing the n observations associated with intraregional flows set to zero in the matrices X_o, X_d, and zeros in the other $N - n$ observations. That is,

$\tilde{X}_o = X_o - X_i$, and $\tilde{X}_d = X_d - X_i$. In addition, a new intercept vector ι_i is introduced that contains ones in the n positions so that $\tilde{\iota}_N = \iota_N - \iota_i$. The adjusted *independence model* now takes the form

$$y = \alpha \tilde{\iota}_N + \alpha_i \iota_i + (X_o - X_i)\beta + (X_d - X_i)\gamma + X_i \psi + \theta d + \varepsilon \qquad (C.3.22a)$$

$$y = \alpha \tilde{\iota}_N + \alpha_i \iota_i + \tilde{X}_o \beta + \tilde{X}_d \gamma + X_i \psi + \theta d + \varepsilon \qquad (C.3.22b)$$

where a corresponding adjustment can be used for the case of the spatial lag model in Eq. (C.3.11) or the spatial error model in Eq. (C.3.13). This model uses the (orthogonal) intercept term ι_i and explanatory variables X_i (and associated Ψ) to capture variation in the vector of flows y across dyads representing intraregional flows and the adjusted variables: $\tilde{\iota}_N, \tilde{X}_d, \tilde{X}_o$ to model variation in interregional flows.

Of course, it is not necessary to rely on the same set of explanatory variables for X_o, X_d, X_i, but this will simplify computation via the moment matrices for models involving large samples n as discussed earlier. LeSage and Pace (2009a) provide expressions for the moment matrices that arise for these adjustments to the model.

As an example, consider that variation in intraregional flows might be explained by variables such as the *area of the regions* or in the case of a migration flow model the *population of the regions*. We would expect that regions having larger population and area should exhibit more intraregional migration. This subset of two explanatory variables could then be used to form the matrix X_i, with corresponding adjustments to these two variables undertaken for the matrices X_o, X_d to produce $\tilde{X}_o, \tilde{X}_d$. Inference regarding the parameter ψ for these two variables would not be of primary interest (since associated with the intraregional control variables) whereas the focus of the model is on the parameters β, γ and θ.

The advantage of this approach is that non-zero intraregional flows can be included in the matrix Y used to form the dependent variable vector y and the spatial lags $W_o y, W_d y, W_w y$. Variation in the flows associated with the large diagonal elements is captured by the embedded model variables ι_i and X_i allowing the coefficient estimates associated with the adjusted explanatory variables $\tilde{X}_o, \tilde{X}_d$ to more accurately characterize variation in interregional flows.

As an illustration of the differences that arise from these adjustments to the model, we use a sample of 1998 commodity flows between the 48 lower U.S. states plus the District of Columbia leading to a sample size of $n = 49$ and $N = 2,401$. The commodity flows were taken from the Federal Highway Administration Freight Analysis Framework State to State Commodity flow Database. As explanatory variables we use the (logged) area of each state and the 1998 Gross State Product (*gsp*). The model was based on a single spatial weight matrix constructed using a row-normalized matrix consisting of $W_d + W_o + W_w$, where the n-by-n matrix W was based on six nearest neighbors. Following convention, the commodity flows were transformed using logs as were the explanatory variables representing area and gsp.

Table C.3.2 shows the coefficient estimates labelled $\hat{\beta}_1$ for the adjusted model along with those from the unadjusted model labeled $\hat{\beta}_0$. In the table, we use the symbol I_gsp and I_area to denote the variables contained in the matrix X_i in the adjusted model expression given by Eq. (C.3.22). A t-test for significant differences between the coefficients $(\hat{\beta}_0 - \hat{\beta}_1)$ common to the two models is presented in Table C.3.3. From the table reporting test results for differences in the two sets of estimates we see evidence of differences that are significant at the 99 percent level in the coefficients on distance and the spatial lag of the dependent variable. There is also a difference between the *origin area* explanatory variable that is significant at the 90 percent level. It is also worth noting that twice the difference in the log-likelihood function values from the two models is 249, which suggests a significant difference between the models. This would be an informal indication since the two models cannot be viewed as formally nested.

Table C.3.2. Unadjusted and adjusted model estimates

Variables	Unadjusted model		Adjusted model	
	Coefficient $(\hat{\beta}_0)$	t-statistic	Coefficient $(\hat{\beta}_1)$	t-statistic
Constants				
$\iota_N / \tilde{\iota}_N$	−19.2770	−38.9	−19.9888	−41.1
ι_i	–	–	−5.2012	−2.2
Origin variables				
$O\_gsp / \tilde{O}\_gsp$	0.3397	15.7	0.3520	17.0
$O\_area / \tilde{O}\_area$	0.5679	27.1	0.4961	23.6
Destination variables				
$D\_gsp / \tilde{D}\_gsp$	0.7374	30.7	0.7021	30.8
$D\_area / \tilde{D}\_area$	0.2806	17.2	0.2608	16.5
I_gsp	–	–	0.6169	4.3
I_area	–	–	0.3738	3.5
Distance	−0.5123	−22.2	−0.3101	−13.1
ρ	0.5219	23.5	0.6429	31.6
σ^2	1.1549		1.0337	
Log-likelihood	−2762.7		−2638.2	

We can also use this model and sample data to illustrate how problems arise when setting the intraregional flows to zero values. For this illustration a spatial weight matrix based on row-normalized $W_d + W_o$ was used, and the unadjusted model was estimated for values of the dependent variable representing intraregional flows flows set to zero as well as the full set of non-zero flows.

Table C.3.3. Test for significant differences between the unadjusted and adjusted model estimates

Variables	$(\hat{\beta}_0 - \hat{\beta}_1)$	t-statistic	t-probability
Constant	0.7118	0.7264	0.4677
Origin variables			
O_gsp	−0.0123	−0.2905	0.7715
O_area	0.0718	1.7139	0.0867
Destination variables			
D_gsp	0.0352	0.7529	0.4516
D_area	0.0198	0.6195	0.5357
Distance	−0.2022	−4.3319	0.0000
ρ	−0.1210	−2.8511	0.0044

The results from this illustration are presented in Table C.3.4 where we see a serious degradation in the log-likelihood function value for the zero-flows model and a dramatic six-fold rise in the noise variance estimate σ^2. A number of problematical coefficient estimates arise, for example the coefficient on distance is negative but not significantly different from zero, contrary to the conventional result. The magnitude of the spatial dependence parameter ρ decreased dramatically, consistent with our admonition that setting the main diagonal elements of the flow matrix to zero will have an adverse impact on the spatial nature of the sample flow data. Finally, given the reported *t*-statistics, we can infer that the coefficient estimates on the origin and destination *gsp* variables are significantly different in the two regressions.

Table C.3.4. Zero intraregional flows versus non-zero intraregional flows

Variables	Zero diagonal flows		Non-zero diagonal flows	
	Coefficient $(\hat{\beta}_0)$	t-statistic	Coefficient $(\hat{\beta}_1)$	t-statistic
Constant	2.1675	2.30	−16.1351	−33.55
Origin variables				
O_gsp	0.3801	7.73	0.2805	13.92
O_area	0.5573	13.42	0.4552	22.72
Destination variables				
D_gsp / $\tilde{D}$_gsp	0.8504	15.35	0.5969	25.77
D_area / $\tilde{D}$_area	0.1801	5.01	0.2341	15.31
Distance	−0.0230	−0.75	−0.4113	−19.15
ρ	0.2979	6.80	0.6449	33.71
σ^2	5.8627		0.9911	
Log-likelihood	−4,707.2		−2,612.1	

The zero flows problem

Another problem that arises involves the presence of a large number of zero flows[1]. This problem arises when analyzing sample data collected using a fine spatial scale. As an example, population migration flows between the largest 50 U.S. metropolitan areas over the period 1995-2000 resulted in only 3.76 percent of the OD-pairs contained zero flows, whereas 9.38 percent of the OD-pairs were zero for the largest 100 metropolitan areas and for the largest 300 metropolitan areas, 32.89 percent of the OD pairs exhibited zero flows.

The presence of a large number of zero flows invalidates use of least-squares regression as a method for estimating the *independence model* and maximum likelihood methods for spatial variants of the interaction model. This is because zero values for a large proportion of the dependent variable invalidate the normality assumption required for inference in the regression model and validity of the maximum likelihood method. Despite this, a number of applications can be found where the dependent variable is modified using log $(1 + y)$ to accommodate the log transformation. This, however, ignores the mixed discrete/continuous nature of the flow distribution. Intuitively, this type of practice should lead to downward bias in the coefficient estimates for the model.

If we can view flows as arising from say positive utility in the case of migration flows or positive profits when considering commodity flows, then the presence of zero flows might be indicative of negative utility or profits. This type of argument is often used to motivate sample censoring models such as in the Tobit regression model. In a non-spatial application to international trade flows, Ranjan and Tobias (2007) treat zero flows using a threshold Tobit model. Their argument is that zero trade flows are indicative of situations where the transportation and other costs associated with trade exceed a threshold making trade unprofitable. A similar argument could be applied to migration flows. Non-zero flows could be viewed as an indication that the origin versus destination characteristics are such that at least one migrant perceives positive utility arising from movement between the origin-destination dyad. In contrast, zero observed migration flows could be interpreted to mean that no individual views destination utility to be greater than utility at the origin for these OD dyads, leading to net negative utility from migration. We note that similar arguments regarding utility from program participation have been used to motivate sample truncation leading to the use of Tobit regression models when evaluating the level of program participation by individuals.

LeSage and Pace (2009a) set forth estimation methods for Tobit models where a spatial lag of the dependent variable is involved. This requires Bayesian MCMC estimation where a set of parameters representing negative utility are introduced for the zero-valued dependent variable observations. Some important caveats are associated with this approach to dealing with zero-valued flows. One is that Tobit

---

[1] Note that zero counts present no serious problem in Poisson regression, but must be handled in the log-normal spatial interaction model case.

models assume the dependent variable follows a truncated normal distribution. This assumption seems reasonable when we are faced with a sample of flows containing less than 50 to 70 percent zero or censored values. However, in situations where we are faced with a very large proportion of zero values, the assumption of a truncated normal distribution seems less plausible.

In the context of modeling knowledge flows between European Union regions, LeSage et al. (2007) note that a large proportion of zero knowledge flows between the sample of European regions should be viewed as indicative that knowledge flows are perhaps a rare event. This view is more consistent with a Poisson distribution for the dependent variable. We will have more to say about this later.

To demonstrate how spatial autoregressive Tobit models can be used to address the issue of zero observations we generated a sample of 2,401 OD flow observations using the explanatory variables *area* and *gsp* from our previous example involving state level commodity flows involving the 48 lower U.S. states and the District of Columbia. A Queen-based spatial contiguity weight matrix was used for W and a single matrix $\tilde{W}$ was generated using a row-normalized version of $W_d + W_o$. The true parameter values for β and γ were set to one and minus one for the *gsp* and *area* variables respectively. Use of both positive and negative coefficient values ensures that the generated flows will include negative values. The parameter θ for distance was set to minus one and that for the intercept to 20. A value of $\rho = 0.65$ was used. This procedure for producing data-generated flows resulted in 1,020 negative flows out of 2,401 observations, or slightly more than 42 percent sample censoring. We should view the dependent variable generated in this fashion as profitability associated with interregional commodity flows, so the magnitude of commodity flows is proportional to profitability. Consistent with this view, we set negative values of the dependent variable to zero, reflecting the absence of commodity flows between dyads where negative profits existed.

Estimates from the set of continuous values for the flows/profitability were constructed using maximum likelihood estimation of the spatial autoregressive model in Eq. (C.3.11). These estimates should of course be close to the true values used to generate the sample data. *A second set of estimates* were based on the sample with zero values assigned for negative values of the generated dependent variable, to explore the impact of ignoring zero flow values and proceeding with conventional maximum likelihood estimation of the spatial autoregressive model. Here we would expect to see downward bias in the coefficient estimates due to the sample truncation.

A third set of spatial autoregressive Tobit model estimates were based on the sample with zero values assigned for negative values of the dependent variable. Ideally, the spatial Tobit model parameters should be close to the true parameter values used to generate the sample of flows, if we have been successful in our spatial econometric treatment of zero valued flows as representing sample truncation. MCMC estimation methods described in LeSage and Pace (2009a) were used to produce estimates for the spatial autoregressive Tobit model.

Results from this illustration are reported in Table C.3.5, where we see coefficient estimates labeled *Uncensored sample* close to the true values used to generate the flow vector *y*. These were based on the sample flow vector that did not impose sample truncation on the negative values of the dependent variable. The estimates labeled *Non-Tobit censored* are those based on ignoring the existence of zero valued flows. The Bayesian spatial autoregressive Tobit model estimates are reported in the columns labeled *Tobit censored*, where the posterior mean reported in the table is based on a sample of 1,000 MCMC draws. The posterior mean was divided by the posterior standard deviation to produce a pseudo *t*-statistic for comparability with these measures of dispersion for the maximum likelihood estimates.

From the table we see that ignoring zero valued flows produces a dramatic downward bias in the coefficient estimates. Most of the estimates are around 50 to 60 percent lower than the true parameters used to generate the sample *y*-vector. In contrast, the spatial autoregressive Tobit estimates produced coefficients very close to the true parameters as well as the benchmark estimates based on the *uncensored sample*. A point worth noting is that use of the spatial autoregressive Tobit model will lead to larger dispersion in the estimates, which from a Bayesian viewpoint reflects greater uncertainty in the posterior means.

Table C.3.5. Spatial Tobit experimental results

Variables	True	Uncensored sample		Non-Tobit censored		Tobit censored	
		Coefficient	*t*-statistic [a]	Coefficient	*t*-statistic [a]	Coefficient	*t*-statistic [a]
Constant	20	19.2933	31.4	15.7547	24.9	19.5794	29.9
Origin variables							
O_gsp	1	1.0309	42.5	0.4746	21.2	1.0519	32.5
O_area	−1	−1.0055	−45.0	−0.6128	−29.3	−1.0169	−45.4
Destination variables							
D_gsp	1	0.9833	41.1	0.4564	20.5	0.9940	31.8
D_area	−1	−0.9691	−44.3	−0.5985	−29.0	−0.9849	−43.2
Distance	−1	−0.9861	−42.8	−0.6016	−27.9	−1.0075	−41.7
ρ	0.65	0.6569	81.9	0.7719	90.8	0.6475	75.1
σ^2	1	0.9654		0.9853		0.9786	

Notes: [a] Pseudo *t*-statistic, posterior mean divided by posterior standard deviation

Some caveats regarding this approach to dealing with zero-valued flows are in order. As already mentioned, this approach is most likely applicable for situations where there is not an excessive amount of zero values. The ability of this approach to produce quality estimates depends on the ability of the spatial Tobit procedure to produce good estimates for the latent parameters introduced in the model (see LeSage and Pace 2009a for a detailed discussion of this). As economists are fond of saying, there is no such thing as a free lunch. This applies to the spatial Tobit model where the cost of censoring is increased uncertainty regarding the posterior estimates. Intuitively, as the proportion of the sample that is censored increases, so does our uncertainty in the estimation outcomes. A final point is that this same approach can be used to deal with zero flow values for the spatially structured effects model set forth in Eq. (C.3.16). LeSage and Pace (2009a) discuss this and LeSage et al. (2008) provide details including an applied example using commuting flows in Toulouse. This involves introducing latent parameters for the zero-valued flows and estimating these using Bayesian MCMC procedures.

As already mentioned, cases where the proportion of zero-valued flows is very large are not amenable to the Tobit model approach. LeSage et al. (2007) provide an extension of the model given by Eq. (C.3.16) that can be used to accommodate this situation. They rely on a variant of the model in Eq. (C.3.16) where the flows are assumed to follow a Poisson distribution, and treat interregional patent citations from a sample of European Union regions as representing knowledge flows. The counts of patents originating in region i that were cited by regions $j = 1, ..., n$ are used to form a *knowledge flows* matrix. Since cross-region patent citations are both counts and rare events, a Poisson distribution seems much more plausible than the normal distribution assumption made for the Tobit model.

The extension of the spatially structured effects model relies on work by Frühwirth-Schnatter and Wagner (2008) who argue that (non-spatial) Poisson regression models (including those with random-effects) can be treated as a partially Gaussian regression model by conditioning on two strategically chosen sequences of artificially missing data. These sequences are similar in spirit to the latent parameters approach described above for estimating the spatial autoregressive Tobit model (LeSage and Pace 2009a). After conditioning on both of these latent sequences, Frühwirth-Schnatter and Wagner (2008) show that the resulting model can be estimated using an MCMC procedure.

The one drawback to the approach pointed out by LeSage et al. (2007) is that one must sample two sets of latent parameters equal to $y_{ij}+1$, where y_{ij} denotes the count for observation i. This can lead to very long sequences of artifically missing data that need to be manipulated during MCMC estimation thousands of times. The authors report that for a sample of $n=188$ regions 23,718 zero values and 199,817 non-zero values, a total of 133,535 latent observations were needed to sample each of the two latent variable vectors. The estimation procedure took over two days to produce estimates for the moderately sized sample based on $n = 188$.

For the spatially structured random effects model from Eqs. (C.3.17) to (C.3.20), let $y = (y_1, ..., y_N)$ denote our sample of $N = n^2$ counts for dyads of flows between regions. The assumption regarding y_i is that $y_i \mid \lambda_i$ follows a Poisson, $P(\lambda_i)$ distribution, where λ_i depends on (standardized) covariates Z_i reflecting the ith row of the explanatory variables matrix Z, with $i = 1, ..., N$. The Poisson variant of this model can be expressed as

$$y_i \mid \lambda_i \sim P(\lambda_i) \qquad \text{(C.3.23a)}$$

$$\lambda_i = \exp(z_i \delta + \delta_{di} \theta_d + \delta_{oi} \theta_o) \qquad \text{(C.3.23b)}$$

where δ_{di} represents the ith row from the matrix Δ_d in Eq. (C.3.16) that identifies region i as a destination region and δ_{oi} identifies origin regions using rows from the matrix Δ_o of Eq. (C.3.16). The insight of Frühwirth-Schnatter and Wagner (2008) was that conditional on the sequences of artifically missing data MCMC samples can be constructed from the posterior distribution of the parameters using draws from a series of distributions that take known forms.

C.3.4 Concluding remarks

In addition to the challenges discussed above that face practitioners interested in empirical implementation of spatial interaction models, there is a need to provide a theoretical justification for the use of spatial lags of the dependent variable (or disturbances) in spatial interaction models. The description provided here motivates the need for these models based on empirically observed spatial dependence in flows.

LeSage and Pace (2008) provide a purely econometric motivation for inclusion of spatial lags of the dependent variable based on missing variables, and LeSage and Pace (2009a) provide a number of additional econometric motivations for use of spatial autoregressive regions models in applied settings not specific to modeling origin-destination flows. Many of these empirical motivations could be extended to the case of flow modeling.

However, a theoretical basis would give the strongest justification for use of these models. Koch et al. (2007) provide a starting point for the special case of international trade flows by extending the work of Anderson and van Wincoop (2004). They rely on a monopolistic competition model in conjunction with a CES (constant elasticity of substitution) utility function to derive a gravity equation for trade flows that contains spatial lags of the dependent variable. A study of theoretical work in the trade literature (Anderson and van Wincoop 2004; Koch et al. 2007) suggests that spatial interaction models may suffer from their focus on bilateral flows between origin-destination dyads. The conclusion drawn from recent theoretical developments in the trade literature is that *bilateral relationships* may not readily extend to a *multilateral world*. Simple relationships based on dyads ignore indirect interactions that link all trading partners. The theoretical work of Koch et al. (2007) leading to a spatial interaction model for trade flows that includes spatial lags of the dependent variable has some important implications for spatial interaction modeling in more general circumstances. One implication is that introducing spatial dependence leads to a situation where dyad relationships are no longer of central importance. In the context of trade flows *and* spatial dependence, price differences between bilateral partners spillover to produce an implicit dependence that quickly encompasses *all* other trading partners. Specifically, the authors argue that when goods are gross substitutes, trade flows from any origin to any destination may depend on the entire distribution of bilateral trade barriers, which reflect prices of substitute goods.

As already motivated, use of spatial regression models that include spatial lags of the dependent variable leads to an implication consistent with the work of Koch et al. (2007). Returning to our example of a ceteris paribus change in labor market opportunities for a single region i, the spatial spillover impacts that arise for these models have the potential to reflect dependence on the entire distribution of regional labor market opportunities available in all regions.

References

Anderson JE, van Wincoop E (2004) Trade costs. J Econ Lit 42(3):691-751
Bailey TC, Gatrell AC (1995) Interactive spatial data analysis. Longman, Harlow
Batty M, Sikdar PK (1982) Spatial aggregation in gravity models. 4. generalisations and large-scale applications. Env Plann A 14(6):795-822
Bolduc D, Laferriere R, Santarossa G (1992) Spatial autoregressive error components in travel flow models. Reg Sci Urb Econ 22(3):371-385
Curry L (1972) A spatial analysis of gravity flows. Reg Stud 6(2):131-147
Curry L, Griffith D, Sheppard E (1975) Those gravity parameters again. Reg Stud 9(3):289-296
Feenstra RC (2002) Border effects and the gravity model: consistent methods for estimation. Scott J Pol Econ 49(5):491-506
Fischer MM (2002) Learning in neural spatial interaction models: a statistical perspective. J Geogr Syst 4(3):287-299
Fischer MM, Griffith DA (2008) Modeling spatial autocorrelation in spatial interaction data: an application to patent citation data in the European Union. J Reg Sci 48(5):969-989
Fischer MM, Reismann M (2002) A methodology for neural spatial interaction modeling. Geogr Anal 34(2):207-228
Fotheringham AS, O'Kelly ME (1989) Spatial interaction models: formulations and applications. Kluwer, Dordrecht
Frühwirth-Schnatter S, Wagner H (2006) Auxiliary mixture sampling for parameter-driven models of time series of counts with applications to state space modelling. Biometrika 93(4):827-841
Getis A (1991) Spatial interaction and spatial autocorrelation: a cross-product approach. Env Plann A 23(9):1269-1277
Griffith D (2007) Spatial structure and spatial interaction: 25 years later. The Rev Reg Stud 37(1):28-38
Griffith D, Jones K (1980) Explorations into the relationships between spatial structure and spatial interaction. Env Plann A 12(2):187-201

Koch W, Ertur C, Behrens K (2007) Dual gravity: using spatial econometrics to control for multilateral resistance, LEG - Document de travail - Economie 2007-03, LEG, Laboratoire d'Economie et de Gestion, CNRS UMR 5118, Université de Bourgogne

Lee M, Pace RK (2005) Spatial distribution of retail sales. J Real Est Fin Econ 31(1):53-69

LeSage JP, Llano C (2007) A spatial interaction model with spatially structured origin and destination effects. Available at SSRN http://ssrn.com/abstract =924603

LeSage JP, Pace RK (2008) Spatial econometric modeling of origin-destination flows. J Reg Sci 48(5):941-967

LeSage JP, Pace RK (2009a) Introduction to spatial econometrics. CRC Press (Taylor and Francis Group), Boca Raton [FL], London and New York

LeSage JP, Pace RK (2009b) Spatial econometric models. In Fischer MM, Getis A (eds) Handbook of applied spatial analysis. Springer, Berlin, Heidelberg and New York, pp.355-376

LeSage JP, Polasek W (2008) Incorporating transportation network structure in spatial econometric models of commodity flows. Spat Econ Anal 3(2):225-245

LeSage JP, Fischer MM, Scherngell T (2007) Knowledge spillovers across Europe: evidence from a Poisson spatial interaction model with spatial effects. Papers in Reg Sci 86(3):393-421

LeSage JP, Rousseau C, Thomas C, Laurent T (2008) Prise en compte de l'autocorrelation spatiale dans l'étude des navettes domicile travail: example de Toulouse, paper presented at INSEE (National Institute for Statistics and Economic Studies), Paris, France

Porojan A (2001) Trade flows and spatial effects: the gravity model revisited. Open Econt Rev 12(3):265-280

Ranjan R, Tobias JL (2007) Bayesian inference for the gravity model. J Appl Econ 22(4):817-838

Roy JR (2004) Spatial interaction modelling. a regional science context. Springer, Berlin, Heidelberg and New York

Sen A, Smith TE (1995) Gravity models of spatial interaction behavior. Springer, Berlin, Heidelberg and New York

Smith TE (1975). A choice theory of spatial interaction. Reg Sci Urb Econ 5(2):137-176

Wilson AG (1967) A statistical theory of spatial distribution models. Transp Res 1(3):253-269

C.4 Spatial Econometric Model Averaging

Oliver Parent and *James P. LeSage*

C.4.1 Introduction

Estimates and inferences that arise from use of empirical models include uncertainty arising from a number of sources. Coefficient estimates produced using statistical regression methods embody uncertainty that we attribute to noise that arises in the process that generated our sample data. There are other sources of uncertainty related to issues of model specification that are typically ignored when we conduct statistical inference regarding model parameters. Uncertainty related to various aspects of model specification is typically excluded from inferential considerations by virtue of the assumption that our models are correctly specified to reflect the true model that generated the sample data. Given this assumption, as well as assumptions regarding the nature of statistical distributions assigned to all random deviates in the data generating process, we can use basic principles from statistical theory to derive distributions for the model parameters that serve as the basis for parameter inference.

An implication that is often ignored in applied practice is that we should consider parameter inference to be *conditional* on the model specification. In this contribution we discuss formal methods that can be used to incorporate model specification uncertainty when making inferences about model parameters. These have been labeled *Bayesian model averaging* and represent one approach to making parameter inference *unconditional* on model specification issues. Instead of selecting a single model, this approach proposes to average estimates across different models. We focus our discussion of model averaging on prominent members of the family of spatial regression models that are widely used by practitioners analyzing spatial data sets. In this setting model uncertainty arises from three sources: (i) the spatial weight or connectivity structure assigned to regions that form the observational basis of spatial data samples, (ii) the type of model employed from the family of models available, and (iii) specific explanatory variables included in the model.

The first source of model uncertainty is unique to spatial regression modeling since conventional regression models assume *independence* between sample observations. The hallmark of spatial regression models that distinguish these from more traditional regression methods is the spatial weight matrix. Uncertainty regarding the spatial weight matrix has long been recognized by practitioners who typically check whether estimates and inferences are *similar* when alternative spatial weight structures are used. As we will see, model averaging represents a more formal approach to this issue. The second and third sources of uncertainty arise in conventional regression models as well as spatial regression models considered here. Again, model averaging provides a formal approach to considering the impact of these two types of uncertainty regarding model specification on the resulting estimates and inferences we draw regarding parameters of interest from our models.

Bayesian inference is based on the posterior distribution of model parameters which refers to an update of the prior parameter distributions that arises from mixing these with sample data. The posterior distribution for our model parameters tells us what we learn about the model parameters from combining our model, prior beliefs and sample data information. This is often referred to as *Bayesian learning*, where the data allows us to update our prior views about the model parameters. The result is the posterior which combines our prior distributions for the model parameters with the data. Non-Bayesian methods such as maximum likelihood focus only on the data distribution arising from random deviates at work in the data generating process to derive statistical distributions for the model parameters. Model averaging extends this approach to estimation and inference by including the model specification in the learning process. This results in a posterior distribution for the model parameters that includes the three sources of uncertainty noted above regarding various aspects of model specification.

This contribution describes details and provides illustrations of how this can be accomplished in the context of spatial regression models. Section C.4.2 sets forth the theory behind Bayesian model averaging with specifics related to prominent members of the family of spatial regression models detailed in Section C.4.3. Implementation issues are taken up in Section C.4.4 with an applied illustration in Section C.4.5.

C.4.2   The theory of model averaging

We consider spatial regression models that involve an n-by-1 dependent variable vector y, where n denotes the number of observations or regions contained in the sample data. Spatial data samples typically consist of a single observation for each region in the sample. Explanatory variables in these models take the form of an n-by-k matrix, where k represents the number of explanatory variables which might include an intercept vector. As noted, a distinguishing feature of spatial regression models is use of a spatial connectivity or weight matrix that describes *neighboring* relationships between the regional observational units. This is usually specified using an *n*-by-*n* matrix that we label W. This matrix contains non-zero entries in row *i*, column *j* to indicate a neighboring relationship between regions/observations *i* and *j*. Zero entries are used to denote the absence of a neighboring relationship, and the main diagonal elements of W are set to zero to prevent a region from being defined as a neighbor to itself. We will have more to say about the spatial weight matrix as it pertains to specific spatial regression models that we consider later.

Given a set $i = 1, \ldots, m$ of Bayesian models, each would be represented by a likelihood function and prior distribution as in Eq. (C.4.1).

$$p(\theta^i | \mathcal{D}, M_i) = \frac{p(\mathcal{D}|\theta^i, M_i) p(\theta^i | M_i)}{p(\mathcal{D}|M_i)} \quad (C.4.1)$$

where $\mathcal{D} = \{y, X, W\}$ represent model data and θ denote model parameters. We note that the posterior distributions for the model parameters in this case are formally *conditional* on the model specification M_i as well as the data, $\mathcal{D}$. Equation (C.4.1) results from application of Bayes' rule. This rule states that for two sets of random variables $\mathcal{D}$ and θ the *joint probability* $p(\mathcal{D}, \theta)$ can be expressed in terms of *conditional probability* $p(\mathcal{D} | \theta)$ or $P(\theta | \mathcal{D})$ and the *marginal probability* $P(\theta)$ as shown in Eqs. (C.4.2) and (C.4.3).

$$p(\mathcal{D}, \theta) = p(\mathcal{D}|\theta) \, p(\theta) \quad (C.4.2)$$

$$p(\mathcal{D}, \theta) = p(\theta|\mathcal{D}) \, p(\mathcal{D}). \quad (C.4.3)$$

Setting these two expressions equal and rearranging gives rise to *Bayes' Rule*

$$p(\theta|\mathcal{D}) = \frac{p(\mathcal{D}|\theta) \, p(\theta)}{p(\mathcal{D})}. \quad (C.4.4)$$

The expression in Eq. (C.4.1) arises from application of Bayes' rule to expand terms like $p(\mathcal{D}|M_i)$ in a fashion similar to that used to arrive at Eq. (C.4.4). A similar approach leads to a set of unconditional posterior model probabilities:

$$p(M_i|\mathcal{D}) = \frac{p(\mathcal{D}|M_i) p(M_i)}{p(\mathcal{D})}. \quad (C.4.5)$$

These posterior model probabilities serve as the basis for inference about different models. As indicated by the notation $p(M_i|\mathcal{D})$, the model probabilities depend only on the sample data. A key point is that the posterior model probabilities are *unconditional* on the model specification. This is unlike the conventional situation motivated in the introduction where inferences regarding the model parameters were considered *conditional* on the model specification which was assumed to be correct. Some discussion of how this is accomplished through the use of Bayes' rule follows (Zellner 1971).

The term $p(\mathcal{D}|M_i)$ that appears on the right-hand-side of Eq. (C.4.5) is called the *marginal likelihood*, and we can solve for this key quantity needed for model comparison finding

$$p(\mathcal{D}|M_i) = \int p(\mathcal{D}|\theta^i, M_i)\, p(\theta^i|M_i)\, d\theta^i. \qquad (C.4.6)$$

An important point is that the model probabilities involve integration over the entire posterior distribution for the parameters in all models, θ^i. This makes these unconditional on any particular values taken by the parameters. Non-Bayesian methods such as maximum likelihood carry out model comparison using mean values of the parameter estimates to evaluate the likelihood function. Models are then compared using scalar values of these parameters that maximize the likelihood function. This means that non-Bayesian inferences about two or more models will depend on particular maximum likelihood parameter estimates used to calculate likelihood function values employed in the comparison. In contrast, Bayesian model comparison constructs model probabilities for comparison purposes by integrating over the entire posterior distribution of possible values that can be taken by the model parameters in all models under consideration. The process of integrating over distributions for unknown quantities such as the model parameters makes our posterior inferences regarding various model specifications unconditional on these quantities.

This suggests the Bayesian approach has advantages, but there is also the computational burden of carrying out integration with respect to the model parameters. Using analytical methods of integration simplifies the task. Unfortunately, analytical methods are not always applicable and if numerical methods are required the task can be computationally demanding.

Assuming we can calculate posterior model probabilities using analytical or numerical methods, Bayesian model averaging proceeds by constructing a linear combination of parameter distributions. The posterior model probabilities are used as weights when forming the linear combination of parameter distributions. As a simple example of this procedure, consider a situation where we are uncertain about the specification used for the spatial weight matrix in our spatial regression model. For simplicity, assume that we assign equal *prior probabilities* to models based on $m = 10$ different models, each based on a different nearest-neighbor weight matrix. Assigning equal prior probabilities to all ten models implies that we believe each of the ten models based on alternative weight matrices to be equally likely a priori. We use the term *a priori* to denote that we have made this assignment without examining the sample data. Further assume that we entertain model specifications based on weight matrices constructed using the single nearest neighboring region, the two nearest neighbors, three neighbors and so on, up to ten nearest neighbors, leading to a set of ten models under consideration. Formally,

$$p(\theta|\mathcal{D}) = \prod_{i=1}^{m} \pi_i p(\theta^i|\mathcal{D}, M_i) \qquad \text{(C.4.7)}$$

where we use π_i to represent the posterior model probabilities. Like all probabilities, these must lie between zero and one, and sum to unity over the set of $m = 10$ models under consideration.

A non-Bayesian approach to inference in this type of situation might be to select a single model based on some criterion such as model fit or likelihood function values. We note however that formal likelihood ratio tests that compare models cannot be applied in this situation because the set of models under consideration is *non-nested*. A set of nested models is such that the simpler models in the set can be expressed as restricted versions of a more elaborate model. For example, a regression model based on two explanatory variable vectors x_1, x_2 nests a model based on the single explanatory variable vector x_1, and a model based on only x_2. These two simpler models can be derived by imposing a zero restriction on the parameters associated with one of the two explanatory variables in the full model involving both variables. An important advantage of Bayesian model comparison methods based on posterior model probabilities such as π_i is that non-nested models can be compared.

As noted in the introduction, it has become conventional non-Bayesian practice to report spatial regression model estimates based on a single selected model and to explore the sensitivity of inferences made when the specification is altered to rely on say the next best model which would have been selected using the selection criterion. Despite this non-Bayesian attempt to explore robustness of inferences to the choice of alternative spatial weight matrices, inferences are drawn from a single model. An implication of this is that uncertainty regarding the choice of weight matrix is not formally incorporated in reported inferences regarding model parameters of interest.

The Bayesian model averaging approach would be to construct a single posterior distribution for the model parameters based on a linear combination of parameter distributions from all ten model specifications based on each of the ten weight matrices. This leads to posterior inferences that incorporate model uncertainty regarding the choice of weight matrix. If the parameter distributions from individual models based on different weight matrices exhibit a great deal of variation and the posterior model probabilities assign large weights to many of the ten models, this will lead to greater dispersion in the posterior parameter distribution arising from model averaging. Intuitively, if the posterior probabilities are dispersed over the set of ten models, this is indicative that the sample data are relatively inconclusive about which weight matrix that should be employed in our model. This aspect of model uncertainty should be taken into account when we draw inferences about model parameters of interest, leading to greater uncertainty in our conclusions. Suppose we are interested in a single model parameter of strategic interest regarding the influence of infrastructure investment on regional economic growth. If model averaged inferences lead us to conclude that infrastructure exerts a positive and significant influence on regional growth, we can be confident that this inference includes model uncertainty regarding the spatial weight matrix used. It might also be the case that after taking into account model uncertainty regarding the spatial weight matrix we find no significant role for infrastructure investment on the regional growth process. In this circumstance, the non-Bayesian approach that produces inferences based on a single model might lead to an erroneous conclusion that is specific to the particular spatial weight matrix employed in the model selected for purposes of inference.

C.4.3 The theory applied to spatial regression models

We wish to consider prominent members of the family of spatial regression models popularized by Anselin (Anselin 1988). Specifically, we focus on the spatial autoregressive (SAR) model

$$y = \rho W y + \alpha \iota_n + X\beta + \varepsilon \tag{C.4.8}$$

$$\varepsilon \sim \mathcal{N}(0, \sigma^2 I_n) \tag{C.4.9}$$

where y is our n-by-1 dependent variable vector, ι_n denotes an n-by-1 vector of ones and α is the associated intercept parameter. The n-by-k matrix X contains non-constant explanatory variables that are assumed exogenous with β being a k-by-1 vector of associated parameters. The n-by-1 vector ε is a disturbance vector that is normally distributed with zero mean and constant scalar variance, σ^2, and zero covariance leading to a variance-covariance matrix $\sigma^2 I_n$, where I_n is an n-dimensional identity matrix. The matrix-vector product, Wy represents a *spatial lag* of the dependent variable vector y. This results from the matrix multiplication because the matrix W consists of non-zero weights reflecting the degree of connectivity between neighboring observations/regions in our sample data. If we assign equal weights to each neighboring observation in the matrix W, the product Wy represents an average of the values taken by the dependent variable in neighboring regions (LeSage and Pace 2009). The weight matrix is normalized to have row-sums of unity to accomplish the task of producing spatial lags that represent linear combinations of values taken by neighboring observations. The scalar parameter ρ in the model reflects the strength of influence or spatial dependence of each observation on values of the dependent variable from neighboring regions. This dependence could be positive or negative, and for stability of the model we require that the parameter ρ takes values less than one. We can assign a Bayesian prior distribution for this parameter that restricts the range to the interval $-1 < \rho < 1$.

It should be clear that the model in Eq. (C.4.8) represents an extension of the conventional regression model when the parameter $\rho \neq 0$, and collapses to the ordinary *independence model* when $\rho = 0$. Since this parameter measures the degree of dependence, a zero value reflects no dependence which is equivalent to independence between observations of the dependent variable.

The results we derive also apply to an extension of this model that has been labeled the *spatial Durbin model* (SDM) by Anselin (1988). This extended variant of the model includes a *spatial lag* of the explanatory variables matrix formed by WX leading to

$$y = \rho Wy + \alpha \iota_n + X\beta + WX\gamma + \varepsilon \qquad (C.4.10)$$

$$\varepsilon \sim \mathcal{N}(0, \sigma^2 I_n) \qquad (C.4.11)$$

where the matrix product WX represents a linear combination, or in the case of equal values assigned to neighbors by the matrix W, an average of the values taken by the explanatory variables from neighboring observations/regions.

Another model that represents a prominent member of the family of spatial regression models is the spatial error model (SEM)

$$y = \alpha \iota_n + X\beta + u \qquad (C.4.12)$$

$$u = \rho Wu + \varepsilon \qquad (C.4.13)$$

$$\varepsilon \sim \mathcal{N}(0, \sigma^2 I_n) \qquad (C.4.14)$$

which models the vector of disturbances u as exhibiting spatial dependence on neighboring region disturbances. This is accomplished by the spatial lag Wu that produces a linear combination (or average) of neighboring disturbances, with the strength of spatial dependence determined by the scalar parameter ρ. As in the case of the SAR model, this model represents a simple extension of the ordinary regression model when the parameter $\rho \neq 0$, and collapses back to a standard regression when $\rho = 0$.

An interesting motivation for use of the SDM model in the presence of model uncertainty regarding use of the SAR or SEM model specification is provided by LeSage and Pace (2009). They make the following observation starting with the data generating processes (DGPs) for the SAR and SEM models shown in Eqs. (C.4.15) and (C.4.16) respectively. We have included the intercept vector ι_n in the matrix of explanatory variables X for notational simplicity in Eqs. (C.4.15) and (C.4.16).

$$y_s = (I_n - \rho W)^{-1} X\beta + (I_n - \rho W)^{-1} \varepsilon \qquad (C.4.15)$$

$$y_e = X\beta + (I_n - \rho W)^{-1} \varepsilon. \qquad (C.4.16)$$

The DGP can be thought of as the process we would use to produce a simulated sample of data observations that obey the model specification. For example, we would use Eq. (C.4.15) to produce an n-by-1 vector of observations y that are consistent with the SAR model statement in Eq. (C.4.8), and parameters ρ, α, β and noise variance σ^2. Using our earlier notation we can use $\mathcal{D} = \{y, X, W\}$ to denote the sample data realization and the vector $\theta = (\alpha, \beta, \rho, \sigma^2)$ for the model parameters. Equations (C.4.15) and (C.4.16) represent DGPs so we are free to assume identical values for the parameter vector θ in both models.

LeSage and Pace (2009) point out that if we entertain only these two models and suppose that posterior model probabilities π_s, π_e have been calculated using the sample data $\mathcal{D}$, model averaging would lead to a linear combination of the SAR and SEM models that could be expressed as

$$y_{avg} = \pi_s y_s + \pi_e y_e \qquad (C.4.17)$$

$$y_{avg} = (I_n - \rho W)^{-1} X\beta\pi_s + X\beta\pi_e + (I_n - \rho W)^{-1} \varepsilon (\pi_s + \pi_e). \qquad (C.4.18)$$

This can be simplified to arrive at (LeSage and Pace 2009)

$$y_{avg} = \rho W y_{avg} + X\beta + WX\gamma + \varepsilon \qquad (C.4.19)$$

$$\gamma = -\rho\beta\pi_e \qquad (C.4.20)$$

which is the SDM model specification set forth in Eq. (C.4.10). Of interest is the fact that if a zero posterior model probability π_e for the SEM model arises, then using Eq. (C.4.20) we have that $\gamma = 0$. Since $\pi_s + \pi_e = 1$, this implies that $\pi_s = 1$, producing the intuitively pleasing result that the SAR model: $y_s = \rho W y_s + X\beta + \varepsilon$, is the appropriate model. On the other hand, if $\pi_e \neq 0$ so there is some posterior probability evidence in favor of an SEM model, we should rely on the SDM model in our empirical application.

In addition to this motivation for use of the SDM model in applied spatial regression work, LeSage and Pace (2009) provide a number of other motivations for this model based on omitted or excluded variables that often arise in applied practice. For this reason, we focus our developments for Bayesian model averaging on the case of the SDM model. To simplify notation we use a slightly altered version of the expression in Eq. (C.4.8) to represent the SDM model by simply re-defining the matrix

$$Z = (X \quad WX) \qquad (C.4.21)$$

$$y = \rho W y + \alpha \iota_n + Z\delta + \varepsilon \qquad (C.4.22)$$

$$\varepsilon \sim \mathcal{N}(0, \sigma^2 I_n). \qquad (C.4.23)$$

We consider two types of model uncertainty that arise in spatial regression modeling. One relates to the specification used to construct the spatial weight matrix W, and the other pertains to which explanatory variables should be included in the matrix X. Given our development here, we ignore model specification issues pertaining to whether we should rely on an SAR or SEM model specification, since the SDM model we work with subsumes both of these models as special cases (LeSage and Pace 2009). It should be clear from our development here that when $\pi_e = 0$ we have an SAR model specification and when $\pi_e \neq 0$ an SDM model arises. The development here obscures the fact that when $\pi_s = 0$, we have that $\pi_e = 1$, leading to the SEM model. This can be seen in the following development, where we apply the fact that $\pi_s = 0$ and $\pi_e = 1$ to Eq. (C.4.19).

$$y_e = \rho W y_e + X\beta + WX(-\rho\beta) + \varepsilon \qquad (C.4.24)$$

$$(I_n - \rho W) y_e = (X - \rho W X) \beta + \varepsilon \qquad (C.4.25)$$

$$(I_n - \rho W) y_e = (I_n - \rho W) X \beta + \varepsilon \qquad (C.4.26)$$

$$y_e = X\beta + (I_n - \rho W)^{-1} \varepsilon \qquad (C.4.27)$$

$$y_e = X\beta + u \qquad (C.4.28)$$

$$u = (I_n - \rho W)^{-1} \varepsilon \qquad (C.4.29)$$

$$u = \rho W u + \varepsilon. \qquad (C.4.30)$$

The fact that the SDM model subsumes both the SAR and SEM model has been overlooked in most applied spatial regression work, leading practitioners to devote a great deal of effort to choosing between the SAR and SEM model specifications. Given that the SDM model subsumes both of these models, there is no need to agonize over this aspect of model uncertainty.

C.4.4 Model averaging for spatial regression models

We consider the two sources of model uncertainty that arise from specification of the spatial weight matrix and selection of explanatory variables separately. There is an important technical difference between these two types of problems that motivates this choice. We consider the weight matrix issue and then turn attention to the variable selection problem.

Model uncertainty associated with spatial weight matrix specification

From our theoretical development we have seen that the key quantity needed to produce posterior model probabilities is the *marginal likelihood*. When we compare models based on a finite set of alternative spatial weight matrices, we are typically considering only a small number of alternative models, say m. Further, each model differs only in terms of the weight matrix specification since we hold the number of explanatory variables used in the model matrix X fixed. Of course, for the SDM model specification changes in the specification for the matrix W imply a change in the explanatory variables constructed using the spatial lag WX. However, an important point is that the *number* of vectors included in the matrix X remain the same. When we turn attention to variable selection the number of vectors in the matrix X change as we consider different explanatory variables. This requires some changes in the approach taken to determining the *marginal likelihood* and accompanying posterior model probabilities.

Determining values for the marginal likelihood for varying weight matrix specifications represents a situation where the *dimension of the model* is fixed because we have the same number of explanatory variables in all models. The models considered differ only in terms of the weight matrices used, allowing us to rely on uninformative prior distributions for the model parameters. This simplifies the task of model specification since we do not need to specify prior distributions for the model parameters.

For the simple case of two models, M_1, M_2 we denote the marginal likelihood of the data given model M_m, $m = 1, 2$ using $p(\mathcal{D}|M_m)$ which can be used to construct a posterior model probability for M_1 which takes the form

$$\pi_1 = p(M_1|\mathcal{D}) = \frac{p(\mathcal{D}|M_1)}{p(\mathcal{D}|M_1) + p(\mathcal{D}|M_2)} \frac{p(M_1)}{p(M_2)} \qquad \text{(C.4.31)}$$

$$p(\mathcal{D}|M_i) = \int p(\mathcal{D}|\theta_i, M_i) p(\theta_i|M_i) d\theta_i \qquad i = 1, 2 \qquad \text{(C.4.32)}$$

where $p(M_1)$ and $p(M_2)$ represent prior probabilities assigned to the two models by the practitioner. If we wish to let the sample data information determine the posterior model probabilities we should rely on a uniform setting for these that assigns equal weight to all models. That is, $p(M_i) = 1/m$, $i = 1, \ldots, m$ for the case of m models. It should be clear that in this case the fraction $p(M_1)/p(M_2) = 1$, eliminating any role for the prior model probabilities in determination of the posterior model probabilities.

A related concept often used to compare two (or more) models is the *posterior odds ratio* for M_1 versus M_2. The odds ratio is constructed using the posterior model probabilities: $O_{1,2} = p(M_1|\mathcal{D}) / p(M_2|\mathcal{D})$, where we use $O_{1,2}$ to denote the odds in favor of model one versus two. There is of course a relationship between the odds ratios and model probabilities

$$\pi_1 = \frac{1}{1 + O_{2,1} + \ldots + O_{m,1}} \qquad \text{(C.4.33)}$$

and in general for the case of m models we have

$$\pi_1 = \frac{p(M_i|\mathcal{D})}{\sum_{j=1}^{m} p(M_j|y)}. \qquad \text{(C.4.34)}$$

For the *independent* regression model where uninformative prior distributions are assigned to the parameters β, σ^2 the marginal likelihood takes the form of a scalar expression:

$$p(\mathcal{D}|M_i) = \Gamma\left(\frac{n-k}{2}\right) \frac{1}{(2\pi)^{(n-k)/2}} \frac{1}{|X^T X|^{1/2}} \frac{1}{(e^T e)^{(n-k)/2}} \quad \text{(C.4.35)}$$

$$e = y - X\hat{\beta} \quad \text{(C.4.36)}$$

$$\hat{\beta} = (X^T X)^{-1} X^T y. \quad \text{(C.4.37)}$$

Hepple (1995a, 1995b) sets forth the expressions needed to calculate the marginal likelihood associated with the SAR model that can be adapted to our case of the SDM model. The development is for the case of uninformative *improper* priors assigned to the model parameters δ, σ that take the form: $p(\delta, \sigma) \propto 1/\sigma$, and a uniform *proper* prior for the parameter ρ having the range D, $p(\rho) = 1/D$. We will have more to say about the role of proper versus improper priors in the next section when we discuss the need for proper priors when carrying out model averaging over models with varying sets of explanatory variables. For now we simply note that assigning these priors used by Hepple (1995a) requires no work on the part of the practitioner. This is because the range D for the uniform prior can be based on the interval $(-1 < \rho < 1)$ and there is no need to think about prior information regarding the parameters δ and σ.

The resulting marginal likelihood is derived from the joint posterior density for the model by analytically integrating over the parameters δ and σ to produce the expression shown in Eq. (C.4.38), where we use the symbol $Z_i = (X \; W_i X)$ to represent a model based on the spatial weight matrix W_i that defines model M_i.

$$p(\mathcal{D}|M_i) = \frac{1}{D}\Gamma\left(\frac{n-k}{2}\right) \frac{1}{(2\pi)^{(n-k)/2}} \frac{1}{|Z_i^T Z_i|^{1/2}} \int |I_n - \rho W_i| \frac{1}{(e^T e)^{(n-k)/2}} d\rho$$

(C.4.38)

$$e = y - Z_i \hat{\delta} \quad \text{(C.4.39)}$$

$$\hat{\delta} = (Z_i^T Z_i)^{-1} (I_n - \rho W_i) y. \quad \text{(C.4.40)}$$

We also note that consistent with our discussion regarding fixing the explanatory variables matrix X we do not place a subscript i on this matrix which remains fixed for all models, M_i, $i = 1, ..., m$.

An important difference arises between the scalar marginal likelihood in Eq. (C.4.35) for the case of the *independent* regression model and Eq. (C.4.38) for the SDM model. In the case of the SDM model we cannot rely on analytical integration methods to completely derive the marginal likelihood. These work to eliminate the parameters δ and σ from the marginal likelihood, but not the spatial dependence parameter ρ. To complete the task of evaluating the marginal likelihood we need to perform numerical integration over the range of the parameter ρ. LeSage and Parent (2007) provide an Appendix that sets forth computationally efficient methods for accomplishing this task.

Uncertainty arising from explanatory variable selection

In the case where we fix the spatial weight matrix W and consider models based on varying numbers of explanatory variables we cannot rely on improper prior distributions for the parameters δ and σ as we did in the previous section. An issue that arises when calculating posterior model probabilities for these models has been labeled the *Lindley paradox* (Lindley 1957). Lindley noted that posterior model probabilities calculated for models based on improper priors resulted in a higher posterior probability always being assigned to the more parsimonious model, that containing fewer parameters. This result arises irrespective of the sample data used. Since we would like the sample data to play a primary role in determining the posterior probabilities for models based on varying sets of explanatory variables, this is a very undesirable result.

The solution to the Lindley paradox is to assign proper prior distributions for the parameters δ and σ in our model. (We have already assigned a proper *uniform* prior for the parameter ρ in the model) There is a trade-off between allowing the sample data to play the only role in determining the explanatory variables which would be the case if we were able to assign uninformative priors and the need to avoid the Lindley paradoxical outcome associated with using this type of prior. We note that this problem arises in the conventional independent regression model as well as the spatial regression models considered here. LeSage and Parent (2007) build on results from the conventional regression literature to devise a *strategic prior*. One implication of the Lindley paradox is that there is no natural way to construct a prior that exerts a total lack of influence on the resulting posterior parameter distributions that arise in Bayesian analysis. Nonetheless, we can devise a prior that exerts a minimal influence on the posterior outcome so the sample data information plays a dominant role in determining the posterior model probabilities. A prior specification that exerts minimal influence on the posterior model probabilities is what we mean when we refer to a strategic prior. We provide specifics regarding our strategic prior later.

There is a second important difference between calculating posterior model probabilities for a finite number of models based on alternative weight matrices and the problem of models based on alternative explanatory variables considered here. In these situations a small set of say 20 candidate explanatory variables will lead to $2^{20} = 1,048,576$ or over one million possible models. That is, if we are interested in entertaining all possible ways of including or excluding combinations of 20 variables we would need to calculate posterior model probabilities for a very large number of models. In general, if we let k denote the number of candidate explanatory variables, there are 2^k possible models to be considered. If $k = 50$, a seemingly realistic number in many applied situations, we have a near infinite 1,000 trillion possible models to consider. Further, determining each model probability requires that we carry out numerical integration of the expression in Eq. (C.4.38) to arrive at the marginal likelihood needed to determine each model probability.

A large literature exists on the topic of Bayesian model averaging for the case of the independent regression model where alternative sets of explanatory variables are the object of interest (Fernandez et al. 2001; Madigan and York 1995). This is perhaps not surprising given the classic trade-off that exists in applied regression modeling between including a sufficient number of explanatory variables to avoid potential omitted variables bias and inclusion of redundant variables that produce a decrease in precision of the estimates. A strategic prior is set forth by (Fernandez et al. 2001) that we rely on to overcome the problem of the Lindley paradox. To address the second issue where a near infinite number of possible models arises when the number of candidate explanatory variables becomes large, we adopt a method that has been labeled *Markov Chain Monte Carlo Model Composition* or MC^3 (Madigan and York 1995). Details regarding these two approaches are provided in LeSage and Parent (2007) as they apply to both the SAR and SEM spatial regression model specifications. We also note there is some more recent literature regarding strategic priors for use with the MC^3 method (Ley and Steel 2009; Liang et al. 2008). For example, Ley and Steel (2009) propose a binomial-beta prior distribution that relaxes the assignment of equal prior probability for each model. This represents an attempt to address a concern that models with more versus fewer variables might be seen as a priori more or less likely.

LeSage and Parent (2007) rely on a normal distribution as a prior for the parameters δ in the SDM model and an inverse gamma prior distribution for the parameter σ. This combination of normal and inverse gamma distribution simplifies analytical integration over these parameters allowing use to arrive at an expression analogous to that in Eq. (C.4.38). As in the case of Eq. (C.4.38), we still require numerical integration over the parameter ρ to complete our evaluation of the marginal likelihood.

The normal prior assigned for the parameters δ is based on a suggestion by Fernandez et al. (2001) that the normal prior distribution from Zellner (1986) known as the g-prior can act as a *strategic prior*. They suggest settings for this prior distribution that they demonstrate to be strategic. By this we mean that the prior settings produce a proper prior that does not exert undue influence on the posterior model probabilities. However, Liang et al. (2008) propose an alternative approach to specifying the Zellner g-prior from Fernandez et al. (2001).

Given the ability to calculate the (logged) marginal likelihood for a single model, LeSage and Parent (2007) suggest using this calculated quantity in the MC^3 method of Madigan and York (1995). The MC^3 method relies on a stochastic Markov Chain process that moves through the near infinite dimensional model space and samples regions of high posterior support. This eliminates the need to consider all possible models. Rather, the Markov Chain process works its way through the model space sampling various models and calculating (log-transformed) marginal likelihoods for each model sampled. These are subjected to a Metropolis-Hastings accept-reject step which steers the sampling process towards regions of the model space with higher posterior probability mass. Specifically, a proposed model M_i is compared to the current model M_j using the Metropolis-Hastings acceptance probability in Eq. (C.4.41).

$$\min\left[1, \frac{p(M_i \mid \mathcal{D})}{p(M_j \mid \mathcal{D})}\right]. \qquad (C.4.41)$$

For details regarding Metropolis-Hastings sampling in the context of spatial regression models [see LeSage and Pace (2009)]. Of note for our purposes is the fact that the fraction in Eq. (C.4.41) is nothing more than our *odds ratio* O_{ij} (see the discussion surrounding Eq. (C.4.31)). There are strict requirements on the procedure used to propose a new model for validity of the MC^3 method. LeSage and Parent (2007) discuss these issues. Basically, if we let M_j denote the current model, a proposed model M_i must contain either one variable more (labeled a *birth step*), or one variable less (a *death step*) than M_j. Of course, birth steps select a variable at random from those not currently included in the model and death steps select at random from the set of variables currently included in the model.

This procedure is not *ad-hoc*. Madigan and York (1995) show that running the Markov Chain process long enough will result in a sample of models that are representative of the true posterior model probabilities. In applied practice, one can select a random set of explanatory variables as a starting point for the sampling procedure and produce a large number of sampled models (say 500,000) along with their posterior model probabilities. Running a second sampling procedure beginning with a different randomly selected set of starting variables to produce another sample of 500,000 models and model probabilities should produce very similar results to the first sample. If similar results do not arise, one should increase the sample size beyond 500,000. This process should be continued until a sample size large enough to produce samples that approximates the true posterior model probabilities arises.

C.4.5 Applied illustrations

We illustrate model averaging for the case of spatial weight matrices based on differing numbers of nearest neighbors using two samples consisting of U.S. counties, one involving 950 counties located in metropolitan areas and the other consisting of 1,754 counties located outside of metropolitan areas. The sample data was taken from the 2002 Census of Governments on county-level spending and some observations were missing resulting in less than 3,108 total county-level observations.

The same example is used to illustrate model averaging over models based on differing sets of explanatory variables. The results reported for these two illustrations do not fully incorporate both sources of model uncertainty. Results from the first illustration are conditional on the explanatory variables matrix X used, and those from the second illustration condition on the spatial weight matrix. A third illustration is used to present posterior inferences that incorporate model specification uncertainty regarding both the spatial weight matrix and explanatory variables.

Weight matrix model averaging

The model was used to explore the impact of population migration on provision of local government services. It is commonly acknowledged that local government service provision and taxes are not *independent*, but rather *spatially dependent*, which means that levels of services and taxes in one county are similar to those of nearby counties. There are a number of theories that provide an explanation for this observed spatial clustering (Tiebout 1956). This suggests an econometric model that takes spatial dependence into account should be used when examining cross-sectional information on county government spending and taxes.

Information on taxes and intergovernmental aid from both state and national sources were used as one explanatory variable in the model along with median household income estimates for the year 2002 taken from Current Population Survey Annual Demographic Supplements. Population for the year 2000 and in- and out-migration were obtained from the year 2000 Census, with the migration magnitudes reflecting cumulative in- and outmigration to each county in the our sample (over the five-year period from 1995-2000) from all other (3,108) counties in the contiguous 48 states.

The model is shown in Eq. (C.4.42), where the dependent variable y represents the (log) marginal tax cost of local government services provision for county i. This is a variable constructed using: $y_i = \ln(s_i P_i^\phi)$, where s_i is the median voter's share of taxes raised from local sources, P_i is the county population and ϕ represents a scalar congestion parameter associated with consumption of local public goods. This parameter reflects the degree of publicness that varies with consumption congestion, with $0 < \phi < 1$. A value of $\phi = 0$ reflects local government services provision that suffer from no consumption congestion effects resulting in a purely public good and a value $\phi = 1$ denotes a private good (Turnbull and Geon 2006). Since this parameter is unknown, we set $\phi = 0.5$ in this illustration, reflecting the midpoint of the zero to one range. This indicates that we view county government services as midway between the extremes of pure public and private goods.

$$y = \alpha_0 \iota_n + \rho W y + \alpha_1 X + \alpha_2 W X + \varepsilon. \tag{C.4.42}$$

Four explanatory variables were used to form the explanatory variables matrix X: intergovernmental grants from state and national sources, which we label A, county government spending (G) (excluding A), and in- and out-migration to the county (I, O). We might expect the effects on marginal tax cost to be positive for G and negative for A. (Intergovernmental grants/transfers essentially act like a reduction in G.) Both the dependent and independent variables were transformed using logs, so we will be able to interpret the coefficient estimates as elasticities.

The effects of in- and out-migration on the marginal tax cost of local government services are less clear. Destination regions should benefit from an inflow of more highly skilled and educated workers, since these are the groups most likely to move. On the other hand, origin regions may suffer from a loss of the more productive members of their communities who are also less dependent on government services. This reasoning has led to the argument that rural-urban migration trends over the past half century have increased the costs of providing local government services in rural areas.

For our illustration here we calculated posterior model probabilities for two models, one based on the sample of 950 metropolitan area counties and the other based on the sample of 1,754 counties located outside of metropolitan areas. These are reported in Table C.4.1 for models based on 15 different spatial weight matrices based on varying the number of nearest neighbors used to construct the matrix W_i over $m = 1$ to $m = 15$. The table also reports the posterior mean estimates for the noise variance parameter σ^2 for the metropolitan area sample which we use later.

From the table, we see high posterior probabilities pointing to a spatial weight matrix based on $m = 8$ and $m = 9$ nearest neighbors in the case of the non-metropolitan county sample and $m = 7, 8, 9$ for the metropolitan area counties. For the U.S. counties, the number of first-order contiguous neighbors (those with borders that touch each county) is around six, so the number of neighbors chosen from the model comparison illustration represents slightly more than just the contiguous counties.

Table C.4.1. Posterior model probabilities for varying spatial neighbors

# Neighbors	Non-metro	Metro	Metro $\hat{\sigma}^2$
$m = 1$	0.0000	0.0000	0.1569
$m = 2$	0.0000	0.0000	0.1397
$m = 3$	0.0000	0.0000	0.1338
$m = 4$	0.0000	0.0000	0.1287
$m = 5$	0.0000	0.0000	0.1286
$m = 6$	0.0000	0.0250	0.1270
$m = 7$	0.0003	0.4305	0.1266
$m = 8$	0.6007	0.2884	0.1272
$m = 9$	0.3299	0.2427	0.1275
$m = 10$	0.0689	0.0102	0.1288
$m = 11$	0.0001	0.0023	0.1297
$m = 12$	0.0000	0.0005	0.1306
$m = 13$	0.0000	0.0005	0.1311
$m = 14$	0.0000	0.0000	0.1318
$m = 15$	0.0000	0.0000	0.1336

To illustrate how model averaging works, we constructed model averaged estimates using the metropolitan area sample of counties and the eight non-zero posterior probability weights to average over models based on weight matrices constructed using nearest neighbors ranging over $m = 6, \ldots, 13$. These model averaged estimates will be compared to estimates based on the $m = 7$ neighbors suggested by the single highest posterior probability model. The single model approach might reflect a conventional approach that selects a single model based on some criterion such as fit. The posterior mean of the parameter σ^2 was indeed a minimum for the model based on $m = 7$ as shown by the values reported in Table C.4.1 for the metropolitan area sample.

Table C.4.2 reports posterior means and standard deviations constructed from a set of 2,500 draws produced using Markov Chain Monte Carlo (MCMC) estimation of the model (LeSage 1997). In addition, we follow conventional MCMC practice and report 0.95 and 0.99 *credible intervals* constructed using the sample of draws from the MCMC sampler. This involves sorting the sampled draws from low to high and finding lower and upper 0.95 and 0.99 points. For example, given a vector of 10,000 sorted draws, we would use the 5,000 − (9,500/2) and 5,000 + (9,500/2) elements of this vector as the lower and upper 0.95 credible intervals. Inferences based on these should correspond to a 95% level of confidence from conventional methods of inference used in regression modeling.

From the table we see that the standard deviation of the model averaged estimates is smaller than that associated with the single model estimates for all variables. This indicates an increase in the posterior precision of the parameters arising from the model averaging procedure. The model averaged standard deviations are around 60 percent of those from the single model.

This increased precision leads to some differences in the inferences that would be drawn based on the two sets of model estimates. In the case of the single $m = 7$ model, we would conclude that neighboring governments expenditures *WG* do not reduce the marginal tax costs of local government service provision using the 0.99 credible interval. In contrast, the model averaged estimates point to a negative impact from *WG* based on the 0.99 credible interval. Another difference arises for the out-migration variable *O* which has a 0.99 lower credible interval near zero (0.0008) for the single model. The positive impact of out-migration on marginal tax costs is much clearer for the model averaged estimates since the lower 0.99 credible interval of (0.0392) is clearly greater than zero.

Another approach to resolving questions regarding the role of the various explanatory variables in explaining variation in the marginal tax costs of local government services would be to rely on model averaging in the context of variable selection. We illustrate this approach next.

Table C.4.2. Metropolitan sample SDM model estimates

Variables[a]	Lower 01	Lower 05	Mean	Upper 05	Upper 01	Std
			Single $m = 7$ model estimates			
Constant	−2.4134	−2.2622	−1.9059	−1.5392	−1.3844	0.2158
G	0.5179	0.5322	0.5718	0.6092	0.6260	0.0235
A	−0.0467	−0.0371	−0.0089	0.0184	0.0291	0.0167
I	−0.3076	−0.2791	−0.2006	−0.1241	−0.0955	0.0465
O	0.0008	0.0321	0.1079	0.1849	0.2158	0.0471
WG	−0.2340	−0.1965	−0.1179	−0.0351	0.0007	0.0492
WA	−0.1715	−0.1531	−0.1088	−0.0653	−0.0470	0.0263
WI	0.1697	0.2239	0.3610	0.4977	0.5519	0.0827
WO	−0.6493	−0.5914	−0.4507	−0.3153	−0.2599	0.0837
			Model averaged estimates			
Variables[a]	Lower 01	Lower 05	Mean	Upper 05	Upper 01	Std
Constant	−2.1247	−2.0319	−1.8252	−1.6205	−1.5326	0.1252
G	0.5495	0.5577	0.5801	0.6028	0.6101	0.0135
A	−0.0357	−0.0296	−0.0130	0.0025	0.0086	0.0096
I	−0.2541	−0.2405	−0.1959	−0.1518	−0.1364	0.0263
O	0.0392	0.0559	0.0996	0.1439	0.1600	0.0266
WG	−0.2126	−0.1931	−0.1449	−0.0978	−0.0768	0.0286
WA	−0.1343	−0.1239	−0.0986	−0.0734	−0.0638	0.0152
WI	0.2369	0.2774	0.3526	0.4314	0.4620	0.0479
WO	−0.5443	−0.5134	−0.4325	−0.3521	−0.3163	0.0487

Notes: [a] *G* is government spending; *A* is intergovernmental revenue; *I* is in-migration; *O* is out-migration; *WG* is the spatial lag of *G*; *WA* is the spatial lag of *A*; *WI* and *WO* are spatial lags of in- and out-migration

Variable selection model averaging

We proceed by allowing each of the four explanatory variables and their spatial lags to enter the model independently, leading to a set of $2^8 = 256$ possible models. The intercept term and spatial lag of the dependent variable are included in all models (LeSage and Parent 2007). There may be modeling contexts where it makes more sense to force an explanatory variable from the matrix *X* to enter the model along with the same variable from *WX*. When implementing the MC^3 procedure, we used only a single spatial weight matrix based on $m = 7$, the highest posterior probability model. This makes the estimates and inferences about the matrix X drawn conditional on the spatial weight matrix W used. In our previous illustration, inferences regarding the spatial weight matrix W were conditional on use of the saturated explanatory variables matrix X containing all explanatory variables. We will have more to say about eliminating the conditional nature of these results later.

Since there are only 256 models we could calculate posterior model probabilities for an enumerative list of these models, but we used the MC^3 procedure to sample the model space. A run of 100,000 sampling draws found 119 unique models, with the top 12 models accounting for 0.9966 of the posterior probability mass determined using all 256 possible models. We note that the MC^3 sampling procedure systematically steered away from around half of the model space where the models exhibited low posterior probabilities. The model probability mass was very concentrated in a few models with the top 5 models accounting for 0.9530 probability and the top 2 models 0.6020 probability.

Results for the top 10 models are shown in Table C.4.3, where the ten columns labeled $m1$ to $m10$ show which variables entered each model using a '1', and a '0' for variables that were not included. The last row of the table shows the posterior model probabilities for each model. These results confirm the earlier uncertainty regarding the influence of neighboring governments expenditures WG on marginal tax costs. The top 2 models ($m1$, $m2$) have probabilities that are roughly equal to 0.30 with the WG variable entering one model and not the other. All other variables are the same for these two models. It is also the case that WG appeared in five of the top 10 models, again pointing to uncertainty about the role of this variable.

Also consistent with our earlier results, the variable A representing intergovernmental aid had posterior credible intervals that spanned zero in Table C.4.2, pointing to a lack of significance. Here we see that this variable did not enter any of the top 5 models.

Table C.4.3. Metropolitan sample MC^3 results for $m = 7$ neighbors

Variables[a]	m10	m9	m8	m7	m6	m5	m4	m3	m2	m1
G	1	1	1	1	1	1	1	1	1	1
A	1	1	0	1	1	0	0	0	0	0
I	1	1	1	1	1	1	1	1	1	1
O	1	1	0	0	0	0	1	1	0	0
WG	1	0	1	0	1	0	1	0	0	1
WA	1	1	1	1	1	1	1	1	1	1
WI	1	1	0	1	1	0	1	1	1	1
WO	1	1	1	1	1	1	1	1	1	1
$p(M_i|\mathcal{D})$	0.005	0.007	0.008	0.010	0.011	.0020	0.132	0.199	0.294	0.308

Notes: [a] G is government spending; A is intergovernmental revenue; I is in-migration; O is out-migration; WG is the spatial lag of G; WA is the spatial lag of A; WI and WO are spatial lags of in- and out-migration

Model averaged estimates were produced using the top 12 models which as noted accounted for 0.9966 of the posterior probability mass determined using all 256 possible models. These estimates are reported in Table C.4.4, where posterior means for the coefficients and 0.95 as well as 0.99 credible intervals are shown. These estimates clearly point toward a zero impact for the variable A, producing a very small coefficient estimate. The posterior mean estimate of -0.0001 produces a sharper inference than the model averaged posterior mean coefficient from Table C.4.2 which was equal to -0.0130. The model averaged estimates resolve the question regarding WG by pointing to a significant negative impact based on the posterior mean and 0.99 credible intervals for this variable reported in the table.

We also report a model averaged coefficient for the spatial dependence parameter associated with the spatial lag of the dependent variable Wy, which was excluded from our previous results to save space. This coefficient points to positive and significant spatial dependence in the marginal tax costs relationship being explored.

Table C.4.4. Model averaged estimates based on the top 12 models

Variables [a]	Lower 0.01	Lower 0.05	Coefficients	Upper 0.95	Upper 0.99
G	0.5460	0.5518	0.5643	0.5777	0.5830
A	−0.0007	−0.0005	−0.0001	0.0004	0.0006
I	−0.1684	−0.1558	−0.1330	−0.1096	−0.0991
O	0.0100	0.0211	0.0400	0.0597	0.0679
WG	−0.0821	−0.0747	−0.0496	−0.0251	−0.0186
WA	−0.1613	−0.1561	−0.1408	−0.1264	−0.1198
WI	0.2065	0.2311	0.2923	0.3521	0.3830
WO	−0.4946	−0.4686	−0.4076	−0.3461	−0.3230
Wy	0.5639	0.5720	0.5941	0.6158	0.6252

Notes: [a] G is government spending; A is intergovernmental revenue; I is in-migration; O is out-migration; WG is the spatial lag of G; WA is the spatial lag of A; WI and WO are spatial lags of in- and out-migration; Wy is a spatial lag of the dependent variable, marginal tax costs of local government services

There is still the question of whether the results reported here fully account for the two aspects of model uncertainty under consideration. Model averaged results that address the weight matrix uncertainty were produced by conditioning on the saturated matrix X containing the full set of explanatory variables. Similarly, the MC^3 results were produced by conditioning on an $m = 7$ neighbors spatial weight matrix. Ideally, we would like to produce posterior inferences that are unconditional on both the weight matrix and explanatory variables employed. These inferences would incorporate all aspects of model uncertainty. We turn attention to this next.

Weight matrix and variable selection model averaging

As noted, the model averaged estimates presented in the previous two sections do not fully incorporate all sources of uncertainty in the posterior inferences. There are applied modeling situations where application of the MC^3 procedure to models based on different spatial weight matrices will produce models and model averaged posterior inferences that do not vary greatly as we change the weight matrix. An examination of results from this approach applied to our model showed this was not the case. For example, the posterior mean model averaged coefficient for the *WG* variable based on a set of 142 unique models identified by the MC^3 procedure applied with an $m = 9$ nearest neighbors weight matrix was equal to -0.1706 which is quite different from the value of -0.0496 reported in Table C.4.4 for these same results based on $m = 7$. The upper 0.99 credible interval for this coefficient in the $m = 9$ procedure was -0.0861 suggesting a significant difference between the posterior mean estimates. There were a number of other differences between the outcomes from the $m = 7$, $m = 8$ and $m = 9$ models, suggesting substantial model uncertainty associated with the particular spatial weight matrix employed.

In the most general case where we are dealing with a near infinite number of possible models, we could adapt our MC^3 procedure to create proposal models based on variation in both the spatial weight matrix as well as the explanatory variables matrix. LeSage and Fischer (2008) discuss this approach and provide an application of the method to European regional growth.

For the relatively small number of models considered here, we can simply average over models produced using the MC^3 procedure three times based on spatial weight matrices involving $m = 7, 8$ and 9 nearest neighbors. As reported in Table C.4.1, models based on these weight matrices account for most of the posterior probability mass. The MC^3 sampling procedure implemented using 100,000 draws produce 119, 126 and 142 unique models for $m = 7, 8, 9$ respectively. Using the log-marginal likelihoods for these models to calculate posterior model probabilities resulted in 31 models that had posterior probabilities greater than 0.0001. Of these 31 models, ten were based on $m = 7$, 9 were associated with $m = 8$ and 12 exhibited $m = 9$. This suggests a relatively uniform distribution of posterior model probabilities with respect to the number of nearest neighbors used to form the spatial weight matrix. Table C.4.5 shows the top 12 models which had posterior model probabilities greater than 0.01 along with the number of neighbors m associated with these models.

Model averaged estimates based on the 31 models having posterior model probabilities greater than 0.0001 are reported in Table C.4.6 along with 0.95 and 0.99 credible intervals. From these estimates we would conclude that all explanatory variables are significant using the 0.95 credible intervals. Based on the 0.99 intervals we see that the variable *A* representing intergovernmental aid does not exert an impact on the marginal tax costs of local government services provision. We see the same small posterior mean coefficient estimate for the variable *A* as in the model averaged estimates results reported in Table C.4.4. Since the variables were transformed using logs, we can interpret the coefficient magnitudes as elasticities. This suggest that despite the statistical significance of the variable *A* based

Table C.4.5. Posterior model probabilities for the top 12 models and associated neighbors

Model	Posterior Probability	# of nearest neighbors
12	0.0123	9
11	0.0131	9
10	0.0193	8
9	0.0235	9
8	0.0383	8
7	0.0396	9
6	0.0630	7
5	0.0736	7
4	0.0944	7
3	0.1455	7
2	0.1599	8
1	0.2257	9

on the 0.95 credible interval, intergovernmental aid is not likely to be economically significant.

All other variables have a statistically significant impact using the 0.99 credible intervals, and their magnitudes are such that we would infer these to be economically significant as well. Another difference between the model averaged estimates from Table C.4.4 and those in Table C.4.6 relates to the estimate for the parameter ρ associated with the spatial lag variable Wy. Averaging over models based on differing spatial weight matrices produces a larger posterior mean estimate for the strength of spatial dependence. The lower 0.01 credible interval for this coefficient equal to 0.5953 is above the posterior mean estimate of 0.5941 reported in Table C.4.4, suggesting a significant increase in spatial dependence.

Table C.4.6. Model averaging over both neighbors and variables

Variables[a]	Lower 0.01	Lower 0.95	Mean	Upper 0.95	Upper 0.99
G	0.5540	0.5581	0.5670	0.5763	0.5797
A	−0.0012	−0.0010	−0.0005	−0.0001	0.0001
I	−0.1418	−0.1354	−0.1226	−0.1099	−0.1050
O	0.0119	0.0155	0.0245	0.0332	0.0363
WG	−0.1529	−0.1416	−0.1149	−0.0872	−0.0750
WA	−0.1369	−0.1312	−0.1183	−0.1049	−0.1002
WI	0.2070	0.2257	0.2711	0.3151	0.3347
WO	−0.4208	−0.4023	−0.3581	−0.3128	−0.2959
Wy	0.5953	0.6031	0.6217	0.6388	0.6467

Notes: [a] G is government spending; A is intergovernmental revenue; I is in-migration; O is out-migration; WG is the spatial lag of G; WA is the spatial lag of A; WI and WO are spatial lags of in- and out-migration; Wy is a spatial lag of the dependent variable, marginal tax costs

An important caveat regarding interpretation of the model averaged coefficient estimates reported in Table C.4.6 is that we *cannot* interpret these in the same fashion as ordinary regression coefficients. Models containing spatial lags of the dependent variable result in a situation where ceteris paribus changes in a single observation *i* associated with any explanatory variable give rise to both *direct and indirect impacts* on the dependent variable *y*. The direct impacts reflect how changes in the *i*th observation of an explanatory variable influence the dependent variable at observation *i*. Indirect impacts indicate how other observations *j* of the dependent variable change in response to this type of ceteris paribus change in the single explanatory variable observation *i*. This is a consequence of allowing for spatial dependence between observations as opposed to the assumption of independence across observations made in conventional regression models (LeSage and Pace 2009).

C.4.6 Concluding remarks

Model specification uncertainty arises in spatial regression models from three sources, (i) the type of model that should be used, (ii) the type of spatial weight matrix, and (iii) the specific explanatory variables to be included in the model. We showed how the first type of uncertainty can be resolved by relying on a spatial Durbin model that subsumes both the spatial lag and spatial error dependence models as special cases.

Bayesian model averaging methods can be used to incorporate uncertainty arising from the other two model specification choices that confront practitioners in applied settings. For prominent members of the family of spatial regression models often used in applied work, the marginal likelihood can be calculated using relatively simple univariate numerical integration. As discussed, this quantity allows calculation of posterior model probabilities that can be used in formal Bayesian model comparison methods. Beyond this, Bayesian model averaging procedures can be used to incorporate uncertainty arising from sources (ii) and (iii) above. This involves constructing a posterior distribution for the model parameters using a linear combination of different models, where the posterior model probabilities are used as weights.

The model averaging approach represents a formal Bayesian solution to the problem of uncertainty regarding various aspects of model specification that arise in applied practice. It can be used with a large number of spatial models, not just those described here (Parent and LeSage 2008; LeSage and Polasek 2008). In more complicated models it may be necessary to produce an approximation or estimate of the marginal likelihood. As discussed, calculating the marginal likelihood requires integration over the model parameters. It is not always possible to use analytical integration of the type illustrated here to reduce the dimensionality of the integration problem. Fortunately, there is a large literature on various approaches to approximating the marginal likelihood (Chib 1995; Chib and Jeliazkov 2001; Newton and Raftery 1994).

Acknowledgements. Olivier Parent would like to acknowledge support of the Charles Phelps Taft Research Center and James LeSage would like to acknowledge support from NSF SES-0729264 and the Texas Sea grant program. Both authors would like to thank Manfred M. Fischer along with an anonymous reviewer for helpful comments on earlier versions of this chapter.

References

Anselin L (1988) Spatial econometrics: methods and models. Kluwer, Dordrecht

Chib S (1995) Marginal likelihoods from the Gibbs sampler. J Am Stat Assoc 90(432):1313-1321.

Chib S, Jeliazkov I (2001) Marginal likelihood from the Metropolis-Hastings output. J Am Stat Assoc, 96(1):270-281

Fernandez C, Ley E, Steel MFJ (2001) Benchmark priors for Bayesian model averaging. J Econometrics 100(2):381-427

Hepple LW (1995a) Bayesian techniques in spatial and network econometrics: 1. Model comparison and posterior odds. Environ Plann A 27(3):447-469

Hepple LW (1995b) Bayesian techniques in spatial and network econometrics: 2. Computational methods and algorithms. Environ Plann A 27(4):615-644

LeSage JP (1997) Bayesian estimation of spatial autoregressive models. Int Reg Sci Rev 20(1&2):113-129

LeSage JP, Fischer MM (2008) Spatial growth regressions: model specification, estimation and interpretation. Spat Econ Anal 3(3):275-304

LeSage JP, Pace RK (2009) Introduction to spatial econometrics. CRC Press (Taylor and Francis Group), Boca Raton [FL], London and New York

LeSage JP, Parent O (2007) Bayesian model averaging for spatial econometric models. Geogr Anal 39(3):241-267

LeSage JP, Polasek W (2008) Incorporating transportation network structure in spatial econometric models of commodity flows. Spat Econ Anal 3(2):225-245

Ley E, Steel M (2009) On the effect of prior assumptions in Bayesian Model Averaging with applications to growth regression. J Econometrics 24(4):651-674

Liang F, Paulo R, Molina G, Clyde MA, Berger JO (2008) Mixtures of g-priors for Bayesian variable selection. J Am Stat Assoc 103(481):410-423

Lindley DV (1957) A statistical paradox. Biometrika, 44(1-2):187-192

Madigan D, York J (1995) Bayesian graphical models for discrete data. Int Stat Rev 63(2):215-232

Newton MA, Raftery AE (1994) Approximate Bayesian inference with the weighted likelihood bootstrap. J Roy Stat Soc B 56(1):3-48

Parent O, LeSage JP (2008). Using the variance structure of the conditional autoregressive specification to model knowledge spillovers. J Appl Econ 23(2):235-256

Tiebout CM (1956) A pure theory of local expenditures. J Polit Econ 64(5):416-424

Turnbull GK, Geon G (2006) Local government internal structure, external constraints and the median voter. Public Choice 129(3-4):487-506

Zellner A (1971) An introduction to Bayesian inference in econometrics. Wiley. New York, Chichester, Toronto and Brisbane.

Zellner A (1986) On assessing prior distributions and Bayesian regression analysis with g-prior distributions. In Goel P, Zellner A (eds.) Bayesian inference and decision techniques: essays in honor of Bruno de Finetti. Elsevier, Amsterdam, pp. 233-243

C.5 Geographically Weighted Regression

David C. Wheeler and *Antonio Páez*

C.5.1 Introduction

Geographically weighted regression (GWR) was introduced to the geography literature by Brunsdon et al. (1996) to study the potential for relationships in a regression model to vary in geographical space, or what is termed parametric non-stationarity. GWR is based on the non-parametric technique of locally weighted regression developed in statistics for curve-fitting and smoothing applications, where local regression parameters are estimated using subsets of data proximate to a model estimation point in variable space. The innovation with GWR is using a subset of data proximate to the model calibration location in geographical space instead of variable space. While the emphasis in traditional locally weighted regression in statistics has been on curve-fitting, that is estimating or predicting the response variable, GWR has been presented as a method to conduct inference on spatially varying relationships, in an attempt to extend the original emphasis on prediction to confirmatory analysis (Páez and Wheeler 2009).

In GWR, a regression model can be fitted at each observation location in the dataset, although the model calibration locations are not restricted to observation locations. The spatial coordinates of the data points, either individual data points or areal centroids, are used to calculate inter-point distances, which are input into a kernel function to calculate weights that represent spatial dependence between observations. For each model calibration location, $i = 1, \ldots, n$, the GWR model is

$$y_i = \beta_{i0} + \sum_{k=1}^{p-1} \beta_{ik} x_{ik} + \varepsilon_i \qquad (C.5.1)$$

where y_i is the dependent variable value at location i, x_{ik} is the value of the kth covariate at location i, β_{i0} is the intercept, β_{ik} is the regression coefficient for the kth covariate, p is the number of regression terms, and ε_i is the random error at location i. We note the distinction between regression terms and regression coefficients, where the number of regression coefficients is np. The obvious difference in this model and the traditional ordinary least squares (OLS) regression model is in regression coefficients estimated at each data location, where they are global, or fixed for the study area, in the OLS model.

C.5.2 Estimation

To facilitate the exposition, it is convenient to express the GWR model in matrix notation

$$y_i = X_i \beta_i + \varepsilon_i \quad (C.5.2)$$

where β_i is a column vector of regression coefficients and X_i is a row vector of explanatory variables at location i. The vector of estimated regression coefficients at location i is

$$\hat{\beta}_i = [X^T W_i X]^{-1} X^T W_i Y \quad (C.5.3)$$

where Y is the n-by-1 vector of dependent variables; $X = [X_1^T, X_2^T, ..., X_n^T]^T$ is the design matrix of explanatory variables, which includes a leading column of ones for the intercept; $W_i = diag[W_{i1}, ..., W_{in}]$ is the n-by-n diagonal weights matrix calculated for each calibration location i; and $\hat{\beta}_i = (\hat{\beta}_{i0}, \hat{\beta}_{i1}, ..., \hat{\beta}_{ip-1})^T$ is the vector of p local regression coefficients at location i for $p-1$ explanatory variables and an intercept. Given Eq. (C.5.3), GWR may be viewed as a locally weighted least squares regression model where the weights associate pairs of data points, and there are weights to associate the model calibration location i with all data points, including the calibration location itself. The weight matrix must be calculated at each location before the local regression coefficients can be estimated with Eq. (C.5.3).

In GWR, the local weights matrix, W_i, is calculated from a kernel function that places more weight on locations that are closer in space to the calibration location than those that are more distant in space. The weighting, therefore, follows the assumption of spatial autocorrelation, which if exists, is expected to result in non-stationary patterns in estimated coefficients. A kernel function in this context takes as input distance between two locations, has a bandwidth parameter that determines the spatial range of the kernel, and returns a weight between two locations that is inversely related to distance. A number of different kernel functions have been proposed for use in GWR. There are two general types of kernel functions, fixed and adaptive, where adaptive kernel functions attempt to adjust for the density of data points and fixed kernel functions do not. As an example of the difference in types of kernels, an adaptive kernel function could use the same number of observations in each local kernel, while a fixed kernel function could use the same spatial range in each local kernel.

Some examples of both fixed and adaptive kernel functions are provided below. In perhaps the simplest case, one could use a binary weighting scheme such as

$$W_{ij} = \begin{cases} 1 & \text{if } d_{ij} \leq d^* \\ 0 & \text{otherwise} \end{cases} \quad (C.5.4)$$

where d_{ij} is the distance between observations i and j, and d^* is a threshold distance that defines the size of the window. This kernel function could result in using fewer observations in the weighted set of a model calibration point located in a sparse area compared to a relatively dense area. Alternatively, the kernel function can be defined as

$$W_{ij} = \begin{cases} 1 & \text{if } y_j \in Y_i(N) \\ 0 & \text{otherwise} \end{cases} \quad (C.5.5)$$

where $Y_i(N)$ is the set of Nth nearest observations to point i, and N is a value to estimate. In this case, the kernel function uses the same number of observations at every point, but these observations may cover a different spatial extent in every case. Despite its simplicity, this kernel function has not been used extensively, perhaps because it does not conform well to established ideas about distance decay that have been strongly flavored by gravity modeling. Most applications of GWR instead have favored continuous functions that produce weights that monotonically decrease with distance, such as the Gaussian kernel function

$$W_{ij} = \exp\left(-\tfrac{1}{2}\left(\frac{d_{ij}}{\gamma}\right)^2\right). \quad (C.5.6)$$

In this function, the weight for observation j relative to observation i changes as a function of the distance d_{ij} and a kernel bandwidth parameter γ that controls the range and decay of spatial correlation. A similar kernel function is the simple exponential function

$$W_{ij} = \exp\left(-\frac{d_{ij}}{\gamma}\right) \tag{C.5.7}$$

which removes the powering and scaling of the Gaussian function. Another continuous, fixed kernel function is the bi-square kernel function

$$W_{ij} = \left(1 - \frac{d_{ij}^2}{\gamma^2}\right). \tag{C.5.8}$$

Several adaptive kernel functions have been proposed to adjust to the density of observations within a region. One such kernel function uses ranks of increasing distance instead of distance to calculate weights

$$W_{ij} = \exp\left(-\frac{R_{ij}}{\gamma}\right) \tag{C.5.9}$$

where R_{ij} is the rank of distance d_{ij} when locations are sorted by increasing distance from model calibration location i. An adaptive kernel function that has a different type of bandwidth parameter is the bi-square nearest neighbor kernel, where, again, the number of nearest neighbors must be determined in order to calculate weights to estimate the local regression coefficients. The kernel specification is

$$W_{ij} = \begin{cases} [1-(d_{ij}/d_{iN})^2]^2 & \text{if } j \text{ is one of the } N\text{th nearest neighbors of } i \\ 0 & \text{otherwise} \end{cases} \tag{C.5.10}$$

where d_{iN} is the distance to the Nth nearest neighbor from location i. This function assigns a weight of zero to points that are beyond the distance to the Nth nearest neighbor and a non-zero weight that decays with distance to points within the threshold distance.

Given the many options for a kernel function, one must first select a type of kernel function before calibrating a GWR model. Furthermore, in all the kernel functions above, there is an unknown kernel bandwidth parameter that must be selected or estimated from the data. Conventional wisdom inherited from the statistical non-parametric roots of GWR holds that selection of a functional form for the kernel is less critical than selection of the kernel bandwidth parameter for estimation results (see Chapter E.2, evidence in an application context), although this is something that has not been thoroughly explored in the geographical literature. Current practice is more concerned with the need for a formal criterion to select the kernel bandwidth or number of nearest neighbors in adaptive specifications. There are currently three different approaches for exogenously estimating the kernel bandwidth in GWR, direct assignment of the bandwidth of number of nearest neighbors (McMillen 1996), cross-validation (Brunsdon et al. 1996; Farber and Páez 2007), and a corrected Akaike Information Criterion (AIC, Fotheringham et al. 2002). In addition, an approach to parameterize the estimation of the kernel bandwidth has been proposed by Páez et al. (2002a). Of these, the most widely used approach by far remains cross-validation.

Cross-validation (CV) is an iterative process that searches for the kernel bandwidth that minimizes the prediction error of all the $y(s)$ using a subset of the data for prediction. If the kernel bandwidth is γ, it is estimated in CV by finding the γ that minimizes the root mean squared prediction error (RMSPE)

$$\hat{\gamma} = \arg\min_{\gamma} \sum_{i=1}^{n} [y_i - \hat{y}_{(i)}(\gamma)]^2 \qquad (C.5.11)$$

where $\hat{y}_{(i)}$ is the predicted value of observation i with calibration location i left out of the estimation dataset. $\hat{\gamma}$ is the kernel bandwidth value that minimizes the RMSPE. There are several search routines available, such as the golden search and the bi-section search, for finding the minimizing kernel bandwidth. Alternatively, one may evaluate the RMSPE over a large range of potential kernel bandwidths. As described, this is leave-one-out CV because only one observation is removed from the dataset for each local model when estimating the kernel bandwidth. The data point i is removed when estimating y_i to avoid estimating it perfectly. In the kernel functions outlined above, the kernel bandwidth is a global parameter. This parameter is applied to all local models individually, both in estimation of the kernel bandwidth and the regression coefficients. Implied in Eq. (C.5.11) is a local model to estimate y_i without using data point i and with the estimated regression coefficients in Eq. (C.5.3) and the current value of γ, and repeating this for each location.

An approach to estimate the kernel bandwidth not based on prediction of the response variable is the corrected AIC, adopted in form from locally weighted regression to GWR. It is instead based on minimizing the estimation error of the response variable. It is a compromise between goodness-of-fit of the model and model complexity, in that there is a penalty in the criterion for the effective number of parameters in the model. The corrected AIC for GWR is

$$AIC_c = 2n\log(\hat{\sigma}) + n\log(2\pi) + n\left(\frac{n+trace(\boldsymbol{H})}{n-2-trace(\boldsymbol{H})}\right) \qquad (C.5.12)$$

where $\hat{\sigma}$ is the estimated standard deviation of the error, $\boldsymbol{H}$ is the hat matrix, and the trace of a matrix is the sum of the matrix diagonal elements. The kernel bandwidth is used in the calculation of $\hat{\sigma}$ and $\boldsymbol{H}$. Each row of the hat matrix is defined by

$$\boldsymbol{H}_i = \boldsymbol{X}_i\,(\boldsymbol{X}^{\mathrm{T}}\,\boldsymbol{W}_i\,\boldsymbol{X})^{-1}\,\boldsymbol{X}^{\mathrm{T}}\,\boldsymbol{W}_i \qquad (C.5.13)$$

which may also be expressed as

$$\boldsymbol{H}_i = \boldsymbol{X}_i\,\boldsymbol{A}_i. \qquad (C.5.14)$$

The estimated error variance is

$$\hat{\sigma}^2 = \frac{\sum_{i=1}^{n}(y_i - \hat{y}_i)^2}{\{n - [2trace(\boldsymbol{H}) - trace(\boldsymbol{H}^{\mathrm{T}}\boldsymbol{H})]\}}. \qquad (C.5.15)$$

As with CV, to estimate the kernel bandwidth one either uses a search algorithm or evaluates the objective function over a range of values of γ. Here, the objective function is the AIC and it is to be minimized.

After estimating the kernel bandwidth with either CV or the AIC, one must calculate the kernel weights at each model calibration location using the estimated kernel function and then estimate the local regression coefficients. Then, one must estimate the response variable by

$$\hat{y}_i = \boldsymbol{X}_i\hat{\boldsymbol{\beta}}_i. \qquad (C.5.16)$$

In many applications of GWR, the spatial analyst maps the estimated regression coefficients and attempts to interpret the spatial pattern of the coefficients in the context of the research problem. Analysts are typically interested in where the estimated regression coefficients are statistically significant, according to some specified significance level. In the frequentist setting of GWR, statistical significance tests of the coefficients use the variance of the estimated regression coefficients. According to Fotheringham et al. (2002, p.55), the variance of the regression coefficients is

$$var[\hat{\beta}_i] = A_i A_i^T \hat{\sigma}^2. \qquad (C.5.17)$$

Technically, this equation is not correct because the Fotheringham et al. (2002) version of GWR is not a formal statistical model with kernel weights that are part of the errors. The equation used for the local coefficient covariance is only approximate with cross-validation because the kernel weights are calculated from the data first before the regression coefficients are estimated from the data. The kernel weights are inherently a function of Y, as are the regression coefficients, and the correct expression for the coefficient covariance would be non-linear.

C.5.3 Issues

While GWR offers the potential of investigating relationships that vary over space between variables in a regression model, there have been several critiques expressed about the methodology that counsel prudence in the application of the method. At a fundamental level, an argument is that GWR does not propose a base model for the source of the variation, and is thus more appropriately seen as a heuristic approach. As a consequence of this, it can be argued that GWR lacks a unified statistical framework since it is in essence an ensemble of local geographical regressions where the dependence between regression coefficients at different data locations is not specified in the model. This results in a fixed effects model with no pooling in estimates.

A second issue is related to the repeated use of data to estimate model parameters at different model calibration locations, which causes a multiple comparisons situation. With an increasing number of local models estimated, the probability that some individual tests will appear significant, even if only by chance, will also increase. The problem in this case is related to the trade-off between amount of information and confidence, since the usual confidence intervals for regression coefficients are no longer reliable. In order to account for multiplicity, each individual test needs to be seen as part of a family of experiments, and its corresponding level of significance needs to be adjusted so that it conforms to a family-wise confidence level. A simple adjustment to achieve this objective is based on the Bonferroni inequality, where the individual (adjusted) significance level is α/m with α being the nominal level of significance and m the number of tests in the family. This adjustment ensures that the family-wise level of confidence will be at most the nominal level. While this simple adjustment does not require any distributional assumptions, the resulting individual tests lack power and are overly conservative when the tests are not independent, and this is certainly the case with GWR because of the use of overlapping subsets of data. Alternative adjustments are available that improve power of the individual tests by introducing multiple-step rejection schemes that adjust the level of significance in a sequential way (see Páez et al. 2002a).

Another issue with GWR that is directly related to the selection of the kernel bandwidth involves high levels of spatial variation and smoothness of estimated regression coefficients. Clearly, if the bandwidth is such as to include a large number of observations, there will be relatively little or no spatial variation in the coefficients, and if the bandwidth is small, there will potentially be large amounts of variation. A natural concern emerges that some variation or smoothness in the pattern of estimated coefficients may be artificially introduced by the technique and may not represent true regression effects. This situation is at the heart of the discussion about the utility of GWR for inference on regression coefficients and is not answered by existing statistical (Leung et al. 2000a) or Monte Carlo (Fotheringham et al. 2002) tests for significant variation of GWR coefficients because these tests do not consider the source of the variation. This is important because one source of regression coefficient variability in GWR can come from collinearity, or dependence in the kernel-weighted design matrix. Collinearity is known in linear models to inflate the variances of regression coefficients (Neter et al. 1996), and GWR is no exception (Griffith 2008). Collinearity has been found in empirical work to be an issue in GWR models at the local level when it is not present in the global linear regression model using the same data (Wheeler 2007). In addition to large variation of estimated regression coefficients, there can be strong dependence in GWR coefficients for different regression terms, including the intercept, at least partly attributable to collinearity. Wheeler and Tiefelsdorf (2005) show in a simulation study that while GWR coefficients can be correlated when there is no explanatory variable correlation, the coefficient correlation increases systematically with increasingly more collinearity.

Inflated regression coefficient variation associated with local collinearity in GWR can lead to overestimates of covariate effect magnitudes and coefficient sign reversals, both of which are likely to lead to incorrect interpretations of relationships in the regression model. Fortunately, collinearity diagnostic tools have been developed for the GWR framework (Wheeler 2007) that detect where estimated regression coefficients are problematic in terms of redundant information and an ill-conditioned variance matrix due to collinearity in the design matrix. As an example, Wheeler (2007) applies the collinearity diagnostic tools to a Columbus (Ohio) crime rate GWR model to clearly link local collinearity to strong GWR coefficient correlation and increased coefficient variation for two economic status covariates at numerous data locations with counter-intuitive positive regression coefficient signs. In any analysis, estimated GWR coefficients from local models that are diagnosed as problematic should be interpreted with extreme caution and additional analysis should be undertaken in these areas to understand the nature of the relationships that are being modeled.

Another issue in GWR is with the standard errors associated with regression coefficient estimates. The standard error calculations in GWR are only approximate due to reusing data for parameter estimation at multiple locations (Congdon 2003; LeSage 2004) and due to using the data to estimate both the kernel bandwidth with cross-validation and the regression coefficients (Wheeler and Calder 2007). In addition, as previously implied, local collinearity can increase variances of estimated regression coefficients in the general regression setting (Neter et al. 1996). This issue with the standard errors indicates that the confidence intervals for estimated GWR coefficients are only approximate and are not exactly reliable for indicating statistically significant covariate effects and model selection.

An open debate about GWR is in the nature of the application of the technique itself. It has been suggested that GWR, given its theoretical origins in local linear regression (developed to estimate a response variable locally), is well suited for estimation and prediction of a response variable but is less useful in the formal statistical inference on spatially varying regression effects (Wheeler 2009). Perhaps a shift in the focus of the utility of GWR towards spatial interpolation would be worthwhile, and there is empirical evidence to support such a move, with GWR producing good comparative results in relation to other interpolation techniques (Páez et al. 2008). One supporting argument for this is that when interpolation of a response variable over space is the sole interest, estimation issues in GWR, such as collinearity, are no longer a major concern.

C.5.4 Diagnostic tools

There are several well-known diagnostic tools available for OLS regression models, including ones to check for autocorrelation, influential observations, and collinearity. In accord with this tradition, use of a more complicated linear regression model such as GWR should be accompanied with diagnostic tools.

Methods to identify spatial residual autocorrelation in GWR models have been developed by Leung et al. (2000b), based on well-established statistics of spatial autocorrelation including Moran's I and Geary's c. The approach proposed by Leung et al. (2000b) compares each local prediction of the dependent variable to its observed value. This provides a set of estimated residuals that can be used to detect map patterns. The application of these statistics is very similar to the application of autocorrelation statistics, and the theory required for hypothesis testing is derived by these authors. A limitation of these statistics is that they are not model based, recalling that the GWR method is a collection of local models that are not part of a unified framework. As a consequence, it is unclear that the source of autocorrelation can actually be identified. A different approach to test for spatial dependencies is proposed in a paper by Páez et al. (2002b) that provides a model-based alternative within a variance heterogeneity model, but it shares the limitation that there is no unified framework to tie in the local models together.

Additional diagnostic tools attend a recognition in the literature (Wheeler and Tiefelsdorf 2005; Waller et al. 2007; Wheeler and Calder 2007; Griffith 2008) of the existence of potentially strong correlation in sets of estimated GWR coefficients, which could come from local collinearity in the model, analysts should strongly consider using diagnostic tools for collinearity when estimating GWR coefficients. Fortunately, there are diagnostic tools one can use to evaluate whether substantial collinearity effects are present in a GWR model. In addition to scatter plots of regression coefficients for pairs of regression terms, maps of approximate local regression coefficient correlations (Wheeler and Tiefelsdorf 2005), local variance inflation factors (VIFs), one can use variance-decomposition proportions and the associated condition indexes (Belsley 1991; Wheeler 2007). An advantage of the variance-decomposition approach over the VIF, which measures how much the estimated variance of a regression coefficient is increased by collinearity, is that it measures and conveys the nature of the collinearity among all regression terms at the same time, including the intercept.

The variance-decomposition proportion and condition index diagnostic tools introduced by Belsley (1991) and modified for GWR by Wheeler (2007) use singular value decomposition of the GWR kernel weighted design matrix to form condition indexes and variance-decomposition proportions of the coefficient covariance matrix. The variance-decomposition proportion is the percentage of the variance of a regression coefficient that is explained with any one component of the variance matrix decomposition. It has an affiliated condition index, which is the ratio of the largest singular value and the smallest singular value of the decomposition. The singular value decomposition (SVD) of the design matrix in the GWR framework is

$$(W_i)^{1/2} X = U D V^T \qquad (C.5.18)$$

where U and V are orthogonal n-by-p and p-by-p matrices respectively; D is a (p-by-p) diagonal matrix of singular values of $(W_i)^{1/2} X$, starting at matrix element (1,1) and decreasing in value down the diagonal; and $(W_i)^{1/2}$ is the square root of the diagonal weight matrix for calibration location i using a kernel function with the GWR estimated kernel bandwidth. Through SVD, the local variance-covariance matrix of the regression coefficients is

$$\mathrm{var}(\hat{\boldsymbol{\beta}}_i) = \sigma^2\, \boldsymbol{V}\boldsymbol{D}^{-2}\boldsymbol{V}^{\mathrm{T}} \qquad (C.5.19)$$

and the variance of the local kth regression coefficient is

$$\mathrm{var}(\hat{\beta}_{ik}) = \sigma^2 \sum_{j=1}^{p} \frac{v_{kj}^2}{d_j^2} \qquad (C.5.20)$$

where the v_{kj} are the elements of the $\boldsymbol{V}$ matrix and the d_j are the singular values. The variance-decomposition proportion for the local kth regression term and the jth component of the decomposition is

$$\pi_{jk} = \frac{\phi_{kj}}{\phi_k} \qquad (C.5.21)$$

where

$$\phi_{kj} = \frac{v_{kj}^2}{d_j^2} \qquad (C.5.22)$$

$$\phi_k = \sum_{j=1}^{p} \phi_{kj}\,. \qquad (C.5.23)$$

The condition index for variance component $j = 1,\ldots,p$ is

$$\eta_j = \frac{d_{\max}}{d_j}. \qquad (C.5.24)$$

Belsley (1991) introduced some relevant guidelines for using the variance-decomposition proportions and condition indexes in the OLS regression setting. Through experimentation results, Belsley (1991) suggests a conservative value of thirty as a threshold for a condition index which indicates collinearity, although the threshold could be as low as ten if there are large variance-decomposition proportions for two or more regression terms for the same variance component. In general, stronger collinearity is indicated by larger condition indexes. Another guideline is that the presence of two or more variance-decomposition proportions greater than 0.5 for the same variance component indicates collinearity existing between those regression terms. One can apply the same guidelines for diagnosing collinearity in GWR. It should be emphasized that the variance-decomposition proportion and condition index diagnostic tools reveal collinearity locally at the GWR model calibration locations and therefore permit one to construct plots of the diagnostic values and link them explicitly to GWR estimated coefficients for visual analysis of any problems that may be present in the model.

C.5.5 Extensions

A number of different models have been proposed to extend the applicability of the concept of geographical weights in regression analysis. Three such extensions are discussed next.

Autoregressive GWR

One of the first extensions to the concept of GWR was to accommodate spatial dependencies in the structure of the model (Brunsdon et al. 1998). One computational challenge faced when working with models that contain spatially autoregressive components is the estimation of the coefficients using leave-one-out cross-validation, since this necessitates the calculation of a determinant of a $(n-1)$-by-$(n-1)$ matrix n times. Brunsdon et al. (1998) suggest conducting cross-validation on a randomly selected sub-sample of points; however, given the existence of influential points in cross-validation, this practice may not be appropriate (see Farber and Páez 2007). A different approach to obtain spatially autoregressive local models based on the concept of geographical weighting is from Páez et al. (2002b) who, by adopting a non-constant variance model, are able to parameterize the estimation of the model coefficients, including the kernel bandwidth. Alternative models have been suggested in the literature, including the spatially autoregressive local estimation (SALE) model of Pace and LeSage (2004) that is based on a decomposition of the estimation matrices, and the ZOOM model of Mur et al. (2008).

Constrained GWR

Issues arising from collinearity can be addressed by constraining the amount of variation in regression coefficients. In the case of GWR, two versions of methods that achieve this objective have been proposed, namely geographically weighted ridge regression (GWRR, Wheeler 2007) and the geographically weighted lasso

(GWL, Wheeler 2009). As the names imply, these techniques are based on ridge regression and the lasso respectively. The methods work by penalizing the regression in order to limit the amount of variation in the coefficients. In both cases, a constraint on the size of the regression coefficients is introduced, but the constraint is slightly different in each. While ridge regression coefficients minimize the sum of a penalty on the size of the squared coefficients and the residual sum of squares

$$\hat{\boldsymbol{\beta}}^R = \arg\min_{\beta} \left\{ \sum_{i=1}^{n} \left(y_i - \beta_0 - \sum_{k=1}^{p} x_{ik}\beta_k \right)^2 + \lambda \sum_{k=1}^{p} \beta_k^2 \right\} \quad \text{(C.5.25)}$$

the lasso coefficients minimize the sum of the absolute value of the coefficients and the residual sum of squares

$$\hat{\boldsymbol{\beta}}^L = \arg\min_{\beta} \left\{ \sum_{i=1}^{n} \left(y_i - \beta_0 - \sum_{k=1}^{p} x_{ik}\beta_k \right)^2 + \lambda \sum_{k=1}^{p} |\beta_k| \right\} \quad \text{(C.5.26)}$$

where λ is the parameter that controls the amount of shrinkage in the regression coefficients. This difference in specification of the two models results in potentially more shrinkage in the lasso regression coefficients, some of which may shrink to zero. In both ridge regression and the lasso, it is common practice to center the response variable, and center and scale the explanatory variables to have unit variances because the methods are scale-dependent. The formula to estimate the GWRR coefficients using centering of the variables is

$$\hat{\boldsymbol{\beta}}_i = (\boldsymbol{X}^{*\text{T}} \boldsymbol{W}_i \boldsymbol{X}^* + \lambda \boldsymbol{I})^{-1} \boldsymbol{X}^{*\text{T}} \boldsymbol{W}_i \boldsymbol{y}^* \quad \text{(C.5.27)}$$

where $\boldsymbol{X}^*$ is the matrix of standardized explanatory variables, $\boldsymbol{y}^*$ is the standardized response variable, and other terms are previously defined. There are options for the type of centering and scaling that can be performed (see Wheeler 2007). The absolute value constraint on the regression coefficients in GWL makes the problem non-linear, but fortunately there are efficient algorithms for estimating the parameters (Wheeler 2009). Cross-validation is employed in estimating the kernel bandwidth in both constrained versions of GWR.

Logistic and probit models with geographical weights

In addition to the linear regression framework, the idea of applying geographical weights has been applied to models for nominal variables, including the geographically weighted logistic model of Atkinson et al. (2003) and the probit model with geographical weights of Páez (2006). These models extend the scope of application of GWR to situations in geomorphology and transportation research that frequently require the analysis of limited dependent variables.

C.5.6 Bayesian hierarchical models as an alternative to GWR

With recent gains in computing power and software availability, it is possible to use Bayesian hierarchical models to estimate spatially varying regression coefficients as an alterative approach to GWR. The Bayesian hierarchical models are hierarchical in that the distribution of the data is specified conditional on unknown parameters, whose distribution is in turn specified conditional on other parameters. In addition, these models can incorporate parameters at different levels of data, for example both at the individual and group level, to model relationships at different scales. There are Bayesian hierarchical models with random effects for both the intercept and covariate effects, where random effects can be specified as independent in the prior and borrow strength across observations globally or be specified to have spatial correlation and borrow strength locally. There are two primary alternatives to GWR within this class of models. One is called the Bayesian spatially varying coefficient (SVC) model, which defines spatial correlation in the regression coefficients through a prior conditional specification of the coefficients that uses only neighboring observations. The other is called the spatially varying coefficient process (SVCP) model which uses a prior joint specification of the coefficients that models correlation in the coefficients as a continuous spatial process. We next outline the two models, beginning with SVC model.

In the Bayesian SVC model, one goal is to describe $E[Y_i \mid f(X(s_i))]$, the expected value of a response variable in location i given a function of the covariates associated with the location. The general Bayesian SVC model is

$$[Y_i \mid \mu_i, \tau] \sim \mathcal{N}(\mu_i, 1/\tau) \qquad (C.5.28)$$

where

$$\mu_i = X_i \boldsymbol{\beta}_i \qquad (C.5.29)$$

specifies the response variable mean at each data location through covariate vector X_i and a vector of spatially varying regression coefficients β_i. The model assumes spatial dependence in the regression coefficients through the prior distribution for the coefficients. The prior for the coefficients is an intrinsic multivariate conditional autoregressive (CAR), or MCAR prior, and is written as $\beta \sim \text{MCAR}(\Omega)$. The MCAR prior for the vector of spatially varying coefficients at each data location i has a multivariate conditional distribution

$$\beta_i | (\beta_{(-i)0}, \beta_{(-i)1}, \ldots, \beta_{(-i)p-1}) \sim \mathcal{N}_p(\overline{\beta}_i, \Omega/m_i) \qquad (C.5.30)$$

where $\overline{\boldsymbol{\beta}}_i = (\overline{\beta}_{i0}, \overline{\beta}_{i1}, \ldots, \overline{\beta}_{ip-1})^T$, $\overline{\beta}_{ik} = \Sigma_{j \in k_i} \beta_{jk}/m_i$, κ_i is the set of neighboring locations for location i, and m_i is the number of neighbors for location i. This prior is a natural one for area-level data where neighborhoods are formed from adjacent areas. The diagonal elements of the variance-covariance matrix Ω are the conditional variances of the β_k. A conjugate prior for this within-area, between-coefficient p-by-p variance-covariance matrix is an inverse Wishart distribution and a conjugate prior for the error precision τ is gamma. Conjugate priors are used for computational convenience. The MCAR prior has an advantage over the prior in the process model discussed next because it is less computationally demanding. It is also a natural extension of the CAR prior that is commonly used for spatial random effects in Bayesian regression models (for more details see Besag et al. 1991; Besag and Kooperberg 1995). For a more thorough introduction to the SVC model, see Banerjee et al. (2004).

As for implementation of this Bayesian SVC model, one can use Markov Chain Monte Carlo (MCMC) simulation in WinBUGS software (Spiegelhalter et al. 2003) to provide samples of model parameter values from the joint parameter posterior distribution for inference. The neighborhood adjacency list required for the MCAR prior can be generated in GeoBUGS software (Thomas et al. 2004). Typically, one uses a 'burn-in' period of samples and then a subsequent number of joint posterior distribution samples in MCMC to calculate posterior mean or median estimates for the model parameters. Statistical inference on the parameters comes from summaries of the posterior distribution, such as credible intervals using certain percentiles of the distribution.

An alternative to the MCAR prior for the regression coefficients is a geostatistical prior specification with a distance-based covariance function (Gelfand et al. 2003). The SVCP model is specified conveniently with matrix notation as

$$\left[Y | \boldsymbol{\beta}^P, \tau^2\right] \sim \mathcal{N}[(X^P)^T \boldsymbol{\beta}^P, \tau^2 I] \qquad (C.5.31)$$

where Y is assumed to be Gaussian conditional on the parameters β^P and τ^2. β^P is a np-by-1 vector of regression coefficient parameters; and $(X^P)^T$ is the n-by-np block diagonal matrix of covariates where each row contains a row from the (n, p) design matrix X, along with zeros in the appropriate places [the covariates from X are shifted p places in each subsequent row in $(X^P)^T$]. The superscript P is meant to indicate the different sizes of the regression coefficient matrix and the design matrix associated with the process model. I is the n-by-n identity matrix and τ^2 is the error variance.

The prior distribution for the regression coefficient parameters is specified as

$$\left[\beta^P \mid \mu_\beta, \Sigma_\beta\right] = \mathcal{N}(I_{n \times 1} \otimes \mu_\beta, \Sigma_\beta) \tag{C.5.32}$$

where the vector $\mu_\beta = (\mu_{\beta_0}, \ldots, \mu_{\beta_p})^T$ contains the means of the regression terms. The Kronecker product operator ($\otimes$), multiplies every element in $I_{n \times 1}$ by μ_β. The prior on the regression coefficients takes into account possible spatial dependence in the coefficients through the covariance, Σ_β, which has a separable form with two distinct components, one for the spatial dependence in the regression coefficients and one for the within site dependence between coefficients. The separable form of the covariance matrix for β^P is

$$\Sigma_\beta = R(\phi) \otimes T \tag{C.5.33}$$

where $R(\phi)$ is the n-by-n correlation matrix that captures the spatial association between the n locations using inter-point distances, ϕ is an unknown spatial dependence parameter, and T is a positive-definite p-by-p matrix for the covariance of the regression coefficients at any spatial location. In contrast to the repeated application of spatial kernel functions in GWR, this np-by-np covariance matrix captures the covariation between all regression coefficients simultaneously. In the separable covariance matrix, each of the p coefficients represented in the covariance is assumed to have the same spatial dependence structure. This aligns with the assumption in GWR of equal spatial ranges for each regression term.

The specification of the Bayesian SVCP model is complete with the specification of prior distributions for the other parameters. A conjugate prior for the coefficient means is Gaussian. A conjugate prior for the within-site covariance matrix is inverse Wishart and a conjugate prior for the error variance is inverse gamma. One can use a uniform or gamma prior for the spatial dependence parameter. Inference on the model parameters is achieved with MCMC by sampling from the joint posterior distribution of the parameters. See Wheeler and Calder (2007) for implementation details of the MCMC for the SVCP model.

C.5.7 Bladder cancer mortality example

To serve as an illustrative example of the recommended way to apply the GWR approach, an analysis of white male bladder cancer mortality rates in the 506 State Economic Areas (SEA) of the contiguous United States for the years 1970 to 1994 is presented. The dataset comes from the Atlas of Cancer Mortality from the National Cancer Institute (Devesa et al. 1999) and contains age standardized mortality rates (per 100,000 person-years). The standardized mortality rates are plotted in Fig. C.5.1 for SEAs. The explanatory variables of interest are population density and lung cancer mortality rate. Population density is used as a surrogate for behavioral and environmental differences with respect to an urban/rural dichotomy. It is expected, as several studies suggest, that with an increase in the population density, there is an increase in the rate of bladder cancer. Lung cancer mortality rates are used as a surrogate for smoking, which is a known risk factor for bladder cancer. There is evidence in public health that an increase in smoking increases bladder cancer risk, therefore, we expect a positive relationship between these variables. There is also evidence to justify the approximation of smoking by lung cancer, as the attributable risk of smoking for lung cancer is greater than 80 percent and the attributable risk of smoking for bladder cancer is greater than 55 percent (Mehnert et al. 1992).

As an initial step in the analysis, a traditional, or global regression model is estimated for bladder cancer mortality. The base model is

$$y(i) = \beta_1 + \beta_2 \, x_1(i) + \beta_3 \, x_2(i) + \varepsilon(i) \tag{C.5.34}$$

where $y(i)$ is the bladder cancer mortality rate for white males for years 1970 to 1994 for the ith SEA, x_1 is lung cancer mortality for time period 1954-1969, and x_2 is the natural log of population density. A smoking surrogate is used from an earlier time period to represent an induction period for bladder cancer given the risk factor. Population density is natural log transformed to linearize the relationship with bladder cancer mortality.

The coefficient of determination for the fitted global model is 0.25 and the root mean square error (RMSE) of the estimated response variable is 1.06. The OLS regression coefficient estimates are $\hat{\beta}_1 = 3.832$, $\hat{\beta}_2 = 0.029$, $\hat{\beta}_3 = 0.277$ and the p-values for all these coefficients are less than 0.001. Both the smoking surrogate risk factor and log population density are significantly positively related to the rate of bladder cancer mortality, as expected. The variance inflation factors for the two explanatory variable coefficients are less than 1.6 and the correlation of the global regression parameters is moderately negative at -0.60, whereas the correlation of the two variables is 0.60. The results from the initial analysis suggest that collinearity is not a significant problem with these data in this type of model.

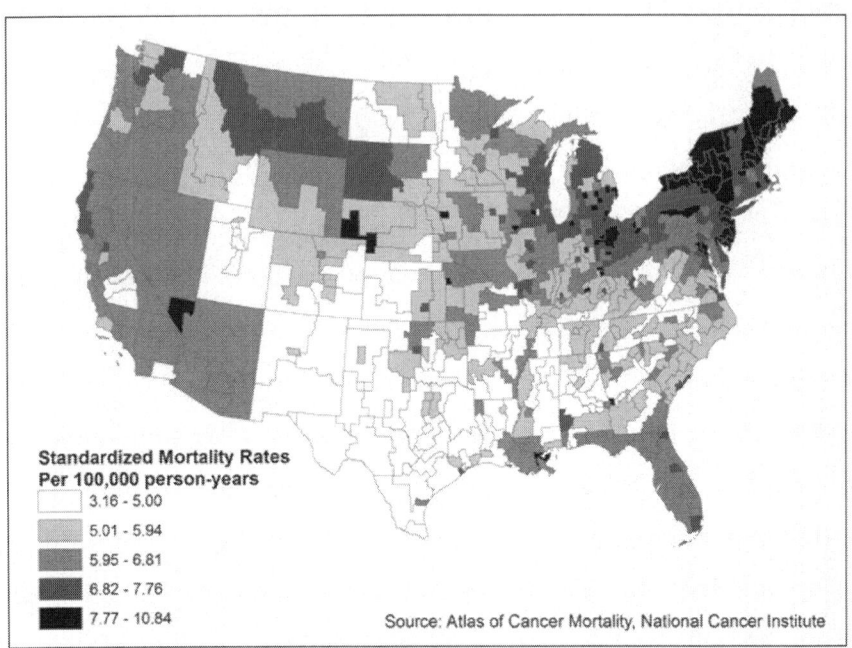

Fig. C.5.1. Standardized mortality rates for bladder cancer among white males from 1970 to 1990 in the State Economic Areas of the contiguous United States

Next, a GWR model is fitted using the bladder cancer mortality data in R software with custom code. Note that there is a free R package for estimating GWR model parameters, spgwr, written by Roger Bivand (see Chapter A.3). We fit the following GWR model,

$$Y(i) = \beta_1(i) + \beta_2(i)\, x_1(i) + \beta_3(i)\, x_2(i) + \varepsilon(i) \tag{C.5.35}$$

where the regression coefficients now vary by SEA. Through cross-validation, the GWR kernel bandwidth is estimated to be $\hat{\gamma} = 1.27$. The RMSE of the estimate of the response variable for the GWR model with this estimated bandwidth and associated estimated regression coefficients is 0.52, which is a marked reduction from the OLS model. Allowing the regression coefficients to vary by SEA provides an improved fit of bladder cancer mortality. Typically, incorporating observation-specific intercepts in a model will improve model fit considerably over a model with a fixed intercept.

The estimated GWR coefficients are graphed in Fig. C.5.2 for the three regression terms. The estimated coefficients exhibit noticeable variation, with counterintuitive negative coefficients in some SEAs for both smoking proxy and population density, although there are more negative coefficients for population density. Wheeler and Tiefelsdorf (2005) recommended using scatter plots of estimated GWR coefficients for pairs of regression terms to visualize the nature of dependence in estimated GWR coefficients. Figure C.5.3 has scatter plots for the three pairs of regression terms in the model. There is discernable correlation in the coefficients in some areas, particularly for the intercept and smoking proxy coefficients and the smoking proxy and population density coefficients. The Pearson correlation coefficients for the GWR coefficients for pairs of regression terms are $r_{12} = -0.36$, $r_{13} = -0.28$, $r_{23} = -0.74$, where the subscripts indicate the regression term. The level of correlation in the coefficients for the smoking proxy and population density terms is notably stronger in the GWR model than in the OLS model. The overall level of correlation in these GWR coefficients could indicate the presence of local collinearity in the GWR model that could lead to problems with inference on the GWR coefficients.

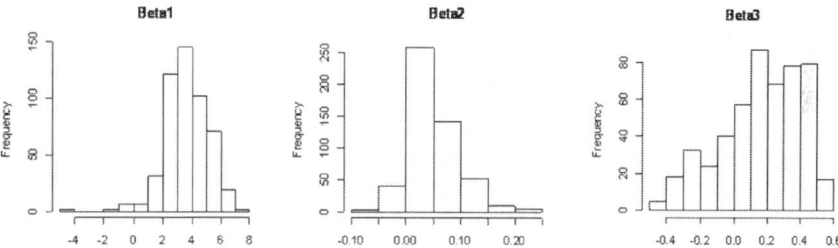

Fig. C.5.2. Estimated GWR coefficients for $\hat{\beta}_1$ (intercept), $\hat{\beta}_2$ (smoking proxy), $\hat{\beta}_3$ (log population density)

To further explore dependence in the regression coefficients, the variance-decomposition proportion and condition index diagnostic tools (Wheeler 2007) described earlier are applied. The GWR estimated bandwidth is used in the variance-decomposition of the kernel weighted design matrix to assess collinearity in the GWR model. Of the 506 SEAs in the dataset, thirteen have a condition index greater than thirty, eighty-five have a condition index greater than twenty, and 500 have a condition index greater than ten for the largest variance component. There are 436 records in the data with large variance proportions (greater than 0.5) for the largest variance component, with the shared component being between the two covariates for some records and between a covariate and the intercept for other records. Of these records, 431 also have a condition index greater than ten for the largest variance component. Overall, the variance-decomposition proportions and condition index values indicate the presence of some substantial local collinearity in the GWR model.

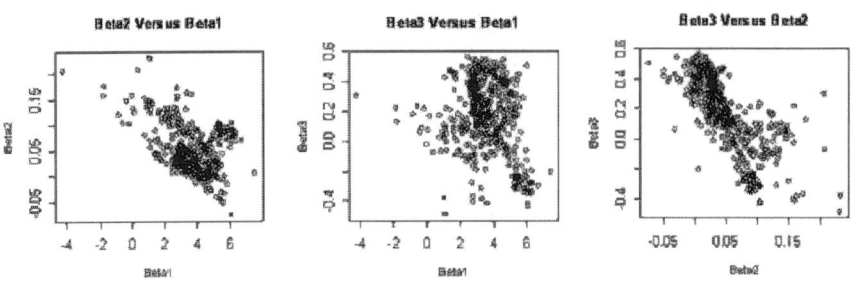

Fig. C.5.3. GWR estimated coefficients $\hat{\beta}_2$ versus $\hat{\beta}_1$ (left), $\hat{\beta}_3$ versus $\hat{\beta}_1$ (middle), $\hat{\beta}_3$ versus $\hat{\beta}_2$ (right)

In addition to looking at summaries of the diagnostic tools, it is useful, particularly for inferential purposes, to visualize the diagnostic values with graphical links to the mapped GWR coefficients to inspect where the especially troublesome coefficients are located (Wheeler 2008). Figure C.5.4 contains maps of the GWR coefficients for the intercept and smoking proxy, a parallel coordinate plot of the condition indexes and variance proportions for the largest variance component, and a histogram of the condition indexes. The lines in the parallel coordinate plot that are highlighted are a selection set of the thirty SEAs with the largest condition indexes. The same selected SEAs are highlighted with a yellow crosshatching on the coefficient maps. Most of the selected SEAs are peripheral ones in the West. It is clear in the parallel coordinate plot that most of the selected SEAs have large proportions for both the intercept and the smoking proxy on the largest variance component.

Table C.5.1. Condition index and variance-decomposition proportions for the largest variance component

η_3	π_{31}	π_{32}	π_{33}
39.5	0.97	0.99	0.27
37.0	0.97	0.99	0.18
36.2	0.97	0.98	0.07
35.8	0.96	0.98	0.05
33.6	0.95	0.99	0.28
33.5	0.97	0.99	0.31
33.0	0.93	0.98	0.18
32.7	0.96	0.98	0.07
32.6	0.99	0.99	0.12
31.1	0.93	0.98	0.18
31.0	0.59	1.00	0.37
30.5	0.73	0.98	0.21
30.3	0.93	0.98	0.20

Notes: η_3 condition index, π_{31}, π_{32} and π_{33} variance-decomposition proportions for intercept, smoking proxy, and population density respectively

The variance-decomposition proportions and condition indexes are listed in Table C.5.1 for records with condition indexes greater than thirty for the largest variance component. As evidenced in the table, most of the records have variance proportions of almost one for the intercept and the smoking proxy, meaning that the variance of these two regression terms is explained by one component in these locations. There is effectively one piece of information from two different sources. Therefore, attempting to separate the two regression coefficients for inference is not possible. This also brings up the important point that while strongly correlated regression coefficients for the intercept and another regression term may not be a general concern in an OLS regression model, strong correlation in local coefficients for these terms is a problem in GWR. The issue with high dependence between terms is that attempting to interpret the spatial pattern in the coefficients for individual terms, i.e. marginal inference, will lead to biased conclusions.

In addition to the shared variance of the intercept and the smoking proxy effect in some SEAs, there are other SEAs with large variance decomposition proportions for both the smoking proxy and population density for the largest variance component. Figure C.5.5 demonstrates this with linked maps of the GWR coefficients for smoking proxy and log population density, a scatter plot of variance decomposition proportions, and a histogram of condition indexes. The selection in all graphics is for SEAs with variance proportions greater than 0.6 for both regression terms. Most of the condition indexes for these SEAs exceed ten. The majority of the selected SEAs with a shared variance component for smoking proxy and population density are located in the Midwest and Northeast. These are areas where the GWR coefficients should be interpreted with caution.

As an alternative to GWR, a Bayesian SVCP model is also fitted. The estimated regression coefficients are plotted in Fig. C.5.6 side-by-side for GWR and the SVCP model as a means of comparison. Certain contrasts are evident, such as more variation in the GWR coefficients. There are both lower and higher coefficients for the GWR model compared to the SVCP model for each regression term, although this is most apparent with the intercept and population density. In addition to being more variable, the GWR coefficients are more spatially smooth for the smoking proxy term than are the SVCP coefficients. The SVCP model better constrains the coefficients in areas where the diagnostic values indicated problems with the GWR local models, such as in California for $\hat{\beta}_1$ and $\hat{\beta}_2$ and in areas of the Midwest for $\hat{\beta}_2$ and $\hat{\beta}_3$. As recommended by Páez and Wheeler (2009), use of complementary approaches provides mutually supporting evidence of non-stationarity in the case of bladder cancer presented in this section.

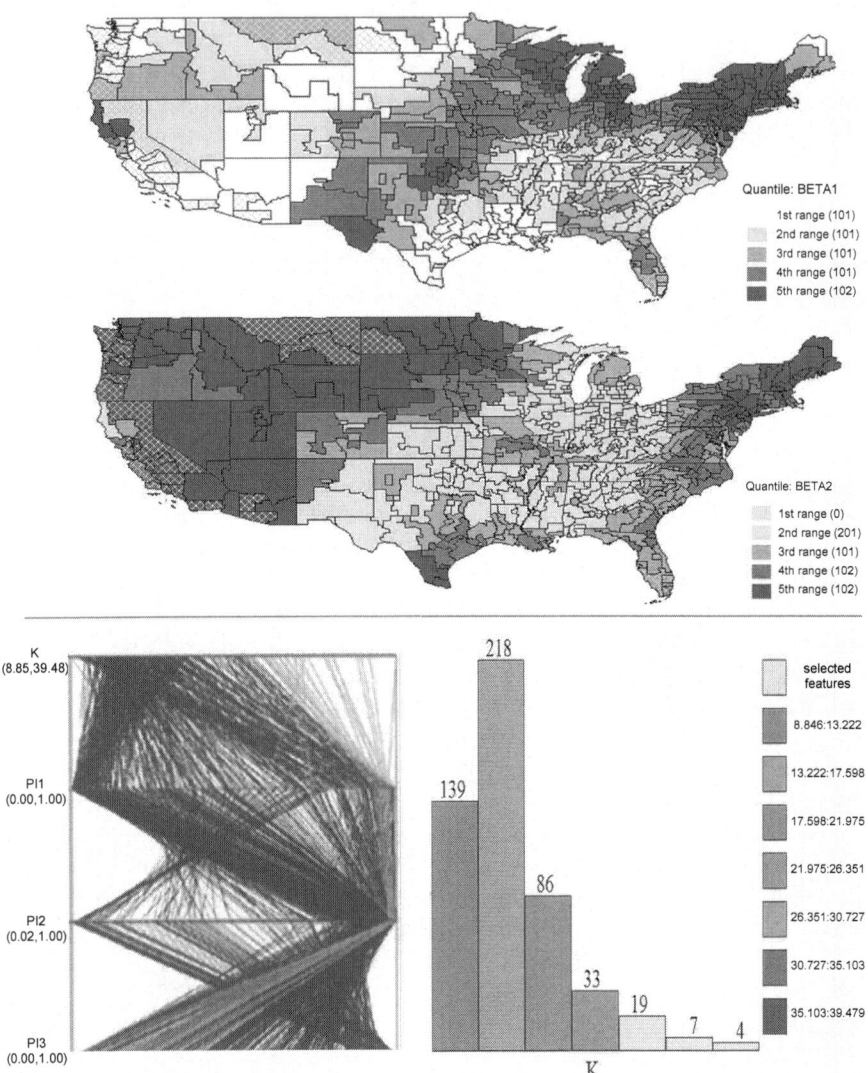

Fig. C.5.4. Estimated GWR coefficients for the intercept (top) and the smoking proxy (middle), parallel coordinate plot for condition indexes and variance decomposition proportions (bottom left), and histogram of condition indexes (bottom right) with a selection set for SEAs with the thirty largest condition indexes for the largest variance component

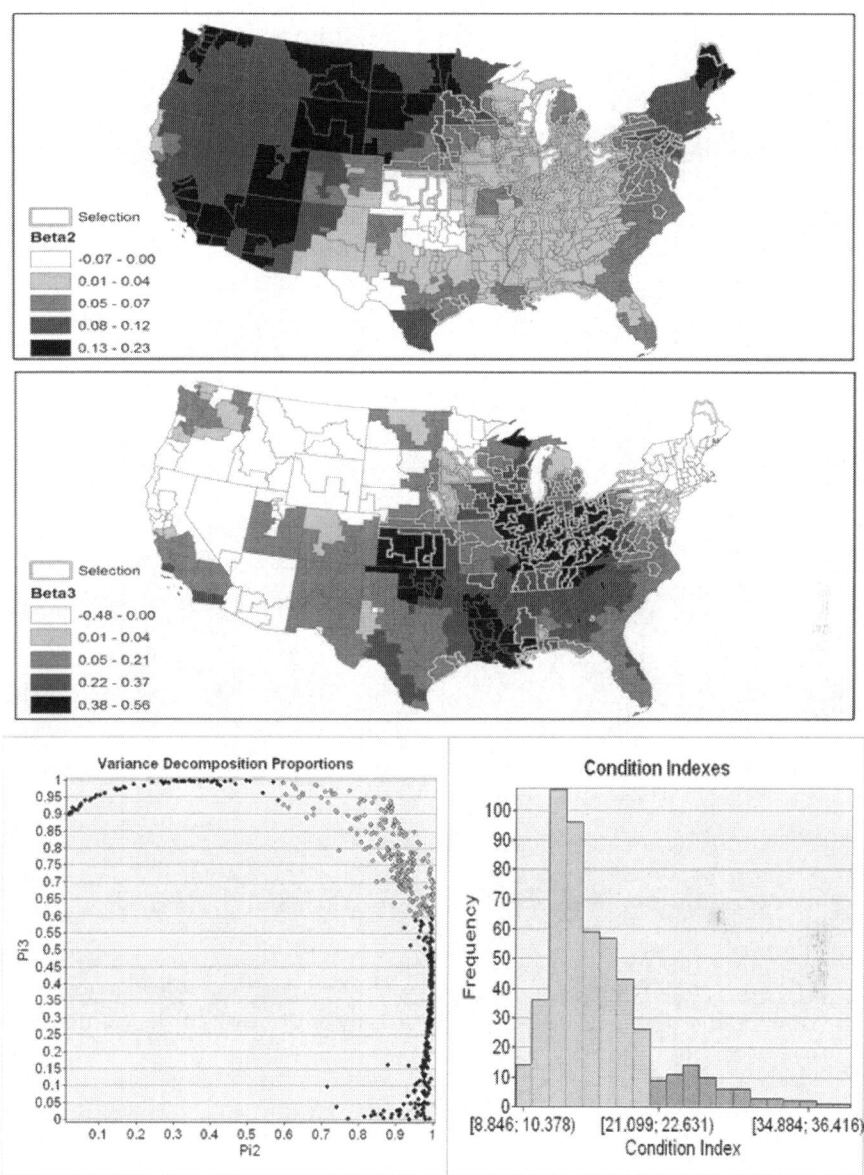

Fig. C.5.5. Estimated GWR coefficients for smoking proxy (top) and population density (middle), scatter plot for variance decomposition proportions for these two regression terms (bottom left), and histogram of condition indexes (bottom right) with a selection set for SEAs with both variance decomposition proportions greater than 0.6 for the largest variance component

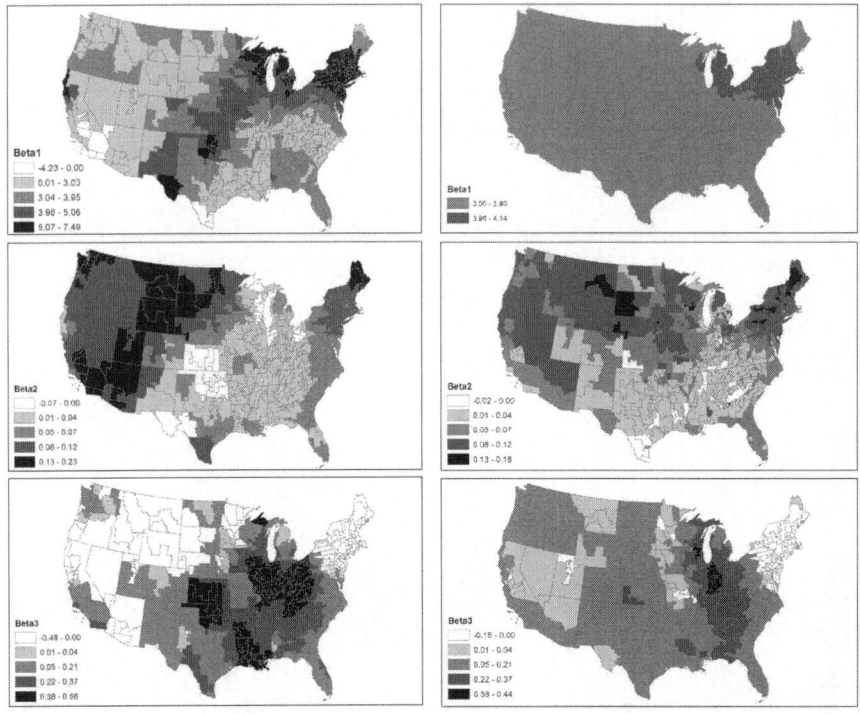

Fig. C.5.6. GWR coefficients (left) and SVCP coefficients (right) for the intercept (top), smoking proxy (middle), and population density (bottom)

References

Atkinson PM, German SE, Sear DA, Clark MJ (2003) Exploring the relations between riverbank erosion and geomorphological controls using geographically weighted logistic regression. Geogr Anal 35(1):58-82

Banerjee S, Carlin BP, Gelfand AE (2004) Hierarchical modeling and analysis for spatial data. CRC Press (Taylor and Francis Group), Boca Raton [FL], London and New York

Belsley DA (1991) Conditioning diagnostics: collinearity and weak data in regression. Wiley, New York, Chichester, Toronto and Brisbane

Besag J, Kooperberg C (1995) On conditional and intrinsic autoregressions. Biometrika 82(4):733-746

Besag J, York J, Mollie A (1991) Bayesian image restoration, with two applications in spatial statistics (with discussion). Ann Inst Stat Math 43(1):1-59

Bivand RS (2009) Spatial econometric functions in R. In Fischer MM, Getis A (eds) Handbook of applied spatial analysis. Springer, Berlin, Heidelberg and New York, pp.53-71

Brunsdon C, Fotheringham AS, Charlton M (1996) Geographically weighted regression: a method for exploring spatial non-stationarity. Geogr Anal 28(4):281-298

Brunsdon C, Fotheringham AS, Charlton M (1998) Spatial non-stationarity and autoregressive models. Environ Plann A 30(6):957-973

Congdon P (2003) Modelling spatially varying impacts of socioeconomic predictors on mortality outcomes. J Geogr Syst 5(2):161-184

Devesa SS, Grauman DJ, Blot WJ, Pennello G, Hoover RN, Fraumeni JF Jr. (1999) Atlas of cancer mortality in the United States, 1950-94. National Cancer Institute, Bethesda. http://www3.cancer.gov/atlasplus/

Farber S, Páez A (2007) A systematic investigation of cross-validation in GWR model estimation: empirical analysis and Monte Carlo simulations. J Geogr Syst 9(4):371-396

Fischer MM, Stumpner (2009) Income distribution dynamics and cross-region convergence in Europe. In Fischer MM, Getis A (eds) Handbook of applied spatial analysis. Springer, Berlin, Heidelberg and New York, pp.599-628

Fotheringham AS, Brunsdon C, Charlton M (2002) Geographically weighted regression: the analysis of spatially varying relationships. Wiley, New York, Chichester, Toronto and Brisbane

Gelfand AE, Kim H, Sirmans CF, Banerjee S (2003) Spatial modeling with spatially varying coefficient processes. J Am Stat Assoc 98(462):387–396

Griffith D (2008) Spatial filtering-based contributions to a critique of geographically weighted regression (GWR). Environm Plann A 40(11):2751-2769

LeSage JP (2004) A family of geographically weighted regression models. In Anselin L, Florax RJGM, Rey SJ (eds) Advances in spatial econometrics. Methodology, tools and applications. Springer, Berlin, Heidelberg and New York, pp.241-264

Leung Y, Mei CL, Zhang WX (2000a) Statistical tests for spatial non-stationarity based on the geographically weighted regression model. Environ Plann A 32(1):9-32

Leung Y, Mei CL, Zhang WX (2000b) Testing for spatial autocorrelation among the residuals of the geographically weighted regression. Environ Planning A 32(5):871-890

Mehnert WH, Smans M, Muir CS, Möhner M, Schön D (1992) Atlas of cancer incidence in the former German Democratic Republic 1978-1982. Oxford University Press, New York

Mur J, López F, Angulo A (2008) Symptoms of instability in models of spatial dependence. Geogr Anal 40(2):189-211

Neter J, Kutner MH, Nachtsheim CJ, Wasserman W (1996) Applied linear regression models. Irwin, Chicago

Pace RK, LeSage JP (2004) Spatial autoregressive local estimation. In Getis A, Mur J, Zoller H (eds) Recent advances in spatial econometrics. Palgrave Macmillan, New York, pp.31-51

Páez A (2006) Exploring contextual variations in land use and transport analysis using a probit model with geographical weights. J Transp Geog 14(3):167-176

Páez A, Wheeler DC (2009) Geographically weighted regression. In International Encyclopedia of Human Geography. Elsevier, Oxford [UK] (in press)

Páez A, Long F, Farber S (2008) Moving window approaches for hedonic price estimation: an empirical comparison of modelling techniques. Urban Stud 45(8):1565-1581

Páez A, Uchida T, Miyamoto K (2002a) A general framework for estimation and inference of geographically weighted regression models: 1. location-specific kernel bandwidths and a test for locational heterogeneity. Environ Plann A 34(4):733-754

Páez A, Uchida T, Miyamoto K (2002b) A general framework for estimation and inference of geographically weighted regression models: 2. spatial association and model specification tests. Environ Planning A34(5):883-894

Spiegelhalter D, Thomas A, Best N, Lunn D (2003) WinBUGS users manual, version 1.4

Thomas A, Best N, Lunn D, Arnold R, Spiegelhalter D (2004), GeoBUGS user manual, version 1.2

Waller L, Zhu L, Gotway C, Gorman D, Gruenewald P (2007) Quantifying geographic variations in associations between alcohol distribution and violence: a comparison of geographically weighted regression and spatially varying coefficient models. Stoch Envir Res Risk Assess 21(5):573-588

Wheeler DC (2007) Diagnostic tools and a remedial method for collinearity in geographically weighted regression. Environ Plann A 39(10):2464-2481

Wheeler DC (2008) Visualizing and diagnosing output from geographically weighted regression models. Technical Report 08-02. Department of Biostatistics, Emory University

Wheeler, DC (2009) Simultaneous coefficient penalization and model selection in geographically weighted regression: the geographically weighted lasso. Environ Plann A 41(3):722-742

Wheeler DC, Calder C (2007) An assessment of coefficient accuracy in linear regression models with spatially varying coefficients. J Geogr Syst 9(2):145-166

Wheeler DC, Tiefelsdorf M (2005) Multicollinearity and correlation among local regression coefficients in geographically weighted regression. J Geogr Syst 7(2):161–187

C.6 Expansion Method, Dependency, and Multimodeling

Emilio Casetti

C.6.1 Introduction

The body of this chapter consists of three sections: expansion method, dependency, and multimodeling. In the first section the expansion method is defined, discussed, illustrated by an example, and pertinent literature items are briefly showcased. In the second section it is shown that when an estimated model's residuals show significant spatial dependency the expansion method can provide a course of action to remedy this dependence. In the third section an expansion based multi-modeling approach to remedying spatial dependence is presented and demonstrated. The themes addressed in these sections are briefly outlined hereafter.

The expansion method encompasses a technique for constructing mathematical models and the rationale concerning its use. As a technique, it involves widening the scope of a simpler initial model by expansion equations that redefine some or all of the initial model's parameters into functions of contextual variables. By replacing the parameters of the initial model with their expansions a terminal model is produced that encompasses both the initial model and a specification of its contextual variation.

The rationale for the expansion method is best clarified by considering that scientific knowledge tends to progress by moving from simple models of realities to more complex ones. The expansion method brings into focus an orderly routine by which the transition from simpler to more complex mathematical models can be carried out. The relevance of the expansion method to the spatial disciplines becomes apparent considering that models born non-spatial can be expanded to encompass spatial contexts, and that models born spatial can be expanded to encompass non-spatial contexts.

The 'dependency' section (see Section 6.3) shows by an example that an 'initial model' that upon estimation and testing displays significant residual spatial autocorrelation can be expanded into 'terminal models' that upon estimation and testing display no significant autocorrelation. Thus, the expansion method may remedy spatial dependency. The reason is quite simple. The econometrics literature has shown that 'omitted variables' can produce autocorrelated residuals. The expansion method generates terminal models that in most cases will contain more independent variables than the initial models from which they originated. Therefore, in these cases, some of the terminal model's variables are 'omitted' from its initial model. If upon estimation and testing, significant autocorrelation is found in the initial model's residuals but not in the terminal model's residuals, it is likely that the variables generated by expansions are what made the difference.

The expansion based multimodeling approach discussed in the 'multimodeling' section (see Section C.6.4) also discusses an omitted variables type of remedy to spatial dependency. The approach involves first assembling a pool of initial models and a pool of expansion variables and then combining the two to create a multimodel pool; specifically, this multimodel pool will contain the initial models from the first pool, and the terminal models generated by expanding each initial model from the first pool by expansion variables from the second pool. Then all the models in the multimodel pool are estimated by OLS and tested so that the models with/without significant spatial dependency are identified. These results can be used to evaluate how prevalent spatial dependency is in a given research, to investigate which properties and attributes of the models are associated with the occurrence of spatial dependency, and finally, also to identify the best among the models with no significant spatial dependency.

C.6.2 Expansion method

Generalities. The expansion method section is divided into three subsections. In this subsection the expansion method and the spatial expansion method are defined and discussed in generalities. In the second, termed Applications, selected literature items are briefly commented. In the third subsection the spatial expansion method is demonstrated via the step-by-step construction of two terminal models.

The expansion method (Casetti 1972, 1986, 1997b) is a technique for constructing mathematical models, bundled with rationales for its use. In a nutshell, it involves widening the scope of a simpler 'initial' model by redefining some or all of its parameters into functions of 'expansion' variables or variates that index a substantively relevant 'context'. The 'terminal model' thus produced will encompass both the initial model and its contextual variation. Let us clarify the difference between the conventional approach to constructing models and the expansion method's.

A mathematical model of any realities links a substantive conceptual frame of reference to an analytical mathematical structure. Mathematics defines analytical structures such as equations, probability distributions, and stochastic processes, in which variables, random variables, and parameters appear. Substantive disciplines define disciplinary-specific conceptual frames of reference. A mathematical model of any realities consists of analytical structures with some or all of their variables, variates, and parameters linked to disciplinary-specific frames of reference.

There are two major approaches to the construction of mathematical models: the conventional modeling and the expansion modeling. The conventional modeling consists in the straightforward linking of an analytical structure to a substantive frame of reference. As example of conventional modeling suppose that a substantive discipline has defined variables and important relations among these variables. A scholar from this discipline with a background in a given inventory of analytical structures selects one of these structures and links it to one such relation. When this link is established, a mathematical model of a segment of reality is born. The expansion modeling consists in the conventional modeling of the parameters of a pre-existing initial model. The mathematical models can be deterministic, stochastic, or mixed, can reflect theory or empirical regularities, and can be intended for solving, maximizing, or estimating. All these types of models can be arrived at by conventional modeling or by expansions.

The relevance of the expansion method to the spatial disciplines becomes readily apparent considering that models born non-spatial can be expanded to encompass spatial contexts, and that models born spatial can be expanded to encompass non-spatial contexts. To clarify this point spatial models and spatial expansions have to be briefly defined and discussed.

A model is spatial if the variables in it are spatial. Examples of spatial variables are spatial 'dummies', distances, directions, spatial coordinates, location quotients, measures of spatial inequality, and spatially referenced measures of phenomena that vary over space or index geographic environments. Spatial models are found not only in geography and regional science, but also in the spatial peripheries of virtually all social science disciplines. A spatial expansion is the redefinition of a parameter of a spatial or non-spatial model in terms of spatial variables.

The expansion method has an especially useful role in spatial modeling. It lends itself to operationalize the integration of complex geographical contexts and non-spatial models. If the initial model is suggested by non-spatial 'theory' and the expansion variables capture pertinent geographical dimensions, the terminal model integrates non-spatial theory and spatial realities. However, the expansion method has been also employed to model the variation of spatial relationships across non-spatial contexts, and the spatial variation of spatial relationships.

The contexts of a model spatial or otherwise are any meaningful dimensions of reality: they can be the dimensions specifying substantively relevant aspects of geographical environments, dimensions of reality that are competing explanators of the same dependent variables, and just about any dimensions external to the model.

Let us sketch a rough typology of spatial expansions constructed from classes of initial models and classes of expansion variables. The initial models can originate from spatial theory, as in the case of spatial interactions models; from mathematical land use theory, as in the case of population density models; from central place theory such as the models relating the distances among centers to the centers' sizes; and so on. Alternatively, the initial models can originate within any non-spatial social science discipline.

The parameters of any of these models can be redefined into functions of non-spatial and spatial variables. Let us confine ourselves to the latter. The spatial expansion variables have included distances from substantively significant geographical reference points, the variables that appear in 'spatial polynomials' such as trend surfaces or two-dimensional Fourier polynomials, indexes of regional characteristics or factors extracted from these by factor analyses, spatial systemic measures such as indexes of accessibility, of metropolitan dominance, or of inequality.

Applications

The bibliographic items referred to in this subsection were selected with the objective of giving the reader a feel for the diversity of the expansion method literature, that encompasses a wide range of research styles, issues, substantive fields, and mathematical techniques. The section includes presentations of the expansion method and of its extensions, discussions of the issues that arise in its implementation, and also touches upon three types of applications that are especially relevant to the themes developed here.

The presentation and discussion of the expansion method in this chapter is a condensed version of some of the themes dealt with at greater length in Casetti (1997b). The article by Casetti (1972) contains the first published definition of the expansion method as a technique for creating or modifying models combined with rationales for its application.

Multicollinearity can be a problem in the estimation of terminal models generated from initial models with many variables, and/or from expansion equations with many variables, and/or from iterated expansions (Kristensen 1997). The use of Bayesian regression and of mixed estimation à la Theil-Goldberger to remedy degrading multicollinearity within an expansion modeling frame of reference are demonstrated respectively in Casetti (1992) and in Casetti (1997c). The application of Bayesian techniques in the estimation of spatial models including spatial expansion models is extensively reviewed in Congdon (2003, pp.273-322). The testing for parametric drift within an expansion modeling frame of reference is the focal point of Casetti (1991).

The dual expansions method (Casetti 1986) is an extension of the expansion method. It builds upon the proposition that an expansion formulation defined as primal implies a dual expansion formulation, much in the same way as dual mathematical programming formulations are implied by primal ones. Specifically, in Casetti (1986) it is shown that a primal linear initial model and linear expansion equations imply a dual linear initial model and linear expansion equations, and that these primal and dual formulations yield the same terminal model. Zhang and Kristensen (1995) used the dual expansions method to combine different gravity-based trade models into a supermodel suitable for estimation and testing.

Jones and Bullen (1994) discuss the relations between the expansion method and multilevel modeling. Jones and Wrigley (1995) extend the scope of the expansion method to the generalized additive models.

The expansion method is well suited to modeling complex spatial non-linear dynamics because it allows the linking of geographical locations to dynamics that vary across space and time. The reasoning and issues involved are discussed in Casetti (1997a). Two examples of applications can be found in Casetti (1989), and Casetti and Pandit (1987). The first of these centers on the spread of economic growth in Europe, and yields estimates of the time when each of 16 European countries transitioned from pre-modern quasi-stagnation to modern economic growth. The second application centers on the favorable and perverse dynamic equilibria that materialized in geographically distinct clusters of countries, and are associated respectively with strong manufacturing growth and with strong growth of the service sector.

Modeling the human-environment interactions via the expansion method is discussed by Gatrell (2005). Reviews of spatial modeling that address the spatial expansion modeling can be found in LeSage (1999), Fotheringham and Brunsdon (1999), and Anselin (1988). A number of spatial expansion articles appeared in a book edited by Jones and Casetti (1992).

The applications clusters that follow are groupings of expansions of the same type. They will be referred to as 'distance cluster', 'trend surface cluster', 'spatial context cluster'. The 'distance cluster' encompasses expansion method contributions in which functions of geographical distance(s) appear either in the initial model or in the expansion equation(s). The modeling of spatial diffusion, of change in urban population densities, of polarized growth and polarization reversal, of temporal change in spatial interactions, to name a few, can be modeled by expanding into distance polynomial(s) the parameters of a time function appearing in an initial model. For example, in Casetti and Semple (1969), and Casetti (1972) a spatial diffusion model was arrived at, from an initial model relating the percent of adopters of an innovation to a logistic function of time, by redefining the logistic parameters as polynomial functions of distance from a diffusion pole. The other artcles selected as illustrations of the 'distance cluster' are by Casetti (1973), Zdorkowski and Hanham (1983), Kellerman and Krakover (1986), Kristensen (1997) and Eldridge and Jones (1991).

Initial models or expansion equations in which trend surfaces appear characterize the 'trend surface cluster'. To exemplify, let us consider the intriguing paper by Eldridge and Jones (1991) that focuses upon the possible spatial instability of the friction of distance parameter in gravity models. They expand this parameter into a quadratic trend surface. The estimated terminal model obtained is then used to produce distorted maps that depict the 'spatial warping' implied by the variation of the estimated friction of distance parameter. Other examples of trend surface expansions are Casetti and Fan (1991); Casetti and Krakover (1990); Tanaka and Casetti (1992), and Aten and Heston (2003).

Initial models or expansion equations including variables that index geographical contextual variation characterize the 'spatial context cluster'. In Can (1990, 1992) initial models relating housing prices to housing attributes is expanded using a 'contextual' neighborhood quality variable. In these articles several variants of the models, some with spatially autoregressive terms, are specified and estimated, and tests for spatial dependence are carried out. Can's work has spawned a growing literature. In Can et al. (1989) and Brown and Goetz (1987) the households' propensity to move is related to households' attributes, and the resulting initial model is expanded using indices descriptive of geographical contexts. Other examples of work fitting the spatial context cluster are by Casetti and Tanaka (1992); Kodras (1986); Jones (1987); Kristensen (1997); Varga (1998, 2000); Theriault et al. (2003); Burford and Zee (2006); Jensen et al. (2004), Jensen et al. (2005).

Example

In this subsection the spatial expansion method is demonstrated by an example. Consider an analyst who wishes to model the spatial distribution of crime in a monocentric city partitioned into neighborhoods. For each neighborhood are available: a measure of crime, CR; average household Income, I; and the Distance, D, between the neighborhood centroid and the central business district. The analyst wishes to specify model(s) of the spatial distribution of crime using these variables.

Crime levels in a neighborhood are likely to be lower in more affluent neighborhoods. A simple initial model that for appropriate parameter values can portray a relation of this type between crime level CR and average household income I is

$$CR = \alpha_0 + \alpha_1 I + \varepsilon \tag{C.6.1}$$

where ε is a well behaved error term. Equation (C.6.1) implies that the spatial distribution of crime is accounted for by the spatial distribution of income. However, since in monocentric cities poverty tends to be greater in the city center than in the suburbs we can hypothesize that Eq. (C.6.1) holds with different parameter values at different distances from the CBD, namely, for different values of the contextual variable D. The contextual variation of the parameters in model in Eq. (C.6.1) in terms of D can be specified by the following linear expansion equations

$$\alpha_0 = \alpha_{00} + \alpha_{01} D \tag{C.6.2}$$

$$\alpha_1 = \alpha_{10} + \alpha_{11} D. \quad (C.6.3)$$

By substituting the right hand sides of Eqs. (C.6.2) and (C.6.3) for the parameters α_0 and α_1 in Eq. (C.6.1) the following terminal model in Eq. (C.6.4) is obtained

$$CR = \alpha_{00} + \alpha_{01} D + \alpha_{10} I + \alpha_{11} I\, D + \varepsilon. \quad (C.6.4)$$

Suppose that the analyst expands only the intercept of Eq. (C.6.1), which is equivalent to set the parameter α_{11} to zero in Eq. (C.6.3), thus creating the terminal model

$$CR = \alpha_{00} + \alpha_{01} D + \alpha_{10} I + \varepsilon. \quad (C.6.5)$$

Let us say that the terminal model in Eq. (C.6.4) was generated by the 'full expansion' of model in Eq.(C.6.1) with respect to the expansion variable D, and that the terminal model in Eq. (C.6.5) was generated by the 'intercept expansion' with respect to D.

The estimation and testing of models in Eqs. (C.6.1), (C.6.4) and (C.6.5), and of related models is discussed in the next sections in connection with the use of the expansion method as one of possible remedies to significant spatial autocorrelation.

C.6.3 Dependency

Generalities. Dependency is a focal point of spatial econometrics and of time series econometrics. The similarities and differences between the two have been discussed by Anselin (1988) and Stern (2000). In the past 20 plus years a rich literature has addressed the effects of the testing for, and the remedies to spatial autocorrelation (Anselin 2006, 2007; Anselin and Bera 1998; Getis 2007).

The remedial options available when a spatial linear model shows significant residual autocorrelation have always included the respecifications of the substantive aspects of the model by adding and/or removing and/or transforming its variables (Miron 1984). However more recently greater attention has been given to the following four approaches: (i) the estimation of spatial lag, spatial autoregressive error, or spatial Durbin models; (ii) the FGLS estimation of models that incorporate spatial structure via a previous parametric estimation of the error's variance-covariance matrix; (iii) the correction of the degrees of freedom to compensate for the partial redundancy of the observations due to the spatial autocorrelation; and (iv) spatial filtering (Getis 1990; Getis and Griffith 2002; Getis and Ord 1995; Griffith 2003; Fischer and Griffith 2008). Reviews of these four approaches to remedy spatial dependency are in Rangel and Diniz-Filho (2006), de Smith (2007), and Dorman et al. (2007).

This section is concerned with the use of the expansion method to remedy residual spatial autocorrelation. The line of reasoning involved is as follows. Omitted variables represent an important determinant of residual spatial autocorrelation A large literature has shown that omitted variables produce temporal and spatial autocorrelation: see for instance Maddala (1992, pp.255-257), Ramanathan (2002 Chapter 9), Mukherjee et al. (1998, pp.23ff and pp.208ff), Cliff and Ord (1981, p.197), McMillen (2003). Most applications of the expansion method generate terminal models with more independent variables than can be found in their respective initial models. Consequently we can say that in these applications the additional variables in a terminal model are 'omitted' from its initial model and that an estimated initial model showing significant spatial autocorrelation 'may' be converted by the expansions into a larger model that does not display any spatial autocorrelation.

Example

Let us show an example of initial model that upon estimation and testing displays a significant spatial autocorrelation, while a terminal model generated from it does not. In the terminal model the variables added by expansions were missing in the initial model. Consequently, the spatial autocorrelation of the initial model is due to omitted variables.

The example in this 'Dependency' section as well as the one in the 'Multimodeling' section are based on Anselin's classical 'Columbus Crime' dataset (Anselin 1988, pp.189) that has been used in a great many spatial analyses, and can be found in Bivand's R-package spdep. The dataset contains 1980 data for 49 neighborhoods in Columbus, Ohio. The variables used in this example and in the other in the 'Multimodeling' section are: the residential burglaries and vehicle thefts per thousand households, CR; the average housing value per household in $1000, H; the average household income in $1000, I; the neighborhoods distances to the CBD, D; and a core-periphery dummy, C. The variable C has a value of one for the 'core' neighborhoods, and of zero otherwise. The variables' names CR, H, I, D, and C are shortened versions of the ones appearing in the dataset's documentation. The spatial dependency tests used here (LM-Err and LM-Lag) are based on the neighborhoods' contiguity matrix available in the Columbus dataset.

The example that follows involves estimating and comparing the initial model and its full expansion specified by Eqs. (C.6.1) to (C.6.4). The results are shown in Table C.6.1.

Upon comparing the estimated models by their performance on the Lagrangian Multiplier tests for spatial lag and spatial error, LM-Lag and LM-Err, it is apparent that the initial model shows a significant LM-Lag, while the terminal model does not show any significant residual spatial autocorrelation.

Table C.6.1. Regression results

	Initial Model	Terminal Model
Intercept	64.46***	92.10***
	13.58	(7.67)
I	−2.04***	−3.06**
	(−6.64)	(−3.25)
D		−13.86***
		(−3.53)
$I*D$		0.56*
		(2.15)
R^2	0.48	0.68
p(LM-Err)	0.109	0.664
p(LM-Lag)	0.009**	0.513

Notes: p(LM-Err) stands for probability of LM-Err, p(LM-Lag) stands for probability of LM-Lag, t values are in parentheses,
***, ** and * indicate significance at the 0.001, 0.01, and 0.05 levels

A significant LM-Lag test can suggests either the respecification of the initial model as a spatial lag model, or an omitted variable solution such as the one that led to the terminal model specified by Eq. (C.6.4), or the use of one of the other remedial courses of action available.

Let us confine ourselves to the respecification of the initial model as a spatial lag model, and to its respecifications by expansions. It can be argued that there is no conflict or incompatibility between these respecifications in the sense that one of them is correct, and the other is not, because the dataset used in the estimation originated from real world measurements and not from an artificial dataset. In the case of an artificial dataset the analyst knows what the true model is, because it was his/hers data generating process the basis on which the dataset was constructed. Instead, in the case of a dataset from real world measurements the analyst does not know whether any of the respecifications is a true model: she/he only knows that they were all effective remedies to the residual spatial dependency observed in the initial model.

Let us restate for emphasis this important point. If the spatial lag model and the terminal model in this demonstration are approximations to the same unknown and unknowable infinitely-complex realities, they will both reflect them, much in the same way that we can have multiple snapshots of the same object taken from different distances and angles of view. Then the issue is not which model or approximation or snapshot is 'true', but rather how they differ, how suited they are for any given purpose, and which model should be preferred when and why. Some aspects of what may cause a preference for one of alternative approaches to remedy residual spatial dependency are briefly discussed in the next few paragraphs.

Ultimately, the analyst's mindset and the specifics of the analysis determine which model among a plurality of acceptable models is preferable. However it is useful to identify a few lines of thought and research situations in which the determinants of such preferences are more apparent.

The commitment to the specific independent variables in an initial model varies greatly across analysts and disciplines. Clearly, though, when a significant spatial dependency in the initial model's residuals is found, a greater commitment to specific independent variables will tend to be associated with a stronger propensity to remedy it by a respecification of the error term rather than by a respecification of the variables.

The rationales for a strong commitment to specific independent variables can be quite compelling. Issues of data availability can force the use of specific independent variables. Data analyses centered upon the estimation of established and well grounded theoretical models will lead to a strong commitment to the theoretical models' variables. Also, when the dependency has theoretical foundations as in the cases discussed in (Anselin 2002), respecification into a spatial lag model is possibly the most straightforward response to it.

However, we can also find analysts inclined to blame significant spatial dependency on the model's inadequate 'complexity' that translates into an 'omitted variables' type of mis-specification. For instance, this is the case when the spatial drift of some or all the parameters in the model has a theoretical basis or a central position in a research.

Between strong preferences for addressing significant residual spatial dependency by respecifying the error term, and strong preferences for addressing it by adding independent variables, there is a vast gray area inhabited by eclectic practitioners. Possibly, empirical studies of the analysts' actual patterns of behavior in matters pertaining to specification and respecification searches have not being carried out to an extent commensurate to their importance. These studies would help considerably to understand which specification strategies are preferred when and why.

C.6.4 Multimodeling

Generalities. The prevailing modeling strategies aim at finding, estimating, testing, and using, a single mathematical model of some phase of reality that is assumed to be 'true', or optimal, or consistent with theory and data, or congruent with estimation assumptions, or useful in some sense, or, in general, that possesses desirable properties and attributes. Let's just note that the strategies involving the iterative stepwise improvement of a starting model by moving from the simple to the complex or vice versa occupy a somewhat intermediate position between the single model strategy and multimodeling.

This section's focus is upon the less common strategies that are here collectively referred to as multimodeling, and that involve generating results by processing simultaneously a plurality of models. Let us briefly touch upon four prominent types of multimodeling: meta-analysis, information theory based multimodeling, Bayesian based multimodeling, and extreme bounds analysis.

In the words of its creator *'Meta-analysis refers to the analysis of analyses ... [Meta-analysis is] the statistical analysis of a large collection of analysis results from individual studies for the purpose of integrating the findings'* (Glass 1976, p.3). The meta-analysis entry in Wikipedia remarks that '... *a meta-analysis combines the results of several studies that address a set of related research hypotheses. ... Because the results from different studies investigating different independent variables are measured on different scales, the dependent variable in a meta-analysis is some standardized measure of effect size. ...* [S]*tudy characteristics such as measurement instrument used, population sampled, or aspects of the studies design ...* [may be] *... used as predictor variables to analyze the ... variation in the effect sizes ... Meta-analysis leads to a shift in emphasis from single studies to multiple studies. It emphasizes the practical importance of the effect size instead of the statistical significance of individual studies. This shift in thinking has been termed meta-analytic thinking'*. Two treatises on meta-analysis are Hunter and Smith (2004) and Sutton et al. (2000). A spatial econometric application is by Florax and de Graaff (2004).

Both the information theory based multimodeling (Anderson 2007; Burnham and Anderson 2002)and the Bayesian based multimodeling (Hilborn and Mangel 1997; Hobbs and Hilborn 2006; McCarthy 2007) involve (a) assembling and estimating a pool of models of same type (such as for instance, linear models with well behaved error terms) that share dependent variable, dataset, and research objectives, and then (b) comparing and ranking the estimated models, and using them as inputs to further processing as in model averaging and multimodel inference.

Kennedy (2003, p.84) characterizes Leamer's extreme bounds and fragility analyses in the following terms: *'Suppose that the purpose of [a] study is to estimate the coefficients of some key variables. The first step of this approach, after identifying a general family of models, is to undertake an extreme bounds analysis, in which the coefficients of the key variables are estimated using all combinations of included/excluded doubtful variables. If the resulting range of estimates is too wide for comfort, an attempt is made to narrow this range by conducting a fragility analysis'*. Key contributions to this type of multimodeling are Leamer (1983, 1985) and Leamer and Leonard (1983). Comments and criticisms on Leamer's extreme bounds and fragility analyses are reviewed in Kennedy (2003, pp. 95-96).

The expansion-based multimodeling discussed here consists in the following sequence of steps. First, a research question is defined, a suitable dataset is secured, a dependent variable is selected, and the potential independent variables in the dataset are divided between candidates for use in the initial model(s), and candidates for use in the expansion equation(s). Second, a pool of initial models is obtained by defining the dependent variable as a linear function of a subset of appropriate candidate variables plus an error term. Third, a pool of terminal models is obtained by expanding some or all the parameters of each initial model into linear functions of expansion variables. Fourth, each initial model and each terminal model are estimated, tested for the attribute(s) of interest, and assigned to one of two groups depending on whether they posses the attribute(s) or not. Fifth, analyses are carried out to identify which characteristics of the estimated terminal models tend to be associated with the presence or absence of the attribute(s) of interest. Sixth, the subset of models possessing the attribute(s) are ranked according to a criterion such AIC, and the best of them is selected for use; alternatively, the 'distance' of the ranked models from the best is computed, and the models within some distance threshold from the best are used for averaging or inference.

Dependency

The illustration of the expansion-based multimodeling that follows is designed to remedy residual spatial autocorrelation, and to gain insights as to which characteristics of a model tend to be associated with significant residual spatial autocorrelation.

The expansion method is very well suited to generate any multiple spatial models. However, here let us focus on the type of expansions emphasized throughout this chapter, whereby a linear initial model born in a substantive social science discipline is expanded in terms of variables that define a geographic context and/or spatial dimensions such as distances or spatial dummies.

A pool of spatial models of this type can be easily generated from a pool of initial models with well behaved error terms and the same dependent variable, and a pool of expansion variables defining geographical contexts or spatial dimensions. To this effect each initial model is expanded in terms of as many subsets of variables from the expansion variables pool as appropriate. All the initial models and all the terminal models generated from them constitute the pool of models to be estimated and tested for spatial dependency.

The estimation and testing of the models in the pool implement the 'Spatial Regression Model Selection Decision Rule' detailed and diagrammed in Anselin (2005, pp.198-200) and Florax et al. (2003, pp.558-559). Essentially, the rule involves estimating a starting linear model by OLS and then testing it by LM-Error and LM-Lag statistics. If neither test rejects the null hypothesis of no-dependency the model is regarded as correctly specified. Otherwise the rule specifies a sequence of steps leading to rerun it either as a spatial error model or as a spatial lag model.

The estimation and testing for spatial dependency of all the models in the final pool is a multimodel implementation of the initial step in Anselin's decision rule, and categorizes these models as having significant spatial dependency or not. This categorization opens the way to analyses aiming at finding which attributes of the models relate how much to spatial dependency. Also it opens the way to rank the models that do not display significant dependency by a criterion such as AIC, in order to select the best model (namely the one with the lowest AIC value) as an end product of the multimodeling. Alternatively, the 'distance' of the ranked models from the best could be used to select an average of the 'best models with no spatial dependency' as an end result. In the next section the approach is demonstrated using the Columbus Crime dataset.

Example

The illustration that follows explores the effectiveness of the expansion based multimodeling approach to remedy spatial autocorrelation. The research question concerns the spatial distribution of crime in Columbus, Ohio, and its determinants.

All the models appearing in the demonstration are linear, share the same dependent 'crime' variable CR, are assumed to have well behaved error terms, and are estimated by OLS. Since all the models share the dependent variable they are defined by their independent variables.

The names of the models were constructed so as to distinguish initial models from terminal models, and the independent variables appearing in the initial models from those added by expansions. In terms of this scheme the initial models are identified just by the independent variables' names. For instance I denotes an initial model with CR as dependent variable, with an intercept and with I as independent variable. Instead HI denotes an initial model with dependent variable CR, with an intercept, and with the two independent variables H and I. Variable names separated by a 'period' denote a terminal model. For instance HI.D identifies a terminal model generated by expanding an initial model with the independent variables H and I, in terms of the expansion variable D. Hence the independent variables in HI.D are $H, I, D, H*D$, and $I*D$. The suffix m as in HI.Dm denotes a terminal model generated by an intercept expansion, so that the independent variables in it are $H, I,$ and D.

Three initial models H, I, and HI relating crime, CR, respectively to mean home values, H, to mean household income, I, and to both H and I, were estimated by OLS, tested by LM-Err and LM-Lag, and found to have all significant residual spatial dependency. The three models have variables that are spatial because are spatially referenced, but the substantive relations that they formalize originate in sociology and economics. The significant spatial autocorrelation indicates that the substantive relations used are inadequate to account for the spatial distribution of crime in Columbus. This illustration explores whether expanding these initial models in terms of the center-periphery dummy, C, or in terms of the distance from the CBD, D, will produce terminal models with no significant spatial dependence, and if so how many.

A pool of fifteen models that include the three initial models plus twelve terminal models were defined, estimated and tested. The terminal models were obtained by the full expansion and the intercept expansion of each initial model in terms of C, and then of D. All the models in the pool were then estimated by OLS, and tested for residuals' spatial dependency by LM-Err and LM-Lag statistics. Relevant results for the fifteen models are shown in Tables C.6.2 and C.6.3.

Table C.6.2. Multimodeling results

Model	R^2	df	AIC	p(LM-Err)	p(LM-Lag)	SSD
H	0.330	3	401	0.000	0.000	1
I	0.484	3	388	0.109	0.009	1
HI	0.552	4	383	0.032	0.005	1
H.C	0.621	5	377	0.045	0.029	1
I.C	0.674	5	369	0.721	0.321	0
HI.C	0.698	7	370	0.539	0.232	0
H.Cm	0.620	4	375	0.044	0.028	1
I.Cm	0.664	4	369	0.719	0.346	0
HI.Cm	0.686	5	367	0.450	0.210	0
H.D	0.664	5	371	0.063	0.289	0
I.D	0.680	5	368	0.664	0.513	0
HI.D	0.721	7	366	0.516	0.865	0
H.Dm	0.609	4	376	0.154	0.419	0
I.Dm	0.647	4	371	0.785	0.737	0
HI.Dm	0.673	5	369	0.375	0.877	0

For each estimated model Table C.6.2 shows the R-square; the degrees of freedom required by the estimation of the intercept, of the independent variables' parameters, and of the variance; the Akaike Information Criterion (AIC); the probability of no dependency due to spatial error p(LM-Err); the probability of no dependency due to spatial lag, p(LM-Lag); and the dummy variable SSD that stands for Significant Spatial Dependency. SSD was defined so that it has a value of one for the models with significant spatial dependency, and a value of zero otherwise. Specifically, SSD = 1 if either p(LM-Err) < 0.05 or p(LM-Lag) < 0.05 or [p(LM-Err) < 0.05 and p(LM-Lag) < 0.05], else SSD = 0.

The table shows that all the initial models H, I, and HI have significant spatial dependency, while only two out of twelve terminal models have significant spatial dependency. This means that in ten out of the twelve terminal models created by expanding the initial models the spatial dependency is no longer found. In the balance, this illustration suggests that when spatial autocorrelation is encountered, the strategy combining the spatial expansion method with multimodeling should be regarded as one of the remedial courses of action available.

However the results in Table C.6.2 allow also identifying which variables tend to be more associated with spatial dependency. Specifically, the variable H appears in ten models, and four of these models show significant spatial dependency.

The variable I also appears in ten models, but of these only two display significant spatial dependency. Clearly, H tends to be associated with dependency more than I. Next, consider the expansion variables C and D. Each of them appears in six models. However, while two of the models in which C appears are spatially dependent, none of the models in which D appears are. Clearly, again, the absence of spatial dependence in a terminal model is more strongly associated with expansions in D than with expansions in C. Finally, note that both H and C appear in the only two terminal models with spatial dependence. This simple counting shows to what extent specific independent variables are associated with the occurrence of spatial dependency.

Table C.6.3. Group means

Groups	Mean R^2	Mean df	Mean AIC
SSD=1	0.52	3.80	385
SSD=0	0.67	5.10	370

Table C.6.3 shows the mean values of R^2, df, and AIC, for the group of five models that show Significant Spatial Dependence labeled SSD=1, and for the group of twelve models that do not show any Significant Spatial Dependence labeled SSD=0. In a capsule, Table C.6.3 shows that the models with no significant spatial dependency have a higher Mean R^2, an higher Mean df, and a lower Mean AIC values than the models with significant spatial dependency.

The Akaike Information Criterion, AIC, rewards goodness of fit and penalizes complexity. In the case of OLS estimation, AIC is an increasing function of both the mean square residuals and the number of estimated parameters. Among a plurality of models the most preferable is the one with the smallest AIC. Table C.6.2 shows that the best, second best, and third best models in terms of AIC are respectively HI.D, HI.Cm, and I.D. Both Tables C.6.2 and C.6.3 suggest that in this illustration the models with more estimated parameters, greater R^2, and smaller AIC values tend to be associated with the absence of spatial autocorrelation.

The analyses presented in this demonstration can be regarded as an example of a broad class of analyses exploring the relationships between/among 'interesting' attributes of a pool of estimated models.

C.6.5 Concluding remarks

The body of this chapter is comprised of three sections. In the first section the spatial expansion method is defined, demonstrated by an example, and illustrated by a brief review of selected items from the expansion method literature.

The possible use of the expansion method as a remedy to residual spatial dependency is discussed in the second section. The key line of reasoning runs as follows. If an initial model upon estimation and testing is found to have significant spatial dependency while a terminal model generated from it does not show any significant dependency, we can say that the variables added to the initial model by expansions are what removed the dependency.

In the third section an expansion based multimodeling approach to remedy spatial dependency is described and illustrated. In some research situations this approach could be of interest as one of the possible responses when spatial dependency strikes.

Acknowledgements. Luc Anselin's 'Columbus Crime' dataset, and the dependency tests LM-Err and LM-Lag used in the examples are from Roger S. Bivand's R-package spdep. Insightful comments and criticisms by an anonymous reviewer are gratefully acknowledged. The errors are mine.

References

Anderson DR (2007) Model based inference in the life sciences: a primer on evidence. Springer, Berlin, Heidelberg and New York
Anselin L (1988) Spatial econometrics: methods and models. Kluwer, Dordrecht
Anselin L (2002) Under the hood: issues in the specification and interpretation of spatial regression models. Agricult Econ 27(3):247-267
Anselin L (2005) Exploring spatial data with GeoDa: a workbook. Spatial Analysis Laboratory, Department of Geography, University of Illinois, Urbana-Champaign
Anselin L (2006) Spatial econometrics. In Mills TC, Patterson K (eds) Palgrave handbook of econometrics. Palgrave Macmillan, New York, pp.901-969
Anselin L (2007) Spatial econometrics in RSUE: retrospect and prospect. Reg Sci Urb Econ 37(4):450-456
Anselin L, Bera AK (1998) Spatial dependence in linear regression models with an introduction to spatial econometrics In: Ullah A, Giles DEA (eds) Handbook of applied economic statistics. Marcel Dekker, New York, pp.255-258
Aten B, Heston A (2003) Regional output differences in international perspective. United Nations University: World Institute for Development Economics Research (WIDER) retrieved from http://wwwwiderunuedu/publications/discussion-papers-2003.htm
Brown LA, Goetz AR (1987) Development-related contextual effects and individual attributes in third world migration processes: a Venezuelan example. Demography 24(4): 497-516
Burford RL, Zee SML (2006) An application of Casetti's expansion method to a variable coefficient regression model of electricity demand: simulation results of alternative estimation methods. Europ J Econ, Fin, Admin Sci (5):176-186
Burnham KP, Anderson DR (2002) Model selection and multimodel inference: a practical information-theoretic approach (2nd edition). Springer, Berlin, Heidelberg and New York

Can A (1990) The measurement of neighborhood dynamics in urban house prices. Econ-Geogr 66(3):254-271

Can A (1992) Specification and estimation of hedonic housing price models. Reg Sci Urb Econ 22(3):453-474

Can A, Anselin L, Casetti E (1989) Spatial variation in the propensity to move of households. Sist Urb 11(1):107-127

Casetti E (1972) Generating models by the expansion method: applications to geographical research. Geogr Anal 4(1):81-91

Casetti E (1973) Testing for spatial-temporal trends: an application to urban population density trends using the expansion method. Canad Geogr 17(2):127-137

Casetti E (1986) The dual expansion method: an application for evaluating the effects of population growth on development. IEEE Transact Syst Man Cybern, SMC-16(1):29-39

Casetti E (1989) The onset and spread of modern economic growth in Europe: an empirical test of a catastrophe model. Environ Plann A 21(11):1473-1489

Casetti E (1991) The investigation of parameter drift by expanded regressions: generalities and a 'family planning' example. Environ Plann A 23(7):1045-1061

Casetti E (1992) Bayesian regression and the expansion method. Geogr Anal 24(1):58-74

Casetti E (1997a) Catastrophe models and the expansion method: a review of issues and an application to the econometric modeling of economic growth. Discr Dyn Nat Soc 1(3):185-202

Casetti E (1997b) The expansion method, mathematical modeling, and spatial econometrics. Int Reg Sci Rev 20(1-2):9-33

Casetti E (1997c) Mixed estimation and the expansion method: an application to the spatial modeling of the AIDS epidemic In Fischer MM, Getis A (eds) Recent developments in spatial analysis. Springer, Berlin, Heidelberg and New York, pp.15-34

Casetti E, Fan CC (1991) The spatial spread of the AIDS epidemic in Ohio: empirical analyses using the expansion method. Environ Plann A23(11):1589-1608

Casetti E, Krakover S (1990) Spatio-temporal dynamics of the US population: estimates and extrapolations. Occasional Paper Series on Socio-Spatial Dynamics 1(3):139-159

Casetti E, Pandit K (1987) The non linear dynamics of sectoral shifts. Econ Geogr 63(3):241-258

Casetti E, Semple RK (1969) Concerning the testing of spatial diffusion hypotheses. Geogr Anal 1(3):254-259

Casetti E, Tanaka K (1992) The spatial dynamic of Japanese manufacturing productivity: an empirical analysis by expanded Verdoorn equations. Papers in Reg Sci 71(1):1-13

Cliff AD, Ord JK (1981) Spatial processes: models and applications. Pion, London

Congdon P (2003) Applied bayesian modelling. Wiley, New York, Chichester, Toronto and Brisbane

De Smith MJ, Goodchild MF, Longley PA (2007) Geospatial analysis. Troubador, Leicester [UK]

Dormann CF, McPherson JM, Araujo MB, Bivand R, Bolliger J, Gudrun C, Davies RG, Hirzel A, Jetz W, Kissling WD, Kuhn I, Ohlemuller R, Peres-Neto PR, Reineking B, Schroder B, Schurr FM, Wilson R (2007) Methods to account for spatial autocorrelation in the analysis of species distributional data: a review. Ecography 30(5):609-628

Eldridge DJ, Jones JPI (1991) Warped space: a geography of distance decay. Prof Geogr 43(4):500-511

Fischer MM, Griffith DA (2008) Modeling spatial autocorrelation in spatial interaction data: an application to patent application data in the European Union. J Reg Sc 48(5):969-989

Florax RJGM, De Graaf T (2004) The performance of diagnostic tests for spatial dependence in linear regression models: a meta-analysis of simulation studies. In Anselin L, Florax RJGM, Rey SJ (eds) Advances in spatial econometrics: methodology, tools and applications. Springer, Berlin, Heidelberg and New York, pp.29-65

Florax RJGM, Folmer H, Rey SJ (2003) Specification searches in spatial econometrics: the relevance of Hendry's methodology. Reg Sci Urb Econ 33(5):557-579

Fotheringham SA, Brunsdon C (1999) Local forms of spatial analysis. Geogr Anal 31(4):340-358

Gatrell JD (2005) Modeling human-environment interactions. In Jensen RR, Gatrell JD, McLean DD (eds) Geo-spatial technologies in urban environments. Springer, Berlin, Heidelberg and New York, pp.47-54

Getis A (1990) Screening for spatial dependence in regression analysis. Papers of the Reg Sci Assoc 69(1):69-81

Getis A (2007) Reflections on spatial autocorrelation. Reg Sci Urb Econ 37(4):491-496

Getis A, Griffith DA (2002) Comparative spatial filtering in regression analysis. Geogr Anal 34(2):130-140

Getis A, Ord JK (1995) Local spatial autocorrelation statistics: distributional issues and an application. Geogr Anal 27(4):287-306

Glass GV (1976) Primary, secondary, and meta-analysis of research. Educ Res 5(10):3-8

Griffith DA (2003) Spatial autocorrelation and spatial filtering: gaining understanding through theory and scientific visualization. Springer, Berlin, Heidelberg and New York

Hilborn R, Mangel M (1997) The ecological detective: confronting models with data. Princeton University Press, Princeton [NJ]

Hobbs NT, Hilborn R (2006) Alternative to statistical hypothesis testing in ecology: a guide to self teaching. Ecol Appl 6(1):5-19

Hunter JE, Smith FL (2004) Methods of meta-analysis: correcting error and bias in research findings (2nd edition). Sage, Thousand Oaks [CA]

Jensen RR, Gatrell JD, Boulton JR, Harper BT (2004) Using remote sensing and geographic information systems to study urban quality of life and urban forest amenities. Ecol Soc 9(5): http://wwwecologyandsocietyorg/vol9/iss5/art5/

Jensen RR, Gatrell JD, Boulton JR, Harper BT (2005) The urban environment, socioeconomic conditions, and quality of life: an alternative framework for understanding and assessing environmental justice. In Jensen RR, Gatrell JD, McLean DD (eds) Geo-spatial technologies in urban environments. Springer, Berlin, Heidelberg and New York, pp.63-72

Jones JPI (1987) Work, welfare, and poverty among black female-headed families. Econ Geogr 63(1):20-34

Jones JPI, Casetti E (eds) (1992) Applications of the expansion method. Routledge, London and New York

Jones K, Bullen N (1994) Contextual models of urban house prices: a comparison of fixed- and random-coefficients models developed by expansion. Econ Geogr 70(3):252-272

Jones K, Wrigley N (1995) Generalized additive models, graphical diagnostics, and logistic regression. Geogr Anal 27(1):1-21

Kellerman A, Krakover S (1986) Multi-sectoral urban growth in space and time: an empirical approach. Reg Stud 20(2):117-129

Kennedy P (2003) A guide to econometrics (5th edition). MIT Press, Cambridge [MA]

Kodras JE (1986) Labor market and policy constraints on the work disincentive effect of welfare. Ann Assoc Am Geogr 76(2):228-246

Kristensen G (1997) Women's economic progress and the demand for housing: theory, and empirical analyses based on Danish data. Urb Stud 34(3):403-418

Leamer EE (1983) Let's take the con out of econometrics. Am Econ Rev 73(1):31-43

Leamer EE (1985) Sensitivity analyses would help. Am Econ Rev 75(3):308-313

Leamer EE, Leonard H (1983) Reporting the fragility of regression estimates. Rev Econ Stat 65(2):306-317

LeSage JP (1999) Spatial econometrics. Regional Research Institute, Morgantown [WV], http://wwwrriwvuedu/WebBook/LeSage/spatial/spatial.html

Maddala GS (1992) Introduction to econometrics (2nd edition). Macmillan, New York

McCarthy MA (2007) Bayesian methods for ecology. Cambridge University Press, Cambridge

McMillen DP (2003) Spatial autocorrelation or model mis-specification? Int Reg Sci Rev 26(2):208-217

Miron JR (1984) Spatial autocorrelation in regression analysis: a beginner's guide. In Gaile GL, Willmott CJ (eds) Spatial statistics and models. Reidel, Dordrecht, pp.201-222

Mukherjee C, White H, Wuyts M (1998) Econometrics and data analysis for developing countries. Routledge, New York and London.

Ramanathan R (2002) Introductory econometrics with applications. Harcourt, Fort Worth [TX}

Rangel TFLVD, Diniz-Filho JAF (2006) Toward the integrated computational tool for spatial analysis in macroecology and biogeography. Glob Ecol Biogeogr 15(4):321-327

Stern DI (2000) Applying recent developments in time series econometrics to the spatial domain. Prof Geogr 52(1): 37-49

Sutton AJ, Jones DR, Abrams KR, TA S, Song F (2000) Methods for meta-analysis in medical research. Wiley, London

Tanaka K, Casetti E (1992) The spatio-temporal dynamics of Japanese birth rates: empirical analyses using the expansion method. Geogr Rev Japan 65 B 1(1):15-31

Theriault M, Des Rosiers F, Villeneuve P, Kestens Y (2003) Modelling interactions of location with specific value of housing attributes. Prop Mgmt 21(1):25-62

Varga A (1998) University research and regional innovation: a spatial econometric analysis of academic technology transfers. Kluwer, Boston

Varga A (2000) Local academic knowledge transfers and the concentration of economic activity. J Reg Sci 40(2):289-309

Zdorkowski TR, Hanham RQ (1983) Two views of the city as a source of space-time trends in economic development and the decline of human fertility. Urb Geogr 4(1):54-62

Zhang J, Kristensen G (1995) A gravity model with variable coefficients: The EEC trade with third countries. Geogr Anal 27(4):307-320

C.7 Multilevel Modeling

S.V. Subramanian

C.7.1 Introduction

Individuals are organized within a nearly infinite number of levels of organization, from the individual up (for example, families, neighborhoods, counties, states), from the individual down (for example, body organs, cellular matrices, DNA), and for overlapping units (for example, area of residence and work environment). It is necessary, therefore, that links should be made between these possible levels of analysis. The term 'multilevel' refers to the distinct levels or units of analysis, which usually, but not always, consists of, individuals (at lower level) who are nested within contextual/aggregate units (at higher level). Multilevel methods consist of statistical procedures that are pertinent when (i) the observations that are being analyzed are correlated or clustered, or (ii) the causal processes is thought to operate simultaneously at more than one level, and/or (iii) there is an intrinsic interest in describing the variability and heterogeneity in the phenomenon, over and above the focus on the average (Diez Roux 2002; Subramanian et al. 2003; Subramanian 2004a, 2004b).

Multilevel statistical models are often used in areas such as image processing and remote sensing (Kolaczyk et al. 2005). Multilevel methods are specifically geared towards the statistical analysis of data that have a *nested* structure. The nesting, typically, but not always, is hierarchical. For instance, a two level structure would have many level-1 units nested within a smaller number of level-2 units. In educational research, the field that provided the impetus for multilevel methods, level-1 usually consists of pupils who are nested within schools at level-2. Such structures arise routinely in health and social sciences, such that level-1 and level-2 units could be, workers in organizations, patients in hospitals, individuals in neighborhoods, respectively. In this chapter, for exemplification, we will consider the structure of individuals nested within neighborhoods (used to reflect one practical realization of place).

The existence of nested data structures is neither random nor ignorable; for instance, individuals differ but so do the neighborhoods. Differences among neighborhoods could either be directly due to the differences among individuals who live in them; or groupings based on neighborhoods may arise for reasons less strongly associated with the characteristics of the individuals who live in them. Regardless, once such groupings are established, even if their establishment is random, they will tend to become differentiated. This would imply that the group (for example, neighborhoods) and its members (for example, individual residents) can exert influence on each other suggesting different sources of variation (for example, individual-induced and neighborhood-induced) in the outcome of interest and thus compelling analysts to consider covariates at the individual *and* at the neighborhood level. Ignoring this multilevel structure of variations not simply risks overlooking the importance of neighborhood effects, but has implications for statistical validity.

To put this in perspective, in an influential study of progress among primary school children, Bennett (1976), using single-level multiple regression analysis, claimed that children exposed to 'formal' style of teaching exhibited more progress than those who were not. The analysis while recognizing individual children as units of analysis ignored their grouping into teachers/classes. In what was the first important example of multilevel analysis using social science data, Aitkin et al. (1981) reanalyzed the data and demonstrated that when the analysis accounted properly for the grouping of children (at lower level) into classes (at higher levels), the progress of formally taught children could not be shown to significantly differ from the others.

What was occurring here was that children within any one class/teacher, because they were taught together, tended to be similar in their performance thereby providing much less information than would have been the case if the same number of children had been taught separately. More formally, the individual samples (for example, children) were *correlated* or *clustered*. Such clustered samples do not contain as much information as simple random samples of similar size. As was shown by Aitkin et al. (1981), ignoring this autocorrelation and clustering resulted in an increased risk of finding differences and relationships where none existed.

Clustered data also arise as a result of sampling strategies. For instance, while planning large-scale survey data collection, for reasons of cost and efficiency, it is usual to adopt a multistage sampling design. A national population survey, for example, might involve a three-stage design, with regions sampled first, then neighborhoods, and then individuals. A design of this kind generates a three-level hierarchically clustered structure of individuals at level-1 nested within neighborhoods at level-2, which in turn are nested in regions at level-3. Individuals living in the same neighborhood can be expected to be more alike than they would be if the sample were truly random. Similar correlation can be expected for neighborhoods within a region.

Much documentation exists on measuring this 'design effect' and correcting for it. Indeed, clustered designs (for example, individuals at level-1 nested in neighborhoods at level-2 nested in regions at level-3) are often a nuisance in traditional analysis. However, individuals, neighborhoods and regions can be seen as distinct structures that exist in the population that should be measured and modeled.

C.7.2 Multilevel framework: A necessity for understanding ecological effects

Figure C.7.1 identifies a typology of designs for data collection and analyses (Blakely and Woodward 2000; Kawachi and Subramanian 2006; Subramanian et al. 2007) where the rows indicate the level or unit at which the outcome variable is being measured [that is, at the individual level (y) or the ecological level (Y)], and the columns indicate whether the exposure is being measured at the individual level (x) or the ecological level (X). The ecological level, in this illustration, relates to the neighborhood level. Study-type (y, x) is most commonly encountered when the researcher aims to link exposure to outcomes, with both being measured at the individual level. Study-type (y, x) typically ignores ecological effects (either implicitly or explicitly).

		Exposure	
		Individual (x) (measured at individual level)	*Ecologic* (X) (measured at ecological level)
Outcome	*Individual* (y)	(y, x) Traditional risk factor study	(y, X) Multilevel study
	Ecologic (Y)	(Y, x)[a]	(Y, X) Ecological study

Notes: [a] This type of study is impossible to specify as it stands. Practically speaking, it will either take the form of (Y, X), that is, ecological study, where X will now simply be central tendency of x. Or, if disaggregation of Y is possible, so that we can observe y, then it will be equivalent to (y, x). Source: (Subramanian et al. 2008)

Fig. C.7.1. Typology of studies (Subramanian et al. 2007)

Conversely, study-type (Y, X) – referred to as an 'ecological study' – may seem intuitively appropriate for research where higher levels (for instances, neighborhoods, regions, states, schools and so on) are the targets of interest. However, study-type (Y, X) conflates the genuinely ecological and the aggregate or 'compositional' (Moon et al. 2005), and precludes the possibility of testing heterogeneous contextual effects on different types of individuals. Ecological effects reflect predictors and associated mechanisms operating primarily at the contextual level. The search for such measures and their scientific validation and assessment is an area of active research (Raudenbush 2003). Aggregate effects, in contrast, equate the effect of a neighborhood with the sum of the individual effects associated with the people living within the neighborhood. In this situation the interpretative question becomes particularly relevant. If common membership of a neighborhood by a set of individuals brings about an effect that is over and above those resulting from individual characteristics, then there may indeed be an ecological effect.

Study-type (y, X) provides a multilevel approach in which an ecological exposure is linked to an individual outcome. A more complete representation would be type (y, x, X) whereby we have an individual outcome, individual confounders (x), and neighborhood exposure reflecting a multilevel structure of individuals nested within neighborhoods. A fundamental motivation for study-type (y, x, X) is to distinguish 'neighborhood differences' from 'the difference a neighborhood makes' (Moon et al. 2005). Stated differently, ecological effects on the individual outcome should be ascertained after individual factors that reflect the composition of the places (and may be potential confounders) have been controlled. Indeed, compositional explanations for ecological variations in health are common. It nonetheless makes intuitive sense to test for the possibility of ecological effects. Besides anticipating their impact on individual outcomes, compositional factors may vary by context. Thus, unless contextual variables are considered, their direct effects and any indirect mediation through compositional variables remain unidentified. Moreover, composition itself has an intrinsic ecologic dimension; the very fact that individual (compositional) factors may 'explain' ecologic variations serves as a reminder that the real understanding of ecologic effects is likely to be complex.

The multilevel framework with its simultaneous examination of the characteristics of the individuals at one level and the context or ecologies in which they are located at another level accordingly offers a comprehensive framework for understanding the ways in which places can affect people (contextual) and/or people can affect places (composition). It likewise allows for a more precise distinction between aggregative fallacy *versus* ecologic effects (Subramanian et al. 2008).

C.7.3 A typology of multilevel data structures

The idea of multilevel structure can be recast, with great advantage, to address a range of circumstances where one may anticipate clustering. Outcomes as well as their causal mechanisms are rarely stable and invariant over time, producing data structures that involve repeated measures, which can be considered a special case of multilevel clustered data structures. Consider the 'repeated cross-sectional design' that can be structured in multilevel terms with neighborhoods at level-3; year/time at level-2 and individuals at level-1. In this example, level-2 represents repeated measurements on the neighborhoods (level-3) over time. Such a structure can be used to investigate what sorts of individuals and what sorts of neighborhoods have changed with respect to the outcome. Alternatively, there is the classic 'longitudinal or panel design' in which the level-1 is the measurement occasion, level-2 is the individual and level-3 is the neighborhood. This time, the individuals are repeatedly measured at different time intervals so that it becomes possible to model changing individual behaviors within a contextual setting of, say neighborhoods.

When different responses/outcomes are correlated this lends itself to a 'multivariate' multilevel data structure in which level-1 are sets of response variables measured on individuals at level-2 nested in neighborhoods at level-3. The 'multivariate responses' could be, for instance, different aspects of, say, health behavior (for example, smoking and drinking). In addition, such responses could be a mixture of 'quality' (do you smoke/do you drink) and 'quantity' (how many/how much) producing 'mixed multivariate responses'. The substantive benefit of this approach is that it is possible to assess whether different types of behavior and whether the qualitative and quantitative aspects of each behavior are related to individual characteristics in the same or different ways. Additionally, we can also ascertain whether neighborhoods that is high for one behavior also high for another and whether neighborhoods with high prevalence of smoking, for instance, also high in terms of the number of cigarettes smoked.

While the previous examples are strictly hierarchical, in that all level-1 units that form a level-2 grouping are always in the same group at any higher level, data structures could be non-hierarchical. For example, a model of health behavior (for instance, smoking) could be formulated with individuals at level-1 and both residential neighborhoods *and* workplaces at level-2 not nested but crossed and are also called as the 'cross-classified structures'. Individuals are then seen as occupying more than one set of contexts, each of which may have an important influence. For instance individuals in a particular workplace may come from different neighborhoods and individuals in a neighborhood may go to several worksites.

A related structure occurs where for a single level-2 classification (for example, neighborhoods), level-1 units (for example, individuals) may belong to more than one level-2 unit and these are also referred as 'multiple membership designs'. The individual can be considered to belong simultaneously to several neighborhoods with the contributions of each neighborhood being weighted in relation to its distance (if the interest is spatial) from the individual. In summary, between some combination of hierarchical structures, cross-classified nesting and multiple membership exhibit a great of complexity that is imprinted either explicitly or implicitly in data can be incorporated via multilevel models.

C.7.4 The distinction between levels and variables

Each of the levels that were discussed in the previous section (for example, neighborhoods) can be considered as variables in a regression equation with an indicator variable specified for each neighborhood. Conversely, why are many categorical variables such as gender, ethnicity/race, social class *not* a level? Critical to treating neighborhoods, for example, as a level is because neighborhoods are treated as a *population* of units from which we have observed one random sample. This enables us to draw generalizations for a particular level (for example, neighborhoods) based on an observed sample of neighborhoods. Further, it is more efficient to model neighborhoods as a random variable given the (likely) large number of neighborhoods. On the other hand gender, for instance is not a level because it is not a sample out of all possible gender categories. Rather, it is an attribute of individuals. Thus, male or female in our gender example are 'fixed' discrete categories of a variable with the specific categories only contribute to their respective means. They are not a random sample of gender categories from a population of gender groupings. Further, we would usually wish to ascribe a fixed-effect to each gender, but not each neighborhood. Rather, we wish to model an ecologic attribute at the neighborhood-level.

It is possible to consider 'levels' as 'variables'. Thus, when neighborhoods are considered as a variable, they are typically reflective of a *fixed* classification. While this may be useful in certain circumstances, doing so robs the researcher of the ability to generalize to all neighborhoods and inferences are only possible for the specific neighborhoods observed in the sample.

C.7.5 Multilevel analysis

There are three constitutive components of multilevel analysis which are now discussed.

Evaluating sources of variation: Compositional and/or contextual. A fundamental application of multilevel methods is disentangling the different sources of variations in the outcome. Evidence for variations in poor health, for example, between different neighborhoods can be due to factors that are intrinsic to, and are measured at, the neighborhood level. In other words, the variation is due to what can be described as *contextual*, or *neighborhood effects*. Alternatively, variations between neighborhoods may be *compositional*, that is, certain types of people who are more likely to be in poor health due to their individual characteristics happen to be clustered in certain neighborhoods. The issue, therefore, is not whether variations between different neighborhoods exist (they usually do), but what is the primary source of these variations. Put simply, are there significant contextual differences in health between neighborhoods, after taking into account the individual compositional characteristic of the neighborhood? The notions of contextual and compositional sources of variation have general relevance and they are applicable whether the context is administrative (for example, political boundaries), temporal (for example, different time periods), or institutional (for example, schools or hospitals).

Describing contextual heterogeneity. Contextual differences may be complex such that it may not be the same for all types of people. Describing such *contextual heterogeneity* is another aspect of multilevel analysis and can have two interpretative dimensions. First, there may be a *different amount* of neighborhood variation, such that, for example, for high social class individuals it may not matter in which neighborhoods they live (thus a lower between neighborhood variation), but it matters a great deal for the low social class and as such shows a large between-neighborhood variation. Second, there may be a *differential ordering*: neighborhoods that are high for one group are low for the other and vice versa. Stated simply, the multilevel analytical question is whether the contextual neighborhood differences in poor health, after taking into account the individual composition of the neighborhood, is different for different types of population groups?

Characterizing and explaining the contextual variations. Contextual differences, in addition to people's characteristics, may also be influenced by the different characteristics of neighborhoods. Stated differently, individual differences may interact with context and ascertaining the relative importance of individual and neighborhood covariates is another key aspect of a multilevel analysis. For example, over and above social class (individual characteristic) health may depend upon the poverty levels of the neighborhoods (neighborhood characteristic). The contextual effect of poverty can either be the same for both the high and low social class suggesting that while neighborhood poverty explains the prevalence of poor health, it does not influence the social class inequalities in health. On the other hand, the contextual effects of poverty may be different for different groups, such that neighborhood poverty adversely affects the low social class, but does the opposite for the high social class. Thus, neighborhood level poverty may not only be related to average health achievements but also shapes social inequalities in health. The analytical question of interest is whether the effect of neighborhood level socioeconomic characteristics on health is different for different types of people?

In the presence of a multilevel data, as described in Section C.7.3, and having motivations as discussed above, there are substantive as well as technical reasons to use multilevel statistical models to analyze such data (Raudenbush and Bryk 2002; Goldstein 2003). We shall not review the basic principles of multilevel modeling here as they have been described elsewhere in the context of health research (Subramanian et al. 2003; Moon et al. 2005; Blakely and Subramanian 2006), but rather provide a brief overview of the type of models invoked for identifying ecologic effects discussed in this section.

C.7.6 Multilevel statistical models

Like all statistical regression equations, multilevel models have the same underlying function, which can be expressed as:

RESPONSE = FIXED/AVERAGE PARAMETERS + (RANDOM/VARIANCE PARAMETERS).

While in a conventional regression model the random part of the model is usually restricted to a single term (called error terms or residuals), in the multilevel regression model the focus is on expanding the random part of a statistical model.

In order to exemplify multilevel models we consider the following example. Suppose we are interested in studying the variation in health score, as a function of certain individual and neighborhood predictors. Let us assume that the researcher collected data on a sample of 50 neighborhoods and, for each of these neighborhoods, a random sample of individuals. We then have a two-level structure where the outcome is a health score (with higher score indicating better health), y, for individual i in neighborhood j. We will restrict this exemplification to one individual-level predictor, poverty, x_{1ij}, coded as zero if not poor and one if poor, for every individual i in neighborhood j; and one neighborhood predictor, w_{1j}, a socioeconomic deprivation index in neighborhood j.

Variance component or random intercepts model. Multilevel models operate by developing regression equations at each level of analysis. In the illustration considered here, models would have to be specified at two levels, level-1 and level-2. The model at level-1 can be formally expressed as

$$y_{ij} = \beta_{0j} + \beta_1 x_{1ij} + e_{0ij} \qquad (C.7.1)$$

where β_{0j} (associated with a constant, x_{0ij}, which is a set of ones, and therefore, not written) is the mean health score for the jth neighborhood for the non-poor group; β_1 is the average differential in health score associated with individual poverty status (x_{1ij}) across all neighborhoods. e_{0ij} is the individual or the level-1 residual term. To make this a genuine two-level model we let β_{0j} become a random variable as

$$\beta_{0j} = \beta_0 + u_{0j} \qquad (C.7.2)$$

where u_{0j} is the random neighborhood-specific displacement associated with the overall mean health score (β_0) for the non-poor group. Since we do not allow, at this stage, the average differential for the poor and non-poor group (β_1) to vary across neighborhoods, u_{0j} is assumed to be same for both groups. Equation (C.7.2) is then the level-2 between-neighborhood model.

It is worth emphasizing that the 'neighborhood effect', u_{0j} can be treated in one of the two ways. One can estimate each neighborhood separately as a fixed-effect (that is, treat them as a variable, with 50 neighborhoods there will be 49 additional parameters to be estimated). Such a strategy may be appropriate if the interest is in making inferences about just those sampled neighborhoods. On the other hand, if neighborhoods are treated as a (random) sample from a population of neighborhoods (which might include neighborhoods in future studies if one has complete population data), the target of inference is the variation between neighborhoods in general. Adopting this multilevel statistical approach makes u_{0j} a random variable at level-2 in a two-level statistical model.

Substituting Eq. (C.7.2) into Eq. (C.7.1) and grouping them into fixed and random part components (the latter shown in brackets) yields the following *random-intercepts* or *variance components* model

$$y_{ij} = \beta_0 + \beta_1 x_{1ij} + (u_{0j} + e_{0ij}). \qquad (C.7.3)$$

We have now expressed the response y_{ij} as the sum of a fixed part and a random part. Assuming a normal distribution with zero mean, we can estimate a variance at level-1 (σ_{e0}^2: the between-individual within-neighborhood variation) and level-2 (σ_{u0}^2: the between-neighborhood variation), both conditional on fixed poverty differences in health score. It is the presence of more than one residual term (or the structure of the random part more generally) that distinguishes the multilevel model from the standard linear regression models or analysis of variance type analysis. The underlying random structure (variance-covariance) of the model specified in Eq. (C.7.3) is

$$var\,(u_{0j}) \sim N(0,\,\sigma_{u0}^2) \qquad (C.7.4a)$$

$$var\,(e_{0ij}) \sim N(0,\,\sigma_{e0}^2) \qquad (C.7.4b)$$

$$cov(u_{0j},\,e_{0ij}) = 0. \qquad (C.7.4c)$$

It is this aspect of the regression model that requires special estimation procedures in order to obtain satisfactory parameter estimates (Goldstein 2003).

The model specified in Eq. (C.7.3) with the above random structure is typically used to partition variation according to the different levels, with the variance in y_{ij} being the sum of σ_{u0}^2 and σ_{e0}^2. This leads to a statistic known as *intra-class correlation*, or *intra-unit correlation*, or more generally *variance partitioning coefficient* (Goldstein 2002), representing the degree of similarity between two randomly chosen individuals within a neighborhood. This can be expressed as

$$\rho = \frac{\sigma_{u0}^2}{\sigma_{u0}^2 + \sigma_{e0}^2}. \qquad (C.7.5)$$

Note that Eq. (C.7.3) estimates a variance based on the observed sample of neighborhoods. While this is important to establish the overall importance of neighborhoods as a unit or level, another quantity of interest may pertain to estimating whether living in neighborhood j_1, as compared to neighborhood j_3, for example, predicts a different health score conditional on compositional influences of covariates. Given Eq. (C.7.3), we can estimate for each level-2 unit

$$\hat{u}_{0j} = E(u_{0j} \mid Y, \hat{\beta}, \hat{\Omega}). \tag{C.7.6}$$

The quantity $\hat{u}_{0j}$ are referred to as 'estimated' or 'predicted' residuals, or using Bayesian terminology, as 'posterior' residual estimates, and is calculated as

$$u_{0j} = r_j \frac{\sigma_{u0}^2}{\sigma_{u0}^2 + \sigma_{e0}^2 / n_j} \tag{C.7.7}$$

where σ_{u0}^2 and σ_{e0}^2 are as defined above, r_j is the mean of the individual-level raw residuals for neighborhood j, and n_j is the number of individuals within each neighborhood j. This formula for $\hat{u}_{0j}$ uses the level-1 and level-2 variances and the number of people observed in neighborhood j to scale the observed level-2 residual r_j. As the level-1 variance declines or the sample size increases, the scale factor approaches one, and thus $\hat{u}_{0j}$ approaches r_j.

These neighborhood-level residuals are 'random variables with a distribution whose parameter values tell us about the variation among the level-2 units' (Goldstein 2003). Another interpretation is that each $\hat{u}_{0j}$ estimates neighborhood j's departure from expected mean outcome. This interpretation is based on the assumption that each neighborhood belongs to a population of neighborhoods, and the distribution of the population provides information about plausible values for neighborhood j (Goldstein 2003). For a neighborhood with only a few individuals, we can obtain more precise estimates by combining the population and neighborhood-specific observations than if we were to ignore the population membership assumption and use only the information from that neighborhood. When the estimated residuals at higher-level units are of interest in their own right, we need to provide standard errors, interval estimates and significance tests as well as point estimates for them (Goldstein 2003).

Modeling places: fixed or random? It is worth drawing parallels between the multilevel or random-effects model given by Eq. (C.7.3) and the conventional OLS or fixed-effects regression model. Consider the fixed-effects model, whereby the neighborhood effect is estimated by including a dummy for each neighborhood, as shown by

$$y_{ij} = \beta_0 + \beta x_{ij} + \beta N_j + e_{0ij} \qquad (C.7.8)$$

where N_j is a vector of dummy variables for $N-1$ neighborhoods. The key conceptual difference between the fixed-effects and the random-effects approach to modeling neighborhoods is that while the fixed part coefficients are estimated separately, the random part differentials (u_{0j}) are conceptualized as coming from a distribution (Goldstein 2003). This conceptualization results in three practical benefits (Jones and Bullen 1994)

(i) *pooling information* between neighborhoods, with all the information in the data being used in the combined estimation of the fixed and random part; in particular, the overall regression terms are based on the information for all neighborhoods;
(ii) *borrowing strength*, whereby neighborhood-specific relations that are imprecisely estimated benefit from the information for other neighborhoods; and
(iii) *precision-weighted estimation*, whereby unreliable neighborhood-specific fixed estimates are differentially down-weighted or shrunk toward the overall city-wide estimate. A reliably estimated within-neighborhood relation will be largely immune to this shrinkage.

The random-effects and the fixed-effects estimates for each neighborhood are related (Jones and Bullen 1994). The neighborhood-specific random intercept (β_{0j}) in a multilevel model is a weighted combination of the specific neighborhood coefficient in a fixed-effects model (β_{0j}^*) and the overall multilevel intercept (β_0), in the following way

$$\beta_{0j} = w_j \beta_{0j}^* + (1-w_j)\beta_0 \qquad (C.7.9)$$

with the overall multilevel intercept being a weighted average of all the fixed intercepts

$$\beta_0 = \frac{\sum w_j \beta_{0j}^*}{\sum w_j}. \qquad (C.7.10)$$

Each neighborhood weight is the ratio of the true between-neighborhood parameter variance to the total variance, which additionally includes sampling variance resulting from observing a sample from the neighborhood. Consequently, the weights represent the reliability or precision of the fixed terms

$$w_j = \frac{\sigma_{uo}^2}{\upsilon_j^2 + \sigma_{uo}^2} \qquad (C.7.11)$$

where the random sampling variance of the fixed parameter is

$$\upsilon_j^2 = \frac{\sigma_e^2}{n_j} \qquad (C.7.12)$$

with n_j being the number of observations within neighborhood j. When there are genuine differences between the neighborhoods and the sample sizes within a neighborhood are large, the sampling variance will be small in comparison to the total variance. As a result, the associated weight will be close to one, with the fixed neighborhood effect being reliably estimated, and the random effect neighborhood estimate will be close to the fixed neighborhood effect. As the sampling variance increases, however, the weight will be less than one and the multilevel estimate will increasingly be influenced by the overall intercept based on pooling across neighborhoods. Shrinkage estimates allow the data to determine an appropriate compromise between specific estimates for different neighborhoods and the overall fixed estimate that pools information across places over the entire sample (Jones and Bullen 1994).

Importantly, the fixed-effects approach to modeling neighborhood differences using cross-sectional data is *not* a choice for a typical multilevel research question, where there is an intrinsic interest in an exposure measured at the level of neighborhood such as the one specified in Eq. (C.7.3). In such instances, a multi-level modeling approach is a necessity. This is because the dummy variables associated with the neighborhoods (measuring the fixed-effects of each neighborhood) and the neighborhood exposure is perfectly confounded and, as such, the latter is not identifiable (Fielding 2004). Thus, the fixed-effects specification to understand neighborhood differences is unsuitable for the sort of complex questions which multilevel modeling can address.

The random coefficient or random slopes model. We can expand the *random structure* in Eq. (C.7.3) by allowing the fixed-effect of individual poverty (β_1) to randomly vary across neighborhoods in the following manner

$$y_{ij} = \beta_{0j} + \beta_{1j} x_{1ij} + e_{0ij}. \qquad (C.7.13)$$

At level-2, there will now be two models

$$\beta_{0j} = \beta_0 + u_{0j} \tag{C.7.14}$$

$$\beta_{1j} = \beta_1 + u_{1j}. \tag{C.7.15}$$

Substituting the level-2 models in Eqs. (C.7.14) and (C.7.15) into the level-1 model in Eq. (C.7.13) gives:

$$y_{ij} = \beta_0 + \beta_1 x_{1ij} + (u_{0j} + u_{1j} x_{1ij} + e_{0ij}). \tag{C.7.16}$$

Across neighborhoods, the mean health score for non-poor is β_0, and $\beta_0 + \beta_1$ is the mean health score for the poor, and the mean poverty-differential is β_1. The poverty differential is no longer constant across neighborhoods, but varies by the amount u_{0j} around the mean, β_1. Such models are also referred to as *random-slopes* or *random coefficient models*. These models have a more complex variance-covariance structure than before

$$\operatorname{var}\begin{bmatrix} u_{0j} \\ u_{1j} \end{bmatrix} \sim \mathcal{N}\left(0, \begin{bmatrix} \sigma_{u0}^2 & \\ \sigma_{u0u1} & \sigma_{u1}^2 \end{bmatrix}\right) \tag{C.7.17}$$

$$\operatorname{var}[e_{0ij}] \sim \mathcal{N}(0, \sigma_{e0}^2). \tag{C.7.18}$$

With this formulation, it is no longer straightforward to think in terms of a summary intraclass correlation statistic ρ as the level-2 variation is now a function of an individual predictor variable, x_{1ij}. In our exemplification when x_{1ij} is a dummy variable, we will have two variances estimated at level-2, one for non-poor which is σ_{u0}^2 and one for poor which is

$$\sigma_{u0}^2 + 2\sigma_{u0u1} x_{1ij} + \sigma_{u1}^2 x_{1ij}^2. \tag{C.7.19}$$

That is, level-2 variation will be a 'quadratic' function of the individual predictor variable when x_{ij} is a continuous predictor. Thus the notion of 'random intercepts and slopes', while intuitive, is not entirely appropriate. Rather, what these models are really doing is modeling variance as some function (constant, quadratic or linear) of a predictor variable (Subramanian et al. 2003).

Building on the above perspective of modeling the variance-covariance function (as opposed to 'random intercepts and slopes'), we can extend the concept to modeling variance function at level-1. It is extremely common to assume that the variance is homoskedastic in the random part at level-1 [σ_{e0}^2; Eq. (C.7.16))], and indeed researchers seldom report whether this assumption was tested or not. One strategy would be to model the different variances for poor and non-poor of the following form:

$$y_{ij} = \beta_0 + \beta_1 x_{1ij} + (u_{0j} + u_{1j} x_{1ij} + e_{1ij} x_{1ij} + e_{2ij} x_{2ij}) \tag{C.7.20}$$

where $x_{1ij} = 0$ for non-poor, one for poor, and the new variable $x_{2ij} = 1$ for non-poor, zero for poor, with $var(e_{1ij}) = \sigma_{e1}^2$ giving the variance for poor, and $var(e_{2ij}) = \sigma_{e2}^2$ giving the variance for non-poor, and $cov(e_{1ij}, e_{2ij}) = 0$. There are other parsimonious ways to model level-1 variation in the presence of a number of predictor variables (Goldstein 2003; Subramanian et al. 2003). With this specification, we do not have an interpretation of the random level-1 coefficients as 'random slopes' as we did at level-2. The level-1 parameters, σ_{e1}^2 and σ_{e2}^2, describe the complexity of level-1 variation, which is no longer homoskedastic (Goldstein 2003). Anticipating and modeling heteroskedasticity or heterogeneity at the individual level may be important in multilevel analysis as there may be cross-level confounding – what may appear to be neighborhood heterogeneity (level-2) to be explained by some ecological variable could be due to a failure to take account of the between individual (within-neighborhood) heterogeneity (level-1).

Modeling the fixed-effect of a neighborhood predictor. An attractive feature of multilevel models – one that is perhaps most commonly used in social science research – is their utility in modeling neighborhood *and* individual characteristics, and any interaction between them, simultaneously. We will consider the underlying level-2 model related to Eq. (C.7.20), which is exactly the same as specified in Eqs. (C.7.14) to (C.7.15), but now including a level-2 predictor w_{1j}, the deprivation index for neighborhood j

$$\beta_{0j} = \beta_0 + \alpha_1 w_{1j} + u_{0j} \tag{C.7.21}$$

$$\beta_{1j} = \beta_1 + \alpha_2 w_{1j} + u_{1j}. \tag{C.7.22}$$

Note that the separate specification of micro and macro models correctly recognizes that the contextual variables (w_{1j}) are predictors of between-neighborhood differences. The extension of Eq. (C.7.20) will now be

$$y_{ij} = \beta_0 + \beta_1 x_{1ij} + \alpha_1 w_{1j} + \alpha_2 w_{1j} x_{1ij} + (u_{0j} + u_{1j} x_{1ij} + e_{1ij} x_{1ij} + e_{2ij} x_{2ij}). \quad (C.7.23)$$

The combined formulation in Eq. (C.7.23) highlights an important feature, the presence of an interaction between a level-2 and level-1 predictor ($w_{1j} x_{1ij}$), represented by the fixed parameter α_2. Now, α_1 estimates the marginal change in health score for a unit change in the neighborhood deprivation index for the non-poor, and α_2 estimates the extent to which the marginal change in health score for unit change in the neighborhood deprivation index is *different* for the poor. This multilevel statistical formulation allows *cross-level effect modification* or *interaction* between individual and neighborhood characteristics to be robustly specified and estimated.

In summary, multilevel models are concerned with modeling both the average and the variation around the average, at different levels. To accomplish this they consist of two sets of parameters: those summarizing the average relationships(s), and those summarizing the variation around the average at both the level of individuals and neighborhoods. Models presented in the preceding section can be easily adapted to other structures with nesting of level-1 units within level-2 units. Additionally, these models can be extended to three or more levels. While the preceding discussion considered a single normally distributed response variable for illustration, multilevel models are capable of handling a wide range of responses. These include: binary outcomes, proportions (for example, logit, log-log, and probit models); multiple categories (for example, ordered and unordered multinomial models); and counts (for example, Poisson and negative binomial distribution models). In essence, these models work by assuming a specific, non-Gaussian distribution for the random part at level-1, while maintaining the normality assumptions for random parts at higher levels. Consequently, the discussion presented in this entry focusing at the neighborhood level would continue to hold regardless of the nature of the response variable, with some exceptions. For instance, determining intra-class correlation or partitioning variances across individual and neighborhood levels in complex non-linear multilevel logistic models is not straightforward (see for details, Browne et al. 2005; Goldstein et al. 2002).

C.7.7 Exploiting the flexibility of multilevel models to incorporating 'realistic' complexity

Current implementations of multilevel models have generally failed to exploit the full capabilities of the analytical framework (Subramanian 2004a; Leyland 2005;

Moon et al. 2005). Much, if not all, of the current research linking neighborhoods and health is cross-sectional, and assumes a hierarchical structure of individuals nested within neighborhoods. This simplistic scenario ignores, for instance, the possibility that an individual might move several times and as such reflect neighborhood effects drawn from several contexts, or that other competing contexts (for example, schools, workplaces, hospital settings) may simultaneously contribute to contextual effects.

Figure C.7.2 provides a visual illustration of one complex, but realistic multilevel structure for neighborhoods and health research, where time measurements (level-1) are nested within individuals (level-2) who are in turn nested within neighborhoods (level-3). Importantly, individuals are assigned different weights for the time spent in each neighborhood. For example, individual 25 moved from neighborhood one to neighborhood 25 during the time period t_1-t_2, spending 20 percent of her time in neighborhood one and 80 percent in her new neighborhood. This multiple membership design would allow control of changing context as well as changing composition. Such designs could be extended to incorporate memberships to additional contexts, such as workplaces, or schools. It can also be extended to enable consideration of weighted effects of proximate contexts (Langford et al. 1998). So, for example, the geographic distribution of disease can be seen not only as a matter of composition and the immediate context in which an outcome occurs, but also a consequence of the impact of nearby contexts with nearer areas being more influential than more distant ones. This is also called spatial autocorrelation and forms an important area of spatial statistical research (Lawson 2001). While such analyses require high-quality longitudinal and context-referenced data, models that incorporate such 'realistic complexity' (Best et al. 1996) are likely to improve our understanding of true neighborhood effects. While the foregoing discussion provides a sound rationale to adopt a multilevel analytic approach for modeling ecologic effects, it obviously does not overcome the limitations intrinsic to any observational study design, single-level or multilevel.

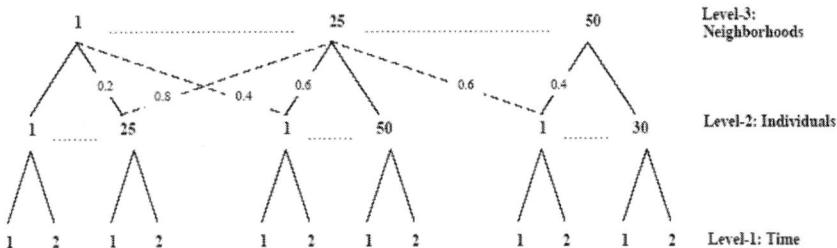

Fig. C.7.2. Multilevel structure of repeated measurements of individuals over time across neighborhoods with individuals having multiple membership to different neighborhoods across the time span. Source: Subramanian (2004b)

C.7.8 Concluding remarks

The multilevel statistical approach – an approach that explicitly models the correlated nature of the data arising either due to sampling design or because populations are clustered – has a number of substantive and technical advantages.

From a substantive perspective, it circumvents the problems associated with ecological fallacy (the invalid transfer of results observed at the ecological level to the individual level), individualistic fallacy (which occurs by failing to take into account the ecology or context within which individual relationships happen), and atomistic fallacy (that arises when associations between individual variables are used to make inferences on the association between the analogous variables at the group/ecological level). The issue common to the above fallacies is the failure to recognize the existence of unique relationships being observable at multiple levels and each being important in its own right. Specifically, one can think of an individual relationship (for example, individuals who are poor are more likely to have poor health), an ecological/contextual relationship (for example, places with a high proportion of poor individuals are more likely to have higher rates of poor health), and an individual-contextual relationship (for example, the greatest likelihood of being in poor health is found for poor individuals in places with a high proportion of poor people). Multilevel models explicitly recognize the level-contingent nature of relationships.

From a technical perspective, the multilevel approach enables researchers to obtain statistically efficient estimates of fixed-effects regression coefficients. Specifically, using the clustering information, multilevel models provide correct standard errors, and thereby robust confidence intervals and significance tests. These generally will be more conservative than the traditional ones that are obtained simply by ignoring the presence of clustering. More broadly, multilevel models allow a more appropriate and realistic specification of complex variance structures at each level. Multilevel models are also precision weighted and capitalize on the advantages that accrue as a result of 'pooling' information from all the neighborhoods to make inferences about specific neighborhoods.

While the advances in statistical research and computing has shown the potential of multilevel methods for health and social behavioral research there are issues to be considered while developing and interpreting multilevel applications. First, it is important to clearly motivate and conceptualize the choice of higher levels in a multilevel analysis. Second, establishing the relative importance of context and composition is probably more apparent than real and necessary caution must be exercised while conceptualizing and interpreting the compositional and contextual sources of variation. Third, it is important that the sample of neighborhoods belong to well-defined population of neighborhoods such that the sample shares exchangeable properties that are essential for robust inferences. Fourth, it is important to ensure adequate sample size at all levels of analysis. In general, if the research focus is essentially on neighborhoods then clearly the analysis requires more neighborhoods (as compared to more individuals within a neighborhood).

Lastly, the ability of multilevel models to make causal inferences is limited and innovative strategies including randomized neighborhood-level research designs (via trials or natural experiments) in combination with multilevel analytical strategy may be required to convincingly demonstrate causal effects of social contexts such as neighborhoods.

References

Aitkin M, Anderson DR, Hinde J (1981) Statistical modelling of data on teaching styles (with discussion). J Roy Stat Soc A 144(4):148-161
Bennett N (1976) Teaching styles and pupil progress. Open Books, London
Best N, Spiegelhalter DJ, Thomas A, Brayne CEG (1996) Bayesian analysis of realistically complex models. J Roy Stat Soc A 159(2):232-342
Blakely TA, Subramanian SV (2006) Multilevel studies. In Oakes M, Kaufman J (eds) Methods for social epidemiology. Jossey Bass, San Francisco, pp.316-340
Blakely TA, Woodward AJ (2000). Ecological effects in multi-level studies. J Epid Comm Health 54(5):367-374
Browne WJ, Subramanian SV, Jones K, Goldstein H (2005) Variance partitioning in multilevel logistic models that exhibit overdispersion. J Roy Stat Soc A168(3):599-613
Diez Roux AV (2002) A glossary for multilevel analysis. J Epid Comm Health 56(8):588-594
Fielding A (2004) The role of the Hausman test and whether higher level effects should be treated as random or fixed. Multil Mode Newsl 16(2):3-9
Goldstein H (2003) Multilevel statistical models. Edward Arnold, London
Goldstein H, Browne WJ, Rasbash J (2002) Partitioning variation in multilevel models. Underst Stat 1(4):223-232
Jones K, Bullen N (1994) Contextual models of urban house prices: a comparison of fixed- and random-coefficient models developed by expansion. Econ Geogr 70(3):252-272
Kawachi I, Subramanian SV (2006) Measuring and modeling the social and geographic context of trauma: a multilevel modeling approach. J Trauma Stress 19(2):195-203
Kolaczyk ED, Ju J, Gopal S (2005) Multiscale, multigranular statistical image segmentation. J Am Stat Assoc 100:1358-1369
Langford IH, Bentham G, McDonald AL (1998) Multilevel modelling of geographically aggregated health data: a case study on malignant melanoma mortality and UV exposure in the European Community. Stat Med 17(1):41-57
Lawson AB (2001) Statistical methods in spatial epidemiology (2nd edition). Wiley, New York, Chichester, Toronto and Brisbane
Leyland AH (2005) Assessing the impact of mobility on health: Implications for life course epidemiology. J Epid Comm Health 59(2):90-91
Moon G, Subramanian SV, Jones K, Duncan C, Twigg L (2005) Area-based studies and the evaluation of multilevel influences on health outcomes. In Bowling A, Ebrahim S (eds) Handbook of health research methods: investigation, measurement and analysis. Open University Press, Berkshire [UK], pp.266-292
Raudenbush SW (2003). The quantitative assessment of neighborhood social environment. In Kawachi I, Berkman LF (eds) Neighborhoods and health. Oxford University Press, New York, pp.112-131
Raudenbush SW, Bryk A (2002) Hierarchical linear models: applications and data analysis methods. Sage, Thousand Oaks [CA]

Subramanian SV (2004a) Multilevel methods, theory and analysis. In Anderson N (ed) Encyclopedia on health and behavior. Sage, Thousand Oaks [CA], pp.602-608

Subramanian SV (2004b) The relevance of multilevel statistical methods for identifying causal neighborhood effects. Soc Sci Med 58(10):1961-1967

Subramanian SV, Glymour MM, Kawachi I (2007) Identifying causal ecologic effects on health: a methodologic assessment. In Galea S (ed) Macrosocial determinants of population health. Springer, New York, pp.301-331

Subramanian SV, Jones K, Duncan C (2003) Multilevel methods for public health research. In Kawachi I, Berkman LF (eds) Neighborhoods and health. Oxford University Press, New York, pp.65-111

Subramanian SV, Jones K, Kaddour A, Krieger N (2009) Revisiting Robinson: the perils of individualistic and ecologic fallacy. Int J Epidem 38 (2):342-360

Part D

The Analysis of Remotely Sensed Data

D.1 ARTMAP Neural Network Multisensor Fusion Model for Multiscale Land Cover Characterization

Sucharita Gopal, Curtis E. Woodcock and *Weiguo Liu*

D.1.1 Background: Multiscale characterization of land cover

Land cover characterization is essential for terrestrial ecosystem and climate modeling, monitoring and prediction. Satellite remote sensing represents the only feasible way to monitor and model many important global change processes by providing repetitive global scale coverage. As a result, typically land cover classification is based on the use of low spatial resolution data such as AVHRR and MODIS (Loveland et al. 2000; Friedl et al. 2002.). But of the global and regional scale, land cover consists of heterogeneous mixtures of different land cover types with exceptions of vast expanses of desert, grassland or forest. Normally, such land cover characterization at regional and global scales are often validated and complemented using fine resolution imagery such as Landsat TM. This line of research has led to multiscale land cover characterization using multisensor fusion of remote sensing data.

Two factors underlie why multisensor fusion of data is becoming increasingly important in land surface characterization. First, there are substantial improvements in land cover characterization using more information from multisensor data sets obtained at various spatial resolutions (from local to regional) as well as at various spectral and angular resolutions. Second, there will be an exponential growth in the availability of data. Under NASA's Earth Science Enterprise (ESE) (formerly called Mission to Planet Earth) initiative, there is a series of satellites and other aircraft-based instruments that provide data relating to Earth. However, the issue of how best to integrate data from different sensors (different spatial resolutions) in order to combine their information and formulate classification and monitoring strategies is not well understood from either a conceptual or a methodological perspective.

The conventional notion of image processing occurring at the scale of pixels is no longer relevant, as there is no single defining image geometry for all sensors. Thus new data models for pixel representation at multiple scales are needed.

Mishandling scale can bias inferences made about the proportions of cover types and the spatial patterns at various scales that have consequences for global and terrestrial models. Both the thematic and proportional accuracy of land cover data are affected by the spatial resolution of the sensor, the interaction between sensor resolution and the spatial characteristics of the phenomenon being mapped, and spatial organization of the classes in the landscape, specifically the interclass adjacencies in the landscape and level of spatial association of classes. There is a scale dependent area bias as there is a general tendency for dominant classes to increasingly dominate the landscape at coarser scales. Conversely, smaller classes will diminish in size at coarser scales (Moody and Woodcock 1994).

D.1.2 Approaches for multiscale land cover characterization

Several approaches have been used to extract the fraction cover at sub-pixel level with remotely sensed data. One approach extracts the sub-pixel information during the image classification procedure using various models including linear mixture models (Adams et al. 1982; Roberts et al. 1993), linear mixture models with multiple end members (Robert et al. 1998; Dennison and Roberts 2003), neural networks (Liu et al. 2004; Carpenter et al. 1999; Foody 1998; Atkinson et al. 1997; Foody et al. 1997), fuzzy classifiers (Foody 1996), regression trees (DeFries et al. 1999; Huang and Townshend 2003), Gaussian mixture discriminant analysis (Ju et al.. 2003), maximum likelihood classifiers (Foody et al. 1992; Schowengerdt 1996; Eastman and Laney 2002), decision trees (McIver 2001) and support vector machines (Brown et al. 1999). Prior research has shown that linear mixture models generate acceptable results. Carpenter et al. (1999) presented a non-linear algorithm for mixture estimation based on an ARTMAP neural network for identifying life form components of the vegetation mixture. Fuzzy ARTMAP architecture achieves a synthesis of fuzzy logic and adaptive resonance theory (ART). Landsat TM imagery was used to estimate the sub-pixel information for life-form components. ARTMAP-based mixture model was able to capture non-linear effects and thus performed better than the conventional linear mixture models. Atkinson et al. (1997) applied Multilayer Perceptron (MLP) based mixture model to decompose AVHRR imagery. The 'unmixture' information from the model was better compared to that generated through a linear mixture model and a fuzzy c-means classifier.

A second approach is to use post-classification calibration procedure to correct biased area estimation (Mayaux and Lambin 1995; Moody and Woodcock 1996; Moody 1998). This method can calibrate the estimation from classification result derived from coarse resolution image with the spatial arrangement of land covers at fine resolution.

In the following, we describe two approaches that are used in this research for multiscale characterization of land cover. The first is based on linear mixture models that have been traditionally used in remote sensing for characterizing mixtures. The second approach is the ARTMAP neural network approach to mixture modeling in remote sensing.

Linear mixture model. Linear mixture models have been widely used in remote sensing for a variety of problems including, estimating crop area (Quarmby et al. 1992), green vegetation, non-photosynthetic vegetation and soil discriminance (Roberts et al. 1993) and measuring land cover change in the Amazon (Adams et al. 1995). The linear mixture model is defined in terms of a set of image end-members, with mixture compositions calculated by linear interpolation within the convex set defined by the end-members. This model is based on an assumption that the reflectance of the pixel is the summation of the component reflectance value weighted by the respective proportion within the pixel

$$R = \sum_{j=0}^{N} r_j F_j + e \qquad \sum_{j=0}^{N} F_j = 1 \qquad (D.1.1)$$

where R is the reflectance value of one pixel, r_j is the reflectance value or 'end-member' of land cover type j, F_j is the proportion of the pixel covered by type j, and e is an error term. One of critical issues for successful application of linear mixture model is the selection of 'end-member' or pure pixels. The 'end-member' can be selected from the field or laboratory measurement (Adams et al. 1995), or based on the result of principle component analysis (Bateson and Curtiss 1996). Recently, researchers have defined methods to select multiple end members that are linear combinations of end members that are allowed to vary in number and type on a per-pixel basis leading to a substantial improvement in accuracy (Dennison and Roberts 2003; Roberts et al. 1998). In this chapter, we are using the conventional linear mixture model to benchmark the performance of the ARTMAP neural network while acknowledging that the non-linear mixture models and multiple end member approaches may provide improved results.

ARTMAP neural network. ART stands for Adaptive Resonance Theory and was introduced by Stephen Grossberg in 1976. The main feature of ART systems is a pattern matching process that compares the current input with a selected learned category representation. ART is capable of developing stable clusters in response to arbitrary sequences of input patterns by self-organization. ARTMAP extends the ART design to include both supervised and unsupervised learning. Fuzzy ARTMAP (Carpenter et al. 1992) incorporates fuzzy logic (Zadeh 1965) in its ART modules and has fuzzy set theoretic operations instead of binary set theoretic operations.

The basic architecture of fuzzy ARTMAP consists of a pair of fuzzy ART modules, ART_a and ART_b, connected by an associative learning network called a Map Field. The other component of this architecture is a controller that uses a minimum learning rule to conjointly minimize predictive error and maximise code compression or predictive generalization (Carpenter et al. 1992). The 'hidden units' in ART_a and ART_b represent learned recognition categories.

In the training phase, ART_a and ART_b modules of the system are presented with a stream of input a^p and desired output pairs b^p respectively. The two modules classify the a^p and b^p vectors into categories and the map field makes the association between ART_a and ART_b categories. A mismatch between the actual b^p and predicted b^p causes a memory search in ART_a. A mechanism called match tracking then raises the ART_a vigilance ρ_a by the minimum amount necessary to trigger a memory search. This can lead to a selection of a new ART_a category that is a better predictor of b^p. Between learning trials, the vigilance relaxes back to its baseline value ρ_a. Match tracking therefore sacrifices only the minimum amount of generalization needed to correct a predictive error. Fast learning and match tracking enable fuzzy ARTMAP to learn to predict novel events while maximizing code compression and preserving code stability.

For the multisensor fusion framework, the ARTMAP neural network model is used to build an association between spectral value and class proportion in pixels (pattern) in any image during the learning process. The trained ARTMAP network can be used to predict class proportion for the test pixels whose spectral values are known but class proportions are unknown. In this study, the spectral values of the coarser resolution (MODIS) pixels form the input vector for ART_a and the land cover class proportions associated with the relative MODIS pixel form the input vector for ART_b. During training, ARTMAP is presented with a stream of ART_a and associated ART_b pairs of inputs. During testing, a stream of ART_a inputs is presented and the ARTMAP neural network model predicts the associated class proportions through ART_b.

D.1.3 Research methodology and data

The objective of the chapter is to describe a multisensor fusion framework for land cover characterization based on ARTMAP neural networks that can help to estimate the proportions of land cover types presented within each coarse resolution pixel. Such an approach will support a wide range of applications within NASA's Earth Science Enterprise including global climate modeling, estimation of photosynthesis, and biophysical parameter estimation (Sellers et al. 1986; Tian et al. 2000; Knyazikhin et al. 1998).

Research methodology. In this research, images of the same region using two different sensors have to be co-registered. The methodology of multisensor fusion framework with an ARTMAP neural network consists of five steps.

- *First*, the Landsat 7 ETM+ image was classified into defined land cover classes with ART-VIP (ART for Visualization and Image Processing) (Liu et al. 2001).
- *Second*, the classified Landsat 7 ETM+ image was registered with the coarser (MODIS) image using GCPs (ground control points) and image-to-image registration with Erdas Imagine software.
- *Third*, the classification map was associated with the MODIS image to estimate the fraction of land cover classes for each pixel at MODIS scale.
- *Fourth*, some training sites were randomly selected from the MODIS image in order to build a training vector with spectral value and the associated class proportion of the training pixel. A series of such training vectors were used to train the ARTMAP multisensor model.
- *Fifth*, the trained network was used to predict class proportions at MODIS scale in regions where Landsat TM imagery is not available.

To evaluate the performance of the ARTMAP neural network multisensor mixture model, results from the model were compared with that from the conventional linear mixture model. The research context is as follows. Multispectral measurements from a low spatial resolution (one K) satellite (MODIS) comprise of a mixture of spectral values from different land cover classes (such as 50 percent water, 25 percent vegetation and 25 percent barren) within one pixel. The sub-pixel information was obtained from a high spatial resolution satellite such as Landsat TM (30m). In this contribution, the multisensor fusion framework was applied to extract the proportion of forest cover for a region of North Central Turkey. We compared the results obtained from the neural network approach with that from a conventional linear mixture model.

Data. The area selected for this study is a portion of north central Turkey. Data consists of a Landsat 7 ETM+ image (path 168 and row 032) for August 19, 1999 and a MODIS/Terra 16 day NBAR global one km image (20 horizontal and 04 vertical (20-4)) for August 27, 2000 and October 31, 2000. The area of MODIS image is 139x146 cells each having a resolution of one km and the corresponding area of the Landsat 7 ETM+ image is 4905x5147 cells each having a resolution of 30m.

The Landsat 7 ETM+ image was classified into the following five land cover classes – *barren*, *water*, *deciduous*, *conifer*, and *grass*. These classes were aggregated into four MODIS classes – *conifer* and *deciduous* were aggregated into a class called *forest*, *barren*, *grass* and *water*. For the purpose of training and testing, a total of 1,884 MODIS pixels were randomly selected from the image. One half of the pixels were used for training while the remaining pixels were used for testing.

For MODIS imagery with one km resolution, most pixels are mixed. Selecting end-member spectra for linear mixture models is problematic. Based on the land cover classification map of Landsat 7 ETM+ image, we selected the spectra of the MODIS pixels deemed to be pure for each class and then calculated the mean spectra value. For the data set of this chapter, it is difficult to find a good spectra example for *forest* class; hence pure spectra examples were selected separately for *deciduous* and *conifer* class. The linear mixture model first predicted the proportion of *deciduous* and *conifer* class respectively for each pixel. Then the summation of the proportion of the two classes was estimated as the fractional cover for class *forest*.

A general practice in many prior studies for the estimation of land cover has been to use simulated coarser scale image by degrading a Landsat TM image (Kalluri et al. 1997). For comparison purposes, we have also used the degraded simulated data generated with an algorithm for simulating MODIS image from Landsat TM imagery (Barket et al. 1992).

D.1.4 Results and analysis

First, this section compares the performance of the ARTMAP approach to the traditional linear mixture model. Second, the performance using a real MODIS image is compared with that with the simulated MODIS data obtained from aggregating a TM image in order to examine the potential sources of error in our model (including registration problems). We expect the aggregated data set to result in better performance compared with the real MODIS data. Third, performance using a single date MODIS image is compared with two dates MODIS image to analyze improvements in classification accuracy with multitemporal images. Three measures of performance are used – an analysis of error bound, root mean square (RMS) errors and analysis of classified land cover maps of the region.

Performance of ARTMAP versus linear mixture model. We compared the performance of the ARTMAP neural network with the linear mixture model. Fig. D.1.1 shows the accuracy in terms of the proportion of error plotted against the percent of total predictions for unseen testing data set. It can be seen that the ARTMAP neural network resulted in a better performance using the real as well as the simulated MODIS data. The best predicted class is *water* while *grass* is the worst predicted class. For each class, the ARTMAP model predicted above 80 percent of pixels within the 20 percent error bound. The linear mixture model did worse for both the simulated and the real MODIS data. The best predicted class is *water* while *grass* is the worst predicted class. The linear mixture model predicted less than 70 percent of the pixels of each class (other than water) within the 20 percent error bound.

Table D.1.1 shows RMS error using the ARTMAP and the linear mixture models for the real and the simulated MODIS testing data sets. First, RMS error results are less for ARTMAP indicating its better performance. Second, RMS error is less for the simulated data set for both the ARTMAP and the linear mixture models.

D.1 ARTMAP neural network multisensor fusion model

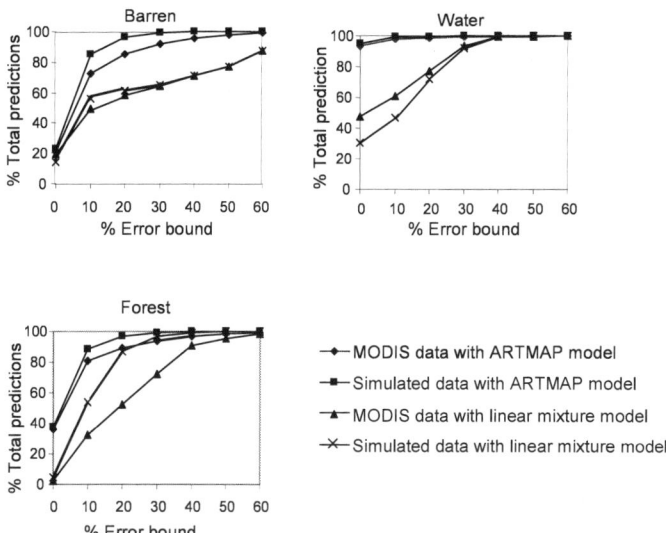

Fig. D.1.1. A comparison of error bound limits of ARTMAP and linear mixture models

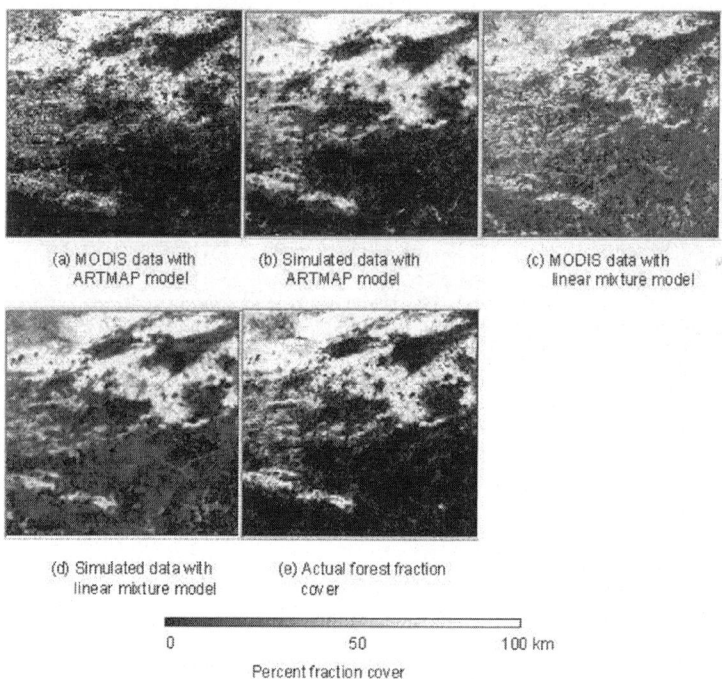

Fig. D.1.2. Predictive and actual forest fraction cover

Table D.1.1. RMS error

Models	Data	Barren	Water	Grass	Forest
Linear Mixture	MODIS	0.34	0.15	0.19	0.26
	Simulated	0.33	0.17	0.18	0.14
ARTMAP	MODIS	0.14	0.05	0.16	0.14
	Simulated	0.09	0.04	0.10	0.08

Figure D.1.2 shows the proportion of land cover classes estimated for the whole image using the two methods for the real MODIS and the simulated MODIS data sets. Performance was compared with ground truth data represented by the aggregated land cover map drawn using the Landsat TM data shown in Fig. D.1.2(e). A visual inspection indicates that maps based on the ARTMAP model estimate similar forest fraction cover to the ground truth map shown in Fig. D.1.2(e). In contrast, the linear mixture models predicted different forest fraction cover. These figures also show that the ARTMAP model has predicted class proportions in the correct areas of the image.

The next stage in analysis involves an estimation of the fraction of *forest* cover. For this purpose, the forest cover shown in Fig. D.1.2(e) was classified into five groups: *sparse* forest cover (0-20 percent), *some* forest cover (20-40 percent), *moderate* forest cover (40-60 percent), *dense* forest cover (60-80 percent) and *very dense* forest cover (80-100 percent). Most pixels of the image (almost 56 percent) have *sparse* forest cover. In other words, other land cover classes dominate the image. *Moderate* to *dense* forest covers about 40 percent of the region, mostly in the north.

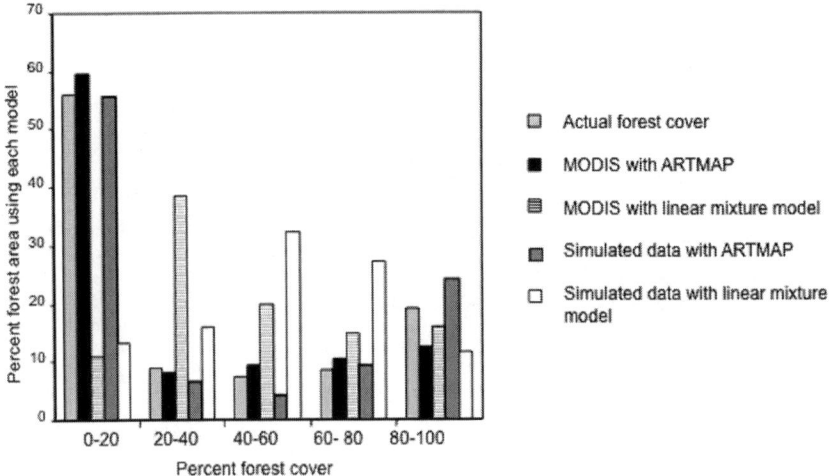

Fig. D.1.3. Predicted forest cover using ARTMAP and linear mixture models

Figure D.1.3 shows the percent forest cover predicted using the two data sets. The ARTMAP model predicts *some*, *moderate* and *dense* forest cover classes reasonably well. It underpredicts the *very dense* forest class and overpredicts the *sparse* forest class. Using the simulated data, the ARTMAP model overpredicts the *dense* forest class and underpredicts *medium* forest cover class. The performance of the ARTMAP model for other forest classes is good. The linear mixture model, in contrast, consistently overpredicts the *some*, *medium* and *dense* forest cover and underpredicts the *sparse* and *very dense* forest cover class. Its poor performance in predicting forest class is reflected in the overall RMS error.

The analysis of forest cover has potential applications in areas where information is needed on the spatial pattern and area of land cover. This is especially important in areas where input maps of sufficient spatial detail are unavailable. The methodology presented in this contribution can utilize widely available coarse spatial resolution imagery to produce a land cover map at the required fine spatial resolution. The framework can also help in the estimation of forest cover from coarse resolution imagery that is useful for forest management and monitoring change.

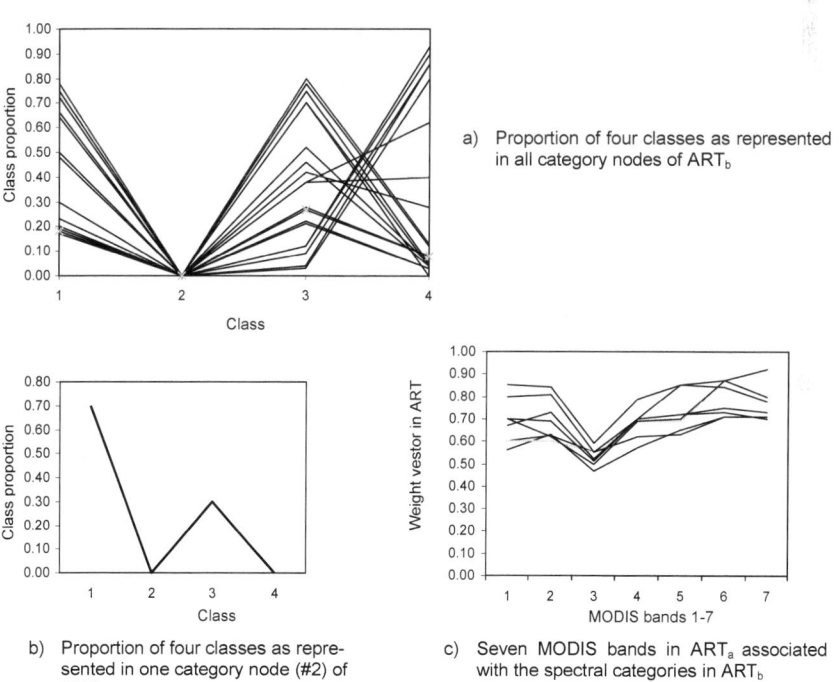

a) Proportion of four classes as represented in all category nodes of ART_b

b) Proportion of four classes as represented in one category node (#2) of ART_b

c) Seven MODIS bands in ART_a associated with the spectral categories in ART_b

Fig. D.1.4. Category visualization in the ARTMAP model

The ARTMAP model generates 453 categories in ART_b module to represent the class proportion information of the training samples. Figure D.1.4(a) shows the distribution of 25 out of all categories, Fig. D.1.4(b) the second category in ART_b with the following class proportions – *barren* (70 percent), *water* (0 percent), *grass* (30 percent) and *forest* (0 percent). At the same time, ART_a produces 1,308 categories to store the spectral information of the training samples. Fig. D.1.4(c) shows the spectral categories in ART_a associated with the class proportion category in ART_b given in Fig. D.1.4(b). The three plots in Fig. D.1.4 illustrate that the learning process of the ARTMAP model is able to capture many different combinations of class proportions. Thus, the ARTMAP model builds an association between the spectral categories and class proportion categories. These plots illustrate the uniqueness of ARTMAP related to its ability to create categories in response different combinations of class proportions. Traditional linear mixture models, on the other hand, lack these capabilities.

Performance comparison of MODIS data versus simulated data. Performance results of the real MODIS data and the simulated MODIS data are shown in Fig. D.1.3. The prediction accuracy of the simulated data is much better for both the ARTMAP model and the linear mixture model. Table D.1.1 shows that the RMS error for the ARTMAP model is superior for the simulated data set since it had the least RMS error. The reasons for such performance differences stems from many sources. Registering a TM image to a MODIS image needs a geometric registration algorithm that may introduce some errors. The simulated data avoids registration error since it simply aggregates spectral values from TM images. Registration error may be an important factor for such multisensor fusion models.

Note that both the simulated and the real MODIS data may have pixels that have similar spectral signatures but characterized by different class proportions. For example, a pixel characterized by 50 percent water and 50 percent barren may look similar to a pixel characterized by 65 percent water and 35 percent barren. This is a potential source of error in our models.

Multitemporal and multisensor image fusion. The prior section described the analysis relating to the performance of the two models using single time period images. In this section, we apply the single date and two dates image respectively to the ARTMAP multisensor fusion model for land cover class fractional estimation in order to analyze if there are differences in performance. The composites of two temporal MODIS images were gathered in August 2000, and October of 2000. The second date image was registered to the first date MODIS image. This ensures that the second date image is co-registered with the Landsat 7 ETM+ image to extract sub-pixel information with the ARTMAP neural network approach.

Performance is compared using the first date, the second date and the two dates image. This analysis shows no significant difference between using single date MODIS image and using two dates MODIS image for sub-pixel information extraction in terms of accuracy of testing data set. For the whole image (see Fig. D.1.5), the predictions of forest cover class are very similar for the three types of temporal data. In general, there is an over prediction for *some*, and *very dense*. But there is an under prediction for other forest cover types in all three date types.

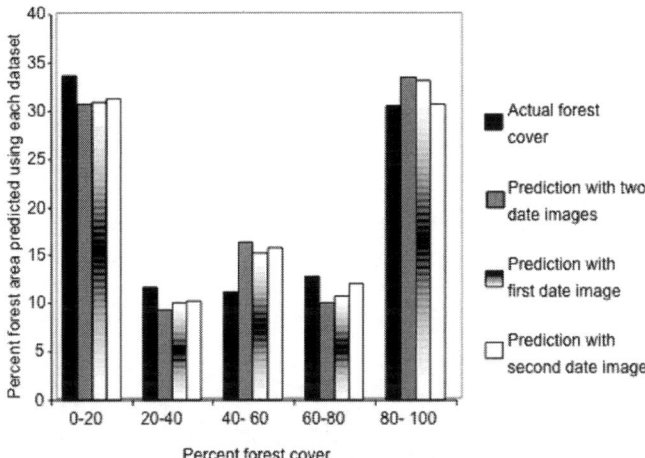

Fig. D.1.5. Predictive and actual forest cover class using multi-temporal images

Figure D.1.4 shows maps of percent forest cover using the scheme presented in Fig. D.1.1. There is no significant difference for the three predictions compared with the ground truth forest cover shown in Fig. D.1.4(d). Comparison of RMS error also provides similar results to support the general finding that there is no significant difference using single date or two date images. All the three prediction results for testing data are similar as shown in Table D.1.2.

Table D.1.2. RMS error of the testing data

	Barren	Water	Grass	Forest
First date image	0.133	0.049	0.180	0.170
Second date image	0.140	0.051	0.199	0.192
Two dates image	0.124	0.040	0.178	0.176

However, a different picture emerges on examining two dates MODIS image for the whole image as shown in Table D.1.3. Compared with the results using one date image, there is two percent improvement for *forest* and *grass* class and one percent improvement for *barren* class. Thus, the introduction of the second date (October of 2000) image is useful in discriminating *forest* from *grass* and similarly *barren* from *grass*.

Table D.1.3. RMS error of the whole image

	Barren	Water	Grass	Forest
First date image	0.145	0.031	0.219	0.216
Second date image	0.140	0.031	0.213	0.213
Two dates image	0.134	0.033	0.197	0.197

D.1.5   Concluding remarks

Results from this study highlight the importance of multisensor modeling in estimating land cover proportions. This is especially critical in large regions where only coarse scale sensor data is available and there is a need to estimate the land cover proportions. This chapter presents a framework for multisensor fusion using an ARTMAP neural network to extract sub-pixel information from coarser resolution imagery for a region of North Central Turkey. Results in terms of analysis of errors and land cover maps demonstrate the utility of the ARTMAP neural network in this context.

The contribution shows that the ARTMAP models are better at estimating land cover class proportions compared with the traditional linear mixture models. They have the potential to predict accurately land cover patterns at the sub-pixel scale. Results in terms of RMS error of both the simulated MODIS data and the real MODIS data using the ARTMAP model are better compared with the linear mixture models. The advantages of the ARTMAP model in this context may be due to its ability to model non-linear relationships and it isn't dependent on the availability of 'end-member'. However, it does require enough high quality training samples containing sub-pixel information in order to accurately predict sub-pixel fractions. The linear mixture models do not require such many training data as the ARTMAP model, but do require an accurate identification of pure 'end-members'. In many contexts, it is difficult to identify pure 'end-member' pixels in the area of interest; in addition, what is pure in a given situation maybe not be so in other regions. The performance of the linear mixture model can be improved if there is a better algorithm for selecting pure 'end-member'. In addition, non-linear mixture models and multiple end members may lead to significant improvements in accuracy.

A major requirement in multisensor framework is the accurate registration of fine and coarse resolution images. The ARTMAP model shows poor performance using the real MODIS data compared with the simulated MODIS data obtained from aggregating Landsat data to MODIS scale may stem from different reasons. The error caused by mis-registration may be one important factor. Geometric registration can be done by a number of methods including matching of image features, correlation in the spatial or frequency domain and phase estimation in the frequency domain. Eastman and Moigne (2001) present a comparison of various gradient descent techniques for image registration and discussed the advantages and disadvantages of such methods. In future research, better image registration method can be used to improve the performance of the ARTMAP model. The linear mixture models offer an advantage in this regard since they do not need image registration. However, this advantage is offset by a need to correctly select endmember spectra for each desired class.

This chapter also demonstrates the performance differences using single date and two dates image. While there is no significant difference using one data or two date imagery for the testing data, there is a difference for the whole image. Performance of the ARTMAP model is improved for certain classes using two dates image. Multidate imagery may be useful when land cover changes seasonally and this change will be useful to distinguish one class from another although the spectral value of two classes are similar in one season.

Acknowledgements. This research project was supported by the National Science Foundation Grants, SES-9973474 and BCS-0079077. We also would like to thank Mutlu Ozdogan of Department of Geography, Boston University for preparing the MODIS data set.

References

Adams J, Sabol D, Kapos V (1995) Classification of multispectral images based on fraction endmembers application to land cover change in the Brazilian Amazon. Remote Sens Environm 52:137-154

Atkinson PM, Cutler M, Lewis H (1997) Mapping sub-pixel proportional land cover with AVHRR imagery. Int J Remote Sens 18(4):917-935

Barket J, Markham B, Burelcach J (1992) MODIS image simulation from Landsat TM imagery In Proceedings of the Global Change and Education ASPRS/ACSM/RT 92, American Society for Photogrammetry and Remote Sensing, Bathesda [MD], pp.156-165

Bateson A, Curtiss B (1996) A method for manual endmember selection and spectal unmixting. Remote Sens Environm 55(3): 229-243

Brown M, Gunn SR, Lewis HG (1999) Support vector machines for optimal classification and spectral unmixing. Ecol Mod 120 (2-3):167-179

Carpenter G, Gopal S, MacomberS, Martens S and Woodcock C (1999) A neural network method for mixture estimation for vegetation mapping. Remote Sensing of Environment 70:138-152

Carpenter G, Grossberg S, Markuzon S, Martens N, Reynolds J, Rosen D (1992) Fuzzy ARTMAP: a neural network architecture for incremental supervised learning of analog multidimensional maps. IEEE Transact Neural Netw 3(5):698-713

DeFries R, Hansen M, Townshend J (1995) Global discrimination of land cover types from metrics derived from AVHRR pathfinder data. Remote Sens Environm 54(3):209-222

DeFries R, Townshend J, Hansen M (1999) Continuous fields of vegetation characteristics at the global scale at 1-km resolution. J Geophys Res 104:16911-16923

Dennison PE, Roberts DA (2003) The effects of vegetation phenology on endmember selection and species mapping in southern California chaparral. Remote Sens Environm 87(2-3):295-309

Eastman JR, Laney RM (2002) Bayesian soft classification for sub-pixel analysis: a critical evaluation. Photog Eng Remote Sens 68(11)1149-1154

Eastman JR, Moigne J (2001) Gradient descent technique for multitemporal and multisensor image registration of remotely sensed imagery. In Proceedings of the 4th Annual Conference on Information Fusion, Montreal Quebec, Canada

Foody G (1996) Approaches for the production and evaluation of fuzzy land cover classifications from remotely-sensed data. Int J Remote Sens 17(7):1317-1340

Foody G (1998) Sharpening fuzzy classification output to refine the representation of subpixel land cover distribution. Int J Remote Sens 19:2593-2599

Foody G, Campbell N, Trodd N, Wood T (1992) Derivation and applications of probabilistic measures of class membership from maximum-likelihood classification. Photog Eng Remote Sens 58(9):1335-1341

Foody G, Lucas R, Curran P, Honzak M (1997) Non-linear mixture modeling without endmembers using an artificial neural network. Int J Remote Sens 18(4):937-953

Friedl MA, McIver DK, Hodges JCF, Zhang XY, Muchoney D, Strahler AH, Woodcock CE, Gopal S, Schneider A, Cooper A, Baccini A, Gao F, Schaaf C (2002) Global land cover mapping from MODIS: algorithms and early results. Remote Sens Environm 83(1-2):287-302

Grossberg S (1976) Adaptive pattern classification and universal recoding. I parallel development and coding of neural feature detectors. Biol Cybern 23(3):121-134

Huang C, Townshend JRG (2003) A stepwise regression tree for non-linear approximation: applications to estimating sub-pixel land cover. Int J Remote Sens 24 (1):75-90

Ju J, Kolaczyk E, Gopal S (2003) Gaussian mixture discriminant analysis and sub-pixel land cover characterization in remote sensing. Remote Sens Environm 84(9):550-560

Knyazikhin Y, Martonchik J, Myneni D, Verstraeter M, Pinty B, Gobron N (1998) Estimation of vegetation canopy leaf area index and fraction of absorbed photosynthetically active radiation from atmosphere-corrected MISR data. J Geophys Res 103: 32239-32256

Liu W, Gopal S, Woodcock C (2001) Spatial data mining for classification visualization and interpretation with ARTMAP neural network. In Grossman R (ed) Data mining for scientific and engineering applications. Kluwer, Dordrecht, pp.205-222

Liu W, Seto K, Wu E, Gopal S, Woodcock C (2004) ART-MMAP: a neural network approach to sub-pixel classification. IEEE Transact Geosci Remote Sens 42(9):1976-1983

Loveland TR, Reed BC, Brown JF, Ohlen DO, Zhu Z, Yang L, Merchant JW (2000) Development of a global land cover characteristics database and IGBP DISCover from 1 km AVHRR data. Int J Remote Sens 1 (6-7):1303-1330

Mayaux P, Lambin E (1995) Estimation of tropical forest area from coarse spatial resolution data a two-step correction function for proportional errors due to spatial aggregation. Remote Sens Environm 53(1):1-15

McIver D (2001) Adapting machine learning approaches for coarser resolution land cover classification. PhD dissertation, Boston University

Moody A (1998) Using landscape spatial relationships to improve estimates of land cover area from coarse resolution remote sensing. Remote Sens Environm 64(2):202-220

Moody A, Woodcock C (1994) Scale-dependent errors in the estimation of land cover proportions implications for global land cover data sets. Photog Eng Remote Sens 60(5):585-594

Moody A, Woodcock C (1996) Calibration-based models for correction of area estimates derived from coarse resolution land cover data. Remote Sens Environm 58(3):225-241

Quarmby NA, Townshend JRG, Settle JJ, White KH, Milnes M, Hindle TL, Silleos N (1992) Linear mixture modeling applied to AVHRR data for crop area estimation. Int J Remote Sens13(3):415-425

Roberts DA, Smith M, Adams J (1993) Green vegetation non-photosynthetic vegetation and soils in AVIRIS data. Remote Sens Environm 44(2-3):255-269

Roberts DA, Gardner M, Church R, Ustin S, Scheer G, Green R (1998) Mapping chaparral in the Santa Monica mountains using multiple endmember spectral mixture models. Remote Sens Environm 65(3):267-279

Schowengerdt R (1996) On the estimation of spatial-spectral mixing with classifier likelihood functions. Patt Recogn Letter 17(13):1379-1387

Sellers P, Mintx Y, Sud Y, Dalcher A (1986) A simple biosphere model (SiB) for use within general circulation models. J Atmosph Sci 43(6):505-531.

Tian Y, Knyaikhin Y, Myneni R, Glassy J, Dedieu G, Running S (2000) Prototyping of MODIS LAI and FPAR algorithms with LASUR and Landsat data. IEEE Transact Geosci Remote Sens 38(4):1-15

Zadeh LA (1965) Fuzzy sets. Inf Contr 8:338-353

D.2 Model Selection in Markov Random Fields for High Spatial Resolution Hyperspectral Data

Francesco Lagona

D.2.1 Introduction

The statistical analysis of high spatial resolution hyperspectral (HSRH) data is a challenging issue in remote sensing because conventional methods may be unmanageable, in terms of computational time, due to the large amount of data used to characterize hyperspectral images. This chapter focuses on efficient methods of model selection in the analysis of HSRH data.

Hyperspectral remote sensing (see, for example, Landgrebe 2000) provides a stack of images of a scene acquired in contiguous bands over spectral range. This is often referred to as an 'image cube', because data can be arranged in a three-dimensional array, where two spatial dimensions give the coordinates of each pixel and the third dimension is wavelength. The general aim in analysis is to assess the local distribution of properties of interest in the image. Major steps in analysis include image restoration, segmentation and classification. Stochastic restoration (Geman and Geman 1984) is concerned with the estimation of the 'true' image from data that are known to have been contaminated by noise. Segmentation (Zhang et al. 1990) is concerned with the partitioning of the map into homogeneous regions (e.g., land cover types, morphological units, rivers, buildings, blocks). Classification (Hoffbeck and Landgrebe 1996) is concerned with assigning classes (e.g., water, bare soil, grass) to pixels (and pixels to classes).

These operations are usually addressed by automatic or semi-automatic algorithms that have the remotely sensed image as an input and provide, as an output, information from the image which can be used to refine maps or other spatial representations. Depending on the format of the output and the level of confidence in the results (accuracy), image-derived information can be directly used to refine maps with a Geographic Information System (GIS).

In practice, algorithms for image analysis may either use training regions of the image where ground data are available (supervised methodology), or they enter no field information and use an unsupervised classification method followed by use of ground survey data to validate the results. Since image data take the form of a multidimensional spatial series, recorded at the vertices (pixels) of a regular square lattice, standard multivariate techniques (e.g., principal component analysis, canonical correlations, multivariate linear and generalized linear models) could be implemented for handling these data and addressing restoration, segmentation and classification. Unfortunately, though, spatial data are not independent, since data at neighboring pixels show similar values, and hence these procedures need to be corrected to account for the natural spatial dependence between observations (Amrhein and Griffith 1997; Wackernagel 1998). Additionally, HSRH datasets are very large, since they are generally composed of about 100–200 spectral bands, whereas the number of bands in a multispectral image is typically about 5–10, and the high spatial resolution means that values are recorded at a resolution of 1m x 1m to 5m x 5m. As a result, the computational complexity of both statistical models and estimation procedures need to be taken into account at each stage of the analysis.

A powerful approach to modeling spatial lattice data is provided by Markov Random Field (MRF) models, introduced by Besag (1974). MRFs have been used in the past for modeling and classifying textures (Cross and Jain 1983). Lately there have been approaches to model whole images with MRFs (Geman and Geman 1984; Kelly et al. 1988; Derin and Howard 1987; Rignot and Chellappa 1992; Smits and Dellepiane 1997). Successful applications include the stochastic restoration of multispectral images (Bennett and Khotanzad 1998), the spatial distribution of sedimentary facies in petroleum reservoirs (Tjelmeland 1996), the classification of coastal environments using synthetic aperture radar (SAR) data (Crawford and Ricard 1998), and unsupervised classification for multi-sensor data (Lee and Crawford 1999).

A major reason for the increasing role played by these models in remote sensing analysis is that they make it easy to incorporate spatial constraints such as region continuity of the spatial scene, by modeling the global (complex) dependence structure between observations through the local (simple) conditional distribution of each observation given the rest of the dataset. What makes the functional form of the conditional distributions simple is the Markov property: a neighborhood is defined around each pixel and the conditional probability that each pixel belongs to a given color (class) depends only on the data at the neighboring pixels. As a result, while MRF distributions were first used in the statistical mechanics literature for modeling molecular interactions in ferromagnetic materials, they allow the development of computationally traceable and robust processing algorithms for what are basically 'noncausal' image models.

Unfortunately, the neighborhood system needs to be elicited *a priori*, but what often happens is that several different neighborhood structures seem viable. Since the definition of an erroneous adjacency can lead to poor quality estimates (Griffith and Lagona 1998) a preliminary screening between candidate models is needed.

As it is well known, model selection is related to a number of trade-offs. *Ceteris paribus*, the efficiency of any estimation procedure increases with the number of the degrees of freedom but a too simple model may give a poor fit of the data. On the other hand, a fully parameterized model is often capable of fitting the data perfectly, but the stability of the estimates is not guaranteed and the chance of a poor data summary is an issue to be addressed. Generally, a parsimonious model is preferable if data are fitted reasonably. Hence, the choice between a model with p parameters and one with $q < p$ parameters is usually driven by both the definition of a measure of their discrepancy (in terms of goodness of fit) and the evaluation of how discrepant they are, because while a small discrepancy might be tolerable a large discrepancy is not.

Limiting the discussion to the works that are directly applicable to HSRH data, a number of discrepancies for model selection are currently available in the statistical literature, including identification criteria (Kashyap and Chellappa 1983; Ji and Seymour 1998; Jona Lasinio and Lagona 2003), the Coding Ratio (Besag 1974; Guyon and Hardoum 1992), the Likelihood Ratio (Guyon 1995) and the Pseudo-Likelihood Ratio (PLR, Lagona 2001) tests. The asymptotic distribution of both the Coding Ratio (CR) and the Likelihood Ratio (LR) tests is known to be the usual Chi Square with $p - q$ degrees of freedom, under standard regularity conditions. However, the Coding Ratio is computed by evaluating the conditional distributions of the field at those observation points belonging to the so-called 'coding set', i.e. a subset D of the observation domain that is constructed in a way that, given any pair of data observed at two pixels belonging to D, these are conditionally independent of each other, given the rest of the image. Hence, CRs can be easily constructed only when the random field is Markovian with respect to a simple neighborhood system and, even in this case, many coding tests can be defined as ways of coding the observation domain. These tests are correlated and it is not generally possible to construct global statistics that summarizes them (Guyon 1995).

On the other hand, the PLR test converges in distribution to a weighted sum of $(p - q)$ Chi Square with one degree of freedom, but in the case of massive datasets (e.g., HSRH data) significant computational difficulties arise for estimating these weights (Lagona 2001). These limitations would lead to consideration of the LR as the method of choice. However, there are a number of computational difficulties with LR, too. Since LR is based on the direct maximization of the likelihood function, a major issue is the presence of a distribution normalizing constant that is intractable (both analytically and numerically) as the sample size increases. To obtain an approximate value of the LR statistic, an MCMC (Markov Chain Monte

Carlo) approximant to the likelihood function (MCMC-MLE, Geyer and Thompson 1992) needs to be constructed.

However, in an HSRH setting, MCMC-MLE can be heavily time consuming for at least two reasons. First, simulation of a multivariate random field through MCMC is not easy in many cases, because parameter estimates are often negatively correlated in real case studies, which leads to an extremely slow mixing (Knorr-Held and Besag 1997). Second, MCMC-MLE requires an initial guess of the parameter value, and its efficiency depends on how close this is to the 'true' value, hence in case of little information about the data generating process, the procedure needs to be repeated for a grid of candidate values to avoid convergence to local maxima.

Finally, ratio tests allow comparison of two different models at a time, this being time consuming when selection is from a battery of alternate models. On the other hand, identification criteria permit selection among several models. These criteria are based on the optimization of either a likelihood or a quasi-likelihood function, coupled with a penalization rate to account for model parsimony. The two most popular penalization rates are referred to as the AIC (Akaike Information Criterion, Akaike 1974) and the BIC (Bayesian Identification Criterion, Schwartz 1978). The basic statistical property required of these procedures is consistency, i.e. the probability that the criterion chooses the wrong model must tend to 0 as the sample size (e.g., the number of pixels of an image) increases, which is a nice property in the case of analysis of massive datasets (e.g., HSRH data).

While AIC is not consistent for spatially dependent data (Kashyap and Chellappa 1983), the BIC-penalized likelihood criterion (BIC-ML, Ji and Seymour 1998) is consistent under very general conditions. Unfortunately, though, the latter procedure relies upon the direct maximization of the likelihood function and the aforementioned issues apply. Only in the case of Gaussian MRFs (e.g., spatial autoregressive models), is the normalizing constant analytically tractable and the penalized likelihood can be evaluated when the image dimension is reasonable (Kashyap and Chellappa 1983): numerical approximations are however necessary for handling high resolution datasets (Griffith 2002). Such computational difficulties can be avoided by using a BIC-penalized pseudo-likelihood (BIC-PL) function. In small samples the performance of the resulting criterion is worse than that of its likelihood counterpart, since the Maximum Pseudo-Likelihood Estimator (MPLE) can overestimate the parameters, but the penalized PL criterion is consistent under very general conditions and can be applied easily even in the framework of multivariate, space-time data series (Jona Lasinio and Lagona 2003).

In this chapter, implementation of MRFs in image restoration, segmentation and classification is briefly reviewed, highlighting the crucial role played by the adjacency definition. Then the practical implementation of the BIC-PL criterion for selecting adjacency is illustrated. Finally, the performance of the BIC-PL procedure and the BIC-ML criterion are compared for detecting spatial structures in a high spatial resolution hyperspectral image for the Lamar area in Yellowstone National Park.

D.2.2 Restoration, segmentation and classification of HSRH images

A HSRH dataset can be viewed as an N-by-B rectangular matrix Y, whose generic element, say y_{ib}, is the wavelength of the bth spectral band, recorded at pixel i. Each row y_i of this matrix is referred to as the hyperspectral signature of the ith pixel. The bth column of Y, say y_b, is the high resolution image of band b. For example, the image considered in the application (Section D.2.4) is characterized by $B = 128$ bands recorded in a lattice of $N=350\times450$ pixels. Classically, the general aim of image analysis is to couple prior field information with data, in order to update each observation y_{ib}, by estimating NB parameters θ_{ib}, and hence obtaining a new N-by-B matrix Θ. Each column θ_b of Θ is usually referred to as *a posteriori* image.

In image restoration, for example, each image y_b is corrupted by noise and θ_b represents the 'true image': here, the prior information is that the components of each θ_b are spatially dependent. In image segmentation, instead, both the wavelength given by each y_b and the true image given by θ_b are discretized in, say, K colors (or grey levels) and the prior information is that the K-colored regions associated to each possible realization of the true image θ_b are geometrically regular, i.e. they exhibit a specific shape. Finally, the prior information in image classification is that there is a number, say H, of classes to which pixels belong; accordingly, each θ_{ib} assumes one of the H values relating to the different classes and image updating reduces to assigning each pixel to its own class.

In this framework, MRFs may easily model prior information. To build a MRF model, we first need the definition of a neighborhood structure between pixels. More precisely, an N-by-N connectivity matrix C needs to be constructed in a way that its (i, j)-th entry is $c_{ij} = 1$ if pixels i and j are neighbors and zero otherwise. As a result, C is an N-by-N, symmetric, binary matrix where diagonal entries are equal to zero. Secondly, we assume that each vector θ_b is sampled from the distribution

$$P(\theta_b) = p(\mathbf{0}) \exp \Phi(\theta_b, C) \qquad (D.2.1)$$

where $p(\mathbf{0}) = p(0, ..., 0)$ is a normalizing constant and

$$\Phi(\theta_b, C) = \sum_{i=1}^{N} \varphi_i(\theta_{ib}) + \sum_{i=1}^{N} \sum_{j:c_{ij}=1} \varphi_{ij}(\theta_{ib}, \theta_{jb}). \qquad (D.2.2)$$

Notice that the functions φ_i and φ_{ij} are log-odds ratios of spatial configurations

$$\phi_i(\theta_{ib}) = \log \frac{p(0, ..., 0, \theta_{ib}, 0, ..., 0)}{p(0)} \quad \text{(D.2.3a)}$$

$$\phi_{ij}(\theta_{ib}, \theta_{jb}) = \log \frac{p(0, ..., 0, \theta_{ib}, 0, ..., 0, \theta_{jb}, 0, ..., 0) p(0)}{p(0, ..., 0, \theta_{ib}, 0, ..., 0) p(0, ..., 0, \theta_{jb}, 0, ..., 0)} . \quad \text{(D.2.3b)}$$

Thus, for example, if $\varphi_{ij} = 0$, identically, then Eq. (D.2.1) reduces to the distribution of a random vector with independent (not necessarily identically distributed) components. If, additionally, $\varphi_i(\theta_{ib}) = \varphi_b(\theta)$ for each pixel i, then θ_b is a simple random sample drawn from one random variable Y, with distribution $p(\theta) = p(0)$ exp $[\varphi_b(\theta)]$. In other words, a unit increase in the function φ_{ij} corresponds to an increase of the log odds of observing the value θ_{ib} to a different value at pixel i, given the values at the neighboring pixels. Moreover, the Markov property of model in Eq. (D.2.1) follows from the computation of the conditional distribution of each value θ_{ib}, given the rest of the prior image, that takes the following form:

$$p_i(\theta_{ib} \mid \theta_{jb}, j \neq i) = p_i(0 \mid \theta_{jb}, j \neq i) \exp\left(\varphi_i(\theta_{ib}) + \sum_{j: c_{ij}=1} \varphi_{ij}(\theta_{ib}, \theta_{jb}) \right). \quad \text{(D.2.4)}$$

In Eq. (D.2.4), the pixel wise conditional distributions do not depend on the values attained by the wavelengths observed outside the neighborhood of each pixel, as defined by matrix C, i.e. model in Eq. (D.2.1) is Markov with respect to the neighborhood structure specified by C.

D.2.3 Adjacency selection in Markov random fields

Depending on the issue to be addressed, parametric MRF models can be elicited by defining a suitable neighborhood structure and assuming that the functions φ in Eq. (D.2.2) are known up to a vector of parameters, say $\beta = (\beta_1, ..., \beta_K)$, to be estimated. As mentioned in Section D.2.1, maximum likelihood (ML) is not the standard method of parameter estimation, because the normalizing constant $p(0)$ in Eq. (D.2.1) becomes unmanageable, both numerically and analytically, as the number of pixels in an image increases. As a result, a number of viable alternative estimation methods have been suggested, using numerically tractable objective functions. In summary, these procedures yield estimators that are less efficient with respect to ML, although they are consistent and asymptotically normal (see Guyon 1995, for a discussion on these issues).

In this chapter, we concentrate on the Besag's (1974) Maximum Pseudo-likelihood Estimator (MPLE), which is defined as

$$\hat{\beta} = \arg\max_{\beta} U_b(\beta) \qquad (D.2.5a)$$

$$U_b(\beta) = \log \prod_{i=1}^{N} p_i(\theta_{ib} \mid \theta_{jb}, j \neq i) = \sum_{i=1}^{N} \log p_i(\theta_{ib} \mid \theta_{jb}, j \neq i) \qquad (D.2.5b)$$

where $U_b(\beta)$ is the pseudo-likelihood (PL) function. Under the specification (D.2.2), the PL function is concave and, as a result, the MPLE is unique. This estimator is moreover consistent and (asymptotically) normally distributed, under very general conditions, although less efficient than its ML counterpart (Guyon 1995).

The computational efficiency of MPLE relies on the form taken by the PL function. Under the specification given by Eq. (D.2.2), optimization of the pseudo-likelihood function reduces to fitting a generalized linear model and can be hence implemented by using conventional software packages, such as R, STATA, SAS or GAUSS. To illustrate, we concentrate on the two examples that will be considered in the application of Section D.2.4.

If the *a priori* image is binary (i.e., θ_{ib} is equal to either zero or one), and isotropic conditional distributions are assumed [(i.e. $\varphi_i(\theta_{ib}) = \beta_1 \theta_{ib}$ and $\varphi_{ij}(\theta_{ib}, \theta_{jb}) = \beta_2 \theta_{ib}\theta_{jb}$)], then Eq. (D.2.1) reduces to the popular spatial auto-logistic model (Besag 1974), as follows:

$$p(\theta_b) = p(\mathbf{0}) \exp\left(\sum_{i=1}^{N}\left(\beta_1 \theta_{ib} + \beta_2 \theta_{ib} \sum_{j:c_{ij}=1} \theta_{jb}\right)\right). \qquad (D.2.6)$$

Under (D.2.6), the conditional distributions at each pixel are given by

$$p_i(\theta_{ib} \mid \theta_{jb}, j \neq i) = \frac{\exp\left(\beta_1 \theta_{ib} + \beta_2 \theta_{ib} \sum_{j:c_{ij}=1} \theta_{jb}\right)}{1 + \exp\left(\beta_1 + \beta_2 \sum_{j:c_{ij}=1} \theta_{jb}\right)} \qquad (D.2.7)$$

motivating the name of the model. For each hyperspectral band b, the PL function of an isotropic auto-logistic model can be written as

$$U_b(\beta_1,\beta_2) = \sum_{i=1}^{N} \log\left[1 + \exp(\beta_1 + \beta_2 c_i \theta_b) - \beta_1 t_1(\theta_b) + 2\beta_1 t_2(\theta_b)\right] \qquad \text{(D.2.8)}$$

where

$$t_1(\theta_b) = \sum_{i=1}^{N} \theta_{ib} \qquad \text{(D.2.9a)}$$

$$t_1(\theta_b) = \tfrac{1}{2} \sum_{i=1}^{N} \theta_{ib} \, c_i^{\mathrm{T}} \, \theta_b. \qquad \text{(D.2.9b)}$$

Function (D.2.8) therefore resembles the log-likelihood function of a logistic regression with intercept β_1 and regression coefficient β_2. Therefore, the MPLEs of β_1 and β_2 can be computed by fitting a logistic regression, using θ_b as a vector of N 'response' observations and the number of neighbors of each pixel, $x_i = \Sigma_j c_{ij}$, as a covariate.

Alternatively, if the *a priori* image is polychromous (i.e., each θ_{ib} takes values in a set of $K+1$ grey levels), a spatial auto-binomial model (Besag 1974) can be specified as follows

$$p(\theta_b) = p(0) \, \exp\left(\sum_{i=1}^{N} \left(\beta_1 \theta_{ib} + \log\binom{K}{\theta_{ib}} + \beta_2 \theta_{ib} \sum_{j:c_{ij}=1} \theta_{jb} \right) \right). \qquad \text{(D.2.10)}$$

Notice that Eq. (D.2.10) can be derived from Eq. (D.2.2) by choosing

$$\varphi_i(\theta_{ib}) = \beta_1 \theta_{ib} + \log\binom{K}{\theta_{ib}} \qquad \text{(D.2.11)}$$

and $\varphi_{ij}(\theta_{ib},\theta_{jb}) = \beta_2 \, \theta_{ib}\theta_{jb}$. In the case of an auto-binomial model, therefore, the pixel wise conditional distributions are given by

$$p_i(\theta_{ib} \mid \theta_{jb}, j \neq i) = \binom{K}{\theta_{ib}} \pi_\beta^{\theta_{ib}} (1 - \pi_\beta)^{K-\theta_{ib}} \quad \text{(D.2.12)}$$

where

$$\pi_\beta = \frac{\exp\left(\beta_1 \theta_{ib} + \beta_2 \theta_{ib} \sum_{j:c_{ij}=1} \theta_{jb}\right)}{1 + \exp\left(\beta_1 + \beta_2 \sum_{j:c_{ij}=1} \theta_{jb}\right)}. \quad \text{(D.2.13)}$$

Accordingly, the MPLEs of β_1 and β_2 can be computed by fitting a binomial regression (with logit link), where β_1 and β_2 are respectively the intercept and the regression coefficient, by using θ_b as a vector of N 'response' observations and the number of neighbors of each pixel, $x_i = \Sigma_j c_{ij}$, as a covariate.

Specification of MRFs is based on the definition of a connectivity matrix C that characterizes the neighborhood structure among image pixels. When different adjacency relationships seem viable, a natural way to handle the specification problem is through the definition of a sequence of non-overlapping neighborhood structures. In other words, we suggest to specify a sequence $C_1, C_2, \ldots, C_H$ of N-by-N connectivity matrices, defined in a way that the corresponding entries $_hc_{ij}\ _kc_{ij} = 0$, as $h \neq k$, $_hc_{ij}$ being the (i,j)-th entry of the hth matrix of the sequence. Under this setting, we obtain a sequence of candidate models

$$p_h(\theta_b) = p(0) \exp\left(\sum_{i=1}^N \phi_i(\theta_{ib}, \beta) + \sum_{h=1}^H \left(\sum_{j \atop c_{ij}=1} \phi_h(\theta_{ib}, \theta_{jb}, \beta_h)\right)\right). \quad \text{(D.2.14)}$$

To choose among the H candidate models, we suggest selecting the model that minimizes the BIC-penalized Pseudo-likelihood (BIC-PL) function

$$U_b(\hat{\beta}_h) + \frac{\log N}{N} |\beta_h| \quad \text{(D.2.15)}$$

where U_b is the PL function evaluated for the ith candidate and $|\beta_h|$ is the number of unknown parameters to be estimated in model h. This criterion chooses the most likely model (with respect to the pseudo-likelihood function), taking into account the parsimony principle, i.e. the model the smallest number of independent parameters is preferred, among equally likely models. The statistical properties of the BIC-PL criterion have been studied by Jona Lasinio and Lagona (2003) who proved its consistency under very general conditions and showed its good performance even in small images (10m×10m), through an extensive simulation study.

D.2.4 A study of adjacency selection from hyperspectral data

To illustrate the BIC-PL selection criterion on a real dataset, we used data collected as part of a Yellowstone Ecosystem Studies project funded through the NASA EOCAP (Earth Observing Commercial Applications Program) program of Stennis Space Flight Center, Mississippi. The Yellowstone HSRH data were collected in August of 1999 using procedures described in detail by Marcus et al. (2000). The area of Lamar River within Yellowstone National Park was captured during clear weather conditions and 128 hyperspectral bands were recorded on 350×450 pixels, with a 5m spatial resolution. Additional details on these data can be found in Jacquez et al. (2002).

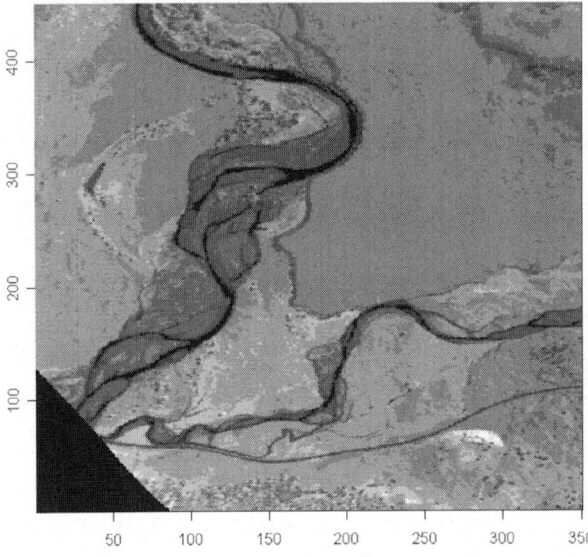

Fig. D.2.1. Grey levels image of the first principal scores of the Lamar imagery

Figure D.2.1 shows the image of the first principal scores that capture the 71 percent of the total variance in the data, as obtained after running an ordinary principal component analysis on the whole HSRH data. For illustration purposes, only bands 7, 19, 24, 55, 62, 65, 66, 79, 109 have been considered for analysis. These bands have been selected as representatives of a number of clusters appearing after mapping bands on the correlations space spanned by the first two principal components of the full Lamar imagery (see Fig. D.2.2).

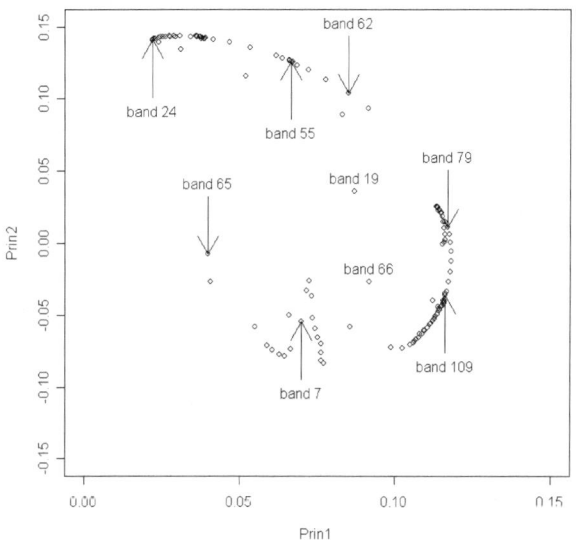

Fig. D.2.2. Lamar imagery scatter plot of the 128 bands mapped onto the correlation space spanned by the first two principal components and representative bands selected for subsequent analysis

High resolution images have been fitted by a number of spatial statistical models. Specifically, each band-specific image has been first transformed to a binary image by using a median cut-off, and fitted by a class of spatial auto-logistic models. In practical applications, a different cut-off might be chosen, based on field information, in a way that wavelength is transformed to one if it belongs to a range of interest. Second, the range of each band has been divided in order to construct four grey levels, according to the three cut-offs specified by the three quartiles of the image distribution, and fitted by a class of spatial auto-binomial models. In case of prior field information, each grey level would correspond to a wavelength class of interest.

Both the auto-logistic and the auto-binomial models were chosen by combining different adjacency definitions (Table D.2.1) and parametric constraints (Table D.2.2). In particular, Table D.2.1 shows the four connectivity matrices that have been considered for analysis: C_1 and C_2 respectively relate to a Bishop- and a

Table D.2.1. Models used in the application

Connectivity matrix	$c_{ij=1}$ if pixel j has coordinates	Number of neighbors
C_1	$(x_i \pm 1, y_i \pm 1)$	4
C_2	$(x_i, y_i \pm 1)$ or $(x_i \pm 1, y_i)$	4
C_3	$(x_i \pm l, y_i \pm m), 2 \leq l+m \leq 4$	16
C_4	$(x_i \pm l, y_i \pm m), 3 \leq l+m \leq 6$	24

Rook-type first-order adjacency structure, while C_3 and C_4 specify a second-order and third-order adjacency relationship, respectively. Matrices C_1 and C_2 have been exploited to specify four different specifications (m1 to m4), according to different parametric constraints. Including matrix C_3 allows for the specification of m5 and m6. Finally, specifications m7 to m9 is defined by combining all the four connectivity matrices of Table D.2.1. We thus obtain nine different auto-logistic models, namely

$$p_h(\theta_b) = p(0)\ \exp\left(\sum_{i=1}^{N}\theta_{ib}\left(\alpha + \sum_{h=1}^{4}\beta_h \sum_{j:\ _hc_{ij}=1}\theta_{jb}\right)\right) \tag{D.2.16}$$

and nine different auto-binomial models, namely

$$p_h(\theta_b) = p(0)\ \exp\left(\sum_{i=1}^{N}\alpha\theta_{ib} + \log\binom{4}{\theta_{ib}} + \sum_{h=1}^{4}\beta_h\,\theta_{ib}\sum_{j:\ _hc_{ij}=1}\theta_{jb}\right). \tag{D.2.17}$$

In Eqs. (D.2.16) and (D.2.17), $_hc_{ij}$ is the (i,j)-th entry of the hth connectivity matrix of Table D.2.1 ($h=1,2,3,4$), while parameters β_1, β_2, β_3, β_4 are constrained according to Table D.2.2. Table D.2.2 also displays the number of independent parameters (model dimension) to be estimated under each specification, i.e. the difference between the number of parameters involved and the number of constraints. Figures D.2.3 to D.2.5 provide a graphical representation of the specifications used for analysis.

Table D.2.2. Parametric specifications used in the application

Model name	Parameter constraints	Model dimension
m1	$\beta_1 = \beta_3 = \beta_4 = 0$	1
m2	$\beta_2 = \beta_3 = \beta_4 = 0$	1
m3	$\beta_1 = \beta_2, \beta_3 = \beta_4 = 0$	1
m4	$\beta_3 = \beta_4 = 0$	2
m5	$\beta_1 = \beta_2 = \beta_3, \beta_4 = 0$	1
m6	$\beta_1 = \beta_2, \beta_4 = 0$	2
m7	$\beta_1 = \beta_2 = \beta_3 = \beta_4$	1
m8	$\beta_1 = \beta_2, \beta_3 = \beta_4$	2
m9	$\beta_1 = \beta_2$	3

In particular, Fig. D.2.3 shows the first order adjacencies upon which specifications m1-m4 are based. Models m1 and m2 assume that the conditional distribution at each pixel is driven by the four nearest neighbors with respect to two different directions. Models m3 and m4 use Queen's case adjacency (also known as the full adjacency in the engineering literature; Rose and Devijver 1984). However, m4 is the anisotropic version of m3.

Figure D.2.4 illustrates both the isotropic (m5) and the anisotropic (m6) versions of a second order neighborhood structure. Finally, Fig. D.2.5 shows the isotropic (m7) and two anisotropic versions (m8 and m9) of third order adjacencies. These second-order and third-order adjacency definitions have been often considered in the statistical analysis of images (Tjelmeland and Besag 1998).

The BIC-PL criterion has been evaluated by considering the product of 147,238 conditional distributions at the pixels belonging to an interior domain resulting after constructing a 3-pixels-wide guard, or buffer, area for the 350×450 pixels image of each band. For comparison purposes, the BIC-ML criterion mentioned in the Introduction was also been evaluated, by simulating a MCMC via block Metropolis steps to improve the mixing rate of the chain, as suggested by Smith and Roberts (1993). A Pentium 800 PC worked for several hours to get BIC-ML results for one band, while BIC-PL results were obtained in a few minutes.

Figures D.2.6 and D.2.7 display the results after fitting the auto-logistic and auto-binomial models to the data. For binary images (Fig. D.2.6), m4 is always the model chosen. This result is in keeping with the remote sensing literature that suggests first order, full adjacency as the neighborhood structure of choice for binary images. On the other hand, m4 is preferred to m3 because of the high level of spatial heterogeneity (e.g., anisotropy) in the full image. Most of the time, the BIC-PL criterion agreed with its pseudo-likelihood counterpart, since the dataset under study is massive and both the procedures are consistent, i.e. they converge to the same model as the sample size increases. This result supports the idea that the pseudo-likelihood approach is to be pursued for massive datasets, since the evaluation of the likelihood function through MCMC was extremely time consuming when implemented for the Lamar imagery, still giving outcomes similar to that given by BIC-PL. Finally, even when the BIC-PL criterion disagreed with the BIC-ML, model m3 was chosen (an exception was band 7), i.e. the isotropic version of the model produced using the pseudo-likelihood approach. This is in line with the known (Geyer and Thompson 1992) behavior of MPLE that tends to overestimate spatial interaction parameters, hence preferring models that are less parsimonious than those chosen by the direct likelihood approach.

For grey levels images, the two criteria were still concordant in most cases although m4 was not always the model chosen. This is as expected, since data disaggregation (from two to four categories) increases the spatial heterogeneity of the Lamar scene and the specific spatial distribution of each band emerges. In other words, bands do not exhibit the same spatial structure and, while this phenomenon

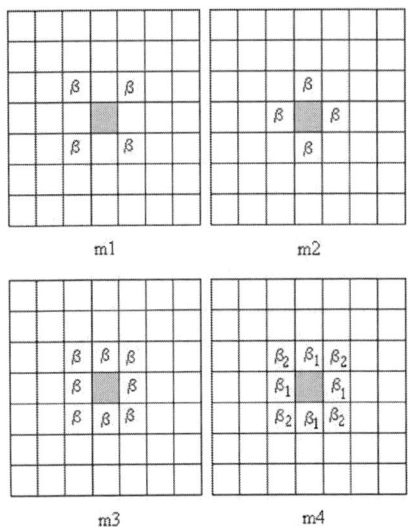

Fig. D.2.3. First-order lattice adjacencies

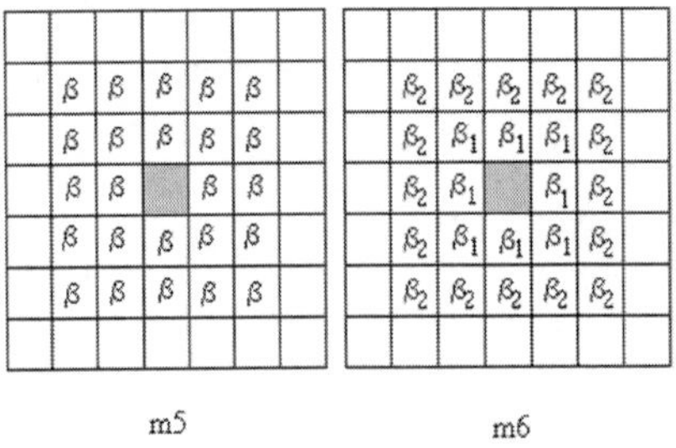

Fig. D.2.4. Second-order lattice adjacencies

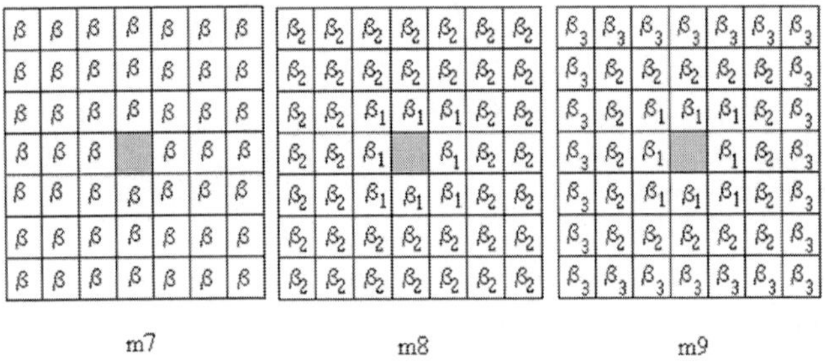

Fig. D.2.5. Third-order lattice adjacencies

was hidden by the binary transformation, the 'best' adjacency definition for the full image simply does not exist and selection from a battery of adjacencies needs to be addressed, especially for multicolored images. Of note, furthermore, is that even when the two criteria disagree, the models chosen are still mostly the isotropic/anisotropic versions of each other. In summary, both the BIC-PL and the BIC-ML criteria seem to agree in selecting the neighborhood structure but yield different results in the choice of model dimension.

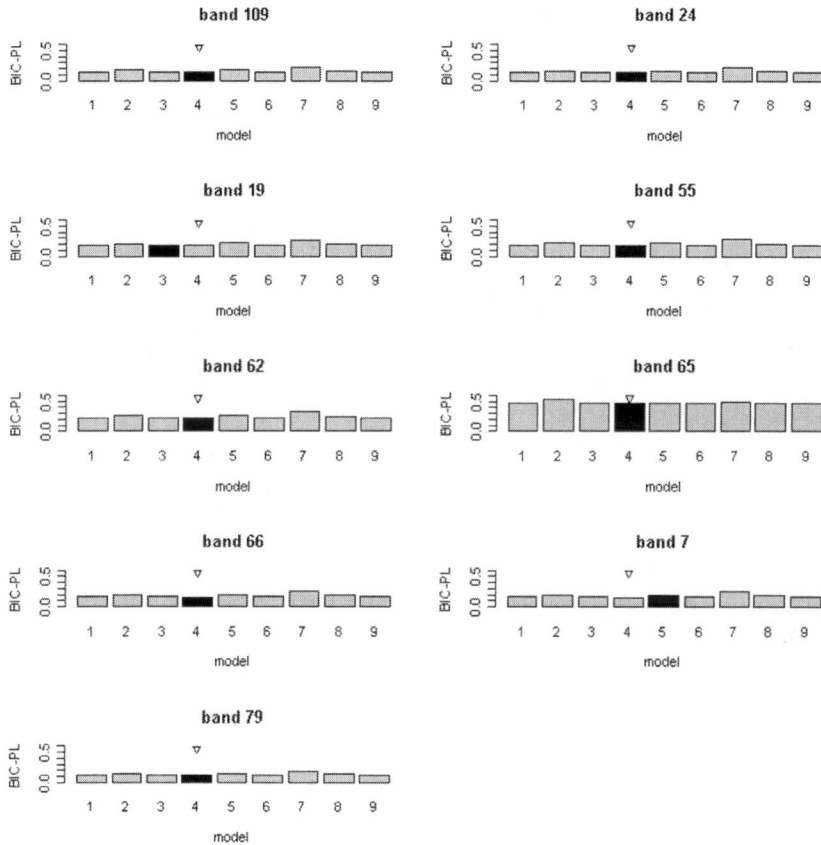

Fig. D.2.6. BIC-PL values of nine spatial auto-logistic models, fitted on a number of hyperspectral bands. Triangles indicate the minimum value attained by the BIC-PL criterion, while black bars indicate the model chosen by the BIC-ML criterion

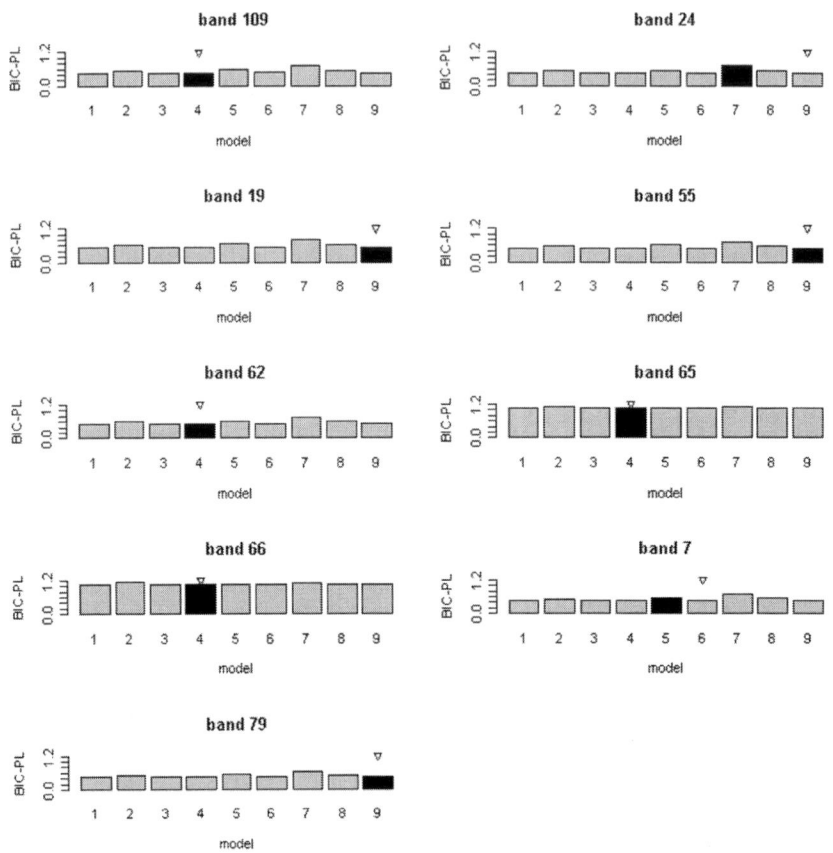

Fig. D.2.7. BIC-PL values of nine spatial auto-binomial models, fitted on further hyperspectral bands. Triangles indicate the minimum value attained by the BIC-PL criterion, while black bars indicate the model chosen by the BIC-ML criterion

D.2.5 Concluding remarks

Typical objectives of statistical procedures for HSRH data are to restore the image by removing stochastic noise, classify and segment the image into areas exhibiting similar spectral properties, determine the composition of a mixture of material within a pixel, and locate signatures of unresolved objects with spatial magnitude less than a single pixel. In the last two decades, an increasing number of research papers have appeared in the Proceedings of the International Society for Optical Engineering, the IEEE Transactions on Pattern Analysis and Machine Intelligence, on Geoscience and Remote Sensing, showing that techniques for optimizing the mismatch between model prediction and sensor data have passed the point of scientific curiosity and are now under active evaluation by researchers in many fields. Within this framework, general purpose statistical techniques can be successfully adapted to be useful tools in the multi-disciplinary analysis of HSHR data, by taking into account the spatial redundancy of the data and the computational difficulties relating to the impressive amount of data to be analyzed.

MRF models play a central role in formalizing aprioristic information about spatial data dependence. Implementation of MRF is not automatic, since the choice of a neighborhood structure is not obvious in most cases. In this contribution a BIC-PL criterion is suggested for choosing both the model dimension (i.e., the number of independent parameters to be estimated) and spatial neighbors (i.e., adjacency).

The discussion presented here highlights how the implementation of BIC-PL leads to selection procedures that match good statistical properties and computational feasibility, while still giving results that are similar to those resulting from pursuing approaches based on the direct maximization of the likelihood function. In particular, the criterion is consistent, i.e., the probability of choosing the wrong model decreases to zero as the sample size increases, which is a nice property for massive dataset (e.g., the Lamar imagery). Furthermore, since evaluation of BIC-PL reduces to maximize objective functions that resemble likelihood functions of Generalized Linear Models (GLM, McCullough and Nelder 1983), advantage may be taken of the large number of efficient numerical GLM algorithms available in the literature. As a result, BIC-PL computation for HSRH datasets is dramatically faster than MCMC approximations of the likelihood function. On the other side, BIC-PL tends to overestimate the dimension of the model, while it seems to produce good results in selecting the neighborhood structure of the image.

Acknowledgements. This work was supported by the Italian Ministry for Research and University, under the Project 'Hierarchical models for spatial and space-time interactions in environmental data series'. The present chapter is a revised version of a paper published in the *Journal of Geographical Systems* 4(1).

References

Akaike H (1974) A new look at the statistical model identification. IEEE Transact Autom-Contr AC-19:716–723

Amrhein C, Griffith DA (1997) Multivariate analysis for geographers. Prentice-Hall, Englewood Cliffs [NJ]

Besag J (1974) Spatial interactions and the statistical analysis of lattice data. J Roy Stat Soc B 36(2):192–225

Bennett J, Khotanzad A (1998) Multispectral random field models for synthesis and analysis of color images. IEEE Transact Patt Anal Machine Intell 20(3):327-332

Crawford MM, Ricard M (1998) Hierarchical classification of SAR data using a Markov random field model. In Proceedings of the Southwest Symposium on Image Processing, Tucson [AZ], pp.81-86

Cross GR, Jain AK (1983) Markov random field texture models. IEEE Transact Patt Anal Machine Intell 5(1):25-39

Derin H, Howard E (1987) Modeling and segmentation of noisy and textured images using Gibbs random fields. IEEE Transact Patt Anal Machine Intell 9(1):39-55

Geman S, Geman D (1984) Stochastic relaxation, Gibbs distributions, and the Bayesian restoration of images. IEEE Transact Patt Anal Machine Intell 6(6):721-741

Geyer CJ, Thompson EA (1992) Constrained Monte Carlo maximum likelihood for dependent data. J Roy Stat Soc 54(3):657-699

Griffith DA (2002) Modeling spatial dependence in high spatial resolution hyperspectral data sets. J Geogr Syst 4(1):43-51

Griffith DA, Lagona F (1998) On the quality of likelihood-based estimators in spatial autoregressive models when the data dependence structure is mis-specified. J Stat Plann Inference 69(1):153-174

Guyon X (1995) Random fields on a network. Springer, Berlin, Heidelberg and New York

Guyon X, Hardoum C (1992) The Chi^2 difference of coding for testing Markov random field hypothesis. In Barone P, Frigessi A (eds) Lecture notes of statistics 74, Springer, Berlin, Heidelberg and New York, pp.165-176

Hoffbeck JP, Landgrebe D (1996) Classification of remote sensing images having high spectral resolution. Remote Sens Environ 57(3):119-126

Jacquez GM, Marcus WA, Aspinall RJ, Greiling DA (2002) Exposure assessment using high spatial resolution hyperspectral (HSRH) imagery. J Geogr Syst 4(1):1-14

Ji C, Seymnour L (1998) A consistent model selection procedure for Markov random fields based on penalized likelihood. Ann Appl Prob 6(2):423-443

Jona Lasinio G, Lagona F (2003) Selection of the neighborhood structure for space-time Markov Random field models. Stat Meth Appl 11(3):293-311.

Kashyap R, Chellappa R (1983) Estimation and choice of neighbors in spatial interaction models of images. IEEE Transact Inform Theory 29(1):60-72

Kelly PA, Derin H, Hartt KD (1988) Adaptive segmentation of speckled images using a hierarchical random field model. IEEE Transact Acoustic, Speech, Sign Proc 36(10):1628-1641

Knorr-Held L, Besag J (1997) Modelling risks from a disease in time and space. NRCSE Technical Report 5:1-24

Lagona F (2001) Parametric restrictions in random fields for binary space-time series. Metron 59(1-2):73-97

Landgrebe D (2000) Information extraction principles and methods for multispectral and hyperspectral image data. In Chen CH (ed) Information processing for remote sensing. World Scientific Publishing, River Edge [NJ], pp.3-37

Lee S, Crawford MM (1999) Unsupervised classification for multi-sensor data in remote sensing using Markov random field and maximum entropy method. In Proceedings of the 1999 International Geoscience and Remote Sensing Symposium, Hamburg, Germany, pp.1200-1202

Marcus WA, Crabtree R, Aspinall RJ, Boardman JW, Despain D, Minshall W, Peel J (2000) Validation of high-resolution hyperspectral data for stream and riparian habitat analysis. NASA EOCAP Final Phase 2 Annual Report. Stennis Space Flight Center

McCullagh P, Nelder JA (1983) Generalized linear models. Chapman and Hall, London

Rignot E, Chellappa R (1992) Segmentation of polarimetric synthetic aperture radar data. IEEE Transact Image Process 1(3):281-300

Rose C, Devijver PA (1984) Connected components in binary images: the detection problem. Wiley, New York, Chichester, Toronto and Brisbane

Smith AFM, Roberts GO (1993) Bayesian computation via the Gibbs sampler and related Markov Chain Monte Carlo methods (with discussion). J Roy Stat Soc B 55(1):3-23

Smits PC, Dellepiane SG (1997) Synthetic aperture radar image segmentation by a detail preserving Markov random field approach. IEEE Transact Geosci Remote Sens 35(4):844-857

Schwartz G (1978) Estimating the dimension of a model. Ann Stat 6(2):461-464

Tjelmeland H (1996) Modeling of the spatial facies distribution by Markov random fields. In Baafi EY, Schofield NA (eds) Proceedings of the 5th International Geostatistical Congress, Wollongong, Australia, September 22-27, 1996. Kluwer, Dordrecht, pp.512-523

Tjelmeland H, Besag J (1998) Markov random fields with higher-order interactions. Scand J Stat 25(3):415-433

Wackernagel H (1998) Multivariate geostatistics: an introduction with applications (2nd edition). Springer, Berlin, Heidelberg and New York

Zhang M, Haralick R, Campbell J (1990) Multispectral image context classification using stochastic relaxation. IEEE Transact Syst Man Cybern 20(1):128-140

D.3 Geographic Object-based Image Change Analysis

Douglas Stow

D.3.1 Introduction

Remote sensing is an effective tool for mapping earth surface objects and phenomena, and provides the sole means for comprehensive monitoring of land surface changes. Normally captured in an image form by sensors mounted on aircraft or satellites, remotely sensed data are spatially contiguous and temporally periodic measurements of the reflected or emitted electromagnetic radiation (EMR) leaving the earth's surface. In order to create or update maps of earth surface objects or phenomena, these image data must be visually interpreted by humans and/or processed by computer routines. For the past 30 years, a major emphasis has been placed on computer-assisted approaches to mapping and monitoring earth surface objects and phenomena, to achieve greater efficiency and objectivity, which are inherent to such approaches.

Object-based approaches to semi-automated mapping and monitoring have been a primary focus area of remote sensing and image processing research in the past ten years conducted in disciplines such as computer science, electrical engineering, and geography, and under interdisciplinary headings such as computer vision, image understanding, biomedical imaging, and forensic science. Relative to single- or per-pixel approaches, object-based image analysis (OBIA) attempts to exploit spatial relationships of groups of image picture elements (called pixels) in order to delineate and identify objects within an imaged scene (Benz et al. 2004). Object-based approaches are most applicable to the high or H-resolution remote sensing situation, where objects or features of interest are larger than the ground resolution element associated with a pixel (Strahler et al. 1986). For geographic object-based image analysis (GEOBIA) (Hay and Castilla 2008), such objects may be related to natural features of land and water surfaces (for example, trees and ocean fronts) or human-made features (for example, buildings or boats). While object-based image processing techniques have been developed to support environmental remote sensing for over 30 years (Ketting and Landgrebe 1976), the greater research emphasis on such techniques in the past several years can be explained by several factors. One of the primary factors is the greater availability of digital remote sensing image data having high spatial resolution; such data are generally not amenable to achieving highly accurate mapping and monitoring results when generated with per-pixel image classification routines. Another factor is the greater availability and affordability of high performance computers and object-based image processing software.

The past five years have seen an initiation of research activity on GEOBIA techniques for detecting, identifying, and/or delineating earth surface changes. Such techniques may be referred to as geographic object-based image change analysis or GEOBICA, and are the focus of this chapter. GEOBICA is based on quantitative spatial analytical methods and generates data sets that can support spatial analysis of geographic areas.

The emphasis of this chapter is on the use of multi-temporal remotely sensed image data to map earth surface changes from an object-based perspective. The chapter starts with a description of the reasons and purposes for conducting GEOBICA, which is followed by treatment of image acquisition and pre-processing requirements, and the types of image data that are input to GEOBICA routines. Next, brief overviews of image segmentation and segment-based classification are provided. These are followed by discussions on approaches to multi-temporal image analysis and GEOBICA strategies. Post-processing techniques are also covered. Finally, fundamentals of accuracy assessment for object-based and land cover change mapping are presented.

D.3.2 Purpose of GEOBICA

There are four general reasons or purposes for conducting GEOBICA: (i) land cover and land use change analysis, (ii) detection, inventory, and/or mapping of specific earth surface object types, (iii) map updating, and (iv) tracking movements and measuring displacements of earth surface features. These reasons or purposes are not necessarily independent and can be complimentary, as is expanded upon below.

Land cover and land use change analysis (LCLUCA) is one of the primary and more successful applications of remote sensing (Gutman et al. 2004; Hayes and Sader 2001; Rogan and Chen 2004). Most remote sensing approaches to land cover and land use analysis have been pixel-based, meaning that changes are detected or identified by comparing corresponding pixels on multi-temporal image data sets, one pixel at a time (Coppin et al. 2004; Mas 1999; Radke et al. 2005). Using an object-based approach, LCLUCA pertains to the delineation of pixel groupings which correspond to land surface units that have experienced a change in land cover (LC) or land use (LU), as well as the identification of the land cover or land use transition states associated with these units. Implicit to the identification of transitions is specification of both the initial ('from') and latter ('to') states of land cover or land use. Land cover pertains to the general surface material composition (for example, vegetation or rock type), while land use pertains to the predominant human activities or utilization of the land. Knowledge of the locations, distributions, and magnitudes of LCLUC is critical to effective urban planning and environmental management, and enables scientific analyses of processes such as urbanization, agricultural expansion, and vegetation response to climate change.

Many earth surface features, particularly those created by humans, tend to be manifested on a remotely sensed image as objects having characteristic geometric shapes and/or patterns (Li and Narayanan 2003). Examples of such human generated features are roads and street networks, trails and trail networks, and buildings, while some natural features that have characteristic image signatures are geological faults and fractures, lakes, ocean frontal boundaries, and clouds and atmospheric storm systems. For example, newly built features are evident by the appearance of the characteristic rectilinear-shaped objects on a multi-temporal difference image, or on the second date of imagery of a bi-temporal image pair, when the building is not evident on the first date. Inventory and mapping of new human-created surface features is primarily conducted to support government and business applications such as transportation planning, real estate development, or law enforcement, but can also be useful for humanitarian efforts such as disaster assessment or studies of informal settlements. For naturally occurring features, the interest is often in tracking and measuring their movements.

GEOBICA also provides a means for updating maps and GIS layers. Maps and GIS layers of dynamic landscapes become out of date because of LCLUC (Walter 1998). Even maps of relatively static themes such as soils types or topography should be updated following human modification of the land surface (for example, covering soil with urban settlements or modifying the terrain during suburban grading). The same GEOBICA procedures that are used to study LCLUC, or monitor new human-created features can be utilized for map updating purposes. The primary difference is that the map or layer requiring revision is normally incorporated into the image analysis procedures.

The fourth reason for implementing GEOBICA is to measure rates of movements or displacements. The strategy of such analyses is to delineate, identify and track over time, a moving earth surface feature. Examples of earth surface features that may be tracked semi-automatically using time sequential remote imaging and GEOBICA are clouds and weather systems, ocean surface features such as fronts and eddies, wildfire flame fronts, earthquake faults and other geologic lineaments associated with seismic activity, vegetation boundaries such as ecotones or tree-limits, expanding urban limits, large animals, automobiles, and marine vessels. The first step is to detect and delineate objects of interest on each image of a registered multi-temporal image sequence through GEOBIA. Next, specific objects or particular components of these objects (for example, leading edge or center of gravity) are located and matched on sequential images. The relative displacement distance between corresponding objects or components is divided by the time interval between sequential imaging to infer rates of movements based on a Lagrangian drift perspective (Stow 1987).

D.3.3 Imagery and pre-processing requirements

The success of GEOBICA depends on the type and quality of multi-temporal image data that are input. Most GEOBICA are performed using a bi-temporal image data set, consisting of two images captured one or more years apart. Ideally, these images are captured by the same remote sensing system, at nearly the same date and time of day, and with a similar viewing geometry. Consistency in imaging systems and environmental conditions tends to reduce artifacts and false detections, and minimize pre-processing requirements.

The spatial, spectral, and radiometric characteristics (resolution in particular) of multi-temporal image data sets will have a major influence on the success of GEOBICA. Image spatial resolution will determine the size of change objects that can be delineated. Spectral resolution and coverage pertains to the width and locations of EMR wavebands, which will influence the magnitude and types of surface changes that can be detected or identified. The highest spatial resolution images tend to be captured in a single broad panchromatic waveband, or in a few broad visible and near infrared wavebands (that is, multi-spectral imagery). Fine or hyperspectral imagery, and sensing in the shortwave and thermal infrared, and microwave part of the spectrum can provide additional levels of discrimination for quantifying or identifying surface changes, but normally at a reduced spatial resolution. High radiometric resolution imagery enables subtle changes in surface reflectance or emittance to be detected, and normally over a broader reflectance/emittance range.

Several types of image processing steps are normally required prior to conducting GEOBICA and therefore, are referred to as pre-processing steps. These can be grouped into geometric and radiometric pre-processing categories. Geometric pre-processing pertains to correction or restoration of the geo-spatial fidelity of multi-temporal images, and radiometric pre-processing pertains to refining the image brightness or color characteristics. Generally speaking, relative consistency in the geometric and radiometric characteristics of two or more multi-temporal images is more critical to effective GEOBICA than absolute geometric or radiometric precision.

The three most common geometric pre-processing requirements are rectification, geographic referencing, and registration (Jensen 2005). Image rectification is the process of correcting geometric errors that result from platform, sensor, and earth distortion factors. Geographic referencing or georeferencing is the process of orienting a digital image grid according to an earth coordinate system and projection, such that each pixel has an associated coordinate location. Image registration is achieved by relative co-alignment of pixels between two or more images, such that corresponding pixels are associated with the same patch of ground surface. These pre-processing steps are not necessarily independent, as rectification, georeferencing, and/or registration can be achieved in an integrated fashion. Image registration is the key requirement for GEOBICA, as misalignment between images can yield false change detections and even omissions of actual changes (Dai and Khorram 1998; Stow, 1999; Stow and Chen 2002). This may be less of an issue for GEOBICA than pixel-based change analyses, as mis-registration artifacts may have characteristic object properties, which can be identified in the classification phase. Georeferencing is required if maps or GIS layers are incorporated in the change analysis, or if map updating is the purpose for performing GEOBICA.

Radiometric pre-processing for GEOBICA normally involves the adjustment of pixel brightness values such that they are relatively consistent within an image frame or strip, between adjacent frames or strips captured at nearly the same time, and between spatially corresponding frames or strips captured at different times. Such relative correction is commonly called radiometric normalization (Du et al. 2002; Schott et al 1988; Yang and Lo 2000). While normalization can be achieved through more absolute radiometric corrections such as radiometric calibration, atmospheric correction, terrain illumination correction, and anisotropic reflectance correction, these more absolute or mechanistic corrections have substantially higher science, technology and data input requirements, and are not always necessary (Song et al. 2001).

Multi-temporal image inputs to GEOBICA routines can take the form of pre-processed panchromatic and/or multi-spectral data, and/or transformed data derived from panchromatic and/or multi-spectral imagery. Texture transforms are commonly applied to panchromatic images as a means for extracting information on local spatial variability and expanding the dimensionality of these single dimensional images. Spectral ratios and other linear algebraic combinations of multi-spectral data are derived to enhance surface materials, reduce or expand dimensionality, and/or normalize terrain and anisotropic reflectance effects. Spectral transformations such as spectral vegetation indices, principal components analysis (PCA), and the tasseled-cap transform (Li and Yeh 1998; Jin and Sader 2005) can be applied to individual dates of imagery to generate useful input images for GEOBICA. Transformations such as PCA (Byrne et al. 1980) and image correlation analysis (Im et al. 2005) can also be applied to multi-temporal data sets to achieve change enhancement and data reduction for subsequent GEOBICA (Im et al. 2008).

D.3.4 GEOBIA principles

GEOBICA is based on the paradigm of GEOBIA, which is that image objects are delineated through segmentation, and then segments are classified and sometimes generalized to identify object types (Hay and Castilla 2008). For GEOBICA, these objects pertain to earth surface changes or new earth surface features. Prior to discussing the approaches and strategies for GEOBICA, it is useful to cover the basics of image segmentation and segment classification.

Segmentation refers to the process of partitioning an image into multiple pixel groups that may represent linear or polygonal features on the ground (Shapiro and Stockman 2001). This may be achieved by delineating linear features (Kaiser et al. 2004) or boundaries of polygonal features that are represented by relatively sharp gradients in image brightness or edges (Hirschmugl et al. 2007; Pouliot et al 2002), or by grouping contiguous pixels that share common image brightness, color or texture characteristics (Ketting and Landgrebe 1976; Ryherd and Woodcock 1996). A myriad of segmentation algorithms have been developed and tested by researchers in such fields as computer vision, image understanding, pattern recognition, and biomedical, forensic and remote sensing image processing (Burnett and Blaschke 2003; Woodcock and Harward 1992). Readers are referred to Haralick and Shapiro (1985) and Pal and Pal (1993), which provide summaries of image segmentation algorithms. Neubert et al. (2008) evaluate and compare commercial and open source segmentation routines.

The second phase of GEOBIA is the identification of the object type or class memberships for image-derived segments. This means that groups of pixels belonging to each segment are classified based on their collective image properties, which differs from per-pixel classification where each pixel is classified one at a time (Lobo 1997). Assignment of class labels to image segments can be achieved by using a per-pixel classification, where a majority or plurality rule is used to assign the most commonly occurring class for pixels composing a segment (Shandley et al 1996; Wang et al. 2004). Segment classification can also be achieved by deriving statistical measures of the frequency distribution of pixels composing a segment and then using those measures in a discriminant analysis or membership function classifier. Measures of central tendency (mean, median, mode) and variability (standard deviation or range) of image brightness, color, texture, spectral transforms, etc. are commonly utilized for this purpose. A potentially promising but relatively unexplored approach is to utilize non-parametric frequency distribution metrics or histogram descriptors as metrics. An advantage of per-object classification approaches is that measures of segment size and shape, sub-object (that is, nested segment) characteristics, and contextual relationships such as adjacency and contiguity can also be incorporated into the classifier (Burnett and Blaschke 2003).

A large number of image classification routines or classifiers exist, which are applicable to both per-pixel and object-based approaches. In the context of GEOBIA, the statistical metrics of each image segment is compared to a template of metrics or signatures for training segments of known class type, or for a user-supplied discriminant or membership function for each object type or class (Gamanya et al. 2007). In the case of supervised classification, the class signature template is based on analyst-selected pixel or segment samples representing classes of interest (that is, training data). For unsupervised classification, the template represents computer-derived cluster classes formed within the statistical feature space defined by the input data. Class labels are assigned by analysts after pixels or segments are classified according to cluster types. Classifiers may be based on parametric or non-parametric statistical assumptions and on hard or soft/fuzzy statistical rules or membership functions. Some of the simpler, yet reliable classifiers are parallelepiped, minimum distance to means (also known as nearest neighbor since an object is classified as the category to which it is nearest its training data in feature space), and maximum likelihood, which is the special case of the Bayesian probability classifier when equal *a priori* probabilities are assigned for all classes. Machine learning classifiers such as classification trees (Laliberte et al. 2007), artificial neural networks, self organizing maps, support vector machines, and genetic algorithms have shown much promise for per-pixel classification, but have received less attention for object-based classification to date. This may be due to the requirement for large training sample sets, which are more costly and challenging to generate for segments than for individual pixels.

D.3.5 GEOBICA approaches

Implicit to GEOBICA is that earth surface change objects are delineated and identified using multi-temporal geo-spatial data, with at least the latter date of data being a remotely sensed image. An analyst may select one of several available approaches to exploiting geo-spatial data when conducting GEOBICA, depending on the purpose or application objective, and the site-specific availability of geo-spatial data. For the bi-temporal case, these approaches can be characterized as: (i) map-to-image, (ii) image-to-image, and (iii) map/image-to-image (Cao et al. 2007). In this context, a map is a digital map or GIS layer, in vector or raster form.

For the map-to-image approach, an older map depicting the theme of interest is compared in a spatially explicit manner to an image captured at some time later than the map's production date. One of the main reasons for taking this approach is map updating. LCLUCA can also be conducted by using the map-to-image approach, but tends to be conducted more effectively when images from two or more dates are utilized. A vector map is more likely to be used in a map-to-image approach for GEOBICA, as it represents earth surface features as objects. A raster map would need to be generalized to form pixel groupings and/or vectorized. First, the image is segmented and then segments are classified as land use and land cover types, or identified as some type of earth surface feature. Parameter selection to determine the size and shape of image-derived segments should be guided by the size and shape of objects represented on the map that is to be updated. Map objects can be used to select training objects for supervised classification, or to determine class labels for clusters derived using unsupervised classification. However, this implies that objects used for training or labeling have been determined to be stable (that is, do not correspond to change areas). Image-generated objects are compared to those on the map, and objects with substantially different spatial characteristics and/or having different class attributes are deemed to be change objects. Change objects supercede previously mapped objects in map updating. They also provide the primary basis for LCLUCA. Walter (2004) is one of the first published works on GEOBICA and describes a map-to-image approach to achieve GIS map updating.

The image-to-image approach is the one most commonly used for GEOBICA, and for LCLUCA in general. A dated image is compared to a more recently captured image to infer and map LCLUC objects, or to detect and delineate new or moving earth surface features. The specific strategies for object comparisons with the image-to-image approach are discussed in the following section.

The map/image-to-image approach is a hybrid of the first two, where a map depicting the theme of interest (for example, land cover and land use) and a remotely sensed corresponding to the first date of the bi-temporal sequence are both utilized, along with a more current image. In many cases, the dated map was derived from the corresponding image. This approach is ideal for map updating, as the older image provides additional spatial information about prior states of earth surface features and in a manner that is more comparable with the recently acquired image.

Another potential use of the map/image-to-image approach is to detect or identify LCLUC within extant map units, rather than strictly on a per-pixel or per-object basis. This is similar to the per-field approach used for image classification (Aplin et al. 1999). Mapped polygons that can provide the basis for summarizing LCLUC are agricultural field boundaries, forest or habitat management units, urban community planning areas, integrated terrain units, and statistical reporting units (for example, census tracts). While these change analysis units are a type of geographic object, changes per map unit could be based on products derived with either per-pixel change detection or GEOBICA methods.

D.3.6 GEOBICA strategies

The crux of GEOBICA is in the various strategies for image-to-image multi-temporal analyses, which are conducted to delineate and identify LCLUC, or newly formed earth surface objects. Two general strategies exist: (i) post-classification comparison and (ii) multi-temporal image object analysis.

Post-classification comparison is the most straightforward strategy, where an initial and one or more subsequently captured images of the same geographic extent are subjected to OBIA and the resultant maps are compared to delineate and classify change objects. Effectively, this strategy transitions from an image-to-image to a map-to-map analysis. Each of the images of the multi-temporal data set should be subjected to the same segmentation and classification routines, using similar processing parameters, decision rules, class schemes, and output formats. This is done to minimize differences in the maps that are not related to actual earth surface changes. The resultant maps can be in either raster object or vector format. The output maps are compared using GIS techniques such as polygon overlay, raster cross-tabulation, or Boolean logic. Change objects occur as the intersection (in the set theory context) of bi-temporal map objects that have different class labels. The disparate class types form the transitional change sequence of the change objects (that is, 'from-to' classes). Spatial and contextual filtering can be applied to remove apparent change objects that are likely artifacts from mis-registration errors, and inconsistencies in illumination, seasonality, segmentation and classification between image dates.

While GEOBICA is just beginning to emerge and only a handful of articles are evident in the earth remote sensing literature, a slight majority of these articles are based on post-classification comparison strategies (Blaschke 2005; Gamanya et al. in press; Laliberte et al. 2004; Lamar et al. 2005; Zhou et al. 2008). A framework for GEOBICA based on post-classification comparison is provided by Blaschke (2005). While no empirical examples with actual image data sets are demonstrated, Blaschke (2005) offers a strong theoretical basis for the comparison of multi-temporal map objects to detect and identify changes. Laliberte et al. (2004) analyzed desert shrub encroachment through a time sequence of detailed vegetation maps generated by applying segmentation and segment classification methods to high spatial resolution aerial and satellite image data. Precise changes in shrub and grass cover were quantified from the resultant map time series.

Lamar et al. (2005) applied segmentation and object-based image classification to a three-date sequence of digital color aerial photographs to monitor patch-level changes in hemlock trees. A challenging component of the research was the reconciliation of false changes and gaps in actual patches, and the development of an object-based accuracy assessment approach. Gamanya et al. (2007) incorporated a bi-temporal data set of moderate spatial resolution imagery from the Landsat TM/ETM+ systems to map and analyze land cover changes in a large African city. Each image was subjected to a hierarchical segmentation routine and then segments were classified based on a large number of segment-based statistical metrics. Land cover changes were derived using objects from the first date of Landsat imagery as the basis. Zhou et al. (2008) also implemented a post-classification comparison strategy to GEOBICA to map land cover changes within an urban setting, only using very high spatial resolution (0.5 m) color infrared digital camera data. They found that the GEOBICA approach achieved a significantly higher accuracy of change identification (approximately 90 percent) relative to a pixel-based approach (approximately 80 percent), when mapping five general land cover change types.

Multi-temporal image object analysis is a strategy of the image-to-image GEOBICA approach that directly applies image segmentation and classification to multi-temporal image data sets. The input data set to segmentation and classification routines is either a multi-temporal composite and/or a multi-temporal transform image data set. A multi-temporal composite image consists of one or more registered panchromatic, multi-spectral waveband, texture, and/or spectral transform images for two or more image dates. In the simplest case, two registered panchromatic images captured on different dates could be combined to form a single two-channel multi-temporal composite image. A multi-temporal transform image is generated by transforming two or more dates of imagery through operations such as like-channel image differencing, change vector analysis (a coordinate transformation of multi-spectral image differencing), multi-temporal correlation (Im et al. 2005), and principal components analysis (Byrne et al. 1980).

An advantage of multi-temporal transform data sets is that they can represent the same spectral-radiometric, texture, or transform data types with fewer channels to process and store (normally half as many) as multi-temporal composite data sets. A disadvantage is that only the change magnitude and direction are captured and not the actual 'from-to' signatures, such that multiple surface transition sequences with disparate reflectance characteristics could have non-unique change signatures. Note that either of the two types of multi-temporal data set could be input to a segmentation routine and the other type used in the classification phase, such as was done by Desclée et al. (2006).

Since two or more dates of imagery are incorporated into the segmentation phase of multi-temporal image object analysis, image-derived segments will generally conform to change and no-change surface features, as well as artifacts that may result from mis-registration and differences in shadowing between dates. In the case of LCLUCA, surface change objects may emerge at several different spatial scales, pertaining to the scales associated with modified land units and their internal characteristic surface material components. For this reason, it may be desirable to implement hierarchical or multi-resolution segmentation. Such a segmentation approach can also be beneficial when mapping new earth surface features that occur over a wide range of sizes (for example, buildings and ephemeral water bodies).

The occurrence of both change and no-change objects also means that the classification or identification phase of multi-temporal image object analysis can be particularly challenging. The classification scheme must consist of the surface change classes or new object identities of interest, plus at least a no-change category. This means that classification signatures consist of both spatial-radiometric and temporal components, which provide a higher-level of complexity than single-date classification. This also means that it can be challenging to delineate training objects for supervised classification or label cluster classes for unsupervised classification routines (Bruzzone and Prieto 2000). While more challenging, advantages of this strategy (relative to post-classification comparison) is that there is not a need to apply map-to-map comparison methods and resultant accuracies are not dependent on the combined accuracy of multiple image-derived maps.

One of the first published research article truly employing multi-temporal object image analysis of Desclée et al. (2006), who did so to map forest cover changes. Three dates of SPOT multi-spectral (20 m) data were combined in a multi-temporal composite data set, which was subjected to image segmentation. Multi-temporal differences were computed for image segments based on several spectral feature types for, and a multivariate iterative trimming technique was used to establish forest change/no-change thresholds. This GEOBICA approach yielded over 90 percent forest change detection accuracy, which was substantially higher than a pixel-based approach. A similar approach was taken by Bontemps et al. (2008) for a four-year time series of SPOT VEGETATION (one km) data to map forest change objects in the Brazilian Amazon region. Conchedda et al. (2008) and Stow et al. (2008) utilized multi-temporal composite images in both segmentation and classification phases to map vegetation change objects. For the Conchedda et al. (2008) study, SPOT multi-spectral (20 m) images captured 20 years apart were utilized to map changes in mangrove forests with greater than 85 percent accuracy. Fine-scale objects associated with shrub cover changes ('to' or 'from' shrub cover) were delineated by Stow et al. (2008) using a multi-temporal airborne digital CIR imagery (one m), enabling an assessment of human and climate disturbance factors on wildlife habitat reserves.

While not truly an example of GEOBICA in the manner in which is defined in this chapter, one of the first published articles that combined GEOBIA and binary change detection was Hall and Hay (2003). The authors tested an object-based change enhancement procedure for analyzing landscape changes. Bi-temporal SPOT panchromatic images were spatially transformed individually to generate multi-scale image sets (Hay et al. 2001), which were then were subjected to a watershed segmentation routine. Per-pixel image differencing was applied to segmented images such that the panchromatic brightness was represented as the segment mean value. Difference thresholds were derived to map 'change' and 'no change' features in a managed forest landscape context.

D.3.7 Post-processing

Upon completion of segmentation and classification phases, post-processing steps may be implemented prior to generating a final product or initiating change analyses. These post-processing steps may be grouped into three categories: (i) automated map generalization and editing, (ii) change reconciliation, and (iii) manual editing.

A common requirement for map generalization is the merging or aggregation of adjacent objects that have been classified as the same category in the single-date classification phase of post-classification comparison, or the classification phase of multi-temporal image object analysis. This is achieved with simple polygon merging techniques used for vector GIS. Another vector GIS generalization and editing technique that is useful for post-processing of GEOBICA products is filtering and removal of objects that are smaller than a specified threshold. Such objects may be smaller than the size of a designated minimum mapping unit for change assessment and/or may be considered to be likely artifacts or noise-related objects rather than true change features. A typical source of noise objects in GEOBICA is mis-registration, where resultant features often have sliver-like shapes. Such objects may be removed or 'filtered out' by utilizing shape (narrow and elongated) and size criteria. Polygon smoothing is another map generalization post-processing technique that may be implemented if the segmentation or post-classification overlay processes generate objects that are deemed to be excessively irregular.

Temporal reconciliation is a post-processing step where initial products from GEOBICA (normally associated with LCLUCA) are evaluated and potentially modified based on the logic or likelihood of particular temporal transitions. For bi-temporal image analysis, certain land LCLUC transition sequences may be deemed illogical (for example, from urban to agriculture). By using hard rules, an object initially classified as 'change' or as a change transition ('from-to') sequence are considered not to have changed, or in map updating, the original map class is preserved. A more probabilistic approach is to incorporate Markov Chain probabilities of LCLU transitions in the classification phase, by using transition probabilities as *a priori* probabilities of a Bayesian classifier (Strahler et al 1980). When the post-classification comparison strategy is applied to a multi-temporal data set consisting of more than two dates of imagery, temporal persistence rules may be used to filter out apparent transitions that likely result from image processing errors. A type of persistence-based filtering is panel analysis, which has been implemented successfully for per-pixel LCLUCA by Crews-Meyer (2002).

D.3.8 Accuracy assessment

The penultimate step in GEOBICA, prior to conducting spatial-temporal analyses, should be the assessment of accuracy of image-derived products. Unfortunately, this step is often avoided for two reasons; it can be very expensive and it is very challenging to generate reference data. In fact, assessing the accuracy of object-based change analysis products may be one the most challenging accuracy assessment tasks in remote sensing, because of both the temporal (Biging et al. 1999; Morisette and Khorram 2000) and the object-oriented nature of the products. Adding to the expense and challenge is that reference data may first and foremost be required for training segmentation and classification routines, with accuracy assessment being a secondary and less essential requirement. Most of the items discussed in this section are pertinent to any approach to geographic image change analysis (that is, not just object-based approaches).

Reference data (sometimes known erroneously as 'ground truth') must be an independent source of data or knowledge about changing states represented on the final GEOBICA product, which are deemed to be more reliable and of greater accuracy than the product being assessed. For most GEOBICA products, this requires knowledge of both the prior and more current states of land cover, or whatever earth surface feature is being examined for change. Since it is impossible to capture reference data for a time in the past (until time machines become reality), some form of archived data must be utilized. This means that a long-term field measurement or finer spatial resolution imaging program has already been implemented, or that one is fortunate to find and access suitable extant data captured for other purposes. As a fallback, GEOBICA products that are semi-automatically generated from a multi-temporal image data set may be assessed for accuracy using reference data generated from visual interpretation of the same image data set (Zhou et al. 2008); the assumption being that the manually-derived data are more accurate and reliable, even if they were not generated in as efficient or repeatable manner.

The two main types of map errors are positional and attribute, where in the case of remote sensing products, the latter pertains primarily to the classification process. For OBIA products, the accuracy of boundary delineation can be influenced by geometric image processing (that is, rectification and georeferencing), and the preciseness and reliability of image segmentation. Complicating matters for boundary delineation accuracy in GEOBICA is the preciseness of image registration and to some degree, the mutual reliability of classifying and overlaying two or more dates of imagery. Image classification or feature identification errors will also influence the attribute representation error of change classes assigned to each change object. While it has been suggested that positional and attribute errors of image-derived maps can be isolated and quantified separately (Pontius 2002), doing so for GEOBICA products is likely intractable.

The spatial unit of analysis for assessing the accuracy of GEOBICA products may be a pixel, an image-derived segment, an independently derived linear or polygonal unit, or an image-derived change object. Using a sample of reference data pixels to assess accuracy is similar to a 'point in polygon' vector GIS approach and has been applied by Conchedda et al. (2008) and Zhou et al. (2008). This enables a general assessment of product accuracy without examining specifically boundary delineation and classification accuracy. Similarly, using image-derived segments of known class attributes will only enable determination of the multi-temporal class accuracy, since such segments will either coincide with or be nested within the final map objects (Stow et al. 2008). This would also hold true if the change objects on the final image-derived map were the spatial unit of assessment.

The only spatial unit that would enable both boundary and class representation accuracy to be considered is to use linear or polygonal objects (depending if image-derived change objects are linear or polygonal) derived independently, to conduct a GIS polygon overlay analysis. For analyses of single change objects such as roads or trees, accuracy assessment can be performed using relative count comparisons between image-derived and reference data for larger reporting units (for example, forest stands or census blocks) (Lamar et al. 2005). Note that if a complete reference change map is available rather than ground reference samples, or if different GEOBICA products are to be compared in a relative, spatially-explicit manner, either a raster or polygon overlay analysis can be conducted to determine relative map agreement or accuracy (Bontemps et al. 2008; Desclée et al. 2006).

Comparisons of GEOBICA products to reference data can yield a variety of different accuracy/error measures on how well change has been detected, or if specific types of changes are identified in a reliable manner. If reference data are sampled randomly or are in the form of a complete change map, then the error in the proportion of area representation (that is, inventory perspective) of change objects can be estimated. Conversely, site-specific accuracy is assessed through spatially-explicit comparison of reference data samples or maps to the GEOBICA product (Jensen 2005). Measures such as overall agreement, kappa index of agreement, producers and users accuracy, and class-specific accuracy can be derived in a change detection (change/no-change classes or presence/absence of specific change features), and/or a change identification (from-to class sequences) context.

D.3.9 Concluding remarks

The 'A' in GEOBICA stands for *analysis* and the ultimate purpose for conducting GEOBICA is to analyze earth surface changes, even if the immediate objective may be to update a map or GIS layer. While the emphasis of this chapter has been on semi-automated, object-based digital image processing for deriving maps of earth surface changes, the analysis of change is ultimately conducted by humans examining statistics, maps, or time series visualizations. To facilitate analyses and provide greater rigor and objectivity, many of the spatial analysis techniques summarized in the other chapters of this book can by applied to image-derived change maps to generate quantitative measures or test statistical relationships between change features and potential causal factors. Products derived through GEOBICA may also be suitable for modeling and forecasting future LCLUC (Raza and Kainz 2001).

While research on GEOBICA techniques has just begun to emerge in the refereed remote sensing and image processing literature over the past five years, it is likely that some experimentation of object-based approaches to mapping earth surface changes have been conducted for many years. Based on the relative proliferation of such article in the past two years and the number of commercially available and open source software packages that support object-based image analysis, it is possible that GEOBICA will become the predominant approach to map updating and LCLUCA. When earth surface changes are manifested as identifiable objects, it makes sense to use object-based approaches that exploit the spatial characteristics of remotely sensed data, and not just the spectral, radiometric, and temporal properties.

References

Aplin P, Atkinson P, Curran P (1999) Per-field classification of landuse using the forthcoming very fine resolution satellite sensors: problems and potential solutions. In Atkinson P, Tate N (eds) Advances in remote sensing and GIS analysis. Wiley, New York, Chichester, Toronto and Brisbane, pp.219-239

Benz UC, Hofmann P, Willhauck G, Lingenfelder I, Heynen M (2004) Multi-resolution, object oriented fuzzy analysis of remote sensing data for GIS-ready information. J Photogramm Remote Sens 58(3-4):239-258

Biging GS, Colby DR, Congalton RG (1999) Sampling systems for change detection accuracy assessment. In Lunetta RS, Elvidge CD (eds) Remote sensing change detection: environmental monitoring methods and applications. Ann Arbour Press, Chelsea, pp.281-308

Blaschke T (2005) Toward a framework for change detection based on image objects. In Erasmi S, Cyffka B, Kappas M (eds) Remote sensing and GIS for environmental studies. [Göttinger Geographische Abhandlungen 113], pp.1-9

Bontemps S, Bogaert P, Titeux N, Defourny P (2008) An object-based change detection method accounting for temporal dependences in time series with medium to coarse spatial resolution. Remote Sens Environm 112(6):3181-3191

Bruzzone L, Prieto DF (2000) Automatic analysis of the difference image for unsupervised change detection. IEEE Transact Geosci Remote Sens 38(3):1171-1182

Burnett C, Blaschke T (2003) A multi-scale segmentation/object relationship modelling methodology for landscape analysis. Ecol Mod 168(3):233-249

Byrne GF, Crapper PF, Mayo KK (1980) Monitoring land cover change by principal component analysis of multitemporal Landsat data. Remote Sens Environm 10(1980):175-184

Cao L, Stow D, Kaiser J, Coulter L (2007) Monitoring cross-border trails using airborne digital multispectrali and interactive image analysis techniques. Geocarto Intern 22(2):107-125

Conchedda G, Durieux L, Mayaux P (2008) An object-based method for mapping and change analysis in mangrove ecosystems. J Photogramm Remote Sens 63(5):578-589

Coppin P, Jonckheere I, Nackaerts K, Muys B, Lambin E (2004) Digital change detection methods in ecosystem monitoring: a review. Int J Remote Sens 25(9):1565-1596

Crews-Meyer K (2002) Characterizing landscape dynamism via paneled-pattern metrics. Photog Eng Remote Sens 68(10):1031-1040

Dai X, Khorram, S (1998) The effects of image mis-registration on the accuracy of remotely sensed change detection. IEEE Transact Geosci Remote Sens, 36(5):1566-1577

Desclée B, Bogaert P, Defourny P (2006) Forest change detection by statistical object-based method. Remote Sens Environm 102(1-2):1-11

Du Y, Teillet PM, Cihlar J (2002) Radiometric normalization of multitemporal high-resolution satellite images with quality control for land cover change detection. Remote Sens Environm 82(1):123-134

Gamanya R, De Maeyer P, De Dapper M (2007) An automated satellite image classification design using object-oriented segmentation algorithms: a move towards standardization. Expert Syst Appl 32(2):616-624

Gamanya R, De Maeyer P, De Dapper M 2009) Object-oriented change detection for the city of Harare, Zimbabwe Expert Systems with Application Expert Systems with Applications 36(1):571-588

Gutman G, Janetos A, Justice CO, Moran EF, Mustard JF, Rindfuss RR, Skole D, Turner II BL, Cochrane MA (2004) Observing monitoring and understanding trajectories of change on the earth's surface series: remote sensing and digital image processing. Springer, Berlin, Heidelberg and New York

Hall O, Hay GJ (2003) A multiscale object-specific approach to digital change detection. Int J Appl Earth Observ Geoinf 4(4):311-327

Haralick RM, Shapiro L (1985) Survey: image segmentation techniques. Comput Vision, Graph Image Proc 29(1):100-132

Hay GJ, Castilla, G (2008) Geographic object-based image analysis (GEOBIA): a new name for a new discipline. In Blaschke T, Lang S, Hay GJ (eds) Object-based image analysis spatial concepts for knowledge-driven remote sensing applications. Springer, Berlin, Heidelberg and New York, pp.75-90

Hay GJ, Marceau DJ, Bouchard A, Dubé P (2001) A multiscale framework for landscape analysis: object-specific up-scaling. Landsc Ecol 16(6):471-490

Hayes DJ, Sader SA (2001) Comparison of change-detection techniques for monitoring tropical forest clearing and vegetation regrowth in a time series. Photog Eng Remote Sens 67(9):1067–1075

Hirschmugl M, Ofner M, Raggam J, Schardt M (2007) Single tree detection in very high resolution remote sensing data. Remote Sens Environm 110(4):533-544

Im J, Jensen JR (2005) A change detection model based on neighborhood correlation image analysis and decision tree classification. Remote Sens Environm 99(3):326-340

Im J, Jensen JR, Tullis JA (2008) Object-based change detection using correlation image analysis and image segmentation. Int J Remote Sens 29(2):399-423

Jensen JR (2005) Introductory image processing: a remote sensing perspective (3rd edition). Prentice-Hall, New Jersey

Jin SM, Sader SA (2005) Comparison of time series tasseled cap wetness and the normalized difference moisture index in detecting forest disturbances. Remote Sens Environm 94(3):364-372

Kaiser J, Stow D, Cao L, Coulter L (2004) Evaluation of remote sensing technologies for mapping trans-border trails. Photog Eng Remote Sens 70(12):1441-1447

Ketting RL, Landgrebe DA (1976) Classification of multispectral image data by extraction and classification of homogeneous objects. IEEE Transact Geosci Electronics 14(1):19-26

Laliberte AS, Fredrickson EL, Rango A (2007) Combining decision trees with hierarchical object-oriented image analysis for mapping arid rangelands. Photog Eng Remote Sens 73(2):197-207

Laliberte AS, Rango A, Havstad KM, Paris JF, Beck RF, McNeely, R (2004) Object-oriented image analysis for mapping shrub encroachment from 1937 to 2003 in southern New Mexico. Remote Sens Environm 93(1-2):198-210

Lamar WR, McGraw JB, Warner T (2005) Multitemporal censusing of a population of eastern hemlock (Tsuga canadensis L) from remotely sensed imagery using an automated segmentation and reconciliation procedure. Remote Sens Environm 94(1):133-143

Li J, Narayanan RM (2003) A shape-based approach to change detection of lakes using time series remote sensing images. IEEE Transact Geosci Remote Sens 41(11):2466-2477

Li X, Yeh AGO (1998) Principal component analysis of stacked multitemporal images for the monitoring of rapid urban expansion in the Pearl River Delta. Int J Remote Sens 19(8):1501-1518

Lobo A (1997) Image segmentation and discriminant analysis for the identification of land cover units in ecology. IEEE Transact Geosci Remote Sens 35(5):1136–1145

Mas JF (1999) Monitoring land-cover changes: a comparison of change detection techniques. Int J Remote Sens 20(1):139-152

Morisette JT, Khorram J (2000) Accuracy assessment curves for satellite-based change detection. Photog Eng Remote Sens 66(7):876-880

Neubert M, Herold H, Meinel G (2008) Assessing image segmentation quality - concepts, methods and application. In Blaschke T, Lang S, Hay GJ (eds) Object-based image analysis spatial concepts for knowledge-driven remote sensing applications. Springer, Berlin, Heidelberg and New York, pp.769-784

Pal NR, Pal SK (1993) A review on image segmentation techniques. Patt Recogn Letter 26(9):1277-1294

Pontius Jr RG (2002) Statistical methods to partition effects of quantity and location during comparison of categorical maps at multiple resolutions. Photog Eng Remote Sens 68(10):1041-1049

Pouliot DA, King DJ, Bell FW, Pitt DG (2002) Automated tree crown detection and delineation in high-resolution digital camera imagery of coniferous forest regeneration. Remote Sens Environm 82(3-4):322-334

Radke RJ, Andra S, Al Kofahi O, Roysam B (2005) Image change detection algorithms: a systematic survey. IEEE Transact Image Process 14(3):294-307

Raza A, Kainz W (2001) An object-oriented approach for modeling urban land use changes. URISA 14(1):37-55

Rogan J, Chen DM (2004) Remote sensing technology for mapping and monitoring landcover and land-use change. Progr Plann 61(4):301-325

Ryherd S, Woodcock CE (1996) Combining spectral and texture data in the segmentation of remotely-sensed images. Photogramm Eng Remote Sens 62(2):181-194

Schott JR, Salvaggio C, Volchok WJ (1988) Radiometric scene normalization using pseudo-invariant features. Remote Sens Environm 26(1):1-16

Shandley J, Franklin J, White T (1996) Testing the Woodcock-Harward image segmentation algorithm in an area of southern California chaparral and woodland vegetation. Int J Remote Sens 17(5):983-1004

Shapiro LG, Stockman GC (2001) Computer vision. Prentice-Hall, New Jersey

Song C, Woodcock CE, Seto KC, Lenney MP, Macomber SA (2001) Classification and change detection using Landsat TM data: when and how to correct atmospheric effects? Remote Sens Environm 75(2):230-244

Stow D (1987) Remotely sensed tracers for hydrodynamic surface flow estimation, Int J Remote Sens 8(3):261-278

Stow D (1999) Reducing mis-registration effects for pixel-level analysis of landcover change. Int J Remote Sens 20(1):2477-2483

Stow D, Chen DM (2002) Sensitivity of multitemporal NOAA-AVHRR data for detecting land cover changes. Remote Sens Environm 80(2):297-307

Stow D, Hamada Y, Coulter L, Anguelova Z (2008) Monitoring shrubland habitat changes through object-based change identification with airborne multi-spectral imagery. Remote Sens Environm 112(3):1051-1061

Strahler AH, Woodcock CE, Smith JA (1986) On the nature of models in remote sensing. Remote Sens Environm 20(2):121-139

Strahler AH, Estes JE, Maynard P, Mertz FC, Stow DA (1980) Incorporating collateral data in Landsat classification and modeling procedures. Proceedings of the 14th International Symposium on Remote Sensing of the Environment, San Jose, Costa Rica, April 1980

Walter V (1998) Automatic classification of remote sensing data for GIS database revision. Int Arch Photogramm Remote Sens 32(4):641-648

Walter V (2004) Object-based classification of remote sensing data for change detection. J Photogramm Remote Sens 58(3-4):225-238

Wang L, Sousa WP, Gong P (2004) Integration of object-based and pixel-based classification for mapping mangroves with IKONOS imagery. Int J Remote Sens 25(24):5655-5668

Woodcock CE, Harward V (1992) Nested-hierarchical scene models and image segmentation. Int J Remote Sens 13(16):3167-3187

Yang X, Lo CP (2000) Relative radiometric normalization performance for change detection from multi-date satellite images. Photogramm Eng Remote Sens 66(8):967-980

Zhou W, Troy A, Grove M (2008) Object-based land cover classification and change analysis in the Baltimore metropolitan area using multitemporal high resolution remote sensing data. Sensors 8(8):1613-1636

Part E

Applications in Economic Sciences

E.1 The Impact of Human Capital on Regional Labor Productivity in Europe

Manfred M. Fischer, *Monika Bartkowska*, *Aleksandra Riedl*, *Sascha Sardadvar* and *Andrea Kunnert*

E.1.1 Introduction

Economists have long stressed the importance of human capital to the process of economic growth. However, recent cross-country studies have shown that economic growth appears to be unrelated to increases in human capital (Griliches 2000). Benhabib and Spiegel (1994), for example, have found only a weak correlation between growth and increases in human capital, measured in terms of educational attainment. Pritchett (2001) has arrived at similar results using a different dataset and more extensive robustness testing. These findings are in contrast to a great deal of evidence indicating high returns to human capital investments in both developing and developed countries (Temple 1999).

Increasing evidence suggests that regional rather than national economies are the decisive units at which growth takes place (Ohmae 1995; Storper 1997; Cheshire and Malecki 2004). Thus, we depart from previous research by shifting attention from countries to regions as objects of the analysis, and focus on levels rather than rates of growth.

The objective of this chapter is to provide empirical evidence on the contribution of human capital to labor productivity differences among regions in Europe. Labor productivity is measured in terms of gross value added per worker. There is no clear-cut definition on how human capital should be represented and measured. As in most previous studies, emphasis is on education rather than on any broader concept of human capital. In this vein, human capital is measured in terms of educational attainment based on data for the active population aged 15 years and older that attained the level of tertiary education. This variable is clearly imperfect: it ignores primary and secondary education, and on-the-job training, and does not account for the quality of education.

The presence of unobserved human capital in conjunction with the educational attainment variable motivates the use of a spatial Durbin model (SDM) to characterize the relationship between human capital and regional labor productivity. This motivation is independent of any economic theoretical justification in that it rests entirely on the plausibility of a conjunction of two circumstances (see LeSage and Fischer 2008). One of these is spatial dependence of the observed and unobserved forms of human capital. The second is that both variables are correlated by virtue of common (correlated) shocks to the spatial autoregressive processes governing these variables[1].

The remainder of this contribution is organized as follows. Section E.1.2 sets forth the spatial Durbin model that describes the basic relationship between regional human capital and labor productivity levels, and outlines the associated methodology for quantifying the direct and indirect impacts of human capital on labor productivity. Section E.1.3 applies the methodology to a sample of 198 NUTS-2 regions, and provides a correct assessment of both direct and indirect effects, in terms of the approach suggested by LeSage and Pace (2009a). A simulation approach is used with 10,000 random draws to produce an empirical distribution of the model parameters that are needed for computing measures of dispersion for the impact estimates. Section E.1.4 offers some closing comments.

E.1.2 Framework and methodology

The spatial Durbin model that describes the relationship between human capital and labor productivity is given by

$$y = \alpha \iota_n + X\beta + \rho W y + W X\gamma + \varepsilon \qquad (\text{E.1.1})$$

with the associated data generating process

$$y = (I_n - \rho W)^{-1}(\alpha \iota_n + X\beta + W X\gamma + \varepsilon) \qquad (\text{E.1.2})$$

and the expectation

$$E(y) = (I_n - \rho W)^{-1}(\alpha \iota_n + X\beta + W X\gamma) \qquad (\text{E.1.3})$$

---

[1] See LeSage and Fischer (2008, 2009) for details on the econometric derivation of the relationship.

$$\boldsymbol{\varepsilon} \sim \mathcal{N}(0, \sigma^2 \boldsymbol{I}_n). \tag{E.1.4}$$

All variables are in log form. $\boldsymbol{y}$ represents an n-by-1 vector of observations on the labor productivity level at the end of the sample period, with n being the number of observations (regions). $\boldsymbol{\iota}$ is an n-by-1 vector of ones with the associated scalar parameter α. $\boldsymbol{X}$ is an n-by-Q matrix of observations on the Q (non-constant) explanatory variables (here: $Q = 2$): labor productivity and human capital at the beginning of the sample period, while $\boldsymbol{\beta}$ is the associated Q-by-1 parameter vector.

$\boldsymbol{W}$ is an n-by-n non-stochastic, non-negative spatial weight matrix that specifies the spatial dependence among observations (regions), or in other words expresses for each row (observation/region) those regions (columns) which belong to its neighborhood[2] set as non-zero elements. We assume $0 \leq W_{ij} \leq 1$ and $W_{ij} = 0$ if $i = j$. The matrix $\boldsymbol{W}$ is row standardized so that $\Sigma_j W_{ij} = 1$ for $i = 1,...,n$. This facilitates the interpretation of operations with the spatial weight matrix as an averaging of neighboring values[3].

The n-by-1 vector $\boldsymbol{Wy}$ is the spatial lag of $\boldsymbol{y}$ that captures spatial effects working through the dependent variable, i.e. labor productivity at the end of the sample period. ρ is the scalar parameter of the first order spatial autoregressive (SAR) process, and is typically referred to as the spatial autoregressive parameter assumed to lie in $(-1, 1)$. This parameter reflects spatial dependence, which is expected to be positive in our model, indicating that regional productivity levels are positively related to a linear combination of neighboring regions' productivity. The presence of the spatial lag variable $\boldsymbol{Wy}$ on the right hand side of the equation will induce a non-zero correlation with $\boldsymbol{\varepsilon}$ that represents an n-by-1 normally distributed, constant variance disturbance term. The spatial lag for an observation (region) i is not only correlated with the error term at i, but also with the error terms at $j \neq i$.

$\boldsymbol{WX}$ is the n-by-Q matrix of the spatially lagged non-constant explanatory variables. $\boldsymbol{\gamma}$ contains the regression parameters associated with these variables. The coefficient estimates on the spatial lag of the explanatory variables capture two types of spatial externalities: spatial effects working through the level of labor productivity and spatial effects working through the level of human capital, both at the beginning of the sample period.

---

[2] Note that the term neighborhood is used here in a more general sense of spatial relatedness, despite that we will use it later in the more restricted sense of map-based first-order contiguity relations.

[3] Row normalizing is also helpful to guarantee that $(\boldsymbol{I}_n - \rho \boldsymbol{W})$ is non-singular for all $\rho \in (-1, 1)$.

This spatial Durbin model occupies an interesting position in the field of spatial regression analysis because it nests many of the models widely used in the literature (see, for example, Abreu et al. 2005; Fingleton and López-Bazo 2006; LeSage and Fischer 2008, 2009):

(a) Imposing the restriction $\gamma = \mathbf{0}$ leads to the spatial autoregressive (SAR) model that includes a spatial lag of labor productivity from neighboring regions, but excludes the influence of the spatially lagged explanatory variables.
(b) The so-called common factor parameter restriction $\gamma = -\rho\beta$ yields the spatial error regression model (SEM) specification that assumes that externalities across regions are mostly a nuisance spatial dependence problem caused by the regional transmission of random shocks[4].
(c) The restriction $\rho = 0$ results in a least-squares spatially lagged X regression model (labeled SLX by LeSage and Pace 2009a, b) that assumes independence between regional productivity levels, but includes characteristics from neighboring regions in the form of spatially lagged explanatory variables.
(d) Finally, imposing the restrictions $\rho = 0$ and $\gamma = \mathbf{0}$ yields the standard least-squares regression model (LSM).

Testing whether the restrictions hold or not implies not much effort. Of particular importance are common factor tests that discriminate between the unrestricted SDM and the SEM specifications, or in other words between substantive and residual dependence in the analysis. The likelihood ratio test proposed by Burridge (1981) is the most popular test in this context (see LeSage and Pace 2009a; Mur and Angulo 2006 for alternative tests and a comparison based on Monte Carlo evidence).

Model estimation

Estimation of the SDM via least-squares can lead to inconsistent estimates of the regression parameters. In constrast, maximum likelihood is consistent for this model. Maximizing the full log-likelihood involves setting the first derivatives with respect to the coefficient vector $\delta = (\alpha, \beta, \gamma)^T$ equal to zero and simultaneously solving these first-order conditions for all the parameters. Equivalent maximum likelihood estimates can be found using the log-likelihood function concentrated (with respect to δ and the noise variance parameter σ^2) which takes the form

---

[4] Note that the spatial Durbin error model (SDEM), which includes a spatial lag of the explanatory variables as well as spatially dependent disturbances, does not nest the SDM and vice versa. It can arise only if there would be no unobserved human capital, or if this unobserved variable would not be correlated with the educational attainment variable, both of which are very unlikely.

$$\ln L(\rho) = \frac{N}{2}\ln 2\pi + \ln|\boldsymbol{I} - \rho \boldsymbol{W}| - \frac{1}{2}\ln(\hat{\boldsymbol{e}}_0 - \rho\hat{\boldsymbol{e}}_L)^{\mathrm{T}}(\hat{\boldsymbol{e}}_0 - \rho\hat{\boldsymbol{e}}_L). \quad (\text{E.1.5})$$

The notation $\ln L(\rho)$ in this equation indicates that the scalar concentrated log-likelihood function value depends on the parameter ρ. $\hat{\boldsymbol{e}}_0$ and $\hat{\boldsymbol{e}}_L$ are the estimated residuals in a regression of y on Z and Wy on Z, respectively, with $Z = [\iota_n \ X \ WX]$.

Optimizing $\ln L(\rho)$ with respect to ρ permits us to find the ML estimate $\hat{\rho}$ and to use this estimate in the closed form expressions for $\hat{\boldsymbol{\beta}}(\hat{\rho})$, $\hat{\boldsymbol{\gamma}}(\hat{\rho})$ and $\hat{\sigma}^2(\rho)$ to produce ML estimates for these parameters. A variety of univariate techniques may be used for optimizing the concentrated log-likelihood function. In this study we used the simplex optimization technique.

Model interpretation

While linear regression parameters have a straightforward interpretation as the partial derivatives of the dependent variable with respect to the explanatory variables, in the SDM specification given by Eq. (E.1.1) interpretation of the parameters becomes more complicated. This comes from the simultaneous feedback nature of the SDM model that originates from the dependence relationships embodied in the spatial lag terms. These lead to feedback effects from changes in the explanatory variables in a region that neighbors i, say region j, that will impact the dependent variable for region i. This feature of the spatial Durbin model enables to quantify spatial spillover effects associated with human capital (see Chapter C.1).

In our spatial Durbin regression setting labor productivity of region i (that we denote by y_i) depends on

- *first*, labor productivity in regions neighboring i, captured by the spatial lag variable $W_i \, y$, where W_i represents the ith row of the spatial weight matrix W,
- *second*, the own-region initial period level of productivity, represented by X_{i1}, the first column of the n-by-Q matrix X,
- *third*, the initial period levels of labor productivity in the neighboring regions, represented by the spatially lagged variable $W_i X_{i1}$,
- *fourth*, the own-region initial period level of human capital, represented by X_{i2}, the second column of X,
- *fifth*, the initial period levels of human capital in the neighboring regions, represented by the spatially lagged variable $W_i X_{i2}$.

Thus, a change in the human capital level in region (observation) i will not only exert a *direct effect* on the productivity level of this region, but also an *indirect effect* on productivity levels in other regions $j \neq i$. This type of impact arises due to the interdependence between observations in the model. To see this, consider the SDM model expressed in Eqs. (E.1.6) to (E.1.7)

$$y = \sum_{q=1}^{Q} S_q(W) X_q + (I_n - \rho W)^{-1} \alpha \iota_n + (I_n - \rho W)^{-1} \varepsilon \qquad (E.1.6)$$

$$S_q(W) = (I_n - \rho W)^{-1} (I_n \beta_q + W \gamma_q) \qquad (E.1.7)$$

where the index q runs from 1 to Q, and X_q is the qth explanatory variable (qth column of X). There are Q explanatory variables. The Q-by-1 vector β contains the regression parameters associated with the non-constant explanatory variables in X, and the Q-by-1 vector γ the regression parameters associated with the spatially lagged variables WX.

To make the role of $S_q(W)$ more transparent, we consider the case of a single dependent variable observation in Eq. (E.1.8)

$$y_i = \sum_{q=1}^{Q} S_q(W)_{i1} X_{1q} + S_q(W)_{i2} X_{2q} + \ldots + S_q(W)_{in} X_{nq}$$

$$+ (I_n - \rho W)^{-1} \iota_n \alpha + (I_n - \rho W)^{-1} \varepsilon \qquad (E.1.8)$$

where $S_q(W)_{ij}$ represents the (i, j)th element from the matrix $S_q(W)$. It follows from Eq. (E.1.8) that

$$\frac{\partial y_i}{\partial X_{jq}} = S_q(W)_{ij}. \qquad (E.1.9)$$

This implies that a change in the human capital level for a single region j can potentially affect the dependent variable in all regions i, with $i \neq j$. This is a consequence of our spatial Durbin model.

The canonical case of own derivative for the ith region shown in Eq. (E.1.10) yields an expression $S_q(W)_{ii}$ that represents the impact on the dependent variable observation i from a change in X_{iq}

$$\frac{\partial y_i}{\partial X_{iq}} = S_q(W)_{ii} \qquad (E.1.10)$$

where $S_q(W)_{ii}$ denotes the (i, i)th element of the matrix $S_q(W)$.

LeSage and Pace (2009a) label $S_q(W)_{ii}$ the *direct* impact that is measured by the (i, i)th element of $S_q(W)$. This includes feedback influences that arise as a result of impacts passing through neighbors, and back to the observation (region) itself[5]. The *indirect* impact that arises from changes in all observations $j = 1, ..., n$, with $j \neq i$, of an explanatory variable X_q are found as the sum of the off-diagonal elements of row i from the matrix $S_q(W)$, for each observation i. Direct plus indirect effects equal the *total* effect from ceteris paribus changes in variable X_q.

Since the impact of changes in an explanatory variable differs over all observations, LeSage and Pace (2009a, b) suggest the following scalar summary measures[6]:

(a) the *average direct* effect constructed as an average of the diagonal elements of $S_q(W)$,
(b) the *average indirect* effect constructed as an average of the off-diagonal elements of $S_q(W)$, where the off-diagonal row elements are summed up first, and then an average of these sums is taken,
(c) the *average total* effect is the sum of the direct and indirect impacts.

We will use these scalar summary measures to quantify the set of non-linear impacts that fall on all regions as a result of changes in the human capital level in a region, and rely on LeSage and Pace's (2009a) approach to calculating measures of dispersion to draw inferences regarding the statistical significance of direct or indirect effects. These are based on simulating parameters from the normally distributed parameters β, γ, ρ and σ^2, using the estimated means and variance-covariance matrix. The simulated draws are then used in computationally efficient formulas to calculate the implied distribution of the scalar summary measures.

---

[5] Despite the fact that the main diagonal of the spatial weight matrix W contains zeros, the main diagonal of higher order matrices W^m (m integer) that arise in the infinite series expansion representation of the matrix inverse are non-zero. W_{ii}^2, for example, is non-zero to reflect the fact that region i is a second-order neighbor to itself, that is a neighbor to its neighbor. This accounts for the feedback effects.

[6] Of course, one could analyze direct and indirect impacts for an individual region i without averaging, but this would take the form of an 1-by-n row vector for each i considered. This type of limited analysis can be found, for example, in Anselin and LeGallo (2006), and Kelejian et al. (2006).

E.1.3 Application of the methodology

The definition of a spatial lag in spatial regression models depends on the choice of a spatial weight matrix that specifies 'neighborhood sets' for each observation[7]. In this study the weight matrix takes the form of a binary first-order contiguity matrix[8], in which only direct interaction between geographically neighboring regions is allowed for. This matrix is constructed on the basis of digital boundary files in a GIS and implemented in row-standardized form. Two regions are defined as neighbors when they show a common boundary.

Our sample includes 198 NUTS-2 regions[9] in continental Europe including 159 regions located in *Western Europe* covering Austria (nine regions), Belgium (11 regions), Denmark (one region), Finland (four regions), France (21 regions), Germany (40 regions), Italy (18 regions), Luxembourg (one region), the Netherlands (12 regions), Norway (seven regions), Portugal (five regions), Spain (15 regions), Sweden (eight regions) and Switzerland (seven regions), and 39 regions in *Central and Eastern Europe* covering the Baltic states (three regions), Czech Republic (eight regions), Hungary (seven regions), Poland (16 regions), Slovakia (four regions) and Slovenia (one region).

All variables are in log form. The dependent variable is labor productivity in 2004, and there are two (non-constant) explanatory variables, labor productivity and human capital in 1995. Human capital is proxied by the skills of the workforce as given by the level of tertiary educational attainment of the active population (aged 15 and over). Labor productivity is measured in terms of gross value added per worker, expressed in euros. Gross value added is the net result of output at basic prices less intermediate consumption valued at purchasers' prices, and measured in accordance with the European system of accounts 1995. Our main data source is Eurostat's Regio database. The data for Norway and Switzerland stem from Statistics Norway (Division for National Accounts) and the Swiss Office Fédéral de la Statistique (Comptes Nationaux), respectively.

The time period from 1995 to 2004 is short due to a lack of reliable figures for the regions in Central and Eastern Europe (see Fischer and Stirböck 2006). The political changes since 1989 have resulted in the emergence of new or re-established states with only a very short history as sovereign national entities. In

---

[7] The specification of the spatial weight matrix is a matter of some arbitrariness. A range of suggestions has been offered in the literature, based on first-order contiguity (thereby precluding islands), (inverse square) distance with or without a critical cutoff, as well as more general metrics. For extensive reviews see Cliff and Ord (1981), Anselin (1988), Anselin and Bera (1998), and Griffith (1995).

[8] This specification is well in line with empirical evidence on the geographic bounding of spatial externalities (see, for example, Fischer et al. 2009)

[9] NUTS-2 regions are defined according to formal rather than functional criteria and thus represent a less satisfactory definition of the region for the purposes of our study. But since data on functionally defined regions is not publicly available we had to make use of the NUTS classification.

most of these states historical series simply do not exist. Even for states such as Hungary and Poland that existed for much longer in their present boundaries, the quality of data referring to the period of central planning imposes serious limitations on a regional analysis. This is closely related to the change in accounting conventions, from the material product system to the European system of accounts 1995. Cross-region comparisons require interregionally comparable data which are not only statistically consistent but also expressed in the same numéraire such as euros. The absence of market exchange rates in the planned economies is seen as a further impediment.

Table E.1.1 reports the parameter estimates, the associated t-statistics and standard deviations not only for the SDM, but also for the SEM specification. A likelihood ratio test rejects the common factor restriction (test statistic: 13.79, $p = 0.001$) and consequently the SEM specification. This indicates that spatial externalities are substantive phenomena rather than random shocks diffusing through space. The parameter estimate of the spatial autoregressive parameter ($\hat{\rho} = 0.664$) provides evidence for the existence of significant spatial effects working through the dependent variable. This result is evidently in accordance with Fingleton and López-Bazo (2006).

Table E.1.1. Parameter estimates from SDM and SEM specifications

Variables	SDM			SEM		
	Parameter	Standard deviation	t-statistic	Parameter	Standard deviation	t-statistic
Constant	1.0831	0.2098	5.1631	3.3926	0.1915	17.7190
Initial labor prod. (β_1)	0.6621	0.0260	25.4287	0.6716	0.0197	34.1043
Human capital (β_2)	0.1476	0.0198	7.4576	0.1365	0.0194	7.0218
W-initial labor prod. (γ_1)	−0.4150	0.0503	−8.2482	–	–	–
W-human capital (γ_2)	−0.1691	0.0247	−6.8577	–	–	–
Spatial autoregressive parameter (ρ)	0.6640	0.0598	11.1002	0.7380	0.0500	14.7480
Sigma squared		0.0064			0.0066	
Log-LIK/n		1.3974			1.3626	

Notes: The dependent variable is labor productivity in 2004, the independent variables are labor productivity and human capital in 1995. The dependent and the independent variables are in log form. Thus, the coefficient estimates can be interpreted on an elasticity scale

As emphasized in the previous section, it is necessary to calculate the direct and indirect effects associated with changes in human capital on regional labor productivity to arrive at a correct interpretation of the model, in terms of the LeSage and Pace (2009a, b) approach. Table E.1.2 presents the corresponding impact estimates, along with inferential statistics. The estimates were produced by simulating parameters using the ML multivariate normal parameter distribution. A series of 10,000 simulated draws was used. The reported means and t-statistics were constructed from the simulation output.

If we consider the direct impacts, we see that these are close to the SDM model coefficient estimates reported in Table E.1.1. The difference between the human capital parameter estimate of 0.1476 and the direct impact estimate of 0.1317 equal to 0.0159 represents feedback effects that arise as a result of impacts passing through neighboring regions and back to the region itself. The discrepancy is negative since the coefficient estimate exceeds the impact estimate, reflecting some negative feedback. Since the difference between the coefficient estimate and the direct impact estimate is rather small, we conclude that feedback effects are small and not likely of economic significance.

The indirect impact estimates are what economists usually refer to as spatial spillovers. The presence or absence of significant spillovers depends on whether the indirect effects that arise result in statistically significant effects. We emphasize that it would be a mistake to interpret the γ-coefficient estimates as representing spatial spillover magnitudes.

To see how incorrect this is, consider the difference between the spatial lag coefficient γ_2 for human capital from the SDM (reported in Table E.1.1) and the indirect impact estimate calculated from the partial derivatives of the model (given in Table E.1.2). The indirect impact estimate for human capital is –0.1968, and significantly different from zero, while the SDM coefficient estimate associated with the spatially lagged human capital variable is –0.1691, and significant. If we would incorrectly view the coefficient γ_2 as reflecting the indirect impact, this would lead us to an inference that the human capital variable exerts a lower negative indirect impact on regional labor productivity. The true impact estimate points to a larger negative indirect impact calculated from the partial derivatives of the model.

Table E.1.2. Direct, indirect and total impact estimates (*t*-statistics in parentheses)

Variables	Spatial Durbin model		
	Mean direct impact	Mean indirect impact	Mean total impact
Initial labor productivity	0.6677	0.0683	0.7361
	(27.5716)	(1.8992)	(26.3921)
Human capital	0.1317	–0.1968	–0.0650
	(6.8644)	(–3.7637)	(–1.1847)

Note: t-statistics based on 10,000 sampled raw parameter estimates of the SDM

The indirect impact estimate can be interpreted in two ways. One interpretation reflects how a change in the human capital level of all regions by some constant would impact the labor productivity of a typical region (observation). The estimate of the indirect impact is equal to –0.1968, so an increase in the initial level of human capital of all other regions would decrease the productivity level of a typical region. This indirect impact takes into account the fact that the change in initial human capital level negatively impacts other regions' labor productivity, which in turn negatively influences our typical region's labor productivity due to the presence of positive spatial dependence on neighboring regions' labor productivity levels (see Table E.1.2).

The second interpretation measures the cumulative impact of a change in region's i initial level of human capital averaged over all other regions. The impact from changing a single region's initial level of human capital on each of the other regions' labor productivity is small, but cumulatively the impact measures – 0.1968. Of course the impact on regions closely related to region i whose initial human capital level has been changed will be greater than the impact on more remotely related regions.

In thinking about these results we note that it may be more intuitive to think about the impact of changes in the human capital variable by taking one or the other of the above two interpretative views. Since the scalar summary magnitudes representing the average over all impacts are numerically equivalent, we are free to do so (LeSage and Fischer 2008). Regarding the lack of impact on labor productivity arising from human capital, it seems more intuitive that raising initial levels of human capital for all regions would likely to have no significant total impact on the labor productivity level of a typical region. This represents the *Average Total Impact to an Observation* view of a change in the human capital levels during the initial period.

The intuition here arises from the notion that it is relative regional advantages in human capital that matter most for labor productivity, so changing human capital across all regions should have little or no total impact on (average) labor productivity levels. This interpretative view is consistent with our finding that the scalar summary measure for total impact of a change in human capital is not sufficiently different from zero (see Table E.1.2). It is interesting to note that the results obtained are consistent with the findings in LeSage and Fischer (2008).

E.1.4 Concluding remarks

The inherent complexity of the spatial Durbin model means that treating the parameter estimates like least-squares parameter estimates is incorrect. A change to a single observation of the human capital variable leads to changes in the dependent variable at each of the n locations. Consequently, our model leads to $n^2 = 39,204$ derivatives to analyze, and this provides too much information to easily digest. To quantify the impact of human capital on regional labor productivity, we used the LeSage and Pace (2009a) approach to calculating scalar summary impact measures. The direct impact is defined as the average impact of a change in the human capital variable at each of the n locations on the dependent variable at the same location, while the indirect impact is summarized using the average impact of a change in the explanatory variable at each location on the dependent variable at different locations.

Since the model is specified using logged levels of labor productivity and human capital, we can interpret the impact estimates obtained as elasticities. Based on the 0.1317 estimate for the direct impact of human capital, we find that a ten percent increase in human capital will on average result in a 1.3 percent increase in the final period level of labor productivity. This positive direct impact is offset by a significant and negative indirect impact producing a negative total impact that is not significantly different from zero.

The inferences were made conditional on the data and the specification of the spatial weight matrix[10]. The assumption that a particular spatial weight matrix specification is correct may be relaxed by treating spatial weight specification as an additional unknown feature, that is, by explicitly incorporating model uncertainty in the statistical analysis. To accommodate this uncertainty issue one may follow LeSage and Fischer (2008) in endorsing the use of Bayesian methods such as Bayesian model averaging in combination with Markov Chain Monte Carlo Model Composition (see also Chapter C.4). Another avenue for future research is to extend our framework to allow not only for geographical, but also for time dependence. This would permit us to study the impact of human capital over time.

Acknowledgements. The authors gratefully acknowledge the grant no. P19025-G11 provided by the Austrian Science Fund (FWF). They thank James LeSage for initiating this research and for helpful suggestions as well as two anonymous referees for useful comments on an earlier draft. All computations were made using LeSage's spatial econometrics library, http://www.spatial-econometrics.com/.

References

Abreu M, de Groot HLF, Florax RJGM (2005) Space and growth: a survey of empirical evidence and methods. Région et Dévelopment 21:12-43

Anselin L (1988) Spatial econometrics: methods and models. Kluwer, Dordrecht

Anselin L, Bera AK (1998) Spatial dependence in linear regression models with an introduction to spatial econometrics. In Ullah A, Giles DEA (eds) Handbook of applied economic statistics. Marcel Dekker, New York, pp.237-289

Anselin L, LeGallo J (2006) Interpolation of air quality measures in hedonic house price models: spatial aspects. Spat Econ Anal 1(1):31-52

Benhabib J, Spiegel MM (1994) The role of human capital in economic development. Evidence from aggregate cross-country data. J Monet Econ 34: 143-173

Burridge J (1981) A note on maximum likelihood estimation for regression models using grouped data. J Roy Stat Soc B43(1):41-45

Cheshire PC, Malecki EJ (2004) Growth, development, and innovation: a look backward and foreward. Papers in Reg Sci 83(1):249-267

Cliff AD, Ord JK (1981) Spatial processes: models and applications. Pion, London

Fingleton B, López-Bazo E (2006) Empirical growth models with spatial effects. Papers in Reg Sci 85(2):177-198

---

[10] It is noteworthy that alternative specifications based on four, five and six nearest neighbors essentially produced the same results.

Fischer MM, Stirböck C (2006) Pan-European regional income growth and club-convergence. Ann Reg Sci 40(4):693-721

Fischer MM, Scherngell T, Reismann M (2009) Knowledge spillovers and total factor productivity. Evidence using a spatial panel data model. Geogr Anal 41(2):204-220

Griffith DA (1995) Some guidelines for specifying the geographic weights matrix contained in spatial statistics models. In Arlinghaus SL, Griffith DA, Arlinghaus WC, Drake WD, Nystrom JD (eds) Practical handbook of spatial analysis. CRC Press (Taylor and Francis Group), Boca Raton [FL], London and New York, pp.65-82

Griliches Z (2000) R&D, education, and productivity. A retrospective. Harvard University Press, Cambridge [MA] and London (England)

Kelejian HH, Tavlas GS, Hondronyiannis G (2006) A spatial modeling approach to contagion among emerging economies. Open Econ Rev 17(4/5):423-441

LeSage JP, Fischer MM (2008) Spatial growth regressions: model specification, estimation and interpretation. Spat Econ Anal 3(3):275-304

LeSage JP, Fischer MM (2009) The impact of knowledge capital on regional total factor productivity. Available at SSRN: http://ssrn.com/abstract=1088301

LeSage JP, Pace RK (2009a) Introduction to spatial econometrics. CRC Press (Taylor and Frtancos Group), Boca Raton [FL], London and New York

LeSage JP, Pace RK (2009b) Spatial econometric models. In Fischer MM, Getis A (eds) Handbook of applied spatial analysis. Springer, Berlin, Heidelberg and New York, pp.355-376

Mur J, Angulo A (2006) The spatial Durbin model and the common factor tests. Spat Econ Anal 1(2):207-226

Ohmae K (1995) The end of the nation state: the rise of regional economies. Free Press, New York

Parent O, LeSage JP (2009) Spatial econometric model averaging. In Fischer MM, Getis A (eds) Handbook of applied spatial analysis. Springer, Berlin, Heidelberg and New York, pp.435-462

Pritchett L (2001) Where has all the education gone? The World Bank Eco Rev 15(3): 367-391

Storper M (1997) The regional world: territorial development in a global economy. Guilford Press, New York

Temple J (1999) A positive effect of human capital on growth. Econ Lett 65: 131-134

E.2 Income Distribution Dynamics and Cross-Region Convergence in Europe

Manfred M. Fischer and *Peter Stumpner*

E.2.1 Introduction

Whether income levels of poorer regions are converging to those of richer is a question of paramount importance for human welfare (Islam 2003). In Europe interest in this question has been enhanced in recent years, with the entry of new countries to the European Union. This chapter looks at evidence for regional income convergence in Europe. By Europe we mean the European Union of 27 member states. The notion of convergence is a fuzzy term that can mean different things (see Quah 1999). In this chapter we understand this notion in the sense of poorer regions catching-up with the richer. The observation units are NUTS-2 regions which the European Commission has chosen as targets for the convergence process and defined as the geographical level at which the persistence or disappearance of inequalities should be measured.

Measuring regional income and the extent to which convergence across regions – or what the European Commission calls regional cohesion – exists is a difficult issue. But per capita gross regional product [GRP] measured in purchasing power units seems like a natural definition if one is interested in an important determinant of average welfare. By focusing upon per capita GRP we are interested in the economic performance of regions and the claims that people living in those regions have over that wealth. Cohesion depends on the degree of equality in the distribution of per capita income and the extent to which there are processes of catch-up, in which less wealthy regions enjoy faster rates of income growth than more developed ones. The data were calculated on the basis of the 1995 European System of Accounts [ESA 95] and refer to the time period from 1995 to 2003. This shorter time span makes apparent the need for a model, before we can speak of the underlying dynamic regularities in these data.

Empirical research on regional income convergence has proceeded in many directions, using different definitions and methodologies[1]. Most research has, however, concentrated on the cross-section regression approach to investigate β-convergence where β is the generic notion for the coefficient on the initial income variable in the growth-initial level regressions. A negative β is interpreted as evidence of convergence in terms of both income level and growth rate. But Quah (1993b), Friedman (1992) and others have emphasized that a negative β can just be an example of the more general phenomenon of reversion to the mean, and, by interpreting it as convergence, growth analysts falling into Galton's fallacy.

This study follows the tradition of the non-parametric approach that views the catching-up question as a question about the evolution of the cross-section distribution of income, and diverts attention from the individual or representative region to the entire distribution as object of interest (see, in particular, Quah 1993a, 1996a, b, 1997a, b, c). The distribution that is relevant here is the distribution of income *across* regions, not that within a given region. Purpose of the analysis is to find the law of motion that describes transition dynamics and implied long-run behaviour of regional income. In the spirit of Quah (1996a, b) we assume that each region's income follows a first-order Markov process with time-invariant transition probabilities. That is, a region's (uncertain) income tomorrow depends only on its income today.

Most of the applications of this approach have worked in a discrete state space set up (see Quah 1996a, b; Fingleton 1997, 1999; Paap and van Dijk 1998; López-Bazo et al. 1999; Magrini 1999; Rey 2001; LeGallo 2004 to mention some). This set up has several advantages, but the process of discretising the state space of a continuous variable is necessarily arbitrary. Experience from the study of income distributions shows that this arbitrariness can matter in the sense that statements on inferred dynamic behaviour of the distribution in question and the apparent long-run implications of that behaviour are sensitive to the choice of the discretisation (Jones 1997; Reichlin 1999). Indeed, it is well known that the Markov property itself can be distorted from inappropriate discretisation (Bulli 2001).

This chapter avoids arbitrary discretisation of the income space and its possible effects on the results by using the stochastic kernel, the continuous equivalent of the transition probability matrix, as a suitable tool to overcome the problem. The remainder of the chapter is divided into two parts. The first, Section E.2.2, provides an empirical framework that extends current research by incorporating two novel techniques into the existing research: kernel estimation and graphical devices for the representation of the stochastic kernel (see Hyndman

---

[1] Recent surveys of the new growth literature in general and the convergence literature in particular can be found in Durlauf and Quah (1999), Temple (1999) and Islam (2003), while Fingleton (2003), Abreu et al. (2004), and Magrini (2004) survey the regional convergence literature, with region denoting a subnational unit.

et al. 1996), and Getis' spatial filtering technique that enables to account for the effects of spatial autocorrelation. The second part of the contribution, Section E.2.3, applies this framework to analyse income distribution dynamics and cross-region convergence in Europe, looking at evolving distributions of purchasing power standardized per capita (relative) gross regional product across 257 NUTS-2 regions in 27 EU-countries from 1995 to 2003. Some concluding remarks are given in the final section.

E.2.2 The empirical framework

A distribution perspective to the study of income dynamics and cross-region convergence directs attention to the evolution of the entire cross-region income distribution, emphasising shape and intra-distribution dynamics, and long-run (ergodic) behaviour. The section introduces a continuous version of the standard model of explicit distribution dynamics, pioneered by Quah (1993a), and argues that the stochastic kernel can be described as a conditional density function. Then we present a product kernel estimator for estimating this transition function, and briefly describe a three-step-strategy for solving the bandwidth selection problem, that appears to be crucial for estimation. Finally, we combine Getis' spatial filtering view with stochastic kernel estimation to account for the issue of spatial autocorrelation that may misguide inferences and interpretations if not properly handled.

A continuous version of the model of distribution dynamics

Let F denote the cross-section distribution of regional incomes at time t, then the simplest scheme for modelling the intra-distribution dynamics of $\{F_t \mid t \text{ integer}\}$ is a first-order Markov process with time-invariant transition probabilities. The distribution evolves according to

$$F_{t+1} = M F_t \qquad (E.2.1)$$

where M maps the distribution from time t to time $t+1$, and tracks where points in F_t end up in F_{t+1}. Iteration of Eq. (E.2.1) gives a prediction for future distributions of the ex-post probabilities

$$F_{t+\tau} = M^\tau F_t \quad \text{for } \tau > 0 \quad \tau = 1, 2, \ldots \qquad (E.2.2)$$

In this framework, there are two goals, the estimation of M will give us information on persistence of regional income inequalities, and the computation of the ergodic (steady-state) distribution. The latter provides information on the limiting behaviour of the regional income distribution. Convergence then might manifest in $\{F_{t+\tau}\}$ tending towards a point mass. A bimodal limit distribution can be interpreted as a tendency towards stratification into two different 'convergence clubs'.

In the discrete version of the model, the operator M can be interpreted as the transition probability matrix of the Markov process. The operator is approximated by partitioning the set of possible income values into a finite number of intervals. These intervals then constitute the states of a (time-homogeneous) finite Markov process, and all the relevant properties of M are described by a Markov Chain transition matrix whose (i, j) entry is the probability that a region in state i transits to state j in income space, in one time step. The inferred dynamic behaviour and the long-run implications of that behaviour are conditional on the discretisation chosen.

Regional income, however, is by nature a continuous variable, and hence discretisation may induce a possible bias. Instead of a state being a fixed interval we let the state be all possible intervals, including the infinitesimal small ones. In this case one may think of the number of distinct cells to tend to infinity and then to continuum. The corresponding transition probability matrix then tends to a matrix with a continuum of rows and columns. In this case, the operator M in Eq. (E.2.1) may be viewed as a stochastic kernel or transition function that describes the (time-invariant) evolution of the cross-section distribution in time. Convergence can then be studied by visualising and interpreting the shape of the income distribution at time $t+\tau$ over the range of incomes observed at time t.

For notational convenience let Y and Z denote the variable (per capita) regional income at times t and $t+\tau$ ($\tau > 0$), respectively. The sample may be denoted then by $\{(Y_1, Z_1), \ldots, (Y_n, Z_n)\}$ and the observations by $\{(y_1, z_1), \ldots, (y_n, z_n)\}$ where n indicates the number of regions. We assume that the cross-region distribution of Y can be described by the density function $f_t(y)$. This distribution will evolve over time so that the density prevailing at $t+\tau$ is $f_{t+\tau}(z)$. If we continue to maintain the assumptions of time-invariance and first-order of the transition process, the relationship between the cross-region income distributions, at time t and τ-periods later, can be written as

$$f_{t+\tau}(z) = \int_0^\infty g_\tau(z|y) f_t(y) \, dy \qquad \text{(E.2.3)}$$

where $g_\tau(z|y)$ is the conditional density function giving the τ-period ahead density of income z, conditional on income y at time t. Evidently, the (first-order) stochastic kernel can be described by a conditional density function assuming that the marginal and conditional income distributions have density functions.

So long as $g_\tau(z|y)$ exists, the long-run (ergodic) density, $f_\infty(z)$, implied by the estimated $g_\tau(z|y)$ function can then be found as solution to

$$f_\infty(z) = \int_0^\infty g_\tau(z|y) f_\infty(y) \, dy. \qquad (E.2.4)$$

In this contribution we will use the solution procedure outlined in Johnson (2004) to estimate this long-run distribution of regional income per capita.

Kernel estimation of the conditional density function

If $f_{t,t+\tau}(y,z)$ denotes the joint density of (Y,Z) and $f_t(y)$ the marginal density of Y, then the conditional density of $Z \,|\, (Y = y)$ is given by

$$g_\tau(z|y) = \frac{f_{t,t+\tau}(y,z)}{f_t(y)} \qquad (E.2.5)$$

Probably, the most obvious estimator of this conditional density function[2] (see Hyndman et al. 1996) is

$$\hat{g}_\tau(z|y) = \frac{\hat{f}_{t,t+\tau}(y,z)}{\hat{f}_t(y)} \qquad (E.2.6)$$

where

$$\hat{f}_{t,t+\tau}(y,z) = \frac{1}{n\, h_y\, h_z} \sum_{i=1}^{n} K\left(\frac{1}{h_y}\|y - Y_i\|_y\right) K\left(\frac{1}{h_z}\|z - Z_i\|_z\right) \qquad (E.2.7)$$

is the kernel estimator of $f_{t,t+\tau}(y,z)$, and

---

[2] For alternative estimators see Hyndman and Yao (2002), and Basile (2006).

$$\hat{f}_t(y) = \frac{1}{n\, h_y} \sum_{i=1}^{n} K\left(\tfrac{1}{h_y}\|y - Y_i\|_y\right) \tag{E.2.8}$$

the kernel estimator of $f_t(y)$ (see Hyndman et al. 1996). h_y and h_z are bandwidth parameters that control the degree of smoothing applied to the density estimate. h_y controls the smoothness between conditional densities in the y-direction, and h_z the smoothness of each conditional density in the z-direction. $\|\cdot\|_y$ and $\|\cdot\|_z$ are distance metrics on the spaces Y and Z, respectively. In this contribution we use the standard euclidean distances, $\|\cdot\|_y = |\cdot|_y$ and $\|\cdot\|_z = |\cdot|_z$.

A multivariate kernel other than the product kernel might be used to define $\hat{g}_t(z \mid y)$. But the product kernel is simpler to work with, leads to conditional density estimators with several nice properties and is only slightly less efficient than other multivariate kernels (Wand and Jones 1995). The kernel $K(x)$, where x is variously y or z, is a real, integrable, non-negative, even function on $\mathbb{R}$ concentrated at the origin so that (Silverman 1986)

$$\int_{\mathbb{R}} K(x)\,dx = 1, \quad \int_{\mathbb{R}} x K(x)\,dx = 0 \quad \text{and} \quad \sigma_K^2 = \int_{\mathbb{R}} x^2 K(x)\,dx = \infty. \tag{E.2.9}$$

Popular choices for $K(x)$ are defined in terms of univariate and unimodal probability density functions. In this contribution we use the Gaussian kernel[3] given by

$$K(x) = \left(\sqrt{2\pi}\right)^{-1} \exp(-\tfrac{1}{2} x^2). \tag{E.2.10}$$

Whatever kernel is being used, bandwidth parameters chosen to minimize the asymptotic mean square error give a trade-off between bias and variance. Small bandwidths yield small bias but large variance, while large bandwidths lead to large bias and small variance. The problem of choosing, how much to smooth, is of crucial importance in conditional density estimation, and the results of the continuous state space approach to distribution dynamics strongly depend on the bandwidth parameters chosen.

---

[3] On the basis of the mean integrated square error criterion, Silverman (1986) has shown that there is very little to choose between alternatives. In contrast, the choice of the bandwidths plays a crucial role. See also Chapter C.5 for this issue.

In this study we follow Bashtannyk and Hyndman (2001) to solve this bandwidth selection problem[4] by a three-step-strategy that combines three different procedures: a Silverman (1986) inspired normal reference rule that has proven useful in univariate kernel density estimation[5], a bootstrap bandwidth selection approach following the approach of Hall et al. (1999) for estimating conditional distribution functions, and a regression-based bandwidth selector[6] (see Fan et al. 1996). *Step* 1 involves finding an initial value for the smoothing parameter h_z using the rule with normal marginal density. Given this value of h_z, *Step* 2 makes use of the regression-based bandwidth selector to find a value for h_y. In *Step* 3 the bootstrap method is used to revise the estimate of h_z by minimising the bootstrap estimator of a weighted mean square error function. *Step* 2 and *Step* 3 may be repeated one or more times.

Spatial autocorrelation and stochastic kernel estimation

Stochastic kernel estimation rests on the implicit assumption that each region represents an independent observation providing unique information that can be used to estimate the transition dynamics of income. In essence, the cross-section observations at one point in time are viewed as a random sample from a univariate distribution, or in other words, X (where X stands variously for Y and Z) is assumed to be univariate and random. If the X_i ($I = 1, ..., n$) are independent, we say that there is no spatial structure. Independence implies the absence of spatial autocorrelation[7]. Spatial autocorrelation reflects a lack of independence between regions. Dependence may arise from a variety of measurement problems, such as boundary mismatches between the NUTS-2 regions and the growth processes. But also interactions or externalities across regions such as, for example, knowledge spillovers, trade as well as commuting and migration flows are likely to be a major source of the violation of the assumption (see Abreu et al. 2004 for a survey of the existing evidence).

A violation of the independence assumption may result in misguided inferences and interpretations (Rey and Janikas 2005). This problem has been

---

[4] It is well known that the selection of the bandwidth parameters rather than the choice between various kernels is of crucial importance in density estimation.

[5] The rule is to assume that the underlying density is normal and to find the bandwidth which could minimise the integrated mean square error function.

[6] For a given h_z and a given value z, finding $\hat{g}(z \mid y)$ is viewed here as a standard non-parametric problem of regressing $h_z^{-1} K(h_z^{-1} \mid z - Z_i \mid)$ on Y_i.

[7] The controverse is not necessarily true (Ord and Getis 1995). Nevertheless, tests for spatial autocorrelation are typically viewed as appropriate assessments of spatial dependence. Moran's I and Geary's c statistics are typical testing tools (see Chapter B.3).

largely neglected in distribution analysis so far. One way[8] to dealing with the problem involves the filtering of the variable X in order to separate spatial effects from the variable's total effects. While insuring spatial independence, this allows us to use the stochastic kernel to properly estimate the underlying regional income distribution and to analyse its evolution over time. The motivation for a spatial filter is simply that a spatially autocorrelated variable can be transformed into an independent variable by removing the spatial dependence embedded in it. The original variable, X, is hence partitioned into two parts, a filtered non-spatial variable, say $\tilde{X}$, and a residual spatial variable L_X. The transformation procedure depends on identifying an appropriate distance δ within which nearby regions are spatially dependent, and examining each individual observation for its contribution to the spatial dependence embedded in the original variable (Getis and Griffith 2002).

There have been several suggestions for identifying δ, but in this contribution we adopt the Getis filtering approach (see Getis 1990, 1995) which is based on the local spatial autocorrelation statistic G_i (Getis and Ord 1992) to be evaluated at a series of increasing distances until no further spatial autocorrelation is evident. As distance increases from an observation (region i), the G_i-value also increases if spatial autocorrelation is present. Once the G_i-value begins to decrease, the limit on spatial autocorrelation is assumed to have been reached, and the associated critical δ identified. The filtered observation $\tilde{x}_i$ is given as

$$\tilde{x}_i = \frac{x_i \left[\frac{1}{n-1} W_i\right]}{G_i(\delta)} \qquad (E.2.11)$$

where x_i is the original income observation for region i, n is the number of observations and

$$W_i = \sum_{j=1}^{n} w_{ij}(\delta) \quad \text{for } j \neq i \qquad (E.2.12)$$

---

[8] Griffith's eigenfunction decomposition approach that uses an eigenfunction decomposition based on the geographic connectivity matrix used to compute a Moran's I statistic provides an alternative way (see Griffith 2006, and Chapter B.5).

with $w_{ij}(\delta) = 1$ if the distance[9] from region i to region j $(i \neq j)$, say d_{ij}, is smaller than the critical distance band δ, and $w_{ij}(\delta) = 0$ otherwise. $G_i(\delta)$ is the spatial autocorrelation statistic[10] of Getis and Ord (1992) defined as

$$G_i(\delta) = \frac{\sum_{j=1}^{n} w_{ij}(\delta) x_j}{\sum_{j=1}^{n} x_j} \quad \text{for } i \neq j. \qquad \text{(E.2.13)}$$

The numerator in Eq. (E.2.13) is the sum of all x_j within δ of i but not including x_i. The denominator is the sum of all x_j not including x_i.

Equation (E.2.11) compares the observed value of $G_i(\delta)$ with its expected value, $(n-1)^{-1} W_i$. $E[G_i(\delta)]$ represents the realisation, $\tilde{X}$, of the variable X at region i when no autocorrelation occurs. If there is no autocorrelation at i to distance δ, then the observed and expected values, x_i and $\tilde{x}_i$, will be the same. When $G_i(\delta)$ is high relative to its expectation, the difference $x_i - \tilde{x}_i$ will be positive, indicating spatial autocorrelation among high observations of X. When $G_i(\delta)$ is low relative to its expectation, the difference will be negative, indicating spatial autocorrelation among low observations of X. Thus, the difference between x_i and $\tilde{x}_i$ represents the spatial component of the variable X at i. Taken together for all i, L_X represents a spatial variable associated, but not correlated, with the variable X. Thus, $L_X + \tilde{X} - X$ (Getis and Griffith 2002).

Combining this spatial filtering approach with stochastic kernel estimation as described in the previous section yields the long-run (ergodic) density, $f_\infty(\tilde{z})$, implied by the estimated $g_\tau(\tilde{z} \mid \tilde{y})$ function

$$f_\infty(\tilde{z}) = \int_0^\infty g_\tau(\tilde{z} \mid \tilde{y}) \, f_\infty(\tilde{y}) \, d\tilde{y} \qquad \text{(E.2.14)}$$

where $\tilde{y}$ and $\tilde{z}$ denote the spatially filtered observations of Y and Z, respectively. To assess the role played by space on income growth and convergence dynamics

---

[9] In this study distances are measured in terms of geodesic distances between regional centers.

[10] Getis and Ord (1992), and Ord and Getis (1995) show that the statistic $G(\delta)$ is asymptotically normally distributed as δ increases. When the underlying distribution of the variable in question is skewed, appropriate normality of the statistic can be guaranteed when the number of j neighbors is large.

across the regions, we consider a specific stochastic kernel[11] that maps the distribution Y to the spatially filtered distribution $\tilde{Y}\,|\,Y$ so that

$$g(\tilde{y}\,|\,y) = \frac{f(y,\tilde{y})}{f(y)} \qquad \text{(E.2.15)}$$

where the stochastic kernel does not describe transitions over time, but transitions from unfiltered to spatially filtered regional income distributions, and, thus, quantifies the effects of spatial dependence. If spatial effects caused by spatial interaction among regions and measurement problems would not matter, then the stochastic kernel would be the identity map.

E.2.3 Revealing empirics

This section applies the above framework to study regional income dynamics and convergence in Europe. In this section we describe the data and the observation units. Kernel smoothed densities and Tukey boxplots are used then to study the shape dynamics of the distribution. Cross-profile plots, continuous stochastic kernels and implied ergodic distributions are taken to investigate intra-distribution dynamics and long-run tendencies in the data. Finally, the section proceeds to the spatial filtering view of the data to gain insights not affected by the spatial autocorrelation problem.

Data and observation units

We use per capita GRP over the period 1995-2003 expressed in Euros. The GRP figures were calculated on the basis of the 1995 European System of Integrated Economic Accounts (ESA 95)[12] and extracted from the Eurostat Regio database.

---

[11] Combining stochastic kernel estimation with the conditioning scheme suggested by Quah (1996b, 1997a) is an alternative way to evaluate the role of spatial interactions among neighboring regions. Conditioning means here normalising each region's observations by the (population weighted) average income of its neighbors. This approach removes substantive, but not nuisance spatial dependence effects.

[12] In order to deal with the widely known problem measuring Groningen's GRP figure we replaced its energy specific gross value added component by the average of the neighboring regions (Drenthe and Friesland).

We use GRP per capita in national PPS (purchasing power standards) as defined by Eurostat.[13]

The time period is relatively short due to a lack of reliable figures for the regions in the new member states of the EU. This comes partly from the substantial change in measurement methods of national accounts in Central and East Europe (CEE) between 1991 and 1995. But more important, even if estimates of the change in the volume of output did exist, these would be impossible to interpret meaningfully because of the fundamental change of production from a centrally planned to a market system. As a consequence, figures for GRP are difficult to compare until the mid-1990s (Fischer and Stirböck 2006).

The observation units of the analysis are NUTS-2 regions[14]. Although varying considerably in size, NUTS-2 regions are those regions that are adopted by the European Commission for the evaluation of regional growth and convergence processes. NUTS is an acronym of the French for 'the nomenclature of territorial units for statistic', which is a hierarchical system of regions used by the statistical office of the European Community for the production of regional statistics. Our sample includes 257 NUTS-2 regions[15] covering the 27 member states of the EU (see the appendix for a description of the regions):

- *the EU-15 member states*: Austria (nine regions), Belgium (eleven regions), Denmark (one region), Finland (five regions), France (22 regions), Germany (40 regions), Greece (thirteen regions), Ireland (two regions), Italy (20 regions), Luxembourg (one region), Netherlands (twelve regions), Portugal (five regions), Spain (16 regions), Sweden (eight regions), UK (37 regions);
- *the twelve new member states*: Bulgaria (six regions), Cyprus (one region), Czech Republic (eight regions), Estonia (one region), Hungary (seven regions), Latvia (one region), Lithuania (one region), Malta (one region), Poland (16 regions), Romania (eight regions), Slovakia (four regions), Slovenia (one region).

---

[13] Figures given in PPPs are derived from figures expressed in national currency by using PPPs as conversion factors. These parities are obtained as a weighted average of relative price ratios in respect to a homogeneous basket of goods and services, both comparable and representative for each individual country. The use of national purchasing power parities is based on the assumption that there are no – or negligible – purchasing power disparities between the regions within individual countries. This assumption may not appear to be entirely realistic, but it is inevitable in view of the data available.

[14] Note that the use of administratively defined regions, such as NUTS-2 regions, can lead to misleading inferences due to the presence of significant nuisance spatial dependence. In the case of Hamburg, for example, the NUTS-2 boundary is very narrowly drawn with respect to the corresponding functional region so that residential areas extend well beyond the boundary and substantial in-commuting takes place. This implies that per capita GRP is overestimated, while in the surrounding NUTS-2 regions underestimated.

[15] We exclude the Spanish North African territories of Ceuta y Melilla, the Portuguese non-continental territories Azores and Madeira, and the French Départements d'Outre-Mer Guadeloupe, Martinique, French Guayana and Réunion.

Shape dynamics of the distribution

When studying income distribution dynamics across regions in Europe, one can consider incomes per region in absolute terms. Alternatively, one can study regional incomes normalized by the European average. Although there are merits to using the absolute income distribution, it is more natural to take relative incomes when considering changes in income distributions over time. Relative incomes allow us to abstract from overall changes in income levels[16]. A natural approach to assess the shape dynamics of the distribution change over the observation period 1995-2003 is to estimate the cross-sectional distributions by using non-parametric kernel smoothing procedures, which avoid the strong restrictions imposed by parametric estimation. In this framework, if there is a bimodal density at a given point in time, indicating the presence of two groups in the population of regions, convergence implies a tendency of the distribution to move progressively towards unimodality.

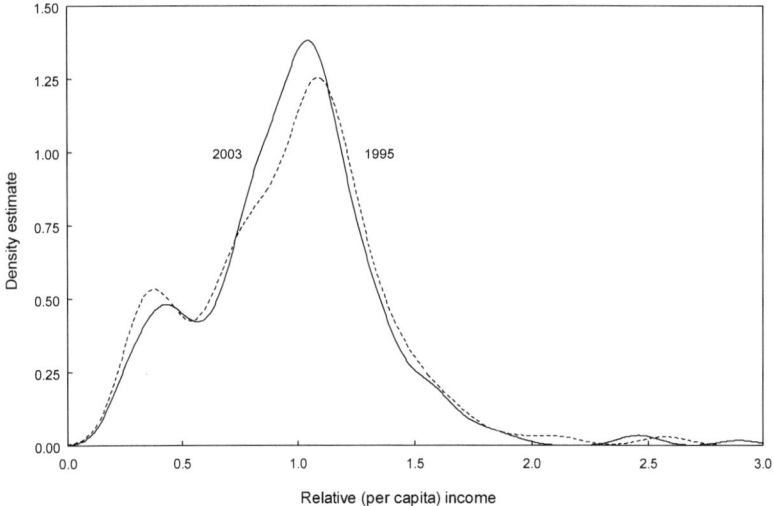

Fig. E.2.1. Distributions of relative (per capita) regional income, 1995 versus 2003

Notes: The plots are densities calculated non-parametrically using a Gaussian kernel with bandwidth chosen as suggested in Silverman (1986), restricting the domain to be non-negative. The *solid line* shows the density for 2003 and the *dashed line* that for 1995

---

[16] This normalization makes it possible to separate the global (European) effects on the cross-section distribution of European forces from the effects from regional-specific effects.

Figure E.2.1 plots the distribution of (per capita) GRP relative to the average of all 257 regions – what we call the Europe relative (per capita) income or simply the relative income. The plots are densities and can be interpreted as the continuous equivalent of a histogram, where the number of intervals has been let tend to infinity and then to the continuum. All densities were calculated non-parametrically using a Gaussian kernel with bandwidths chosen as suggested in Silverman (1986), restricting the range to the positive interval. The solid line shows the distribution in 2003, and the dashed line that in 1995. To read this type of figure, note that 1.0 on the horizontal axis indicates the European average of regional income, 2.0 indicates twice the average, and so on. The height of the curve over any point gives the probability that any particular region will have that relative income. Since the height of the curve at any particular point gives the probability, the area under the curve between, say 0.0 and 1.0, gives the total likelihood that a region will have a relative income that is between 0.0 and 1.0.

The figure shows a distribution with twin-peaks – to use the appellation coined by Quah (1993a) – in 1995, one corresponding to low income regions and the other to middle-income ranges, and a long tail with two smaller bumps at the upper end of the distribution. Technically, the income distribution is said to show a bimodal shape. The main mode[17] is located at about 110 percent of the European average, and the second mode at about 38 percent. The estimated densities reveal several changes over the observation period. The kernel estimated median value decreases by two percent, while the level of dispersion exhibits a small reduction. The kernel estimated standard deviation decreases by 3.3 percent from 0.393 in 1995 to 0.380 in 2003.

Perhaps most remarkable is the change in the shape of the distributions. By 2003, the peaks have become closer together, and the richer peak has risen moderately at the expense of the poorer. We see this by noting that the area under the 2003 curve, that is between 0.5 and 1.1, is greater than the corresponding area under the 1995 curve, while the area that is to the left of 0.5 is smaller. The smaller peak seems to progressively collapse over time. This finding may suggest an improvement in economic conditions of the poorest regions and reflect a trend, in some sense, of catching-up.

Figure E.2.2 gives a sequence of Tukey boxplots for the 257 NUTS-2 regions. Recall that the units of income are PPS units scaled to the EU-27 average. Time appears on the horizontal axis, while the vertical axis maps relative per capita income values. To understand these pictures, recall the construction of a Tukey boxplot. Each boxplot includes a box bounded by Q_1 and Q_3 denoting sample quartiles. Thus, the box contains the middle 50 percent of the distribution. The thick line in the box locates the median. The upwards and downwards distances from the median to the top and bottom of the box provide information on the shape of the distribution. If these distances differ, then the distribution is asymmetric. Thin dashed vertical lines emanating from the box both upwards and

---

[17] A mode is defined as a point at which the gradient changes from positive to negative.

downwards, reach upper and lower adjacent values, respectively. The upper adjacent value is the largest value observed that is not greater than the top quartile plus 1.5 times $(Q_3 - Q_1)$. The lower quartile is similarly defined, extending downwards from the 25th percentile. Dots indicate upper and lower outside values, that is, observations that lie outside the upper and lower adjacent values, respectively. These denote regions which have performed extraordinarily well or extraordinarily poorly relative to the set of other regions. Of course, upper and lower outside values might not exist. The adjacent values might already be the extreme points in a specific realisation.

There are no extraordinarily poorly performing regions, more accurately when regions performed especially badly, they were not alone. On the upside, by contrast, the figure shows several outstanding performers. At the beginning of the sample, five regions showed upper outside values, and by the end of the sample six outside values. The spreading apart in the regional income distribution has one distinct source, the pulling away of the upper outside values – representing Inner London, Brussels, Luxembourg, Hamburg, Île-de-France and Vienna – from the rest of the regions. The figure, moreover, makes clear that the interquartile range is decreasing by more than 15 percent, and this falling is due to a decrease of Q_3 rather than Q_1.

The matching counterparts in Fig. E.2.1 and Fig. E.2.2 use exactly the same data. But they emphasize different empirical regularities. The bimodal shape is striking in Fig. E.2.1, but is far from obvious in Fig. E.2.2. The spreading out of the upper tail of the distribution is apparent in Fig. E.2.2. It appears in form of two smaller bumps in Fig. E.2.1.

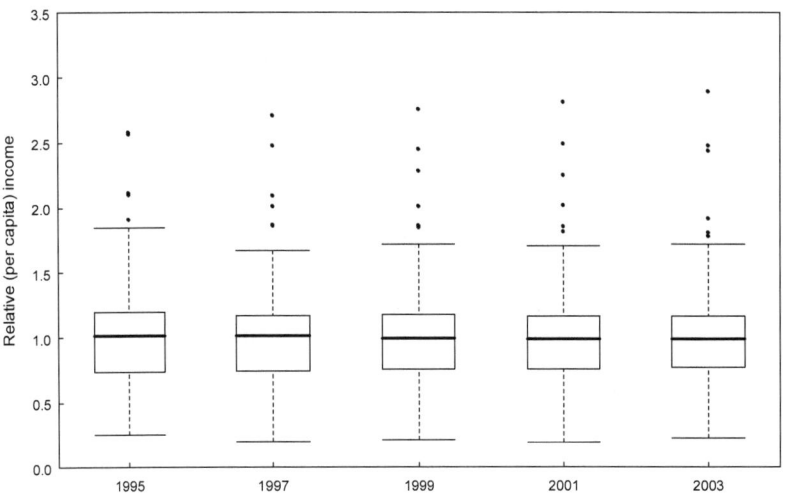

Fig. E.2.2. Tukey boxplots of relative (per capita) regional income across 257 European regions

Intra-distribution dynamics and long-run tendencies

Thus far, we have considered only point-in-time snapshots of the income distribution across the regions. This section takes the next step in the analysis, and looks at the intra-distribution dynamics and then at the long-run (ergodic) tendencies. We start with Fig. E.2.3 showing cross-profile dynamics[18]. The vertical axis is the log of relative (per capita) incomes. Each curve in the figure refers to the situation at a given point in time. The lowest curve gives the cross-section of regions at time 1995 in increasing order. This ordering is then maintained throughout the time periods considered. Proceeding upwards, we see curves for 1999 and 2003. The character of the upper plots, thus, depends on 1995 when the ordering is taken.

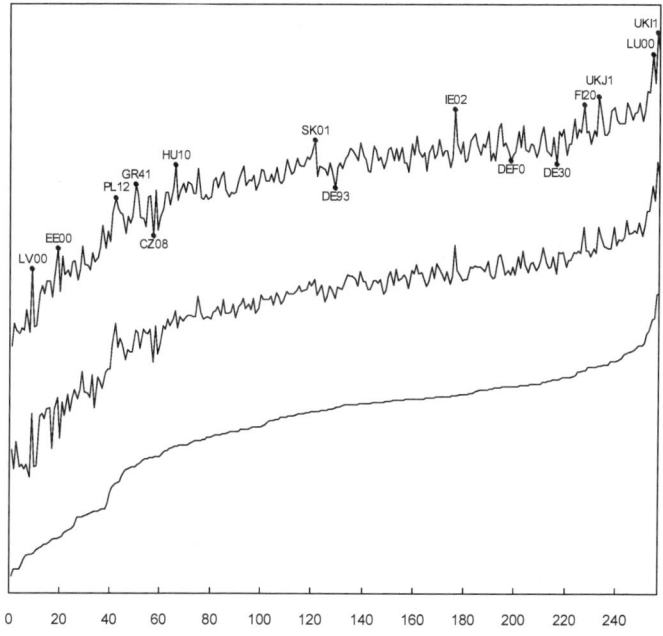

Fig. E.2.3. Cross-profile dynamics across 257 European regions, retaining the ranking fixed at the initial year, relative (per capita) income, advancing upwards: 1995, 1999 and 2003 (a guide to region codes can be found in the appendix)

In the plots, increasing jaggedness indicates intra-distribution mobility. In contrast, if each cross-profile would always monotonically increase over time, then income rankings were invariant. The most striking feature of Fig. E.2.3 is not this comparative stability through time. It is the change in choppiness through

---

[18] The idea for this picture comes from Quah (1997a), and López-Bazo et al. (1999).

time in the cross-profile plots indicated by local peaks. By 2003, we observe local peaks, for example, at the lower end of the distribution around regions ranked 9th, 19th, 42nd and 66th poorest in 1995, and at the upper end around regions ranked second and fourth richest. These turn out to be Latvia, Estonia, Mazowieckie (Warszawa) and Közép-Magyarország (Budapest), and Inner London and Luxembourg, respectively. By contrast, Moravskoslezko (57th poorest in 1995) in the Czech Republic, Lüneburg (129th poorest) and Berlin (the 41st richest region) experienced economically significant relative declines by 2003. The cross-profile dynamics are informative. They illustrate when regions overtake one another, fall behind, or pull ahead. But they do not identify underlying dynamic regularities in the data. We thus turn to the stochastic kernel representation of intra-distribution dynamics next.

Figure E.2.4 shows the conditional kernel density estimate $\hat{g}_\tau(z \mid y)$ with fixed bandwidths ($h_y = 0.036$, $h_z = 0.023$) [19] that describes the stochastic kernel across the 257 regions, averaging over 1995 through 2003. The stochastic kernel has been estimated for a five-year transition period, setting $\tau = 5$. The figure displays the estimate, using Hyndman's (1996) visualisation tools. Figure E.2.4(a) presents the stochastic kernel in terms of a three-dimensional stacked conditional density plot in which a number of conditional densities are plotted side by side in a perspective plot. For any point y on the period t axis, looking in the direction parallel to the $t+5$ time axis traces out a conditional probability density. The graph shows how the cross-section income distribution at time t evolves into that at time $t+5$. Just as with a transition probability matrix in a discrete set up, the 45-degree diagonal in the graph indicates persistence properties. When most of the graph is concentrated along this diagonal, then the elements in the cross-section distribution remain where they started. As evident from Fig. E.2.4(a), a large portion of the probability mass remains clustered along the main diagonal over the five-year horizon, and most of the peaks lie along this line indicating a low degree mobility and modest change in the regional income distribution.

The highest density regions (HDRs) boxplot, given in Fig. E.2.4(b), makes this clearer. A highest density region is the smallest region of the sample space containing a given probability. Figure E.2.4(b) shows a plot of the 50 percent and 99 percent highest density regions[20], computed from the density estimates shown in Fig. E.2.4(a). Each vertical strip represents the conditional density for one y value. The darker shaded region in each strip is a 50 percent HDR, and the lighter shaded region is a 99 percent HDR. The mode for each conditional density is shown as a bullet •.

---

[19] The bandwidths for the estimator were chosen according to Bashtannyk and Hyndman's three-step-strategy. See Section E.2.2 for more details.

[20] An HDR boxplot replaces the box bounded by the interquartile range with the 50 percent HDR, the region bounded by the upper and lower adjacent values is replaced by the 99 percent HDR that roughly reflects the probability coverage of the adjacent values on a standard boxplot for a normal distribution. In keeping with the emphasis on highest density, the mode rather than the median is marked.

(a)

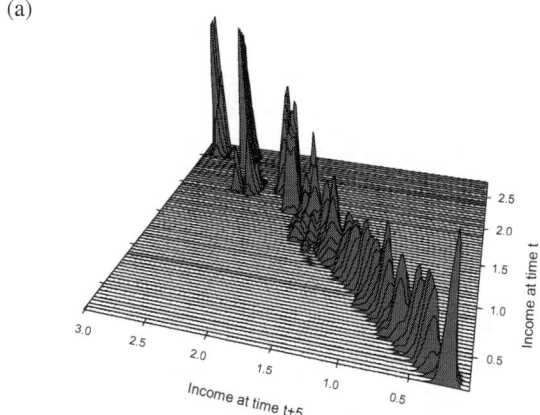

(b)

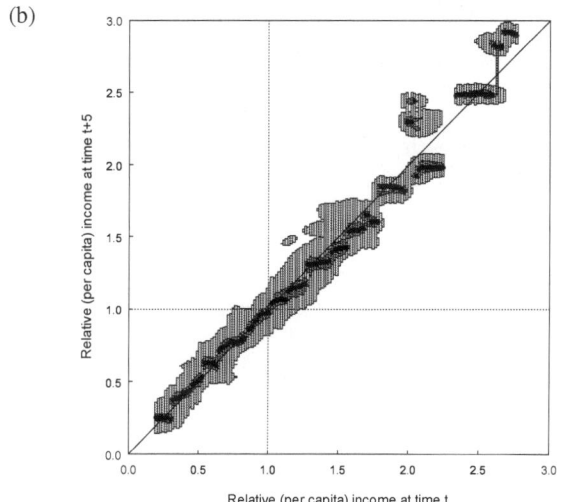

Fig. E.2.4. Relative income dynamics across 257 European regions, the estimated $g_5(z|y)$, see Eq. (E.2.6): (a) stacked density plot, and (b) highest density regions boxplot

Notes: ad (b) The *lighter shaded regions* in each strip is a 99% HDR, and the *darker shaded regions* a 50% HDR. The mode for each conditional density is shown as a bullet •. *Technical notes*: The conditional density $g_\tau(z|y)$ is estimated over a 5-year transition horizon $\tau = 5$ between 1995-2003. Estimates are based on a Gaussian product kernel density estimator with bandwidth selection ($h_y = 0.036$, $h_z = 0.023$) based on the three-step-strategy suggested by Bashtannyk and Hyndman (2001). The stacked conditional density plot and the high density region boxplot were estimated at 70 and 150 points, respectively. Calculations of the plots were performed using the R package HRDCDE, provided by Rob Hyndman

The vertical dashed line at 1.0 marks regions with income equal to the European average at time t, and the horizontal dashed line at 1.0 those with income equal to the average at $t+5$. The 45-degree diagonal indicates intra-distribution persistence over the five-year transition horizon.

To read this type of boxplot note that strong persistence is evidenced when the main diagonal crosses the 50 percent HDRs. It means that most of the elements in the distribution remain where they started. There is a low persistence and more intra-distribution mobility if that diagonal crosses only the 99 percent HDRs. Strong (weak) global convergence towards equality would manifest in fifty percent (99 percent) HDRs crossed by the horizontal line at 1.0. Fifty percent HDRs consisting of two disjoint intervals would indicate a two-peaks property of the distribution.

The plot not only reveals persistence, but also mobility and polarisation features. Regions with an income range of 0.8–1.2 times the European average show strong persistence. Some mobility occurs at the extremes of the distribution, more at the upper extreme than at the lower. Some portions of the cross-section in the income range below 0.8 times the average tend to slightly increase their relative position over the five-year transition horizon, indicating a process of catching-up of the poorest regions with the richer ones. In contrast, portions in the income range above 1.2–1.8 times the average lose out their relative position, becoming relatively poorer. The boxplot also shows signs of polarisation, the opposite of catching-up. This is indicated by the disjoint intervals of the 50 and 99 percent HDRs at the upper extreme of the income range. We see that regions starting with an income of 2.0–2.3 times the European average at time t are unlikely to remain there. Most see their European relative income fall and others rise, with the result that this income class appears to vanish. The position of a small very rich group around 2.3–2.6 times the average remains either unchanged or shifting away.

The evidence of Fig. E.2.4 is corroborated by the ergodic density function that is obtained by solving Eq. (E.2.4). Figure E.2.5 plots the estimated long-run (ergodic) density[21], $\hat{f}_\infty(z)$, implied by the estimated $g_\tau(z|y)$ function for $\tau = 5$, along with the initial income distribution. The solid line shows the point estimate of the ergodic distribution and the dashed line the initial income distribution. Comparing these two distributions we see that the ergodic distribution is wider, both at the top and at the bottom. This reflects a shift in the mass of the distribution away from the lower end to the middle, and from the middle to the upper end. In particular, the peak in the initial distribution between 20 and 50

---

[21] It is well known that the shape of the estimated ergodic density is sensitive to the bandwidths chosen in computing the underlying estimated joint density functions. Wider bandwidths tend to obscure detail in the shapes while narrower bandwidths tend to increase it but possibly spuriously so. It is important to note that smaller equi-proportionate decreases and increases in bandwidths do not remove the tendency to bimodality in the ergodic density.

percent of the European relative per capita income has shifted upward into the 60 to 100 percentage range and shows a tendency to disappear.

The stationary distribution across the 257 regions, plotted in Fig. E.2.5, is distinctively bimodal. The dominant peak[22] represents regions clustered just below the European average income, while a small group of relatively rich regions gathers around three times of the average European (per capita) income. The bimodal nature of the ergodic distribution in comparison with the initial income distribution provides indication for two types of processes at work over time: a gradual and slow catching-up of the poorest regions which turn out to be – with very few exceptions – regions in Central and Eastern Europe, and simultaneously a tendency towards polarisation – a small group of richer regions separating from the rest of the cross-section.

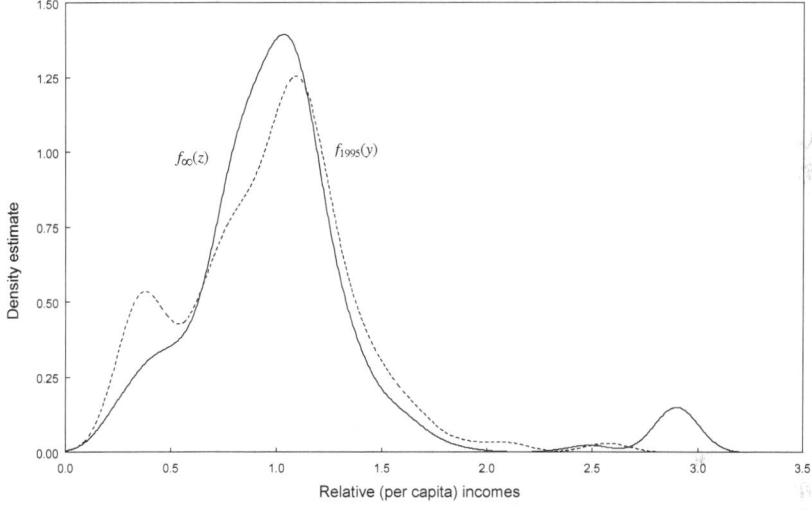

Fig. E.2.5. The ergodic density $f_\infty(z)$ implied by the estimated $g_5(z|y)$ and the marginal density function $f_{1995}(y)$

Notes: The *solid line* shows the point estimate for $f_\infty(z)$ and the *dashed line* the estimate for the marginal density $f_{1995}(y)$. The ergodic function $f_\infty(z)$ has been found as solution to Eq. (E.2.4)

The bimodal shape of the ergodic distribution contradicts with Quah's (1996a) unimodal ergodic solution found in a discrete state space set up with a largely reduced set of 78 European regions over 1980-1989. The observation, however, is

---

[22] The upper peak, however, is imprecisely estimated. Only few observations were actually made there, and the precision of the estimate is low.

in line with Pittau and Zelli's (2006) findings, obtained for a set of 110 regions covering twelve EU member countries over the time period from 1977 to 1996.

To sum up this first pass through the data, we conclude that the data show a wide spectrum of intra-distribution dynamics. Overtaking and catching-up occur simultaneously with persistence and polarisation. Polarisation manifests itself in the emergence of a twin-peak structure in the long-run regional income distribution.

The spatial filtering perspective

Large significant and positive values of Moran's I reveal the presence of spatial association of similar values of neighbouring European regions in relative (per capita) income[23]. This motivates a spatial filtering pass[24] through the data to avoid inferences and interpretations, misguided by the violation of the independence assumption in the previous analysis.

Figure E.2.6 presents the spatially filtered counterpart of Fig. E.2.1. Comparing these densities with those in Fig. E.2.1 indicates that the mode, which was situated at around 38 percent of the European average, has disappeared. Consequently, the economic performance of the regions is well explained by the neighbouring regions' performances, except may be for regions with very high relative (per capita) income.

The filtered distributions in this figure are tighter and more concentrated than those in Fig. E.2.1. The boxplots in Fig. E.2.7 make this particularly clear. Upper and lower outliers exist here, but the 25th and 75th percentiles are located close to the average income. Lower and upper adjacent values are compactly situated within about 0.5 and 1.5 times average income levels. The filtered distribution has a kernel estimated standard deviation of 0.262 in 1995, which increases to 0.283 in 1999, and then to 0.310 in 2003. The increase over the time 1995-2003 is 15 percent. The estimated standard deviations of the unfiltered data were found to be 0.393 in 1995 and 0.380 in 2003, indicating a slight decline by 3.3 percent.

From this, it is clear that the evidence for σ-convergence found above is caused by spatial dependence embedded in the income data[25].

---

[23] Using Moran's I, the spatial autocorrelation latent in each of the income variables ranges from $z(MI)=8.86$ for the 1995 income variable to $z(MI)=8.06$ for the 2003 income variable where $z(MI)$ denotes the z-score value of Moran's I. From this, it is clear that there is a strong spatial autocorrelation, and hence the assumption of spatial independence does not hold.

[24] Rather than use an individual δ for each observation, the modal value for δ was chosen for each income variable as recommended by Getis and Griffith (2002).

[25] See Rey and Dev (2006) for appropriate inference methods of σ-convergence in the presence of spatial effects.

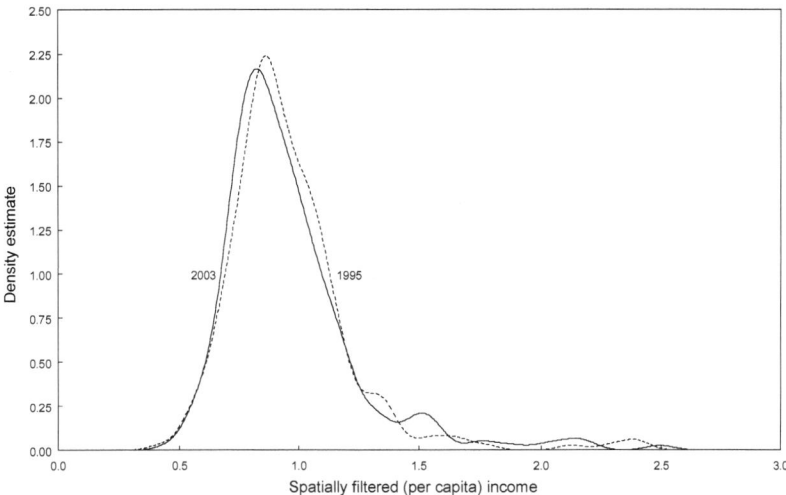

Fig. E.2.6. Densities of relative (per capita) income, 1995 versus 2003: the spatial filtering view

Notes: The plots are densities calculated non-parametrically using a Gaussian kernel with bandwidth chosen as suggested in Silverman (1986), restricting the domain to be non-negative. The *solid line* shows the density for 2003 and the *dashed line* that for 1995

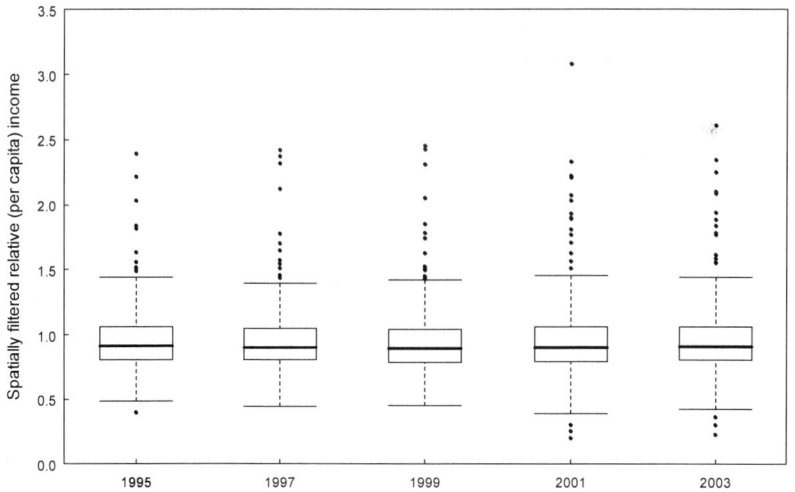

Fig. E.2.7. Tukey boxplots of relative (per capita) income, across 257 European regions: the spatial filtering view

(a)

(b)

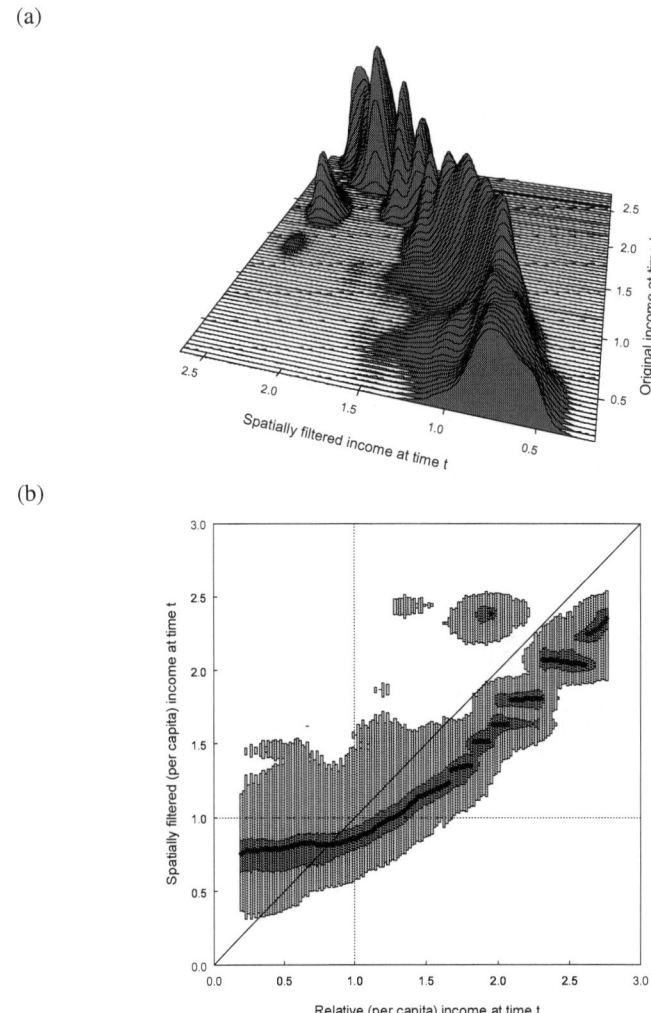

Fig. E.2.8. Stochastic kernel mapping from the original to the spatially filtered distribution, the estimated g $(\tilde{y}|y)$: (a) the stacked conditional density plot, and (b) the highest density regions boxplot

Notes: ad (b) The *lighter shaded region* in each strip is a 99 percent HDR, and the *darker shaded region* a 50 percent HDR. The mode for each conditional density is shown as a bullet •. *Technical notes*: The conditional density $g(\tilde{y}|y)$ is estimated over a 5-year transition horizon $\tau = 5$ between 1995 and 2003. Estimates are based on a Gaussian product kernel density estimation with bandwidth selection ($h_y = 0.103$, $h_{\tilde{y}} = 0.052$) based on the three-step-strategy suggested by Bashtannyk and Hyndman (2001). The stacked conditional density plot and the high density region boxplot were estimated at 70 and 150 points, respectively. Calculations of the plots were performed using the R package HRDCDE, provided by Rob Hyndman, and spatial filtering, using the PPA package, provided by Arthur Getis

(a)

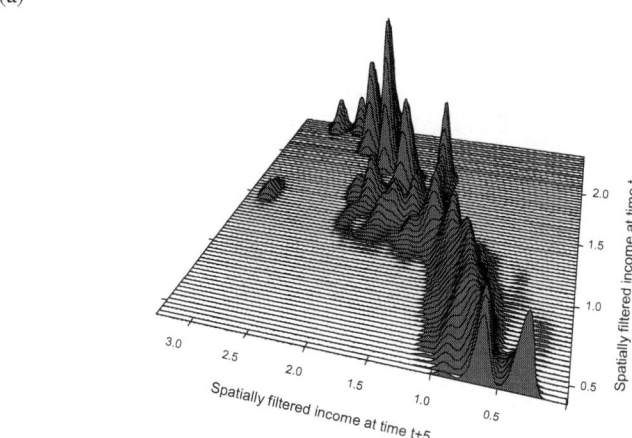

(b)

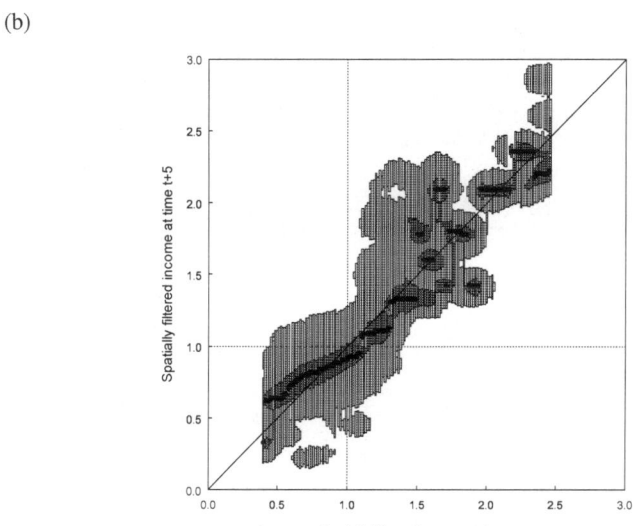

Fig. E.2.9. The spatial filter view of relative income dynamics: the estimated $g_5(\tilde{z} \mid \tilde{y})$, (a) stacked density plot, and (b) highest density regions boxplot

Notes: ad (b) The *lighter shaded region* in each strip is a 99% HDR, and the *darker shaded region* a 50% HDR. The mode for each conditional density is shown as a bullet •. *Technical notes*: The conditional density $g_\tau(\tilde{z}|\tilde{y})$ is estimated over a 5-year transition horizon $\tau = 5$ between 1995-2003. Estimates are based on a Gaussian product kernel density estimator with bandwidth selection ($h_{\tilde{y}} = 0.061$, $h_{\tilde{z}} = 0.047$) based on the three-step-strategy suggested by Bashtannyk and Hyndman (2001). The stacked conditional density plot and the high density region boxplot were estimated at 70 and 150 points, respectively. Calculations of the plots were performed using the R package HRDCDE, provided by Rob Hyndman, and spatial filtering using the PPA package, provided by Arthur Getis

More information on the role of spatial effects becomes evident when looking at the stochastic kernel in Fig. E.2.8 that shows how the original (unfiltered) relative (per capita) income distribution is transformed into the spatially filtered one. Figure E.2.8(a) displays the conditional kernel density estimate $\hat{g}(\tilde{y}|y)$ with fixed bandwidths ($h_y = 0.103$, $h_{\tilde{y}} = 0.052$) in terms of a three-dimensional stacked conditional plot as given in Fig. E.2.8(a), and an HDR boxplot in Fig. E.2.8(b).

If spatial effects account for a substantial part of the distribution, then the stochastic kernel mapping from the original (unfiltered) to the spatially filtered distribution would depart from the identity map. Indeed, Fig. E.2.8(a) precisely conveys this message. The graph shows the kernel mapping the original to the filtered distribution in the same year. The evident clockwise reversal on the lower, but also on the higher part of the distribution indicates that spatial effects do account for a large part of income dynamics in Europe. Figure E.2.8(b) reinforces this interpretation. The dominant feature in this figure appears to be intra-distribution mobility rather than persistence. Regions with an income less than 0.7 times the European average show a clear tendency towards cohesion. There are strong indications that the probability of the poorest regions to move up is negatively affected by the presence of spatial dependence effects. This is evidenced by the 99 percent HDRs crossing the horizontal line at 1.0 and by the 50 percent HDRs coming much closer to this line. However, while this is happening, the very highest parts of the income distribution show tendencies away from cohesion, and provide evidence for emerging twin peaks.

Figure E.2.9 provides stochastic kernel representations of five-year transition dynamics in the spatially filtered income space, using again a stochastic kernel estimator with fixed bandwidths ($h_{\tilde{y}} = 0.061$, $h_{\tilde{z}} = 0.047$). This figure is the counterpart to Fig. E.2.4 for spatially filtered relative (per capita) regional incomes. Figure E.2.9(a) presents the stochastic kernel in terms of a three-dimensional stacked conditional density plot, and Fig. E.2.9(b) in terms of a highest density regions boxplot.

The picture that emerges from the estimates here is that of a substantial degree of intra-distribution mobility at the upper and lower tails of the income distribution. The remarkably different dynamics that emerge – in comparison to the unfiltered regional income case – suggest that – if we are to evaluate growth and convergence dynamics across regions correctly – the use of spatially filtered data is pretty much essential to avoid misleading interpretations.

E.2.4 Concluding remarks

The study follows the tradition of the non-parametric approach studying both the shape and mobility dynamics of cross-sectional distributions of relative (per capita) income that appears to be generally more informative about the actual patterns of cross-sectional growth than convergence empirics within the β-

convergence regression approach. It differs from most of the previous work by going for a continuous kernel route which is more informative than research with discretely-defined income cells.

This contribution incorporates two novel techniques into the continuous analysis: kernel estimation and more powerful graphical devices for the representation of the stochastic kernel, and Getis' spatial filtering technique to explicitly account for the spatial dimension of the growth process. The chapter illustrates that the use of spatially filtered data is pretty much essential to evaluate growth and convergence dynamics across regions. The lack of an appropriate inferential theory, however, restricts the study to a descriptive stage.

The study has produced some interesting results. *First*, there is no development trap in the long-run into which the poorer Central and Eastern European regions will be permanently condemned. *Second*, the findings suggest a tendency of the cross-section distribution of regional per capita income to split up into two separate groups, where a small group of richer metropolitan regions is growing away from the rest of the European regions. This evidence is coherent to Pittau and Zelli's (2006) stationary distribution estimated on a sample of 110 EU-12 regions over the period 1977-1996. *Third*, spatial effects explain a substantial part of the income distribution, but not the emergence of the two-club regional world in the long-run. Growth theories now need to explain these facts. The distribution dynamics analysis carried out in this chapter does not help further in this respect.

Acknowledgements. The authors gratefully acknowledge the grant no. P19025-G11 provided by the Austrian Science Fund (FWF). They also thank two anonymous referees for their comments which improved the quality of the contribution. The calculations were done using a combination of the R package HDRCDE, provided by Rob Hyndman, and the PPA package, provided by Arthur Getis. Special thanks to Roberto Basile for providing the original stimulus to carry out this study.

References

Abreu M, Groot HLF, Florax RJG (2004) Space and growth: a survey of empirical evidence and methods. Tinbergen Institute, Working Paper TI04-129/3. http://ssrn.com/abstract=631007

Bashtannyk DM, Hyndman RJ (2001) Bandwidth selection for kernel conditional density estimation. Comput StatData Anal 36(3):279-298

Basile R (2006) Intra-distribution dynamics of regional per capita income in Europe: evidence from alternative conditional density estimators. Paper presented at the 53rd North-American Meetings of the RSAI, November 2006, Toronto

Bulli S (2001) Distribution dynamics and cross-country convergence: a new approach. Scott J Pol Econ 48(2):226-243

Durlauf SN, Quah DT (1999) The new empirics of economic growth. In Taylor JB, Woodford M. (eds) Handbook of macroeconomics, volume 1. Elsevier, Amsterdam, pp.235-308

Fan J, Yao Q, Tong H. (1996) Estimation of conditional densities and sensitivity measures in nonlinear dynamical systems. Biometrika 83(1):189-206

Fingleton B (1997) Specification and testing of Markov Chain models: an application to convergence in the European Union. Oxford Bulletin of Econ Stat 59(3):385-403

Fingleton B (1999) Estimates of time to economic convergence: an analysis of regions of the European Union. Int J Reg Sci Rev 22(1):5-34

Fingleton B (ed) (2003) European regional growth. Springer, Berlin, Heidelberg and New York

Fischer MM, Stirböck C (2006) Pan-European regional income growth and club-convergence. Ann Reg Sci 40(4):693-721

Friedman M (1992) Do old fallacies ever die? J Econ Lit 30(4):2129-2132

Getis A (1990) Screening for spatial dependence in regression analysis. Papers of the Reg Sci Assoc 69:69-81

Getis A (1995) Spatial filtering in a regression framework: Examples using data on urban crime, regional inequality, and government expenditures. In Anselin L, Florax R (eds.) New directions in spatial econometrics. Springer, Berlin, Heidelberg and New York, pp.172-188

Getis A (2009) Spatial autocorrelation. In Fischer MM, Getis A (eds) Handbook of applied spatial analysis. Springer, Berlin, Heidelberg and New York, pp.255-278

Getis A, Griffith DA (2002) Comparative spatial filtering in regression analysis. Geogr Anal 34(2):130-140

Getis A, Ord JK (1992) The analysis of spatial association by use of distance statistics. Geogr Anal 24(3):189-206

Griffith DA (2006) Spatial autocorrelation and spatial filtering. Springer, Berlin, Heidelberg and New York

Griffith DA (2009) Spatial filtering. In Fischer MM, Getis A (eds) Handbook of applied spatial analysis. Springer, Berlin, Heidelberg and New York, pp.301-318

Hall P, Wolff RC, Yao Q (1999) Methods for estimating a conditional distribution function. J Am Stat Assoc 94(445):154-163

Hyndman RJ (1996) Computing and graphing highest density regions. The Am Stat 50(2):120-126

Hyndman RJ, Yao Q (2002) Nonparametric estimation and symmetry tests for conditional density functions. J Nonparametric Stat 14(3):259-278

Hyndman RJ, Bashtannyk DM, Grunwald GK (1996) Estimating and visualizing conditional densities. J Comput Graph Stat 5(4):315-336

Islam N (2003) What have we learnt from the convergence debate? J Econ Surv 17(3):309-362

Johnson PA (2004) A continuous state space approach to 'Convergence by parts'. Vassar College Economics Working Paper 54, Department of Economics, Vassar College

Jones CI (1997) On the evolution of the world income distribution. J Econ Persp 11(3):19-36

LeGallo J (2004) Space-time analysis of GDP disparities among European regions: a Markov Chains approach. Int J Reg Sci Rev 27(2):138-163

López-Bazo E, Vaya E, Mora AJ, Surinach J (1999) Regional economic dynamics and convergence in the European Union. Ann Reg Sci 33(3):343-370

Magrini S (1999) The evolution of income disparities among the regions of the European Union. Reg Sci Urb Econ 29(2):257-281

Magrini S (2004) Regional (di)convergence. In Henderson JV, Thisse J-F (eds) Handbook of regional and urban economics. Elsevier, Amsterdam, pp.2741-2796

Ord JK, Getis A (1995) Local spatial autocorrelation statistics: distributional issues and an application. Geogr Anal 27(4):286-306

Paap R, van Dijk HK (1998) Distribution and mobility of wealth of nations. Europ Econ Rev 42(7):1269-1293

Pittau MG, Zelli R (2006) Empirical evidence of income dynamics across EU regions. J Appl Econ 21(5):605-628

Quah DT (1993a) Empirical cross-section dynamics in economic-growth. Europ Econ Rev 37(2-3):426-434

Quah DT (1993b) Galtons fallacy and tests of the convergence hypothesis. Scand J Econ 95(4):427-443

Quah DT (1996a) Regional convergence clusters across Europe. Europ Econ Rev 40(3-5):951-958

Quah DT (1996b) Empirics for economic growth and convergence. Europ Econ Rev 40(6):1353-1375

Quah DT (1997a) Regional cohesion from local isolated actions: I. Historical outcomes. Centre for Economic Performance, Discussion Paper No. 378. School of Economics, London

Quah DT (1997b) Regional cohesion from local isolated actions: II. Conditioning. Centre for Economic Performance, Discussion Paper No. 379. School of Economics, London

Quah DT (1997c) Empirics for growth and distribution: stratification, polarization, and convergence clubs. J Econ Growth 2(1):27-59

Quah DT (1999) Regional cohesion from local isolated actions – historical outcomes. In Study of the Socio-economic Impact of the Projects Financed by the Cohesion Fund – A Modelling Approach, volume 2. Office for Official Publications of the European Communities, Luxembourg

Reichlin L (1999) Discussion of 'convergence as distribution dynamics' by D Quah. In Baldwin R, Cohen D, Sapir A, Venables A (eds) Market, integration, regionalism, and the global economy. Cambridge University Press, Cambridge, pp.328-335

Rey SJ (2001) Spatial empirics for economic growth and convergence. Geogr Anal 33(3):195-214

Rey SJ, Dev B (2006) σ-convergence in the presence of spatial effects. Papers in Reg Sci 85(2):217-234

Rey SJ, Janikas MV (2005) Regional convergence, inequality and space. J Econ Geogr 5(2):155-176

Silverman BW (1986) Density estimation for statistics and data analysis. Chapman and Hall, London and New York

Temple J (1999) The new growth evidence. J Econ Lit 37(1):112-156

Wand MP, Jones MC (1995) Kernel smoothing. Chapman and Hall, London

Wheeler D, Páez A (2009) Graphically weighted regression. In Fischer MM, Getis A (eds) Handbook of applied spatial analysis. Springer, Berlin, Heidelberg and New York, pp.461-486

Appendix

NUTS is an acronym of the French for the 'nomenclature of territorial units for statistics", which is a hierarchical system of regions used by the statistical office of the European Community for the production of regional statistics. At the top of the hierarchy are NUTS-0 regions (countries) below which are NUTS-1 regions and then NUTS-2 regions. The sample is composed of 257 NUTS-2 regions located in 27 EU member states (NUTS revision 1999, except for Finland NUTS revision 2003). We exclude the Spanish North African territories of Ceuta and Melilla, and the French Départements d'Outre-Mer Guadeloupe, Martinique, French Guayana and Réunion, and the Portuguese non-continental territories Azores and Madeira. Thus, we include the NUTS-2 regions listed in the table.

Country	ID Code	Region	Country	ID Code	Region
Austria	AT11	Burgenland	Finland	FI13	Itä-Suomi
	AT12	Niederösterreich		FI18	Etelä-Suomi
	AT13	Wien		FI19	Länsi-Suomi
	AT21	Kärnten		FI1A	Pohjois-Suomi
	AT22	Steiermark		FI20	Åland
	AT31	Oberösterreich	France	FR10	Île de France
	AT32	Salzburg		FR21	Champagne-Ardenne
	AT33	Tirol		FR22	Picardie
	AT34	Vorarlberg		FR23	Haute-Normandie
Belgium	BE10	Région de Bruxelles-Capitale		FR24	Centre
	BE21	Prov. Antwerpen		FR26	Bourgogne
	BE22	Prov. Limburg (B)		FR30	Nord-Pas-de-Calais
	BE23	Prov. Oost-Vlaanderen		FR41	Lorraine
	BE24	Prov. Vlaams Brabant		FR42	Alsace
	BE25	Prov. West-Vlaanderen		FR43	Franche-Comté
	BE31	Prov. Brabant Wallon		FR51	Pays de la Loire
	BE32	Prov. Hainaut		FR52	Bretagne
	BE33	Prov. Liège		FR53	Poitou-Charentes
	BE34	Prov. Luxembourg (B)		FR61	Aquitaine
	BE35	Prov. Namur		FR62	Midi-Pyrénées
Bulgaria	BG11	Severozapaden		FR63	Limousin
	BG12	Severen tsentralen		FR71	Rhône-Alpes
	BG13	Severoiztochen		FR72	Auvergne
	BG21	Yugozapaden		FR81	Languedoc-Roussillon
	BG22	Yuzhen tsentralen		FR82	Provence-Alpes-Côte d'Azur
	BG23	Yugoiztochen		FR83	Corse
Cyprus	CY00	Kypros / Kibris	Germany	DE11	Stuttgart
Czech	CZ01	Praha		DE12	Karlsruhe
Republic	CZ02	Střední Čechy		DE13	Freiburg
	CZ03	Jihozápad		DE14	Tübingen
	CZ04	Severozápad		DE21	Oberbayern
	CZ05	Severovýchod		DE22	Niederbayern
	CZ06	Jihovýchod		DE23	Oberpfalz
	CZ07	Střední Morava		DE24	Oberfranken
	CZ08	Moravskoslezko		DE25	Mittelfranken
Denmark	DK00	Danmark		DE26	Unterfranken
Estonia	EE00	Eesti		DE27	Schwaben

cont.

Country	ID Code	Region	Country	ID Code	Region
Germany	DE30	Berlin	Italy	ITD5	Emilia-Romagna
	DE40	Brandenburg (Südwest and Nordost)		ITE1	Toscana
	DE50	Bremen		ITE2	Umbria
	DE60	Hamburg		ITE3	Marche
	DE71	Darmstadt		ITE4	Lazio
	DE72	Gießen		ITF1	Abruzzo
	DE73	Kassel		ITF2	Molise
	DE80	Mecklenburg-Vorpommern		ITF3	Campania
	DE91	Braunschweig		ITF4	Puglia
	DE92	Hannover		ITF5	Basilicata
	DE93	Lüneburg		ITF6	Calabria
	DE94	Weser-Ems		ITG1	Sicilia
	DEA1	Düsseldorf		ITG2	Sardegna
	DEA2	Köln	Lithuania	LT00	Lietuva
	DEA3	Münster	Luxembourg	LU00	Luxembourg (Grand-Duché)
	DEA4	Detmold	Latvia	LV00	Latvija
	DEA5	Arnsberg	Malta	MT00	Malta
	DEB1	Koblenz	Netherlands	NL11	Groningen
	DEB2	Trier		NL12	Friesland
	DEB3	Rheinhessen-Pfalz		NL13	Drenthe
	DEC0	Saarland		NL21	Overijssel
	DED1	Chemnitz		NL22	Gelderland
	DED2	Dresden		NL23	Flevoland
	DED3	Leipzig		NL31	Utrecht
	DEE1	Dessau		NL32	Noord-Holland
	DEE2	Halle		NL33	Zuid-Holland
	DEE3	Magdeburg		NL34	Zeeland
	DEF0	Schleswig-Holstein		NL41	Noord-Brabant
	DEG0	Thüringen		NL42	Limburg (NL)
Greece	GR11	Anatoliki Makedonia, Thraki	Poland	PL11	Lódzkie
	GR12	Kentriki Makedonia		PL12	Mazowieckie
	GR13	Dytiki Makedonia		PL21	Malopolskie
	GR14	Thessalia		PL22	Slaskie
	GR21	Ipeiros		PL31	Lubelskie
	GR22	Ionia Nisia		PL32	Podkarpackie
	GR23	Dytiki Ellada		PL33	Swietokrzyskie
	GR24	Sterea Ellada		PL34	Podlaskie
	GR25	Peloponnisos		PL41	Wielkopolskie
	GR30	Attiki		PL42	Zachodniopomorskie
	GR41	Voreio Aigaio		PL43	Lubuskie
	GR42	Notio Aigaio		PL51	Dolnoslaskie
	GR43	Kriti		PL52	Opolskie
Hungary	HU10	Közép-Magyarország		PL61	Kujawsko-Pomorskie
	HU21	Közép-Dunántúl		PL62	Warminsko-Mazurskie
	HU22	Nyugat-Dunántúl		PL63	Pomorskie
	HU23	Dél-Dunántúl	Portugal	PT11	Norte
	HU31	Észak-Magyarország		PT15	Algarve
	HU32	Észak-Alföld		PT16	Centro (P)
	HU33	Dél-Alföld		PT17	Lisboa
Ireland	IE01	Border, Midlands and Western		PT18	Alentejo
	IE02	Southern and Eastern	Romania	RO01	Nord-Est
Italy	IT31	Bolzano-Bozen e Trento		RO02	Sud-Est
	ITC1	Piemonte		RO03	Sud
	ITC2	Valle d'Aosta/Vallée d'Aoste		RO04	Sud-Vest
	ITC3	Liguria		RO05	Vest
	ITC4	Lombardia		RO06	Nord-Vest
	ITD3	Veneto		RO07	Centru
	ITD4	Friuli-Venezia Giulia		RO08	Bucuresti

cont.

Country	ID Code	Region	Country	ID Code	Region
Slovakia	SK01	Bratislavský kraj	United Kingdom	UKE1	East Riding and North Lincolnshire
	SK02	Západné Slovensko		UKE2	North Yorkshire
	SK03	Stredné Slovensko		UKE3	South Yorkshire
	SK04	Východné Slovensko		UKE4	West Yorkshire
Slovenia	SI00	Slovenija		UKF1	Derbyshire and Nottinghamshire
Spain	ES11	Galicia		UKF2	Leicestershire, Rutland and Northants
	ES12	Principado de Asturias			
	ES13	Cantabria		UKF3	Lincolnshire
	ES21	País Vasco		UKG1	Herefordshire, Worcestershire and Warks
	ES22	Comunidad Foral de Navarra			
	ES23	La Rioja		UKG2	Shropshire and Staffordshire
	ES24	Aragón		UKG3	West Midlands
	ES30	Comunidad de Madrid		UKH1	East Anglia
	ES41	Castilla y León		UKH2	Bedfordshire, Hertfordshire
	ES42	Castilla-La Mancha		UKH3	Essex
	ES43	Extremadura		UKI1	Inner London
	ES51	Cataluña		UKI2	Outer London
	ES52	Comunidad Valenciana		UKJ1	Berkshire, Bucks and Oxfordshire
	ES53	Illes Balears			
	ES61	Andalucía		UKJ2	Surrey, East and West Sussex
	ES62	Región de Murcia		UKJ3	Hampshire and Isle of Wight
Sweden	SE01	Stockholm		UKJ4	Kent
	SE02	Östra Mellansverige		UKK1	Gloucestershire, Wiltshire and North Somerset
	SE04	Sydsverige			
	SE06	Norra Mellansverige		UKK2	Dorset and Somerset
	SE07	Mellersta Norrland		UKK3	Cornwall and Isles of Scilly
	SE08	Övre Norrland		UKK4	Devon
	SE09	Småland med öarna		UKL1	West Wales and The Valleys
	SE0A	Västsverige		UKL2	East Wales
United Kingdom	UKC1	Tees Valley and Durham		UKM1	North Eastern Scotland
	UKC2	Northumberland, Tyne and Wear		UKM2	Eastern Scotland
	UKD1	Cumbria		UKM3	South Western Scotland
	UKD2	Cheshire		UKM4	Highlands and Islands
	UKD3	Greater Manchester		UKN0	Northern Ireland
	UKD4	Lancashire			
	UKD5	Merseyside			

E.3 A Multi-Equation Spatial Econometric Model, with Application to EU Manufacturing Productivity Growth

Bernard Fingleton

E.3.1 Introduction

Recently major new advances have occurred in urban economics and economic geography (Huriot and Thisse 2000; Fujita et al. 1999; Brakman et al. 2001) which provide a formal, general equilibrium, theory of economic geography and urban agglomeration. One of the most significant aspect of this new theory is that it accommodates increasing returns, which is more or less universally agreed to be the 'sine qua non' of urban and regional economics (Fingleton and McCombie 1998; Fingleton 2003), but which has previously restricted the integration of regional economics into the economic mainstream. Also the theory goes some way towards realism by explicitly incorporating a monopolistic competition market structure (Dixit and Stiglitz 1977) with each agent solving a clearly defined economic problem, and by introducing externalities.

One of the criticisms that has been made in the past of the new economic geography theory is that hitherto there has been little or no empirical testing of its assumptions and predictions. One reason for this is that the theory as presented by Fujita et al. (1999) is largely deductive and abstract, designed to provide a window on the real world by logical argument clarified by simplifying assumption. It is not particularly designed for a confrontation with data. However the situation is changing rapidly, and latterly there has been an outpouring of work which takes the new theory as its starting point, but which strives to estimate some of the fundamental parameters, or in various ways to operationalize various version and extensions of the new theory (Combes and Lafourcade 2001, 2004; Combes and Overman 2003; Forslid et. al. 2002; Head and Mayer 2003; Redding and Venables 2004; Rice and Venables 2003; Fingleton 2005a, 2005b, 2006).

In this chapter, the focus is on a model with the same provenance as the foregoing theory, but which comes from the urban economics wing of the literature (Rivera Batiz 1988; Abdel-Rahman and Fujita 1990; Ciccone and Hall 1996), since this has tended to be somewhat overshadowed and yet in many ways it offers a more straightforward route to modeling urban and regional disparities. Implementing new economic geography models can be difficult, partly because of problems of obtaining realistic indicators of transport costs and market potential (McCann 2005; Fingleton 2005a; Fingleton and McCann 2007). Moreover there is some evidence that, at the regional level, models based on urban economic theory compare favourably with those with a basis in new economic geography (Fingleton 2006).

The econometric approach adopted is spatial seemingly unrelated regression (SUR) with spatially lagged dependent variables (see Anselin 1988; Fingleton 2001b). While the current chapter uses the same data and underlying theory as in this earlier paper (see also Fingleton 2000, 2001a), it builds on the earlier analysis with new findings based on data (taken from Cambridge Econometrics' European regional database) allowing the division of the 178 EU regions into two groups, namely a core group and a periphery group (see Appendix). The regions are confined to the UK, Ireland, France, Italy, Germany, Spain, Portugal, Austria, Netherlands, Belgium, Luxembourg, Denmark, Greece. With space partitioned in this way, and time divided into three periods, 1975-81, 1981-89, 1989-95, the systems modelling approach allows testing for parameter homogeneity across space and time. While more recent data are available, this period of history is one of considerable interest, since it was an era of rapid adjustment by new entrants to the EU. The chapter begins by setting out the theoretical basis of the reduced form which lies at the heart of the empirical model. It then introduces spatial heterogeneity and spatial interaction between regions thus allowing geographically-conditioned technology growth. The kernel of the chapter is dedicated to the quest for a preferred model, which is a simplification of an initial unrestricted spatial SUR model. The chapter concludes by interpreting the results in a policy context, and by calling for a new layer of theory, in which the constant parameter assumptions, which have largely dominated the literature, are replaced by a more realistic approach.

E.3.2 Theory

This section sketches the theory behind the econometric model, following Fingleton (2001b). In order to save space, and because this has been set out in detail elsewhere, it is necessarily brief. It is well known (see for instance Abdel-Rahman and Fujita 1990) that assuming a two–sector characterization of the economy of each region, divided between a competitive 'manufacturing' sector and a monopolistically competitive 'producer services' sector, we obtain the following reduced form

$$Q = \phi N^\gamma \tag{E.3.1}$$

in which competitive sector output (Q) relates nonlinearly to the intensity of activity in a unit area (given by the total labour per unit area N), and in which the elasticity γ depends on the exogenous parameters α, β and μ, and is a measure of external increasing returns.

$$\gamma = \alpha[1 + (1-\beta)(\mu-1)] \tag{E.3.2}$$

The level of monopoly power in the imperfectly competitive sector is given by the exogenous parameter $\mu \geq 1$. As μ increases, we see rising monopoly power and falling elasticity of substitution. The parameter β determines the relative importance of labour in the competitive sector (M) versus services (in the form of a composite index I), as indicated by the Cobb-Douglas production function

$$Q = (M^\beta I^{1-\beta})^\alpha L^{1-\alpha} \tag{E.3.3}$$

in which L is land area used in production. However L is redundant as a factor affecting the growth of productivity, which is the subject of this contribution, since it is constant across time and therefore is eliminated on differentiation with respect to time (see Eq. (E.3.15)). In order to simplify the notation, we set $L = 1$ at this juncture so that we have production per unit area, hence $Q = (M^\beta I^{1-\beta})^\alpha$. This means that the output level Q is reduced by restricting land area, and the parameter $\alpha < 1$ can be seen to represent the impact of congestion which, to some extent, will offset increasing returns, as indicated by Eq. (E.3.2) (Ciccone and Hall 1996).

For increasing returns, simultaneously, producer services (I) have to be relevant ($\beta < 1$) to competitive sector output, service firms have to exert a degree of monopoly power under monopolistic competition ($\mu > 1$), and the effect of congestion ($\alpha < 1$) has to be sufficiently small, in order to ensure that $\gamma < 1$. There are a priori reasons why the value of the estimable returns to scale parameter γ may not be constant between core and (typically) later EU entrants in the periphery, and vary between different time periods. For instance the importance (β) of producer services for competitive sector output may change as economic structure changes in response to the exposure of formerly protected markets to EU and global competition. Likewise congestion (α), considered in the widest sense, may reflect variations in infrastructural investment over time and between core and periphery, due perhaps to the impact of EU investment programmes in the periphery and slower infrastructure investment in the well-established infrastructure of the core. The degree of monopoly power (μ) is also unlikely to be a constant, as a result of institutional and structural differences and changes over time.

Although of less direct consequence for the empirical analysis, it is worth noting that ϕ is also a function of constants given by

$$\phi = \left[\beta^\beta (1-\beta)^{\mu(1-\beta)} \alpha^{-(1-\beta)} s^{-(1-1/\mu)(1-\beta)\mu} (1/\mu)^{1-\beta} (1-1/\mu)^{(1-1/\mu)(1-\beta)\mu}\right] \quad (E.3.4)$$

Equation (E.3.4) extends Eq. (E.3.18) in Abdel-Rahman and Fujita (1990) by incorporating the congestion-induced diminishing returns of the current set up.

In order to move closer to a convenient reduced form, we linearize Eq. (E.3.1) by taking natural logarithms and rearranging to give an expression in terms of the level of manufacturing output per labour unit as a function of the level of manufacturing output, thus

$$\ln(Q/N) = \ln(\phi)/\gamma + [(\gamma-1)/\gamma]\ln(Q) \quad (E.3.5)$$

Also, standard competitive equilibrium theory (see for example Fingleton 2003) allows us to assume $M/N = \beta$, the preferred expression is in terms of the level of manufacturing output per unit of manufacturing labour, thus

$$\ln(Q/M) = \ln(\phi)/\gamma + [(\gamma-1)/\gamma]\ln(Q) - \ln(\beta). \quad (E.3.6)$$

E.3.3 Incorporating technical progress variations

Technological externalities involving information spillovers within and between regions and cities are increasingly recognized (e.g. Fujita and Thisse 1996) as an important contributor to the spatial concentration of economic activity, reflecting the earlier work of Jacobs (1969) among others. In the current empirical modeling context, their presence is compelled by the need to avoid estimation bias. In what follows, attention focuses on the rate of technical progress as a manifestation of technological externalities. We see that technical progress is not confined by artificial region boundaries but influences and is influenced by technical progress in other regions, so that it varies by region rather than being an unmodeled constant.

In order to model the technical progress rate, the approach again follows that in Fingleton (2001b) and Fingleton (2000). First we consider the labour units referred to above to be labour efficiency units, hence

$$M_t = E_t A_t = E_t A_0 \exp(\lambda t) \qquad (E.3.7)$$

where E_t is the level of manufacturing employment and A_t is the efficiency level at time t that is determined by the initial level A_0 and the rate of technical progress λ. It then become possible to re-express the model in terms of the level of manufacturing productivity per unit area by substitution, hence

$$\ln(Q/E) = \ln(\phi)/\gamma + [(\gamma-1)/\gamma]\ln(Q) - \ln(\beta) + \ln(A_0) + \lambda t \qquad (E.3.8)$$

The submodel for the technical progress rate λ proposes that

$$\lambda = \lambda^* + \rho W \lambda \qquad (E.3.9)$$

so that the technical progress rate depends partially on the rate of technical progress due to within-region factors as represented by λ^*, and on the rate of technical progress in 'nearby' regions $(W\lambda)$.

Consider first the determinants of technical progress which exist within the region. The basic assumption is that λ^* is determined by H, the stock of human capital within a region, and the 'technology gap' G which is (a function of) the level of technology within a region, hence

$$\lambda^* = \pi G + \nu H. \qquad (E.3.10)$$

In Eq. (E.3.10), G reflects the ubiquitous public good dimension of knowledge. There is no distance effect involved and some components of knowledge are free to diffuse to any region, irrespective of geography. However, the hypothesis is that the impact will be spatially variegated. The reason is differential acceptance of knowledge because of its varying usefulness. Some regions will gain great benefit from the diffusion of knowledge from the technological frontier region, others that are close to the technological frontier will see only minor gains. We therefore envisage that there will be a positive relationship $(\pi > 0)$ between the technical progress rate and the initial technology gap, although this is a testable hypothesis and not an assumption of the model. A positive relationship indicates catching-up due to the faster growth of the initially lower level technology regions, but a negative relationship implies regional divergence.

There are several proxies for initial level of technology and thus the technology gap (see Fingleton and McCombie 1998), but here we work with the (initial) level of technology gap $G = 1 - P_{1975} / P_{*1975}$ in which P_{1975} is the level of manufacturing productivity in 1975, and P_{*1975} is the level of the leading technology (manufacturing productivity) region. Since it is defined for the initial year, it is treated as exogenous in the analysis that follows.

Independently of catching-up, it is hypothesized that regions with substantial human capital assets will make faster technical progress $(v > 0)$ since human capital facilitates the research and development on which technical progress depends. Human capital level (H) is proxied by two exogenous variables, namely the start-of-period population density (U) and remoteness (L). In high population density regions, which are invariably highly urbanized, we envisage dense social networks and institutions in which, compared to more rural regions, there is more scope for the exchange of (embodied and disembodied) information leading to technical progress. This type of knowledge spillover is seen as a technological externality. The assumption that remote regions tend to possess less human capital than more central regions is based on their traditionally comparatively large agricultural and non-market services sectors, which partly because of additional transactions costs have not fostered a strong manufacturing research and development and skills base. Combining the dual effects of remoteness and urbanization gives

$$H = \varepsilon + \theta L + \tau U \qquad (\text{E.3.11})$$

in which remoteness variable (L) is measured by km distance between each region's centre and Luxembourg, and population density (U) relates to 1975.

Next we turn to spatially impeded information flows. The hypothesis is that regions with fast technical progress occurring in 'neighbouring' regions will see faster than otherwise technical progress, and regions with slower technical progress neighbours will see a technical progress slowdown. 'Who your neighbours are' matters in this context, because of the impedance to information flow across space. Working with the so-called NUTS-2 regional system of the EU means that we are likely to see flows across region boundaries. This is because this regionalizing does not define self-contained economic regions but rather regions are largely defined on a formal or administrative basis. In addition, national barriers have been progressively reduced within the EU with the consequence that there are minimal barriers to remote cross-region spillover. Hence we anticipate that the strength and significance of spillover effects will strengthen over time, as indeed is evident from the empirical analysis reported later.

Two cross-region spillover mechanisms are envisaged. One is based on the sharing of the same labour pool in a common local labour market area straddling regional boundaries, the thesis being that the rate of productivity growth occurring in one region will be transmitted to other nearby regions as workers embodying technical progress switch jobs within local labour market (i.e. journey to work) areas. The second mechanism involves inter-firm interaction across region boundaries. Because of proximity, firms locally and in nearby regions may be competitors for the same local markets, or collaborate as part of a localized production chain. In either case, fast technical progress in neighbouring regions will tend to induce technical progress and thus fast productivity growth locally.

The assumptions set out above result in $W\lambda$ in Eq. (E.3.9) being specified as the weighted average of technical progress in 'surrounding' regions, with surrounding broadly defined so as to acknowledge that even the remotest regions interact and that the EU comprises a more or less integrated economy. Size also is considered relevant, and to a degree offsets remoteness, because of the extensive trade and labour market that a large diverse local economy naturally generates. This is apparent in the definition of the absolute (conditional) level of interaction between regions i and j,

$$W_{ij}^* = Q_{j1975}^\eta d_{ij}^{-\delta}. \tag{E.3.12}$$

In Eq. (E.3.12), the term Q_{j1975} represents the economy size proxy, the 1975 level of output in region j, so that given the size of region i, the interaction with region j is likely to be stronger if region j possesses a larger economy. The definition of W_{ij}^* ignores the size of region i as an influence. The size of i could be represented by Q_{i1975}, so that the unconditional interaction is then $W_{ij}^* = Q_{i1975}^\vartheta Q_{j1975}^\eta d_{ij}^{-\delta}$ in which ϑ is an additional parameter, but the outcome with respect to W is the same irrespective of whether conditional or unconditional specifications are used. Proximity is represented by the great circle distance (d_{ij}) between the centres of regions i and j. Given a presumably negative parameter (δ), increasing distance reduces the absolute conditional interaction between i and j. The importance of the size of the economy is controlled by the parameter η. It is assumed that $\delta = 2$ and $\eta = 1$ as a result of trials of different values reported in Fingleton (2001b). Likewise, comparison between the power function of Eq. (E.3.12) and the negative exponential function $W_{ij}^* = Q_{j1975}^\eta \exp(-\delta d_{ij})$ provide empirical support for Eq. (E.3.12).

The standardized matrix W has cell entries for regions i and j equal to

$$W_{ij} = \frac{W_{ij}^*}{\sum_j W_{ij}^*} \tag{E.3.13}$$

which sum to unity across rows with zeros on the main diagonal, so that the matrix product $W\lambda$ is a vector of weighted averages. Note that because it has been standardized, W is asymmetrical with the consequence that i's effect on j may not equal j's effect on i, since j may dominate i but the reverse may not be true, hence the spillover effects are conditional on the spatial context of each region. This means that the parameter ρ in Eq. (E.3.9) is the change in technical progress per unit change in the 'average' in 'surrounding' regions.

Combining the earlier equations, we obtain

$$\lambda = \rho W \lambda + \pi G + v(\varepsilon + \theta L + \tau U) \qquad (E.3.14)$$

and differentiating Eq. (E.3.8) with respect to time, we obtain

$$p = (\gamma-1)q/\gamma + \rho W \lambda + \pi G + v(\varepsilon + \theta L + \tau U) \qquad (E.3.15)$$

The rate of technical progress is unknown but if we treat it as a 'residual', $\lambda = p - (\gamma-1)q/\gamma$ and $Wp = W\lambda + (\gamma-1)Wq/\gamma$. In fact it is simpler for the purposes of estimation to assume that $\lambda = p$ and therefore $W\lambda = Wp$. Otherwise we need to include both Wp and Wq in the specification, and estimation involves a constraint. Therefore we specify the model as

$$p = (\gamma-1)q/\gamma + \rho Wp + \pi G + v(\varepsilon + \theta L + \tau U) \, . \qquad (E.3.16)$$

An advantage of this specification is that it is easily estimated by maximum likelihood in the single equation context. Given a standardized W matrix, estimated ρ is automatically less than an upper bound equal to one, thus facilitating interpretation (see Fingleton 2000). A similar outcome is produced by scaling all values by the maximum eigenvalue of the matrix. The emphasis given by row standardization is on relative rather than absolute distance. With the equation system defined here, ρ is also typically below one, although alternative estimation techniques are employed as outlined below. Writing this in matrix terms, simplifying the parameterization and adding an error term gives

$$p = \rho Wp + Xb + u \qquad (E.3.17)$$

in which p is the exponential growth rate of manufacturing productivity by EU region, X is the matrix of regressors q, G, L, U and a constant with composite parameters b, and $u \sim \mathcal{N}(0, \sigma^2)$ is an error term capturing other unmodeled effects.

E.3.4 The econometric model

The model set out in Eq. (E.3.17) is a conventional spatial lag model as commonly found in the spatial econometrics literature. From the point of view of estimation, the main feature of the model is the endogenous term Wp (it is assumed that the remaining variables are exogenous), so that for consistent estimation in the single equation context, maximum likelihood or two-stage least squares (2sls) is required.

The subsequent analysis focuses on the question of whether the basic model in Eq. (E.3.17) has constant parameters across different regions and times. In order to test these hypotheses, we adopt a systems modeling approach, using three stage least squares estimation via LIMDEP, with results replicated by PcFIML. The equations are linked by error covariance, in other words we apply three stage least squares to spatial SUR with spatially lagged dependent variables (see Anselin 1988). While this preserves the basic specification given by Eq. (E.3.17), it improves the efficiency of estimation giving estimates with smaller asymptotic variance, and captures potential omitted variables that different equations might have in common. Most importantly, it also facilitates nested hypothesis tests of parameter equality across equations. This approach was first used in the current context by Fingleton (2001b), but here is extended to both space and time contrasts.

Given the possible existence of structural instability over both space and time, and assuming $T = 3$ time periods, 1975-81, 1981-89, and 1989-95 and $R = 2$ areas (core and periphery), the consequence is $2RT = 12$ equations. There are six equations for the rate of manufacturing productivity growth p, collectively the six equations are

$$p_{rt} = \rho_{rt} \tilde{W} p_{rt} + X_{rt} b_{rt} + \eta_{rt} . \tag{E.3.18}$$

In Eq. (E.3.18), the matrix X_{rt} denotes the exogenous variables specific to time t and area r, and b_{rt} denotes parameter vectors (with elements b_{0rt}, b_{1rt}, etc.). The parameter set ρ_{rt} allows space and time varying effects of the endogenous lags derived from the matrix products Wp_{rt}. Since these involve the matrix W which applies to all regions irrespective of whether they are core or periphery, this product by itself gives a vector with non-zero elements for all regions. Zeros are introduced for non-core regions by multiplying the vector by a dummy variable with ones indicating core regions and zeros indicating periphery regions. Similarly zeros are introduced for non-periphery regions by multiplying by the periphery region dummy (see below). Therefore $\tilde{W}p_{rt}$ denotes these modified variables, with zeros in place as appropriate.

The other six equations relate to the set of endogenous vectors $\tilde{W}p_{rt}$. It is assumed that the lagged causal variables matrices $\tilde{W}X_{rt}$ (treated as above with multiplication by the appropriate area dummy) are the exogenous determinants of $\tilde{W}p_{rt}$ and there is also feed back from p_{rt} to $\tilde{W}p_{rt}$. This is summarized by

$$\tilde{W}p_{rt} = \tilde{W}X_{rt}m_{rt} + p_{rt}n_{rt} + \xi_{rt} \qquad (E.3.19)$$

with parameter sets m_{rt} and n_{rt}. Subsequently, while restrictions are imposed on b_{rt} and ρ_{rt}, m_{rt} and n_{rt} remain unrestricted in all models fitted and are unreported in the tables of estimates given below.

The matrix of variances and covariances of the error terms η_{rt} and ξ_{rt} is of dimension twelve and so is omitted here. The main diagonal of this matrix is

$$(\sigma^2_{\eta_{c1}}, \sigma^2_{\eta_{c2}}, \sigma^2_{\eta_{c3}}, \sigma^2_{\xi_{c1}}, \sigma^2_{\xi_{c2}}, \sigma^2_{\xi_{c3}}, \sigma^2_{\eta_{p1}}, \sigma^2_{\eta_{p2}}, \sigma^2_{\eta_{p3}}, \sigma^2_{\xi_{p1}}, \sigma^2_{\xi_{p2}}, \sigma^2_{\xi_{p3}}). \qquad (E.3.20)$$

Off the main diagonal, the elements of the matrix relating to core regions are given by $E(\eta_{ct}, \eta_{ct}) = \sigma_{\eta_{ct}\eta_{ct}}$, $E(\xi_{ct}, \xi_{ct}) = \sigma_{\xi_{ct}\xi_{ct}}$ and $E(\xi_{ct}, \eta_{ct}) = \sigma_{\xi_{ct}\eta_{ct}}$ with $t = 1, 2, 3$. For periphery regions the same equations apply, but with suffix c replaced by p. Non-zero covariance is therefore confined to equations for different times, but there is no covariance between the equations for the core and periphery. The assumption is that unmodelled effects in the core are confined to the core but carry across time giving non-zero covariance, and likewise for unmodelled effects specific to the periphery. Hence $E(\eta_{ct}, \eta_{pt}) = 0$, $E(\xi_{ct}, \xi_{pt}) = 0$, $E(\xi_{ct}, \eta_{pt}) = 0$ and $E(\eta_{ct}, \xi_{pt}) = 0$, giving the twelve by twelve covariance matrix which is block-diagonal with zeros in the top right and bottom left quadrants. From the computational perspective, this is achieved by defining dummy variables for core and periphery areas, and multiplying the variables by these dummies to give area-specific variables, as described above and as exemplified in the appendix. These covariances are set to zero by using the appropriate area-specific variables in each of the twelve equations, together with the appropriate dummy variable as an additional regressor.

E.3.5 Model restriction

Table E.3.1 gives the unrestricted 3sls parameter estimates which are the starting point for the process of producing a parsimonious final model in which parameters across location and time are the subject of simplifying restrictions. On the whole the parameter estimates in Table E.3.1 are similar to those reported in Fingleton (2001b), the difference being that in this chapter core and periphery regions are separated. There is a positive parameter on q indicating increasing returns which is highly significant regardless of location or of time, although there are variations in magnitude. Likewise the initial technology gap variable G parameter is consistently positive and largely significant, reaffirming the importance attributed earlier to innovation diffusion. As one would anticipate from the previous study, the effect of distance from Luxembourg (L) is negative throughout, although in this case it is not significant for all times and locations. The effect of population density (U) is significant and positive across all periods in core regions, but is only significant in the earliest period for the peripheral regions. There are also core-periphery differences in the strength and significance of the effect of 'neighbouring' productivity growth captured by the endogenous variable Wp.

Table E.3.1. Unrestricted model estimates

Variable	Core Parameter	t-ratio	Goodness-of-fit	Periphery Parameter	t-ratio	Goodness-of-fit
1975-81						
Constant	–0.24930E–01	–2.862	RSS= 3.4576E–02	0.28396E–03	0.042	RSS= 3.6658E–02
q	0.83893	9.960	Adj R-sq= 0.68785	0.79776	25.439	Adj R-sq = 0.84175
G	0.11286	8.183	F = 79.01	0.69615E–01	5.151	F = 189.30
L	–0.54168E–03	–3.973		–0.12850E–01	–2.795	
U	0.64439E–02	3.365		0.51569E–02	2.550	
Wp	0.13541	0.621		–0.22766	–1.762	
1981-89						
Constant	–0.10269E–01	–2.047	RSS= 1.3157E–02	–0.34152E–01	–3.753	RSS= 4.4630E–02
q	0.82019	19.692	Adj R-sq = 0.86339	0.44494	7.585	Adj R-sq = 0. 69873
G	0.35955E–01	4.332	F = 224.73	0.65461E–01	4.189	F = 83.10
L	–0.20727E–03	–2.550		–0.90314E–04	–1.655	
U	0.29026E–02	2.639		0.33179E–02	1.408	
Wp	0.47948	4.560		0.86547	4.978	
1989-95						
Constant	–0.11683E–01	–1.389	RSS= 4.8366E–02	0.47115E–02	0.446	RSS= 5.2430E–02
q	0.53722	5.618	Adj R-sq = 0.28533	0.68412	9.699	Adj R-sq = 0. 60963
G	0.30547E–01	1.979	F = 15.13	0.11012E–01	0.661	F = 56.28
L	–0.39709E–04	–0.244		–0.14016E–03	–2.327	
U	0.82493E–02	3.982		–0.32429E–02	–1.273	
Wp	0.53137	2.758		0.79522	3.383	

Across-space restrictions. The above preliminary interpretation takes no account of random variation which may be partly responsible for apparent trends and differences in parameter estimates across space and time. This section imposes various restrictions on the parameters of the unrestricted model reported in Table E.3.1 to produce restricted models which indicate, via the loss-of-fit incurred, whether the across-space restrictions are consistent with the data or whether we do actually need separate estimates for core and periphery. The analysis commences with the joint test (see Table E.3.2) in which the thirty-six parameters of the unrestricted model relating to q, G, L, U, Wp and the constant are replaced by eighteen parameters. For example, the restriction $b_{1ct} - b_{1pt} = 0$ is imposed equalizing the core and periphery q parameters for each of $t = 1,2,3$, with similar restrictions nullifying core-periphery differences for the other variables. The joint test is therefore a test of the overall significance of core-periphery effects, given the presence of time-period effects. It turns out that this restricted model produces a significant loss-of-fit compared to the unrestricted model, since the χ^2 test statistic of 86.58 is significantly large when referred to the χ^2_{18} distribution. Clearly there are some, as yet unspecified, core-periphery differences required to be present in a final model, and this motivates the subsequent analysis.

The lack-of-fit of the model nullifying core-periphery differences may be attributable to only a subset of variables. More insight regarding the sources of lack-of-fit is provided in Table E.3.2 which summarizes conditional tests in which the unrestricted model (Table E.3.1) is compared to models with variable-specific restrictions. For example, in order to test whether there are core-periphery differences in the effect of q, the six parameters differentiating the q effect by time and space are replaced by three (one per time period) thus giving three degrees of freedom for this test conditional on the existence of time and space differences in all other parameters. Table E.3.2 indicates that, given time differences, we also need parameters reflecting locational differences across the range of variables, although for Wq the test is only significant at the 10 percent level.

Table E.3.2. Conditional tests of core-periphery contrasts

Variable	Chi-square test statistic	df	Prob
q	28.64	3	0.00000
G	11.381	3	0.00984
L	11.458	3	0.00949
U	13.372	3	0.00392
Wp	6.906	3	0.07497
Constant	12.408	3	0.00611
Joint test	86.577	18	0.00000

Across-time restrictions. Given the evidence suggesting the need to allow for core-periphery contrasts, we now reconsider the case for time homogeneity. Previously (Fingleton 2001b), it appeared that there was a strong case for differences across time period in the strength and significance of all the model parameters, with the possible exception of U which was only significant at the 10% level. This existence of some time inhomogeneity is reaffirmed in the current analysis, as shown by the joint test in Table 3. This indicates the consequence in terms of loss-of-fit of imposing restrictions on the thirty-six parameters (six per endogenous variable, $p_{1c},...,p_{3p}$) in order to completely eliminate time-inhomogeneity. For example, the three parameters reflecting time differences in the effect of q in core regions are replaced by a single parameter via the restrictions $b_{1c1} - b_{1c2} = 0$ and

$b_{1c1} - b_{1c3} = 0$ with equivalent restrictions ($b_{1p1} - b_{1p2} = 0$ and $b_{1p1} - b_{1p3} = 0$) on the effects of q in peripheral regions. These restrictions are repeated across all variables simultaneously to create a restricted model in which twelve parameters (two per endogenous variable) replace the thirty-six of the unrestricted model. The resulting Wald test statistic (163.69) is highly significant when referred to the χ^2_{24} distribution. Conditional on regional inhomogeneity, there is significant evidence for time inhomogeneity.

The question now arises as to whether these time inhomogeneities are a feature of all variables, or apply to a specific subset. Table E.3.3 also shows the outcome of variable-specific tests in which the models being compared are the unrestricted model (Table E.3.1) and a model restricted to time homogeneity across the parameters of the specific variable, conditional on time inhomogeneity for the other variables. Hence, the three parameters needed to pick up time differentiated effects of the variable are replaced by a single parameter, thus giving two degrees of freedom. The results show that this restriction produces a significant loss-of-fit for the majority of variables, implying that a successful final model will include parameters controlling time variation in location-specific parameters. The exceptions are that for core regions there is no strong evidence that the effect of Wp varies across time, and for peripheral regions the effect of L is apparently constant over time.

Table E.3.3. Conditional tests of time homogeneity

Variable	Chi-square test statistic	df	Prob
Core			
q	7.904	2	0.01922
G	42.836	2	0.00000
L	11.088	2	0.00391
U	6.066	2	0.04818
Wp	2.790	2	0.24789
Constant	3.098	2	0.21249
Periphery			
q	31.437	2	0.00000
G	11.486	2	0.00321
L	0.464	2	0.79292
U	10.954	2	0.00418
Wp	28.933	2	0.00000
Constant	10.987	2	0.00411
Joint test	163.690	24	0.00000

Space-time restrictions. An additional question is whether the parameter differences that exist are evidence that peripheral regions lag core regions, so that the parameters for peripheral regions equal core region parameters for an earlier period. Thus across-region inhomogeneity may conceal across-region homogeneity across time. Broadly speaking peripheral regions tend to be more recent EU members or have lagged economic development or moves to closer economic integration which have occurred in the core, so it seems reasonable to hypothesize that peripheral regions may go through stages of development which were experienced earlier by core regions.

In order to test this hypothesis, peripheral region parameters are first of all jointly restricted to equal core region parameters for an earlier period. Hence the six parameters for 1981-89 for peripheral regions and for 1975-81 for core regions are constrained to be equal. This provides six degrees of freedom and a Wald test statistic equal to 27.17 which has a p-value of 0.00013 in the chi-squared distribution with six degrees of freedom. A similar test jointly restricting the 1989-95 and 1975-81 parameters (since the time lag may be longer) produces a test statistic equal to 50.38. It is clearly evident that the set of parameters in peripheral regions do not simply mimic the values of the earlier periods in the core.

The joint test conceals possible variable-specific across-region across-time equalities. The equalization of a subset of the parameters could be the outcome of peripheral regions partially replicating at a later stage the process determining productivity growth in the core, while other determinants of productivity growth possess a significantly different effect in the periphery. An initial investigation of this involves constraining to equality of the six parameters in turn, while allowing the remaining parameters to vary. This produces a conditional test with one degree of freedom. In all cases, equating periphery parameters to core parameters for an earlier period produced a significant loss-of-fit, with the Wald test statistic exceeding the upper 5% point of the chi-squared distribution with one degree of freedom. The exceptions[1] to this rule were the equalization of the periphery q parameter for 1989-95 and the core q parameter for 1975-81 (Wald test statistic equal to 1.99), and the 1975-81 core U parameter and constant and 1981-89 periphery U parameter and constant (1.06 and 0.54 respectively).

E.3.6 The final model

The evidence presented here indicates that on the whole there is core-periphery and temporal parameter heterogeneity in parameters. On possible exception to this relates to the variable Wp, which according to Table E.3.3 is evidently time homogeneous in effect for core regions. In contrast Wp is time heterogeneous for periphery regions. However there is also some weak evidence for core-periphery homogeneity for Wp in Table E.3.2. The dominant feature of the Wp estimates from Table E.3.1 is however their insignificance for 1975-81 and significance subsequently. Therefore as an initial simplifying step we opt to eliminate spillover effects between 'neighbouring' regions in the earliest period. This is assumed to reflect the barriers that existed between countries and regions in this early period of the EU's development, prior to the much more complete liberalization that exists now.

---

[1] However, given the number of tests carried out, the probability of wrongly failing to reject a hypothesis that the parameter values are equal is going to be larger than the nominal 0.05 one would associate with a single test.

Another feature of the Table E.3.1 estimates is the much diminished significance (as apparent from the *t*-ratios) of the level of technology gap (G) in the most recent period. This is to be anticipated, as the initial level of technology in 1975 will undoubtedly have become increasingly irrelevant to productivity growth as time increases. Therefore it seems appropriate to constrain the model by restricting the initial period Wp parameters and the final period G parameters to 0. The Wald test statistic as a result of imposing these four restrictions is equal to 7.9549, which has a *p*-value equal to 0.0932 in the χ_4^2 distribution, showing that there is no significant loss of fit.

However Table E.3.1, and the estimates of the model with these four restrictions, point to additional simplifying restrictions, namely setting to zero the L and U coefficients for periphery regions in 1981-89, and also, for 1989-95, setting core L to zero and periphery U to zero. Table E.3.4 is the final model with all eight restrictions imposed, and again these can be sustained by observing that the Wald test statistic is equal to 15.2167, which just fails to be significant in χ_8^2, with a *p*-value equal to 0.0551. In Table E.3.4 all the variables are significant, although several goodness-of-fit indicators, the adjusted R-squared values, the sum of squared residuals and the F statistic (which is referred to F_{172}^5), highlight the relatively poor performance of this restricted model in the most recent period. This however is not a consequence of these additional restrictions, as is evident from comparison with Table E.3.1.

Table E.3.4. Final model estimates

Variable	Core Parameter	*t*-ratio	Goodness-of-fit	Periphery Parameter	*t*-ratio	Goodness-of-fit
1975-81						
Constant	−1.72385E−02	−2.677	RSS= 3.5564E−02	−5.10838E−03	−1.004	RSS= 3.6387E−02
q	0.860237	11.319	Adj R-sq = 0.67892	0.777821	26.220	Adj R-sq = 0.84292
G	0.102871	8.433	F = 75.85	6.09888E−02	5.104	F = 190.96
L	−5.51187E−04	−4.618		−1.04896E−04	−2.641	
U	6.83510E−03	3.794		5.72922E−03	3.466	
Wp	–	–				
1981-89						
Constant	−9.84770E−03	−1.966	RSS= 1.3159E−02	−3.13665E−02	−3.506	RSS= 4.6068E−02
q	0.818766	19.679	Adj R-sq = 0.86337	0.459335	8.033	Adj R-sq = 0. .68902
G	3.50972E−02	4.235	F = 224.70	4.32023E−02	4.278	F = 79.43
L	−2.06585E−04	−2.545		–	–	
U	2.89906E−03	2.636		–	–	
Wp	0.477992	4.546		0.885518	5.237	
1989-95						
Constant	−4.93244E−04	−0.114	RSS= 4.8953E−02	7.66117E−03	0.822	RSS= 5.3924E−02
q	0.513071	5.550	Adj R-sq = 0.27666	0.707763	10.174	Adj R-sq = 0.59851
G	–	–	F = 14.54	–	–	F = 53.77
L	–	–		−1.05706E−04	−2.558	
U	8.15701E−03	3.946		–	–	
Wp	0.556147	3.083		0.743493	3.192	

Table E.3.4 shows by virtue of the significant and positive parameters on q that increasing returns to scale is a realistic hypothesis that can be applied to both core and periphery, and for each time period. The parameter estimate heterogeneity points to variation in the underlying determining parameters, since we know that according to our underlying theory, the parameter for q is equal to $(\gamma - 1)/\gamma$, with $\gamma = \alpha[1 + (1 - \beta)(\mu - 1)]$. The shrinking of the parameter on q towards zero, as occurs in the core for the most recent period, is commensurate with γ moving towards one and could be due to increased congestion (α goes towards zero), less relevant intermediate services (β moves towards one), or reduced producer service variety (μ goes towards one). It is possible also that we are seeing the net effect of all three parameters changing, although estimating their individual values is an exercise beyond the scope of this contribution.

The effect of G is mainly significant and positive, indicating that there is an overall tendency for regions with initially lower technology levels to see faster manufacturing productivity growth. However, the G effect is constrained to zero in the most recent period with no significant loss of fit. It appears that by 1989-95, the stimulus to productivity growth of innovation diffusion differentially benefiting the lower technology regions had disappeared.

The distance from Luxembourg (L) coefficient is appropriately negative throughout (see Table E.3.1), but we find that we can constrain some of the coefficients to zero. For the most recent 1989-95 period, since we have used L as a proxy for human capital, the inference is that by L no longer reflects human capital differentials in core regions. For peripheral regions, it is apparent that L regains its significance so that the remoter regions in the periphery continue to have significantly slower productivity growth, suggesting that the non-manufacturing legacy and consequently the comparative lack of appropriate human capital remains a factor. The other proxy for human capital is population density/urbanization (U). For core regions, this remains consistently significant over time, but for the periphery it is only significant for the earliest period. This suggests that the concentration of human capital in cities is of continuing importance for core productivity growth, but in the periphery urban concentration is not a factor from 1981.

E.3.7 Concluding remarks

There are two main conclusions that are drawn from the above analysis, the first relating to policy implications and the second to the implications for the development of new theory. Regarding policy, the model estimated in this chapter provides further evidence of the importance of increasing returns to scale for regional economic growth, which implies divergent productivity levels. While there was a catching-up effect due to innovation diffusion as reflected in the positive coefficients on the variable G, the effect had become less significant by period 1989-95, leaving the EU regions more exposed to divergence effects due to increasing returns to scale. Likewise, the heterogeneous effects of regionally differentiated human capital also point to divergent growth rates and levels. Most notably, for core regions, there is a significant positive impact of urban density on productivity growth which is not replicated for regions in the periphery. This suggests a need for additional and different infrastructures and institutions promoting the growth of human capital appropriate to the industrial composition of peripheral regions. The significance of cross-regional spillovers suggests that the impact of policy instruments on the productivity growth of one region may have had effects on productivity growth in other 'nearby' regions, with unintended beneficial consequences of structural or other improvements. Equally, there may be unintended negative effects across regions due to the relative lack of policy intervention. For the most recent periods, spillover effects are stronger for peripheral regions compared with core regions, a factor that should be taken into account in evaluating the likely impact of policy intervention.

Parameter instability is largely excluded from current theory which on the whole tends to be written in terms of fixed, exogenous parameters. For instance, in the very simplest new economic geography theory of Fujita et al. (1999), the emergence of structure is illustrated by altering the parameter for trade costs while holding constant parameters μ and β the fixed and marginal labour requirements and the allocation of an immobile constant returns sector across regions. Hence we see the effect of economic integration conditional on an absence of other changes. These simplifications, while useful as illustrations, are however at odds with the empirical evidence. This shows clearly that when we focus on the estimated γ which embodies the principal parameters of interest coming directly from our urban economics theory, namely α, β and μ, which reflect in turn the impact of congestion, the importance of intermediate services for manufacturing output and the degree of monopoly power and differentiation in the service sector. It is clear from estimated γ that there exist significant variation in time and space in one or more of these underlying parameters. This has implications for the development of new theory, since it suggests that they should be endogenised, since they clearly are dependent on higher causes that have been omitted from the current model.

To summarize, working from a reduced form derived from the urban economics and new economic geography tradition, we have shown the existence of spatiotemporal parameter instability. Rather than being assumed away, some progress could be made by acknowledging the reality of these variations, exploring their causes and ultimately developing theory to account for them.

References

Abdel-Rahman H, Fujita M (1990) Product variety, Marshallian externalities, and city sizes. J Reg Sci 30(2):65-183
Anselin L (1988) Spatial econometrics: methods and models. Kluwer, Dordrecht
Brakman S, Garretsen H, van Marrewijk C (2001) An introduction to geographical economics. Cambridge University Press, Cambridge
Ciccone A, Hall RE (1996) Productivity and the density of economic activity. Am Econ Rev 86(1):54-70

Combes P-P, Lafourcade M (2001) Transportation costs decline and regional inequalities: evidence from France. CEPR DP 2894

Combes P-P, Lafourcade M (2004) Trade costs and regional disparities in a model of economic geography: structural estimations and predictions for France. Unpublished paper available from http://www.enpc.fr/ceras/combes/.

Combes P-P, Overman H (2003) The spatial distribution of economic activity in the EU. CEPR DP 3999

Dixit A, Stiglitz JE (1977) Monopolistic competition and optimum product diversity. Am Econ Rev 67(3):297-308

Fingleton B (2000) Spatial econometrics, economic geography, dynamics and equilibrium: a third way? Environ Plann A 32(8):1481-1498

Fingleton B (2001a) Equilibrium and economic growth: spatial econometric models and simulations. J Reg Sci 41(1):117-148

Fingleton B (2001b) Theoretical economic geography and spatial econometrics: dynamic perspectives. Econ Geogr 1(2): 201-225

Fingleton B (2003), Increasing returns: evidence from local wage rates in Great Britain. Oxford Economic Papers 55(4):716-739

Fingleton B (2005a), Towards applied geographical economics: modelling relative wage rates, incomes and prices for the regions of Great Britain. J Appl Econ 37(21):2417-2428

Fingleton B (2005b), Beyond neoclassical orthodoxy: a view based on the new economic geography and UK regional wage data. Papers in Reg Sci 84(3):351-375

Fingleton B (2006), The new economic geography versus urban economics: an evaluation using local wage rates in Great Britain. Oxford Economic Papers 58(3):501-530

Fingleton B, McCann P (2007) Sinking the iceberg? On the treatment of transport costs in new economic geography. In Fingleton B (ed) New directions in economic geography. Edward Elgar, Cheltenham, pp.168-203.

Fingleton B, McCombie JSL (1998) Increasing returns and economic growth: some evidence for manufacturing from the European Union regions. Oxford Economic Papers 50(1):89-105

Forslid R, Haaland J, Midelfart Knarvik K-H (2002) A U-shaped Europe? A simulation study of industrial location. J Int Econ 57(2):273-297

Fujita M, Thisse, J-F (1996) Economics of agglomeration. J Jap Int Econ 10(4):339-378

Fujita M, Krugman P, Venables, A J (1999) The spatial economy. MIT Press, Cambridge [MA]

Head K, Mayer T (2003) The empirics of agglomeration and trade. CEPR DP 3985

Huriot J-M, Thisse J-F (eds) (2000) Economics of cities. Cambridge University Press, Cambridge

Jacobs J (1969) The economy of cities. Jonathan Cape, London

McCann P. (2005), Transport costs and new economic geography. J Econ Geogr 5(3):305-318

Redding S, Venables AJ (2004) Economic geography and international inequality. J Int Econ 62(1):53-82

Rice P, Venables AJ (2003) Equilibrium regional disparities: theory and British evidence. Reg Stud 37(6-7):675-686

Rivera-Batiz F (1988) Increasing returns, monopolistic competition, and agglomeration economies in consumption and production. Reg Sci Urb Econ 18(1):125-153

Appendix: Core and periphery values of G, L and U in each region

Country	NUTS-2region	Core	G	L	U	Periphery	G	L	U
Germany	Schleswig	0	0.00	0.00	0.0	1	0.25	0.56	164.3
Germany	Hamburg	1	0.45	0.49	2296.7	0	0.00	0.00	0.0
Germany	Braunschweig	1	0.28	0.38	206.8	0	0.00	0.00	0.0
Germany	Hannover	1	0.27	0.35	231.3	0	0.00	0.00	0.0
Germany	Luneberg	1	0.62	0.46	91.9	0	0.00	0.00	0.0
Germany	Weser_Ems	1	0.38	0.36	139.7	0	0.00	0.00	0.0
Germany	Bremen	1	0.18	0.42	1792.1	0	0.00	0.00	0.0
Germany	Dusseldorf	1	0.33	0.18	1064.5	0	0.00	0.00	0.0
Germany	Koln	1	0.40	0.12	488.5	0	0.00	0.00	0.0
Germany	Munster	1	0.32	0.25	354.5	0	0.00	0.00	0.0
Germany	Detmold	1	0.38	0.30	278.4	0	0.00	0.00	0.0
Germany	Arnsberg	1	0.19	0.20	466.4	0	0.00	0.00	0.0
Germany	Darmstadt	1	0.40	0.19	457.8	0	0.00	0.00	0.0
Germany	Giessen	1	0.57	0.21	179.7	0	0.00	0.00	0.0
Germany	Kassel	1	0.00	0.26	144.9	0	0.00	0.00	0.0
Germany	Koblenz	1	0.36	0.10	170.0	0	0.00	0.00	0.0
Germany	Trier	1	0.50	0.04	97.0	0	0.00	0.00	0.0
Germany	Rheinhessen_Pfalz	1	0.17	0.13	268.6	0	0.00	0.00	0.0
Germany	Stuttgart	1	0.34	0.27	329.8	0	0.00	0.00	0.0
Germany	Karlsruhe	1	0.36	0.21	346.4	0	0.00	0.00	0.0
Germany	Freiburg	1	0.39	0.25	199.4	0	0.00	0.00	0.0
Germany	Tuebingen	1	0.53	0.31	166.2	0	0.00	0.00	0.0
Germany	Oberbayern	1	0.49	0.46	203.1	0	0.00	0.00	0.0
Germany	Niederbayern	1	0.56	0.49	96.1	0	0.00	0.00	0.0
Germany	Oberpfalz	1	0.33	0.44	100.7	0	0.00	0.00	0.0
Germany	Oberfranken	1	0.35	0.37	148.1	0	0.00	0.00	0.0
Germany	Mittelfranken	1	0.24	0.34	211.3	0	0.00	0.00	0.0
Germany	Unterfranken	1	0.62	0.27	140.8	0	0.00	0.00	0.0
Germany	Schwaben	1	0.33	0.36	151.7	0	0.00	0.00	0.0
Germany	Saarland	1	0.43	0.07	429.3	0	0.00	0.00	0.0
Germany	Berlin	0	0.00	0.00	0.0	1	0.19	0.58	4216.7
France	Ile_De_France	1	0.30	0.29	822.3	0	0.00	0.00	0.0
France	Champagne_Ardenne	1	0.39	0.15	52.2	0	0.00	0.00	0.0
France	Picardie	1	0.43	0.24	86.7	0	0.00	0.00	0.0
France	Haute_Normandie	1	0.26	0.38	129.7	0	0.00	0.00	0.0
France	Centre	1	0.46	0.41	54.9	0	0.00	0.00	0.0
France	Basse_Normandie	1	0.47	0.49	74.3	0	0.00	0.00	0.0
France	Bourgogne	1	0.46	0.32	49.9	0	0.00	0.00	0.0
France	Nord_Pas_De_Calais	1	0.40	0.24	315.2	0	0.00	0.00	0.0
France	Lorraine	1	0.41	0.12	98.7	0	0.00	0.00	0.0
France	Alsace	1	0.38	0.20	183.6	0	0.00	0.00	0.0
France	Franche_Comte	1	0.44	0.30	65.5	0	0.00	0.00	0.0
France	Pays_De_La_Loire	0	0.00	0.00	0.0	1	0.42	0.58	86.3
France	Bretagne	0	0.00	0.00	0.0	1	0.42	0.68	95.4
France	Poitou_Charentes	0	0.00	0.00	0.0	1	0.43	0.61	59.2
France	Aquitaine	0	0.00	0.00	0.0	1	0.25	0.78	61.7
France	Midi_Pyrenees	0	0.00	0.00	0.0	1	0.45	0.76	49.9
France	Limousin	0	0.00	0.00	0.0	1	0.54	0.57	43.7
France	Rhone_Alpes	0	0.00	0.00	0.0	1	0.39	0.51	109.8
France	Auvergne	0	0.00	0.00	0.0	1	0.47	0.51	51.2
France	Languedoc_Roussillon	0	0.00	0.00	0.0	1	0.35	0.72	69.0
France	Provence_Alpes_Cote	0	0.00	0.00	0.0	1	0.25	0.64	117.1
France	Corse	0	0.00	0.00	0.0	1	0.40	0.88	25.2
Italy	Piemonte	0	0.00	0.00	0.0	1	0.48	0.52	176.4
Italy	Valle_Daosta	1	0.40	0.47	34.0	0	0.00	0.00	0.0
Italy	Liguria	0	0.00	0.00	0.0	1	0.37	0.64	341.8
Italy	Lombardia	0	0.00	0.00	0.0	1	0.45	0.54	365.6
Italy	Trintio_Alto_Adige	0	0.00	0.00	0.0	1	0.23	0.55	63.1
Italy	Veneto	0	0.00	0.00	0.0	1	0.59	0.62	230.3
Italy	Friuli_Venezia_Giul	0	0.00	0.00	0.0	1	0.53	0.66	156.9
Italy	Emilia_Romagna	0	0.00	0.00	0.0	1	0.44	0.70	176.7
Italy	Toscana	0	0.00	0.00	0.0	1	0.56	0.81	153.6
Italy	Umbria	0	0.00	0.00	0.0	1	0.59	0.91	93.3
Italy	Marche	0	0.00	0.00	0.0	1	0.70	0.89	142.7
Italy	Lazio	0	0.00	0.00	0.0	1	0.33	1.01	280.8
Italy	Campagna	0	0.00	0.00	0.0	1	0.54	1.21	384.2
Italy	Abruzzi	0	0.00	0.00	0.0	1	0.64	1.03	110.1
Italy	Molise	0	0.00	0.00	0.0	1	0.61	1.11	73.0
Italy	Puglia	0	0.00	0.00	0.0	1	0.53	1.29	191.0
Italy	Basilicata	0	0.00	0.00	0.0	1	0.59	1.29	60.9
Italy	Calabria	0	0.00	0.00	0.0	1	0.54	1.45	134.0

cont.

Italy	Sicilia	0	0.00	0.00	0.0	1	0.57	1.51	186.1
Italy	Sardegna	0	0.00	0.00	0.0	1	0.46	1.11	63.3
Netherlands	Groningen	1	0.38	0.38	180.5	0	0.00	0.00	0.0
Netherlands	Friesland	1	0.56	0.37	103.7	0	0.00	0.00	0.0
Netherlands	Drenthe	1	0.38	0.35	149.6	0	0.00	0.00	0.0
Netherlands	Overijssel	1	0.54	0.30	273.3	0	0.00	0.00	0.0
Netherlands	Gelderland	1	0.50	0.26	315.1	0	0.00	0.00	0.0
Netherlands	Flevoland	1	0.38	0.31	34.0	0	0.00	0.00	0.0
Netherlands	Utrecht	1	0.49	0.26	612.0	0	0.00	0.00	0.0
Netherlands	Noord_Holland	1	0.24	0.32	625.0	0	0.00	0.00	0.0
Netherlands	Zuid_Holland	1	0.16	0.27	900.6	0	0.00	0.00	0.0
Netherlands	Zeeland	1	0.24	0.25	107.6	0	0.00	0.00	0.0
Netherlands	Noord_Brabant	1	0.46	0.20	381.9	0	0.00	0.00	0.0
Netherlands	Limburg	1	0.49	0.16	472.6	0	0.00	0.00	0.0
Belgium	Brabant	1	0.58	0.16	361.6	0	0.00	0.00	0.0
Belgium	Antwerpen	1	0.43	0.19	542.4	0	0.00	0.00	0.0
Belgium	Limburg	1	0.55	0.14	280.3	0	0.00	0.00	0.0
Belgium	Oost_Vlaanderen	1	0.58	0.21	444.0	0	0.00	0.00	0.0
Belgium	West_Vlaanderen	1	0.62	0.26	341.4	0	0.00	0.00	0.0
Belgium	Hainaut	1	0.61	0.17	349.4	0	0.00	0.00	0.0
Belgium	Liege	1	0.57	0.08	264.1	0	0.00	0.00	0.0
Belgium	Luxembourg	1	0.61	0.05	49.3	0	0.00	0.00	0.0
Belgium	Namur	1	0.63	0.10	106.1	0	0.00	0.00	0.0
Belgium	Bruxelles_Brussel	1	0.45	0.17	6552.8	0	0.00	0.00	0.0
Luxembourg	Luxembourg	1	0.42	0.00	138.1	0	0.00	0.00	0.0
United Kingdom	Cleveland_Durham	0	0.00	0.00	0.0	1	0.46	0.75	397.2
United Kingdom	Cumbria	0	0.00	0.00	0.0	1	0.46	0.82	70.0
United Kingdom	Northumberland_Tyne	0	0.00	0.00	0.0	1	0.47	0.82	265.1
United Kingdom	Humberside	0	0.00	0.00	0.0	1	0.50	0.64	240.3
United Kingdom	North_Yorkshire	0	0.00	0.00	0.0	1	0.50	0.70	79.6
United Kingdom	South_Yorkshire	0	0.00	0.00	0.0	1	0.50	0.66	863.5
United Kingdom	West_Yorkshire	0	0.00	0.00	0.0	1	0.50	0.69	1018.1
United Kingdom	Derbyshire_Nottigha	0	0.00	0.00	0.0	1	0.50	0.64	395.4
United Kingdom	Leicestershire_Nort	0	0.00	0.00	0.0	1	0.50	0.58	272.6
United Kingdom	Lincolnshire	0	0.00	0.00	0.0	1	0.50	0.58	88.8
United Kingdom	East_Anglia	1	0.51	0.48	142.6	0	0.00	0.00	0.0
United Kingdom	Bshire_Hertfordshir	0	0.00	0.00	0.0	1	0.44	0.51	530.5
United Kingdom	Berks_Bucks_Oxfords	0	0.00	0.00	0.0	1	0.44	0.55	309.7
United Kingdom	Surrey_East_Westsus	1	0.44	0.46	430.9	0	0.00	0.00	0.0
United Kingdom	Essex	1	0.44	0.44	404.4	0	0.00	0.00	0.0
United Kingdom	Greater_London	1	0.53	0.48	4302.1	0	0.00	0.00	0.0
United Kingdom	Hampshire_Isle_Of_W	0	0.00	0.00	0.0	1	0.44	0.55	382.2
United Kingdom	Kent	1	0.44	0.41	399.4	0	0.00	0.00	0.0
United Kingdom	Avon_Gloucestershir	0	0.00	0.00	0.0	1	0.49	0.62	259.0
United Kingdom	Cornwall_Devon	0	0.00	0.00	0.0	1	0.49	0.75	129.4
United Kingdom	Dorset_Somerset	0	0.00	0.00	0.0	1	0.49	0.64	162.5
United Kingdom	Hereford_Worcs_Wraw	0	0.00	0.00	0.0	1	0.54	0.64	182.2
United Kingdom	Shropshire_Stafford	0	0.00	0.00	0.0	1	0.54	0.68	219.6
United Kingdom	West_Middlands_Coun	0	0.00	0.00	0.0	1	0.54	0.63	3054.5
United Kingdom	Cheshire	0	0.00	0.00	0.0	1	0.48	0.71	393.5
United Kingdom	Greater_Manchester	0	0.00	0.00	0.0	1	0.48	0.72	2095.6
United Kingdom	Lancashire	0	0.00	0.00	0.0	1	0.48	0.75	449.9
United Kingdom	Merseyside	0	0.00	0.00	0.0	1	0.48	0.75	2440.2
United Kingdom	Clwyd_Dyfed_Gwynedd	0	0.00	0.00	0.0	1	0.50	0.77	60.8
United Kingdom	Gwent_Mid_S_W_Glamo	0	0.00	0.00	0.0	1	0.50	0.71	483.3
United Kingdom	Borders_Central_Fif	0	0.00	0.00	0.0	1	0.52	0.94	102.1
United Kingdom	Dumfries_Galloway_S	0	0.00	0.00	0.0	1	0.52	0.96	129.4
United Kingdom	Highlands_Islands	0	0.00	0.00	0.0	1	0.52	1.14	8.5
United Kingdom	Grampian	0	0.00	0.00	0.0	1	0.52	1.01	54.0
United Kingdom	Northern_Ireland	0	0.00	0.00	0.0	1	0.51	1.03	107.9
Ireland	Ireland	0	0.00	0.00	0.0	1	0.48	1.07	45.3
Denmark	Hovedstadsregionen	0	0.00	0.00	0.0	1	0.47	0.77	617.7
Denmark	Ost_For_Storebaelt_	0	0.00	0.00	0.0	1	0.52	0.71	81.8
Denmark	Vest_For_Storebaelt	0	0.00	0.00	0.0	1	0.51	0.75	81.7
Greece	Anatoliki_Makedonia	0	0.00	0.00	0.0	1	0.96	1.77	39.6
Greece	Kentriki_Makedonia_	0	0.00	0.00	0.0	1	0.81	1.67	76.0
Greece	Dytiki_Makedonia_We	0	0.00	0.00	0.0	1	0.75	1.59	28.7
Greece	Thessalia_Thessaly	0	0.00	0.00	0.0	1	0.84	1.72	45.0
Greece	Ipeiros_Epirus	0	0.00	0.00	0.0	1	0.85	1.61	33.3
Greece	Ionia_Nisia	0	0.00	0.00	0.0	1	0.30	1.57	75.9
Greece	Dytiki_Ellena_Weste	0	0.00	0.00	0.0	1	0.76	1.82	52.3
Greece	Sterea_Ellena_Centr	0	0.00	0.00	0.0	1	0.80	1.83	32.0
Greece	Peloponnisos	0	0.00	0.00	0.0	1	0.67	1.91	35.4
Greece	Attiki	0	0.00	0.00	0.0	1	0.76	1.91	809.9
Greece	Voreio_Aigaio	0	0.00	0.00	0.0	1	0.88	1.97	50.1

cont.

Greece	Notio_Aigaio	0	0.00	0.00	0.0	1	0.85	2.32	40.1
Greece	Kritti	0	0.00	0.00	0.0	1	0.68	2.22	55.1
Spain	Galicia	0	0.00	0.00	0.0	1	0.70	1.34	91.3
Spain	Asturias	0	0.00	0.00	0.0	1	0.61	1.17	102.9
Spain	Cantabria	0	0.00	0.00	0.0	1	0.58	1.07	90.0
Spain	Pais_Vasco	0	0.00	0.00	0.0	1	0.53	1.01	289.9
Spain	Navarra	0	0.00	0.00	0.0	1	0.58	1.00	45.2
Spain	Rioja	0	0.00	0.00	0.0	1	0.66	1.07	46.1
Spain	Aragon	0	0.00	0.00	0.0	1	0.71	1.08	23.6
Spain	Madrid	0	0.00	0.00	0.0	1	0.62	1.30	545.6
Spain	Castilla_Leon	0	0.00	0.00	0.0	1	0.62	1.23	26.5
Spain	Castilla_La_Mancha	0	0.00	0.00	0.0	1	0.58	1.35	18.3
Spain	Extremadurra	0	0.00	0.00	0.0	1	0.82	1.53	23.7
Spain	Cataluna	0	0.00	0.00	0.0	1	0.63	0.97	178.6
Spain	Communidad_Valencia	0	0.00	0.00	0.0	1	0.68	1.29	144.5
Spain	Baleares	0	0.00	0.00	0.0	1	0.66	1.17	117.9
Spain	Andalucia	0	0.00	0.00	0.0	1	0.53	1.63	66.5
Spain	Murcia	0	0.00	0.00	0.0	1	0.46	1.45	74.4
Portugal	Norte	0	0.00	0.00	0.0	1	0.84	1.41	148.4
Portugal	Centro	0	0.00	0.00	0.0	1	0.85	1.53	71.5
Portugal	Lisboa_E_Vale_De_Te	0	0.00	0.00	0.0	1	0.77	1.67	253.6
Portugal	Alentejo	0	0.00	0.00	0.0	1	0.76	1.68	21.0
Portugal	Algarve	0	0.00	0.00	0.0	1	0.76	1.81	59.7
Austria	Wien	0	0.00	0.00	0.0	1	0.43	0.77	3867.5
Austria	Niederosterreich	0	0.00	0.00	0.0	1	0.34	0.72	74.5
Austria	Burgenland	0	0.00	0.00	0.0	1	0.61	0.81	68.6
Austria	Steiermark	0	0.00	0.00	0.0	1	0.48	0.70	73.2
Austria	Karnten	0	0.00	0.00	0.0	1	0.48	0.67	56.1
Austria	Oberosterreich	0	0.00	0.00	0.0	1	0.39	0.60	104.8
Austria	Salzburg	0	0.00	0.00	0.0	1	0.41	0.57	59.3
Austria	Tirol	1	0.39	0.50	44.8	0	0.00	0.00	0.0
Austria	Vorarlberg	1	0.44	0.40	113.6	0	0.00	0.00	0.0

Part F

Applications in Environmental Sciences

F.1 A Fuzzy *k*-Means Classification and a Bayesian Approach for Spatial Prediction of Landslide Hazard

Pece V. Gorsevski, Paul E. Gessler and *Piotr Jankowski*

F.1.1 Introduction

The increasing availability of geospatial data and the rapid advances in the Geographic Information Systems (GIS) technology for statistical and mathematical modeling and simulation have led to a variety of applications and a growing spatial literature. Spatial statistical methods and techniques have been widely used in a number of discipline-specific applications, some of which are described in this handbook and result from those rapid advances in GIS tools, techniques, and literature.

In particular, this chapter presents an application of spatial analyses which link spatial correlations of environmental attributes and landslide datasets of known landslide locations initiated from roads (presence of human interactions) and known landslide location outside of roads (absence of human interactions) for predicting landslide hazard. The chapter pays special attention to the automated spatial extraction analyses of fuzzy *k*-means classification, the computation of an optimal number of classes and their overlap, the implementation of Mahalanobis distances to extrapolate the continuous classification to a broad region, and the derivation of Bayesian predictive models using relationships of landslide locations and continuous landform classes. The spatial approach is demonstrated through a regional case study in the U.S. Pacific Northwest where substantial landslide impacts to the environment occurred in the winter of 1995/96.

The present chapter is largely taken from a paper published in the Journal of Geographical Systems 5(3), 223-251, but additions to this work have been also extended and applied using a number of spatial approaches (Gorsevski et al. 2004, 2005, 2006a, 2006b, 2006c; Gorsevski and Jankowski 2008). An article that closely follows this modeling approach by Gorsevski et al. (2005) is based on integration of the fuzzy *k*-means classification and the Dempster-Shafer (*D-S*) theory of evidence. The *D-S* theory is an extension of the Bayesian theory and it is more flexible in the sense that it waives the need for complete knowledge of prior or conditional probabilities before modeling can take place. Also the *D-S* theory introduces the representation of ignorance, which represents the lack of evidence. For instance, absence of a landslide in the database may suggest that the landslide was not identified through the aerial photo interpretation for various reasons. In addition, in Bayesian theory evidence is used to support individual hypothesis whereas in *D-S* theory a single piece of evidence can support multiple hypotheses. Another distinction is that the *D-S* approach outputs uncertainties through consideration of lower and upper probability intervals induced by multi-valued mapping, rather than explicit probability values as with the approach that follows.

Landslides are natural geologic processes that cause different types of damage and affect people, organizations, industries, and the environment (Glade 1998). Globally, landslides cause billions of dollars in damage and thousands of deaths and injuries each year. Developing countries suffer the most, where 0.5 percent of the gross national product has been lost due to landslides, and 95 percent of landslide disasters have been recorded in developing countries (Chung et al. 1995). In the U.S. alone an estimated annual average cost of $1.5 billion dollars due to landslides has been reported (Glade 1998).

Human activities, such as deforestation and urban expansion, accelerate the process of landslides (Chung et al. 1995). Landslides initiated from roads and forest harvesting are considered to be a significant sediment source (Dyrness 1967). Landslides contribute to decreased water quality, loss of fish spawning habitat and organic matter, and debris jams that may break during peak flows, thereby scouring channels and destroying riparian vegetation. Concern about landsliding especially from forest roads (road related landslides) calls for improved forest management practices in all forestlands where humans are active as well as forestlands where roads are absent. Road related (RR) landslides are defined as landslides that occur or initiate within the road right-of-way, while non-road related (NRR) landslides are landslides that occur outside the road right-of-way. Road right-of-way includes the clearing width and the roadway with its elements (cut slope, ditch, shoulder, travel way, and fill slope). McClelland et al. (1997) define a road related landslide as a landslide originating between the top of a road cut and 100 feet below the base of the fill. A method for predicting RR or NRR landslide hazard may be a valuable decision support tool for future planning and management of forestlands. A robust method for predicting RR and NRR landslides may provide information important for road-siting and forest management practices in areas prone to landslide occurrence.

In recent years the use of GIS for landslide hazard modeling has increased because of the development of commercial and noncommercial systems, such as ArcGIS (ESRI), IDRISI, and GRASS and the quick access to data obtained through the Global Positioning System (GPS) and remote sensing tools. GIS technology has made it possible to derive surface morphometry from a digital elevation model (DEM) (Moore et al. 1993; Hengl and Reuter 2008), which may be used in landslide hazard modeling. Primary and secondary attributes are derived from a DEM, which reduces the high cost of collecting detailed field data. A GIS allows rapid combination and assessment of terrain attributes. Landslide hazard areas may then be identified based on spatial correlation between the terrain and landslide occurrences. Such predictive correlations have been applied to various disciplines including: hydrology, soil-landscape modeling, wildlife studies, climatology, and geohazard assessment (Carrara 1983; Grayson et al. 1992a, 1992b; Moore et al. 1993; Gessler et al. 1995, 2000; Mladenoff et al. 1995; Mladenoff and Sickley 1998; McKenzie et al. 2000; Ryan et al. 2000, Chamran et al. 2002; Gorsevski et al. 2003, 2004, 2005, 2006a, 2006b; Gorsevski and Jankowski 2008; Hengl and Reuter 2008).

Numerous slope stability studies using GIS technology acknowledge that topography, soil thickness, hydrologic processes, and vegetation surcharge influence landslide initiation (Montgomery and Dietrich 1994; Wu and Sidle 1995; Gorsevski et al. 2006c). Some studies have also used landform variables as predictors for modeling landslide hazard (Carrara et al. 1995; McClelland et al. 1997; Robison et al. 1999). Other studies have shown that landslide hazard is a function of steep hillslope gradients originating in areas of topographic convergence (Reneau and Dietrich 1987; Ellen et al. 1988; Montgomery and Dietrich 1994). However, delineating useful morphological units (landforms) and representing the uncertainties inherent in classifying these continuously varying landforms is lacking in most studies. This study presents a new approach to predicting landslide hazard that combines modeling of landforms with modeling landslide hazard probabilities. The approach is based on developing continuous landform classifications (i.e., identifiable elements) on a watershed scale using fuzzy k-means methods and Bayes' theorem to generate landslide hazard probability maps associated with road related and non-road related landslide hazard. The fuzzy k-means approach will ensure that derived landform classifications are reproducible and objectively applied (Burrough et al. 2000), while extrapolation to a broader-scale area using a distancing technique is possible. Therefore, this contribution proposes to improve upon current approaches by explicit incorporation of uncertainty through the fuzzy k-means approach and building models capable of predicting RR and NRR landslide hazard that may enable forestland management to avoid critical areas or provide information that suggests modified practices in areas prone to landslide occurrence.

F.1.2 Overview of current prediction methods

Landslide hazard models have been developed in various ways and include: (i) landslide inventory (Wieczorek 1984; Wright et al. 1974), (ii) statistical modeling (Carrara 1983; Carrara et al. 1991, 1992; Chung et al. 1995; Mark and Ellen 1995;

Chung and Fabbri 1999; Dhakal et al. 2000; Gorsevski et al. 2003, 2004, 2005, 2006b), (iii) heuristic methods (McClelland et al. 1997; Gorsevski et al. 2006a), (iv) process-based modeling (Okimura and Ichikawa 1985; Montgomery and Dietrich 1994; Wu and Sidle 1995; Gorsevski et al. 2006c), (v) probabilistic or stochastic modeling (Hammond et al. 1992), and (vi) artificial intelligence (Gorsevski and Jankowski 2008).

An example of a heuristic method is the current Forest Service method (McClelland et al. 1997) specifically developed for the Clearwater National Forest (CNF) in northcentral Idaho. The method (FSmet) uses a heuristic rule that defines high hazard areas as those locations with slopes greater than 60 percent with parent materials of schist or granitics at elevations below 1,400 m. This heuristic has been developed based on data collected specifically within the CNF. The output of FSmet can be a binary map delineating hazard and no hazard areas.

Another method used for assessing landslide hazard is the SHALSTAB model developed by Dietrich and Montgomery (1998). SHALSTAB is a physically-based model based on a combination of the infinite slope equation of the Mohr-Coulomb failure law and a hydrological component based on steady-state shallow subsurface flow (O'Loughlin 1986). SHALSTAB was implemented as an ArcView extension to generate landslide susceptibility output using a DEM, soil bulk density, and friction angle as inputs. SHALSTAB's one-step option (Dietrich and Montgomery 1998) was used to calculate the critical value of the ratio of steady-state effective precipitation (rain minus evapotranspiration; q to transmissivity, (the ground's subsurface ability to convey water downslope; T needed to generate a landslide. The q/T ratio has dimensions of $(L/T)/(L^2/T)$ or L^{-1}.

A large q/T ratio implies that the soil approaches saturation, and a high susceptibility of slope failure. Because q/T is always less than one, $\log(q/T)$ is reported. SHALSTAB classifies landslide susceptibility as: 'unconditionally stable', 'potentially unstable', and 'unconditionally unstable'. Unconditionally stable elements are predicted not to fail even when saturated, while unconditionally unstable are predicted to fail even when dry. Potentially unstable elements are associated with values of $\log(q/T)$ ranging from -1.9 to -3.4 incremented by -0.3 where divisions within the range of values are user-imposed. Areas with large absolute values of $\log(q/T)$ represent the least stable areas, whereas areas with small absolute values of $\log(q/T)$ represent the most stable areas. We applied arbitrary cut-off values of greater than -2.2 to represent 'unconditionally stable' areas, greater than -2.8 to represent the midpoint of 'potentially unstable' areas, and greater than -3.1 to represent 'unconditionally unstable' areas. The landslide database from 1995-96 used by McClelland et al. (1997) was also used for comparison. However, outputs generated from FSmet and SHALSTAB are associated with overall landslide hazard and not associated with RR or NRR landslide hazard. Also, while outputs are difficult to compare, neither of the methods incorporates uncertainty associated with the parameters used for the modeling.

Uncertainties are usually introduced by soil-related parameters that have different properties and vary over space in different ways. Burrough et al. (2000) and others (Fisher and Pathirana 1990; Butler 1982; Webster and Oliver 1990; McSweeney et al. 1994) argue that soil spatial variation captured in a soil map is often generalized because of many uncertainties involved in soil mapping. Odeh et al. (1992) suggested that soil variation is more continuous than discrete and therefore an approach that models the soil as a continuum is more appropriate.

Modeling uncertainty may be handled through probability theory or fuzzy set theory (Zadeh 1965, 1978). Probability theory uses probabilistic models to quantify the uncertainty associated with the prediction of the phenomenon, and measures incomplete knowledge through objective modeling (Chen and Hwang 1992). Fuzzy set theory, on the other hand, models uncertainty based on expert knowledge, and measures incomplete knowledge through subjective modeling. Therefore, methods of fuzzy classification may be used to account for uncertainty and replace crisp classification methods by providing class overlap that is more realistic for modeling continuous landscape patterns. Fuzzy set theory extends the crisp, unambiguous theory and deals with continuous classification of entities. Irvin et al. (1997) suggest that fuzzy classification of landforms may be a preferred way to encapsulate key variation of complex land attributes for describing and understanding landscape processes. Fuzzy classes may be helpful in defining management areas for dealing with problems such as soil drainage, erosion, or landsliding (Burrough et al. 2000).

Fuzzy methods, such as the fuzzy k-means, which is analogous to traditional cluster analysis but allows class overlap, has been implemented in soil-landscape studies (McBratney and deGruijter 1992; Odeh et al. 1992; Irvin et al. 1997; Ventura and Irvin 2000; Burrough et al. 2000; MacMillan et al. 2000; Hengl and Reuter 2008; Evans et al. 2008). These studies explore the relationship between various earth surface processes, topography, and soil development and test hypotheses about the spatial distribution of soil attributes. Irvin et al. (1997) demonstrated that soil spatial variability could be modeled using six topographic attributes (elevation, slope, profile curvature, tangent curvature, compound topographic index, and solar radiation) that characterize landform shape. However, Irvin et al. (1997) did not take into account the size of the area to be classified and the computational issues associated with the derivation of the secondary attributes (i.e., compound topographic index, and solar radiation). Burrough et al. (2000) suggested an approach to overcome some of the limitations associated with computational issues from the previous approach by using spatial sampling methods, statistical modeling of the derived stream topology, and fuzzy k-means classification using a distance metric.

Other methods, such as Bayes' theorem, provide a potential means of converting knowledge of predictive correlations from a fuzzy classification of landforms in combination with landslide location data, to landslide hazard probabilities. Bayes' theorem has been applied for geologic hazard prediction, geologic soil-landscape modeling, and more recently in environmental science or wildlife studies (Spiegelhalter 1986; Chung and Fabbri 1999; Skidmore et al. 1996; Aspinall 1992; Aspinall and Veitch 1993). The Bayesian approach is a mathematical method used for decision-making under conditions of uncertainty (Aspinall 1992). This method could be used to link prior (known) probabilities of landslide hazard with the fuzzy k-means classes, and consequently assign conditional probabilities. These probabilities at first may or may not be accurate. However, after corroborating the probabilities with additional information from landslide hazard monitoring or more extensive datasets, the accuracy may be improved to acceptable levels (Malczewski 1999).

F.1.3 Modeling theory

In the proposed modeling theory, the fuzzy k-means approach is used to organize the complex multivariate data derived from a DEM into continuous landform classes. The approach is applied to a watershed scale landform classification (training set) that is extrapolated to a broader-scale area using a Mahalanobis metric. The Bayesian theorem follows to quantify the relationships between landslides and the extrapolated classes. The following sections will detail the individual components for the integrated fuzzy k-means classification and a Bayesian approach.

The fuzzy set theory

Fuzzy logic (Zadeh 1965) is a superset of conventional (Boolean) logic that has been extended to handle the concept of partial truth-values between 'completely true' and 'completely false'. In conventional logic the degree to which an individual z is a member or is not a member of a given set A is expressed by the membership function MF^B. The membership function MF^B can take the value zero or one shown in Eqs. (F.1.1) and (F.1.2).

$$MF^B(z)=1 \quad \text{if } b_1 \leq z \leq b_2 \tag{F.1.1}$$

$$MF^B(z)=0 \quad \text{if } z < b_1 \text{ or } z > b_2 \tag{F.1.2}$$

where b_1 and b_2 define the exact boundaries of set A. For instance, if the boundaries b_1 and b_2 for 'steep' slope were defined between 45 percent and 70 percent, then the conventional set theory would assign value one for each individual belonging to the set and zero otherwise. On the other hand, the idea behind fuzzy logic is to describe the vagueness of entities in the real world, where belonging to a set is really a matter of degree (Malczewski 1999). For instance, linguistic terms and qualitative data such as 'gentle', 'moderate', 'steep', and 'very steep' land can be translated into fuzzy sets.

A fuzzy set is a class of elements or objects without well-defined boundaries between objects that belong to the class and those that do not. Fuzzy logic allows objects to belong partially to multiple sets and it is multivalued logic that allows intermediate values to be formulated mathematically. The fuzzy set is specified by a membership function. The function represents any elements on a continuous scale from one (full membership) to zero (full-non-membership). Mathematically a fuzzy set A is defined as follows: If Z denotes a space of objects, then the fuzzy set A in Z is the set of ordered pairs

$$A = \{z,\ MF_A^F(z)\} \qquad z \in Z \tag{F.1.3}$$

where the membership function $MF_A^F(z)$ is known as the 'degree of membership of z in A'. The higher the membership value of $MF_A^F(z)$, the more it belongs to the set. Fuzzy sets for developing spatial decision support systems can be used to represent geographical entities that imprecisely define boundaries as fuzzy objects or fuzzy regions. Fuzzy regions can be conceptualized as a set of pixels between the pixels of full membership values. For example, many researchers (Carrara et al. 1995; McClelland et al. 1997; Robison et al. 1999) have used a correlation between landform attributes and landslide locations to predict landslide hazard. Although, linking landform attributes and landslide location is a valid technique, establishing a clear boundary between different types of landforms is a difficult task. Landforms are more continuous than discrete which calls for a continuous approach to landform classification. Thus, fuzzy classification can be expected to provide such an approach by assigning landforms to continuous classes for determining the strength of the relationship between landform attributes and landslide locations.

There are two basic methods for building membership functions: the fuzzy semantic import (*SI*) model and fuzzy *k*-means clustering (Burrough and McDonnell 1998; MacMillan et al. 2000). The use of the *SI* model for classification purposes depends on the existence of a well-defined and functional classification based on expert knowledge. On the other hand, fuzzy *k*-means clustering is an unsupervised classification method and is not dependent on prior knowledge. Fuzzy *k*-means cluster analysis was used herein and the theory is discussed below.

Fuzzy k-means approach

The fuzzy *k*-means clustering approach, also known as *c*-means (Bezdek 1981; Fisher and Pathirana 1990; McBratney and deGruijter 1992; Odeh et al. 1992; Burrough and McDonnell 1998; Burrough et al. 2000; 2001; Evans et al. 2008) is analogous to traditional cluster analysis. Cluster analysis or clustering is a method that groups patterns of data that in some sense belong together and have similar characteristics. The clustering technique uses a repetitive procedure by selecting a set of random cluster points and building clusters around each seed. This is accomplished by assigning every point in the data set to its closest seed, using distance measures such as: Euclidean, Mahalanobis or Diagonal distance. The iteration stops when a stable solution is reached meaning that the objects in each cluster are similar to one another while those in different clusters are not similar to one another.

The idea of fuzzy clustering was introduced first by Ruspini (1969) as an alternative to the traditional cluster analysis by applying membership values to points between clusters (i.e., difficult to classify) as an inverse function of distance from the cluster centers. This led to further refinement and development of additional algorithms of fuzzy clustering (Bezdek 1981; McBratney and deGruijter 1992). The fuzzy k-means clustering is the most commonly used technique and has been applied to various disciplines such as climatic modeling (McBratney and Moore 1985), geologic modeling (Bezdek 1981), suburban environment modeling (Fisher and Pathirana 1990), and soil-landscape modeling (McBratney and deGruijter 1992; Odeh et al. 1992; Irvin et al. 1997; Burrough et al. 2000; 2001; MacMillan et al. 2000; Ventura and Irvin 2000; Iwahashi and Pike 2007; Hengl and Reuter 2008; Evans et al. 2008). McBratney and deGruijter (1992) refer to the fuzzy k-means clustering term as a 'continuous classification' where each data point is not required to be an exclusive member of one and only one class. The membership value is assigned through the class centroid concept for each data point in each class. The final membership values with fuzzy k-means range between zero and one for each data point, while the sum of values for a particular data point across all classes equals to one. The fuzzy k-means clustering with extragrades is another technique used for continuous classification, which provides better representations of outliers (i.e., data that have low membership in most or all of the classes) (McBratney and deGruijter 1992).

Fuzzy k-means algorithms. For a set of n individuals classified into c classes with conventional (Boolean) classification the membership function equals $M = \mu_{ij} = 1$, where individual i belongs to class j, and $M = \mu_{ij} = 0$, when individual i does not belong to class j. Three conditions ensure that conventional sets are exclusive and jointly exhaustive

$$\sum_{j=1}^{c} \mu_{ij} = 1 \qquad 1 \leq i \leq n \qquad (F.1.4)$$

$$\sum_{i=1}^{n} \mu_{ij} > 0 \qquad 1 \leq j \leq c \qquad (F.1.5)$$

$$\mu_{ij} \in \{0,1\} \qquad 1 \leq i \leq n; \ 1 \leq j \leq c. \qquad (F.1.6)$$

Equation (F.1.4) indicates that the sum of membership of an individual across all classes is one. Equation (F.1.5) ensures that at least one individual belongs to each class, so the classes are not empty. Finally, Eq. (F.1.6) suggests that an individual belongs to a class or does not belong at all. This equation institutes the difference between hard and fuzzy classes. Fuzzy set theory relaxes Eq. (F.1.6) so that class memberships are allowed to be partial and can take on any value between and including zero and one (see Eq. (F.1.7)).

$$\mu_{ij} \in [0,1] \qquad 1 \leq i \leq n; 1 \leq j \leq c. \qquad (F.1.7)$$

Several algorithms are used for computing fuzzy k-means (Bezdek 1981; Burrough and McDonnell 1998; McBratney and deGruijter 1992). Fuzzy c-means (ordinary k-means), as used here, is the best-known classification (Bezdek et al. 1984). The optimal fuzzy classification is achieved by minimization of the objective function to satisfy the conditions in Eqs. (F.1.4), (F.1.5) and (F.1.6). The generalized objective function is given in Eq. (F.1.8).

$$J_F(M,c) = \sum_{i=1}^{n} \sum_{j=1}^{c} \mu_{ij}^{\phi} d_{ij}^2 \qquad (F.1.8)$$

where μ is the membership of the ith object to the jth class; $\mathbf{C} = (\mathbf{C}_{jv})$ is a c-by-p matrix of class centroids with $\mathbf{C}_{jv}$ denoting the centroid of class j for variable v; p is the number of attributes; d_{ij}^2 is the square of the distance between the individual i and the class center j; and ϕ determines the amount of fuzziness or overlap and is called the fuzzy exponent. For example, when ϕ equals one no overlap is allowed and there is no fuzziness (a 'hard class' is generated), for large ϕ there is complete overlap and the clusters are identical. With ϕ greater than one, minimization of $J_F(M,c)$ is achieved by Langrangian differentiation of Eq. (F.1.8) using Picard's iteration (McBratney and deGruijter 1992) with Eqs. (F.1.9) and (F.1.10)

$$\mu_{ij} = \frac{[(d_{ij})^2]^{-1/(\phi-1)}}{\sum_{k=1}^{c}[(d_{ik})^2]^{-1/(\phi-1)}} \qquad 1 \leq i \leq n; 1 \leq j \leq c \qquad (F.1.9)$$

and

$$C_{cj} = \frac{\sum_{i=1}^{n}(\mu_{ij})^{\phi} X_{ij}}{\sum_{i=1}^{n}(\mu_{ij})^{\phi}} \qquad 1 \leq j \leq c \qquad (F.1.10)$$

where C is the cluster center of the cth cluster for the jth attribute, and X is the vector representing the individual data value i for the jth attribute. Equation (F.1.8) is capable of assigning intermediate memberships and solving the problem of intergrades, which are data points between two classes. The solution of the Eqs. (F.1.9) and (F.1.10) is obtained by an iterative procedure.

Size of database to be classified. One approach to extrapolating from a small-scale (i.e., watershed scale) landform classification of identifiable elements using fuzzy k-means to a broader-scale is by defining training areas upon which the classification is derived (Burrough et al. 2000). Such training areas should be representative of the broader area for which the classification will be implemented (i.e., similar domains). A probability density function (PDF) for the smaller area that closely matches the shape of overall regional dataset PDF's can be used to determine appropriate training areas. The PDF's describe the univariate data for individual environmental attributes by determining a reasonable distributional model for the data. The training areas are used to determine the optimal number of classes required for classification of the area, to decide the optimal values of the performance parameters, and to calculate class centroids for interpretation of similarities and differences between classes. After the class centroids, attribute variances, and attribute variance-covariances have been calculated, Burrough et al. (2000) suggests that a membership value for each cell in the area can be computed based on the distance measure from the class centroids. The distances between attributes are calculated by the following equations

$$(d_{ij})^2 = \sum_{j=1}^{c} [(x_{ic} - C_{cj})]^2 \tag{F.1.11}$$

$$(d_{ij})^2 = \sum_{j=1}^{c} [(x_{ic} - C_{cj})/s_j]^2 \tag{F.1.12}$$

$$(d_{ij})^2 = (x_{ic} - C_{cj})^T \sum_{j=1}^{c} (x_{ic} - C_{cj}) \tag{F.1.13}$$

where Eq. (F.1.11) calculates the Euclidean, Eq. (F.1.12) the Diagonal, and Eq. (F.1.13) the Mahalanobis distance respectively. In the equations d_{ij}^2 is the square of the distance between an individual i and a class center j; x_{ic} is an attribute for individual i and the class c; C_{cj} denotes the centroid of class c for attribute j; s_j is the attribute variance. The membership value for each class and each individual cell is then calculated using Eq. (F.1.9).

Choice of distance-dependent measures. The distance function is used to measure the similarity or dissimilarity between two individual observations and then later the similarity or dissimilarity between two clusters. The simplest measure of distance that gives equal weight to all measured variables is Euclidean distance and it is insensitive to statistically dependent variables (Bezdek 1981; Odeh et al. 1992). This measure of distance is useful for uncorrelated variables on the same scale when attributes are independent and the clusters have the general shape of spherical clouds. Euclidean distance should not be used where different attributes have widely varying average values and standard deviations, since large numbers in one attribute will prevail over smaller numbers in another (McBratney and Moore 1985; Minasny and McBratney 2000). The diagonal distance is also insensitive to statistically dependent variables but compensates for distortions in the assumed spherical shape caused by disparities in variances among the measured variables (Odeh et al. 1992). Measuring the distances between a pair of points often requires standardization or transformation before distances are computed. Diagonal distance measurement is useful for uncorrelated variables that are on different scales because it transforms the dataset to one in which all attributes have equal variances (Bezdek 1981; McBratney and Moore 1985). The third possibility is to compute Mahalanobis-type distances, as used here. This type of distance measurement also compensates for distortions like the diagonal measurement and requires initial data transformation, and accounts for statistically dependent variables (Odeh et al. 1992). Mahalanobis distance is used for correlated variables on the same or different scales (Bezdek 1981; McBratney and Moore 1985). Mahalanobis distance transforms the dataset to one in which all attributes have zero mean and unit variances while correlations between variables are taken into account.

Performance measurement. The optimal number of classes or the degree of fuzziness should be chosen based upon the required degree of detail (Burrough et al. 2000). User's knowledge of the data is often used in choosing the optimal number of classes or the degree of fuzziness. Another approach in choosing the optimal number of classes for fuzzy *k*-means is done by repeating the classification for different numbers of classes and applying different degrees of fuzziness (McBratney and Moore 1985; Odeh et al. 1992). For each generated classification, analyses need to be performed and the results should be validated. Two validity functions, the fuzzy performance index (FPI), and the normalized classification entropy (MPE) (Modified Partition Entropy) are used to evaluate the effects of varying the number of classes. The FPI as per Minasny and McBratney (2000) is defined in Eqs. (F.1.14) and (F.1.15).

$$FPI = 1 - \frac{(F - 1/c)}{(1-1)c} \qquad \text{(F.1.14)}$$

where F is the partition coefficient

$$F = \tfrac{1}{n}\sum_{i=1}^{n}\sum_{j=1}^{c}(\mu_{ij})^2 \qquad (F.1.15)$$

The *MPE* as per Odeh et al. (1992) is defined in Eqs. (F.1.16) and (F.1.17):

$$MPE = \frac{H}{\log c} \qquad (F.1.16)$$

where *H* is the entropy function

$$H = -\tfrac{1}{n}\sum_{i=1}^{n}\sum_{j=1}^{c}\mu_{ij}\log(\mu_{ij}). \qquad (F.1.17)$$

The fuzzy performance index function estimates the degree of fuzziness generated by a specified number of classes, while the normalized classification entropy estimates the degree of disorganization created by a specified number of classes (Minasny and McBratney 2000). After *FPI* and *MPE* are calculated the optimum number of continuous and structured classes can be established on the basis of minimizing these two measures (McBratney and Moore 1985).

In the fuzzy *k*-means clustering algorithm the fuzzy exponent ϕ controls the degree of fuzziness. As the fuzzy exponent ϕ approaches one the degree of fuzziness diminishes and the clustering becomes harder. The value of two for ϕ often has been used in previous studies and a quasi-physical justification is provided by Bezdek (1981) for using this value. For instance, if ϕ is too low the classes are discrete and the membership value approaches zero or one, but if ϕ is too high the classes will not discriminate and classification may fail to converge. McBratney and Moore (1985) suggested that the rate of change is not constant by changing ϕ although, the objective function value decreases monotonically by increasing ϕ and increasing the number of classes. Since the goal of the objective function value is to find an optimal balance between structure and continuity, they also argue by choosing a value of ϕ that maximizes the objective function ($\delta J_E/\delta \phi$) generates the 'hardest' fuzzy clustering solution. McBratney and Moore (1985) devised the measure of fuzziness for determining the objective function value by obtaining $-[(\delta J_E/\delta \phi)c^{0.5}]$. Their method plots a series of ϕ after the objective function is determined versus a given class where the best value of ϕ for that class is at the maximum of the curve (Odeh et al. 1992). The function $\delta J_E/\delta \phi$ is defined by Bezdek (1981) in Eq. (F.1.18).

$$\frac{\delta J_E}{\gamma \phi} = \sum_{i=1}^{n}\sum_{j=1}^{c}\mu_{ij}^{\phi}\log(\mu_{ij})\,d_{ij}^2\,. \qquad (F.1.18)$$

Therefore choosing an optimal combination of classes and fuzzy exponent is established on the basis of minimizing *FPI* and *MPE* as described above and choosing the curve with the least maximum of $-[(\delta J_E/\delta\phi)c^{0.5}]$ (Odeh et al. 1992).

Class overlap, confusion index and defuzzification. The confusion index (*CI*) is a measure of the degree of class overlap in attribute space (Burrough and McDonnell 1998; Hengl et al. 2004; Shi et al. 2005). The concept of a 'confusion index' is a measure of how well each individual observation has been classified. The *CI* is used to translate combined maps of fuzzy memberships into easy to understand crisp zones. The *CI* is calculated by Eq. (F.1.19) where MF_{max} denotes the dominant membership value, and MF_{max2} is the subdominant membership value for each observation:

$$CI = 1 - (MF_{max} - MF_{max2}). \tag{F.1.19}$$

If the calculated *CI* approaches zero, then the observation is more likely to belong to the dominant class, while if the *CI* approaches one, the difference between the dominant and subdominant classes are negligible which creates confusion in classification of that particular observation. After the membership values have been calculated, defuzzification (Burrough et al. 2000) is applied to get a crisp numeric output value. Each observation is assigned to a 'hard class' when membership is high (i.e, $\mu \geq 0.7$), to an 'intragrade' when membership is intermediate, and to an 'extragrade' when membership is low. For example, high membership means that observation is more likely to belong to one class, intermediate membership means that observation might belong to two or more classes, and low membership means that observation belongs equally to all classes. Thus, the continuous landform fuzzy *k*-means classification may be codified into classes (but not restricted) for examining the relationship between landslide locations and fuzzy classes. The relationship strength therefore may be used to establish a *priori* probabilities for each of the classes to be used in conjunction with the Bayes' theorem.

Bayes' theorem

The Bayesian approach or Bayes' theorem (Malczewski 1999; Skidmore et al. 1996; Aspinall and Veitch 1993; Aspinall 1992) is a method used for decision-making under uncertainty. The method is a framework for combining subjective probability (of being true or false) with conditional probability (of being true or false). Subjective probability is an expression of the degree of belief in an event occurring based on a person's experience, prejudices, optimism, etc. (Malczewski 1999). Conditional probability is the knowledge about the likelihood of the hypothesis to be true given a piece of evidence. For example, one cannot be certain whether landslides always occur in areas of topographic convergence. The knowledge might be expressed as the user being 90 percent certain (i.e., probability equals 0.9) that landslides will occur in areas of topographic convergence.

Implementing the Bayesian approach requires the following information: (i) prior knowledge or *a priori* probabilities that a particular hypothesis is true, and (ii) the probability that current evidence is true given that the hypothesis is true (Skidmore et al. 1996; Malczewski 1999). Thus, calculated conditional probabilities from the relative frequency of association between the knowledge of presence and absence of landslide locations and categorized membership values of fuzzy *k*-means classes (i.e., 0.0 – 0.1, 0.1– 0.2, 0.2 – 0.3, 0.3 – 0.4, 0.4 – 0.5, 0.5 – 0.6, 0.6 – 0.7, 0.7 – 0.8, 0.8 – 0.9, 0.9 – 1.0) are combined in Bayes' theorem to generate a probability map of landslide hazard. The equation for the Bayesian calculation is shown in Eq. (F.1.20):

$$p_p = \frac{pp \prod_{i=1}^{n} cp_i}{pp \prod_{i=1}^{n} cp_i + pa \prod_{i=1}^{n} ca_i} \qquad (F.1.20)$$

The p_p is the Bayesian probability for presence, *pp* is *a priori* probability for presence, *pa* is *a priori* probability for absence/random, $\prod_{i=1}^{n} cp_i$ is the product of conditional probabilities for presence for attributes $i = 1, ..., n$ of predictor datasets, $\prod_{i=1}^{n} ca_i$ is the product of conditional probabilities for absence/random for attributes $i = 1, ..., n$ of predictor datasets. Statistical assumptions that must be met for the Bayes' theorem (Aspinall 1992) include: subjective probabilities adequately represent uncertainty over a particular event (*a priori* probabilities for presence and absence); conditional probabilities are adequately expressed as relative frequencies of occurrence; conditional probabilities are orderly expressions of relationships between datasets (this assumption is tested through chi-square analysis of discriminatory significance); Bayes' theorem provides an optimal (rational and normative) method for modifying probabilities; and predictor datasets are conditionally independent. The simplicity of this approach is that after the classes with similar characteristics are grouped together by fuzzy *k*-means, the probabilities can always be revised with Bayes' theorem when additional information is obtained.

F.1.4 Application of the modeling approach

The modeling approach to estimating landslide hazard probability in Idaho's CNF was tested and compared against the other approaches previously described (FSmet and SHALSTAB) for the study area. The approach is discussed in the following.

Study area

The study area is within the CNF, located on the western slopes of the Rocky Mountains in north central Idaho (115°46'W, 46°07'N, 114°19'W, 47°00'N), USA. The CNF is located west of the Montana state border and is bounded on three sides by four other National Forests; the Lolo National Forest in Montana; the Bitterroot National Forest in Montana and Idaho; the Nez Perce National Forest in Idaho; and the Panhandle National Forests in Idaho. The CNF map is shown in Fig. F.1.1.

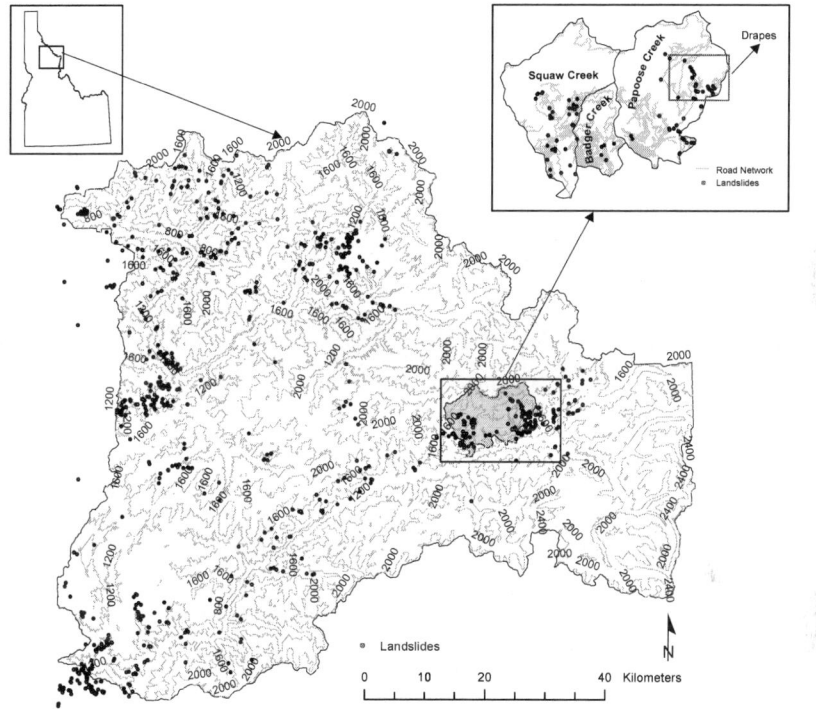

Fig. F.1.1. Distribution of landslides over the Clearwater National Forest drainage during the winter 1995/96 storm events (training area is the shaded area)

The training area of 111.8 km^2 used for classification includes the Papoose, Badger and Squaw Creek watersheds (Fig. F.1.1) located northwest of Lowell, Idaho in the Lochsa Basin of the CNF. The training area was chosen because of the high landslide density initiated by landslide events in November 1995 and February 1996 and the similarity of the topographic attribute distributions with the overall CNF. The highly dissected mountainous topography of the training area is typical for Idaho's Clearwater River Basin. Elevation in the training area watershed ranges from 966 m to 2,154 m and slopes vary between zero and 45 degrees. The climate is characterized by dry and warm summers, and cool wet winters.

Precipitation averages about 1,320 mm annually, which changes significantly across the elevational gradient. Most of the annual average precipitation falls as snow during winter and spring, while peak stream discharge occurs in late spring and early summer. The highly variable steep soils are well drained and are primarily derived from parent materials such as granitics, metamorphic rocks, quartzites, and basalts or surface erosion and depositions. Vegetation includes Grand fir (*Abies grandis*), Douglas fir (*Pseudotsuga menziesii*), Subalpine fir (*Abies lasiocarpa*), Western red cedar (Thuja plicata), Western white pine (*Pinus Monticola*), and various other shrubs and grasses that have short growing seasons, particularly at the higher elevations.

Historically, the CNF in Idaho has experienced periodic floods and landslide events. Major floods occurred in 1919, 1933, 1948, 1964, 1968, and 1974. All of these floods were documented through streamflow records (McClelland et al. 1997). Although the approximate frequency of major landslide events is known for the last century, accurate mapping of individual landslides through time does not exist.

Methods

The data used for this study were derived or obtained from airphoto interpretation, field inventory, and DEMs. The landslides were assessed through aerial reconnaissance flights acquired in July 1996 in conjunction with field inventory following the landslide events that occurred in the CNF in November 1995 and February 1996. The landslide dataset therefore, is limited to a specific set of circumstances (storm events) over a limited area in space and time. A total of 865 landslides were recorded, of which 55 percent were RR and 45 percent were NRR landslides. The presence/absence of landslide data was represented on (30 m) grid coverages with a value of one for presence and zero for absence. The initiation area of each landslide (i.e. the area where the main scarp of the landslide occurred) was used to represent presence of the landslides. The RR landslides, which are associated with forest roads, were separated from the NRR landslides. Furthermore, the RR and the NRR grid coverages were separated so the data from one area (subwatershed) were used to develop the quantitative models, and the data from the other area were used to test the quantitative models.

A total of six environmental attributes [elevation, slope, profile curvature, tangent curvature, compound topographic index (CTI), and solar radiation] were derived from 30 m DEMs using TAPES-G (Gallant and Wilson 1996) and HEMI (Fu and Rich 2000) software. The continuous fuzzy k-means landform classification was performed using these six environmental attributes as the input to FuzME a PC Windows-based program (Minasny and McBratney 2000). In the program, the fuzzy *k*-means algorithm with Mahalanobis distance was applied to the training area to determine the optimal number of classes required to classify the area, to find the optimal values of the performance parameters, and to calculate class centroids for interpretation of similarities and differences between classes. After finding the optimal number of classes, plus the overlaps between the classes and the class centroids, a custom built program using the Arc/Info GRID module was used to extrapolate the fuzzy *k*-means elements from the training area to the entire study area. Along with the derived fuzzy *k*-means classes, the maximum 'Maxclass' (most dominant membership value) and 'Minclass' (least dominant) classes were calculated. The Bayesian modeling followed using an ArcView (ESRI) extension (Aspinall 2000) by correlating derived fuzzy *k*-means classes and the landslide occurrence datasets. This step required fuzzy *k*-means classes to be categorized so landslide occurrences can be linked to the classes.

Results from fuzzy k-means classification

The fuzzy *k*-means were computed for two to nine classes with different fuzzy exponents (ϕ). Using the performance measurement criteria of *FPI* and *MPE*, the optimal number of classes was computed for the lowest ϕ (1.15) and for the highest ϕ (1.50). The fuzzy performance measures are shown in Figs. F.1.2(a) and F.1.2(b) for the initial implementation with a fuzzy exponent $\phi = 1.15$ and $\phi = 1.50$ respectively. Figure F.1.2 suggests that the optimum number of classes for both fuzzy exponents is six using the minimization criteria of *FPI* and *MPE* measures.

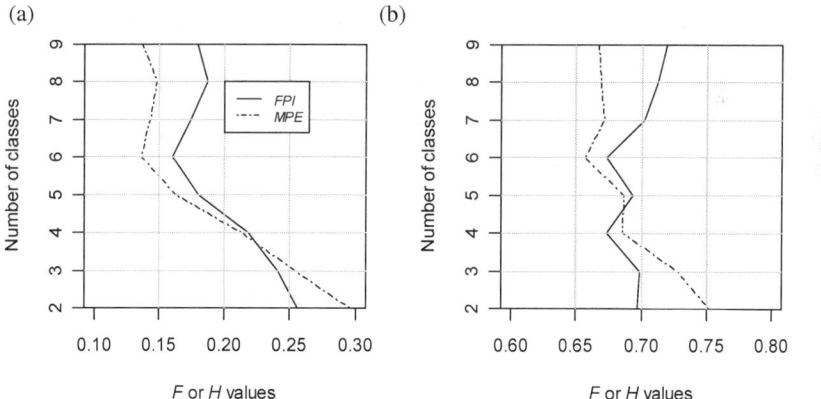

Fig. F.1.2. Fuzziness performance index (*F*) and normalized classification entropy (*H*) versus number of classes: (a) fuzziness exponent $\phi = 1.15$, and (b) fuzziness exponent $\phi = 1.50$

After determining the optimal number of classes, and assuming that no deviation will occur in the number of classes between the lowest and the highest fuzzy exponents, the optimal ϕ [1.15, 1.50] was computed for $c = 6$ at an increment of $\phi = 0.05$. In Fig. F.1.3 derivation of the optimal fuzzy exponent is shown. The best value of ϕ for $c = 6$ is at the maximum of the curve ($\phi = 1.40$).

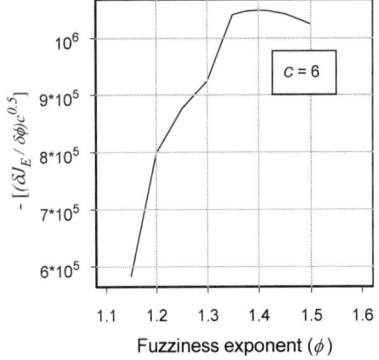

Fig. F.1.3. Plot of $-[(\delta J_E / \delta \phi) c^{0.5}]$ versus ϕ for $c = 6$

The class centroids given in Table F.1.1 can be used to interpret similarities and dissimilarities between classes. The centroid is the average point in the multidimensional space and represents, in a sense, the center of gravity for the respective cluster. For example, class b has the highest wetness index, and the lowest (negative) profile and tangent curvature values suggesting that this class represents convergent areas with high soil moisture. The high solar radiation of class a and the high profile and tangent curvature indicates convex areas. Thus, from the cluster centers in Table F.1.1 the analyst can interpret landscape pattern differences captured by the fuzzy classes. The fuzzy k-means approach has distinguished potentially useful classes defining multivariate patterns and clusters within the area of interest based on the input variables.

Table F.1.1. Cluster centers for six classes

Input data	a	b	c	d	e	f
Elevation	1445.56	1412.93	1784.37	1362.64	1398.89	1606.85
Slope	32.94	44.31	49.17	66.77	58.60	50.12
Wetness Index	9.32	11.61	9.37	8.96	8.93	9.24
Solar Radiation	1108.65	943.71	1260.90	510.01	1101.07	691.26
Profile Curvature	0.10	-0.45	0.04	0.02	0.03	0.01
Tangent Curvature	0.31	-1.35	0.05	0.16	0.15	0.04

Notes: a represents mid elevation, gentle convex slopes with high solar radiation, b is mid elevation, concave drainages, c is high elevation, high solar radiation (southerly slopes) locations, d is low elevation, steep low solar insolation locations (northerly slopes), e is low elevation, steep high solar insolation locations (southerly slopes), and f is high elevation, low solar insolation (northerly slopes)

Figure F.1.4 shows the spatial patterns of these classes to compliment the cluster centers. For example, class a represents mid elevation, gentle convex slopes with high solar radiation, class b is mid elevation, concave drainages, class c is high elevation, high solar radiation (southerly slopes) locations, class d is low elevation, steep low solar insolation locations (northerly slopes), class e is low elevation, steep high solar insolation locations (southerly slopes), and class f is high elevation, low solar insolation (northerly slopes).

In Fig. F.1.5(a) the confusion index (CI) map is shown. The CI map illustrates areas where spatial boundaries between classes exist. High values of CI represent areas with fuzzy boundaries where a grid cell belongs to two or more classes, while low values represent areas with sharp boundaries where a grid cell is more likely to belong to one class. The map of maximum class (Fig. F.1.5(b)) shows the map classification based on the dominant class membership.

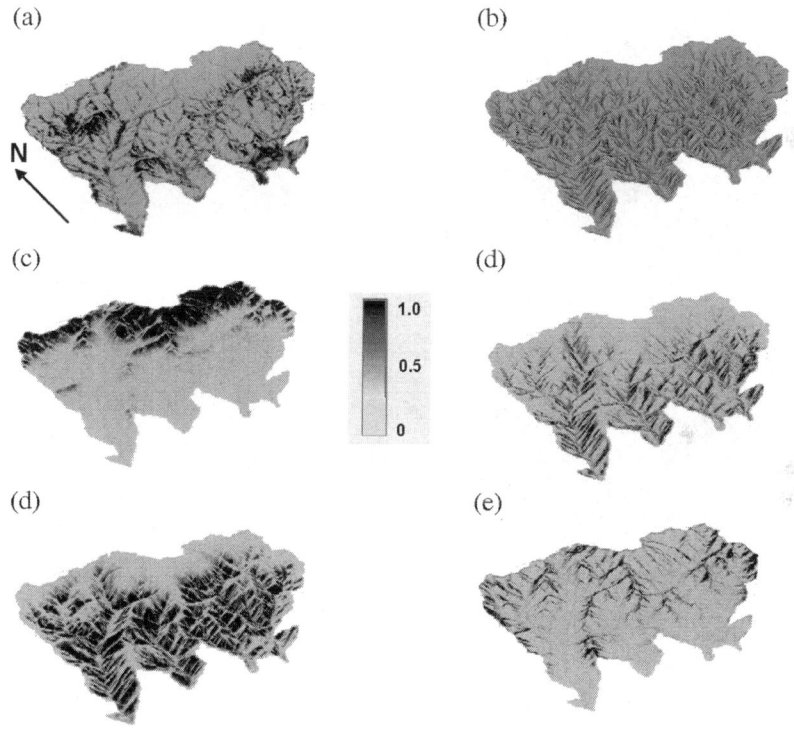

Fig. F.1.4. Drapes of fuzzy k-means classification of training area with six classes

These fuzzy classes may now be assessed for landslide hazard prediction. The modeling approach is based on calculating conditional probabilities from the relative frequency of statistical association between attributes of a dataset to be modeled (i.e., presence and absence of landslides at a given spatial location) and attributes of predictor datasets (i.e., fuzzy k-means classes). The predictor datasets represent classes of continuous landforms with membership values between zero and one. As previously described these classes are a function of several topographic attributes (Table F.1.1) extrapolated for the entire study area using the class centroids and Mahalanobis distance and subsequently categorized into sub-classes for implementing the Bayes' theorem.

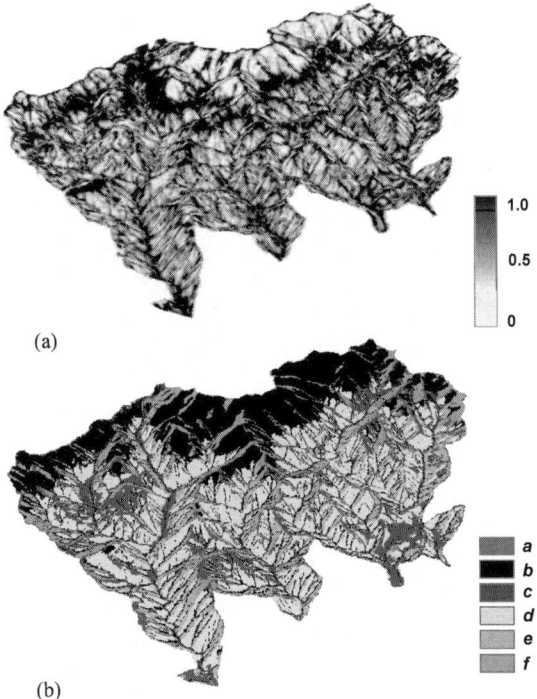

Fig. F.1.5. Drapes of the confusion index (*CI*) and the most dominate class

Results from model testing

The presented methodology was used to accomplish the following three objectives: (i) to evaluate if the predictor fuzzy k-means classes used for RR and NRR landslide hazard models are the same; (ii) to evaluate if predictive Bayesian models of landslide hazard for RR and NRR landslides are different; and (iii) to evaluate if spatial prediction of landslide hazard using the integrated Fuzzy/Bayesian approach is better than spatial prediction using existing models (FSmet, SHALSTAB).

Comparison of predictor datasets. In order to evaluate the first objective the Bayes' theorem was used to provide a framework for combining predictor datasets. This modeling approach is based on calculating conditional probabilities from relative frequencies of datasets to be modeled (presence or absence of NRR or RR landslides) and categorized predictor datasets (fuzzy k-means classes). The significance of the conditional probabilities for discriminating between distribution classes for each predictor dataset is tested using chi-square (see Table F.1.2). The table shows the degrees of freedom, chi-square values, and the level of significance (* less than 0.05; ** less than 0.01; *** less than 0.005) for discriminating between fuzzy k-means classes and landslides for the NRR and RR landslides. The assumptions presented above guided the selection of appropriate predictor datasets for the NRR and RR landslides.

Table F.1.2. Conditional probabilities of fuzzy k-means predictor datasets for development of the Bayesian models for the CNF

Predictors	df	NRR landslides		RR landslides	
		Chi-square	p	Chi-square	p
Minclass	5	64.748	***	23.144	***
Maxclass	5	168.114	***	144.589	***
Cl	9	13.249	NS	8.478	NS
Ca	9	11.859	NS	27.661	***
Cb	9	97.364	***	48.139	***
Cc	9	39.286	***	52.578	***
Cd	9	37.05	***	28.129	***
Ce	9	156.594	***	118.918	***
Cf	9	66.196	***	64.176	***

Notes: NS denotes not significant, * significant at the 0.05, ** at the 0.01 and *** at the 0.005 level

Figure F.1.6 illustrates the spatial implementation of the Bayes' theorem for NRR landslides derived from five-predictor datasets. The five-predictor datasets that were statistically significant (using chi-square) without violating other statistical assumptions included the maximum and minimum class (based on most and least dominant class membership); mid elevation, concave drainages; low elevation, steep low solar insolation locations (northerly slopes); and low elevation, steep high solar insolation locations (southerly slopes).

Figure F.1.7 illustrates the spatial implementation of the Bayes' theorem for RR landslides derived from four-predictor datasets. The four-predictor datasets include the maximum class; mid elevation, gentle convex slopes with high solar radiation; low elevation, steep low solar insolation locations (northerly slopes); and low elevation, steep high solar insolation locations (southerly slopes). The difference between the two outputs (Figs. F.1.6 and F.1.7) is shown in Fig. F.1.8. The figure is an estimate of the uncertainty associated with the derivation of the Bayesian model using RR and NRR landslide predictor subsets. Figures F.1.9 and F.1.10 show the implementation of the Bayes' theorem for a smaller area within the CNF for the NRR and RR landslides draped over a DEM with 50 m surface elevation contours indicating relief.

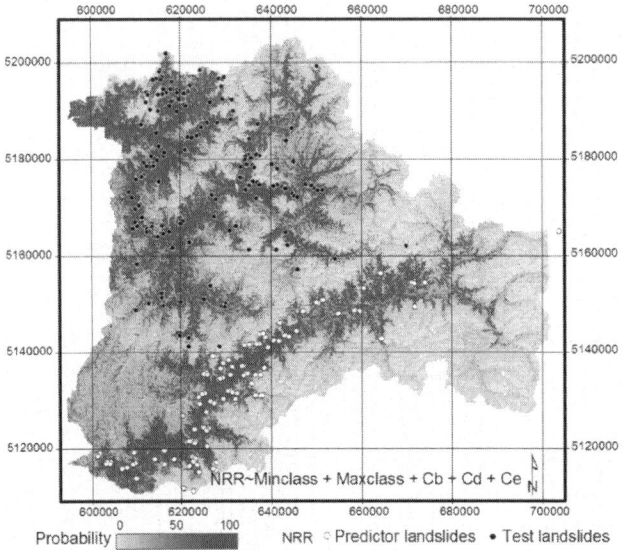

Fig. F.1.6. Bayesian model output. Probabilities of the occurrence of non-road related landslides in the CNF (The units are represented as probability on a scale of zero to one, but have been rescaled to zero to 100)

The prediction surface differences between the RR and NRR landslides are shown in Fig. F.1.11. The legend in the figure suggests that high hazard for NRR landslides is associated with negative values, while high hazard for RR landslides is associated with positive values. Values close to zero represent areas of agreement between both landslide hazards. The 'blocky' appearance of predicted landslide hazard in the figures is likely associated when fuzzy k-means classes were simplified for the Bayesian modeling by using 10 sub-classes (i.e., 0-0.1, 0.1-0.2, 0.2-0.3, 0.3-0.4, 0.4-0.5, 0.5-0.6, 0.6-0.7, 0.7-0.8, 0.8-0.9, 0.9-1) and from the statistical significance associated with each sub-class of significant attributes used for the modeling.

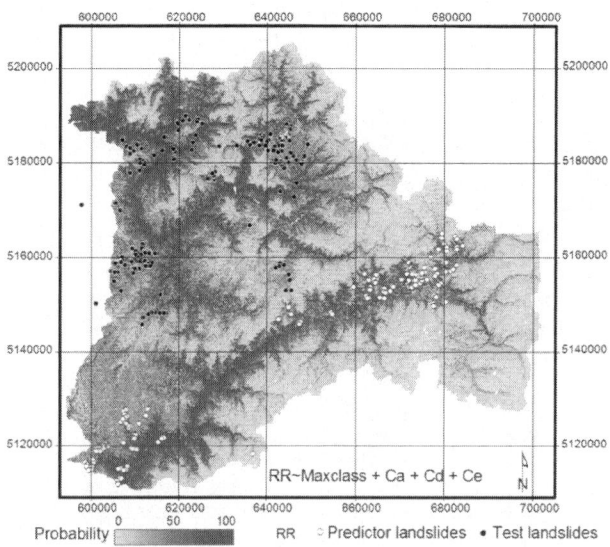

Fig. F.1.7. Bayesian model output. Probabilities of the occurrence of road related landslides in the CNF (The units are represented as probability on a scale of zero to one, but have been rescaled to zero to 100)

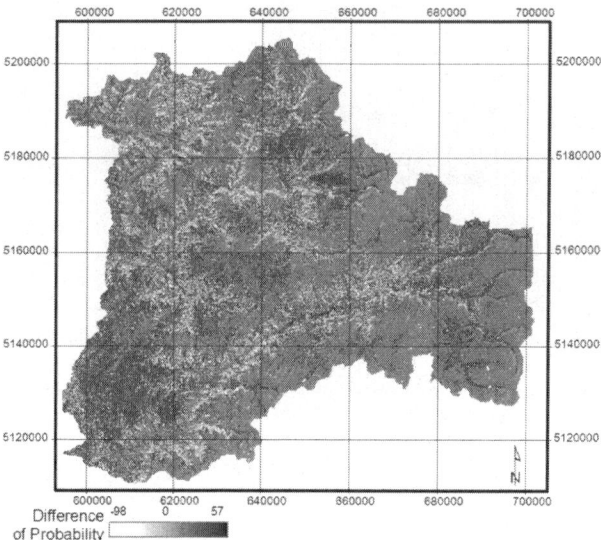

Fig. F.1.8. Difference of probabilities of the occurrence of non-road related versus road related landslides in the CNF

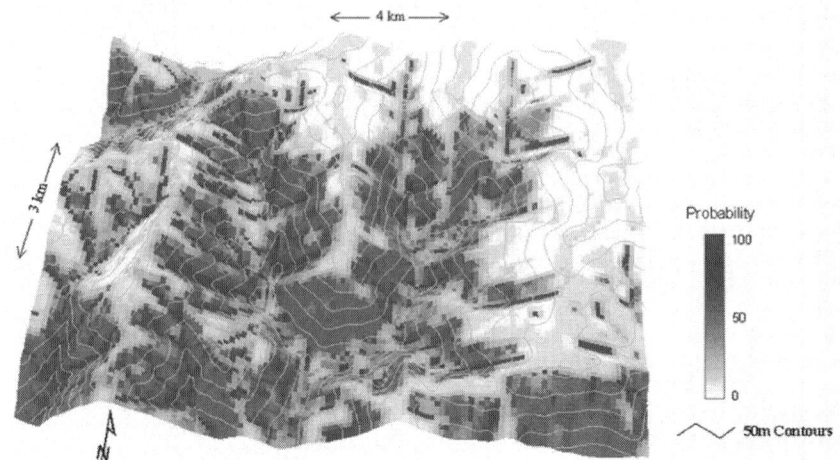

Fig. F.1.9. Drape of predicted landslide hazard for non-road related landslides using Bayesian modeling

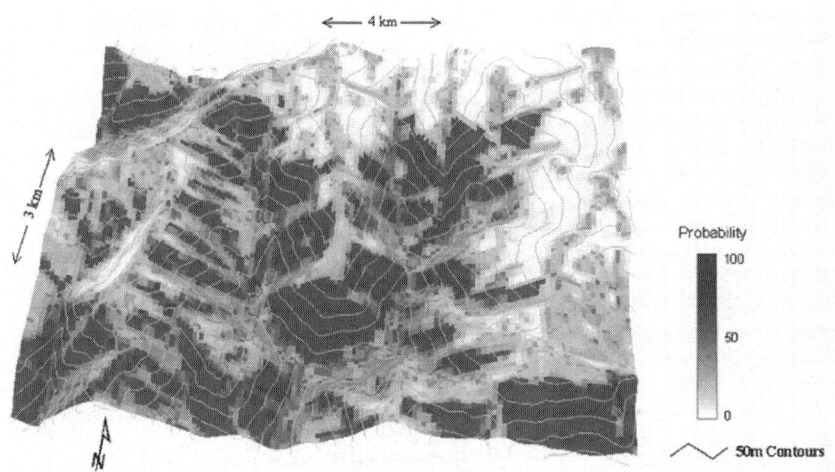

Fig. F.1.10. Drape of predicted landslide hazard for road related landslides using Bayesian modeling

Figure F.1.11 illustrates the difference in the spatial prediction between the RR and NRR landslides. NRR landslides appear more likely associated with steeper concave drainages while the RR landslides are associated with both steep and gentle slopes (more random across the landscape). RR landslides exclude the concave drainages and include gentle slopes and convex areas. The average solar radiation from the class centers (Table F.1.1) for the predictor datasets appear to be lower for the NRR landslides than for the RR landslides. This suggests that cooler and moister locations are likely associated with NRR landslides. These areas also exhibit a higher average wetness index based on the class centers. The class center averages of the profile and tangent curvatures also have negative values for the NRR landslides, which suggest areas of converging flow. Conversely, the class center averages of the profile and tangent curvatures have positive values for the RR landslides, which suggests areas of diverging flow. The average slope from the class centers is lower for the RR landslides, while the elevation is approximately the same. Thus, predictor fuzzy k-means classes used in the development of the RR and NRR models are different.

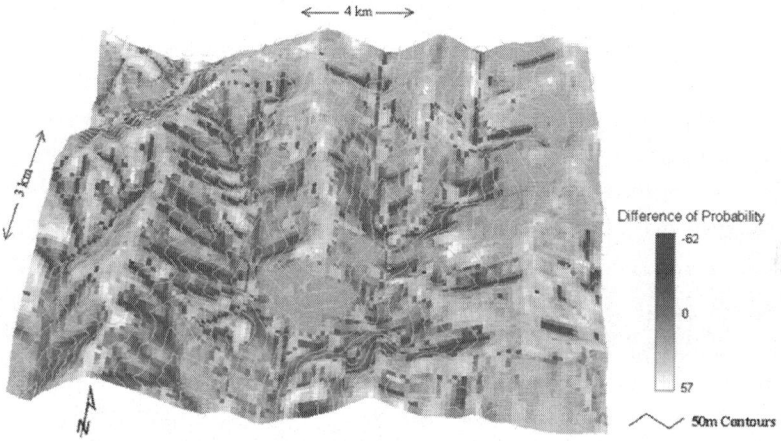

Fig. F.1.11. Drape of predicted landslide hazard difference for road related and non-road landslides using Bayesian modeling

Evaluation of hazard maps. To evaluate the second objective, independent test data for the RR and NRR landslides from another sub-basin within the CNF was used to test the goodness-of-fit between the models and the independent test data. The evaluation of the accuracy of the hazard maps yielded significant χ^2 values for both models. The RR model yielded $\chi^2 = 376.52$ and NRR model yielded $\chi^2 = 507.01$ which in both cases was greater than the critical value ($\chi^2 = 140.16$ with 100 degrees of freedom, and with 95 percent confidence interval) while other assumptions were met. Table F.1.3 shows cross-tabulation of the independent test data for the RR and NRR landslides against the results from the Bayesian model. The results from Table F.1.3 suggest that the NRR model provides a better overall fit with better discrimination between high and low hazard areas than the RR model. Thus the Bayesian prediction for NRR landslides seems to have a higher level of certainty than the prediction for RR landslides. This may be because the model prediction for NRR landslides is better suited to identify natural processes associated with landsliding. For instance, the predictor datasets from the first objective suggest that the NRR landslides are occurring mostly on steep slopes, while the RR landslides can occur on steep slopes as well as gentle slopes. Also, the predictor datasets appear to suggest that roads increase the likelihood of landslides occurring in many places where they likely wouldn't occur in a natural or un-roaded situation. Hence, NRR landslides may be restricted to a more narrow set of environmental conditions than RR landslides. Therefore, the evaluation of this objective suggests that predictive Bayesian models of landslide hazard for RR and NRR landslides are different.

Table F.1.3. Proportion of presence/absence associated with probabilities for non-road and road related landslides

Probability (x 100)	Non-road related		Road related	
	Presence (%)	Absence (%)	Presence (%)	Absence (%)
0 – 20	11.05	60.83	8.56	45.58
20 – 40	5.26	6.05	8.11	11.43
40 – 60	4.21	4.47	20.72	19.64
60 – 80	17.37	10.26	26.13	10.97
80 – 100	62.11	18.39	36.49	12.38

Comparison of models. Finally, the third objective is evaluated through the comparison of new and existing models applied to the CNF (Tables F.1.4 and F.1.5). The comparison of each modeling technique is based on arbitrary cut-off values that measure the proportions of correctly identified landslides and the corresponding proportions of areas at risk. The ratio values in the tables measure the accuracy, which is obtained by dividing the correctly identified landslides by the area classified to be unstable. Higher ratio values suggest better prediction of landslide hazard. The ratio values in both tables show that the derived predictions from the new (fuzzy k-means/Bayes' theorem) models for the RR and NRR landslides are better than the existing models. For example, the p value associated with the fuzzy k-means/Bayes' theorem model corresponds to probability of landslide hazard greater than 50 ($p > 50$) on scale 0 to 100. Tables F.1.4 and F.1.5 show that the fuzzy k-means/Bayes' theorem model does not correctly identify as many known landslide locations as the SHALSTAB modeling techniques, but this method also does not classify as large an area as unstable. Conversely the SHALSTAB model has the lowest ratio values for both NRR (1.31) and the RR (1.22) landslides. This model, in order to correctly identify more than 80 percent of the landslides, classifies a large area (69.26 percent) as unstable. Therefore, the ratio value suggests that spatial predictions from the proposed modeling methodology are better than spatial predictions from existing methods.

Table F.1.4. Evaluation of new and existing modeling techniques in the CNF derived by arbitrary cut-off values for the NRR landslides

Modeling techniques	Area classified to be unstable		Correctly identified landslides		Ratio
	% (1)	km²	% (2)	km²	(2)/(1)
Fuzzy k-means/Bayes' theorem ($p > 50$)	30.96	1996.03	63.70	0.18	2.06
Fuzzy k-means/Bayes' theorem ($p > 70$)	25.89	1669.16	57.28	0.21	2.21
Fuzzy k-means/Bayes' theorem ($p > 90$)	8.05	518.99	20.49	0.06	2.54
SHALSTAB (cut-off > −2.2)	69.26	4465.28	90.88	0.33	1.31
SHALSTAB (cut-off > −2.8)	62.74	4044.92	87.42	0.25	1.39
SHALSTAB (cut-off > −3.1)	32.39	2088.22	62.89	0.23	1.94
Current forest service method	5.54	1996.03	10.06	0.18	1.82

Table F.1.5. Evaluation of new and existing modeling techniques in the CNF derived by arbitrary cut-off values for the RR landslides

Modeling techniques	Area classified to be unstable		Correctly identified landslides		Ratio
	% (1)	km²	% (2)	km²	(2)/(1)
Fuzzy k-means/Bayes' theorem ($p > 50$)	34.32	2212.65	73.83	0.21	2.15
Fuzzy k-means/Bayes' theorem ($p > 70$)	19.57	1261.70	46.91	0.17	2.40
Fuzzy k-means/Bayes' theorem ($p > 90$)	4.41	284.32	12.10	0.03	2.74
SHALSTAB (cut-off > −2.2)	69.26	4465.28	84.20	0.31	1.22
SHALSTAB (cut-off > −2.8)	62.74	4044.92	78.02	0.22	1.24
SHALSTAB (cut-off > −3.1)	32.39	2088.22	45.19	0.16	1.40
Current forest service method	5.54	2212.65	9.63	0.21	1.74

F.1.5 Concluding remarks

The presented work demonstrates that robust landslide hazard predictions can be achieved through an integration of GIS, fuzzy k-means, and Bayesian modeling techniques. The potential of this spatial approach for landslide hazard is high. In the modeling approach the optimal number of classes is derived by iterative classification for a range of classes or from expert knowledge (desired degree of detail). The continuous fuzzy k-means classification provides a significant amount of information about the character and variability of data, and seems to be a useful indicator of landscape processes relevant for predicting landslide hazard. For the study area the continuous classification identified various areas, which were described by combinations of a few or more attributes and their correlations to landform regions. The fuzzy k-means small-scale landform classification (training set) was effectively extrapolated to a broader-scale area for spatial prediction of landslide hazard. Thus, a broad range of explanatory variables useful for landslide hazard prediction were integrated through this continuous fuzzy k-means classification.

The evaluation of objective one demonstrated that predictor fuzzy k-means classes used for RR and NRR landslide hazard models were not the same. The fuzzy k-means classes showed different predictive power for useful landslide hazard relationships as well as different statistical significance for the RR and NRR models. This suggests that different landscape processes are associated with NRR and RR landslides. The model prediction of NRR landslides is more comprehensive in interpreting landslide processes while the model prediction of RR incorporates elements of randomness in describing processes associated with landslide hazard. The explanatory variables suggest that the NRR landslides are occurring mostly on steep slopes and concave drainages, while the RR landslides can occur on steep slopes as well as gentle slopes and convex areas. Also, NRR landslides are associated with wet concave areas, whereas RR landslides are associated with dryer and convex areas. Thus, for management planning strategies and land-use activities such as road construction, decision-makers should consider landscape processes associated with RR landslides in order to avoid vulnerable areas prone to landslide hazard.

The second objective confirmed that predictive Bayesian models of landslide hazard for RR and NRR landslides are different. The test data further demonstrate that the goodness-of-fit for the NRR landslides is better than for RR landslides, and that prediction with higher levels of certainty is possible for NRR landslides. This justifies the need for development of two independent RR and NRR models that may be used for addressing different questions for future planning and management of natural resources in predicting landslide hazard.

Objective three demonstrated that comparison of spatial prediction using the integrated Fuzzy/Bayesian approach is better than spatial prediction using existing models (Fsmet, SHALSTAB). Therefore, this suggests that the integrated Fuzzy/Bayesian approach is appealing for management applications and decision-making when RR and NRR landslides are considered.

Each individual model may be analyzed in greater detail than demonstrated here to improve the understanding between processes driven by RR and NRR landslides. However, the intention was to demonstrate a quantitative methodology for landslide hazard prediction that is explicit, consistent and repeatable. The Fuzzy/Bayesian methodology provides a basis for understanding landslide processes initiated from RR and NRR landsides and may be used to provide forest land managers with information on landslide hazard that can be integrated with other spatial techniques or to focus limited resources toward the prevention of future landslides.

References

Aspinall R (1992) An inductive modelling procedure based on Bayes' theorem for analysis of pattern in spatial data. Int J Geogr Inform Syst 6(2):105-121

Aspinall R (2000) Bayesian modeling with ArcView GIS. Paper presented at the GeoSpatial New West Intermountain GIS Conference, Kalispell [MA]

Aspinall R, Veitch N (1993) Habitat mapping from satellite imagery and wildlife survey data using a Bayesian modelling procedure in a GIS. Photogramm Eng Remote Sens 59(4):537-543

Bezdek JC (1981) Pattern recognition with fuzzy objective function algorithms. Plenum Press, New York

Bezdek JC, Ehrlich R, Full W (1984) FCM: the fuzzy c-means clustering algorithm. Comp Geosci 10(2):191-203

Burrough PA, McDonnell RA (1998) Principles of geographic information systems. Oxford University Press, Oxford

Burrough PA, van Gaans PFM, MacMillan RA (2000) High-resolution landform classification using fuzzy k-means. Fuzzy Sets and Systems 113(1):37-52

Burrough PA, Wilson JP, van Gaans PFM, Hansen AJ (2001) Fuzzy k-means classification of topo-climatic data as an aid to forest mapping in the Greater Yellowstone Area, USA. Landsc Ecol16(6):523-546

Butler BE (1982) A new system for soil studies. J Soil Sci 33(4):581-595

Carrara A (1983) Multivariate models for landslide hazard evaluation. Math Geol 15(3):403-426

Carrara A, Cardinali M, Guzzetti F (1992) Uncertainty in assessing landslide hazard and risk. ITC Journal 2:172-183

Carrara A, Cardinali M, Guzzetti F, Reichenbach P (1995) GIS technology in mapping landslide hazard. In Carrara F, Guzzetti A (eds) Geographical information systems in assessing natural hazards. Kluwer, Dordrecht, pp.135-175

Carrara A, Cardinali M, Detti R, Guzzetti F, Pasqui V, Reichenbach P (1991) GIS techniques and statistical models in evaluating landslide hazard. Earth Surf Proc Landforms 16(5):427-445

Chamran F, Gessler PE, Chadwick OA (2002) Spatially explicit treatment of soil-water dynamics along a California catena. Soil Sci Soc Am J 66:1571-1583

Chen SJ, Hwang CL (1992) Fuzzy multiple attribute decision making. Springer, Berlin, Heidelberg and New York

Chung CF, Fabbri AG (1999) Probabilistic prediction model for landslide hazard mapping. Photogramm Eng Remote Sens 65(12):1389-1399

Chung CF, Fabbri AG, van Westen CJ (1995) Multivariate regression analysis for landslide hazard zonation. In Carrara F, Guzzetti A (eds) Geographical information systems in assessing natural hazards. Kluwer, Dordrecht, pp.107-133

Dhakal AS, Amada T, Aniya M (2000) Landslide hazard mapping and its evaluation using GIS: an investigation of sampling schemes for a grid-cell based quantitative method. Photogramm Eng Remote Sens 66(8):981-989

Dietrich WE, Montgomery DR (1988) A digital terrain model for mapping shallow landslide potential (SHALSTAB) [accessed 11 October 2008, http://socrates.berkeley.edu/~geomorph/shalstab/]

Dyrness T (1967) Mass soil movements in the H.J. Andrews experimental forest. Pacific Northwest Forest and Range Experiment Station, USDA Forest Service: Research Paper PNW-42

Ellen SD, Cannon SH, Reneau SL (1988) Distribution of debris flows in Marin County. In Ellen SD, Wieczorek GF (eds) Landslides, floods, and marine effects of the storm of January 3-5, 1982, in the San Francisco Bay region, California. U.S. Geol Surv Prof Paper 1434

Evans IS, Hengl T, Gorsevski PV (2008) Geomorphometry: applications in geomorphology. In Hengl T, Reuter HI (eds) Geomorphometry: concepts, software, applications. [Development in Soil Science 33]. Elsevier, Amsterdam, pp. 497-525

Fisher PF, Pathirana S, (1990) The evaluation of fuzzy membership of land cover classes in the suburban zone. Remote Sens Environm 34(2):121-132

Fu P, Rich PM (2000) The solar analyst users manual. Helios Environmental Modeling Institute (HEMI), USA [accessed 11 October 2008, http://www.fs.fed.us/informs/solar-analyst/solar_analyst_users_guide.pdf]

Gallant JC, Wilson JP (1996) TAPES-G: a grid-based terrain analysis program for environmental sciences. Comp Geosci 22(7):713-722

Gessler PE, Moore ID, McKenzie NJ, Ryan PJ (1995) Soil-landscape modelling and spatial prediction of soil attributes. Int J Geogr Inform Syst 9(4):421-432

Gessler PE, Chadwick OA, Chamran F, Althouse L, Holmes K (2000) Modeling soil-landscape and ecosystem properties using terrain attributes. Soil Sci Soc Am J 64(6):2046-2056

Glade T (1998) Establishing the frequency and magnitude of landslide-triggering rainstorm events in New Zealand. Environm Geol 35(2-3):160-174

Gorsevski PV, Jankowski P (2008) Discerning landslide susceptibility using rough sets. Comp Environ Urban Syst 32(1):53-65

Gorsevski PV, Gessler PE, Jankowski P (2003) Integrating a fuzzy k-means classification and a Bayesian approach for spatial prediction of landslide hazard. J Geogr Syst 5(3):223-251

Gorsevski PV, Gessler PE, Jankowski P (2004) Spatial prediction of landslide hazard using fuzzy k-means and Bayes' theorem. In Widacki W, Bytnerowicz A, Riebau A (eds) A message from the Tatra: geographical information systems and remote sensing in mountain environmental research. Jagiellonian University Press, Krakow, pp.159-172

Gorsevski PV, Jankowski P, Gessler PE (2005) Spatial prediction of landslide hazard using fuzzy k-means and Dempster-Shafer theory. Transactions in GIS, 9(4): 455-474

Gorsevski PV, Jankowski P, Gessler PE (2006a) An heuristic approach for mapping landslide hazard by integrating fuzzy logic with analytic hierarchy process. Control and Cybernetics 35(1):121-146

Gorsevski PV, Gessler PE, Foltz RB, Elliot WJ (2006b) Spatial prediction of landslide hazard using logistic regression and ROC analysis. Trans GIS 10(3):395-415

Gorsevski PV, Gessler PE, Boll J, Elliot WJ, Foltz RB (2006c) Spatially and temporally distributed modeling of landslide susceptibility. Geomorphology 80(3-4):178-198

Grayson RB, Moore ID, McMahon TA (1992a) Physically based hydrologic modeling 1. A terrain-based model for investigative purposes. Water Resources Research 28(10):2639-2658

Grayson RB, Moore ID, McMahon TA (1992b) Physically based hydrologic modeling 2. Is the concept realistic? Water Resources Research 28(10):2659-2666

Hammond C, Hall D, Miller S, Swetik P (1992) Level I stability analyses (LISA) Documentation for Version 2.0. Gen. Tech. Rep. INT-285, Ogden, UT: U.S. Department of Agriculture, Forest Service, Intermountain Research Station

Hengl T, Reuter HI (eds) 2008 Geomorphometry: concepts, software, applications. Developments in Soil Science 33, Elsevier, Oxford

Hengl T, Walvoort D, Brown A, Rossiter D (2004) A double continuous approach to visualisation and analysis of categorical maps. Int J Geogr Inform Sci 18(2):183-202

Irvin BJ, Ventura SJ, Slater BK (1997) Fuzzy and isodata classification of landform elements from digital terrain data in Pleasant Valley, Wisconsin. Geoderma 77(2-4):137-154

Iwahashi J, Pike RJ (2007) Automated classifications of topography from DEMs by an unsupervised nested-means algorithm and a three-part geometric signature. Geomorphology 86(3-4):409-440

Malczewski J (1999) GIS and multicriteria decision analysis. Wiley, New York, Chichester, Toronto and Brisbane

Mark RK, Ellen SD (1995) Statistical and simulation models for mapping debris-flow hazard, In Carrara F, Guzzetti A (eds) Geographical information systems in assessing natural hazards. Kluwer, Dordrecht, pp.93-106

MacMillan RA, Pettapiece WW, Nolan SC, Goddard TW (2000) A generic procedure for automatically segmenting landforms into landform elements using DEMs, heuristic rules and fuzzy logic. Fuzzy Sets and Systems 113(1):81-109

McBratney AB, deGruijter JJ (1992) A continuum approach to soil classification by modified fuzzy k-means with extragrades. J Soil Sci 43(1):159-175

McBratney AB, Moore AW (1985) Application of fuzzy sets to climatic classification. Agricult Forest Met 35(1-4):165-185

McClelland DE, Foltz RB, Wilson WD, Cundy TW, Heinemann R, Saurbier JA, Schuster RL (1997) Assessment of the 1995 & 1996 floods and landslides on the Clearwater National Forest: landslide assessment. Report to the regional Forester Northern Region U.S. Forest Service, December 1997

McKenzie NJ, Gessler PE, Ryan PJ, O'Connell DA (2000) The role of terrain analysis in soil mapping. In Wilson JP, Gallant JC (eds) Terrain analysis principles and applications. Wiley, New York, Chichester, Toronto and Brisbane, pp.245-265

McSweeney K, Gessler PE, Slater B, Hammer RD, Bell J, Petersen GW (1994) Towards a new framework for modelling the soil-landscape continuum. In Factors of soil formation: a fiftieth anniversary retrospective. Madison [WI] SSSA Special Publ. 33:127-145

Minasny B, McBratney AB (2000) FuzME version 2.1. Australian Centre for Precision Agriculture. The University of Sydney, NSW 2006 [accessed 11 October 2008, http://www.usyd.edu.au/su/agric/acpa]

Mladenoff DJ, Sickley TA (1998) Assessing potential gray wolf restoration in the northeastern United States: a spatial prediction of favorable habitat and potential population Levels. J Wildlife Mgmt 62(1):1-10

Mladenoff DJ, Sickley TA, Haight RG, Wydeven AP (1995) A regional landscape analysis and prediction of favorable gray wolf habitat in the northern Great Lakes regions. Conservation Biology 9(2): 279-294

Montgomery DR, Dietrich WE (1994) A physically based model for the topographic control on shallow landsliding. Water Resources Research 30(4):1153-1171

Moore ID, Gessler PE, Nielsen GA, Peterson GA (1993) Soil attribute prediction using terrain analyses, Soil Sci Soc Am J 57:443-452

Odeh IOA, McBratney AB, Chittleborough DJ (1992) Soil pattern recognition with fuzzy c-means: application to classification and soil-landform interrelationship. Soil Sci Soc Am J 56:505-516

Okimura T, Ichikawa R (1985) A prediction method for surface failures by movements of infiltrated water in a surface soil layer. Natural Disaster Science 7(1):41-51

O'Loughlin EM (1986) Prediction of surface saturation zones in natural catchments by topographic analysis. Water Resources Research 22(5):794-804

Reneau SL, Dietrich WE (1987) Size and location of colluvial landslides in a steep forested landscape. IAHS Publ 165: 39-49

Robison EG, Mills K, Paul J, Dent L, Skaugset A (1999) Storm impacts and landslides of 1996: Final Report, Forest Practices Technical Report No. 4, Oregon Department of Forestry

Ruspini EH (1969) A new approach to clustering. Inf. Control 15(1):22-32

Ryan PJ, McKenzie NJ, O'Connell D, Loughhead AN, Leppert PM, Jacquier D, Ashton L (2000) Integrating forest soils information across scales: spatial prediction of soil properties under Australian forests. Forest Ecology Mgmt 138(1):139-157

Shi X, Zhu A-X, Wang R-X (2005) Deriving fuzzy representations of some special terrain features based on their typical locations. In Cobb M, Petry F, Robinson V (eds.) Fuzzy modeling with spatial information for geographic problems. Springer, Berlin, Heidelberg and New York, pp. 233-252

Skidmore AK, Watford F, Luckananurug P, Ryan PJ (1996) An operational GIS expert system for mapping forest soils. Photogramm Eng Remote Sens 62(5):501-511

Spiegelhalter DJ (1986) A statistical view of uncertainty in expert systems. In Gale WA (ed) Artificial intelligence and statistics. Addison-Wesley, Reading [MA], pp.17-55

Ventura SJ, Irvin BJ (2000) Automated landform classification methods for soil-landscape studies. In Wilson J P, Gallant J C (eds) Terrain analysis principles and applications. Wiley, New York, Brisbane, Toronto and Brisbane, pp.267-294

Webster R, Oliver MA (1990) Statistical methods in soil and land resource survey. Oxford University Press, Oxford

Wieczorek GF (1984) Preparing a detailed landslide inventory map for hazard evaluation and reduction. Bulletin of the Association of Engineering Geologists 21:337-342

Wright RH, Campbell RH, Nilsen TH (1974) Preparation and use of isopleth maps of landslide deposits. Geology 2(10):483-485

Wu W, Sidle RC (1995) A distributed slope stability model for steep forested basins. Water Resources Research 31(8):2097-2110

Zadeh LA (1965) Fuzzy sets. Inf Contr 8:338-353

Zadeh LA (1978) Fuzzy sets as a basis for a theory of possibility. Fuzzy Sets and Systems 1(1):3-28

F.2 Incorporating Spatial Autocorrelation in Species Distribution Models

Jennifer A. Miller and *Janet Franklin*

F.2.1 Introduction

Species distribution models, based on ecological niche theory and gradient analysis, require digital maps of environmental factors that influence species distributions, such as topography and climate, as well as spatial information on the species attribute of interest (for example, presence/absence, type, abundance), typically sampled directly or compiled from existing datasets such as museum records. Reflecting the fact that their use spans several disciplines, these types of models have been referred to previously as 'predictive vegetation mapping' (Franklin 1995), 'predictive habitat distribution modeling' (Guisan and Zimmermann 2000), and 'niche modeling' (Stockwell 2007). The terminology seems to be converging on 'species distribution modeling', which is used here. It should be noted that it is technically the environmental habitat suitability that is produced (mapped) from these models, which renders them appropriate for studying the distribution of communities/assemblages (Ferrier et al. 2002) as well as species, in addition to a number of related biogeographical variables, such as species richness (Rangel et al. 2006), invasive species (Richardson and Thuiller 2007), and disease transmission (Peterson 2006).

There is an increasingly wide variety of statistical methods from which to choose, ranging from more traditional generalized regression to artificial neural networks and genetic algorithms (for review, see Franklin 1995; Guisan and Zimmermann 2000; Elith et al. 2006). The choice of method to use is based upon, among other things, data characteristics (known vs. unknown parameters, distribution, measurement level), model use (prediction vs. inference), and intended final product (categorical map, abundance map).

These models are probabilistic and generally static, spatially and temporally, assuming that the species is in equilibrium with its environment. Many of the commonly used statistical methods such as generalized regression further assume that the observations of the species attribute of interest are random, an assumption which casual observation proves unrealistic. In fact, there are very few things distributed randomly in nature or elsewhere, leading to what Waldo Tobler (1970) termed the 'First law of geography', to describe the regularity that near things are more related than distant things (also see Sui 2004). A more specific ecological manifestation of this acknowledges that elements of an ecosystem close to one another are more likely to be influenced by the same generating process and will therefore be similar (Legendre and Fortin 1989).

This property of spatial autocorrelation in species data is borne out by the fact that two plot locations nearby are more likely to have similar characteristics than two more distant plots. Some of this can be explained by spatially structured environmental predictor variables, such as precipitation, temperature and elevation. The remaining spatial autocorrelation can result from either unmeasured or unobservable variables, or biotic processes, such as competition, predation or dispersal, that cause spatial patterning.

When spatially autocorrelated data are used with traditional (non-spatial) statistical methods, the result is poorly specified models in general and inflated significance estimates for predictor variables (resulting in increased type I errors) in particular (Legendre 1993). Several studies have indicated the importance of including spatial structure in models as a way of clarifying the influence of environmental predictor variables (Wu and Huffer 1997; Hubbell et al. 2001; Keitt et al. 2002). The potential predictive ability of spatial autocorrelation in species distribution models has only recently been explored (see Miller et al. 2007, for review).

Spatial autocorrelation has been identified as an important area of future research in species distribution models (Franklin 1995; Guisan and Zimmermann 2000). Many of these studies that acknowledge it attempt to eliminate it by manipulating the sampling strategy to avoid observations within a certain distance of each other (Legendre and Fortin 1989; Davis and Goetz 1990; Borcard et al. 1992; Smith 1994). Borcard et al. (1992) and Legendre and Legendre (1998) used partial regression to separate the explanatory ability (of vegetation distribution) of environment from spatial factors (see also Lobo et al. 2002; Lobo et al. 2004; Graae et al. 2004; Nogués-Bravo and Martinez-Rica 2004 for recent examples). In a recent review of species distribution models that have incorporated spatial autocorrelation, Miller et al. (2007) conclude that limited availability of sample data at appropriate and varying spatial resolutions has been a limiting factor in the ability of models to describe it well enough to include it.

This study expands upon previous research that focused on incorporating spatial structure explicitly in species distribution models. In Miller (2005), the model residuals were used as a proxy for spatially structured local variation and were interpolated and added to the model predictions. Miller and Franklin (2002) used kriging to calculate *autocovariates,* additional predictor variables based on observed locations of the response variable to represent spatial autocorrelation in models for four vegetation alliances. We extend this work here by using stochastic simulation in addition to kriging to calculate autocovariates for eleven vegetation alliances. Two types of statistical models are used, generalized linear models (GLM) and classification trees (CT), to predict presence/absence of eleven vegetation alliances of varying distributions, in a section of the Mojave Desert, California, USA. Model accuracy was compared using receiver-operating characteristic (ROC) plots, a threshold- and prevalence-independent metric that gauges how well a binary outcome has been classified.

F.2.2   Data and methods

Data. The study area for this research is a 50,369 km^2 portion of the Mojave Desert Ecoregion within California, referred to as the Eastern California Subsection (see Fig. F.2.1). The predictor variables used represent both broad-scale climate and fine-scale topographic variables, as well as two categorical geomorphic landform and surface composition variables (see Table F.2.1).

Eleven vegetation alliances (see Table F.2.2) were selected for modeling, with a goal of achieving a representative variety of distribution types in the Mojave Desert (for example, rare, common; weak and strong environmental relationships). The dataset consisted of 3,819 observations and was partitioned into 75:25 train:test portions following a heuristic suggested by Fielding and Bell (1997) for presence/absence data with more than ten predictor variables. The data are described in more detail in Miller and Franklin (2006).

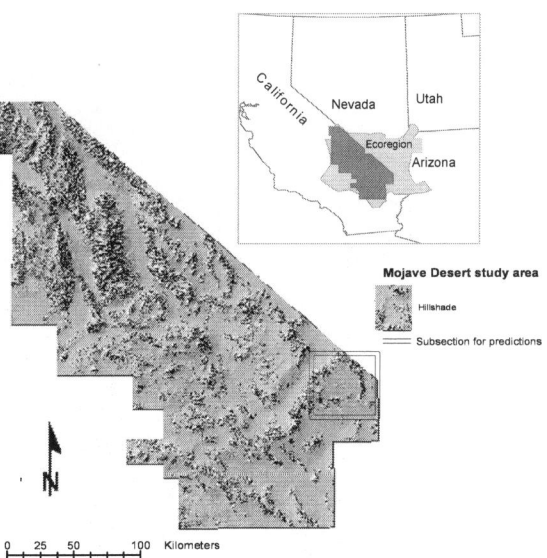

Fig. F.2.1. Mojave Desert study area. The square highlights the section used for predictions in Figs. F.2.3 and F.2.4

Table F.2.1. Environmental variables used in this study. Climate variables are one km resolution; all others are 30m resolution

Variable	Description	Range of values
Sumprecip	Average summer precipitation	11 to 146 mm
Winprecip	Average winter precipitation	45 to 579 mm
Jantemp	Minimum January temperature	−11.3 to 4.8° C
Jultemp	Maximum July temperature	16.6 to 44.4° C
Elevation	From USGS 7.5' DEM	−85 to 3390 m
Slope	Derived from DEM	0 to 78
Swness	Cosine (aspect - 225°) (Franklin et al. 2000)	−1 to 1
Lpos4	Landscape position; Average difference between cell and 4 neighbors (positive in valleys, neutral in mid-slope position, negative on ridges) (Fels 1994)	−1732 to 2311
Solrad	Potential solar radiation (Dubayah 1994)	0 to 383 W/m^2
TMI	Topographic moisture index; Number of cells draining into a cell divided by the tangent of slope (Beven and Kirkby 1979)	0 to 22.6
Landform	Geomorphic landform (Dokka 1999)	29 nominal classes
Landcomp	Surface composition (Dokka 1999)	six aggregated nominal classes

Table F.2.2. Vegetation alliances modeled (Thomas et al. 2004), and the proportion of the full (test and train, $n = 3,819$) dataset in which they are present. Species abbreviations comprise the first two letters of genus and specific epithet of indicator species. Number of observations of present (P) are given for test and train data (which consisted of 960 and 2,859 observations respectively)

Label, (proportion)	Alliance name	P test	P train
ATCA (0.006)	*Atriplex canascens* Shrubland Alliance	7	16
ATCO (0.028)	*Atriplex confertifolia* Shrubland Alliance	34	73
CORA (0.034)	*Coleogyne ramosissima* Shrubland Alliance	21	110
EPNE (0.006)	*Ephedra nevadensis* Shrubland Alliance	5	17
GALL (0.011)	*Pleuraphis rigida* Herbaceous Alliance	9	34
LATR (0.158)	*Larrea tridentata* Shrubland Alliance	145	460
LATR-AMDU (0.427)	*Larrea tridentata - Ambrosia dumosa* Shrubland Alliance	417	1214
MESP (0.007)	*Menodora spinescens* Dwarf-shrubland Alliance	10	17
PIMO (0.013)	*Pinus monophylla* Woodland Alliance	12	38
YUBR (0.092)	*Yucca brevifolia* Wooded Shrubland Alliance	87	265
YUSC (0.047)	*Yucca schidigera* Shrubland Alliance	49	132

Spatial autocorrelation variables. Two geostatistical interpolation methods (kriging and simulation) were used to calculate spatial autocorrelation variables (autocovariates) based on the distribution of presence/absence in the training data, and these terms were included with other environmental variables in GLM and CT models. Kriging methods result in one set of predicted values that are optimized based on the variogram and the spatial configuration of the data, but the result is overly smooth. Rather than one optimal prediction, stochastic simulation generates a series of equally probable predictions, allowing the 'roughness' of the data to be maintained (Burrough and McDonnell 1998). Although multiple realizations are often used to describe the range of variation in values at unsampled locations (see Rossi et al. 1993), here a single simulation realization is used to represent one possible scenario of the spatial distribution of a vegetation alliance.

An autocovariate has been explicitly incorporated along with environmental variables in logistic models, formally called autologistic models, following work by Besag (1972) and subsequent modifications by Augustin et al. (1996). However, including an autocovariate term requires complete information on the distribution of the response variable, which is rarely available. Gibbs sampler and Markov Chain methods (Augustin et al. 1996, 1998) and Markov Chain Monte Carlo methods (Gumpertz et al. 1997; Wu and Huffer 1997) have been used to 'fill in the blanks' of the sample data, but these are computationally intensive and in some cases unstable or intractable (Wu and Huffer 1997). Here we use geostatistical interpolation methods, specifically indicator kriging and conditional simulation, as a less computationally intense way to generate complete surfaces based on spatial structure of and values of the original observations.

Indicator kriging is the non-linear form of kriging used with binary response data and its product is a surface with the probability that the condition coded '1' (for example, 'presence') will occur (Burrough and MacDonnell 1998). Similarly, indicator simulation is used with binary sample data (Burrough and MacDonnell 1998). The result is a layer with values of one and zero based on the variogram and the proportion of one and zero in the sample data.

An indicator variogram was fit to the training data for each alliance. All variograms were fit using the common heuristic 'by eye' approach (Gotway and Hartford 1996), and are therefore highly subjective. Three of the most commonly used variogram models were tested: spherical, exponential, and Gaussian. Only spherical, which describes a clear range and sill, and exponential, which describes a more gradual approach to the range (Burrough and MacDonnell 1998) were used. For comparison purposes, a variogram was fit to all alliances, even when positive spatial autocorrelation was not apparent. This way, the effects of incorporating it in models when it is not empirically detected could be explored. Indicator kriging of the variograms and sample data was used to calculate a layer of probability values for each alliance. Similarly, indicator simulation was used to calculate a layer with values of zero and one that mirrored the sample data proportions. Based on the resolution of the sample data and the environmental variables, an approximation of the 'average' spatial structures among the alliances studied here, and potential processing time, 500 m was selected as the resolution of the autocovariate layers.

This resulted in eleven maps (one for each alliance) with values that represented the probability that a specific alliance would be present in each 500 m grid cell based on indicator kriging (referred to as 'K'); eleven maps with values of one and zero indicating whether an alliance is predicted to be present or absent based on indicator simulation ('1sim'); and eleven layers with values that represented the mean of ten simulations ('Msim'). Generally, the mean of 100 simulations should approximate the kriged result (Burrough and MacDonnell 1998) – the mean of ten simulations should retain some characteristic roughness of the data, with the flexibility of having non-binary values.

To represent the neighborhood around each cell, the values of each of the interpolated variables for the eight surrounding grid cells of each observation were summed using ArcInfo GIS and added to the modeling datasets as the autocovariate (Auto) term (for K, Msim, or 1sim):

$$\text{Auto term} = \sum_{i=1}^{8} P(pres)_i$$

where $P(pres)_i$ represents the kriged or simulated prediction for each cell (i) and can range from zero to one, therefore the autocovariate term representing the neighborhood sum, K/Msim, can range from zero, indicating no observations of presence nearby, to eight, indicating a cluster of observations of presence (Besag 1974; Augustin et al. 1996).

Models. The models that form the basis of our comparison were developed using generalized linear models (GLMs) and classification trees (CTs). GLMs are one of the most commonly used methods (see Guisan et al. 2002). Model specification with GLMs is fairly subjective, and as a result they are less data-driven and exploratory as non-parametric models such as CTs.

We used logistic regression because the response data were binary. The GLMs were developed based on a combination of stepwise and subjective, iterative, variable addition and subtraction methods with a goal of minimizing the AIC statistic (Akaike 1973; Hastie et al. 2001). Pairwise interaction terms based on biophysical principles (for example, elevation/aspect) or observed in the CT structure (described below) were also tested for significance. Generalized additive models were used in an exploratory way to identify higher order relationships (polynomial, piecewise linear) between the environment variables and response variable (see Brown 1994; Franklin 1998; Miller and Franklin 2002 for similar methods) that were then specified in the GLM. Once a subset of variables was selected for the non-spatial model, the autocovariate term was added (always as a linear term) and any subsequently non-significant variables were removed.

CT models are non-parametric and use a rule-based structure developed by partitioning data into subsets that are increasingly homogeneous with respect to the response variable (Breiman et al. 1984). Partitioning continues until either the resulting branches are homogeneous or a minimum number of observations remains in the subset. Associated with each terminal node is the number of points in the training data that were observed in locations that met the environmental criteria, as well as the number that are correctly classified. This proportion can be interpreted as the 'suitability' (Pontius and Schneider 2001) for a class to occur, analogous to the probability that results from GLMs.

Each CT model was given all predictor variables (twelve environmental variables for non-spatial models, one additional spatial variable for each of the three spatial models), then was pruned (based on cross-validation, see Breiman at al. 1984) to sizes that ranged from six to thirty-one terminal nodes for the non-spatial CT models, and three to twenty-seven nodes for the spatial CT models.

Model assessment. Classification accuracy was the focus of model assessment in this work, and receiver-operating characteristic (ROC) plots were used as the accuracy metric as they are threshold- and prevalence-independent (Fielding and Bell 1997). A ROC plot is obtained by plotting all sensitivity values (true positive fraction) on the y-axis against their equivalent (one – specificity) (false positive fraction) values on the x-axis. The area under the curve (AUC) of the resulting plot provides a measure of overall accuracy at all available thresholds. Swets (1988) provides a rough guide for classifying the accuracy as: 0.50–0.60 = fail; 0.60–0.70 = poor; 0.70–0.80 = fair; 0.80–0.90 = good; 0.90–1.00 = excellent. Based on plots of sensitivity and specificity for a range of probability values, a threshold was selected to produce the binary present/absent maps.

To summarize, a total of eight models for each of the eleven alliances were developed: (i) a non-spatial CT model based on the environmental variables; (ii) the same (non-spatial) CT model to which the kriged autocovariate term was added (these models will be referred to as 'CT_K'); (iii) the same CT model to which the mean simulation autocovariate was added (referred to as 'CT_Msim'); (iv) the same CT model to which the simulation autocovariate was added ('CT_1sim'); (v) a (non-spatial) GLM based on the environmental variables; (vi) the same (non-spatial) GLM to which the kriged autocovariate was added ('GLM_K' models); (vii) the same GLM to which the mean simulation autocovariate was added ('GLM_Msim'); and (viii) the same GLM to which the simulation autocovariate was added ('GLM_1sim'). The classification accuracy of the models was compared using ROC plots based on predictions on the test data.

F.2.3 Results

The variograms that were calculated from the binary training data visually indicated that spatial autocorrelation in the training data ranged from negligible (ATCA, EPNE, MESP) to moderate (ATCO, GALL) to quite recognizable (CORA, LATR, LATRAMDU, YUBR, YUSC, and PIMO). There is an obvious association between sample proportion and evident spatial structure – the noisiest variograms belonged to the rarest alliances. This is not surprising, as the variograms are dependent upon data quality and data quantity. The exception, PIMO occurs in only 0.013 of the sample, but its distribution is notably spatially clustered. The lag distance intervals for the alliance variograms varied from 1.8 km (ATCA) to 14 km (YUBR).

As spatial autocorrelation in data is an important component in statistical analysis of spatial data (Cressie 1993), the spatial structure of predictor variables used in the models was also examined. Variograms were fit to the ten continuous predictor variables (*landform* and *landcomp* excepted) using the values contained in the training data ($n = 2,859$), so that the sampling scale would be consistent with the alliance data. The spatial structure in predictor variables can explain some of the spatial pattern in the response variables (Aspinall and Pearson 1996). As expected, the climate variables and *elevation* had a great deal of positive spatial autocorrelation. The climate variables were derived using universal kriging (see Franklin et al. 2001 for methods), and it is understandable that their values are autocorrelated. *Elevation* also displays positive autocorrelation – the range occurs at a distance of approximately thirty km, which corresponds to the difference in basins and ranges in this study area. *Slope* and *swness* also show a moderate amount of autocorrelation, which could be related to the basin and range physiography. The three other complex topographic variables, *TMI*, *solrad*, and *lpos4* show no discernible spatial structure, as their correlations are on a much finer scale than was captured in the sample of locations for the vegetation alliance data. The lag distance intervals for the ten environmental variograms ranged from 3.3 km (*swness*) to 7.5 km (*elevation*).

Although there are scale effects (broad-scale variables tend to be selected before fine-scale variables), and some degree of arbitrariness (when two variables result in equally good splits, one is chosen arbitrarily), a variable is selected in a tree because it results in the two most homogeneous subgroups, therefore, the order in which it is used is an approximate indication of its relative importance.

When included, the autocovariates were always used in the first split, and often again in subsequent splits. When spatial CT models were pruned, few of the environmental variables remained. For example, the non-spatial YUBR CT model used eight environmental variables, but when the kriged autocovariate was included, only *elevation* was retained, and when the mean simulation autocovariate was included, only *elevation* and *landform* were used. The YUBR CT model with the single simulation autocovariate was more complex, using nine environmental variables, but the simulation autocovariate was used often in early splits. The classification accuracies were similar among these CT spatial models, and all were better than the accuracy for the non-spatial model.

As expected, the non-spatial models required more environmental variables than the spatial models. *Elevation* and *landform* were used most often in the CT_K models; *landform*, *TMI*, *swness* and *elevation* were used most often in the CT_Msim models; *landform* and *elevation* were used most often with the

CT_1sim models. The CT_1sim model is generally closer in complexity and variables used to the non-spatial model. The non-spatial CT models used climate, *elevation*, *slope* and *landcomp* more often than the spatial models. With the exception of *landcomp*, these variables all showed a substantial amount of spatial autocorrelation (a categorical variable, *landcomp* was not tested). The spatial structure in these environmental variables may be confounding the actual effects of the environmental gradients they represent (see Lennon 2000). The spatial models used *lpos4*, *solrad* and *TMI* (the three environmental variables that showed no evidence of spatial autocorrelation at the scale of the training data), as much or more than the non-spatial models.

Using the CT models for YUBR as an example, the tree using the single simulation autocovariate was more similar to the non-spatial tree with respect to the variables used. The first split divides the data into a majority side (2,332 observations) that are not near other observations and a smaller portion (527 observations) that are. The *sumprecip* split (at 36.5 mm) results in two very homogeneous terminal nodes – all 1791 observations were correctly classified absent at lower precipitation, and 537 of 541 observations were correctly classified as absent at higher precipitation. That terminal node included four present observations, even though they occurred where the autocovariate term was zero.

The single simulation autocovariate is less forgiving in terms of resolution differences (overlaying sample data points with predictor variable grids) and data anomalies. These four observations occurred at least five km away from other observations of presence, and while the smoothing effects of the kriged or mean simulation variables resulted in autocovariates with values greater than zero, this was not the case with the single simulation term model. This resulted in fairly robust results with the test data as well: 776 observations (of 960) were associated with values of zero for the autocovariate term, 600 of 602 observations were classified correctly as absent below the 37.5 mm precipitation threshold and 165 of 174 observations were correctly classified as absent with higher precipitation. The autocovariate is likely acting as a proxy for a combination of environmental variables that are suitable for YUBR occurrence.

Use of the autocovariates always decreased the AIC statistic in the GLMs. In most cases (except MESP, PIMO and YUBR), each autocovariate explained more variance than all of the other predictor variables combined. The simulation autocovariate explained the least of the three terms, and the kriged term usually explained the most. While the climate variables were often used in the non-spatial CTs, they were usually replaced by the autocovariate in the spatial models. In contrast, the only variables removed from the spatial GLMs were typically interaction terms, allowing more retention of the original environmental variables, although their effects were considerably overshadowed by the spatial terms.

Landform, *TMI*, *swness*, *lpos4*, and *solrad* were all used with greater frequency in CT models than in GLMs – these variables were often important in the spatial CT models, as the autocovariates replaced the broad-scale climate variables and the previously important *elevation* and *slope* variables. *Elevation*, *slope*, *jantemp*, and *sumprecip* were used more often in the GLMs. *Elevation* was expected to be important, as were the climate variables, but the frequency with which *slope* was used was surprising. As a topographic variable, it may describe a combination of features that are both moisture- and landform-related, both very important to desert vegetation.

Model accuracy was assessed using test data. For CTs, all three spatial terms resulted in significantly higher accuracy compared to the non-spatial model for two alliances (LATR and YUBR). Five other alliances had improved accuracy relative to the non-spatial models with at least one of the spatial models (ATCA, ATCO, CORA, LATRAMDU, and PIMO). The four remaining alliances (EPNE, GALL, MESP, and YUSC) all had either higher or similar non-spatial model accuracy compared to the spatial models. Among the autocovariate terms, CT_K and CT_Msim models usually had higher accuracies than the CT_1sim models (with the exception of CORA and YUSC).

For GLMs, all three spatial models had higher accuracy than the non-spatial model for four alliances (ATCO, CORA, LATR, and YUBR). Eight of the eleven non-spatial models were improved by inclusion of at least one of the autocovariates. None of the GLM spatial models improved upon the non-spatial model accuracy for GALL, MESP, and PIMO.

Two alliances had higher accuracy with the non-spatial model than with any of the spatial models for both GLMs and CT models (GALL, and MESP). Additionally, the non-spatial GLM of PIMO had higher accuracy than any of the spatial GLM models and the non-spatial CT models of EPNE and YUSC were better than any of their respective spatial CT models. Of these, MESP and EPNE had variograms that did not indicate any obvious spatial structure. The variogram for GALL indicated a gradual increase towards a sill, but the spatial structure was not very distinct, and it is a relatively rare alliance. PIMO and YUSC both had variograms that showed positive spatial autocorrelation. PIMO distribution is highly correlated with *elevation* and climate, and even though there is strong spatial autocorrelation among the observations, it is most likely a result of the environmental variables, which result in more robust models. While its habitat preferences are not as specific as PIMO, YUSC had very high accuracy with the non-spatial CT model (AUC = 0.917). The CT_1sim model for YUSC had the highest accuracy of the three spatial models, and the variables it used most closely approximated those used in the non-spatial model.

GLM_K models resulted in higher model accuracies slightly more often than the other spatial GLMs, but all had very similar effects. When the mean of all model AUC are compared, the non-spatial GLM and GLM_1sim share the highest value (0.84), while CT_1sim has the lowest value (0.74) (see Fig. F.2.2).

It was surprising that GLM_1sim (EPNE) and CT_1sim (CORA) were very high compared to the other spatial models, but most likely a factor of the difference in what the single simulation term represents (binary presence/absence based on one simulation, rather than the smoother mean simulation term). This illustrates the relatively inconsistent results when the single simulation term was used.

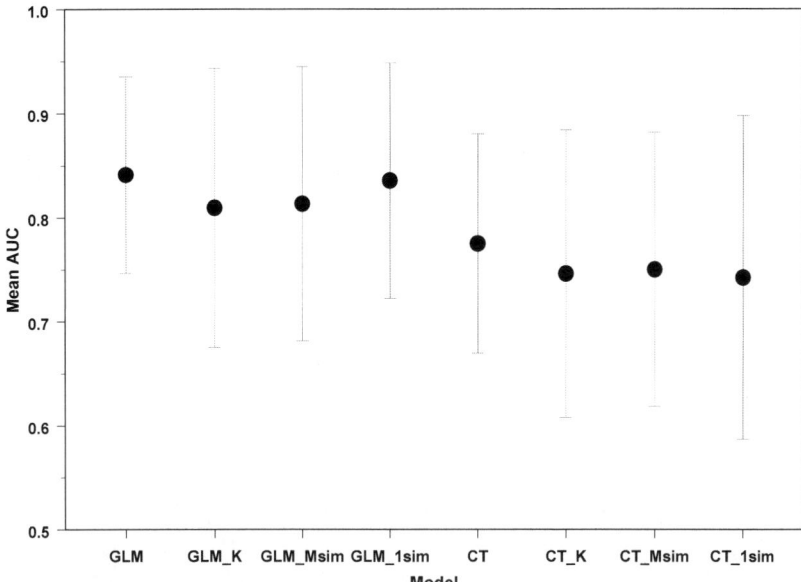

Fig. F.2.2. Mean AUC values (with standard deviation) from all alliances for each model

Model predictions. Mapped predictions were generated for all models. The implementation of the GLMs for predictions is straightforward – each predictor variable is multiplied by its model coefficient, then summed to provide the linear predictor (LP) for the alliance. In order to obtain probability values between zero and one, a logistic transformation of the LP is used. For the CT models, the decision rules are combined with the environmental variable layers to produce maps of suitability of presence. ROC plots are threshold-independent and therefore less subjective than threshold-dependent measures of model accuracy, but in order to produce a map of presence/absence, a threshold must be selected. Sensitivity, specificity, and total accuracy were plotted for a range of probabilities, and a threshold was selected that was near the intersection of the three lines (see Miller 2005 for example). Maps of YUBR presence based on these thresholds are shown in Figs. F.2.3 and F.2.4, along with test presence/absence data. While these maps are only in a subsection of the study area (see Fig. F.2.1) and for only one alliance (YUBR), they illustrate the potential for improving predictive ability by including spatial autocorrelation.

While the GLM has no false negative (present in test data but predicted absent) errors in this section, quite a few false positive (absent in test data but predicted present) errors are produced (Fig. F.2.3a). All of the spatial models reduce the false positive errors at no cost to the false negative errors [see Figs. F.2.3b) to F.2.3d)]. The CT models of the same area have few false positive errors, but more false negative errors [see Fig. F.2.4a)]. The addition of the autocovariate term reduces the false negative errors, although there is an increase in the false positive errors [Figs. F.2.4b) to F.2.4d)].

The main difference among the predictions from the spatial models is visual. The kriged results appear unrealistically smooth [see Figs. F.2.3b) and F.2.4b)], the single simulation autocovariate results in very patchy predictions [Figs. F.2.3d) and F.2.4d)], and as expected, the mean simulated autocovariate results in predictions that are somewhat intermediate between the single simulation and kriged results [Figs. F.2.3c) and F.2.4c)].

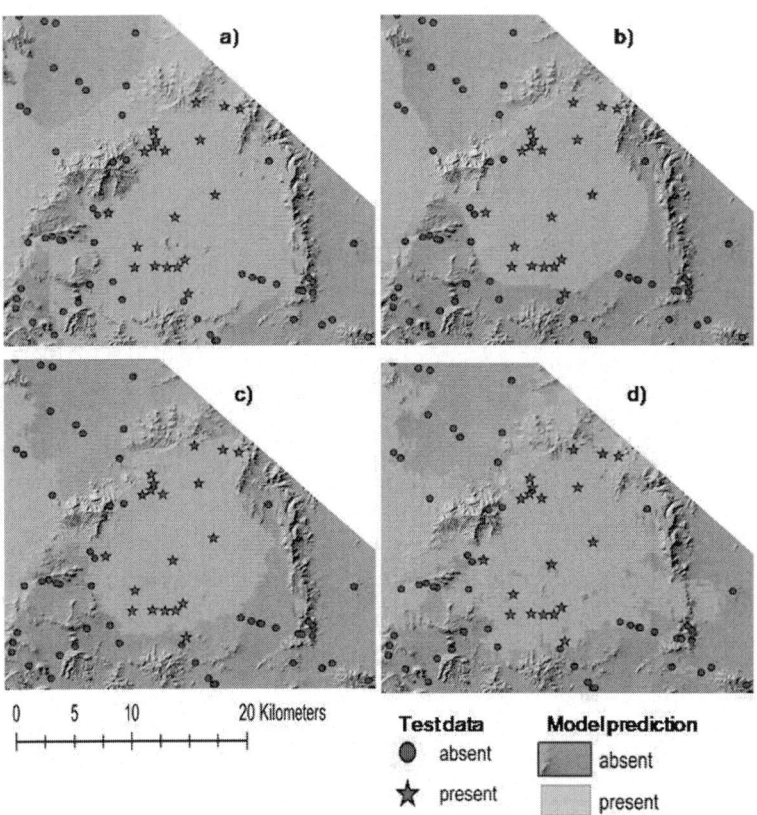

Fig. F.2.3. Generalized linear model predictions of Yucca brevifolia (YUBR) presence in the section shown in Fig. F.2.1: a) non-spatial model (0.15); b) model with kriged autocovariate 0.2); c) model with mean simulation autocovariate (0.2); d) model with single simulation autocovariate (0.12). Probability thresholds used to dichotomize continuous map predictions are shown in parentheses

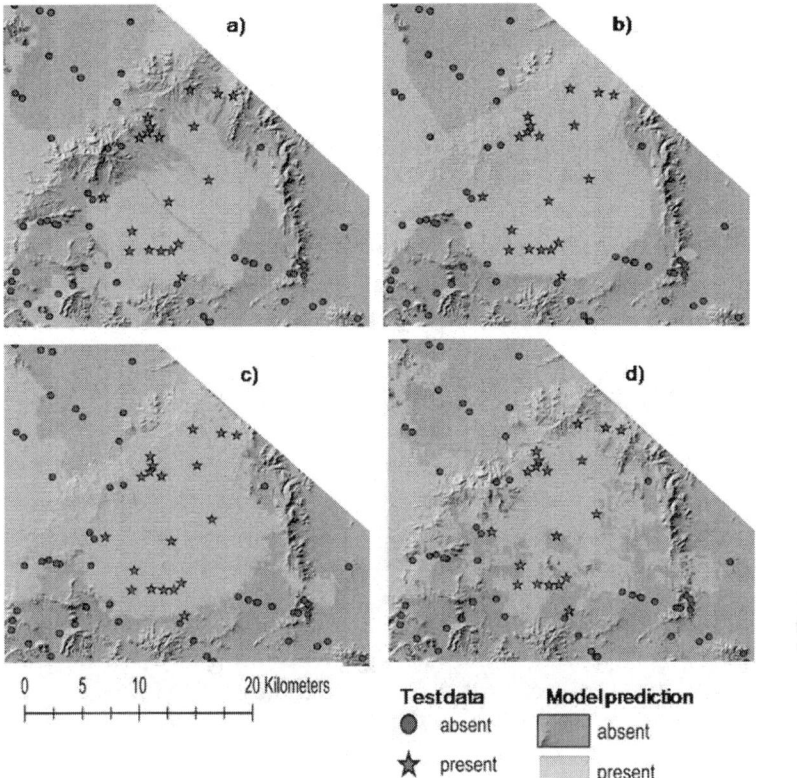

Fig. F.2.4. Classification tree model predictions of Yucca brevifolia (YUBR) presence in the section shown in Fig. F.2.1: a) non-spatial model (0.19); b) model with kriged autocovariate (0.2); c) model with mean simulation autocovariate (0.2); d) model with single simulation autocovariate (0.2). Probability thresholds used to dichotomize continuous map predictions are shown in parentheses

F.2.4 Concluding remarks

The aim of this study was to explore methods of incorporating spatial autocorrelation in species distribution models in order to increase classification accuracy. The emphasis was on prediction rather than inference, and it was expected that the variables representing autocorrelation here could actually be representing biotic effects, abiotic effects or a combination of both.

The methods used here represent an extension of methods that have been used previously in similar studies (see Miller et al. 2007). Autoregressive models are relatively new to species distribution modeling, and allow other predictor variables to be included, but involve computationally intensive methods to calculate the autocovariate term and require all data forms and functions to be specified *a priori*. Geostatistical interpolation methods incorporate spatial structure explicitly, but are not appropriate for including many additional predictor variables and model form is also not flexible. CT models have resulted in higher classification accuracy than GLMs in some modeling studies (Franklin 1998; Vayssières et al. 2000; Thuiller et al. 2003) but they have not been used to incorporate spatial autocorrelation explicitly. The issues of interest here were whether classification accuracy of the test data increased when spatial structure was included (and for which alliance types), which model (GLM or CT) had better results, and which interpolation method calculated the most suitable autocovariate term.

Some GLMs showed improved accuracy with an autocovariate term, some had decreased accuracy and the same inconsistent effect was observed with CT models. Similarly, there was little congruity with respect to the effects of each of the autocovariate terms. However, the three alliances whose models were improved more consistently by incorporating spatial autocorrelation were LATR, LATRAMDU, and YUBR, also the three most prevalent alliances. More information on alliance presence generally increases both model performance and the ability of the variogram to adequately describe the spatial structure in the training data.

Two of the rarest alliances (GALL and MESP) had consistently lower accuracy with the spatial models. The two rarest alliances had either poor models in general (ATCA), or had decreased accuracy with most of the spatial models (EPNE). ATCA, EPNE, and MESP had variograms that indicated very little spatial structure. As these were also rare, it is difficult to determine whether these alliances are actually less spatially clustered, or whether the data are insufficient to model it. CORA, LATR, LATRAMDU, PIMO and YUSC all had variograms that indicated clear spatial structure and were all fairly prevalent alliances, or had very specific environmental requirements within the study area, resulting in spatial clustering (PIMO).

Lack of sufficient data affects the variogram calculation, resulting in an uninformative or even confounding autocovariate term. Additionally, spatial structure is apparent in different degrees among the alliances studied here. Although the variogram for YUSC showed positive spatial autocorelation, the non-spatial models had particularly high accuracy, indicating any spatial clustering was likely represented by the environmental variables.

Sample proportion and distribution were factors in determining the effect spatial autocorrelation had on the models. LATRAMDU and PIMO have different types of distributions—LATRAMDU is one of the most ubiquitous alliances in the Mojave Desert (present in 1,631 of 3,819 observations). PIMO is found only in high elevations with higher precipitation (present in only fifty observations). Each of these alliances had curious results when spatial structure was incorporated in the GLMs. The classification accuracy of LATRAMDU_GLM increased, while PIMO_GLM decreased. It is difficult to make generalizations from these alliances, but the autocovariate seems to have a greater effect on model accuracy when the alliance is more common, acting more as a weighting of the sample data, rather than as compensation for missing environmental variables.

Incorporating spatial autocorrelation can be an efficient way to increase classification accuracy by 'getting more' out of the sample data as well as other models that involve data with spatial structure. In addition, including spatial autocorrelation explicitly in the models can clarify the influence of other predictor variables whose spatial structure may produce spurious correlations (Lichstein et al. 2002). However, when calculated by interpolation and simulation methods and added as explanatory variables in the models, the autocovariate term tends to overpower the environmental variables, resulting in less generalizable models. As the autocovariate terms were derived from the training data, their importance in the models resulted in less robust predictive models – rather than being based on environmental relationships that might be consistent throughout the study area, they are based primarily on sample presence in neighboring sites in the training data.

As opposed to GLMs, for which equation coefficients represent global averages for the entire dataset, the hierarchical partitioning nature of CT models could provide a more appropriate structure for incorporating local variation as represented by the autocovariate terms. Unfortunately when the autocovariates were selected as the first split, as they consistently were here, this is less feasible.

A focus of future work will be to explore different methods for calculating spatial structure. Although we used only a single simulation realization (and the mean of ten simulations) for each model here, any of the other single simulation layers could have produced different model results. An analysis of the range of model results based on incorporating 50 or 100 different simulation layers could provide important insight into differences with respect to sample data prevalence and strength of habitat associations.

Although mixed, the results here still show that spatial autocorrelation *can be* an important and efficient way to increase model accuracy based on available data. However the sampling strategy of the data used is more important than ever. Gradsect sampling approaches are often used in species distribution modeling to maximize floristic variation in the sample, but this approach may not effectively characterize spatial structure in the sample. In order for adequate models to be developed, and for spatial structure to be appropriately quantified, sampling should take place at several different spatial scales. Spatial autocorrelation in rare distributions can be modeled (assuming it exists) if the samples are taken at a more appropriate (and variable) scale.

References

Akaike H (1973) Information theory and an extension of the maximum likelihood. Paper presented at the 2nd International Symposium of Information Theory, Budapest

Aspinall RJ, Pearson DM (1996) Data quality and spatial analysis: analytical use of GIS for ecological modeling. In Goodchild MF, Steyart L, Parks B, Crane M, Johnston C,

Maidment D, Glendinning S (eds) GIS and environmental modelling: progress and research issues. GIS World, Ft. Collins [CO], pp.35-38

Augustin N, Mugglestone M, Buckland S (1996) An autologistic model for the spatial distribution of wildlife. J Appl Ecol 33(2):339-347

Augustin N, Mugglestone M, Buckland S (1998) The role of simulation in modelling spatially correlated data. Environmetrics 9(2):175-196

Besag J (1972) Nearest-neighbour systems and the autologistic model for binary data. J Roy Stat Soc B 34(1):75-83

Besag J (1974) Spatial interaction and the statistical analysis of lattice systems. J Roy Stat Soc B 36(2):192-236

Beven K, Kirkby M. (1979) A physically based variable contributing area model of basin hydrology. Hydrol Sci Bull 24(1):43-69

Borcard D, Legendre P, Drapeau P (1992) Partialling out the spatial component of ecological variation. Ecology 73(3):1045-1055

Breiman L, Freedman J, Olshen R, Stone C (1984) Classification and regression trees. Wadsworth International Group, Belmont [CA]

Brown D (1994) Predicting vegetation types at treeline using topography and biophysical disturbance variables. J Veget Sci 5(5):641-656

Burrough PA, McDonnell RA (1998) Principles of geographical information systems. Oxford University Press, Oxford

Cressie NAC (1993) Statistics for spatial data (revised edition). Wiley, New York, Chichester, Toronto and Brisbane

Davis F, Goetz S (1990) Modeling vegetation pattern using digital terrain data. Landsc Ecol 4(1):69-80

Dokka R, Christenson C, Watts J. (1999) Geomorphic landform and surface composition GIS of the Mojave Desert ecosystem in California. Paper presented at the 4th International Conference on GeoComputation, Fredericksburg [VA]

Dubayah R (1994) Modeling a solar radiation topoclimatology for the Rio Grande River Basin. J Veget Sci 5(5):627-640

Elith J, Graham C, Anderson R, Dudik M, Ferrier S, Guisan A, Hijmans RJ, Huettmann F, Leathwick JR, Lehmann A, Li J, Lohmann LG, Loiselle BA, Manion G, Moritz C, Nakamura M, Nakazawa Y, Overton JM, Peterson AT, Phillips SJ, Richardson K, Scachetti-Pereira R, Schapire RE, Soberon J, Williams S, Wisz MS, Zimmermann NE (2006) Novel methods improve prediction of species' distributions from occurrence data. Ecography 29(2):129-151

Fels J. (1994). Modeling and mapping potential vegetation using digital terrain data. Ph.D. North Carolina State University, Raleigh [NC]

Ferrier S., Drielsma M., Manion G, Watson G. (2002) Extended statistical approaches to modelling spatial pattern in biodiversity in north-east New South Wales. II. Community-level modelling. Biodiv Cons 11(12):2309-2338

Fielding AH, Bell JF (1997) A review of methods for the assessment of prediction errors in conservation presence/absence models. Environ Cons 24(1):38-49

Franklin J (1995) Predictive vegetation mapping: geographic modeling of biospatial patterns in relation to environmental gradients. Progr Phys Geogr 19(4):474-499

Franklin J (1998) Predicting the distributions of shrub species in California chaparral and coastal sage communities from climate and terrain-derived variables. J Veget Sci 9(5):733-748

Franklin J, McCullough P, Gray C (2000) Terrain variables used for predictive mapping of vegetation communities in Southern California. In Wilson J, Gallant J (eds) Terrain analysis: principles and applications. Wiley, New York, Chichester, Torono and Brisbane, pp.331-353

Franklin J, Keeler-Wolf T, Thomas K, Shaari D, Stine P, Michaelsen J, Miller J (2001) Stratified sampling for field survey of environmental gradients in the Mojave Desert ecoregion. In Millington A, Walsh S, Osborne P (eds) GIS and remote sensing applications in biogeography and remote sensing. Kluwer, Dordrecht, pp.229-251

Gotway CA, Hartford AH (1996) Geostatistical methods for incorporating auxiliary information in the prediction of spatial variables. J Agricult Biol Environm Stat 1(1):17-39

Graae BJ, Økland RH, Petersen PM, Fritzbøger B (2004) Influence of historical, geographical and environmental variables on understorey composition and richness in Danish forests. J Veget Sci 15(4):465-474

Guisan A, Zimmermann N (2000) Predictive habitat distribution models in ecology. Ecol Mod 135(2-3):147-186

Guisan A, Edwards T, Hastie T (2002) Generalized linear and generalized additive models in studies of species distributions: setting the scene. Ecol Mod 157(2-3):89-100

Gumpertz M, Graham J, Ristaino J (1997) Autologistic model of spatial pattern of Phytophthora epidemic in bell pepper: effects of soil variation on disease presence. J Agricult Biol Environ Stat 2:131-156

Hastie TJ, Tibshirani R, Friedman J (2001) The elements of statistical learning: data mining, inference and prediction. Springer, Berlin, Heidelberg and New York

Hubbell SP, Ahumada JA, Condit R, Foster RB (2001) Local neighborhood effects on long-term survival of individual trees in a neotropical forest. Ecol Res 16(5):859-875

Keitt TH, Bjornstad ON, Dixon PM, Citron-Pousty S (2002) Accounting for spatial pattern when modeling organism-environment interactions. Ecography 25(5):616-625

Legendre P (1993) Spatial autocorrelation: problem or new paradigm? Ecology 74(6):1659-1673

Legendre P, Fortin MJ (1989) Spatial pattern and ecological analysis. Vegetatio 80(2):107-138

Legendre P, Legendre L (1998) Numerical ecology (2nd edition). Elsevier, Amsterdam

Lennon, JJ (2000) Red-shifts and red herrings in geographical ecology. Ecography 23(1):101-113

Lichstein JW, Simons TR, Shriner SA, Franzreb K (2002) Spatial autocorrelation and autoregressive models in ecology. Ecol Monogr 72(3):445-463

Lobo JM, Jay-Robert P, Lumaret J-P (2004) Modelling the species richness distribution for French Aphodiidae (Coleoptera, Scarabaeoidea). Ecography 27(2):145-156

Lobo JM, Lumaret J-P, Jay-Robert P (2002) Modelling the species richness of French dung beetles (Coleoptera, Scarabaeidae), and delimiting the predictive capacity of different groups of explanatory variables. Glob Ecol Biogeogr 11(4):265-277

Miller JA (2005) Incorporating spatial dependence in predictive vegetation models: residual interpolation methods. Prof Geogr 57(2):169-184.

Miller JA, Franklin J (2002) Modeling the distribution of four vegetation alliances using generalized linear models and classification trees with spatial dependence. Ecol Mod 157(2-3):227-247

Miller JA, Franklin J (2006) Explicitly incorporating spatial dependence in predictive vegetation models in the form of explanatory variables: a Mojave Desert case study. J Geogr Syst 8(4):411-435

Miller JA, Franklin J, Aspinall RJ (2007) Incorporating spatial dependence in predictive vegetation models. Ecol Mod 202(3-4):225-242

Nogués-Bravo D, Martinez-Rica JP (2004) Factors controlling the spatial species richness pattern of four groups of terrestrial vertebrates in an area between two different biogeographic regions in northern Spain. J Biogeogr 31(4):629-640

Peterson AT (2006) Ecological niche modeling and spatial patterns of disease transmission. Emerg Infect Dis 12(12):1822-1826

Pontius RG, Schneider LC (2001) Land-cover change model validation by an ROC method for the Ipswich watershed, Massachusetts, USA. Agricult Ecosyst Environm 85(1-3):239-248

Rangel TFL, Diniz-Filho J, Bini L (2006) Towards an integrated computational tool for spatial analysis in macroecology and biogeography. Global Ecol Biogeogr 15(4):321-327

Richardson DM, Thuiller W (2007) Home away from home-objevtive mapping of high-risk source areas for plant introduction. Divers Distrib 13(3):299-312

Rossi RE, Borth PW, Tollefson JJ (1993) Stochastic simulation for characterizing ecological spatial patterns and appraising risk. Ecol Appl 3(4):719-735

Smith PA (1994) Autocorrelation in logistic regression modeling of species' distributions. Global Ecol Biogeogr Lett 4:47-61

Stockwell D (2006) Niche modeling: predictions from statistical distributions. CRC Press (Taylor & Francis Group), Boca Raton [FL], London and New York

Sui D (2004) Tobler's first law of geography: a big idea for a small world? Ann Assoc Am Geogr 94(2):269-277

Swets JA (1988) Measuring the accuracy of diagnostic systems. Science 240(4857):1285-1293

Thomas K, Franklin J, Keeler-Wolf T, Stine P. (2004). Mojave Desert ecosystem program central Mojave vegetation mapping project. USGS-BRD, DoD, Flagstaff [AZ]

Thuiller W, Araújo M, Lavorel S (2003) Generalized models vs. classification tree analysis: predicting spatial distributions of plant species at different scales. J Veget Sci 14(5):669-680

Tobler WR (1970) A computer movie simulating urban growth in the Detroit region. Econ Geogr 46:234-240

Vayssières MP, Plant RE, Allen-Diaz BH (2000) Classification trees: an alternative non-parametric approach for predicting species distributions. J Veget Sci 11(5):679-694

Wu H, Huffer FW (1997) Modelling the distribution of plant species using the autologistic regression model. Environ Ecol Stat 4(1):49-64

F.3 A Web-based Environmental Decision Support System for Environmental Planning and Watershed Management

Ramanathan Sugumaran, James C. Meyer and *Jim Davis*

F.3.1 Introduction

Many of the largest problems faced by local governments such as cities and counties involve issues associated with urbanization. The increased rate of urbanization has led to haphazard growth, increased infrastructure costs, deterioration of living conditions and worsening of the environment. This phenomenon places a heavy burden on local planners and managers, who struggle to balance competing demands for residential, commercial, and industrial development with imperatives to minimize environmental degradation. In order to effectively manage this development process on a sustainable basis, local government planners increasingly rely on the use of information technologies, spatial modeling techniques (Dragicevic et al. 2000) and Spatial Decision Support Systems (SDSS) (Fedra 1995; Sugumaran et al. 2000; Sugumaran and Sugumaran, 2007; Sugumaran, and Bakker, 2007; Zhang et al. 2008). These technologies and techniques include combinations of remote sensing, GIS, spatial modeling, Multi-Criteria Evaluation (MCE), computational neural networks, and Internet technology.

Recently, Internet technology has been widely used for application development because of advantages such as platform independency, reductions in distribution costs and maintenance problems, ease of use, ubiquitous access and sharing of information by the worldwide user community (Abel et al. 1998; Doyle et al. 1998; Peng and Nebert 1997; Peng and Tsou 2003; Shriram et al. 2007). This has resulted in research studies demonstrating the suitability of the Web as a medium for implementation of planning tools. For example, several researchers demonstrated the use of the Internet and GIS for improved decision-making (Peterson 1997; Doyle et al. 1998; Peng 1999; Dragicevic et al. 2000; Pandey et al. 2000;

Sugumaran et al. 2000; Peng and Tsou 2003; Sugumaran and Sugumaran 2007), and environmental modeling (Carver 1996; Al-Sabhan 2000; Zhang and Wang 2001). Although there has been significant progress in the use of the Web as a medium for environmental data sharing and data visualization (Dragicevic et al. 2000; Houle et al. 2000; Sugumaran et al. 2003), not as many studies have focused on developing a Web-based environmental planning tool using Spatial Decision Support System (SDSS) for local level planning. There is now increased interest in pursuing the development of SDSS on the Web to support better decision-making and policy formulation. Examples include: HYDRA – a Spatial Decision Support System for water quality management in urban rivers (Taylor 2002), development of a decision support system for a fish and wildlife assessment in the Columbia river (Parsley et al. 2000), Agricultural Farm Analysis (Vernon 1999), emergency planning (Carver and Myers 1996; Carver 1999; Carver et al. 2001) environmental decision-making (Kingston et al. 2000), Web-based urban prediction modeling and visualization tool (Compas and Sugumaran 2004) and snow removal (Shriram et al. 2007). The research center CARES at the University of Missouri – Columbia has developed a number of WWW-based SDSS applications addressing such topics as habitat suitability for bird species, hydrologic response to land use decisions, site selection for livestock feeding operations, management of woody draws in agricultural fields, and more (Center for Agricultural, Resource and Environmental Systems 2002).

This chapter presents a Web-based Environmental Decision Support System (WEDSS) prototype to prioritize local watersheds on the basis of environmental sensitivity using multiple criteria. The Internet was chosen as a development platform in this study mainly because of openness and interactivity, which allows public access to and participation in the analysis (Dragicevic et al. 2000; Sugumaran et al. 2000).

F.3.2 Study area

Columbia (Missouri) is a quality city that has been rated as one of the best places in the USA in which to live, work, raise a family and retire. As a result, Columbia is one of the fastest growing Metropolitan Statistical Areas in Missouri with a population increase of over 20.5 percent between 1990 and 2000 (Missouri Census Data Center 2001) (see Fig. F.3.1). In response to this development pressure, local planners, managers, and stakeholders desire assistance in developing smart growth policies that allow for growth while preserving water quality, reducing storm water runoff problems and protecting local natural areas. To do this, it is necessary to evaluate a wide range of information and to analyze alternative development strategies. The present study attempts to address issues facing local planners as they confront storm water management issues by developing a Web-based Environmental Decision Support System (WEDSS) to identify and prioritize local watersheds using multiple environmental criteria. Figure F.3.1 shows the location of all 23 local watersheds involved in the study. These watersheds encompass the most rapidly developing areas of Boone County surrounding the city of Columbia. New development is occurring around the fringes of the current city limits of Columbia and to a lesser extent around the smaller city of Ashland to the south. Broad, gently sloping, upland ridges with a significant amount of row crop production and livestock pasture characterize the eastern part of the study area. Most of the rest of the study area is comprised of narrower ridges with moderate to steep slopes; heavily wooded rolling hills and river bluffs dominate this terrain. There are significant areas of karst topography west and south of Columbia, featuring many caves, sinkholes, losing streams, and springs. Because of these features, the effects of storm water runoff on ground water quality are of concern. A number of state and local parks, conservation areas, and a national forest occupy the sparsely populated southern part of the study area. The adverse impacts posed to these public lands and to the streams that flow through them by potential development are issues of local concern. Most of the streams that currently exhibit superior biological health are located in this area.

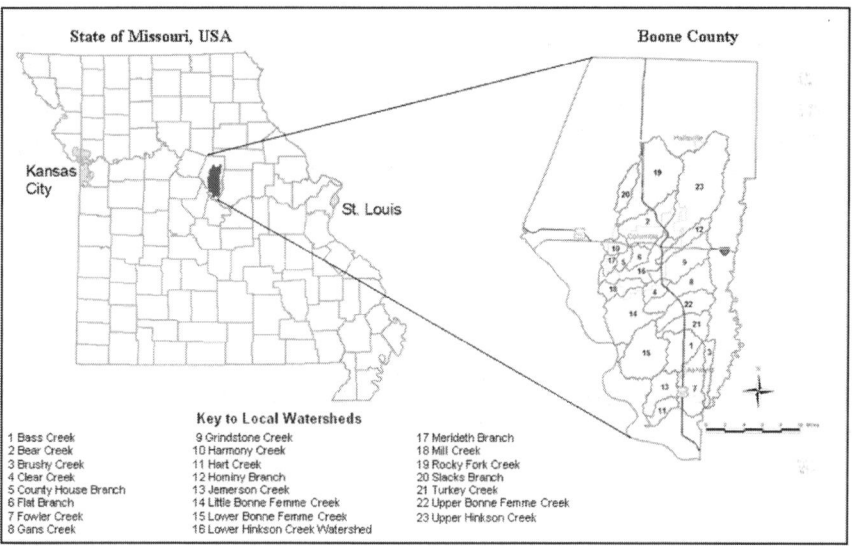

Fig. F.3.1. The study area

F.3.3 Design and implementation of WEDSS

The WEDSS design is based on the client/server model in which clients send requests to services running on a server and receive appropriate information in response (see Fig. F.3.2). Client-server architecture was used because it facilitates maintenance of the application and its data layers. In addition, the functionality of the application can be upgraded or replaced at any time without affecting the user's computer system. The client/server used in this study has a three-tiered configuration consisting of: Tier 1 a WWW client, Microsoft Internet Explorer (IE) or Netscape Navigator; Tier 2 a WWW server, Microsoft Internet Information Server (IIS); and Tier 3 a WWW-based GIS server, ArcView Internet Map Server (Arc-IMS). Figure F.3.2 shows the flow of information in the client-server transaction. The user initiates a request by manipulating tools and buttons in the Web browser (see arrow indicated by 1). The IIS Web server passes the request to the appropriate instance of ArcIMS (see arrow indicated by 2). Avenue Scripts within the Model Management System make calls to the DBMS to perform the analysis (see arrow indicated by 3 and 4). ArcIMS creates map images and data tables of the results and passes them to the IIS Web server (see arrow indicated by 5). IIS formats the output into HTML pages and serves the content to the calling client's Web browser (see arrow indicated by 6). The Web browser on the client machine displays the results and supports further user interaction, which spawns additional requests (see arrow indicated by 7). The Display and Report generator described below is implemented as a set of frames integral to the Web browser's display. There are several techniques are available for the development of the Web-based data visualization and decision support, such as Common Gateway Interface (CGI), Browser Plug-Ins, and ActiveX controls.

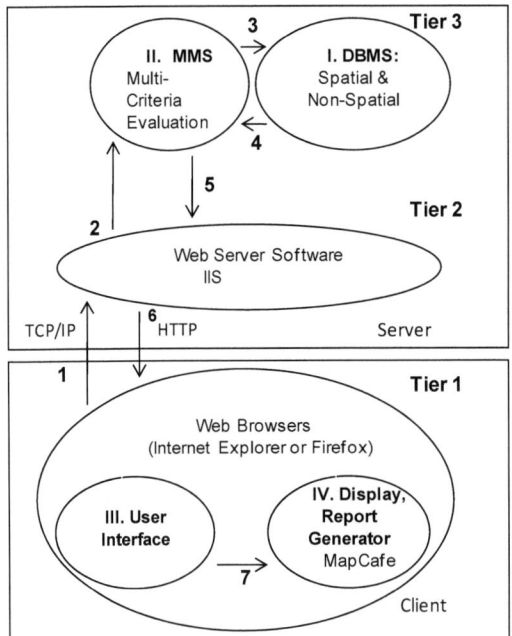

Fig. F.3.2. Overall client server transaction

This application utilizes Java applet-based ArcIMS developed by ESRI. A Java-based programming environment is desirable because it is object-oriented and supports the development of portable, interpreted, system-independent, and distributed applications served on the WWW. The Java functionality used here is integral to the Web browser's Java Virtual Machine and does not require the user to install any additional components. This is particularly helpful to users in large organizations and government institutions, who often do not possess user access permissions to install software or Web browser plug-ins on their own workstations.

A Spatial Decision Support System (SDSS) requires five key components including (a) a Data Base Management System (DBMS); (b) a Model Management System (MMS), implementing the analysis algorithms; (c) a display generator; (d) a report generator and (e) a user interface (Armstrong and Densham 1990). Figure F.3.2 demonstrates the components involved in the conceptual design of the WEDSS. The DBMS and MMS components reside on the server side and the other three components are located on the client side. The next section explains the key components of WEDSS.

Server side processes

The server side environment is where most of the actual functionality was implemented. This includes IIS with the ESRIMap Web server extension (esrimap.dll) and ArcView GIS with the ArcIMS extension (see Fig. F.3.2). ArcIMS is an extension to ArcView GIS that provides WWW communication facilities. IIS transfers data between the client side Web browser and ArcIMS via the ESRIMap Web server extension and vice versa. Connections between IIS and ArcIMS are achieved utilizing sockets (see Fig. F.3.2). Sockets are a common method used to pass messages between processes on the same or different hosts using the TCP/IP protocol.

Data base management system (DBMS). The DBMS component manages spatial and non-spatial data to support the MMS. The development of the DBMS involved two stages, the first was to establish the relevant watershed level environmental criteria by collecting input from members of the storm water steering committee, a local advisory committee comprised of a half dozen staff members from local government agencies charged with addressing storm water management issues. These criteria were chosen to address public health, surface water quality and groundwater quality concerns. The committee selected thirteen environmental criteria to use in twenty-three local watersheds comprising an area of about 177,000 acres (Table F.3.1). These criteria were selected based on the committee's assessment that they quantified aspects of the watersheds that were important from a water quality perspective taking into account both physical processes and human health exposure. Individual committee members and their agencies approached water quality from a variety of perspectives and had differing priorities. The broad areas of concern that emerged from their discussions included human health exposure to contaminants and bacteria suspended in surface water, particularly as related to recreational uses of the streams; the potential for increased soil erosion and sedimentation under development; the potential for surface water to contact and contaminate groundwater; and the potential impacts of degraded water quality on wildlife. Coming into compliance with USEPA Phase II Storm Water Regulations was a concern. The next step was to develop GIS data layers to address each criterion. Relevant GIS layers were developed from a variety of sources including remotely sensed imagery, digital soils surveys, the National Elevation Dataset, field sampling, and other sources. Table F.3.2 shows the base data layers collected, sources, and scales. There is not a one to one correspondence between the source datasets in Table F.3.2 and the criteria in Table F.3.1, since some criteria required interim data layers to be generated from several of the base data layers listed.

In performing this analysis, the goal was to compare the watersheds by considering each of the criteria individually and then generating an index number that would summarize the relative rankings. To make meaningful watershed comparisons, it was necessary to focus on base datasets that were collected in a consistent fashion across the study area. The explicit trade-off made in data collection was to trade contemporaneous data acquisition and spatial resolution/scale for consistency in data collection methodology. One example of this trade off was in the stream network layer. The National Hydrography Dataset (NHD) at a scale of 1:100,000 was readily available for the study area. This is a smaller scale than most of the other vector GIS layers used in the analysis. 1:24,000 scale stream networks are being developed for the State of Missouri, but were not available at the time of this analysis and generating them was outside the scope and budget of this project. It is well known that smaller scale hydrology line work will exhibit less sinuosity and shorter stream lengths when comparing the same features depicted in a dataset collected at a larger scale. However, in this analysis, the most important factor was the relative difference in stream mileage between the watersheds, not their absolute magnitudes. The 1:100,000 scale NHD was the largest scale dataset available that had been collected in a uniform way for the entire study area. The authors deem this consistency from watershed to watershed to be vital and chose to use this dataset rather than other possible sources that varied in scale and collection methodology from location to location within the study area.

For some of the criteria it is recognized that the best data currently available has a sampling bias. For example the endangered species observations from the Missouri Department of Conservation's Natural Heritage Dataset are likely biased based on where agency personnel have done fieldwork and does not represent a concerted countywide sampling effort. The committee deemed the presence of a confirmed endangered species sighting in a watershed to be of great enough significance that it determined to include the best available data in the analysis, recognizing the possible sampling bias.

Table F.3.1. Criteria used in the analysis and their weights

	Criteria	Weight (in %)
1	Portion of watershed area with slope greater than 15%	6.25
2	Portion of watershed area with slope between 7% and 15%	1.75
3	Relative erosion potential as measured by the mean K-factor from the soils survey	3.00
4	Portion of stream miles designated for Whole Body Contact or Boating and Canoeing	9.25
5	Portion of watershed area draining to losing streams	8.50
6	Portion of stream miles running thru or within 50 meters of recreational land	6.25
7	Relative abundance of Endangered Species	14.50
8	Portion of watershed area within a FEMA designated 100-year floodplain	1.75
9	Portion of 50 meter buffer around streams that is wooded	6.25
10	Portion of watershed area that is wooded	7.50
11	Relative stream biological health	18.25
12	Portion of watershed area that is designated as karst in the soils survey	9.00
13	Portion of watershed area that is designated as Palustrine wetland in the National Wetlands Inventory	7.75

Table F.3.2. Base data layers collected and their sources

	Base Data Layer / Source	Resolution / Scale
1	USGS 30 Meter Digital Elevation Model (DEM)	30m raster
2	National Hydrography Dataset (NHD)	1:100,000
3	10 CSR 20-7 Missouri Code of State Regulations	Stream segments defined by mileage
4	Missouri Department of Natural Resources (MDNR) Parks Dataset	1:24,000
5	City of Columbia Recreational Land Dataset	1:24,000
6	Boone County parcel dataset – selected parcels of recreational land	1:4,800
7	Missouri Department of Conservation, Natural Heritage Program, Missouri Natural Heritage dataset	1:24,000
8	Federal Emergency Management Agency (FEMA) Q3 Flood Dataset	1:24,000
9	U.S. Fish & Wildlife Service (USFWS) National Wetlands Inventory	1:24,000
10	Natural Resources Conservation Service (NRCS) Digital Soils Survey	1:24,000
11	CARES 1999 30 meter remotely sensed land cover dataset from Landsat ETM	30m raster
12	Macro invertebrate field sampling performed by the University of Missouri Department of Fisheries and Wildlife	Field observations near watershed outlet

After collecting these base datasets, several interim data sets were developed using ArcView GIS and ArcInfo GIS for the final analysis. For example, the watershed boundaries used as the units of analysis were created from the 30m USGS Digital Elevation Model (DEM). There are roughly 1,500 small sub-watersheds were created by the software using the DEM and different parameters such as flow direction, flow accumulation and stream segmentations. These were combined to form the 23 watersheds used in the analysis. The basis for the designated stream uses was Missouri Statute 10 CSR 20-7, which is a tabular list of designations based on stream mileage from fixed points. In order to create a spatial dataset for the analysis, individual stream segment features in the NHD were selected and attributed using a manual GIS editing process.

Model management system (MMS). After having several meetings with the steering committee, a simple Multi-Criteria Evaluation (MCE) analysis using weighting method particularly weighted linear combination was adopted for the prototype. Weighted linear combination is the most direct means of obtaining weighting information from the decision-maker and least amount of operations to transform information supplied by the decision-maker (Hajkowicz et al. 2002). The reader is referred to Carver (1991), Keller and Strapp (1993), Heywood et al. (1994), Eastman et al. (1995), Jankowski (1995), Carsjens et al. (1996), Hajkowicz et al. (2002) for a detailed description about different Multi-Criteria Evaluation (MCE) methods and their strengths and weaknesses. The primary advantage of weighted linear combination is that it forces decision-makers to make trade-offs in a decision problem (Hajkowicz et al. 2002). In addition, this method allows for a simple, straightforward user interface that is easy to explain and can be understood and operated by members of the general public over the Web without face-to-face instruction. This was an important user-specified requirement. In contrast to pair-wise comparison or rating, it also allows users to make explicit trade-off decisions and to exclude some criteria entirely by assigning them zero weight. The downside to this method is that it requires careful consideration of the relative importance of each criterion and does not provide feedback about the consistency of choices as the pair-wise method does. In order to collect weights for each criterion, the steering committee assigned weights to each according to their importance relative to one another. The weights assigned by individual committee members were averaged and the resulting weights served as the basis for discussion during a meeting where a consensus set of weight values was adopted. A detailed description of the sensitivity calculations is given next. The overall MCE models implementation and process using client server approach is explained in detail at the result section.

ESI calculations: *Standardization of criterion values.* The rating number for each criterion for each watershed is constructed to be a decimal number in the interval [0,1] inclusive mainly because of easy interpretation of the user input values. The criteria were defined in such a way that larger rating values correspond to higher environmental sensitivity. Each rating value is intended to be an interval scale number that reflects the relative magnitude of a watershed for that criterion.

Not only the rank order but also the magnitude of the differences between values is significant.

For most of the criteria, this number is generated as described below from the relevant proportion. For two of the criteria, 'Relative abundance of Endangered Species' and 'Relative stream biological health,' the rating numbers are handled differently since they are not based on proportions of the watershed, but are essentially measures of the number of times an organism was observed. The rating for 'Relative abundance of endangered species' is binary having a value of one if an endangered species was observed anywhere in the watershed or a value of zero if none were observed. It was intended that this rating be interval scale; however, the data had no more than one occurrence in any watershed with the majority of watersheds having zero occurrences so a binary distribution resulted. The rating for 'Relative stream biological health' is based on the number of species of environmentally sensitive macro-invertebrates (i.e. Ephemeroptera, Plecoptera, and Trichoptera) that were observed for each watershed. The number observed ranged from one to thirteen. This raw number was divided by thirteen to normalize these values, resulting in fractions ranging from 0.0769 to 1.0000.

For the remaining eleven criteria, a decimal number indicating the proportion of the watershed meeting the definition of the criterion was calculated. For example, proportion of recreational stream length was calculated by dividing the total length of recreational streams by the total length of all streams in a given watershed. For each proportion value, the corresponding *Standardized score* was calculated as follows, where *median* is the median value of the set, excluding zero values:

$$Standardized\ score = 1 - [1 + proportion\ value\ (median)^2]^{-1} \quad (F.3.1)$$

This formula was chosen to fix the median of the rating values at 0.5. This was necessary because some of the distributions of values were highly skewed. Simply using the proportions directly as ratings would have acted to defeat the weighting scheme applied in the next step. Once rating numbers were calculated using the methods outlined above, they were used to generate an overall Environmental Sensitivity Index by computing a weighted sum. For each watershed, the ratings for all thirteen criteria were summed using the weights selected by the steering committee as shown in Table F.3.1.

$$ESI_{watershed} = \sum_{i=1}^{13} rating_i - weight_i \quad (F.3.2)$$

Client side processes

The client side user interface was developed using a Java, JavaScript, HTML, and Active Server Pages (ASP). JavaScript was used to format URLs for communication with IIS and ESRIMAP.DLL on the server side using user input from the HTML pages (see Fig. F.3.2). The client front-end allows users to interact with the application. The application uses a simple Graphical User Interface (GUI) to dynamically create 'user-friendly' HTML pages. Microsoft's Active Server Pages (ASP) technology was used because it is simple to implement, and allows the creation of dynamic pages to collect user input and session information required by the server-side processes.

The map display and report generator was developed by customizing the MapCafé Java applet that is a standard part of ArcIMS. MapCafé constructs and sends requests corresponding to user button and tool manipulations to the server-side applications. Java was used to develop the visualization and decision support tools that were placed on the MapCafé toolbar. This modification significantly extended the capability of the MapCafé application. The ArcView application running on the Web server responds to the URL encoded requests received from the client by processing them and sending updated map images, text, and tabular data back to the requesting user's MapCafé applet, which renders the results graphically to the user (see Fig. F.3.2).

F.3.4 The WEDSS in action

The WEDSS homepage is available to the general public at http://maproom.missouri.edu/analysis/esi/index.asp. The homepage explains the goals and describes the project. Clicking the 'start program' located at the top of the homepage starts WEDSS. Upon entering the interactive page, the user can display the data layers and also perform analysis (see Fig. F.3.3a). WEDSS uses standard image/GIS data browsing tools such as zoom in, zoom out, pan, feature info etc. built into MapCafé to allow users to interact with the map display. For example, the extent buttons allow the user to move between different views of the map display, drawn at different scales and centered at different points. These buttons function analogous to a Web browser's forward, back, and home buttons. The pan, zoom in and zoom out buttons allow the user to 'move around' in the map display and to change the map scale. The feature info button allows access to information about features on the map display (Fig. F.3.3a). When used to query a watershed in the 'User ESI' GIS layer, the tool brings up a table showing the model results and other attribute information for the selected watershed. The measure distance tool allows the user to measure of the lengths of linear map features or the distance between locations on the map. The help button (right side frame) provides a description of each buttons.

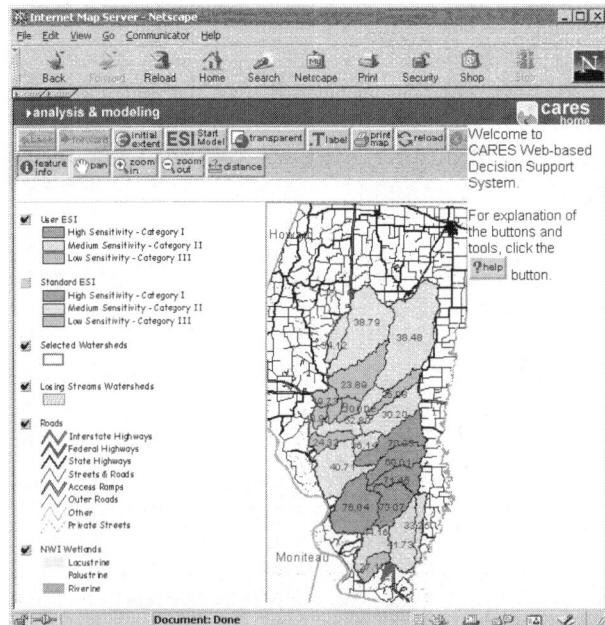

Fig. F.3.3a. Web-based user interface for multi-criteria analysis

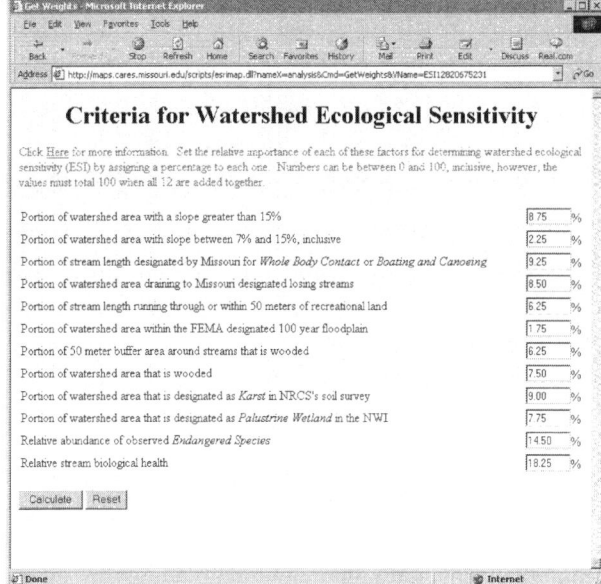

Fig. F.3.3b. User input window for the model

In addition to the aforementioned built-in functions, we have developed several customized tools such as transparent layers, label, print and 'start ESI model' to provide WEDSS specific functions. The transparent tool is extremely useful for visualizing overlapping GIS layers. The level of transparency of any GIS layer can be set so that features or imagery below the layer are made partially visible through the layer. This option gives the user a very good idea of extent of the GIS layer in terms of landmarks from a base layer, such as an aerial photo, and also visually highlights the overlap of map features in different layers. A label tool allows the user to turn labels on and off for each GIS layer. This allows roads to be labeled with their names, watersheds to be labeled with their ESI values and so forth for easier map navigation and increased comprehension. The print map tool allows the user to specify a title for their map, and print the current map display, including visible GIS layers, images, labels, etc. The map is formatted for an 8.5' by 11' page at the scale the user was viewing and includes cartographic elements such as a legend, scale bar, north arrow and user specified title.

The 'start ESI model' button allows the user to perform the analysis with their set of criteria weights. When 'start ESI model' button is pressed, a new window opens to provide the user with the list of criteria and to accept weight values input by the user (see Fig. F.3.3b). The overall process of this ESI models for both client side and server side are described in Fig. F.3.4. The default values displayed initially are the weights set by the storm water steering committee (see Table F.3.1).

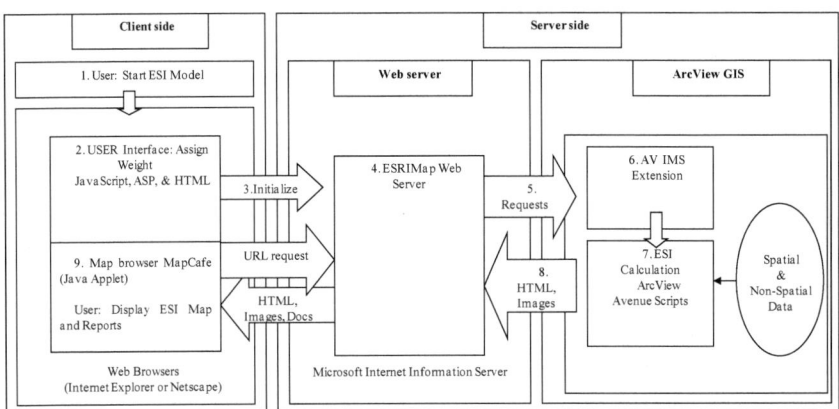

Fig. F.3.4. Overall ESI model process

Interested parties may access the underlying data and conduct their own analyses using their own assessments of the relative importance of the environmental criteria. If a user wants to discard one or more of the criteria completely, (s)he may set the weight value to zero for that criterion. The input form performs validation to verify that the sum of the weight values is 100 before submitting this input to the model. All these user interface forms are written in JavaScript, ASP and HTML (see Fig. F.3.4). When the user is satisfied with criteria and its weight, (s)he can calculate the ESI by clicking the 'Calculate' button which is located at the bottom left of the user interface (see Fig. F.3.3a). The user information is then sent to the server, where ArcView Avenue scripting language processes the user information and sends the result back to client-side (see Fig. F.3.4). Figure F.3.5 is a screen shot of portion of the ESI calculation which was written in ArcView avenue script. After the successful completion of the model calculation, the result will be displayed in the client side (Fig. F.3.3a). Loading and execution time of this model is based on the user's Internet bandwidth. In order to run ESI model successfully, users must be running Internet Explorer or Netscape version version 4.x or higher, browser must have Cookies enabled, Java scripts must be enabled and browser must allow for Pop-up windows. There is also a tutorial with detailed instructions for browser-side setup to the users and it is available now in the ESI website.

```
CalculateRatingValues
'****************************************************************
'Create Sorted Lists of Values for Each Criterion, Strip off Values in Ascending O
'****************************************************************
strFormat = "d.dddd"

ftbWatersheds.SetEditable(TRUE)
For Each fieldItem in 0 .. ((lstCalcRatingCriteriaFieldNames.Count) - 1)
    lstValues = {}
    For Each rec in ftbWatersheds
        portionVal = ftbWatersheds.ReturnValue(ftbWatersheds.FindField(lstCalcRati
        lstValues.Add(portionVal)
    End
    lstValues.Sort(TRUE) 'Ascending Order

    maxNum = lstValues.Get(recCount - 1)
    minNum = lstValues.Get(0)
    rangeNum = maxNum - minNum

    lstNoZeros = lstValues - {0}
    lstNoZeros.Sort(TRUE) 'Ascending Order
    medianElement = ((lstNoZeros.Count)/2).Ceiling
    medianIndex = medianElement - 1
    median = lstNoZeros.Get(medianIndex)
    'msgbox.listAsString(lstValues,"Median = "+median.AsString,fieldItem.GetName)
    a = maxNum / median

    For Each rec3 in ftbWatersheds
        portionVal = ftbWatersheds.ReturnValue(ftbWatersheds.FindField(lstCalcRati
        newRating = 1-(1/(1+((a*(portionVal)/(maxNum))^2)))
        ftbWatersheds.SetValue(lstCalcRatingFields.Get(fieldItem),rec3,newRating)
    End
End

ftbWatersheds.SetEditable(FALSE)
```

Fig. F.3.5. A screen shot of portion of the ESI calculation which was written in ArcView avenue script

F.3.5 Concluding remarks

The Web based system developed for this project provides the user with a simple decision support tool to identify and prioritize local watershed environmental sensitivity using a simple point source MCE method. This system can be used by planners and mangers within local government, the general public, real estate developers, environmental analysts, and other interested parties. The present prototype of the WEDSS is limited to thirteen environmental criteria. In the future, based on feedback received from users, we hope to enhance the WEDSS by collecting improved data sets and addressing additional environmental criteria identified by users. Since WEDSS was developed, discussions and decisions about zoning and permitted uses for some tracts of land have revolved around environmental sensitivity in general and the criteria used by WEDSS in particular. While this particular implementation is focused on the Boone County, Missouri study area, many of the GIS layers used were taken 'off the shelf' from standardized United States federal datasets. This fact allows the model to be quickly adapted to other locations within the United States where the same GIS datasets provide coverage. In addition, the model's framework and coding is abstracted to permit the inclusion of additional criteria and the substitution of more locally relevant criteria for the ones of local significance used in this particular study area. For example, the karst criteria used in this example would not be highly appropriate for areas with differing geomorphology. Other more locally relevant criteria could be substituted without modifying the underlying model framework. Future work will also involve migrating WEDSS from ArcIMS to ESRI's new Internet Map Server technology, ArcIMS. The advantages of upgrading to ArcIMS include greater scalability and stability for WWW applications.

Acknowledgement. This work was supported by NASA/Raytheon/STX Corporation, Subcontract #012100MJ-3. We also wish to thank of George Montgomery and Roy Dudark of city of Columbia, Missouri; and Bill Florea and David Nichols of Boone County, Missouri for their ideas and co-operation in this work. We also thank two anonymous reviewers and Professor Arthur Getis for their comments on improving this chapter.

References

Abel D, Kerry T, Ackland R, Hungerford S (1998) An exploration of GIS architectures for internet environments. Comp Environ Urban Syst 22(1):7-23

Al-Sabhan W (2000) Database-powered web applications for real time watershed simulation model. Paper presented at the GIS 2000 Conference, March 13-16, Toronto, Canada, CD-ROM

Armstrong MP, Densham PJ (1990) Database organization strategies for spatial decision support systems. Int J Geogr Inform Syst 4(1):3-20

Carsjens GJ, Wim GM, Knaap VD (1996) Multi-criteria techniques integrated in GIS applied for land use allocation problems. Paper presented at the Second Joint European Conference and Exhibition on Geographical Information, March 1996, Barcelona, Spain

Carver SJ (1991) Integrating multi-criteria evaluation with geographic information systems. Int J Geogr Inform Syst 5(3):321-339

Carver SJ (1996) Open spatial decision-making on the internet: collaborative spatial decision-making on the internet [accessed Oct 21, 2008, http://www.geog.leeds.ac.uk/people/s.carver/hpcsdm.htm #discussion]

Carver SJ (1999) Developing web-based GIS/MCE: improving access to data and spatial decision support tools. In Thill JL (ed) Multi-criteria decision-making and analysis: a geographic information sciences approach. Ashgate, Aldershot, pp.49-75

Carver SJ, Myers A (1996) Developing a geographical information systems based decision support system for emergency planning in response to hazardous gas releases. In Barham R (ed) Fire engineering and emergency planning. Chapman and Hall, London, pp.301-315

Carver SJ, Evans A, Kingston R, Turton I (2001) Public participation, GIS, and Cyberdemocracy: evaluating on-line spatial decision support systems. Environ Plann A 28(6):907-921

Center for Agricultural, Resource and Environmental Systems (CARES) (2002) Analysis and modeling application page [accessed Oct 21, 2008, http://maproom.missouri.edu/assessment.html]

Compas E, Sugumaran R. (2004) Urban growth modeling on the web: a decision support tool for community planners, In Papers of the Applied Geography Conferences, St.Louis [MO], pp.255-269

Doyle S, Dodge M, Smith A (1998) The potential of web-based mapping and virtual reality technologies for modeling urban environments. Comp Environ Urban Syst 22(2):37-55

Dragicevic S, Balram S, Lewis J (2000) The role of Web GIS tools in the environmental modeling and decision-making process. Proceedings of the 4th International Conference on Integrating GIS and Environmental Modeling (GIS/EM4): Problems, Prospects and Research Needs, September 2 - 8, 2000, Alberta, Canada [accessed Oct 21, 2008, http://www.colorado.edu/ research/cires/banff/pubpapers/134/]

Eastman JR, Jin W, Kyem AK, Toledano J (1995) Raster procedures for multi-criteria / multi-objective decisions. Photogramm Eng Remote Sens 61(5):539-547

Fedra K (1995) Decision support for natural resources management: models, GIS and expert systems. AI Applications 9(1):3-19

Hajkowicz SA, McDonald GT, Smith PN (2000) An evaluation of multiple objective decision support weighting techniques in natural resource management. J Environ Plann Mgmt 43(4):505-518

Heywood I, Oliver J, Tomilson S (1994) Building an exploratory multi-criteria modeling environment for spatial decision support. Proceedings of GISRUK'94, Leicester, pp.327-336

Houle M, Dragicevic S, Boudreault F (2000) A web-based GIS tool for accessing spatial environmental data on the St. Lawrence River. Proceedings of the 4th International Conference on Integrating GIS and Environmental Modeling (GIS/EM4): September 2 to 8, 2000 Banff, Alberta, Canada [accessed Oct 21, 2008, http://www.colorado.edu/ research/cires/banff/pubpapers/ 231/]

Jankowski P (1995) Integrating geographical information systems and multiple criteria decision-making methods. Int J Geogr Inform Syst 9(3):251-273

Keller CP, Strapp JD (1993) Design of multi criteria decision support system for land reform using GIS and API. Paper presented at the Second International Conference on Integrating GIS and Environmental Modeling, Breckenridge [CO]

Kingston R, Carver SJ, Evans A, Turton I (2000) Web-based public participation geographical information systems: an aid to local environmental decision-making. Comp Environ Urban Syst 24(2):109-125

Malczewski J (2004) GIS-based land-use suitability analysis: a critical overview. Progr Plann 62(1):pp.3-65

Missouri Census Data Center (2001) Public law 94-171 trend reports [accessed Oct 21, 2008, http://mcdc2.missouri.edu/webrepts/pl94trend/Missouri_MSAS.pdf]

Pandey S, Gunn R, Lim K, Engel B, Harbor J (2000) Developing a web-enabled tool to assess long-term hydrologic impact of land use change: information technologies issues and a case study [accessed Oct 21, 2008, http://www.urisa.org/Journal/Vol12%20No4/pandey/modeling_the_long.htm]

Parsley MJ, Korschgen CE, Guyton J (2000) Development of a decision support system for a fish and wildlife assessment in the Columbia river: a prototype for the john day reservoir [accessed Sept 12, 2003, http://biology.usgs.gov/wfrc/jddss.htm]

Peng Z (1999) An assessment framework for the development of internet GIS. Environ Plann B 26(1):117-132

Peng Z, Nebert D (1997) An internet-based GIS data access system, J Urb Reg Inf Syst Ass 9(1):20-30

Peng Z, Tsou MH (2003) Internet GIS: distributed geographic information services for the internet and wireless networks. Wiley, London

Peterson MP (1997) Trends in internet map use. Proceedings of the 18th International Cartographic Conference, ICA Stockholm, Sweden, pp.1635-1642

Shriram, I, Sugumaran R, Sugumaran V (2007) Development of a web-based intelligent spatial decision support system (WebSDSS): a case study with snow removal operations. In Hilton BN Emerging spatial information systems and applications. Idea Group, Hershey [PA], pp. 184-202

Sugumaran R, Bakker B (2007) GIS-based site suitability decision support system for planning confined animal feeding operations in Iowa. In Hilton BN (2006) Emerging spatial information systems and applications. Idea Group, Hershey [PA], pp.218-238

Sugumaran R, Davis CH, Meyer J, Prato T, Fulcher C (2000) Web-based decision support tool for floodplain management using high resolution DEM. J Photogram Eng Remote Sens 66(10),1261-1265

Sugumaran R, Meyer J, Barnett Y, Fulcher C, Prato T (2003) A web-enabled spatial data visualization and decision support system. Proceeding of the 7th World Multiconference on Systemics, Cybernetics and Informatics, Orlando [USA] July 27-30, CD-ROM

Sugumaran V, Sugumaran R (2007) Web-based spatial decision support systems (WebSDSS): evolution, architecture, and challenges. J Comm Ass Inform Syst 19(1): 844-875

Taylor K (2002) Hydra 5: catchment management on the web [accessed Oct 21, 2008, http://www.cmis.csiro.au/]

Vernon LT, Aiken RM, Waltman W (1999) Agri-FACTs: agricultural farm analysis and comparison tool [accessed: Oct 21, 2008, http://gis.esri.com/library/userconf/proc99/proceed/papers/pap385/p385.htm]

Zhang X, Wang YQ (2001) Web based spatial decision support for ecosystem management. Proceedings of the ASPRS 2001, April 23-27, St.Louis [MO]

Zhang Y, Barten PK, Sugumaran R, DeGroote J (2008) Evaluating forest harvesting to reduce its hydrologic impact with a spatial decision support system. Applied GIS 4(1):1-16

Part G

Applications in Health Sciences

G.1 Spatio-Temporal Patterns of Viral Meningitis in Michigan, 1993 - 2001

Sharon K. Greene, Mark A. Schmidt, Mary Grace Stobierski and *Mark L. Wilson*

G.1.1 Introduction

Viral meningitis, also known as aseptic or nonpurulent meningitis, is an infection of the meninges that may result in severe systemic disease and neurological damage, particularly in the very young (Rotbart 2000; ProMED-mail 2003). Viral meningitis results in an estimated 26 to 42 thousand hospitalizations in the U.S. each year (CDC 2003). The incidence of this and other diseases can be successfully understood and controlled by examining cases in terms of person, place, and time and exploring spatio-temporal patterns. Areas with high incidence may be targeted for heightened surveillance, education, and prevention efforts. In this contribution, we applied spatial analytical techniques to investigate viral meningitis incidence in Michigan and clarify disease patterns. Specifically, viral meningitis cases from 1993 to 2001 were analyzed using standard epidemiological methods, mapped with a GISystem, and then further analyzed using spatial and temporal cluster statistics.

Enteroviruses are the most common cause of viral meningitis in the U.S., accounting for 85–95 percent of all cases with an identified cause (Rotbart 2000). A total of 64 enterovirus serotypes are recognized, including 61 non-polio enteroviruses (CDC 2002). Of these, echoviruses (types 2, 5, 6, 7, 9, 10, 11, 14, 18, and 30) cause about half of cases, and coxsackieviruses (group B, types 1–6) are responsible for another third of cases (Chin 2000). Sporadic disease may be due to coxsackievirus group A, arboviruses, measles virus, herpes simplex virus, varicella virus, lymphocytic choriomeningitis virus, adenovirus, and others (Chin 2000). Outbreaks can be caused by any one of these viruses or a combination thereof, and laboratory typing is seldom routine, so the causes behind periods of elevated incidence are often left unexplained. Infection is common although clinical illness is rare, as fewer than one in 500 enterovirus infections actually results in viral meningitis (CDC 2000a).

Enteroviruses causing viral meningitis are transmitted via various routes, including fecal-oral, water, food, air, inoculation, and blood. Fecal-oral transmission is the classic means of enterovirus spread, and is common wherever hygiene and sanitation are inadequate, as is sometimes the case among children in day-care centers. Air-borne transmission may predominate in other settings (Morens and Pallansch 1995). Common symptoms include malaise, fever, headache, stiff neck, abdominal pain, nausea, and vomiting. Patients may also develop sore throat, chest pain, photophobia, or a maculopapular rash (Huether and McCance 2000). Cerebrospinal fluid of patients with viral meningitis is characterized by the absence of bacteria that cause these symptoms, and typically the presence of signs such as pleocytosis, increased protein, and normal sugar. Active illness usually lasts less than ten days, and recovery is often complete within a few weeks, although irritability and fatigue may persist (Chin 2000). Treatment is symptomatic (CDC 2003). Five to ten percent of patients experience complications including febrile disorders, movement disorders, lethargy, complex seizures, or coma (Modlin 2004).

Viral meningitis cases are highly seasonal, although the mechanisms underlying this observation are unknown. In temperate climates, enteroviral disease incidence generally increases during late summer and early autumn (Chin 2000). This is potentially because fecal-oral transmission of enteroviruses is aided by warm weather and sparse clothing, especially among children (Rotbart 2000). The predominant serotypes of enteroviruses circulating in a community during a given year cycle with varying periodicity. Outbreaks coincide with the availability of new susceptible hosts (e.g., children not previously exposed to a particular virus). Serotypes responsible for elevated incidence frequently cause disease in the young. A serotype that was absent from a community for several years and reintroduced will affect older children and adults who have never been exposed (Rotbart 2000). Thus, analyses of the temporal pattern of incidence should shed light on risk factors for symptomatic disease.

Young age and immunodeficiency are considered risk factors for viral meningitis. For unknown reasons, higher infection rates appear to occur among males, as well as among those who are of lower socioeconomic status and live in crowded areas, due to the increased potential for fecal contamination (Modlin 2004; Morens and Pallansch 1995).

Viral meningitis ceased to be a nationally notifiable disease in 1995 (CDC 2001), thereby limiting investigations into its national distribution. United States case data through 1994 (CDC 1995) indicate that viral meningitis is present throughout the country, but the incidence in Michigan historically far exceeded that of most of the nation (Fig. G.1.1). The Michigan Department of Community Health (MDCH) continues to collect data on viral meningitis cases.

Various investigations have applied spatial analytic techniques to other diseases at the county-level in efforts to improve understanding of transmission dynamics in a state. For example, methods such as the Ederer-Myers-Mantel and

Moran's *I* procedures have been applied to human brucellosis in California, identifying Hispanics as the subpopulation most at risk for infection (Fosgate et al. 2002). In addition, the spatial autocorrelation of Lyme disease has been investigated in New York state at the county scale, helping to determine a characteristic spatial scale for infection patterns, to quantify the extent and intensity of clustering about disease foci, and to suggest a scale for control efforts (Glavanakov et al. 2001). Such approaches, if applied to ongoing data collection, could help lay the foundation for a Space-Time Information System (STIS) that jointly considers the spatial and temporal patterns of infectious disease, aids in the determination of etiology and risk factors, guides outbreak investigations, and, ultimately, may reduce disease incidence. Accordingly, we undertook a retrospective spatio-temporal analysis of viral meningitis in Michigan to better understand historical disease patterns.

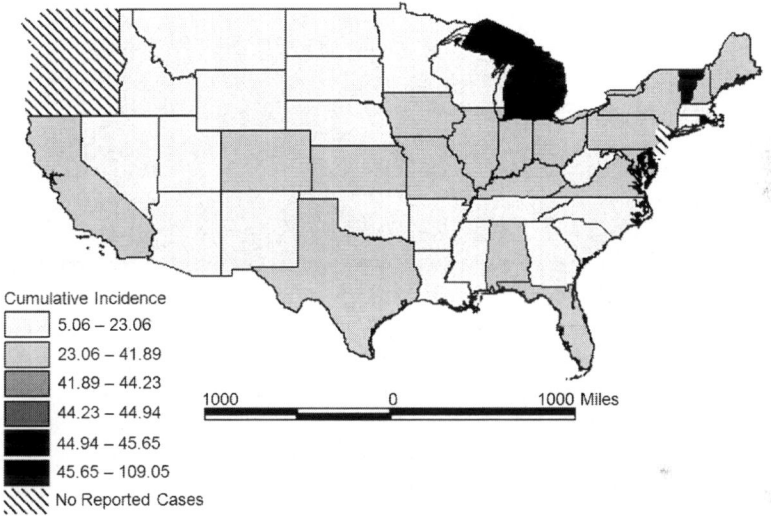

Fig. G.1.1. Cumulative incidence per 100,000 of viral meningitis by state, continental United States, 1989-1994 (CDC 1995)

G.1.2 Materials and methods

Study site and population. Roughly ten million people lived in Michigan in the year 2000. The state's 56,804 square miles are divided by the Great Lakes into two main landmasses: the sparsely populated Upper Peninsula (UP) has a humid climate with long severe winters and short mild summers, while Lower Michigan has more moderate weather, with higher summer temperatures, less severe winter temperatures, and less snowfall. Generally, the Great Lakes serve to moderate cold temperatures (UniversalMAP 2000). Michigan's population is most dense in the southeast; 9.6 percent of the state's residents live in Detroit (U.S. Census). For this study, age-specific population counts for each Michigan county were drawn from the 1990 U.S. Census, and total counts for each county were obtained from the 2000 U.S. Census. Annual viral meningitis incidence in children less than ten years old ($n = 3,031$ cases) was compared using 1990 versus intercensal county-specific population denominators, and the incidences were highly correlated ($r^2 = 0.99$, $p < 0.0001$). Based on these data, it was determined that age-adjusting incidence rates to the 1990 Census population was sufficient.

Surveillance data. The MDCH routinely collects data on all diagnoses of reportable diseases or conditions, as defined by the Michigan Public Health Code (Michigan Compiled Laws 1978). The MDCH provided information on 8,803 viral meningitis cases reported though the state's passive disease surveillance system from 1993 to 2001. A confirmed viral meningitis case was defined as 'a clinically compatible case diagnosed by a physician as aseptic meningitis, with no laboratory evidence of bacterial or fungal meningitis' (CDC 1997, p.44). This is essentially a diagnosis of exclusion. The only city in Michigan that reported cases independently from its county was Detroit, which is within Wayne County. Eleven of 83 counties did not report any cases over the study period, seven of which are in the UP and four in Lower Michigan. Reporting was nearly complete for most case characteristics, including date of report (100 percent), county of residence (100 percent), age (99.9 percent), and sex (99.2 percent), but less so for race (69.4 percent).

Prior to the time frame of this study, total annual counts of viral meningitis in Michigan with no individual-level case information were available from MDCH (1984–1992). To enable a comparison of incidence with nearby states, annual counts were obtained for Illinois from 1990– 2000 (Illinois Department of Public Health website: http://www.idph.state.il.us/health/infect/communicabledisease.htm) and for Indiana from 1990– 2001 (personal communication via electronic mail, Julia Butwin, March 2002).

Virus type identification was obtained through the MDCH Bureau of Laboratories for isolates of some viral meningitis cases during June–December 2001. Samples or specimens from 114 such patients were obtained from cerebrospinal fluid (most commonly), bronchial fluid, throat, stool, or vagina. Antibodies against enteroviral proteins were used as presumptive evidence of the presence of enteroviruses in the isolates. Typing was performed using monoclonal fluorescent antibody testing and pooled antisera. Isolates not typeable by these methods were sent to the Centers for Disease Control and Prevention (CDC) for PCR sequencing. An enterovirus subtype was identified from 43 (37.7 percent) of these samples.

Statistical analyses. Cumulative incidences and relative risks were calculated using SAS for Windows v8 (SAS Institute, Cary, NC). Disease incidence was mapped using ArcView GIS v3.2a (Environmental Systems Research Institute, Inc., Redlands, CA). Time series analysis techniques were applied to the data using R v1.7.1 (R Foundation for Statistical Computing, Vienna, Austria). Kulldorff's Scan test (Kulldorff 1997) was applied to monthly, county-level incidence using ClusterSeer (TerraSeer Inc., Ann Arbor, MI) to search for spatio-temporal clusters. All non-reporting counties were excluded from the spatio-temporal analysis, as were the counties in the UP because Kulldorff's Scan test could not appropriately adjust for the space across the water of the Great Lakes within which no one was resident. A total of 8,743 cases remained for inclusion in the spatio-temporal analysis.

G.1.3 Results

Demographics of cases. A total of 8,803 viral meningitis cases were reported during the nine years of study. Overall, 4,402 (50.4 percent) cases were female and 4,329 (49.6 percent) were male, proportions that were representative of the population of Michigan in 2000 (U.S. Census). Cases ranged in age from less than one to 93 years (mean = 21.8 years, SD = 31.3 years). The median age of cases (18.0 years) was less than that of the Michigan population in 2000 (35.5 years, U.S. Census). Indeed, 1,345 cases (15.2 percent) involved infants less than one year old, representing a cumulative nine-year incidence of 1,040.6 per 100,000. Cumulative incidence for the rest of the population was 81.4 per 100,000 (relative risk [RR] for less than one year old is equal to 12.9; 95 percent confidence interval [CI] 12.2–13.7).

Blacks represented 14.2 percent of Michigan's population in 2000 (U.S. Census), yet they were disproportionately represented among viral meningitis cases, comprising 2,044 (33.5 percent) of cases. The cumulative incidence of viral meningitis among Whites was 51.1 per 100,000, but for Blacks it was 158.2 per 100,000 (RR for Blacks is equal to 3.1; 95 percent CI 2.9–3.3).

Temporal trends. Cases were seasonally distributed, with more than two-thirds reported during July through October. The cumulative statewide incidence of viral meningitis during these four months, summed over the nine study years and calculated using county-specific 2000 Census population denominators, was 60.2 cases per 100,000, compared with 28.4 cases per 100,000 from November through June (RR of July–October is equal to 2.1; 95 percent CI is equal to 2.0 - 2.2). Despite strong annual seasonality, considerable inter-annual variability in epidemic magnitude was apparent (Fig. G.1.2(a)).

Periodic extremes in reported cases of viral meningitis occurred roughly every three years for this time series. The largest epidemics were in 1995, 1998, and (most strikingly) 2001. The autocorrelation function (ACF) revealed a strong autocorrelation of 0.43 at three years and 0.16 at its multiple of six years (Fig. G.1.2(b)). However, peak epidemic years in Michigan prior to the beginning of this study in 1993 were 1987 and 1990, suggesting that a regular three-year cycle may vary somewhat (data not shown). The ACF pattern did not change appreciably when applied to high-risk subsets of the case population such as children less than ten years-olds and Blacks (results not shown).

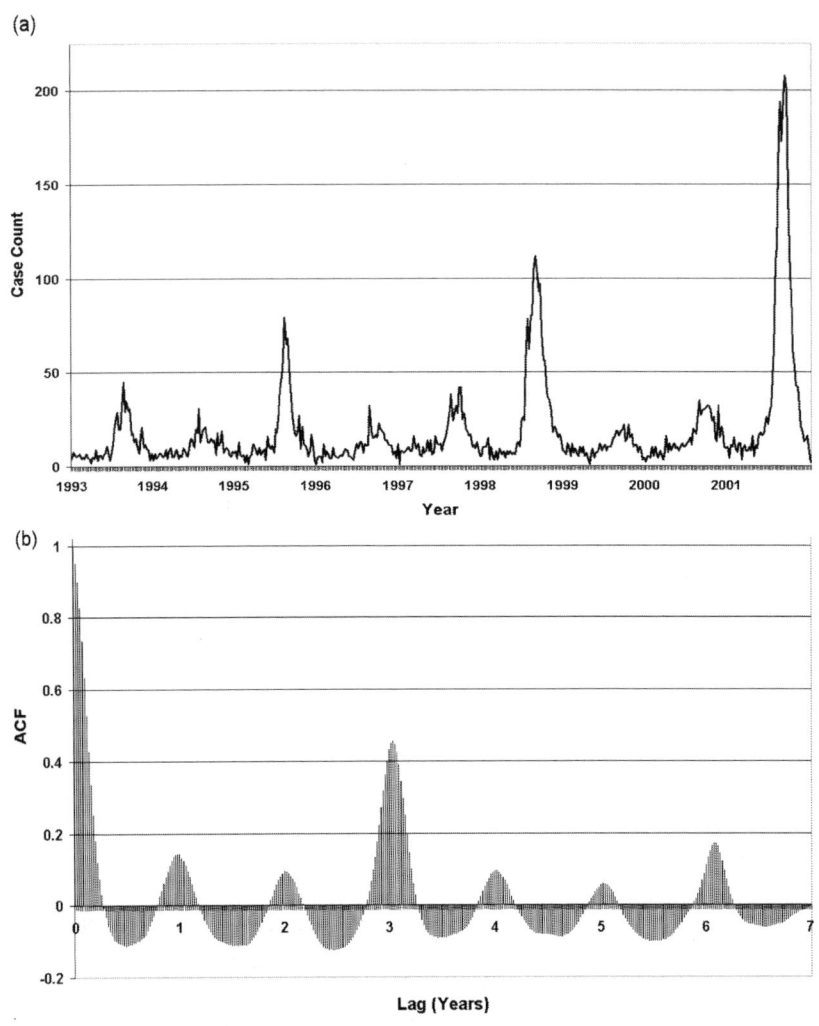

Fig. G.1.2. (a) Weekly number of viral meningitis cases in Michigan, 1993-2001; (b) the autocorrelation function (ACF) from a zero to a seven-year lag

Spatial trends. Age-adjusted cumulative incidence tended to be highest in southern Michigan, particularly in the southeast, near and around Detroit (154.9 cases per 100,000; Fig. G.1.3). Other areas of notably high age-adjusted cumulative incidence relative to the surrounding area include the counties of Muskegon (211.8 cases per 100,000) and Newaygo (121.8 cases per 100,000) in the western central part of Lower Michigan, Kalamazoo in the southwest (185.8 cases per 100,000), and Marquette in the UP (57.3 cases per 100,000).

County-specific cumulative incidence (Fig. G.1.3) was similar to the spatial pattern of county-level population density (i.e., number of residents / sq. mi.; results not shown). The Pearson correlation coefficient for cumulative incidence and population density was positive and significant ($r = 0.45, p = 0.0003$). This suggests that counties with greater human density tended to have more transmission.

To determine whether there was a spatial pattern in inter-annual variability of incidence, the coefficient of variation (CV denotes standard deviation/mean) for nine years of annual incidence in each county was mapped (Fig. G.1.4), with the understanding that counties with low incidence may yield unstable CV estimates. Interestingly, Figs. G.1.3 and G.1.4 show essentially opposite patterns, such that the areas of highest incidence (generally Lower Michigan) were less variable, whereas the areas of lower incidence (northern Lower Michigan and the UP) showed greater year-to-year variability.

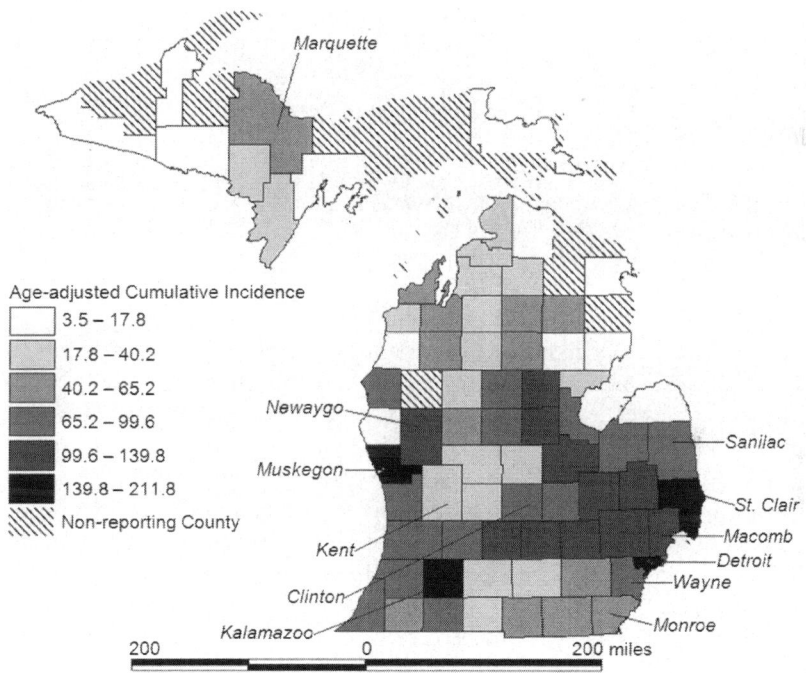

Fig. G.1.3. Cumulative incidence per 100,000 of viral meningitis in Michigan by county, 1993-2001, age-adjusted to the 1990 population. Counties mentioned in text are labeled

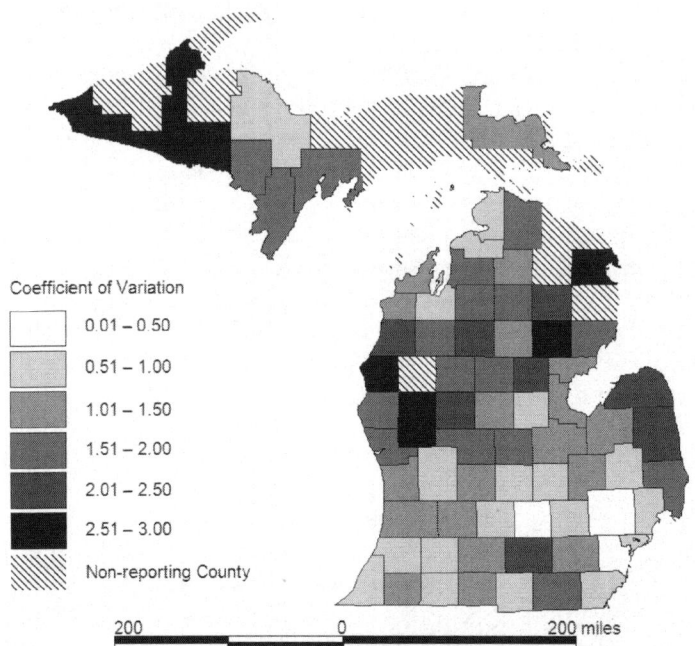

Fig. G.1.4. Coefficient of variation (CV) of annual incidence of viral meningitis cases in Michigan by county, 1993-2001. The CV for each county was calculated by dividing the standard deviation of that county's annual incidence by its mean annual incidence

Spatio-temporal trends. Kulldorff's Scan test identified the three most likely spatio-temporal clusters in the 65 reporting sites of Lower Michigan (Fig. G.1.5). The first most likely cluster was a 42-county region centered on Kent County during four months in late summer/early fall 2001. The second most likely cluster was located around St. Clair County during the same period in 1998, and the third most likely was centered on Monroe County in that period in 2001. Each space-time cluster was significant at $p = 0.01$. This test was also run separately for Blacks and for children aged less than ten years old. For Blacks, there were only two significant spatio-temporal clusters ($p = 0.01$ for both): a 37-county region centered on Clinton County during July–September 2001, and solely St. Clair County from June–September 1998.

The most likely cluster for less than ten year-olds was the same as that found for all cases (42-county region centered on Kent County), but with a slight temporal shift to July–September 2001 ($p = 0.01$). The second and third most likely clusters encompassed five counties plus Detroit centered on Macomb County during July–September 1998 ($p = 0.01$) and solely Sanilac County during July–September 1998 ($p = 0.03$).

Within Cluster #1 (Fig. G.1.5), children less than ten years old comprised 533 (42 percent) of 1,268 cases. Of all 8,743 cases included in the spatio-temporal analysis, 3,014 (34 percent) cases were in this age group. Thus, children less than ten years old were at significantly greater risk for viral meningitis in the first most likely spatio-temporal cluster than they were over the entire region and study period (chi-square for specified proportions is equal to 36.5, df = 1, $p < 0.0001$). This information could be beneficial in determining the reasons behind the increased number of cases in this cluster.

Annual incidences of viral meningitis in Michigan and in Illinois were strongly correlated from 1990–2000 ($r = 0.88$, $p = 0.0004$). Both states experienced maximum incidence over this period in 1998 (yet incidence in Michigan was nearly two times greater than that in Illinois). The correlation between annual incidences of Michigan and the southern border state of Indiana was nonsignificant ($r = 0.49$, $p = 0.11$).

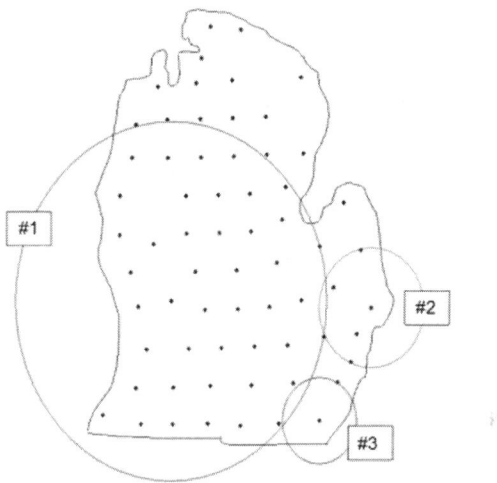

Fig. G.1.5. Spatio-temporal clusters of viral meningitis cases by county in Lower Michigan, 1993-2001. The three most-likely overall clusters as determined by Kulldorff's Scan test include [#1] a 42-county region during July to October 2001, [#2] a five-county region and Detroit from July to October 1998, and [#3] a two-county cluster from August to October 2001

Laboratory results. Of the 43 subtyped enteroviral isolates obtained from viral meningitis patients in June-December 2001, 28 (65.1 percent) were from children less than ten years old. There were differences between this age group and cases at least ten years old in the proportions of viral subtypes isolated (Table G.1.1). Although not representative of all viral meningitis infections, these data suggest that younger cases were more likely to be infected with coxsackie B-3 or echovirus 13, while only older cases harbored echovirus 30. This apparent pattern might be explained by true differences in enterovirus activity, age-specific clinical syndromes of specific enteroviruses, or differences in ease of isolation (Strikas et al. 1986).

Table G.1.1. Relative presence of identified causative viruses associated with viral meningitis cases in Michigan, June to December 2001

Virus	Number of isolates (%)	
	Cases < 10 years	Cases ≥ 10 years
Coxsackie B-2	3 (10.7%)	1 (6.7%)
Coxsackie B-3	8 (28.6%)	1 (6.7%)
Coxsackie B-4	0 (0.0%)	1 (6.7%)
Echovirus 4	3 (10.7%)	1 (6.7%)
Echovirus 6	1 (3.6%)	1 (6.7%)
Echovirus 9	0 (0.0%)	1 (6.7%)
Echovirus 11	4 (14.3%)	1 (6.7%)
Echovirus 13	9 (32.1%)	3 (20.0%)
Echovirus 30	0 (0.0%)	5 (33.3%)
Total	28 (100.0%)	15 (100.0%)

G.1.4 Concluding remarks

To our knowledge, this represents the first study that examines spatio-temporal patterns of viral meningitis cases at a resolution as fine as county-level. Our results confirm the existence of certain high-risk groups and disease clustering in both space and time within Michigan.

Several findings from previous studies of viral meningitis are supported, including risk factors related to seasonality, age, race, and crowding. Temporally, our results are consistent with the seasonal pattern of viral meningitis typically reported from temperate climates, that is, a peak in the summer and autumn (Rotbart 2000; CDC 2000a, 2001). In a study of enteroviral isolates from across the U.S. collected through the Enterovirus Surveillance Program during 1970–1979, 82 percent of all enteroviral isolates were submitted from June to October, and the average number of isolates for each of these months was 6.6 times higher than the monthly average for the other seven months (Moore 1982). Generally, viruses carried in the digestive tract tend to peak during July and August (ProMED-mail 2003).

Young children are the primary reservoir of human enteroviruses (Moore 1982), with infants under one year of age accounting for 45 percent of all reported U.S. enteroviral isolates from 1997–1999 (CDC 2000b). Thus, youth is a predisposing factor for viral meningitis, with several studies showing age-specific incidence to be greater among young infants and school-aged children five to ten years old (Rotbart 2000). The annual incidence of viral meningitis in children under one year-old in a large Finnish cohort was 219/100,000 children (Rotbart

2000). In our study, children less than five years old constituted 21 percent of all cases, and 23 percent of all cases were children aged five to 14 years.

Viral meningitis incidence among Nonwhites in the United States has previously been shown to be higher than that for Whites, a finding attributed to socioeconomic inequality and other confounding variables (Morens and Pallansch 1995). We too found elevated relative risk for viral meningitis among Blacks in Michigan. In other studies, infection rates have been reported to be higher in areas of crowding and among people of lower SES (Rotbart 2000), consistent with our finding of a significant correlation between cumulative incidence and population density. There was high viral meningitis incidence in Detroit, an area with higher population density, a large proportion of Blacks, and many residents of low SES. Being Black and of low SES are strongly autocorrelated in the Detroit area (Zenk 2005), and any separate effects they may have on incidence cannot be teased apart using our study design. These comparisons suggest that our case reports were consistent with those from other locations.

Our study departs from previous reports that found proportionately more enterovirus cases among males than females. Others have noted a male-to-female sex ratio ranging from 1.3:1 (Rotbart 1995) to 2.5:1 (Morens and Pallansch 1995). In a viral meningitis outbreak in Romania, males were more often found to be ill ($p = 0.04$) (CDC 2000a). However, in Michigan, males and females were diagnosed with viral meningitis in equal proportions.

Our finding that there may be a roughly triennial cycle of elevated incidence has some support in the literature. Interannual trends have been reported in poliovirus cases every three years in southern India (Morens and Pallansch 1995). Highly transmissible agents may rapidly circulate in a population, quickly building high levels of population immunity. The observed pattern may also reflect the different periodicities of predominant enteroviral serotypes (CDC 2002). Cyclic epidemics may also occur as the population reaches a threshold level of newly introduced susceptibles (e.g., very young children or immigrants), who reduce herd immunity and permit reappearance of intense transmission. This may be consistent with the large 2001 spatio-temporal cluster of viral meningitis cases (#1, Fig. G.1.5) that was dominated by children less than ten years of age. Further studies of seasonal trends could improve understanding of factors that increase risk, hence providing an early warning of intense transmission periods.

Regional environmental factors also may play a role in driving incidence cycles, as evidenced by the strong correlation in incidence between Michigan and Illinois. Strikas and colleagues (1986) grouped these two states together in their analysis of enterovirus isolates and identified a regional pattern of incidence. Other studies have reported geographical clustering of enteroviral strains in the U.S. and in Belgium (CDC 2003). However, any regional drivers are likely to be influenced by local factors, as the correlation between the annual incidences of Michigan and Indiana during the same time period was poor. Latitude, weather effects, and other environmental differences may be at play.

The most commonly identified virus isolated during the 2001 outbreak in Michigan was echovirus 13 (Table G.1.1). This outbreak occurred in the context of increased detection of viral meningitis cases throughout the U.S. and across the globe. Michigan, along with Illinois, Tennessee, and Wisconsin, reported the most echovirus 13 isolates in 2001 in the country (CDC 2002), and epidemics of echovirus 13-associated viral meningitis took place in Louisiana, Mississippi, Montana, and Tennessee from April to July 2001, prompting outbreak investigations in these states (CDC 2001). Indeed, widespread circulation of echovirus 13 in the U.S. occurred for the first time in 2001 (CDC 2002, Noah and Reid 2002). Echovirus 13 was also very prevalent during 2000–2001 in Australia (ProMED-mail 2001, Quirk 2001) and Europe (Noah and Reid 2002).

Geography is represented using centroids when applying Kulldorff's Scan test, in this case the centroids of Michigan counties. The presence of the Great Lakes therefore prevented the scan statistic from assessing spatio-temporal clustering for the state as a whole. Furthermore, the geographic template employed in this test is a space-time cylinder with a maximum radius and duration. It is consequently capable of reporting clusters only as space-time cylinders, despite whatever the true shape of a particular cluster may be. The test could not precisely pinpoint, for instance, a cone-shaped spatio-temporal cluster, with an outbreak spreading out from an initial focus. If there is not enough power to detect a cluster of this shape, the spatial scan may continue to expand until it contains a large enough number of cases to declare a spurious cluster. This may explain why Cluster #1 in this study encompassed such a large number of counties, and implies that, as always, caution is necessary when drawing statistical inferences. Kulldorff's Scan test also assumes that risk within the space-time cylinder is uniform despite the potential for variability in disease incidence rates within a cluster. It is therefore possible for the test to return false positives, as geographic areas with comparatively low incidence rates may be declared part of a cluster with high rates (Jacquez 2007). Other tests for space-time clustering could also be included in a future STIS for viral meningitis surveillance, to assess the sensitivity of identified clusters to the assumptions of the Kulldorff's Scan test.

Our study analyzed case data from a passive surveillance system and may not be a complete evaluation of all Michigan residents who had viral meningitis during the study period. It has been suggested that viral meningitis cases occur at more than ten times the number reported to the CDC (Rotbart 1995). There could have been reporting variability among counties or regions due to unequal human and monetary resources or heightened concern and greater efforts at detection. For example, a ten-county district in the central region of Michigan (District #10) was known to have intensified viral meningitis case finding during 2001 (personal communication via electronic mail, Dr. Patricia Somsel, August 2002). Another possible source of reporting bias might have involved the August 2001 detection of West Nile Virus (WNV) in birds in Michigan (Calisher 2001), which may have sensitized physicians to diagnose and report viral meningitis. WNV and enteroviruses have a similar seasonal pattern and both are associated with neurological signs and symptoms. However, WNV-associated meningitis tends to occur in adults (median age 46 years, CDC 2003), yet children are at highest risk for enteroviral meningitis. Furthermore, temporal analysis was based on date of report, not date of disease onset, and there may have been variation across counties in the timeliness of record submission. Since the residence of each case was defined at the spatial scale of county, we were unable to track transmission or to draw conclusions about individual risk factors.

Despite these limitations, we identified groups at elevated risk and larger spatio-temporal patterns that would be useful in hypothesis generation for future studies on epidemic patterns. However, our results suggest that improved knowledge of the causative agent may be essential to understanding transmission dynamics. To that end, more reliable and representative laboratory identification of specific pathogens is needed to determine whether outbreaks are caused by one or more viral strains. Indeed, the MDCH is embarking on an enterovirus laboratory surveillance program in reaction to recent outbreaks and this study's findings, in order to better characterize future enterovirus epidemiology. Clinical laboratories throughout the state are being encouraged to submit samples for typing on a weekly basis. Evidence suggests that numerous viruses and serotypes commonly cocirculate (CDC 2001) and that outbreaks of specific echovirus types are more widespread than previously believed (Noah and Reid 2002). Temporal trends in viral meningitis incidence stratified by virus type would have more interpretability than undifferentiated, amalgamated disease incidence.

The identification of spatial and temporal clusters in this study should encourage further research aimed at identifying local and socio-demographic influences on infectious disease agent transmission. We determined that counties with high incidence were consistently high (i.e., low coefficient of variation), suggesting intransient risk factors within these sites. There are no specific prevention or control measures known to reduce the transmission of non-polio enteroviruses beyond good hygienic practices, including hand washing, disinfecting contaminated surfaces, and not sharing utensils or drinking containers (CDC 2003). Lacking specific options for preventing enterovirus epidemics, the incorporation of data and methods such as are described here into a STIS could prospectively focus surveillance activities and allocate limited resources for response to the districts with greatest need. Clinicians could thereby be better informed and equipped to assist health departments in responding to outbreaks by emphasizing preventive measures (Noah 1989). Spatial analytic techniques provide public health practitioners an opportunity to enhance early warning and response activities for viral meningitis and other diseases.

Acknowledgments. The authors gratefully acknowledge Dr. Duane Newton, former manager of the Virology Section, Michigan Department of Community Health Bureau of Laboratories, for his provision of data on viral isolates. We also appreciate the contributions of

Drs. Leah Estberg and Dunrie Greiling for ClusterSeer support, Scott Swan for his assistance with ArcView GIS, and Dr. Edward L. Ionides for his advice on time series analysis. We are grateful to Dr. Pierre Goovaerts and two anonymous reviewers for their insightful comments on an earlier draft of this chapter. This project was supported in part by a grant from the National Oceanic and Atmospheric Administration's Joint Program on Climate Variability and Human Health, a consortium including the EPA, NASA, NSF, and EPRI (NA16GP2361) to MLW. This chapter was lightly edited from Greene et al. (2005).

References

Calisher C (2001) West Nile virus surveillance 2001 – USA. ProMED-mail 20010825.2017. www.promedmail.org

CDC (1995) Summary of notifiable diseases, United States, 1994. Morbidity and Mortality Weekly Report 43(53)

CDC (1997) Case definitions for infectious conditions under public health surveillance. MMWR Recommendations and Reports, 46(RR-10):1-55

CDC (2000a) Outbreak of aseptic meningitis associated with multiple enterovirus serotypes – Romania, 1999. Morbidity and Mortality Weekly Report 49(29):669-671

CDC (2000b) Enterovirus surveillance - United States, 1997-1999. Morbidity and Mortality Weekly Report 49(40):913-916

CDC (2001) Echovirus type 13 – United States, 2001. Morbidity and Mortality Weekly Report 50:777-780

CDC (2002) Enterovirus surveillance – United States, 2000-2001. Morbidity and Mortality Weekly Report 51(46):1047-1049

CDC (2003) Outbreaks of aseptic meningitis associated with echoviruses 9 and 30 and preliminary surveillance reports on enterovirus activity – United States, 2003. Morbidity and Mortality Weekly Report 52(32):761-764

Chin J (ed) (2000) Control of communicable diseases manual. American Public Health Association, Washington, DC

Fosgate GT, Carpenter TE, Chomel BB, Case JT, DeBess EE, Reilly KF (2002) Time-space clustering of human brucellosis, California, 1973–1992. Emerging Infectious Diseases 8(7):672-678

Glavanakov S, White DJ, Caraco T, Lapenis A, Robinson GR, Szymanski BK, Maniatty WA (2001) Lyme disease in New York State: Spatial pattern at a regional scale. Am J Trop Med Hyg 65(5):538-545

Greene SK, Schmidt MA, Stobierski MG, Wilson ML (2005) Spatio-temporal pattern of viral meningitis in Michigan, 1993-2001. J Geogr Syst 7(1):85-99

Huether S, McCance K (2000) Understanding pathophysiology. Mosby Inc, St. Louis

Kulldorff M (1997) A spatial scan statistic. Communications in Statistics – Theory and Methods 26:1481-1496

Jacquez GM (2007) Spatial cluster analysis. In Wilson JP, Fotheringham AS (eds) The handbook of geographic information science. Blackwell, London, pp.395-416

Michigan Compiled Laws 333.5111 (1978). Public Acts, 368

Modlin JF (2004) Enteroviruses. In Gershon AA, Hotez PJ, Katz SL (eds) Krugman's infectious diseases of children (11th edition). Mosby, Philadelphia, pp.117-133

Moore M (1982) Enteroviral disease in the United States, 1970–1979. J Infect Dis 146(1):103-107

Morens D, Pallansch M (1995) Epidemiology. In Rotbart HA (ed) Human enterovirus infections. ASM Press, Washington, DC, pp.3-24
Noah N (1989) Cyclical patterns and predictability in infection. Epidemiology and Infection 102(2):175-190
Noah N, Reid F (2002) Recent increases in incidence of echoviruses 13 and 30 around Europe. EuroSurveillance Weekly 6(7). Available at http://www.eurosurveillance.org/ ViewArticle.aspx?ArticleId=2029
ProMED-mail (2001) Meningitis, echoviruses 13 & 30 - Australia (Western). www.promedmail.org
ProMED-mail (2003) Meningitis, Viral – USA (Georgia). www.promedmail.org
Quirk M (2001) Echovirus to be considered in meningitis diagnosis. Lancet Infect Dis 1:220
Rotbart HA (1995) Meningitis and encephalitis. In Rotbart HA (ed) Human enterovirus infections. ASM Press, Washington, DC, pp.271-290
Rotbart HA (2000) Viral meningitis. Sem Neurol 20(3):277-292
Strikas RA, Anderson LJ, Parker RA (1986) Temporal and geographic patterns of isolates of nonpolio enterovirus in the United States, 1970–1983. J Infect Dis 153(2):346-351
UniversalMAP (2000) Michigan county atlas. Williamston, MI
U.S. Census, http://www.census.gov
Zenk SN, Schulz AJ, Israel BA, James SA, Bao S, Wilson ML (2005) Neighborhood racial composition, neighborhood poverty, and the spatial accessibility of supermarkets in metropolitan Detroit. Am J Publ Health 95(4):660-667

G.2 Space-Time Visualization and Analysis in the Cancer Atlas Viewer

Dunrie A. Greiling, Geoffrey M. Jacquez, Andrew M. Kaufmann and *Robert G. Rommel*

G.2.1 Introduction

For chronic diseases such as cancer, which have long latency and can display significant spatial pattern, atlases of health data are an important resource. Atlases allow researchers and the public alike to formulate and evaluate hypotheses about geographic variation, such as clustering (Jacquez 1998; Moore and Carpenter 1999; Rushton et al. 2000; Jacquez and Greiling 2003). The identification of spatial pattern in mortality has stimulated research to elucidate causative relationships such as the association between snuff dipping and oral cancer (Winn et al. 1981); the association between shipyard asbestos exposure and lung cancer (Blot et al. 1980) and others.

Mortality atlases are available in print form, such as the *Atlas of United States Mortality* (Pickle et al. 1996) and the *Atlas of Cancer Mortality of the United States 1950-1994* (Devesa et al. 1999) and in web format (see Table G.2.1). Web atlases are increasingly available at the state and national levels as the technology for online mapping has matured. Both print and web atlases provide a considerable amount of data and statistics in an easy to understand visual format, but online atlases offer a level of interactivity not available in printed books, as the user can change the colors, zoom and pan, click through to data tables, customize the maps to address a question or purpose not envisioned by the print map's creators, and share maps and collaborate (Gao et al. 2008).

While online atlases provide greater flexibility and customization than print atlases, their use may be hindered by performance limitations that result from Internet communication between the user's computer and the mapping engine. An alternative to web mapping is to download a local version of the data for mapping and interaction on a desktop computer. Once downloading has occurred, the time for map rendering is minimal and the user can explore the data more quickly. The

Cancer Atlas Viewer is an example of a Space-Time Intelligence System (STIS) (described in Chapter A.6). This architecture supplies animated, interactive maps that allow researchers to visualize and explore the dynamic nature of cancer mortality patterns.

In this chapter, we demonstrate the Cancer Atlas Viewer by exploring colon cancer patterns for African-American and white females and males using NCI (National Cancer Institut) data. Among cancers, the highest mortality for men is from lung, prostate, and colon cancers respectively; for women it is lung, breast, and colon cancers, all of which demonstrate spatial patterns (Devesa et al. 1999).

Table G.2.1. A listing of a few online atlas projects

Initiative	Description
Washington State's Epidemiologic Querying and Mapping System (EpiQMS) http://app2.health.state.pa.us/epiqms/	Death certificate data (cause of death) by county along with population information. Users can map the data or prepare graphs, or view tables.
New York State's Cancer Surveillance Improvement Initiative (NY CSII) http://www.health.state.ny.us/nysdoh/cancer/csii/nyscsii.htm	Information on breast, colorectal, lung and prostate cancer diagnoses by ZIP code in New York State. Users can view prepared PDF maps, or view the data for individual ZIP codes by county.
Reproductive Health Atlas http://www.cdc.gov/reproductivehealth/GIS Atlas	Information on variables such as infant mortality, pregnancy outcomes, infant health, and maternal risks by demographic groups and different geographic aggregation. No data is currently available, but the website talks about distributing starter shapefiles for policymakers and service providers.
Cancer Mortality Maps and Graphs (NCI) http://www3.cancer.gov/atlasplus/	Data on mortality for 40 site-specific cancers by county, state economic area, and states from 1950-1994. The data can be mapped online or downloaded for viewing and manipulation.

One challenge in exploring patterns for multiple groups is that there are low populations of African-Americans in rural areas of the midwest and western states. Because of low population numbers, the counts used to calculate the mortality rates are based on small samples and are therefore unstable, subject to fluctuations that may be due to chance. The NCI print and online atlas masks data based on few counts (fewer than six deaths in the five year time period). We focus on the southeastern United States and Gulf Coast, including part of eastern Texas, Mississippi, Louisiana, Alabama, Georgia, Florida, South Carolina, and North Carolina. This region has high enough populations of African-Americans to avoid most rural areas becoming masked out, as geographies with a lot of missing (masked) data are unsuitable for spatial analysis. The southeastern US has been identified as a region of persistently high mortality (Cossman et al. 2003), though it is not the highest mortality region for colon cancer in the US. For colon cancer mortality rates, the southeast is exceeded by the northeastern states (Devesa et al. 1999).

Specifically, we assess the spatial patterns of mortality from colon cancer in the Southeast, using descriptive data visualization of the cancer data for state economic areas (SEAs) and tools and spatial autocorrelation measures as discussed in Chapter B.3. Emphasis is laid on assessing the changes in spatial patterns by examining trends in the local Moran and the Getis-Ord statistics, and the persistence of patterns over time.

G.2.2 Data and methods

Data description. The National Cancer Institute has released age-adjusted cancer mortality rates for US counties, state economic areas (SEAs), and states, for 40 site-specific cancers, four groups (African-American females, African-American males, white females, white males), and for several time periods from 1950-1994. The rates are the number of cancers per 100,000 person-years, age-adjusted to the 1970 US population standard age distribution. We focus here on the SEA datasets. SEAs are aggregations of counties within state boundaries that were similar according to 1960 socioeconomic data (US Bureau of the Census 1966). The SEA data has better temporal resolution than the counties (counties have 20 or 25 year times only) and finer spatial resolution than the state datasets. Data for African-American males and females starts in 1970, while data for white males and females begins in 1950. More information on this data is available from the National Cancer Institute cancer mortality maps and graphs website (http://www3.cancer.gov/atlasplus/) and in the printed atlas (Devesa et al. 1999). The National Atlas has compiled metadata for this dataset, available at http://www.nationalatlas.gov/mld/cancerp.html. We focus on colorectal cancer rates for African-American and white females and males for SEAs in five-year time intervals from 1970 through 1994. We use the age-adjusted rates produced by the NCI: these rates are for 100,000 person-years and are adjusted to the 1970 age-classes (Devesa et al. 1999). We repeated this analysis for the county-level rates data, but did not include it in this write-up because of space constraints. The comparison between I_i and G^* conclusions for the counties geography was similar to the SEA results we present.

Software description. The Cancer Atlas Viewer is the first implementation of the more general Space-Time Information System described in Chapter A.6. Currently, the Cancer Atlas software loads text files downloaded from the NCI website. This version of the graphical user interface works on Windows operating systems, although the underlying architecture is cross-platform and can be compiled for other operating systems. At the time of this writing, the Cancer Atlas Viewer software can be accessed through the Internet (http://biomedware.com/software/Atlas _download.html) free of charge. A more general version of this software that accepts other datasets is available from TerraSeer, called the TerraSeer Space-Time Intelligence System (STIS).

The Cancer Atlas Viewer and STIS software contain several statistical methods, from data transformations, such as the Z-score standardization, over the creation of difference datasets, to the calculation of Moran's I, local Moran, and local G^* clustering statistics, described in Chapter B.3. The cluster statistics are evaluated with Monte Carlo randomization-based hypothesis testing.

The Cancer Atlas Viewer (and its STIS counterpart) has time as a dimension of the data. The spatial relationships among the observed objects (whether point objects or polygons) and the attribute data can be brought in as separate pieces. For instance, in the case of the NCI Atlas data, there is only one geography of the polygons (at the county, SEA, or state level) for the entire analysis. Although the outline of some US counties has changed over time, the NCI standardized it as one static geography for representation in GIS.

Z-score. The Cancer Atlas Viewer uses Z-score standardization to prepare the data for the Moran analysis. The Z-score standardizes the mortality rates by taking the observed rate, subtracting the mean rate for the entire region, and then dividing by the standard deviation. The Z-score is only one of several possible epidemiologically relevant standardizations of mortality data, including the standardized mortality rate or ratio (observed cases/expected). It is a required step for Moran's I and I_i analyses. A Z-score standardizes the mortality rate for area i, m_{it}, by its mean and standard deviation, creating a new variable $\hat{m}_{it}$

$$\hat{m}_{it} = \frac{m_{it} - \overline{m}_t}{s_{mt}} \qquad (G.2.1)$$

where $\overline{m}_t$ is the mean mortality at time t, and s_{mt} is the standard deviation at time t. After Z-score transformation, all variables in a larger dataset have equal means (transformed mean is equal to zero) and standard deviations (transformed $s = 1$), but different ranges. Negative Z-scores indicate the location is below the mean of the data, positive that it is above the mean. The magnitude of the Z-score is the distance in standard deviation units away from the mean.

Difference datasets. Cancer Atlas also calculates difference datasets, to allow the user to view change maps. Absolute change in cancer mortality, m_i for area i between times t and $t + 1$ is calculated as:

$$\Delta \alpha_{it} = m_{it+1} - m_{it} \qquad (G.2.2)$$

Significance and multiple testing. We calculate p-values for I_i and G^* using 999 conditional Monte Carlo permutations of the data values. I_i and G^* statistics calculated for a given study area are not independent of one another, and hence their p-values should be corrected for multiple testing.

Lack of independence arises in two ways: Monte Carlo distributions under the null hypothesis, and in the test statistics. The reference distributions for two different regions i and j are not independent since they will be constructed from repeated drawings from the same population. Lack of independence also arises in the test statistics for two regions that are neighbors of one another. Because they are neighbors, the local Moran statistics for i, j and k will each use the values associated with one another when calculating I_i, I_j, and I_k. The test statistics therefore are correlated, and their *p*-values should be adjusted accordingly.

We use the Simes (1986) method to adjust the *p*-value; the Simes correction is not as conservative as the Bonferroni correction. It is calculated as in Eq. (G.2.3). Assume three *p*-values p_i, p_j, and p_k – suppose they are (0.002, 0.001, 0.036). Rank the *p*-values from lowest to highest, obtaining the vector (0.001, 0.002, 0.036). We wish to calculate the 'Simed' *p*-value for $p_i = 0.002$, the second element in this vector. This is done as

$$p'_i = (n+1-a)p_i \qquad (G.2.3)$$

where n is the number of *p*-values, and a is the index (starting at one) indicating the location in the sorted vector of p_i. The Simed *p*-value is then 0.002 = 0.004.

Classification. After the software calculates the G^* and the local Moran for each location, it classifies all of the SEAs in the geography. For the local Moran analysis, it classifies all of the SEAs as being the center of low-low clusters, high-high clusters, a significant high outlier (high-low), a significant low outlier (low-high), or nonsignificant. It compares the Simed *p*-values to a pre-specified alpha level (in this case $\alpha = 0.05$) and then assigns the classes based on the sign of the local Moran (positive indicates cluster, negative indicates outlier) and its Z-score (high or low). This treatment is parallel to the treatment of the local Moran in other software products, such as ClusterSeer, the SpaceStat extension for ArcView, and GeoDa. For G^*, the software compares Simed *p*-values to a pre-specified alpha level, in this case $\alpha = 0.05$, and then assigns the classes based on the sign of the G^* (positive indicates high cluster, negative indicates low cluster).

We then assessed each set of maps for similarity of classifications and for cluster persistence. For each race-gender subgroup, we examined cluster classifications resulting from the I_i and the G^* analyses at a particular time interval (for example, African-American female mortality rate 1990-1994). We considered an I_i high-high cluster equivalent to a high G^* cluster, similarly matching an I_i low-low cluster with a low G^* cluster, and a finding of nonsignificant in both analyses matched. All matching classifications were considered concordant. Non-concordant situations occurred when the outcomes differed from this matched pairing.

For cluster persistence, we compared sets of maps over time within one classification, for example white male G^* mortality rate classes in 1985-1989 and 1990-1994. Clusters identified in the last time period could not be scored for persistence as we had no information about clustering after 1994. As most of the locations were classed as non-significant and stayed that way, we did not count transitions from the non-significant class, just transitions from a cluster class to another class (such as outlier or non-significant).

G.2.3 Results

In this Section, we describe the Atlas software and a comparison of the two clustering statistics' conclusions about the patterns in colon cancer mortality from 1970 through 1994 rates in five-year time intervals for SEAs. The data for SEAs are efficiently represented as time slices in five-year time intervals (1970-1974, 1975-1979, etc.) with a static geography. Thus, what changes when data are animated in a map, graph, or table are only the attributes, rather than the shapes and positions of the geographic units.

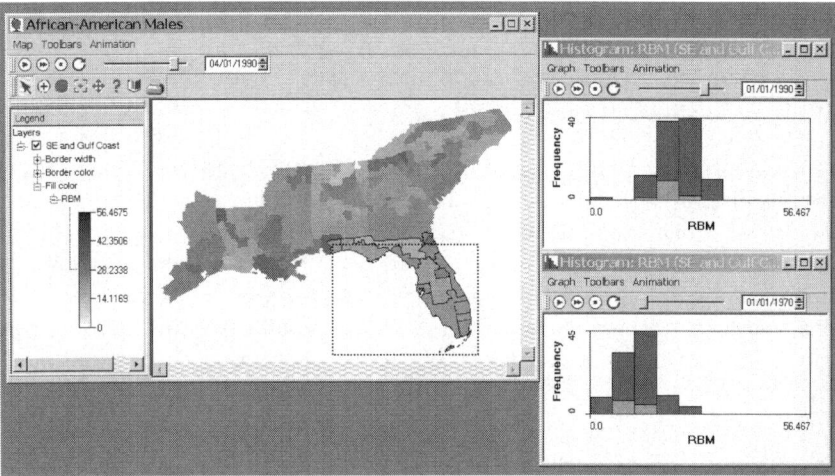

Fig. G.2.1. Time enabled visualization in the Cancer Atlas Viewer. The SEAs in Florida have been selected on the map. The histogram views of the data are also linked, and so Florida's contribution to the histograms is also highlighted in grey. The bottom histogram displays the distribution of 1970-1974 African-American male colon cancer mortalities. The top histogram and the map illustrate 1990-1994 rates. Each view of the data (maps, graphs, tables) has a time slider and media play buttons, so the data can be played through time

Figure G.2.1 is a screenshot illustrating the software's time-enabled views. Notice that each view (maps and graphs as shown, but also scatterplots, boxplots, and tables) have time sliders and media buttons that allow the viewer to pan through time and animate the data. The software also has linked views, a feature common to spatio-temporal visualization software (Haslett et al. 1991, reviewed in Andrienko et al. 2003). In Fig. G.2.1, SEAs in Florida have been selected on the map. Since all data views are linked together, the selected items in the map are also selected (highlighted in grey) in both histograms. The bottom histogram displays the distribution of 1970-1974 African-American male colon cancer mortalities. The top histogram and the map illustrate 1990-1994 rates.

Fig. G.2.2. Difference maps for colon cancer mortality rates between the time intervals 1990-1994 and 1970-1974. The classification of differences in rates is diverging, with white indicating no change from 1970-1974, a grayscale gradient indicating increased mortality rates from colon cancer, and hatching indicating decreased rates

For all but a few SEAs, the count of deaths from colon cancer has increased since 1970 for all gender-race combinations. Similarly, the mortality rates from colon cancer increased for African-American females and males and white males. The differences in rates are largest for African-American males, who experienced the greatest increase in mortality rates from 1970 to 1990. White females, however, experienced decreasing rates in most of the study area. The differences are illustrated for all four groups in Fig. G.2.2, and the distribution of rates for African-American males is illustrated in the histograms in Fig. G.2.1. The rates in 1990-1994 follow a similar pattern to the differences (Fig. G.2.3), with the rates for African-American males being highest and white females lowest, with African-American females and white males intermediate.

Fig. G.2.3. State Economic Areas shaded by their rates for colon cancer mortality in the years 1990-1994

The global spatial pattern in these rates has been somewhat variable over time, but typically increasing. Table G.2.2 shows the I statistics and p-values. White males show the strongest patterns, with significant global autocorrelation in four of five SEA time periods. The white male SEA pattern is variable, with the I statistic between 0.14 and 0.21. For the other groups, there are few SEA time intervals with significant spatial correlation with only one or two time periods showing any significant pattern (with a significant Moran's I). While like values are clustered near each other, these patterns are not consistently strong through time.

Table G.2.2. Moran's I statistics and p-values (in parentheses) calculated from 999 Monte Carlo randomizations. Values in bold are significantly below $\alpha = 0.01$ for SEAs

Group	1970-1974	1975-1979	1980-1984	1985-1989	1990-1994
African-American Females	0.1170	-0.0670	**0.2228**	0.0477	0.0640
	(0.021)	(0.142)	(0.001)	(0.084)	(0.021)
African-American Males	0.0015	0.1008	**0.1758**	0.0347	0.1136
	(0.428)	(0.019)	(0.001)	(0.196)	(0.010)
White Females	**0.1750**	0.0141	0.0066	0.0118	**0.1749**
	(0.001)	(0.333)	(0.370)	(0.312)	(0.001)
White Males	**0.2019**	**0.1355**	**0.1811**	0.0795	**0.1519**
	(0.001)	(0.008)	(0.001)	(0.066)	(0.004)

Table G.2.3. Classification of 1980 locations (99 regions over five time intervals by four population subgroups) by the local Moran and the Getis-Ord statistic G^*. The cells that hold the totals for agreement between the two statistics are shaded grey. Other cells show differences between the clustering results

	Local G^*			Total
	High	Low	NS	not concordant
Local Moran				
High-High	20	0	6	6
Low-Low	0	16	15	15
NS	1	3	1901	4
Low-High	2	0	4	6
High-Low	0	3	9	12
Total not concordant	3	6	34	43

Notes: NS means not significant

Because there are four race-gender combinations and five time intervals by two cluster tests (that is forty cluster maps to compare), we will not detail the results of any single cluster test at any particular time. Instead, we present the pattern of results across all race-gender subgroups over all time intervals. Table G.2.3 compares the results from the local Moran and the local G^* tests. Overall the two local statistics were in concordance. Over 97 percent of the time, the two statistics agreed on the status of a location. Both agreed that there were twenty significant clusters of high values, sixteen significant clusters of low values, and 1,901 non-significant areas. There is no case where the Local Moran finds a cluster of high values and the G^* finds a cluster of low values, or the reverse. They appear to be drawing similar conclusions about these data.

Yet, there are some differences between the results. These differences could be caused by differences in the search pattern of each statistic (the geographic alternative hypothesis to which the statistic is sensitive) or because of the random nature of Monte Carlo probability assessment. We classed the forty-three non-concordant results into four categories for convenience: no comment on outlier, outlier disagreement, significance disagreement, and marginal significance disagreement. These categories are summarized in Table G.2.4.

Table G.2.4. Some characteristics of the non-concordant classes from Table G.2.3

Category	Local Moran (I_i)		Getis-Ord statistic (G^*)	
	Mean (p-value)	Standard deviation	Mean (p-value)	Standard deviation
No comment on outlier (13 cases)	−1.171 (0.018)	0.965	1.577 (0.405)	0.457
Outlier disagreement (13 cases)	−0.354 (0.020)	0.134	3.198 (0.028)	0.437
Marginal disagreement (13 cases)	1.104 (0.042)	0.654	2.683 (0.059)	0.518
Significance disagreement (13 cases)	0.364 (0.038)	0.284	1.895 (0.284)	0.222

No comment on outlier was a category we expected in the beginning—that the G^* may have 'no comment' on locations identified as significant spatial outliers by I_i. Since G^* is not designed to detect outliers, and the local Moran is, we expected outliers by the Moran analysis not to show up as clusters in the G^* analysis. This occurred thirteen times (where the local Moran was Low-High or High-Low and the G^* was not significant). What was unexpected, however, was the five times that the local G^* called something a cluster and the local Moran called it an outlier. We will discuss two examples, Columbus (GA) and Greenville (SC).

The Columbus (GA) SEA is considered the center of a cluster of low mortality rates for white males in 1970-1974 by the G^* but a high outlier among low values by I_i. As shown in the left side of Fig. G.2.4, Columbus is surrounded by several low SEAs, with Z-scores between –4.6 to –0.69. Columbus is near the dataset mean, its Z-score is 0.09, and its southern neighbor is also close to the mean. In this case, the description of Columbus as a significant spatial outlier, specifically a higher outlier among low neighbors, does not correspond to the map pattern. Columbus is an average SEA with several low neighbors. Thus, the G^* is a better descriptor of the local area – the group of SEAs is lower than the regional average. The G^*, however, does not describe Columbus itself very well, it is not low, but it does connect the low group of SEAs. This cluster of low values continues to the northwest, as the Auburn (AL) SEA is classed as a significant cluster of low values by I_i and a marginally significant low cluster by G^* (shown in Fig. G.2.6).

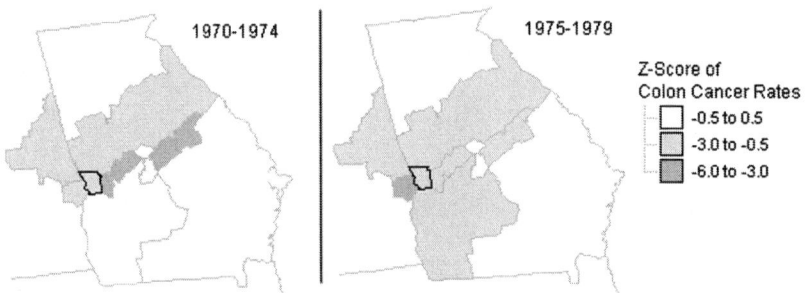

Fig. G.2.4. Statistical disagreement about the significance of the colon cancer mortality rate for white males (RWM) in Columbus (GA). This map shows the location of the SEA (with the dark outline) and its neighbors in Georgia and Alabama. The SEAs shown are colored by their Z-scores for the mortality rate for white males in the period, with SEAs within half a standard deviation of the mean shown as white and below the mean shown hatched. The local Moran and Getis-Ord statistics disagree about the classification in 1970-1974, but both call Columbus the center of a cluster of low values in 1975-1979

The Greenville (SC) SEA is considered the center of a cluster of high mortality rates for African-American females in 1980-1984 by G^* but a low outlier among high neighbors by I_i. Yet, as shown in Fig. G.2.5, neither classification provides an entirely adequate description of the local spatial pattern. Greenville is average, neither especially high nor low, but it does have some very high neighbors, specifically Easley (SC) and Waynesville (NC). These high neighbors seem to be driving both classifications – the high neighbors result in a high local mean which is deemed a high cluster by G^*, and they cause Greenville to be declared a low outlier by I_i. As shown in Fig. G.2.5, it is not a cluster of high values but only two locations with high values, and there is nothing particularly extreme about Greenville itself. The other location that neighbors both Easley and Waynesville (Cornelia SEA in northeastern Georgia) is also the center of a significant cluster of high values, but this time both tests agree. As its rate is also high, this result makes more sense. So both tests found a strong signal in the vicinity of Greenville, but it is not correct to say that Greenville is significantly low or even surrounded by high neighbors or part of a cluster of high values. Neither classification provides a fully accurate description of the local pattern.

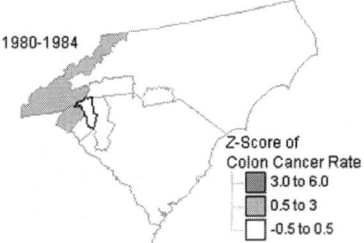

Fig. G.2.5. Statistical disagreement about the significance of the colon cancer mortality rate for African-American females (RBF) in Greenville (SC). This map shows the location of the Greenville SEA (with the dark outline) and its five neighbors. The disjoint polygon on the North Carolina border to the east of the group is actually part of a polygon set that does border the Greenville SEA. The SEAs shown are colored by their Z-scores for the mortality rate for African-American females in the period 1980-1984, with darker grey indicating a higher mortality rate in the period

The twenty-five other cases of difference between G^* and local Moran results occurred when only one of the two tests called a location the center of a significant cluster. In all cases, the statistics agreed about the pattern, both I_i and the G^* showed clustering of high or low values for each location, but their results disagreed about the significance of the pattern. G^* called four locations clusters that the local Moran called not significant, while the local Moran called 21 locations clusters that G^* called not significant. Overall, the local Moran finds clusters more often than G^* does. Whether either is more accurately reporting the 'true' number of clusters in the region cannot be determined with this dataset, but we can examine those cases where the two tests differ to see what triggers each statistic.

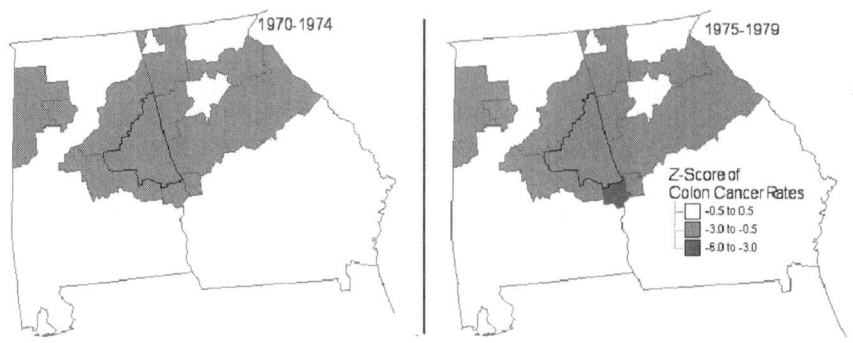

Fig. G.2.6. Persistent clustering of low white male colon cancer mortality rates (RWM) centered on Auburn (AL). The Auburn SEA has the darker outlined in black, and its neighbors are shown with grey borders. The map shading is from the Z-score of the rate, with negative Z-scores shown as hatched. Auburn has three neighbors made up of more than one polygon; this is why some polygons are shown as neighbors but do not share a border with Auburn

Of the twenty-five disagreements about the significance of the clustering, twelve occur when there is a marginal difference in the p-values of the two statistics. For all items in this category, the p-value for both I_i and G^* were smaller than 0.10. For example, for white males in 1970, Auburn (AL) was the center of a cluster of low values according to the local Moran ($I_i = 0.67$, $p = 0.049$); its G^* was marginally significant ($G^* = -2.72$, $p = 0.056$). Both statistics are in agreement about the pattern and its strength, but the Moran statistic happens to be just below the decision criterion ($\alpha = 0.05$) and the G^* above, so they provide different answers. The mean p-value for each statistic in this class was low (mean I_i; $p = 0.042$; mean G^*, $p = 0.059$, see Table G.2.4). Hence this lack of concordance reflects the arbitrariness of the $\alpha = 0.05$ decision threshold. Because of the random nature of the conditional Monte Carlo randomization used to assess significance for both statistics, it is entirely possible that the significance for Auburn (AL) would be the same for both statistics (either a significant low cluster or non-significant) or the pattern of significance reversed (with G^* being below the threshold and I_i above) if the analysis was re-run. Also, we could have chosen a higher number of Monte Carlo randomizations (such as 9,999) to get a more precise p-value from the software. More p-value precision could alleviate these minor p-value disagreements.

The other thirteen cases of disagreement about the significance of a cluster are more interesting. In these cases, the difference in the p-value is large. For example for African-American females in 1970, Sumter (SC) was the center of a significant low cluster according to I_i ($I_i = 0.291$, $p = 0.040$) but not close to significant by G^* ($G^* = -1.68$, $p = 0.500$). Similarly, Prattville (AL) was the center of a local Moran cluster of low mortality for African-American females in 1970 ($I_i = 0.05361$, $p = 0.024$) but not for the G^* ($G^* = -1.83$, $p = 0.500$). In these and other cases where the Moran p-value is much lower than the p-value of G^* it is often the case that the I_i statistic is low (mean of this class is 0.364, Table G.2.4) though significant. There is a significant but weak correlation between the values in the local neighborhood. Overall, the range of I_i values was from about negative two to seven. Positive I_i values less than 0.5, while significant, do not indicate strong clusters of extreme values, and correspond instead to clusters of values slightly lower or higher than the mean. The findings for G^* and I_i differ because G^* is considering divergence from the mean rather than correlation between the values. In many of these cases, G^* provides a more reasonable interpretation of the pattern in terms of what we are looking for in the study of cancer mortality rates – researchers are understandably more concerned about clusters of extremely high or low mortality rates than clusters of rates within one standard deviation of the mean.

The differences in the results of the two cluster detection statistics stem from two factors. The marginal significance disagreement arises from the use of the 0.05 decision criterion. The outlier disagreement, no comment on outlier, and significance disagreement arise from differences in the search pattern of each statistic. In some cases, such as Columbus (GA) G^* provides a better description of the local pattern but not the ego location, while in others, such as Greenville (AL), neither explanation fits.

For cluster persistence, the results were clear: most of the significant clusters do not persist to the next time period, as detailed in Table G.2.5. For sixty-two clusters or outliers identified by the local Moran from 1970 through 1989, sixty were no longer significant in the next time period. For thirty-six clusters identified by the Getis-Ord statistic, thirty-five were no longer significant in the next time period. Only one I_i cluster persisted into the next time period, a cluster of low mortality around Auburn (AL) in 1970-1974 and 1975-1979. The area around Auburn in both time intervals is illustrated in Fig. G.2.6. G^* found this cluster to be marginally significant in 1970-1974 ($p = 0.056$). The I_i classification change occurred around Columbus (GA). Columbus is shown in Fig. G.2.4. In 1970-1974, Columbus was classified as a high outlier among low neighbors by I_i, and then in 1975-1979 it was classified as the center of a low cluster. The change in the data that drives this change in class seems to be that the rate in Columbus went down in this interval. So, it was more similar to its low neighbors in the second time period. Columbus was a significant low cluster according to G^* in both time periods, and is the only persistent G^* cluster in the times studied.

Table G.2.5. Cluster persistence over time, measured in terms of number of cases. Totals are from all race-gender subgroups over all pairs of two sequential times

Transitions between categories	Local Moran I_i	Getis-Ord statistic G^*
From cluster to not significant	60	35
From outlier to cluster	1	n/a
No change	1	1
Total	62	36

G.2.4 Concluding remarks

The lack of persistent clustering in the data suggests that the clusters detected may be ephemeral. There is not persistent clustering of high values indicating a stable environmental exposure or a stable social or genetic contributing factor to colon cancer. Ephemeral clusters can be explained in several ways. They might result from unstable mortality rates generated by small populations-at-risk, population migration, or short term factors such as geographic differences in treatment or screening that do not persist. This seems to be a positive conclusion – there is no area at high for risk colon cancer that persists through time. There are some disparities in colon cancer mortality, as shown in Fig. G.2.3, with males having higher mortality rates than females, and African-Americans having higher mortality than whites. This does not seem to be a local phenomenon, but instead applies across the study geography. And, the mortality rates for African-American males and females are increasing more steeply than they are for whites (see Fig. G.2.2). Researchers have suggested that the increasing rates of colon cancer may be due to changing diets and increasing rates of obesity (Murphy et al. 2000). Other research has shown that patterns in colon cancer mortality in African-Americans may in part be due to socioeconomic factors, specifically differential access to health care. Freeman and Alshafie (2002) found that poorer individuals are diagnosed with more advanced cancers and die more frequently.

This contribution finds that the results of the two statistics are similar, with agreement on their classifications over 97 percent of the time. Because of the large amount of concordance between the two cluster statistics, there seems to be little additional value gained from applying both cluster statistics to a dataset. Because of the differences in reporting of significance, with the local Moran reporting some clusters quite near the mean, the Getis-Ord statistic may be more sensitive to clusters of extreme values. The cluster classifications produced by the statistics need to be further examined to be interpreted well. Although several software products, including the Cancer Atlas Viewer, produce crisp maps classifying locations into clusters, outliers, and non-significant areas, the simplicity of these maps can obscure the complexity of the observed data. The differences we found in the classifications of G^* and I_i pointed out a few locations where the interpretation for either classification requires careful examination of the mapped data. These are instances in which the local map pattern do not correspond to the search image of either cluster statistic.

The few cases they disagree on the status of an outlier seem to come from limitations on the shape of the cluster to be detected imposed by the first-order neighbor relationships considered. Disagreement emerged from a situation where there was variability among the neighbor set, so the 'true cluster', if it existed at all, was likely a subset of the neighbor set, rather than the whole group of first-order neighbors. This is shown in Figs. G.2.4 to G.2.6, where significant clusters or sets of outliers contain individuals that are close to the mean and those that are more extreme. These locations would be better described by a different type of cluster statistic, one that connects areas into sinuous shapes that reflect similarity of values rather than searches for clusters matching a preexisting shape pattern (such as first-order neighbors). This argument is similar to one we have made before about the limitations of centroid-based cluster statistics that search for circular clusters (Jacquez and Greiling 2003).

The analysis reported in this chapter was performed using the free Cancer Atlas Viewer software, providing researchers with sophisticated visualization and statistics tools for the exploration of patterns in mortality from 40 site-specific cancers. It can act as a quicker and more interactive way to explore the data made available by the National Cancer Institute, cutting out the delays inherent in web mapping that may hinder exploration. The statistical analysis presented here may be beyond the interest and commitment of a casual user, but the software provides a means for researchers to assess cancer mortality patterns, examine these patterns over time in animated maps and graphics, and to assess the persistence of clustering.

Acknowledgements. This project was funded by grant CA92669 from the National Cancer Institute to BioMedware, Inc. The positions espoused in this chapter are those of the authors and do not necessarily represent the official views of the National Cancer Institute. Constructive criticism from Heidi Durbeck of BioMedware, Peter Rogerson of SUNY-Buffalo, and three anonymous reviewers helped us improve the interpretation and presentation of these results.

References

Andrienko NV, Andrienko GL, Gatalsky P (2003) Exploratory spatio-temporal visualization: an analytical review. J Vis Lang Comp 14(6): 503-541

Blot WJ, Morris LE, Stroube R, Tagnon I, Fraumeni JF Jr (1980) Lung and laryngeal cancers in relation to shipyard employment in coastal Virginia. J Nat Cancer Inst 65(3): 571-575

Cossman RE, Cossman JS, Jackson R, Cosby A (2003) Mapping high or low mortality places across time in the United States: a research note on a health visualization and analysis project. Health and Place 9(4): 361-369

Devesa SS, Grauman DG, Blot WJ, Pennello G, Hoover RN, Fraumeni JF Jr. (1999) Atlas of cancer mortality in the United States, 1950-94. US Govt Print Off, Washington [DC] [NIH Publ No. (NIH) 99-4564]

Freeman HP, Alshafie TA (2002) Colorectal carcinoma in poor blacks. Cancer 94(9): 2327-2332

Gao S, Mioc D, Anton F, Yi X, Coleman DJ (2008) Online GIS services for mapping and sharing disease information. Int J Health Geographics 7(1): 8

Haslett J, Bradley R, Craig P, Unwin A, Wills G (1991) Dynamic graphics for exploring spatial data with application to locating global and local anomalies. The American Statistician 45(3):234-242

Jacquez GM (1998) GIS as an enabling technology. In Gatrell A, Loytonen M (eds) GIS and health. Taylor and Francis, London, pp.17-28

Jacquez GM, Greiling DA (2003) Local clustering in breast, lung and colorectal cancer in Long Island, New York. Int J Health Geographics 2(1): 3

Moore DA, Carpenter TE (1999) Spatial analytical methods and geographic information systems: use in health research and epidemiology. Epidem Rev 21(2): 143-161

Murphy TK, Calle EE, Rodriguez C, Kahn HS, Thun MJ (2000) Body mass index and colon cancer mortality in a large prospective study. Am J Epidemiol 152(9): 847-854

Pickle LW, Mungiole M, Jones GK, White AA (1996) Atlas of United States mortality. US Department of Health and Human Services, Centers for Disease Control and Prevention, National Center for Health Statistics, Hyattsville [MD.] DHHS Publication No. (PHS) 97-1015

Rushton G, Elmes G, McMaster R (2000) Considerations for improving geographic information system research in public health. J Urb Reg Inf Syst Ass 12(2): 31-49

Simes RJ (1986) An improved Bonferroni procedure for multiple tests of significance. Biometrics 73(3): 751-754

US Bureau of the Census (1966) Census of population: 1960, number of inhabitants, United States summary, final report PC (1)-1A. US Government Printing Office, Washington, [DC]

Winn DM, Blot WJ, Shy CM, Pickle LW, Toledo A, Fraumeni JF (1981) Snuff dipping and oral cancer among women in the southern United States. New England J Med 304(13): 745-749

G.3 Exposure Assessment in Environmental Epidemiology

Jaymie R. Meliker, Melissa J. Slotnick, Gillian A. AvRuskin, Andrew M. Kaufmann, Geoffrey D. Jacquez and *Jerome O. Nriagu*

G.3.1 Introduction

A key component of environmental epidemiologic research is the assessment of historic exposure to environmental contaminants. The continual expansion of space-time databases, coupled with the recognized need to incorporate mobility histories in environmental epidemiology, has highlighted the deficiencies of current software to visualize and process space-time information for exposure assessment (Mather et al. 2004; Pickle et al. 2005). This need is most pressing in retrospective studies or large studies where collection of individual biomarkers is unattainable or prohibitively expensive, and models and software tools are required for exposure reconstruction. In diseases of long latency such as cancer, exposure may need to be reconstructed over the entire life-course, taking into consideration residential mobility, occupational mobility, changes in risk behaviors, and time-changing maps generated from models of environmental contaminants. Even for outcomes of short latency such as asthma attacks, exposure reconstruction may involve daily mobility/activity patterns and temporally-varying maps of contaminants. These types of datasets, for example, mobility histories and time-changing maps of environmental contaminants, are almost always characterized by spatial, temporal, and, spatio-temporal variability. While current state-of-the-art methods can integrate datasets that contain either spatial or temporal variability, datasets exhibiting both spatial and temporal variability have proven largely unmanageable until now, and researchers have been forced to simplify the dynamic nature of their datasets by reducing or eliminating the spatial or temporal dimension.

Despite modern computer technologies for storing and managing temporal and spatiotemporal datasets, surprisingly few tools are available for visualizing the 'what, where, and when' of events (Andrienko et al. 2003; Chittaro et al. 2003). One visualization tool, GISystem software, enables users to visualize what happened, and where; and has augmented assessment of exposure to environmental contaminants. For example, researchers have geocoded locations of industries, industrial waste sites, pollution plumes, as well as homes, schools and jobs of study participants. From these geocoded features, disease maps have been created and spatial analyses performed (Brauer et al. 2003; Maantay et al. 2002; Meliker et al. 2001; Reif et al. 2003; Swartz et al. 2003). A frequent criticism of GIS, however, is its inability to support temporal data structures (Beaubroef and Breckenridge 2000; Dragicevic and Marceau 2000). This can be problematic if, for example, investigators wish to explore whether residences of cancer cases cluster at any time in the past fifty years; or whether living in close proximity to a chemical industry at any point in time is associated with subsequent cancer development. With GISystems, spatial patterns at different isolated moments can be examined, using animation tools. However, each map, or snapshot, must be created independently, requiring a substantial amount of effort and introducing greater likelihood of data entry error. Furthermore, information about change is not available in the interval between two consecutive snapshots. Visualization of a map that displays smooth, continuous changes over time can generate additional insights about spatial disease patterns.

Time-GIS technology has recently been developed that takes advantage of the space-time variability inherent in many datasets (AvRuskin et al. 2004; Greiling et al. 2005; Jacquez et al. 2005; Meliker et al. 2005). Time-GIS support evaluation and query of spatio-temporal datasets, and also can enrich analysis of temporal datasets that are devoid of geographic coordinates. A backbone of these tools is their temporal data structure, which enables non-geographic attributes, such as temporally-variant exposure estimates, to be visually examined.

In this chapter, space-time methods are illustrated using preliminary data from a bladder cancer case-control study in Michigan. Established causes of bladder cancer include cigarette smoking and exposure to aromatic amines in occupational settings; however, many cases of bladder cancer remain unexplained. One possible cause of bladder cancer is exposure to arsenic in drinking water. Concentrations of arsenic in drinking water exceeding World Health Organization (WHO) and US Environmental Protection Agency (EPA) guidelines (10 µg/L) have been identified in ground-water supplies of eleven counties in southeastern Michigan: Genesee, Huron, Ingham, Jackson, Lapeer, Livingston, Oakland, Sanilac, Shiawassee, Tuscola, and Washtenaw (Kim et al. 2002; Kolker et al. 2003; Slotnick et al. 2006). Previous individual-level studies of arsenic in drinking water and bladder cancer have been criticized for imprecise exposure assessments (Cantor 2001) which failed to account for (i) changes in arsenic concentration in water over time, (ii) individual residential mobility patterns, and (iii) behavioral changes in drinking water consumption. These shortcomings are familiar to many investigations of environmental exposures and cancer; Time-GIS are essential for alleviating some of these shortcomings.

G.3.2 Data and methods

A sample of 39 cases and 39 controls from a bladder cancer case-control study in Michigan was selected to demonstrate applications of Time-GIS for exposure assessment. This size of the dataset allows for straightforward manipulation and visualization of complex exposure scenarios, enhancing communication of the intricacies of these visualization tools. Cases were recruited from the Michigan State Cancer Registry and diagnosed in the year 2000. Controls were frequency matched to cases by age (± five years), race, and gender, and recruited using a random digit dialing procedure from an age-weighted list. To be eligible for inclusion in the study, participants must have lived in the eleven county study area for at least the past five years and had no prior history of cancer (with the exception of non-melanoma skin cancer). Participants were offered a modest financial incentive and research was approved by University of Michigan IRB-Health Committee. Participants answered a telephone questionnaire concerning drinking water habits, smoking, and medical history, and completed a written questionnaire describing residential mobility history. Information obtained from these questionnaire instruments was used to create spatiotemporal datasets.

This section describes the key functionalities of Time-GIS software being developed to facilitate space-time exposure reconstruction. This software, STIS^{TM} (TerraSeer, Ann Arbor), supports spatio-temporal datasets but does not yet provide all of the functionalities described here (see Chapter A.6 for a description of TerraSeer).

Temporal and spatio-temporal datasets on residential mobility. Participants provided written residential histories of each home lived-in for at least one year for a total of 519 homes. The duration of residence and exact street address were obtained, otherwise the closest cross streets were provided. Each residence in the study area was geocoded and assigned geographic coordinates in ArcGIS; residences outside the study area were not geocoded. Participants resided at 288 homes within the study area, with time spent averaging 66 percent of their lifetimes. Street files were downloaded from Michigan Center for Geographic Information website, and were part of the Michigan Geographic Framework. Michigan Geographic Framework datasets use the Michigan GeoRef System, based on an Oblique Mercator projection. Residences within the study area were successfully geocoded: 78 percent automatically matched using ArcGIS settings of spelling sensitivity equal to 80, minimum candidate score equal to ten, and a minimum match score equal to 60. Unmatched addresses were manually matched using cross streets with the assistance of internet mapping services (seventeen percent). If cross streets were not provided, best informed guess placed the address on the road (three percent), and as a last resort, residence was matched to town centroid (two percent).

Water supply history. Participants provided written information about primary water supply and any changes in water supply at each residence (for example, public surface, public well, private well, or bottled water). Managers of 135 public water supplies in the study area answered questions about quality of drinking water, source of water, changes in water supply, changes in extent of coverage of water supply, and changes in treatment procedures. Each residence was classified by its primary water supply for a span of time based on accounts provided by participants. At approximately three percent of the addresses, participants did not assign a water supply. Time-GIS was used in those cases to assign a water supply.

Arsenic data. The water sample that provided current arsenic exposure was collected from the kitchen tap, or primary source of water for drinking and cooking at each participant's current home. All plasticware was acid-washed for trace metals determination following modification of a previously described protocol (Nriagu et al. 1993). Blanks and replicates were collected for at least ten percent of the samples. Water samples were stored on ice, acidified with 0.2 percent trace metal grade nitric acid, and refrigerated until analysis. Water samples were subsequently analyzed for arsenic using an inductively coupled plasma mass spectrometer (ICP-MS, Argilent Technologies Model 7500c).

Historic databases were used to estimate arsenic concentrations at past residences. Michigan Department of Environmental Quality (MDEQ) maintains a database of arsenic measurements (1993–2002) in private (N = 11,615 arsenic measurements) and public well water supplies (N = 1675 arsenic measurements) in the study area, analyzed in a state laboratory with graphite furnace atomic absorptions spectrometry (GF/ AAS) (1993–1995), hydride flame (quartz tube AAS) (1993–1995), and an ICP-MS (1996–2002). Private well water measurements from MDEQ database were utilized to generate city or township averages (means) of arsenic concentrations for past private well waters not monitored for arsenic (a geostatistical model is also being developed to predict arsenic concentrations in past private wells; Goovaerts et al. 2005). The MDEQ database of public well water supplies was used to calculate a mean value of arsenic for each community's ground-water supply. Community supplies relying on surface water were assigned an arsenic concentration equal to 0.3 µg/L, the mean level detected in tap water samples that rely on surface water in the area. Residences outside the study area were similarly assigned an arsenic concentration of 0.3 µg/L. Arsenic concentrations in private and public water supplies were assumed not to change over time, since evidence suggests limited temporal variability (Slotnick et al. 2006; Steinmaus et al. 2005; Ryan et al. 2000).

Water consumption patterns. Estimates of water consumption (liters/day) were calculated based on answers to a series of questions from a telephone interview. Participants were asked to self-report the number of glasses of water and beverages made with water drank at home during the past year (the year prior to cancer diagnosis for cases), the previous ten years, and changes in drinking water consumption over the course of a lifetime.

G.3.3 Features and architecture of Time-GIS

Data structure. Time-GIS support datasets in which time is a principal feature. For example, the data structure requires that: (i) each row in a dataset represents a space-time intersection and a variable of interest and (ii) when a geographic location or a variable of interest changes value, a new row is created. Data tables can be linked to spatial features, and the following information must be specified: unique ID, start date, end date, and attributes during that time window. In the example of drinking water history and residential mobility of participants, each home and water source occupy a unique row with a start year, end year, geographic coordinate, and participant ID number. Any change in location of residence or source of drinking water is characterized by a new row with a defined duration, using the same participant ID number (see Table G.3.1). Other variables, such as water consumption patterns, use of home water treatment systems, and concentrations of drinking water contaminants, are stored in separate datasets, including rows with defined durations and ID numbers. Despite different durations for variables, participant datasets can be joined together, using participant ID numbers. Using Time-GIS, a participant's mobility history can be visualized by displaying specified attributes of a participant. In effect, changes in water consumption patterns, water supply, contaminant concentrations in water, or other variables can be illustrated as participants' move through time.

Table G.3.1. Spatio-temporal dataset format for STIS point features: residential mobility history and water consumption history for two participants

Residential mobility history						Estimated water consumption			
Sample ID	Year moved in	Year moved Out	Address*	X-Coord*	Y-Coord*	Sample ID	Start period	End period	Liters/day
001366	04/08/1951	01/01/1963	Address #1	694980	264132	001366	04/08/1951	01/01/1995	1.50
001366	01/01/1963	01/01/1971	Address #2	687299	268878	001366	01/01/1995	01/01/2004	2.75
001366	01/01/1971	01/01/1972	Address #3	694161	272042	001397	04/08/1933	01/01/1949	0.25
001366	01/01/1972	01/01/1975	Address #4	680421	278791	001397	01/01/1949	01/01/1982	1.00
001366	01/01/1975	01/01/2004	Address #5	649645	275342	001397	01/01/1982	01/01/2004	0.60
001397	01/01/1933	01/01/1937	Address #1	692980	168978				
001397	01/01/1937	01/01/1950	Address #2	687699	174042				
001397	01/01/1950	01/01/1953	Address #3	692161	176791				
001397	01/01/1953	01/01/1957	Address #4	660421	177342				
001397	01/01/1957	01/01/1964	Address #5	684656	274665				
001397	01/01/1967	01/01/1969	Address #6	694766	278743				
001397	01/01/1969	01/01/1993	Address #7	686910	274183				
001397	01/01/1993	01/01/1998	Address #8	692830	280704				
001397	01/01/1998	01/01/2004	Address #9	685618	270049				

Notes: *Address and geographic coordinates are altered to protect participant confidentiality

From a data input perspective, datasets of point, line, or polygon features are structured similarly. As with point features, visualizing changes in community attributes requires geographic coordinates, a duration, unique community ID number, and characteristics of a community. These variables are recorded in unique rows for each space-time intersection. The ID number remains constant, even if the geographic area of the community changes – analogous to a unique ID number for each participant, even though residences change. In this manner, polygons can change shape and attributes over time.

Space-time maps. Smooth, temporally continuous, space-time maps are created in Time-GIS using the data structure described above. To illustrate the changes in the source of drinking water, a series of snap-shots are presented, representing the region's water supply status in 1932, 1964, and 1993 (see Fig. G.3.1). The slider bar is dragged to the left (distant past) or right (more recent past) to display different years. With limitations of the printed page, only static images can be presented here. However, Time-GIS produces continuous space-time animations. Valuable information can be gained by visualizing water supply changes. Water supply is designated as: private wells, public ground-water, public surface water, purchased surface water, or mainly private wells (i.e., some small residential developments in the community have public ground-water supplies). From 1932 through 1993, most of the region was served by private wells (white color in Fig. G.3.1). In 1932, only a few communities were served by surface water; by 1993, several communities purchased surface water from the city of Detroit water system. Visualization of changes shows that over time, small communities developed public ground-water systems and some public ground-water systems changed over to public surface water distribution systems.

In addition to attributes changing, variations in town boundaries are displayed, as when new towns become incorporated, communities expand their borders, and when communities merge (see Fig. G.3.2). Between 1950 and 1992, several new communities were incorporated in Oakland County. In the database, each community is assigned a unique ID number that remains the same; any other variable, including geographic coordinates, is permitted to change.

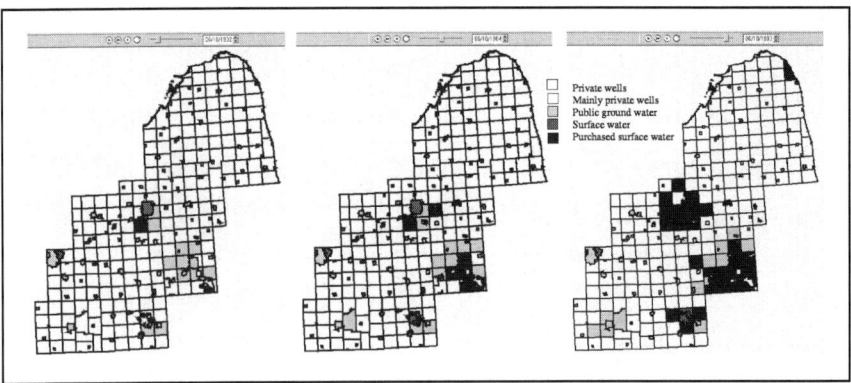

Fig. G.3.1. Public and private water supply status in Southeastern Michigan in 1932, 1964, and 1993. Slider Bar controls the year displayed

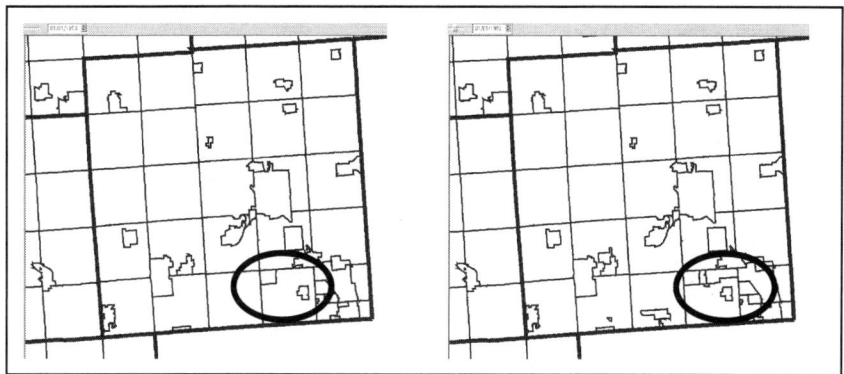

Fig. G.3.2. Feature of Time-GIS. Town boundaries change with time (1950, 1992)

G.3.4 Application

Arsenic exposure reconstruction. Given space-time maps, a natural extension is for the software to process space-time joins. This procedure reduces data entry time and error associated with manually assigning an attribute to a space-time object. For example, in the bladder cancer study, participants were unable to recall their drinking water source at approximately three percent of their lifetime residences. The public water supply map (see Fig. G.3.1) from 1920–2003, was overlaid with the residential history map, and each residence without an assigned drinking water source was visually overlaid with the public water supply map and assigned a drinking water source. The space-time join functionality of Time-GIS makes this assignment automatic.

As a further example, space-time join tools can be used to assign an arsenic concentration to drinking water at each residence. Distinct space-time maps of arsenic concentration are created for residences on public water supplies and private wells. A space-time join procedure can differentiate between residences on private well water and those on public water supply, and assign arsenic to each residence, at all points in time. This automated procedure is flexible, allowing efficient recalculation of lifetime exposure when a different exposure metric or route is selected, or when the underlying models of environmental dispersion are refined. This improvement is substantial and highlights the broad application of Time-GIS in facilitating exposure calculations and therefore improving exposure assessment.

Intake was then calculated by multiplying arsenic concentration (μg/L) by volume of water drank at home and used for making beverages at home (L/day). Each change in water consumption and change in arsenic concentration was used to estimate exposure for a particular time-window. Results are presented as average exposure to arsenic in units of μg/day. Exposure calculations were performed for

587 unique space-time periods, with each space-time period defined by a unique combination of residential location, water source, water treatment, and water consumption behavior. Joining data across these space-time intervals is automated in Time-GIS; additional evidence of its benefit for exposure reconstruction.

Exposure life-lines. One of the most commonly used temporal visualization tools is the time series graph. A time series of arsenic exposure for each case and control is shown in Fig. G.3.3. Each line represents a different participant's arsenic exposure trajectory. The thick line depicts average arsenic exposure for cases and controls, in respective graphs. Individual participants' trajectories are difficult to follow, visually, because the lines intersect. The average arsenic exposure trajectory for cases and controls may generate insights but information is lost by simply averaging participants' exposures. Other traditional tools, such as histograms and scatter plots, display variables at slices of time in Time-GIS. Similar to space-time maps, these tools allow for scanning smooth, continuous, temporally-variant figures for relationships between variables at any moment in time. Histograms display one variable, and scatter plots display two variables, at any time slice. For example, histograms for cases and controls were compared for drinking water source. In 1965, fifteen cases and eleven controls were drinking well water, while twenty-four cases and twenty-eight controls were drinking surface water (see Fig. G.3.4). Scatter plots were used to compare arsenic exposure and cigarette smoking for cases and controls (see Fig. G.3.5). In 1972, controls with arsenic exposure exceeding twenty µg/L smoked fewer than twenty-one cigarettes/day. Cases with arsenic exposure exceeding twenty µg/L, in comparison, smoked greater than thirty cigarettes/day. Limited evidence suggests that cigarette smokers, exposed to elevated levels of arsenic 30–40 years ago (Bates et al. 1995) or 40 or more years ago (Steinmaus et al. 2003) are at an increased risk for bladder cancer. But the temporal relationship between cigarette smoking and arsenic exposure has not been well documented. For example, when does smoking cigarettes interact with arsenic exposure: if exposure is simultaneous; if heavy smoking precedes a period of elevated arsenic exposure; or if heavy smoking occurs following a period of elevated arsenic exposure? Is there a critical time when cigarette smoking and arsenic intake interact to increase risk of bladder cancer? Scatter plots in Time-GIS help researchers address these types of questions.

While scatter plots, histograms, time graphs, and space-time maps, can each be employed to generate insights about space-time variability, these tools do not display the entire temporal dataset in a straightforward manner. The tools rely either on time slices of continuous maps or crisscrossing trajectories of participants, in which the relationships between participants, their exposure, and how exposure changes with time, are difficult to visually comprehend. One solution is exposure life-lines, which display participants on the horizontal axis, time on the vertical axis, and measures of exposure in the life-lines' color or thickness.

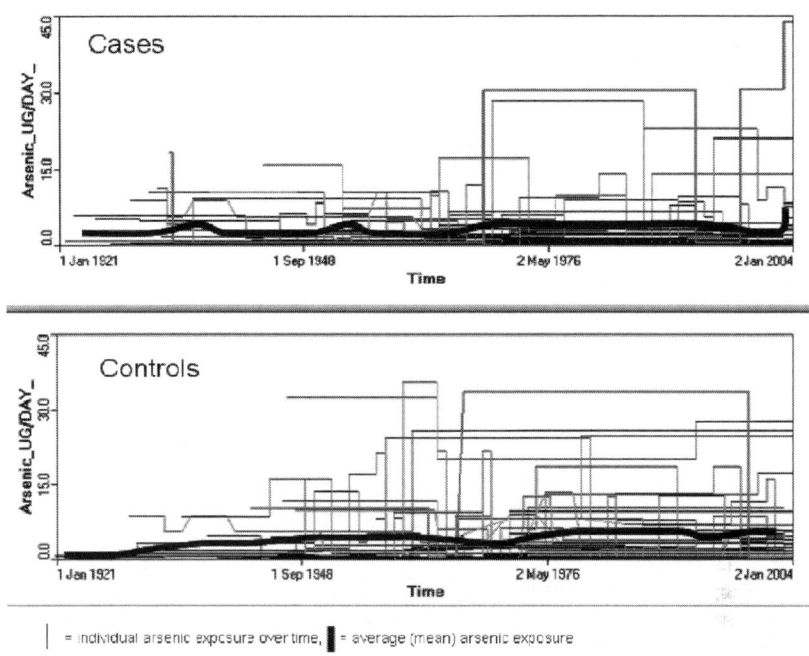

Fig. G.3.3. Time series graphs of arsenic exposure: cases and controls. Each line represent a participant's average daily exposure to arsenic (μg/day) over his/her lifetime

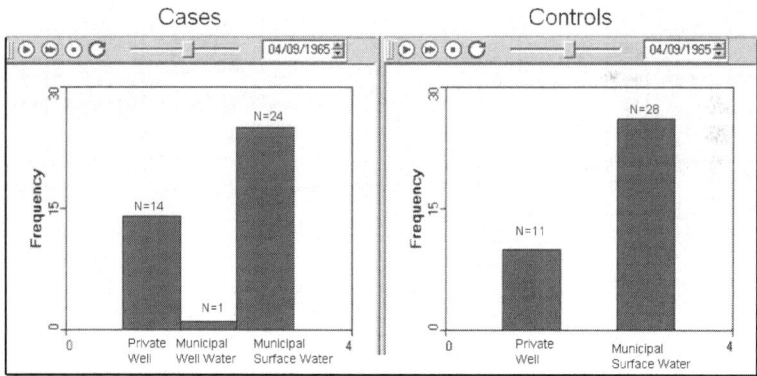

Fig. G.3.4. Histogram of source of drinking water in 1965: cases and controls. Slider Bar controls the year displayed

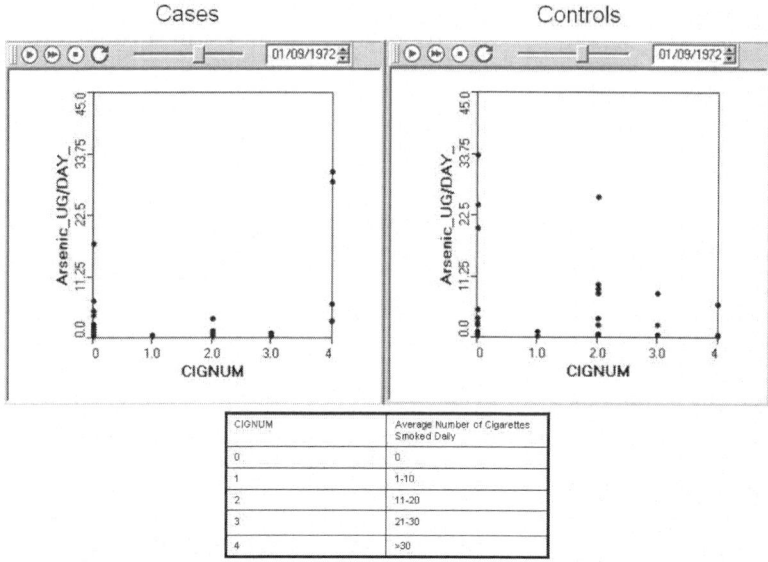

Fig. G.3.5. Scatter plot of arsenic exposure and cigarettes smoked in 1972: cases and controls. Two cases exposed to elevated arsenic exposure and more than 30 cigarettes/day. Slider Bar controls the year displayed

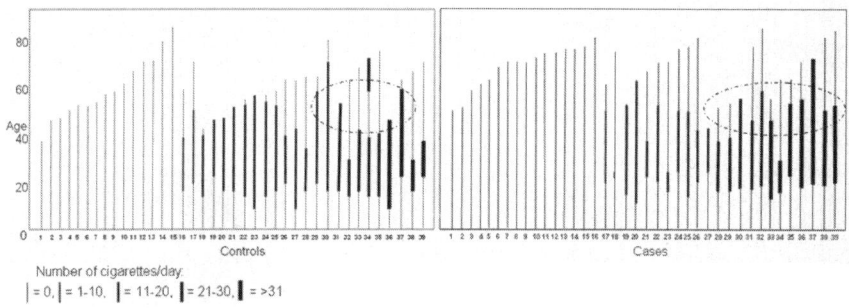

Fig. G.3.6. Exposure life-lines for cigarette smoking: cases and controls. Thickness of life-line increases with higher frequency of cigarettes smoked. There appear to be more heavy smokers around 40-50 years old among cases, compared with controls

Exposure life-lines (see Fig. G.3.6) are presented to illustrate changes in cigarette smoking exposure among cases and controls. The thickness of the life-line reflects the average number of cigarettes smoked each day, with thicker lines indicating heavier smoking. Using these life-lines, investigators can compare cases and controls to evaluate whether cases began smoking at a younger age, quit at an older age, or whether there was a particularly vulnerable period in which heavy smoking appears to be strongly associated with bladder cancer. Similar numbers of cases (sixteen) and controls (fifteen) never smoked cigarettes, but seven cases smoked more than thirty cigarettes/day, in comparison with four controls. In addition, nine of the twelve cases who smoked more than twenty cigarettes/day, were smoking in their 40s and 50s, in comparison with just three of the nine controls who smoked more than a pack a day. While cigarette smoking is an established risk factor for bladder cancer, the timing of when cigarette smoking might exacerbate bladder cancer risk is not well understood. The apparent temporal cluster of heavy smokers among cases in their 40s and 50s could shed light on the temporal relationship between cigarette smoking and bladder cancer.

As Fig. G.3.6 illustrates, exposure life-lines may be used to investigate temporal clusters of high exposure at any point in time. They are particularly useful for investigating the latency period of exposure-disease relationships. These life-lines are a substantial improvement over such methods as cumulative exposure estimates or pre-defined time windows of exposure, which rely on temporally-aggregated exposure estimates. Exposure life-lines are helpful to investigate whether any years of high exposure are more prominent in cases compared with controls.

Bivariate exposure life-lines can be constructed to represent historic exposure levels from two variables with shades of grey for the first variable, and thickness for the second variable. Exposure life-lines depicting arsenic exposure and cigarette smoking history for cases and controls are shown in Fig. G.3.7. A life-line's thickness increases with frequency of cigarettes smoked; increases in darkness correspond to higher levels of arsenic exposure. It is seen that more controls, compared with cases, were exposed to elevated levels of arsenic in drinking water over the course of their lives. Only one control, however, experienced simultaneous exposure to drinking water arsenic more than twenty-five μg/day, and more than twenty cigarettes/day, and that was only for three years, from 1953-1956. In comparison, two cases were simultaneously exposed to arsenic exposure more than twenty-five μg/day, and more than thirty cigarettes/day in the 1970s and 1980s, suggesting the possibility of interaction between arsenic and cigarette smoking, twenty-to-thirty years prior to cancer diagnosis.

Without exposure life-lines, an a priori hypothesis with a specified point in time is required to investigate when multiple variables interact. For example, researchers could not visualize and evaluate the relationships between arsenic exposure at any point in time, cigarette packs smoked at any point in time, and subsequent disease development. Exposure life-lines can be used to evaluate these relationships and shed light on interactions between variables, and when those interactions may be occurring.

In comparison to Fig. G.3.6 where age was used to track time, Fig. G.3.7 uses calendar years. Displaying different temporal orientations, such as calendar years and age, is beneficial in highlighting distinct trends in the data. In some situations, calendar years of exposure may be important, since they indicate key changes in the external environment. In other situations, age at exposure may be critical, because individuals may be more vulnerable to toxic agents when they are at a specific age. Exposure life-lines can be flexible to present data using either of these temporal orientations, or others, including years prior to diagnosis or interview (Meliker and Jacquez 2007).

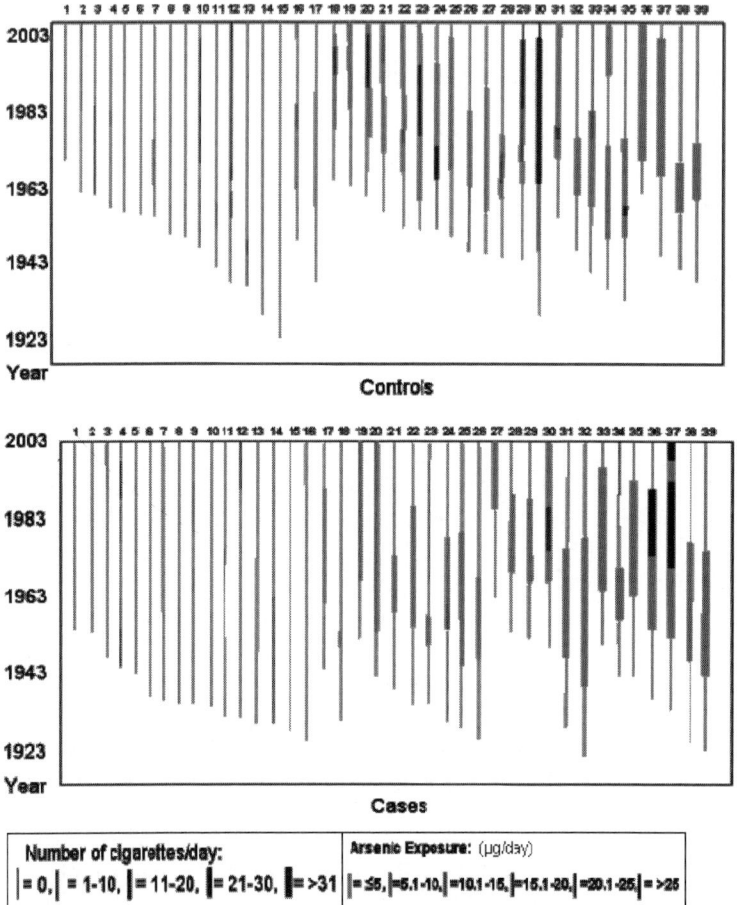

Fig. G.3.7. Exposure life-lines for arsenic exposure and cigarette smoking: cases and controls. Thickness of life-line increases with frequency of cigarettes smoked. Darkness of life-line increases with higher arsenic exposure. There appears to be more heavy smoking in 1970s among cases. A pair of heavy smokers was exposed to arsenic greater than 25 μg/day in 1970s and 1980s

G.3.5 Concluding remarks

Application of Time-GIS for exposure assessment and environmental epidemiology is in its infancy. Processing and supporting visualization of spatio-temporal datasets is essential for exploring patterns in data. Beyond visualization, however, statistical analyses are necessary to identify significant relationships between timing or location of exposure and subsequent disease development. Statistical procedures that appear to be important include spatial and spatio-temporal clustering techniques; focused cluster tests to examine if a cluster is associated with a point source, such as an industry or landfill; and temporal epidemiologic analyses to investigate if exposure at a point in time is associated with subsequent disease development. Another area of future research is the propagation of uncertainty associated with calculation of exposure profiles. Specific to the bladder cancer example presented here, uncertainty in exposure assessment arises at several levels in data collection and manipulation, including water consumption estimates, measured arsenic concentrations, estimated arsenic concentrations, and geographic location. Efforts are underway to assess sensitivity of arsenic exposure estimates to different sources of uncertainty, and to incorporate techniques for propagating uncertainty in Time-GIS.

Despite the common perception that cancers are often caused by environmental contaminants, limited evidence exists to support widespread associations. Reports of weak, non-significant associations between environmental agents and cancer (Gammon et al. 2002; Bates et al. 1995; Steinmaus et al. 2003) may be attributed to exposure assessments that inadequately incorporate temporal variability. A recent study of exposure to organochlorines and breast cancer suggests that biomarkers collected in younger women decades prior to diagnosis reveal an association not seen in studies using more recent measures of exposure (Cohn et al. 2007). Access to historical biomarkers, however, is rare, substantiating the need for improved historical exposure reconstruction methods using Time-GIS which enable spatially-and temporally-explicit exposure assessment and the investigation of ensuing disease development.

Acknowledgments. This study was supported by grants R01 CA96002, R43 ES10220, R01 CA92669, and R43 CA132341 from the National Institutes of Health. Access to cancer case records was provided by Michigan Cancer Surveillance Program within the Division for Vital Records and Health Statistics, Michigan Department of Community Health. The authors thank Michigan Public Health Institute for conducting the telephone interview and Stacey Fedewa and Lisa Bailey for entering written surveys into a database. The authors thank also reviewers for their helpful comments.

References

Andrienko NV, Andrienko GL, Gatalsky P (2003) Exploratory spatio-temporal visualization: an analytical review. J Vis Lang Comp 14(6):503-541

AvRuskin GA, Jacquez GM, Meliker JR, Slotnick MJ, Kaufmann AM, Nriagu JO (2004) Visualization and exploratory analysis of epidemiologic data using a novel space time information system. Int J Health Geographics 3(1):26

Bates MN, Smith AH, Cantor KP (1995) Case-control study of bladder-cancer and arsenic in drinking water. Am J Epidemiol 141(6):523-530

Beaubroef T, Breckenridge J (2000) Real-world issues and applications for real-time geographic information systems (RT-GIS). J Navig 53(1):124-131

Brauer M, Hoek G, Van Vliet P, Meliefste K, Fischer P, Gehring U, Heinrich J, Cyrys J, Bellander T, Lewne M, Brunekreef B (2003) Estimating long-term average particulate air pollution concentrations: application of traffic indicators and geographic information systems. Epidemiology 14(2):228-239

Cantor KP (2001) Invited commentary: arsenic and cancer of the urinary tract. Am J Epidemiol 153(5):419-421

Chittaro L, Combi C, Trapasso G (2003) Data mining on temporal data: a visual approach and its clinical application to hemodialysis. J Visual Lang Compu 14(6):591-620

Cohn BA, Wolff MS, Cirillo PM, Sholtz RI (2007) DDT and breast cancer in young women: new data on the significance of age at exposure. Environ Health Persp 115(10):1406-1414.

Dragicevic S, Marceau DJ (2000) A fuzzy set approach for modeling time in GIS. Int J Geogr Inf Sci 14(3):225-245

Gammon MD, Wolff MS, Neugut AI, Eng SM, Teitelbaum SL, Britton JA, Terry MB, Levin B, Stellman SD, Kabat GC, Hatch M, Senie R, Berkowitz G, Bradlow HL, Garbowski G, Maffeo C, Montalvan P, Kemeny M, Citron M, Schanbel F, Schuss A, Hajdu S, Vinceguerra V, Niguidula N, Ireland K, Santella RM (2002) Environmental toxins and breast cancer on Long Island. II. Organochlorine compound levels in blood. Cancer Epidemiol Biomarkers Prev 11:686-697

Goovaerts P, AvRuskin G, Meliker JR, Slotnick M, Jacquez GM, Nriagu J (2005) Geostatistical modeling of the spatial variability of arsenic in groundwater of Southeast Michigan. Water Resources Research 41(7):W07013

Greiling DA, Jacquez GM, Kaufmann AM, Rommel RG (2005) Space time visualization and analysis in the Cancer Atlas Viewer. J Geograph Syst 7(1):67-84

Jacquez GM, Greiling DA, Kaufmann AM (2005) Design and implementation of space time information systems. J Geograph Syst 7(1):7-24

Kim MJ, Nriagu J, Haack S (2002) Arsenic species and chemistry in groundwater of southeast Michigan. Environ Pollut 120(2):379-390

Kolker A, Haack SK, Cannon WF, Westjohn DB, Kim MJ, Nriagu J, Woodruff LG (2003) Arsenic in southeastern Michigan. In Welch AH, Stollenwerk KG (eds) Arsenic in groundwater: geochemistry and occurrence. Kluwer, Dordrecht, pp.281-294

Maantay J (2002) Mapping environmental injustices: pitfalls and potential of geographic information systems in assessing environmental health and equity. Environ Health Persp 110(2):161-171

Mather FJ, Whited LE, Langlois EC, Shorter CF, Swalm CM, Shaffer JG, Harley WR (2004) Statistical methods for linking health, exposure and hazards. Environ Health Persp 112(14):1440-1445

Meliker JR, Jacquez GM (2007) Space-time clustering of case-control data with residential histories: Insights into empirical induction periods, age-specific susceptibility, and calendar year specific effects. Stoch Environ Res Risk Assess 21(5):625-634

Meliker JR, Nriagu JO, Hammad AS, Savoie KL, Jamil H, Devries JM (2001) Spatial clustering of emergency department visits by asthmatic children in an urban area: southwestern Detroit, Michigan. Amb Child Health 7(3-4):297-312

Meliker JR, Slotnick MJ, AvRuskin GA, Kaufmann AM, Jacquez GM, Nriagu JO (2005) Improving exposure assessment in environmental epidemiology: application of spatiotemporal visualization tools. J Geograph Syst 7(1):49-66

Nriagu JO, Lawson G, Wong HKT, Azcue JM (1993) A protocol for minimizing contamination in the analysis of trace metals in Great Lakes waters. J Great Lakes Res19:175-182

Pickle LW, Szczur M, Lewis DR, Stinchcomb DG (2006) The crossroads of GIS and health information: a workshop on developing a research agenda to improve cancer control. Int J Health Geog 5(1):51

Reif JS, Burch JB, Nuckols JR, Metzger L, Ellington D, Anger WK (2003) Neurobehavioral effects of exposure to trichloroethylene through a municipal water supply. Environ Res 93(3):248-258

Ryan PB, Huet N, MacIntosh DL (2000) Longitudinal investigation of exposure to arsenic, cadmium, and lead in drinking water. Environ Health Persp 108(8):731-735

Slotnick MJ, Meliker JR, Nriagu JO (2006) Effects of time and point-of-use devices on arsenic levels in Southeastern Michigan drinking water, USA. Sci Total Environ 369 (1-3):42-50

Steinmaus C, Yuan Y, Smith AH (2005) The temporal stability of arsenic concentrations in well water in western Nevada. Environ Res 99(2):164-168

Steinmaus C, Yuan Y, Bates MN, Smith AH (2003) Case-control study of bladder cancer and drinking water arsenic in the western United States. Am J Epidemiol 158(12):1193-1201

Swartz CH, Rudel RA, Kachajian JR, Brody JG (2003) Historical reconstruction of wastewater and land use impacts to groundwater used for public drinking water: exposure assessment using chemical data and GIS. J Exp Anal Env Epid 13(5):403-441

List of Figures

PART A   GI Software Tools

A.1   Spatial Statistics in ArcGIS

Fig. A.1.1	Right click on a script tool and select *Edit* to see the Python source code	28
Fig. A.1.2	Weighted mean center of population, by county, 1910 through 2000	29
Fig. A.1.3	Core areas for five gangs based on graffiti tagging	30
Fig. A.1.4	Relative per capita income for New York, 1969 to 2002	32
Fig. A.1.5	Components of the K function graphical output	32
Fig. A.1.6	An analysis of poverty in Ecuador using local Moran's I	34
Fig. A.1.7	An analysis of vandalism hot spots in Lincoln, Nebraska using G_i^*	34
Fig. A.1.8	Traffic conditions or a barrier in the physical landscape can dramatically change actual travel distances, impacting results of spatial analysis	35
Fig. A.1.9	GWR optionally creates a coefficient surface for each model explanatory variable reflecting variation in modeled relationship	36
Fig. A.1.10	Default output from the regression tools is a map of model over- and underpredictions	38
Fig. A.1.11	Geoprocessing Resource Center Web page	39

A.2   Spatial Statistics in SAS

Fig. A.2.1	Moran's I workflow implemented in SAS	46
Fig. A.2.2	Visual Basic interface inside ArcGIS, the Load Data and Model Tabs	47
Fig. A.2.3	Eigenvector spatial filtering work flow implemented in SAS	49

A.3   Spatial Econometric Functions in **R**

Fig. A.3.1	Boston tract log median house price data: plots of spatial autoregressive error model fit components and residuals for all 506 tracts	65
Fig. A.3.2	Comparison of model prediction root mean square errors for four models divided north/south, Boston house price data	66
Fig. A.3.3	Comparison of model prediction root mean square errors means and standard deviations for 100 random samples of 250 in-sample tracts and 256 out-of-sample tracts, for five models, Boston house price data	67

A.4 GeoDa: An Introduction to Spatial Data Analysis

Fig. A.4.1	The opening screen with menu items and toolbar buttons	78
Fig. A.4.2	Linked box maps, box plot and cartogram, raw and smoothed prostate cancer mortality rates	79
Fig. A.4.3	Multivariate exploratory data analysis with linking and brushing	81
Fig. A.4.4	LISA cluster maps and significance maps	83
Fig. A.4.5	Maximum Likelihood estimation of the spatial error model	85

A.5 STARS: Space-Time Analysis of Regional Systems

Fig. A.5.1	STARS in GUI mode	95
Fig. A.5.2	STARS in command line interface mode	96
Fig. A.5.3	STARS in CLI+GUI mode	97
Fig. A.5.4	Multiple views of U.S. per capita income data	98
Fig. A.5.5	Linking multiple views	99
Fig. A.5.6	Brushing multiple views	100
Fig. A.5.7	Roaming a map with brushing	101
Fig. A.5.8	Spatial traveling with brushing	102
Fig. A.5.9	Time roaming	103
Fig. A.5.10	Space-time instabilities	104
Fig. A.5.11	Scatter plot generated time path	105
Fig. A.5.12	Map generated time series	106
Fig. A.5.13	Distributional mixing	107
Fig. A.5.14	Spatial and temporal covariance networks	108
Fig. A.5.15	Spider graph of temporal networks	109

A.6 Space-Time Intelligence System Software for the Analysis of Complex Systems

Fig. A.6.1	Visualization and exploration of space-time patterns in daily beer sales at Dominick's stores in the greater Chicago area in 1990	117
Fig. A.6.2	STIS visualizes spatial weights by outlining the selected location (centroid or polygon), and the localities to which it is connected	118
Fig. A.6.3	Automatic variogram model fitting of soil Cadmium concentrations in the Jura mountains, France	119
Fig. A.6.4	A-spatial regression analysis of breast cancer in the northeastern United States. A local Moran analysis found significant clusters of high and low residuals (map top center) and a global Moran's I of 0.18 ($p < 0.001$)	121

A.8 GeoSurveillance: GIS-based Exploratory Spatial Analysis Tools for Monitoring Spatial Patterns and Clusters

Fig. A.8.1	Structure of GeoSurveillance	137
Fig. A.8.2	Statistical analysis procedures in GeoSurveillance	138

Fig. A.8.3	Result tables linked to the map in Fig. A.8.4	*143*
Fig. A.8.4	Map of the local score statistic for North Carolina SIDS data	*144*
Fig. A.8.5	Results for the local score (upper) and M (lower) statistics	*145*
Fig. A.8.6	Maps of adjusted local score statistic for different bandwidths	*146*
Fig. A.8.7	Linked windows of cusum map, tables and charts	*146*
Fig. A.8.8	Enlarged image (tables and parameter control panels)	*147*
Fig. A.8.9	Maximum cusum charts and 1998 map when $\sigma = 0$	*147*
Fig. A.8.10	Maximum cusum charts and 1998 maps when $\sigma = 1.5$	*148*

A.9 Web-based Analytical Tools for the Exploration of Spatial Data

Fig. A.9.1	Basic Geotools architecture (original)	*159*
Fig. A.9.2	Geotools interface	*159*
Fig. A.9.3	Extended Geotools architecture	*160*
Fig. A.9.4	Customized interface	*161*
Fig. A.9.5	Welcome screen and general options	*163*
Fig. A.9.6	Excess Rate map, St. Louis region homicides (1979-84)	*164*
Fig. A.9.7	Empirical Bayes smoothing, colon cancer, Appalachia (1994-98). Two box maps with smoothed map on top and original raw rate on bottom	*165*
Fig. A.9.8	Empirical Bayes subset smoothing, colon cancer, Appalachia (1994-98). Two box maps with smoothed map on top and original raw rate on bottom	*166*
Fig. A.9.9	Spatial smoothing, colon cancer, Appalachia (1994-98)	*167*
Fig. A.9.10	Moran scatterplot, St. Louis region homicide rate (1979-84)	*168*
Fig. A.9.11	Moran scatterplot, East subregion homicide rate (1979-84)	*169*

A.10 PySAL: A Python Library of Spatial Analytical Methods

Fig. A.10.1	PySAL components	*178*
Fig. A.10.2	Nearest neighbor graphs	*180*
Fig. A.10.3	Spatial time paths	*181*
Fig. A.10.4	Spider and temporal contiguity graphs	*183*
Fig. A.10.5	Spatial smoothing of ACN county prostate rates	*184*
Fig. A.10.6	Regionalization of State incomes using AZP	*185*
Fig. A.10.7	Spatial regression model specification	*187*
Fig. A.10.8	Spatial regression model object attributes	*188*
Fig. A.10.9	Spatial two stage least squares with HAC error variance	*188*
Fig. A.10.10	Architecture of spatial weights Web service	*189*
Fig. A.10.11	Weights Web service user interface	*190*
Fig. A.10.12	Weights in XML format	*190*

PART B Spatial Statistics and Geostatistics

B.1 The Nature of Georeferenced Data

Fig. B.1.1	Binary and standardized connectivity matrices based on area adjacencies	*198*
Fig. B.1.2	Stages in the construction of the spatial data matrix	*200*
Fig. B.1.3	Examples of attributes by levels of measurement and types of space	*202*
Fig. B.1.4	From geographical reality to the spatial data matrix	*204*
Fig. B.1.5	Cressie's (1993) typology of spatial data and the two conceptualizations of geographic reality	*210*

B.2 Exploratory Spatial Data Analysis

Fig. B.2.1	Displays of the reported Medicaid program quality score values 1986: a) stem and leaf display; b) stripchart with jittered points; c) boxplot with standard whiskers; d) histogram with overplotted density curves for selected bandwidths	*222*
Fig. B.2.2	Medicaid program quality scores 1986: a) empirical cumulative distribution function, and b) dotchart	*223*
Fig. B.2.3	Medicaid program quality scores, 1986: thematic cartography as a method for statistical display	*226*
Fig. B.2.4	North Carolina Freeman-Tukey transformed SIDS rates by county for 1974-1978 conditioned on four shingles of the Freeman-Tukey transformed nonwhite live birth rates	*227*
Fig. B.2.5	Choropleth maps of population per crime against property, rank wealth and percentage literacy, France	*228*
Fig. B.2.6	Choropleth maps of population per crime against property, conditioned on ranked wealth and percentage literacy, France	*229*
Fig. B.2.7	Seismic events near Fiji since 1964, conditioned on depth	*230*
Fig. B.2.8	Kernel density plots of seismic events near Fiji; three increasing bandwidth settings	*231*
Fig. B.2.9	Exploratory geostatistical display of Swiss precipitation data from the 1997 Spatial Interpolation Comparison contest: a) precipitation quartiles; b) plot of precipitation by northings; c) plot of precipitation by eastings; d) histogram and density of precipitation	*233*
Fig. B.2.10	Influence plots for trend surfaces, Swiss precipitation data: a) quadratic trend surface; b) cubic trend surface	*233*
Fig. B.2.11	h-scatterplots: scatterplots of pairs of observed values conditioned on distance	*235*
Fig. B.2.12	Swiss precipitation data – binned classic and robust variogram values: a) variogram cloud display; b) variogram values	*235*
Fig. B.2.13	Detecting directionality in the variogram of Swiss precipitation data: a) variogram map showing binned semivariance values by direction and distance; b) classical variograms for four axes at 0°, 45°, 90° and 135°	*236*
Fig. B.2.14	Median polish for North Carolina SIDS data – the Freeman-Tukey transformed SIDS rates and fitted smoothed values are mean-centred to use the same scale as the residuals	*238*

Fig. B.2.15	Local G_i and G_i^* statistics: population per crime against property, France	240
Fig. B.2.16	Local I_i statistics for the null model, the residuals of the simultaneous autoregressive model, and the residuals of the linear model including literacy and wealth: population per crime against property, France	242
Fig. B.2.17	Conditioned choropleth LISA map: Moran's I_i for the null model conditioned on the LISA quadrant	242
Fig. B.2.18	Moran scatterplots for a) null; b) simultaneous autoregressive; c) linear model with covariates; and d) influence map for the three models	243
Fig. B.2.19	APLE plot and local APLE values for the population per crime rate: a) approximate profile-likelihood estimator plot, showing observations with influence; b) local APLE values, with observations with influence marked by asterisks	244
Fig. B.2.20	Six eigenvector maps for eigenvectors: null model	245
Fig. B.2.21	Two eigenvector maps for eigenvectors: linear model with covariates	245
Fig. B.2.22	Population per crime against property: a) population per crime against property; b) geographically weighted means; and c) geographically weighted standard deviations	246
Fig. B.2.23	Conditioned choropleth map of the geographically weighted standard deviation on the inverted crime rate, conditioned on population size	247
Fig. B.2.24	Maps of geographically weighted regression coefficients: a) intercept; b) percent literacy; c) rank wealth; and d) the coefficient of determination	248

B.4 Spatial Clustering

Fig. B.4.1	The Moran scatter plot	291

B.5 Spatial Filtering

Fig. B.5.1	Geographic distributions across the Cusco Department of Peru: a) transformed population density; (b) transformed elevation standard deviation	302
Fig. B.5.2	The Cusco Department of Peru: (a) geographic distribution of areal unit centroids; (b) a three-parameter gamma distribution description of the d_i values for $G_i(d)$; (c) four selected areal unit trajectories for identifying the d_i values for transformed population density	306
Fig. B.5.3	Geographic distributions across the Cusco Department of Peru of $G_i(d)$-based spatial variates: (a) extracted from the transformed population density; (b) extracted from the transformed elevation standard deviation	306
Fig. B.5.4	The Cusco Department of Peru; magnitude in the choropleth maps is directly related to gray tone darkness: (a) the minimum spanning tree connecting the areal unit centroids; (b) E_1, $MC/MC_{max} = 1$; (c) E_3, $MC/MC_{max} = 0.78$; (d) E_9, $MC/MC_{max} = 0.52$; (e) E_{14}, $MC/MC_{max} = 0.25$	308
Fig. B.5.5	Typology-based spatial filters for the Cusco Department of Peru: (a) for transformed population density; (b) for transformed elevation standard deviation	311

Fig. B.5.6 Generalized linear model (GLM) results: (a) the population density GLM spatial filter; (b) scatterplot of the predicted versus the observed *pd*; (c) scatterplot of the predicted versus the observed *pd* with the four largest values set aside   *313*

Fig. B.5.7 Geographically varying coefficients for the GLM population density model: (a) spatially varying intercept term; (b) spatially varying slope coefficient   *314*

Fig. B.5.8 Generalized linear model (GLM) imputation results: (a) scatterplot of the imputed versus the observed population densities (*pd*); (b) scatterplot of the imputed versus the observed population densities (*pd*) with the four largest values set aside   *316*

B.6 The Variogram and Kriging

Fig. B.6.1 Discretization of the lag into bins for irregularly scattered data   *322*

Fig. B.6.2 Three idealized variogram forms: (a) unbounded; (b) bounded; and (c) is the spatially correlated component   *326*

Fig. B.6.3 (a) Directional variogram computed on the raw data (230 points) from the Yattendon Estate, and (b) directional variogram computed on the residuals from the class means   *332*

Fig. B.6.4 Experimental variograms computed by the method of moments (MoM) estimator for lag distances of 20 m, 15 m and 40 m, and for the complete data set of 230 sites [(a), (b) and (c), respectively], subset of 94 data [(d), (e), (f)] and subset of 47 data [(g), (h), (i)] for topsoil K on the Yattendon Estate   *333*

Fig. B.6.5 Directional variogram computed on the pH data from Broom's Barn Farm (433 sampling points): (a) with the best fitting isotropic model, and (b) with an isotropic exponential function   *335*

Fig. B.6.6 Variogram of yield 2001 for the Yattendon Estate: (a) experimental variogram, and (b) the experimental variogram with the fitted double spherical model   *336*

Fig. B.6.7 Maps of block kriged predictions of topsoil potassium at the Yattendon Estate for: (a) complete set of 230 data, (b) subset of 94 data, and (c) subset of 46 data   *345*

Fig. B.6.8 Maps of block kriged kriging variances for topsoil potassium at the Yattendon Estate for: (a) total of 230 data, (b) subset of 94 data, and (c) subset of 46 data   *346*

Fig. B.6.9 Map of punctually kriged kriging variances for topsoil potassium at the Yattendon Estate for the complete set of 230 data   *347*

Fig. B.6.10 Map of punctually kriged predictions for topsoil pH at the Broom's Barn Farm   *348*

Fig. B.6.11 Maps of wheat yield for 2001 at the Yattendon Estate for: (a) ordinary kriged predictions, (b) predictions of the long-range component of the variation, and (c) predictions of the short-range component of the variation   *349*

Part C   Spatial Econometrics

C.5 *Geographically Weighted Regression*

Fig. C.5.1	Standardized mortality rates for bladder cancer among white males from 1970 to 1990 in the State Economic Areas of the contiguous United States	478
Fig. C.5.2	Estimated GWR coefficients for $\hat{\beta}_1$ (intercept), $\hat{\beta}_2$ (smoking proxy), $\hat{\beta}_3$ (log population density)	479
Fig. C.5.3	GWR estimated coefficients $\hat{\beta}_2$ versus $\hat{\beta}_1$, $\hat{\beta}_3$ versus $\hat{\beta}_1$, $\hat{\beta}_3$ versus $\hat{\beta}_2$	480
Fig. C.5.4	Estimated GWR coefficients for the intercept and the smoking proxy, parallel coordinate plot for condition indexes and variance decomposition proportions, and histogram of condition indexes with a selection set for SEAs with the thirty largest condition indexes for the largest variance component	482
Fig. C.5.5	Estimated GWR coefficients for smoking proxy and population density, scatter plot for variance decomposition proportions for these two regression terms, and histogram of condition indexes with a selection set for SEAs with both variance decomposition proportions greater than 0.6 for the largest variance component	483
Fig. C.5.6	GWR coefficients and SVCP coefficients for the intercept, smoking proxy, and population density	484

C.7 *Multilevel Modeling*

Fig. C.7.1	Typology of studies	509
Fig. C.7.2	Multilevel structure of repeated measurements of individuals over time across neighborhoods with individuals having multiple membership to different neighborhoods across the time span	522

Part D   The Analysis of Remotely Sensed Data

D.1 *ARTMAP Neural Network Multisensor Fusion Model for Multiscale Land Cover Characterization*

Fig. D.1.1	A comparison of error bound limits of ARTMAP and linear mixture models	535
Fig. D.1.2	Predictive and actual forest fraction cover	535
Fig. D.1.3	Predicted forest cover using ARTMAP and linear mixture models	536
Fig. D.1.4	Category visualization in the ARTMAP model	537
Fig. D.1.5	Predictive and actual forest cover class using multi-temporal images	539

D.2 *Model Selection in Markov Random Fields for High Spatial Resolution Hyperspectral Data*

Fig. D.2.1	Grey levels image of the first principal scores of the Lamar imagery	554

Fig. D.2.2	Lamar imagery scatter plot of the 128 bands mapped onto the correlation space spanned by the first two principal components and representative bands selected for subsequent analysis	555
Fig. D.2.3	First-order lattice adjacencies	558
Fig. D.2.4	Second-order lattice adjacencies	558
Fig. D.2.5	Third-order lattice adjacencies	558
Fig. D.2.6	BIC-PL values of nine spatial auto-logistic models, fitted on a number of hyperspectral bands	559
Fig. D.2.7	BIC-PL values of nine spatial auto-binomial models, fitted on further hyperspectral bands	560

Part E   Applications in Economic Sciences

E.2   Income Distribution Dynamics and Cross-Region Convergence in Europe

Fig. E.2.1	Distributions of relative (per capita) regional income, 1995 versus 2003	610
Fig. E.2.2	Tukey boxplots of relative (per capita) regional income across 257 European regions	612
Fig. E.2.3	Cross-profile dynamics across 257 European regions, retaining the ranking fixed at the initial year, relative (per capita) income, advancing upwards: 1995, 1999 and 2003	613
Fig. E.2.4	Relative income dynamics across 257 European regions, the estimated $g_5(z\|y)$: (a) stacked density plot, and (b) highest density regions boxplot	615
Fig. E.2.5	The ergodic density $f_\infty(z)$ implied by the estimated $g_5(z\|y)$ and the marginal density function $f_{1995}(y)$	617
Fig. E.2.6	Densities of relative (per capita) income, 1995 versus 2003: the spatial filtering view	619
Fig. E.2.7	Tukey boxplots of relative (per capita) income, across 257 European regions: the spatial filtering view	619
Fig. E.2.8	Stochastic kernel mapping from the original to the spatially filtered distribution, the estimated $g(\tilde{y}\|y)$: (a) the stacked conditional density plot, and (b) the highest density regions boxplot	620
Fig. E.2.9	The spatial filter view of relative income dynamics: the estimated $g_5(\tilde{z} \mid \tilde{y})$, (a) stacked density plot, and (b) highest density regions boxplot	621

Part F   Applications in Environmental Sciences

F.1   A Fuzzy k-Means Classification and a Bayesian Approach for Spatial Prediction of Landslide Hazard

Fig. F.1.1	Distribution of landslides over the Clearwater National Forest drainage during the winter 1995/96 storm events	667
Fig. F.1.2	Fuzziness performance index (F) and normalized classification entropy (H) versus number of classes: (a) fuzziness exponent $\phi = 1.15$, and (b) fuzziness exponent $\phi = 1.50$	669

Fig. F.1.3	Plot of $-[(\delta J_E/\delta\phi)c^{0.5}]$ versus ϕ for $c = 6$	670
Fig. F.1.4	Drapes of fuzzy k-means classification of training area with six classes	671
Fig. F.1.5	Drapes of the confusion index (CI) and the most dominate class	672
Fig. F.1.6	Bayesian model output. Probabilities of the occurrence of non-road related landslides in the CNF	674
Fig. F.1.7	Bayesian model output. Probabilities of the occurrence of road related landslides in the CNF	675
Fig. F.1.8	Difference of probabilities of the occurrence of non-road related versus road related landslides in the CNF	675
Fig. F.1.9	Drape of predicted landslide hazard for non-road related landslides using Bayesian modeling	676
Fig. F.1.10	Drape of predicted landslide hazard for road related landslides using Bayesian modeling	676
Fig. F.1.11	Drape of predicted landslide hazard difference for road related and non-road landslides using Bayesian modeling	677

F.2 Incorporating Spatial Autocorrelation in Species Distribution Models

Fig. F.2.1	Mojave Desert study area. The square highlights the section used for predictions in Figs. F.2.3 and F.2.4	687
Fig. F.2.2	Mean AUC values (with standard deviation) from all alliances for each model	695
Fig. F.2.3	Generalized linear model predictions of Yucca brevifolia (YUBR): a) non-spatial model; b) model with kriged autocovariate; c) model with mean simulation autocovariate; d) model with single simulation autocovariate	696
Fig. F.2.4	Classification tree model predictions of Yucca brevifolia (YUBR): a) non-spatial model; b) model with kriged autocovariate; c) model with mean simulation autocovariate; d) model with single simulation autocovariate	697

F.3 A Web-based Environmental Decision Support System for Environmental Planning and Watershed Management

Fig. F.3.1	The study area	705
Fig. F.3.2	Overall client server transaction	706
Fig. F.3.3a	Web-based user interface for multi-criteria analysis	713
Fig. F.3.3b	User input window for the model	713
Fig. F.3.4	Overall ESI model process	714
Fig. F.3.5	A screen shot of portion of the ESI calculation which was written in ArcView avenue script	715

778   List of figures

Part G   Applications in Health Sciences

G.1   Spatio-Temporal Patterns of Viral Meningitis in Michigan, 1993-2001

Fig. G.1.1	Cumulative incidence per 100,000 of viral meningitis by state, continental United States, 1989-1994	723
Fig. G.1.2	(a) Weekly number of viral meningitis cases in Michigan, 1993-2001; (b) the autocorrelation function (ACF) from a zero to a seven-year lag	726
Fig. G.1.3	Cumulative incidence per 100,000 of viral meningitis in Michigan by county, 1993-2001, age-adjusted to the 1990 population	727
Fig. G.1.4	Coefficient of variation (CV) of annual incidence of viral meningitis cases in Michigan by county, 1993-2001	728
Fig. G.1.5	Spatio-temporal clusters of viral meningitis cases by county in Lower Michigan, 1993-2001	729

G.2   Space-Time Visualization and Analysis in the Cancer Atlas Viewer

Fig. G.2.1	Time enabled visualization in the Cancer Atlas Viewer	742
Fig. G.2.2	Difference maps for colon cancer mortality rates between the time intervals 1990-1994 and 1970-1974	743
Fig. G.2.3	State Economic Areas shaded by their rates for colon cancer mortality in the years 1990-1994	744
Fig. G.2.4	Statistical disagreement about the significance of the colon cancer mortality rate for white males in Columbus (GA)	746
Fig. G.2.5	Statistical disagreement about the significance of the colon cancer mortality rate for African-American females (RBF) in Greenville (SC)	747
Fig. G.2.6	Persistent clustering of low white male colon cancer mortality rates (RWM) centered on Auburn (AL)	748

G.3   Exposure Assessment in Environmental Epidemiology

Fig. G.3.1	Public and private water supply status in Southeastern Michigan in 1932, 1964, and 1993	758
Fig. G.3.2	Feature of Time-GIS. Town boundaries change with time (1950, 1992)	759
Fig. G.3.3	Time series graphs of arsenic exposure: cases and controls	761
Fig. G.3.4	Histogram of source of drinking water in 1965: cases and controls	761
Fig. G.3.5	Scatter plot of arsenic exposure and cigarettes smoked in 1972: cases and controls	762
Fig. G.3.6	Exposure life-lines for cigarette smoking: cases and controls	762
Fig. G.3.7	Exposure life-lines for arsenic exposure and cigarette smoking: cases and controls	764

List of Tables

PART A   GI Software Tools

A.1   Spatial Statistics in ArcGIS

Tab. A.1.1	Tools in the measuring geographic distributions toolset	29
Tab. A.1.2	A summary of the tools in the analyzing patterns toolset	31
Tab. A.1.3	A summary of the tools in the mapping clusters toolset	33
Tab. A.1.4	A summary of the tools in the modeling spatial relationships toolset	35
Tab. A.1.5	A variety of potential applications for regression analysis	37

A.2   Spatial Statistics in SAS

Tab. A.2.1	Neighbor file format	48
Tab. A.2.2	Eigenvector spatial filtering for ArcGIS and SAS output file descriptions	50

A.4   GeoDa: An Introduction to Spatial Data Analysis

Tab. A.4.1	GeoDa functionality overview	77

A.5   STARS: Space-Time Analysis of Regional Systems

Tab. A.5.1	Geocomputational methods contained in STARS	93
Tab. A.5.2	Visualization capabilities in STARS	94

A.6   Space-Time Intelligence System Software for the Analysis of Complex Systems

Tab. A.6.1	Summary of STIS functionality	114

A.7   Geostatistical Software

Tab. A.7.1	List of main geostatistical software with the corresponding reference	126
Tab. A.7.2	List of functionalities for main geostatistical software	131

A.10 PySAL: A Python Library of Spatial Analytical Methods

Tab. A.10.1 PySAL functionality by component   *179*

PART B Spatial Statistics and Geostatistics

B.5 Spatial Filtering

Tab. B.5.1 Transformed population density and elevation variability:
           spatial autocorrelation in terms of *MC* and *GR*   *303*
Tab. B.5.2 Spatial autocorrelation contained in the 30 PCNM eigenvectors with
           non-zero eigenvalues   *309*
Tab. B.5.3 Eigenvector spatial filter regression results using a 10 percent level of
           significance selection criterion   *310*
Tab. B.5.4 Geographically varying coefficients: spatial autocorrelation in terms of
           MC and *GR*   *315*

B.6 The Variogram and Kriging

Tab. B.6.1 Summary statistics   *331*
Tab. B.6.2 Variogram model parameters   *334*

Part C Spatial Econometrics

C.2 Spatial Panel Data Models

Tab. C.2.1 Two goodness-of-fit measures of the four spatial panel data models   *401*
Tab. C.2.2 Prediction formula of the four spatial panel data models   *403*

C.3 Spatial Econometric Methods for Modeling Origin-Destination Flows

Tab. C.3.1 Data organization convention   *414*
Tab. C.3.2 Unadjusted and adjusted model estimates   *425*
Tab. C.3.3 Test for significant differences between the unadjusted and adjusted
           model estimates   *426*
Tab. C.3.4 Zero intraregional flows versus non-zero intraregional flows   *426*
Tab. C.3.5 Spatial Tobit experimental results   *429*

C.4 Spatial Econometric Model Averaging

Tab. C.4.1 Posterior model probabilities for varying spatial neighbors   *452*
Tab. C.4.2 Metropolitan sample SDM model estimates   *453*
Tab. C.4.3 Metropolitan sample MC^3 results for $m = 7$ neighbors   *454*

Tab. C.4.4	Model averaged estimates based on the top 12 models	455
Tab. C.4.5	Posterior model probabilities for the top 12 models and associated neighbors	457
Tab. C.4.6	Model averaging over both neighbors and variables	457

C.5 Geographically Weighted Regression

Tab. C.5.1	Condition index and variance-decomposition proportions for the largest variance component	480

C.6 Expansion Method, Dependency, and Multimodeling

Tab. C.6.1	Regression results	495
Tab. C.6.2	Multimodeling results	500
Tab. C.6.3	Group means	501

Part D   The Analysis of Remotely Sensed Data

D.1 ARTMAP Neural Network Multisensor Fusion Model for Multiscale Land Cover Characterization

Tab. D.1.1	RMS error	536
Tab. D.1.2	RMS error of the testing data	539
Tab. D.1.3	RMS error of the whole image	540

D.2 Model Selection in Markov Random Fields for High Spatial Resolution Hyperspectral Data

Tab. D.2.1	Models used in the application	556
Tab. D.2.2	Parametric specifications used in the application	556

Part E   Applications in Economic Sciences

E.1 The Impact of Human Capital on Regional Labor Productivity in Europe

Tab. E.1.1	Parameter estimates from SDM and SEM specifications	593
Tab. E.1.2	Direct, indirect and total impact estimates	594

E.3 A Multi-Equation Spatial Econometric Model, with Application to EU Manufacturing Productivity Growth

Tab. E.3.1	Unrestricted model estimates	639

Tab. E.3.2	Conditional tests of core-periphery contrasts	*640*
Tab. E.3.3	Conditional tests of time homogeneity	*641*
Tab. E.3.4	Final model estimates	*643*

Part F Applications in Environmental Sciences

F.1 *A Fuzzy k-Means Classification and a Bayesian Approach for Spatial Prediction of Landslide Hazard*

Tab. F.1.1	Cluster centers for six classes	*670*
Tab. F.1.2	Conditional probabilities of fuzzy k-means predictor datasets for development of the Bayesian models for the CNF	*673*
Tab. F.1.3	Proportion of presence/absence associated with probabilities for non-road and road related landslides	*678*
Tab. F.1.4	Evaluation of new and existing modeling techniques in the CNF derived by arbitrary cut-off values for the NRR landslides	*679*
Tab. F.1.5	Evaluation of new and existing modeling techniques in the CNF derived by arbitrary cut-off values for the RR landslides	*679*

F.2 *Incorporating Spatial Autocorrelation in Species Distribution Models*

| Tab. F.2.1 | Environmental variables used in this study | *688* |
| Tab. F.2.2 | Vegetation alliances modeled, and the proportion of the full (test and train, $n = 3,819$) dataset in which they are present | *688* |

F.3 *A Web-based Environmental Decision Support System for Environmental Planning and Watershed Management*

| Tab. F.3.1 | Criteria used in the analysis and their weights | *709* |
| Tab. F.3.2 | Base data layers collected and their sources | *709* |

Part G Applications in Health Sciences

G.1 *Spatio-Temporal Patterns of Viral Meningitis in Michigan, 1993-2001*

| Tab. G.1.1 | Relative presence of identified causative viruses associated with viral meningitis cases in Michigan, June to December 2001 | *730* |

G.2 *Space-Time Visualization and Analysis in the Cancer Atlas Viewer*

| Tab. G.2.1 | A listing of a few online atlas projects | *738* |
| Tab. G.2.2 | Moran's I statistics and p-values calculated from 999 Monte Carlo randomizations | *744* |

Tab. G.2.3 Classification of 1980 locations (99 regions over five time intervals by four population subgroups) by the local Moran and the Getis-Ord statistic G^* 745

Tab. G.2.4 Some characteristics of the non-concordant classes from Tab. G.2.3 745

Tab. G.2.5 Cluster persistence over time, measured in terms of number of cases. Totals are from all race-gender subgroups over all pairs of two sequential times 749

G.3 *Exposure Assessment in Environmental Epidemiology*

Tab. G.3.1 Spatio-temporal dataset format for STIS point features: residential mobility history and water consumption history for two participants 757

Subject Index

A

Accuracy assessment, in remote sensing, 576-578
AIC. *See* Akaike information criterion
Akaike information criterion (AIC), 120, 454, 500, 401, 548, 690
    corrected for GWR, 466
AMOEBA, 295
Anisotropy, 326
ArcGIS, 274
    spatial statistical tools in, 8, 27-39
Area data, 197, 208, 280
ART (adaptive resonance theory), 531
ARTMAP neural network multisensor fusion model, 17, 529-541
Asymptotic variance matrices, 388, 392, 394, 395, 398

B

Bandwidth. *See* Kernel
Bayes' theorem, 437, 658, 666
Bayesian
    hierarchical models, 474
    spatially varying coefficient (SVP) model, 474-475
    spatially varying coefficient process (SVCP) model, 475-476
Bayesian Markov Chain Monte Carlo (MCMC) estimation, spatial regression models, 363-364
Bayesian MCMC estimation. *See* Bayesian Markov Chain Monte Carlo estimation
Bayesian model averaging for spatial regression models, 458,
    applied illustration, 450-458
    models based on different spatial weight matrices, 444-447
    models based on different variables, 447-449
    theory, 436-440
    weight matrix and variable selection model averaging, 455-457
Bayesian posterior distribution, 363-364
Bayesian spatial autoregressive Tobit model estimates, 429,
Besag and Newell's method, 293, 294
Box-Cox transformation, 301
Brushing. *See* STARS

C

Cancer Atlas Viewer, 21, 737-740
Casetti's expansion method, 488-493
Change of support problem, 212, 213
Classification techniques
    fuzzy k-means, 659-665
    machine learning classifiers, 571
    supervised classifiers, 570
Classification trees (CT), 690-691
ClusterSeer 2, 274
Common factor parameter restriction, 588
Complete spatial randomness (CSR), 286
Concentrated log-likelihood function, 365
    fixed effects SEM, 394
    fixed effects spatial lag model, 391
    random effects SEM, 397
    random effects spatial lag model, 395
    SDM, 588
Conditional density function, 603
Conditional negative-semi-definite (CNSD), 327-329
Confirmatory spatial data analysis, 209-210
Connectivity structure of regions, 357
    See also Spatial weight matrix
Contextual heterogeneity, 512-513
Conventional regression model, 357
Cross-product statistic, 12, 259

786 *Subject index*

Cross-profile dynamics, across regions, 613
Cusum test, 141

D

Data generating process (DGP), 355
    SAR model, 358
    SDM, 359
    SEM, 358
Database management system (DBMS), 707
Decision support system. *See* WEDSS
Dempster-Shafer (D-S) theory of evidence, 653-654
Destination-based spatial dependence. *See* Spatial dependence in origin-destination (OD) flows
Direct impact estimates of human capital, 594
DGP. *See* Data generating process
Distance deterrence functions, 412-413
Distribution dynamics, 18, 106
    continuous model of, 601-603
    intra-distribution dynamics, 613-616
    shape dynamics, 610-612

E

Ecological fallacy, 213-214
Ecological inference problem, 213
Eigenfunction spatial filtering and geographical interpolation, 315-316
    GLM, 312-313
    GWR, 313-315
    spatial interaction data, 316-317
Empirical Bayes' smoother, 155, 166
Environmental epidemiology, 21
    exposure assessment, 753-764
Ergodic density function, 603
Error propagation, 208
ESDA. *See* Exploratory spatial data analysis
European regions, 592, 626-628, 647-649
Expansion method. *See* Casetti's expansion method
Expansion-based multimodeling, 498-501
Exploratory data analysis (EDA). 76-77, 80, 219, 219-221, 225
    plotting and EDA, 220-224

Exploratory space-time data analysis, 92-97, 117-118, 181
Exploratory spatial data analysis (ESDA), 3-4, 11-12, 116, 128, 209, 219, 220-221, 225, 232, 236
    areal data, 236-249
    extension to space-time, 92-97, 117-118, 181
    outliers in maps, 152-157
    point patterns, 229-236
    Tukey boxplots, 611-612, 619
    Web-based tools, 151-163
Exploring colon cancer patterns, 737-749

F

Factorial kriging. *See* Kriging
Filtering. *See* Spatial filtering
First law of geography. *See* Tobler's first law of geography
Fixed effects spatial panel data models, 389-394
Flow data. *See* Spatial data, spatial interaction data
Fuzzy
    ARTMAP, 530
    k-means classification, 19, 659-665, 669
    performance index (FPI), 665-666
    set theory, 660-661

G

Gaussian product kernel density estimator, 604
Geary ratio, 44, 303, 315
Geary's c, *44, 55, 95,* 265, 269-270, 284-285, 470, 605
Generalized linear model (GLM), 312, 690
GEOBICA (geographic object-based image change analysis, 17
    accuracy assessment, 576-578
    approaches, 569-572
    goal, 566-567
    image change analysis, 17
    imagery and pre-processing requirements, 568-569
    principles and post-processing, 575-576
    strategies, 572-575

GeoDa, 9
   functionality, 76-77
   goal, 74-75
   mapping and geovisualization, 76-80
   spatial autocorrelation analysis, 82-83
   spatial regression, 84-85
Geographical analysis machine. *See*
   Openshaw's geographical analysis
   machine
Geographical interpolation. *See* Spatial
   interpolation
Geographically weighted lasso, 472-473
Geographically weighted regression
   (GWR), 5-6, 114, 271-272
   applied illustrations, 477-484
   autoregressive, 472
   constrained, 472-473
   diagnostic tools, 469-472
   estimation, 462-467
   local weights matrix, 462
Geographically weighted ridge regression, 473
Geographic space,
   continuous (field) space, 202
   discrete (object) space, 202
Geographical user interface (GUI), 641
Geometric pre-processing, 568
Georeferenced data. *See* Spatial data
Georeferencing, 568-569
Geostatistical data. *See* Spatial data
Geostatistical software, 10
   affordability and user-friendliness, 131-132
   main functionalities, 128-130
   main packages, 126
   opensource code vs. black box software, 127
Geostatistics, 6
   applied illustrations, 344-350
   geostatistical prediction, 337-344
   kriging. *See* Kriging
   software. *See* Geostatistical software
   theory, 319-321
   variogram analysis, 321-336
GeoSurveillance (tools for monitoring spatial patterns and clusters), 10
   applied illustrations, 142-148
   goal, 136
   spatial statistical tests, 138-142
   structure, 137-138
GeoVista Studio, 75, 91, 176

Geovisualization and mapping, 77-80
Getis-Ord statistics. *See* Local G statistics
Getis' spatial filtering technique, 304-306
   and stochastic kernel estimation, 606-607
Gibb's sampling, 364
GISystem (GIS), 6, 10, 16, 19, 21, 567, 721, 253-754
GLM. *See* Generalized linear model
Global Moran. *See* Moran's I
Goodness-of-fit measures, 14, 378, 399-401, 466, 677
G statistics, 95, 118, 135, 142
   global, 285
   local, 33, 114, 269, 290
GWR. *See* Geographically weighted regression

H

Hausman's specification test, 399
HDR plots. *See* Highest density regions plots
Hessian matrix, 365
Heterogeneity, across regions, 356
Highest density regions (HDR) plots, 615, 620-621
High spatial resolution hyperspectral (HSRH) data, 17, 545
Hot spot analysis, 33-34. *See also* Getis-Ord local statistics
HSRH data. *See* High spatial resolution hyperspectral data
Human capital, 18, 585, 634

I

Image processing and remote sensing techniques
   classification techniques, 570-571
   pre-processing, 568-569
   rectification of images, 568
Impact estimates,
   applied illustration, 592-595
   calculating summary measures of, 369-370
   direct and indirect, 370
Indirect (spatial spillover) impact estimate of human capital, 594

Instrumental variables or generalized method of moments (IV/GMM) estimators, 380
Intra-distribution dynamics, 613-616

J

Join-count statistics, 263, 282

K

K function. *See* Ripley's K function
Kernel
   adaptive function, 464
   bandwidth, 465-466, 605
   fixed function, 464
   Gaussian function, 463, 604
k-means. *See* Classification techniques, *and* Fuzzy k-means classification
Kriging, 6, 13, 337
   applied illustrations, 344-349
   factorial, 342-343. 348
   ordinary, 338-341, 344
   types of, 338
   weights, 341
Kronecker product, 390, 414
Kulldorff's Scan test, 211, 728

L

Labor productivity, 583
Lagrange multiplier (LM) test, 365, 384-385
Land cover, 567
Land cover and land use change analysis, 566-567
Landslide, 656
Landslide hazard models, 657-658
Lawson-Waller score test, 296
Least-squares spatially lagged X regression model, 588
LeSage's spatial econometrics toolbox, 274
Likelihood ratio (LR) test, 365, 386, 547
LIMDEP, 637
Lindley paradox, 447
LISA. *See* Local indicators of spatial association
Local G statistics. *See* G statistics, local

Local indicators of spatial association (LISA), 76, 82-83, 93, 114, 269-270, 290-291
Local Moran. *See* Local indicators of spatial association
Local weights matrix, 462
Log-determinant, 360
Log-likelihood function, for the
   fixed effects SEM, 393
   fixed effects spatial lag model, 389
   random effects SEM, 395
   random effects spatial lag model, 394
   SAR model, 360
LSDV estimator, 383

M

Mahalanobis distance, 662
Markov Chain Monte Carlo (MCMC), 363, 452, 547
Markov Chain Monte Carlo estimation, 363, 452
Markov Chain Monte Carlo Model Composition (MC^3), 448-449, 454, 456
Markov random field (MRF) models, 17, 546, 549-556
   adjacency selection in, 550-554
   applied illustration, 554-560
Matheron's estimator. *See* Matheron's method of moments (MoM) estimator
Matheron's method of moments (MoM) estimator, 321-323
Matrix exponential spatial specification (MESS), 361
MAUP. *See* Modifiable areal units problem
Maximum likelihood estimation
   computational challenges, 360-361, 396
   spatial econometric interaction models, 418
   spatial lag models, 360
   spatial panel data models, 389-398
MCMC. *See* Markov Chain Monte Carlo
Metropolis-Hastings acceptance probability, 499
Metropolis-Hasting sampling, 364
Model comparison, spatial panel models, 399-401
Modifiable areal units problem (MAUP), 213
Modified partition entropy (MPE), 665

Moran scatter plots, 156-157
Moran's *I*, 6, 8, 31, 33, 34, 43-46, 55, 75, 77, 82, 83, 93, 122, 135, 142, 157, 162, 169, 189, 239-245, 257, 263-265, 290-292, 470, 605, 606, 618, 722, 740, 744, 749
Moran's *I* regression residual test, 264
Multilevel modeling, 15-16
    contextual heterogeneity, 512-513
    data structures, 510-511
    random coefficient models, 518-519
    random intercepts models, 514-515
Multimodeling, 496-498
Multisensor fusion models
    applied illustration, 534-539
    ARTMAP neural network, 531-532
    linear mixture model, 531

N

Nearest neighbor analysis, 286-287
Neighbor structure of the sample observations, 357-358
New economic geography, 630-632
Non likelihood-based estimation methods, 361

O

Object-based image analysis, remote sensing, 565
OD flows. *See* Origin-destination flows
Omitted variables, 356
Openshaw's geographical analysis machine (GAM), 211
Ordinary kriging. *See* Kriging
Origin-based spatial dependence. *See* Spatial dependence in origin-destination (OD) flows
Origin-centric ordering of the spatial interaction matrix, 414
Origin-destination (OD) flows, 417
    See also Spatial data, spatial interaction data
Origin-to-destination based spatial dependence. *See* Spatial dependence in origin-destination (OD) flows
Outlier maps, 153

P

Point data, 209, 286
Point pattern. *See* Point data
Poisson regression, 15, 114, 115, 120-122, 244, 427, 431,
Polygon, 9, 35, 48-50, 76, 77, 82, 94, 99, 107, 114-115, 118, 120, 130, 180, 200, 202, 206-207, 210, 221, 225, 280, 337, 575-578, 740, 747-748, 757-758
Posterior model probabilities, 445
Posterior odds ratio, 445
PPA(point pattern analysis), 274, 621
Product kernel estimator, 603-604
Pseudo-likelihood ratio test, 547
PySAL (a Python library of spatial analytical methods), 11
    applied illustrations, 180-191
    components and functionality, 177-179
    goal, 175-176

Q

Quadrat analysis, 286
Quantitative revolution in geography, 205
Queen-based spatial contiguity, 356

R

R, 53-55, 74
Radiometric pre-processing, 569
Random effects spatial panel data models, 394-398
Rank correlation statistics, 181
Rate smoothing, 154-156, 182
Regionalization, 184
Regression. *See* Conventional regression model, *and* Spatial regression models
Remote sensing, 16, 529, 565, 581
    hyperspectral, 545
Remotely sensed data, 565, 566
Residual spatial dependence, 358, 379
Ripley's *K* function, 5, 31-33. 266-267, 287
Robust Lagrange multiplier test, 385
Rook-based spatial contiguity, 281, 357
Root mean square error (RMSE), 66-67
Root mean square prediction error (RMSPE), 465

S

SANET, 274
SAR model. *See* Spatial autoregressive model
SAS, 8, 43-51, 56, 126, 131, 275, 551, 724
SaTScan procedure, 136, 137, 294
Scan test. *See* Kulldorff's Scan test
SDM. *See* Spatial data matrix *and* Spatial Durbin model
Segmentation, image objects, 569-570
SEM. *See* Spatial error model
Shape dynamics, 610-612
Smoothing of rates, 154-156, 182
Sources of model uncertainty arising from
    explanatory variable selection, 447-448
    spatial weight matrix specification, 444-446
Space-time data cube, 198
Space-time lag, 367
Space-time visualization, in the Cancer Atlas Viewer, 20, 742-748
Sparse matrix, 418
Spatial autocorrelation, 12, 82-83, 206
    attributes and uses, 257-258
    concept and definition, 256
    cross-product statistic, 259
    global measures and tests, 262-268
        Gamma, 262
        Geary's c, 265, 284-285
        Getis-Ord G, 285
        join count, 263
        Moran's I, 31, 46, 264, 282-283
    local measures and tests, 268-272
        Getis-Ord local statistics, 33, 114 269, 290
        LISA, 76, 82-83, 93, 114, 269-270, 290-291
        local Moran, 82-83, 157, 240, 244 739, 741, 745-749
    problems in dealing with, 272-273
    in OD flows, 417-423
    software, 274-275
    visualization, 156-157
Spatial autoregressive (SAR) model,
    DGP, 358
    expectation, 358
Spatial clusters, 12
    definition, 279
    general tests of clustering, 280-288
    local clustering statistics, 289-297

Spatial connectivity matrix. *See* Spatial weight matrix
Spatial data
    area (or lattice) data, 197, 208, 236, 280, 357
    geostatistical data, 208-209
    implications for data analysis, 208-211
    nature of, 197-214
    object data, 409
    point (pattern) data, 209, 236, 286, 357
    properties of, 204-208
    spatial data matrix, 199-201
    spatial interaction (or flow) data, 409-411
    spatial panel data, 377
    typology of, 210
Spatial data matrix (SDM), 198, 201-206
    properties of spatial data in, 204-208
    stages in the construction, 200
Spatial dependence in origin-destination (OD) flows
    destination-based, 418
    origin-based, 417
    origin-to-destination based, 418
Spatial Durbin model (SDM), 359, 382, 441, 586
    applied illustration, 585-596
    DGP, 359, 586
    interpretation of estimated parameters, 589-591
    model estimation, 588-589
Spatial econometric functions in R, 8
    classes and methods, 57-60
    illustrations, 65-68
Spatial econometric interaction models. *See* Spatial interaction models, spatial econometric extensions to
Spatial econometrics, 5, 13-15, 409
Spatial error dependence, 358
Spatial error model (SEM), 441
    DGP, 358
    expectation, 359
Spatial filtering, 12-13
    eigenfunction spatial filtering, 312-317
    Getis' G_i specification, 304-306, 606-607
    types of, 303-312
Spatial heterogeneity models, 372
    Casetti's expansion method, 487-502
    GWR, 461-484
Spatial interaction models, 14
    adjustments for addressing the zero flows problem, 427-431

classical (unconstrained) model, 411
data organization convention, 414
distance deterrence functions, 412-413
efficient computation, 416-417
large diagonal flow matrix elements, 423-424
log-normal model, 415
origin and destination variables, 412
spatial econometric extensions to, 417-431
Spatial interpolation, 13, 127, 129-130, 132, 252, 315, 469
Spatial lag models, 359
estimation, 360-365, 588-589
interpreting parameter estimates, 366-371, 589-591
SAR model, 358
SDM model, 359, 586
Spatial lag of the dependent variable, 359
Spatial panel data models, 377-404
Elhorst's Matlab routines, 377
fixed effects spatial error model, and estimation, 393-394
fixed effects spatial lag model, and estimation, 389-393
model comparison and prediction, 399-403
random effects spatial lag model, and estimations, 394-395
random effects spatial error model, and estimations, 395-398
standard models for spatial panels, and estimations, 378-388
Spatial regression models, cross-section data, 355-374
Bayesian MCMC estimation, 363-364
bias of least-squares, 362-363
estimates of parameter dispersion and inference, 365
inconsistent parameter estimations with least-squares, 359
interpreting parameter estimates, 366-371
LeSage's toolbox, 274
MESS model, 361
ML estimation, 361
SAR model, 358
SDM model, 359, 441, 586-588
SEM model, 358-359, 441
simultaneous feedback, 366
spatial heterogeneity models, 461, 487

Spatial regression models, panel data. *See* Spatial panel data models
Spatial sampling, 16, 21, 657
Spatial seemingly unrelated regression (SUR) with spatially lagged dependent variables, 630, 637-638
Spatial statistics, 4-5, 319
in ArcGIS, 27-39, 274
in R, 57-60, 274
in SAS, 44-51, 275
Spatial statistics in SAS, 8
eigenvector spatial filtering, 46-50
See also Spatial filtering
Moran scatterplot, 46
spatial analysis built-ins, 44-45
Spatial weight matrix, 260-261
distance decay formulations, 260, 281
first-order contiguity, 366
higher-order contiguity, 366
Queen-based contiguity, 281, 357
Rook-based contiguity, 281, 357
row-standardization, 260, 281
Spatial weights. *See* Spatial weight matrix
Spatially structured effects parameters, spatial econometric interaction models, 422
spdep package, 55
Species distribution models, 19
spatial autocorrelation in, 19-20, 685-699
Spectral transformations, 569
Stacked density plots, 615, 620-621
STARS (space-time analysis of regional systems), 9, 274
components and built-in methods, 93
goal, 91-92
illustrations, 98-103
visualization capabilities, 34
STIS (space-time intelligence system software), 9
analysis and modeling, 113-122
exploratory space-time analysis, 117-119
functionality, 114
goal, 115
visualization, 116-117

T

Time-GIS, 754
application for environmental

epidemiology, 759-764
features and architectures, 757-758
Tobler's first law of geography, 205, 256
Tukey boxplots of regional income, 612-619

V

Variance-covariance matrix, 14, 16, 186, 323, 360, 364-365, 372, 378, 404, 440, 470, 475, 494, 591, 662
Variogram, 13, 16, 122, 128-129, 132, 234-236, 265-266, 320-336, 689-691, 694, 697-698
  applied illustrations, 331-336
  features of, 325-327
  method of moments (MoM) estimator, 322-323
  residual maximum likelihood (REML) estimator, 323-325
  model types, 328-329,334
  modeling the variogram, 327-330

Viral meningitis, 20
  spatial-temporal clusters, 728-729
  spatial trends, 726-272
Voronoi polygons, 5

W

Wald test, 268, 365, 640-643
Web-based environmental decision support system (WEDSS), 20
  design and implementation issues, 705-712
Web-based spatial data analysis, 10-11
  outlier maps, 152-154
  smoothing procedures, 154-156
  spatial autocorrelation analysis, 156-157
WEDDS. *See* Web-based environmental decision support system

Z

Zero flows problem, spatial interaction models, 427

Author Index

A

Abdul-Rahman A, 27
Abdel-Rahman H, 630, 632
Abel D, 703
Abreu M, 588, 600, 605
Adams J, 530, 531
Aitkin M, 508
Akaike H, 548, 690
Aldstadt J, 12, 35, 258, 260, 261, 274, 279, 295
Allers MA, 379
Al-Sabhan W, 704
Alshafie TA, 750
Amrhein C, 546
Anderson DR, 497
Anderson JE, 431
Andrienko GL, 151, 177, 209, 224, 743, 754
Andrienko NV, 151, 177, 209, 224, 743, 753
Angulo A, 588
Anselin L, 5, 8, 10, 35, 55, 56, 73, 74, 75, 76, 78, 82, 84, 86, 91, 93, 98, 118, 151, 152, 153, 156, 157, 175, 176, 181, 182, 186, 189, 210, 211, 224, 228, 238, 239, 241, 243, 255, 257, 258, 265, 268, 269, 270, 281, 290, 291, 292, 361, 374, 379, 380, 382, 384, 387, 389, 390, 391, 392, 393, 395, 440, 441, 491, 493, 494, 496, 498, 591, 592, 630, 637
Aplin P, 572
Arbia G, 208, 258
Armstrong MP, 707
Aspinall RJ, 658, 665, 666, 669, 692,
Assunção RM, 82, 157, 207, 284, 295
Aten B, 492
Atkinson PM, 474, 530
Augustin N, 689, 690
AvRuskin GA, 21, 115, 126, 753, 754

B

Baddeley A, 230, 274
Bailey TC, 78, 152, 155, 219, 225, 267, 283, 287, 288, 411
Baker RD, 292
Bakker B, 703
Balestra P, 387
Baller R, 84, 86, 186
Baltagi BH, 386, 387, 392, 395, 396, 398, 399, 402, 403
Banerjee S, 207, 209, 210, 212, 475
Bao S, 273
Barket J, 534
Barry R, 273, 360, 361, 365, 391, 396
Bartkowska M, 585
Bartlett MS, 206
Bashtannyk DM, 605, 615, 620, 621
Basile R, 603
Bates DM, 68
Bates MN, 765
Bateson A, 531
Batty M, 409
Beaubroef T, 754
Beck N, 387
Becker RA, 54, 78, 221, 223, 226, 227
Bell JF, 687
Belsley DA, 470, 471
Benhabib J, 585
Benjamini Y, 273, 292
Bennett J, 546
Bennett N, 508
Bennett RJ, 60, 197, 508
Benz UC, 565
Bera AK, 84, 380, 382, 392, 493, 592
Berry BJL, 256
Besag J, 118, 135, 266, 283, 287, 293, 475, 546, 547, 548, 551, 552, 557, 689, 690
Best N, 522
Beven K, 688

Bezdek JC, 659, 661. 662, 663, 664
Biehl K, 213
Bierkens MFP, 212
Biging GS, 576
Bithell JF, 297
Bivand RS, 8, 11, 39, 53, 54, 55, 74, 177, 219, 225, 229, 230, 232, 234, 235, 239, 240, 249, 274, 374, 478
Blakely T, 509, 513
Blaschke T, 570, 573
Blot WJ, 737
Bogaert P, 126
Bolduc D, 410
Bontemps S, 575, 578
Boots BN, 46, 230, 265, 271, 283, 286, 292
Borcard D, 303, 686
Boudriault G, 45
Brakman S, 629
Brauer M, 754
Breckenridge J, 754
Breiman L, 691
Breusch TS, 387
Brewer CA, 225
Brooks-Gunn J, 213
Brown A, 151
Brown D, 690
Brown LA, 492
Brown M, 530
Browne WJ, 521
Brueckner JK, 379
Brunsdon C, 69, 206, 220, 221, 224, 246, 247, 461, 465, 472, 491
Bruzzone L, 574
Bryk A, 513
Bullen N, 491, 517, 518
Bulli S, 600
Burford RL, 492
Burnett C, 570
Burnham KP, 497
Burridge J, 588
Burrough PA, 199, 655, 656, 657. 659, 661, 662, 663, 665, 689, 690
Butler BE, 657
Butler H, 176
Buttenfield B, 151
Byrne GF, 569, 574

C

Caldas de Castro M, 292

Calder C, 469, 470, 476
Calisher C, 732
Camara G, 126
Can A, 261, 492
Cantor KP, 754
Cao L, 571
Carlino GA, 92
Carpenter G, 530, 531, 532
Carpenter TE, 737
Carr DB, 78, 226
Carrara A, 655, 659
Carroll ZL, 330, 334
Carsjens GJ, 710
Carver SJ, 704, 710
Casetti E, 15, 372, 487, 490, 491, 492
Castilla G, 565, 569
Ceccato V, 239
Chainey SP, 27
Chambers JM, 54, 57, 219, 228
Chamran F, 655
Charlton ME, 294
Chellappa R, 546, 547, 548
Chen DM, 566, 569
Chen J, 365
Chen SJ, 657
Cheshire, 585
Chetwynd AG, 289
Chib S, 457
Chin J, 721, 722
Chittaro L, 754
Choynowski M, 212
Christakos G, 92, 126
Chung CF, 655, 657
Ciccone A, 630, 631
Clark PJ, 286
Clayton D, 78, 155
Cleveland WS, 74, 153, 221
Cliff AD, 4, 48, 56, 61, 205, 207, 255, 256, 257, 263, 264, 282, 283, 285, 494, 592
Clifford P, 210
Cohn BA, 765
Coles J, 176
Combes PP, 629
Compas E, 704
Conchedda G, 575, 577
Congdon P, 469, 490
Conley J, 293
Cook D, 74, 221, 223
Coppin P, 566
Cossman RE, 738

Cox NJ, 57, 219, 220, 236
Crawford MM, 546
Cressie NAC, 6, 56, 201, 208, 209, 210,
    219, 226, 228, 229, 231, 234, 237,
    265, 266, 319, 336, 387, 692
Crews-Meyer K, 576
Crighton EJ, 239
Cromley EK, 27
Cross GR, 546
Curry L, 410
Curtiss B, 531
Cuzick J, 288

D

Dacey MF, 205, 257
Dai X, 569
Dalgaard P, 54
Dall'erba S, 81, 370
Davis J, 19
Davis F, 686, 703
de Castro MC, 238, 273
de Cesare L, 126, 129
de Graaff T, 497
de Smith MJ, 494
Dearing BE, 57
deFries R, 530
deGruijter JJ, 657, 659, 661
Dellepiane SG, 546
Dennison PE, 530, 531
Densham PJ, 707
Derin H, 546
Desclée B, 574, 578
Deutsch CV, 126, 127
Dev B, 110, 618
Devesa SS, 477, 737, 738, 739
Devijver PA, 557
Devine O, 156
Dhakal AS, 656
Dietrich WE, 655, 656
Diez Roux AV, 507
Diggle PJ, 56, 67, 126, 230, 232, 287,
    288, 289, 297
Diniz-Filho JAF, 494
Dixit A, 629
Dokka R, 688
Dorling D, 78, 209, 225
Dorman CF, 494
Doyle S, 703
Dragicevic S, 703, 704, 754
Dray S, 303, 307

Du Y, 569
Duan N, 113
Dubayah R, 688
Dubin R, 255
Duczmal L, 295
Duque JC, 185
Durham H, 225
Durlauf SN, 600
Dykes JA, 75, 91, 220, 221, 224, 247
Dyrness T, 654

E

Eastman JR, 530, 540, 710
Edwards R, 288
Egenhofer MJ, 92, 113
Eldridge DG, 491
Elhorst JP, 14, 377, 379, 380, 382, 385,
    392, 395, 396, 398, 204
Elith J, 685
Ellen SD, 655
Englund E, 125, 126
Ertur C, 355, 382
Evans FC, 286
Evans IS, 657, 659, 661

F

Fabbri AG, 656, 657
Fan J, 492, 605
Farber S, 465, 472
Fedra K, 703
Feenstra RC, 421, 422
Fels J, 688
Fernandez C, 448
Ferrier S, 685
Fieguth PW, 212
Fielding AH, 518, 687
Fingleton B, 18, 255, 404, 588, 593,
    600, 629, 630, 632, 634, 635,
    636, 637, 638, 640
Fischer MM, 1, 14, 17, 18, 73, 151, 176,
    197, 317, 372, 409, 410, 411, 417,
    419, 456, 459, 494, 585, 586, 588,
    593, 596, 599, 609
Fisher R, 205
Fisher PF, 657, 659, 661
Florax RJGM, 382, 497, 498
Folmer H, 382
Foody G, 530
Forslid R, 629

Forster BC, 208
Fortin MJ, 259, 686
Fosgate GT, 723
Fotheringham AS, 5, 36, 73, 120, 151, 152, 246, 258, 271, 294, 372, 411, 412, 465, 467, 468, 491
Frank AU, 197
Franklin J, 19, 685, 686, 687, 688, 690, 692, 698
Franzese RJ, 381
Freeman HP, 750
Freisthler B, 239
Freret S, 382, 392
Friedl MA, 529
Friedman M, 600
Friendly M, 220, 227, 228, 241
Frogbrook ZL, 335
Froidevaux R, 125
Frühwirth-Schnatter S, 430
Fu P, 668
Fujita M, 629, 630, 632, 645

G

Gabrosek J, 212
Gahegan M, 75, 91, 151, 177, 224
Gallant JC, 668
Galli A, 343, 344
Gamanya R, 570, 573
Gammon MD, 765
Gandin LS, 319
Gao S, 737
Gatrell AC, 78, 152, 155, 219, 225, 267, 283, 287, 288, 411
Gatrell JD, 491
Geary RC, 205, 256, 284
Gebhardt A, 39, 54, 74, 177
Gehlke CE, 213
Gelfand AE, 364, 372, 475
Gelman A, 207
Geman D, 364, 545, 546
Geman S, 364, 545, 546
Gentleman R, 54
Geon G, 450
Gessler PE, 19, 653, 655
Getis A, 1, 11, 27, 33, 35, 65, 69, 73, 118, 151, 176, 238, 239, 255, 258, 259, 260, 261, 267, 269, 272, 273, 281, 285, 286, 290, 292, 295, 303, 304, 410, 493, 494, 605, 606, 607, 618
Geyer CJ, 548, 557

Gillies S, 176
Gilley OW, 65
Glade T, 654
Glass GV, 497
Glavanakov S, 723
Godambe VP, 201
Goetz AR, 492, 686
Goldberger AS, 402
Goldstein H, 513, 515, 516,517, 520, 521
Golledge RG, 92, 259
Gomez-Hernandez JJ, 126
Gómez-Rubio V, 230
Goodchild MF, 73, 113, 151, 175, 197, 198, 199, 255
Goovaerts P, 9, 118, 125, 126, 130, 209, 292, 319, 756
Gopal S, 16, 529
Gorsevski PV, 19, 653, 655, 656
Gotway CA, 44, 56, 135, 203, 208, 212, 220, 239, 284, 296, 297, 298, 689
Gould P, 57
Graae BJ, 686
Granger CWJ, 206
Grayson RB, 655
Greene SK, 20, 721
Greene WH, 383, 384, 386
Greig-Smith P, 286
Greiling DA, 20, 35, 115, 118, 737, 750, 754
Griffith DA, 8, 12, 43, 44, 45, 48, 49, 50, 56, 60, 62, 69, 151, 210, 244, 245, 255, 259, 268, , 301, 303, 309, 317, 387, 396, 410, 411, 417, 419, 420, 468, 473, 494, 546, 547, 548, 592, 606, 607, 618
Griliches Z, 585
Grossberg S, 531
Guisan A, 685, 686, 690
Gumpertz N, 689
Guptill SC, 202
Gutman G, 566
Guyon X, 547, 551

H

Hägerstrand T, 113
Haining R, 11, 56, 57, 60, 61, 62, 68, 73, 151, 197, 198, 201, 202, 203, 206, 207, 208, 209, 210, 211, 228, 236, 255, 258, 284

Hajkowicz SA, 710
Hall AD, 319
Hall O, 575
Hall P, 605
Hall RE, 630, 631
Hammond C, 656
Han D, 113
Hanham RQ, 491
Haralick RM, 570
Hardoum C, 547
Hargitai P, 202
Hartford AH, 689
Hartwig F, 57
Harward V, 570
Haslett J, 74, 224, 743
Hastie TJ, 54, 57, 690
Hastings WK, 364
Hawkins DM, 141
Hay GJ, 565, 569, 575
Hayes DJ, 566
Hays JC, 381
Head K, 629
Hendry DF, 382
Hengl T, 655, 657, 661, 665
Henry M, 273
Hepple LW, 265, 446
Herzog A, 151
Heston A, 492
Heywood I, 710
Hilborn R, 497
Hinsen K, 95
Hipp JR, 213
Hirschmugl M, 570
Hoaglin D, 220
Hobbs NT, 497
Hochberg Y, 273, 292
Hoffbeck JP, 545
Hornsby K, 113
Houle M, 704
Howard E, 546
Hsaio C, 377, 384, 392
Huang B, 151
Huang C, 530
Huang L, 294
Hubbell SP, 686
Hubert LJ, 256, 259
Hudak S, 390, 391, 393
Huether S, 722
Huffer FW, 686, 689
Huijbregts CJ, 319, 337
Hunneman A, 401

Hunter JE, 497
Huriot JM, 629
Hwang CL, 657
Hyndman RJ, 600, 603, 604, 605, 614, 615, 620, 621

I

Ichikawa R, 656
Ihaka R, 54
Illian J, 33
Im J, 569, 574
Irvin BJ, 657, 661
Isaaks E, 126
Isard W, 5
Ishizawa H, 239
Islam N, 599, 600
Iwahashi J, 661

J

Jackson C, 214
Jacobs J, 632
Jacoby WG, 221
Jacquez GM, 9, 20, 21, 35, 113, 115, 118, 123, 275, 289, 292, 554, 732, 737, 750, 753, 754, 764
Jain AK, 546
Janikas MV, 7, 9, 27, 75, 91, 92, 175, 258, 274, 605
Jankowski P, 19, 151, 653, 655, 656, 710
Jaquet O, 344
Jarque CM, 380
Jeliazkov I, 457
Jelinski DE, 206
Jennrich R, 365
Jensen JR, 568, 578
Jensen RR, 492
Ji C, 547, 548
Jin SM, 569
Johnson PA, 603
Johnston CA, 27, 62
Jona Lasinio G, 547, 548, 554
Jones CI, 600
Jones JPI, 491, 492
Jones K, 57, 219, 220, 237, 410, 491, 517, 518
Jones MC, 604
Journel AG, 126, 127, 319, 337
Ju J, 530

Jung I, 294

K

Kafadar K, 78, 156
Kähkonen J, 151
Kainz W, 578
Kaiser J, 570
Kaldor J, 78, 155
Kalluri R, 534
Kaluzny SP, 56, 61, 62, 228
Kangas A, 239
Kapoor M, 381
Kashyap R, 547, 548
Kaufmann AM, 21, 737, 753
Kawachi I, 509
Keitt TH, 686
Kelejian HH, 186, 189, 268, 361, 370, 381, 382, 404, 591
Keller CP, 710
Kellerman A, 491
Kelly PA, 546
Kelsall JE, 67
Kennedy P, 497
Kerry R, 323
Ketting RL, 565, 570
Kho Y, 8, 73
Kholodilin KA, 404
Khorram S, 569, 576
Khotanzad A, 546
Kim MJ, 754
Kim YW, 10, 136, 151, 213
King G, 213
Kingston R, 704
Kirkby M, 688
Kitanidis PK, 324
Knorr-Held L, 548
Knox EG, 279
Knyazikhin Y, 532
Koch W, 355, 382, 423, 431
Kodras JE, 492
Kolaczyk ED, 507
Kolker A, 754
Kolmogorov AN, 319, 336
Kooijman S, 261
Koop G, 363
Kooperberg C, 475
Koopman JS, 116
Korniotis GM, 404
Kraak MJ, 151
Krakover S, 491

Krige DG, 6, 319, 336
Krishna-Iyer PV, 205, 256
Kristensen G, 490, 491, 492
Krivoruchko K, 249
Kulldorff M, 136, 211, 212, 294, 295, 297, 724
Kunnert A, 585
Kwan MP, 113

L

Lafourcade M, 629
Lagona F, 16, 387, 545, 547, 548, 554
Lahiri SN, 387
Laliberte AS, 571, 573
Lam NSM, 199
Lamar WR, 573, 577
Lambin E, 530
Landgrebe DA, 545, 565, 570
Lane RW, 324
Laney RM, 530
Langford IH, 522
Langtangen HP, 95, 176
Lark RM, 337
Lauridsen J, 237
Lautaportii K, 176
Lawson AB, 136, 138, 152, 154, 296,
Lawson G, 522
Layne LJ, 56
Leamer EE, 497
Lee G, 10, 135, 142
Lee LF, 360, 379, 382
Lee M, 417
Lee S, 546
Leenders RTAJ, 380
LeGallo J, 81, 370, 404, 591, 600
Legendre P, 255, 259, 273
Legendre L, 686
Lennon JJ, 693
Leonard H, 497
Lery B, 239
LeSage JP, 13, 14, 39, 177, 275, 316, 355, 356, 359, 360, 361, 363, 364, 365, 367, 369, 370, 371, 372, 374, 391, 401, 404, 409, 410, 414, 416, 417, 418, 419, 420, 421, 422, 423, 424, 427, 428, 429, 430, 431, 435, 441, 442, 443, 447, 448, 449, 452, 453, 456, 457, 469, 472, 491, 586, 587, 588, 591, 593, 595, 596
Leung Y, 258, 272, 468, 469

Levine N, 30, 33, 75, 230
Lewis B, 176
Ley E, 448
Leyland AH, 521
Li D, 402, 403
Li H, 243
Li J, 567
Li X, 569
Liang F, 448
Lichstein JW, 699
Lin H, 151
Lindley DV, 447
Liu W, 16, 529, 530, 533
Livingstone D, 113
Llano C, 421, 422
Lloyd CD, 228, 232
Lo CP, 569
Lobo A, 570
Lobo JM, 686
Loeber R, 213
López-Bazo E, 588, 593, 600
Louks OL, 213
Loveland TR, 529
Lozano-Gracia M, 186

M

Maantay J, 27, 754
MacDonald H, 27
MacDonnell RA, 689, 690
MacEachren A, 224
MacMillan RA, 657, 659, 661
Maddala GS, 494
Madigan D, 448, 449
Magnus JR, 388, 392
Magrini S, 600
Malczewski J, 658, 665, 666
Malecki, 585
Mangel M, 497
Manski CF, 381
Mantel N, 280
Marble DF, 256
Marceau DJ, 213, 754
Marcus WA, 554
Mark D, 113
Mark RK, 655
Marsh TL, 361
Marshall RJ, 78, 155
Martin RJ, 60, 62, 63
Martinez-Rica JP, 686
Mas JF, 566

Matérn B, 319
Mather FJ, 753
Matheron G, 6, 201, 257, 319, 321, 324, 336, 337
Mattheis DJ, 110
Mayaux P, 530
Mayer T, 629
McBratney AB, 657, 659, 661, 663, 664, 668
McCance K, 722
McCann P, 630
McCarthy HH, 213
McCarthy MA, 497
McClelland DE, 654, 656, 659, 668
McCombie JSL, 629, 634
McCullough B, 561
McDonnell RA, 199, 659, 661, 665, 689
McGill M, 74
McIver D, 530
McKenzie NJ, 655
McLafferty SL, 27
McMillen DP, 494
McSweeney K, 657
Mehlich A, 319
Mehnert WH, 477
Meinel G, 570
Meliker JR, 21, 113, 115, 123, 753, 754, 764
Mercer WB, 319
Messner SF, 84, 86, 152, 163, 170
Metropolis N, 364
Meyer JC, 19, 703
Miller JA, 19, 27, 113, 685, 686, 687, 697
Mills LO, 92
Minasny B, 126, 663, 664, 668
Miron JR, 255, 493
Missouri Census Data Center, 704
Mitchell A, 27, 30
Mittelhammer RC, 361
Mladenoff DJ, 655
Modlin JF, 722
Moigne J, 540
Monmonier MS, 74, 209, 224
Montgomery DR, 655, 656
Montouri BD, 92
Mood AM, 398
Moody A, 530
Moon G, 509, 510, 513, 522
Moore DA, 737
Moore ID, 655, 661, 663, 664

Moore M, 730
Moran PAP, 46, 205, 256, 282
Morens D, 722, 731
Moriera W, 192
Morisette JT, 576
Morrison JL, 202
Moser EB, 44
Mountain D, 225
Mukherjee C, 494
Müller W, 232
Mur J, 236, 472, 588
Murphy DK, 750
Murrell P, 221, 231
Myers A, 704

N

Narayanan RM, 567
Nebert T, 703
Nelder JA, 561
Nerlove M, 387
Neter J, 468, 469
Neubert M, 570
Neudecker H, 392
Newell J, 118, 135, 283, 293
Newton MA, 457
NiDardo J, 62
Nijkamp P, 73, 151
Noah N, 732, 733
Nogués-Bravo D, 686
Nriagu JO, 21, 747, 753, 756

O

O'Kelly ME, 136, 411, 412
Odeh IOA, 657, 659, 661, 662, 663, 664
Oden N, 284
Odland J, 255
Ohmae, 585
Okabe A, 27, 258, 274
Okimura T, 656
Olea RA, 203
Oliphant T, 38
Oliver M, 12, 319, 322, 323, 327, 330, 331, 334, 335, 341, 343, 657
O'Loughlin M, 656
Olsson G, 57
Olwell DH, 141
Omre H, 338
Openshaw SM, 136, 211, 213, 261, 293

Ord JK, 4, 33, 35, 48, 56, 61, 69, 84, 118, 205, 207, 238, 239, ,256, 257, 258, 263, 264, 269, 272, 273, 282, 283, 285, 290, 292, 304, 360, 380, 493, 494, 592, 605, 606, 607
Overman H, 629

P

Paap R, 600
Pace RK, 13, 65, 273, 316, 355, 356, 359, 360, 361, 363, 365, 367, 369, 370, 371, 372, 374, 391, 396, 404, 410, 414, 416, 417, 418, 419, 420, 421, 423, 424, 427, 428, 429, 430, 431, 441, 442, 443, 449, 457, 472, 586, 588, 591, 593, 595
Paelink JHP, 5
Páez A, 14, 272, 431, 461, 465, 468, 469, 470, 472, 474
Pal NR, 570
Pal SK, 570
Pallansch M, 722, 731
Pandey S, 703
Pandit K, 491
Pannatier Y, 126, 128
Pardo-Igúzquiza E, 323, 325
Parent O, 14, 435, 447, 448, 449, 453, 457
Parsley MJ, 704
Partridge MD, 386
Patacchini E, 239
Pathirana S, 657, 659, 661
Patterson HD, 324
Payne RW, 126, 129, 332
Pearson DM, 692
Pebesma EJ, 126, 234
Pedroni S, 192
Peng Z, 151, 704
Peres-Neto, 303
Pérez F, 192
Peters A, 27
Peterson AT, 685
Peterson MP, 703
Pettit C, 27
Peuquet DJ, 27, 92, 113
Pick JB, 27
Pickle LW, 225, 737, 753
Pike RJ, 661
Pinheiro JC, 68
Pittau MG, 618, 623

Plewe P, 151
Polasek W, 419, 457
Pontius RG, 577, 691
Porojan A, 417
Portnov BA, 55, 74, 240
Pouliot DA, 570
Price PN, 207
Prieto DF, 574
Pritchett L, 585
ProMED-mail, 721, 730, 732
Prucha IR, 186, 361, 382, 404

Q

Quah DT, 599, 600, 601, 608, 611, 613, 617
Quarmby NA, 531
Quirk M, 732

R

Radke RJ, 566
Raftery AE, 457
Ramanathan R, 494
Rangel TFLVD, 238, 494, 685
Ranjan R, 427
Raper J, 113
Rappin N, 192
Ratcliffe JH, 27
Räty M, 239
Raudenbush SW, 509, 513
Ravenstein EG, 256
Raza A, 578
Redding S, 629
Reichlin R, 600
Reid F, 732, 733
Reif JS, 754
Reis EA, 82, 157, 207, 284
Reismann M, 410
Remy N, 126
Reneau SL, 655
Reuter HI, 655, 657, 661
Rey SJ, 9, 10, 75, 80, 91, 92, 93, 97, 110, 175, 176, 181, 185, 189, 258, 274, 600, 605, 618
Ribeiro PJR, 126, 127, 232
Ricard M, 546
Rice P, 239, 629
Rich PM, 668
Richardson S, 210, 685
Riedl A, 585

Rignot E, 546
Ripley BD, 4, 54, 55, 56, 58, 127, 266, 287
Rivera-Batiz F, 630
Roberts DA, 530, 531
Roberts GO, 557
Robertson GP, 126
Robinson DP, 268
Robinson WS, 213
Robison EG, 655, 659
Rogan J, 566
Rogerson P, 10, 73, 135, 136, 139, 140, 141, 142, 151, 287, 296
Rommel RG, 20, 737
Rose C, 557
Rosenberg MS, 266
Rossi RE, 689
Rotbart H, 721, 722, 730, 731, 732
Rowlingson BS, 297
Roy JR, 410
Rura MI, 8, 43, 44, 48, 49
Rushton G, 737
Ruspini EH, 659
Ryan PB, 756
Ryan PJ, 655
Ryherd S, 570

S

Sabel CE, 115
Sader SA, 566, 569
Saenz J, 95
Sampson RJ, 213
Sardadvar S, 585
Sarkar D, 221, 226, 231
Schabenberger O, 220
Schaerstrom A, 113
Schliep A, 95
Schmertmann CP, 239
Schmidt MA, 20, 721
Schneider LC, 691
Schott JR, 569
Schowengerdt R, 530
Schwartz G, 548
Scott LM, 7, 27, 30
Sellers P, 532
Semple RK, 491
Sen A, 410, 411, 412
Seymour L, 547, 548
Shandley J, 570
Shapiro LG, 570
Shaw SL, 27

Shi X, 665
Shriram I, 703, 704
Sickley DA, 655
Sidle RC, 655, 656
Siegmund D, 141
Sikdar BK, 409
Silverman PW, 604, 605, 610, 611, 619
Simes RJ, 741
Singer BH, 238, 273, 292
Sinha G, 113
Skidmore AK, 657, 665, 666
Slocum TA, 225
Slotnick MJ, 21, 753, 754, 756
Smirnov O, 84, 361
Smith AFM, 364, 557
Smith FL, 497
Smith MJ, 27
Smith PA, 686
Smith TE, 410, 411, 412
Smits PC, 546
Sokal RR, 239, 258, 271
Song C, 297, 569
Sparks A, 125, 126, 585
Spiegel MM, 585
Spiegelhalter DJ, 475, 657
Sridharan S, 239
Srivastava RM, 126
Steel M, 448
Stein ML, 337
Steinmaus C, 756, 760, 765
Stern DI, 493
Stevens G, 239
Stewart MB, 62
Stiglitz JE, 629
Stirböck C, 593, 609
Stobierski MG, 20, 721
Stockman GC, 570
Stockwell D, 685
Stone R, 296
Storper, 585
Stow D, 17, 565, 568, 569, 575, 577
Strahler AH, 565, 576
Strapp JD, 710
Strikas RA, 730, 731
Student, 205
Stuetzle W, 74
Stumpner P, 18, 585, 599
Su L, 404
Subramanian Sv, 15, 507, 509, 510, 513, 520, 521, 522
Sugumaran R, 19, 703, 704

Sugumaran V, 703, 704
Sui D, 686
Sullivan J, 338
Sun Y, 136, 141, 142
Sutton AJ, 497
Swartz CH, 754
Swayne DF, 221, 223
Swets JA, 691
Syabri I, 8, 10, 73, 151
Symanzik J, 74, 91, 151, 223

T

Takahashi K, 295, 297
Takatsuka M, 75, 91, 151, 177, 224
Tanaka K, 492
Tango T, 140, 284, 292, 295, 297
Taylor K, 704
Taylor PJ, 202, 213
Temple J, 585, 600
Theil H, 93
Theriault M, 492
Theus M, 221, 224
Thill JC, 27
Thisse JF, 629, 632
Thomas E, 256, 475
Thomas K, 688
Thompson EA, 548, 557
Thompson ME, 201
Thompson R, 324
Thomson B, 239
Thuiller W, 685, 698
Tian Y, 532
Tiebout CM, 450
Tiefelsdorf M, 46, 239, 244, 245, 248, 261, 265, 272, 281, 283, 309, 468, 470, 479
Tjelmeland H, 546, 557
Tobias JL, 427
Tobler WR, 213, 256, 303, 686
Tomlin DC, 27
Townshend JRG, 530
Tsou MH, 151, 704
Tukey JW, 3, 116, 220
Turnbull BW, 118, 136
Turnbull GK, 450
Turner R, 230, 274

U

Unwin A, 74, 228

Unwin DJ, 232, 233
Upton GJ, 255

V

van Dijk HK, 600
van Wincoop E, 431
Varga A, 492
Vayssières MP, 698
Veitch N, 657, 665
Velleman P, 220
Venables AJ, 629
Venables WN, 54, 58
Ventura SJ, 657, 661
Verbeek M, 399
Veregin H, 202
Vernon LT, 704
von Thünen JH, 256
Voss PR, 239
Vrijburg H, 404

W

Wackernagel H, 327, 546
Wagner H, 430
Wall P, 156
Waller LA, 44, 135, 136, 138, 220, 239, 284, 296, 297, 298, 470
Wallis KF, 62
Walter SD, 284
Walter V, 567, 572
Wand MP, 604
Wang F, 45
Wang L, 570
Wang YQ, 704
Warmerdam N, 30
Warner RM, 292
Warnes G, 192
Wartenberg D, 258
Webster R, 319, 322, 327, 331, 335, 341, 344, 657
Wesseling CG, 126
Wheeler D, 14, 33, 247, 272, 461, 468, 469, 470, 472, 473, 479, 480
White H, 186
Wie YD, 239
Wieczorek GF, 655
Wiener N, 336
Wikle CK, 212
Wikström PO, 213
Wilson AG, 197, 410

Wilson ML, 20, 721
Wilson JP, 668
Wingle WL, 126
Winn DM, 737
Wise S, 74, 91
Wold H, 336
Wong DWS, 30, 258
Wood J, 225
Woodcock C, 16, 529, 530, 570
Woodward AJ, 509
Woolridge JM, 36
Worboys MF, 113, 151
Wright RH, 655
Wrigley N, 232, 233, 491
Wu H, 686, 689
Wu J, 206, 213
Wu W, 655, 656

Y

Yamada I, 10, 135, 141, 287
Yamamoto D, 239
Yang X, 569
Yang Z, 397, 404
Yao Q, 603
Yearsley C, 113
Yeh AGO, 569
Yiannakoulias N, 44
York J, 448, 449
Youden WJ, 319
Young LJ, 56, 203, 208, 212
Yu D, 239
Yu J, 379, 404

Z

Zadeh LA, 531, 658
Zdorkowski TR, 491
Zee SML, 492
Zelli R, 618, 623
Zellner A, 438, 448
Zenk SN, 731
Zenou Y, 239
Zhan FB, 294
Zhang C, 35, 151
Zhang J, 490, 491
Zhang M, 545
Zhang X, 704
Zhand Y, 703
Zhou W, 573, 577
Ziegler J, 27

Zimmermann N, 685, 686
Zipf GK, 256

Contributing Authors

Jared Aldstadt is Assistant Professor in the Department of Geography, University at Buffalo, State University of New York, Buffalo NY, USA
geojared@buffalo.edu

Luc Anselin is Foundation Professor and Director of the School of Geographical Sciences and Urban Planning; Director of the GeoDa Center for Geospatial Analysis and Computation, Arizona State University, Tempe AZ, USA; Editor, *International Regional Science Review*
luc.anselin@asu.edu

Gillian A. AvRuskin is Geographic Information Scientist at BioMedware Inc., Ann Arbor MI, USA
avruskin@biomedware.com

Monika Bartkowska is a PhD Student in the Institute for Economic Geography and GIScience, Department of Social Sciences, Vienna University of Economics and Business, Vienna, Austria
monika.bartkowska@wu.ac.at

Roger S. Bivand is Professor in the Economic Geography Section, Department of Economics, Norwegian School of Economics and Business Administration, Bergen, Norway
Roger.Bivand@nhh.no

Emilio Casetti is Professor Emeritus in the Department of Geography, Ohio State University, Columbus OH, USA
casetti.1@osu.edu

Jim Davis is former research faculty, Rural Sociology Department, University of Missouri, Columbia MO, USA

J. Paul Elhorst is Associate Professor in the Department of Economics and Econometrics, University of Groningen, Groningen, The Netherlands
J.P.Elhorst@rug.nl

Bernard Fingleton is Professor of Economics in the Department of Economics, Scottish Institute for Research in Economics (SIRE); Strathclyde University, Glasgow, UK; Managing Editor, *Spatial Economic Analysis*
bernard.fingleton@strath.ac.uk

Manfred M. Fischer is Professor and Director of the Institute for Economic Geography and GIScience, Department of Social Sciences, Vienna University of Economics and Business, Vienna, Austria; Editor, Springer book series on *Advances in Spatial Science*; and Editor-in-Chief, *Journal of Geographical Systems*
manfred.fischer@wu.ac.at

Janet Franklin is Professor in the School of Geographical Sciences and School of Life Sciences, Arizona State University, Tempe AZ, USA
Janet.Franklin@asu.edu

Paul E. Gessler is Associate Professor and Co-Director of the Geospatial Laboratory for Environmental Dynamics, College of Natural Resources, University of Idaho, Moscow ID, USA
paulg@uidaho.edu

Arthur Getis is Distinguished Professor Emeritus in the Department of Geography, San Diego State University, San Diego CA, USA; Honorary Editor, *Journal of Geographical Systems*
arthur.getis@sdsu.edu

Pierre Goovaerts is Chief Scientist, BioMedware Inc. and courtesy Associate Professor in the Soil and Water Science Department, University of Florida, Gainesville FL, USA
goovaerts@biomedware.com

Sucharita Gopal is Professor in the Department of Geography and Environment, Center for Remote Sensing, Center for Cognitive and Neural Systems, Boston University, Boston MA, USA
suchi@bu.edu

Pece V. Gorsevski is Associate Professor, Geospatial Sciences, School of Earth, Environment and Society, Bowling Green State University, Bowling Green OH, USA
peterg@bgnet.bgsu.edu

Sharon K. Greene is an epidemiologist in the Department of Ambulatory Care and Prevention Care, and Department of Epidemiology, University of Michigan School of Public Health, Ann Arbor MI, USA
skgreene@umich.edu

Dunrie Greiling is Director of Happiness, Pure Visibility Inc., Biomedware Inc., Ann Arbor MI, USA
dunrie@biomedware.com

Daniel A. Griffith is Ashbel Smith Professor in the School of Economic, Political and Policy Sciences, University of Texas, Dallas TX, USA; Editor, *Geographical Analysis*
dagriffith@utdallas.edu

Robert P. Haining is Professor of Human Geography, Department of Geography, University of Cambridge, Cambridge, UK
bob.haining@geog.cam.ac.uk

Geoffrey M. Jacquez is President of Biomedware, Inc., Ann Arbor MI, USA
Jacquez@Biomedware.com

Mark V. Janikas is Spatial Statistics Product Engineer, ESRI, Redlands CA, USA
janikas@rohan.sdsu.edu

Piotr Jankowski is Professor in the Department of Geography, San Diego State University, San Diego CA, USA
piotr@typhoon.sdsu.edu

Andrew M. Kaufmann is Software Engineer, Biomedware Inc., Ann Arbor MI, USA
kaufmann@Biomedware.com

Younghin Kho is a PhD Student, Department of Computer Science, University of Illinois, Urbana-Champaign IL, USA
yhkho@ uiuc.edu

Yong Wook Kim is Research Programmer, Land Use Evolution and Impact Assessment Model (LEAM) University of Illinois, Urbana-Champaign IL, USA
ywkim@ uiuc.edu

Andrea Kunnert is former Research Associate, Institute for Economic Geography and GIScience, Department of Social Sciences, Vienna University of Economics and Business, Vienna, Austria

Francesco Lagona is Associate Professor of Statistics in the Department of Public Institutions, Economy and Society, University Roma Tre, Rome, Italy
lagona@uniroma3.it

Gyoungju Lee is Associate Research Fellow in National Infrastructure & GIS Research Division, Korea Research Institute for Human Settlements (KRIHS), Seoul, Korea
lgjracer@gmail.com

James P. LeSage is Jerry D. and Linda Gregg Fields Chair in Urban and Regional Economics, Finance and Economics Department, McCoy College of Business Administration, Texas State University, San Marcos TX, USA; Editor, *Papers in Regional Science*
jlesage@spatial-econometrics.com

Weiguo Liu is with the Department of Geography and Planning, The University of Toledo, Toledo OH, USA

Jaymie R. Meliker is Associate Professor in the Department of Environmental Health Sciences, School of Public Health, University of Michigan, Ann Arbor MI, USA
jmeliker@umich.edu

James Charles Meyer is former Senior GIS Scientist, University of Missouri, Columbia MO, USA

Jennifer A. Miller is Assistant Professor in the Department of Geography and the Environment, University of Texas, Austin TX, USA
jennifer.miller@austin.utexas.edu

Jerome O. Nriagu is Professor for Environmental Health Sciences and Research Scientist, Center for Human Growth and Development, University of Michigan School of Public Health, University of Michigan, Ann Arbor MI, USA; Editor, *Science of the Total Environment*
jnriagu@umich.edu

Margaret A. Oliver is Visiting Professor in the Department of Soil Science, The University of Reading, Reading, UK; Co-editor, *Precision Agriculture*
m.a.oliver@reading.ac.uk

R. Kelley Pace is Distinguished Professor in the Department of Computer Science, Louisiana State University, Baton Rouge LA, USA
kpace@lsu.edu

Antonio Páez is Associate Professor in the School of Geography and Earth Sciences, McMaster University, Hamilton, Ontario; Editor-in Chief, *Journal of Geographical Systems*
paezha@mcmaster.ca

Olivier Parent is Assistant Professor in the Department of Economics, University of Cincinnati, Cincinnati OH, USA
Olivier.Parent@uc.edu

Sergio J. Rey is Professor in the School of Geographical Sciences, Arizona State University, Tempe AZ, USA; Editor, *International Regional Science Review*
serge@rohan.sdsu.edu

Aleksandra Riedl is Teaching and Research Associate in the Institute for Economic Geography and GIScience, Department of Social Sciences, Vienna University of Economics and Business, Vienna, Austria
aleksandra.riedl@wu.ac.at

Peter Rogerson is Professor in Geography and Biostatistics, University of Buffalo, State University of New York, Buffalo NY, USA
rogerson@buffalo.edu

Robert G. Rommel is Software Engineer, Biomedware Inc., Ann Arbor MI, USA
rommel@Biomedware.com

Melissa J. Rura is PhD student in the Department of Geospatial Information Science, University of Texas, Dallas TX, USA
melissa.rura@utdallas.edu

Sascha Sardadvar is former Teaching and Research Associate, Institute for Economic Geography and GIScience, Department of Social Sciences, Vienna University of Economics and Business, Vienna, Austria

Mark A. Schmidt is Surveillance Officer for the Active Bacterial Core Surveillance Component of the Emerging Infections Program at the Oregon Public Health Division, and Michigan Department of Community Health, Lansing MI, USA
schmidtma@michigan.gov

Lauren M. Scott is Product Engineer on the Geoprocessing Team, ESRI, Redlands CA, USA
lscott@esri.com

Melissa J. Slotnick is PhD student in the Department of Environmental Health Sciences, School of Public Health, University of Michigan, Ann Arbor MI, USA
slotnick@umich.edu

Mary Grace Stobierski is State Public Health Veterinarian and Manager of the Infectious Disease Epidemiology Section, Michigan Department of Community Health, Lansing MI, USA
stobierskim@michigan.gov

Douglas Stow is Professor in the Department of Geography and Co-Director of the Center for Earth Systems Analysis Research at San Diego State University, San Diego CA, USA
stow@mail.sdsu.edu

Peter Stumpner is former Research Associate, Institute for Economic Geography and GIScience, Department of Social Sciences, Vienna University of Economics and Business, Vienna, Austria

S.V. Subramanian is Associate Professor of Society, Human Development and Health, Department of Society, Human Development and Health, Harvard School of Public Health, Harvard University, Boston MA, USA
svsubram@hsph.harvard.edu

Ramanathan Sugumaran is Associate Professor and Director of the GeoTREE Center, University of Northern Iowa and Department of Society, Human Development and Health, Harvard School of Public Health, Boston MA, USA
sugu@uni.edu

Ibnu Syabri is with the Spatial Analysis Laboratory, Department of Agricultural and Consumer Economics, University of Illinois, Urbana-Champaign IL, USA
syabri@uiuc.edu

David C. Wheeler is a National Cancer Institute Prevention Fellow and attending Harvard University in Public Health; Postdoctoral Research Fellow at the Department of Biostatistics, Rollins School of Public Health, Emory University, Atlanta GA, USA
wheelerdc@mail.nih.gov

Mark L. Wilson is Professor in the Department of Epidemiology and the Department of Ecology and Evolutionary Biology, University of Michigan, School of Public Health, Ann Arbor MI, USA
wilsonml@umich.edu

Curtis E. Woodcock is Professor in the Department of Geography and Environment, Center for Remote Sensing, Center for Cognitive and Neural Systems, Boston University, Boston MA, USA
curtis@bu.edu

Ikuho Yamada is Associate Professor in the Geography Department, University of Utah, Salt Lake City UT, USA
ikuho.yamada@geog.utah.edu